W9-BUP-958

Useful Equations

Vector: $\mathbf{A} = A\mathbf{e}_A$

Unit vector: $\mathbf{e}_A = \cos\theta_x\mathbf{i} + \cos\theta_y\mathbf{j} + \cos\theta_z\mathbf{k}$

Scalar product: $\mathbf{A}\cdot\mathbf{B} = AB\cos\theta$

Vector product: $\mathbf{A}\times\mathbf{B} = AB\sin\theta\,\mathbf{e}_c$

Derivative of a vector: $\dfrac{d\mathbf{A}}{dt} = \dfrac{dA}{dt}\mathbf{e}_A + A\dfrac{d\mathbf{e}_A}{dt}$

Rectilinear motion of a particle:

$$ds = v\,dt \qquad dv = a\,dt \qquad v\,dv = a\,ds$$

Angular motion of a line:

$$d\theta = \omega\,dt \qquad d\omega = \alpha\,dt \qquad \omega\,d\omega = \alpha\,d\theta$$

Derivative of a unit vector **e** rotating at a rate of ω:

$$\frac{d\mathbf{e}}{dt} = \boldsymbol{\omega}\times\mathbf{e}$$

General expression for the absolute velocity:

$$\mathbf{v} = \dot{\mathbf{R}} + \dot{\mathbf{r}}_r + \boldsymbol{\omega}\times\mathbf{r}$$

General expression for the absolute acceleration:

$$\mathbf{a} = \ddot{\mathbf{R}} + \ddot{\mathbf{r}}_r + \boldsymbol{\alpha}\times\mathbf{r} + \boldsymbol{\omega}\times(\boldsymbol{\omega}\times\mathbf{r}) + 2\boldsymbol{\omega}\times\dot{\mathbf{r}}_r$$

Velocity and acceleration representation using moving polar coordinates:

$$\mathbf{v} = \dot{r}\mathbf{e}_r + r\dot{\theta}\mathbf{e}_\theta$$

$$\mathbf{a} = (\ddot{r} - r\dot{\theta}^2)\mathbf{e}_r + (2\dot{r}\dot{\theta} + r\ddot{\theta})\mathbf{e}_\theta$$

Velocity and acceleration representation using moving normal and tangential components:

$$\mathbf{v} = v\mathbf{e}_\tau$$

$$\mathbf{a} = \dot{v}\mathbf{e}_\tau + \frac{v^2}{\rho}\mathbf{e}_n$$

Gravitational force: $\mathrm{F} = \dfrac{K}{r^2}\mathbf{e}_r = -\dfrac{Gm_Am_B}{r^2}\mathbf{e}_r$

Weight: $w = mg_e$

Acceleration of gravity: $g = g_e\dfrac{R^2}{(R+h)^2}$

Newton's law: $\mathbf{F} = m\mathbf{a}$

$$\mathbf{F} = m\frac{d\mathbf{v}}{dt}$$

Work: $W = \displaystyle\int_A^B \mathbf{F}\cdot d\mathbf{r}$

Potential Energy:

Spring: $V = \frac{1}{2}kx^2$

Gravity: $V = -mgR^2/(R+h)$

Gravity: $V = mgh$ (close to earth)

Kinetic energy of a particle: $T = \frac{1}{2}mv^2$

Kinetic energy of a plane rigid body:

$$T = \tfrac{1}{2}mv_c^2 + \tfrac{1}{2}I_c\omega^2$$

Conservation of energy: $T_A + V_A = T_B + V_B$

Work-energy: $W_{\mathrm{NC}} = \Delta T + \Delta V$

Linear momentum: $\mathbf{p} = m\mathbf{v}$

Angular momentum: $\mathbf{H} = \mathbf{r}\times m\mathbf{v}$

Linear impulse: $\mathcal{I} = \displaystyle\int_{t_1}^{t_2}\mathbf{F}\,dt$

Angular impulse: $\mathcal{M} = \displaystyle\int_{t_1}^{t_2}\mathbf{M}\,dt$

Impulse-momentum

Linear: $\mathcal{I} = \Delta\mathbf{p}$

Angular: $\mathcal{M} = \Delta\mathbf{H}$

Potential energy: $V = \dfrac{K}{r}$

Momentum in orbit: $r^2\dot{\theta} = h$ a constant

Orbit equation: $r = \dfrac{h^2/\mu}{1 + \epsilon\cos\theta}$

$$r = \frac{(1-\epsilon^2)a}{1+\epsilon\cos\theta}$$

Velocity in orbit: $v^2 = \mu\left(\dfrac{2}{r} - \dfrac{1}{a}\right)$

Period of orbit: $\tau = \xi a^{3/2}$

Conservation of linear momentum:

$$m_1v_1 + m_2v_2 = m_1u_1 + m_2u_2$$

Definition of the coefficient of restitution:

$$e(v_1 - v_2) = -(u_1 - u_2)$$

Conservation of angular momentum–eccentric impact:

$$m_1v_1y + I_c\omega_1 = m_1u_1y + I_c\omega_2$$

Coefficient of restitution–eccentric impact:

$$e_1(v_1 - v_2 - y\omega_1) = -(u_1 - u_2 - y\omega_2)$$

Lagrange's equations:

$$\frac{d}{dt}\left(\frac{\partial\mathcal{L}}{\partial\dot{q}_i}\right) - \frac{\partial\mathcal{L}}{\partial q_i} = Q_i$$

$$\mathcal{L} = T - V \qquad Q_i = \frac{\delta W}{\delta q_i}$$

Euler's equations:

$$M_x = I_x\dot{\omega}_x + (I_z - I_y)\omega_y\omega_z$$

$$M_y = I_y\dot{\omega}_y + (I_x - I_z)\omega_x\omega_z$$

$$M_z = I_z\dot{\omega}_z + (I_y - I_x)\omega_x\omega_y$$

ENGINEERING MECHANICS

Statics and Dynamics

D. K. ANAND
Professor

P. F. CUNNIFF
Professor

Department of Mechanical Engineering
University of Maryland
College Park

ALLYN AND BACON, INC.
Boston London Sydney Toronto

To Our Wives

ASHA PATRICIA

This book is part of the
ALLYN AND BACON SERIES IN ENGINEERING
Consulting Editor: Frank Kreith
University of Colorado

Copyright © 1984 by Allyn and Bacon, Inc., 7 Wells Avenue, Newton, Massachusetts 02159. All rights reserved. No part of the material protected by this copyright notice may be reproduced or utilized in any form or by any means, electronic or mechanical, including photocopying, recording, or by any information storage and retrieval system, without written permission from the copyright owner.

Portions of this book appeared in *Engineering Mechanics: Statics and Dynamics* by D. K. Anand and P. F. Cunniff, copyright © 1973 by Allyn and Bacon, Inc.

LIBRARY OF CONGRESS CATALOGING IN PUBLICATION DATA

Anand, D. K. (Davinder K.), 1939–
Engineering mechanics.

(Allyn and Bacon series in engineering)
Includes index.
1 Statics. 2. Dynamics. I. Cunniff, Patrick F., 1933– . II. Title III. Series.
TA351.A52 1984 620.1'05 82-22638
ISBN 0-205-07810-9

Printed in the United States of America

10 9 8 7 6 5 4 3 2 1 88 87 86 85 84

Contents

3 Rigid Body Equilibrium *73*

4 Structures *101*

5 Distributed Forces *147*

6 Moments of Inertia *199*

7 Beam Analysis 239

8 Friction 267

9 Virtual Work and Energy 293

Dynamics

1 Basic Concepts 1

2 Kinematics of a Particle *27*

3 Kinetics of a Particle *77*

4 Central Force Motion *117*

5 Planar Kinematics of Rigid Bodies *135*

6 Planar Kinetics of Rigid Bodies *167*

7 Impact *207*

8 Advanced Topics *231*

9 Some Special Problems *253*

Preface

This book contains a complete two-volume *Statics* and *Dynamics* set. We have arranged the material in a way designed to make engineering mechanics an interesting and challenging subject for college students. However, the basic format of each chapter also facilitates self-study. The theoretical material is carefully developed, and supported by a generous selection of examples with worked-out solutions, to assist the reader in acquiring problem-solving capabilities. As an aid to the student, each chapter concludes with a summary of the important ideas and equations covered.

After each major topic there are numerous exercise problems arranged approximately in order of increasing complexity. The instructor may choose to assign even-numbered problems for which the answers can be found at the back of the book. Also, we note that the International System of units is included with the standard English gravitational system of units throughout the book. Consequently, the examples and assigned problems incorporate both sets of units.

In the statics volume, Chapter 1 reviews the subject of vectors and the fundamental laws useful for the study of statics. The concepts of force and couple are introduced in Chapter 2, where we also consider the equilibrium of a particle. The equilibrium of rigid bodies is then covered in Chapter 3. Although two- as well as three-dimensional equilibrium is considered, stress is placed on the former since the salient points can be grasped without introducing algebraic complexity. The equilibrium of structures and frames is presented in Chapter 4, where the graphical approach has been adopted in light of current developments in computer graphics.

Centroids and distributed loads are considered in Chapter 5, which also includes the analysis of the cable. Following up on the concept of the first moment, we introduce the area moment of inertia in Chapter 6. Here we have purposely eliminated a lengthy treatment of three-dimensional integration procedures. Area and mass moments of inertia are discussed early in the text to ensure that the student is exposed to the concept of inertia before studying dynamics.

All of Chapter 7 is devoted to beam analysis. Friction is then treated in Chapter 8. We have not included fluid friction, which from our experience, can be more adequately covered in subsequent coursework. The concept of virtual work is introduced in Chapter 9. We believe that it is important to include this chapter for two reasons. First, it provides an alternate method for deriving equations of static equilibrium, and second, it serves as a logical introduction to more advanced problems in mechanics.

In the dynamics volume, Chapter 1 reviews vectors and principles essential for the study of dynamics. Chapter 2 is devoted to the kinematics of planar motion. Moving and fixed reference frames are also introduced here. General expressions of velocity and acceleration are derived for a moving coordinate system and then applied to many specific examples. The kinetics of a particle is then considered in Chapter 3, where attention is focused on the behavior of a single particle.

Chapter 4 is devoted entirely to the motion of a particle in a central force field, since the particle dynamics studied earlier in the book can be used in a very effective way to introduce orbital motion. This in particular, we have found, engages the interest of the student, especially with the continued commitment to the space program by our government.

The kinematics and kinetics of rigid body motion are introduced in Chapters 5 and 6. Here again we have restricted the treatment to planar motion. This simplifies the examples while helping to clarify the salient points. All of Chapter 7 is devoted to direct and oblique impact. We feel this emphasis is warranted, since the subject matter merits special attention. All solutions up to this point in the book are based on the newtonian viewpoint, while the lagrangian and eulerian approaches are introduced in Chapter 8. Finally in Chapter 9, we solve a variety of interesting problems. The treatment of these problems has a twofold benefit. First, it shows the student how fundamental concepts acquired in dynamics may be employed to solve complex problems. Second, and perhaps more important, it extends the student's view of the subject to a variety of physical problems such as electrical circuits and mechanical vibrations.

It is assumed that while studying statics the student is at least concurrently enrolled in the second-semester calculus course of a four-year engineering program. Students will thus have had some exposure to limits, continuity, derivatives, definite and indefinite integrals, and be currently learning techniques of integration, improper integrals, and applications of integration. For dynamics, it is assumed that the student has completed at least the first two semesters of calculus including integral calculus.

All of the material in the statics volume is suitable for a four-credit semester course. In the case of a three-credit semester course, the instructor may wish to omit all of Chapter 7 along with special topics such as the equilibrium of space trusses in Chapter 4 (Section 4.5), cables in Chapter 5 (Section 5.8), principal axes in Chapter 6 (Section 6.4), and potential energy and stability of equilibrium in Chapter 9 (Sections 9.4 and 9.5). The material in the dynamics book is designed so that the first seven chapters can be covered in a three-credit semester course in dynamics. The last two chapters are included for those situations where time and student interest warrant the inclusion of more advanced material.

The material in these volumes has been used in many schools, in addition to the University of Maryland. This revised edition was reviewed by Professors T. J. Zilka and John T. Tielking, to whom we are most grateful. We also wish to acknowledge those students, teachers, and colleagues who have used the earlier edition and have made many valuable suggestions for improvement.

D. K. A.
P. F. C.

List of Symbols

Statics

A	Area
$\mathbf{A}$	Vector with magnitude A
A_x, A_y, A_z	Components of **A** in the direction of X, Y, and Z
a	Constant or length
$\mathbf{B}$	Vector with magnitude B
B_x, B_y, B_z	Components of **B** in the direction of X, Y, and Z
b	Constant or length
b	Distributed load
C	Signifies a compression force
$\mathbf{C}$	Vector of magnitude C
c	Constant or length
C_x, C_y, C_z	Components of **C** in the direction of X, Y, and Z
$\mathbf{D}$	Vector of magnitude D
D	Distance or length
D_x, D_y, D_z	Components of **D** in the direction of X, Y, and Z
d	Distance or length
$\bar{d}$	Centroidal distance
$\mathbf{e}$	Unit vector
$\mathbf{e}_A$	Unit vector in the direction of vector **A**
$\mathbf{E}$	Vector of magnitude E
E_x, E_y, E_z	Components of **E** in the direction of X, Y, and Z
$\mathbf{F}$	Force vector of magnitude F
F_x, F_y, F_z	Components of force in the direction of X, Y, and Z
F_{AB}	Force magnitude along truss member AB
f_k	Kinetic friction force
f_m	Static friction force
G	Universal gravitational constant
g	Gravitational acceleration
h	Distance or constant
I_x	Area or mass moment of inertia about X axis
$\bar{I}_{xy}$	Area product or mass product moment of inertia about X-Y axes

$\bar{I}_x$	Centroidal area or mass moment of inertia about X axis
$\bar{I}_{xy}$	Centroidal area or mass moment of inertia about X-Y axes
$\mathbf{i}$	Unit vector along X axis
J	Polar moment of inertia
$\bar{J}$	Centroidal polar moment of inertia
$\mathbf{j}$	Unit vector along Y axis
k	Radius of gyration
k	Constant
$\mathbf{k}$	Unit vector along Z axis
k	Spring modulus
L	Length
ℓ, l	Length
M	Bending moment
$\mathbf{M}$	Moment vector of magnitude M
$\mathbf{M}_o$	Moment vector about point o
m	Mass
N	Normal reaction
p	Pressure
P	Load or force magnitude
p_a	Absolute pressure
p_o	Atmospheric pressure
Q	Shear force or magnitude of force
$\mathbf{R}$	Resultant vector of magnitude R
R_x, R_y, R_z	Components of vector $\mathbf{R}$ in X, Y, and Z directions
r	Radius or distance
$\mathbf{r}$	Position vector
s	Distance or length
T_o	Tension in cable evaluated at origin
$\mathbf{T}$	Tension vector of magnitude T
t	Plate thickness
u	Derivative of y with respect to x
V	Volume
V	Potential energy
W	Work
W	Watts
w	Magnitude of weight
w	Distributed weight or load
$\mathbf{w}$	Weight vector
X, Y, Z	Cartesian coordinate axes
$\bar{x}$, $\bar{y}$, $\bar{z}$	Centroidal distance
α	Angle related to pitch and radius of thread
δ	Signifies virtual displacement of virtual work
ω	Angular velocity
θ	Angle
ϕ	Angle
ϕ_0	Angle of repose
μ	Coefficient of kinetic friction
μ_0	Coefficient of static friction
π	Constant

Dynamics

A	Constant
$\mathbf{A}$	Vector with magnitude A
A_x, A_y, B_z	Components of $\mathbf{A}$ in the direction of X, Y, and Z
a	Semimajor axis
$\mathbf{a}$	Acceleration vector with magnitude a
a_x, a_y, a_z	Components of $\mathbf{a}$ in the direction of X, Y, and Z
a_{av}	Average acceleration
B	Constant
B	Magnetic field intensity
$\mathbf{B}$	Vector with magnitude B
B_x, B_y, B_z	Components of $\mathbf{B}$ in the direction of X, Y, and Z
b	Constant
b	Semiminor axis
C	Constant
C	Capacitance
$\mathbf{C}$	Vector with magnitude C
C_x, C_y, C_z	Components of $\mathbf{C}$ in the direction of X, Y, and Z
c	Damping constant
c	Constant
D	Rayleigh dissipation function
d	Distance or constant
E	Total energy per unit mass
E	Electric potential
$\mathbf{E}$	Electrostatic field
e	Coefficient of restitution
$\mathbf{e}_A$, $\mathbf{e}_B$	Unit vectors in the direction of $\mathbf{A}$ and $\mathbf{B}$
$\mathbf{e}_n$	Normal unit vector
$\mathbf{e}_r$, $\mathbf{e}_\theta$	Cylindrical unit vectors
$\mathbf{e}_\phi$	Spherical unit vector
$\mathbf{e}_b$	Binomial unit vector
$\mathbf{e}_\tau$	Tangential unit vector
F_r	Magnitude of force in radial direction
F_θ	Magnitude of force in θ direction
$\mathbf{F}$	Force vector with magnitude F
$\mathbf{F}_e$	Electrostatic field force vector
$\mathbf{F}_m$	Magnetic field force vector
$\mathbf{F}_{NC}$	Nonconservative force vector
f_{ij}	Internal force on particle i due to particle j
G	Universal constant
g	Acceleration due to gravity
$\mathbf{H}$	Angular momentum vector
H_o	Angular momentum magnitude about point o
h	Height or constant
$\mathcal{J}$	Impulse
I	Mass moment of inertia
I_o	Mass moment of inertia about point O
$\mathbf{i}$	Unit vector in X direction
$\mathbf{j}$	Unit vector in Y direction
K	Constant
k	Spring modulus
k	Constant
k_t	Torsional spring modulus
$\mathbf{k}$	Unit vector in Z direction
L	Length
$\mathcal{L}$	Lagrangian function
l	Length
l	Electrical inductance
M	Mass of earth
M	Total mass

$\mathbf{M}$	Moment vector with magnitude M
M_o	Magnitude of moment about point O
$\mathcal{M}$	Angular impulse
m	Mass
m_i	Mass of ith particle
N	Normal reaction
O	Origin of fixed reference
O'	Origin of moving reference
P	Power
P	Force magnitude
P	Pressure
$\mathbf{p}$	Linear momentum of vector of magnitude p
Q	Generalized force
q	Electrostatic charge
q	Generalized coordinate
R	Earth radius
R	Electrical resistance
$\mathbf{R}$	Position vector
R_x, R_y, R_z	Components of $\mathbf{R}$ in X, Y, and Z directions
$\mathbf{r}$	Position vector
r_a	Apogee radius
r_c	Distance from origin to center of mass
r_p	Perigee radius
r_r	Relative distance
s	Distance
T	Magnitude of tension
T	Kinetic energy
t	Time
u	Magnitude of velocity before or after impact
V	Potential energy
V_A	Potential energy at point A
$\mathbf{v}$	Velocity vector
v_{av}	Average velocity
v_e	Escape velocity
W	Work
W_C	Work of conservative forces
W_{NC}	Work of nonconservative forces
w	Weight
w	Equal to $\frac{1}{r}$ in the orbital equation
x	Amount spring compresses or stretches
X, Y, Z	Cartesian coordinate directions
β	Constant
θ, ϕ, λ, γ	Angle or constant
τ	Period of orbit
ξ	Constant in orbital motion
ζ	Damping ratio
μ	Constant in orbital motion
ε	Eccentricity in orbital motion
δW	Virtual work
δx	Virtual displacement
$\boldsymbol{\rho}$	Position vector
ρ	Radius of curvature
$\boldsymbol{\omega}$	Angular velocity vector of magnitude ω
ω_n	Natural frequency
ω_d	Damped natural frequency
ω_r	Angular velocity of a body relative to another
Ω	Magnitude of angular velocity
$\boldsymbol{\alpha}$	Angular acceleration vector of magnitude α
π	Constant

List of Abbreviations

Statics

cos	cosine
cosh	hyperbolic cosine
da	derivative of variable *a*
ft	foot
J	joule or N·m
kg	kilogram
kip	kilopound
kN	kilonewton
lb	pound
ln	natural logarithm
m	meter
N	newton
Pa	pascal or N/m^2
PW	pitch of wrench
rad	radian
s, sec	second
sin	sine
sinh	hyperbolic sine
t	1000 kg
tan	tangent
tanh	hyperbolic tangent

Dynamics

CCW	counterclockwise
cos	cosine
CW	clockwise
$\frac{d}{dt}$	first derivative with respect to time
$\frac{d^2}{dt^2}$	second derivative with respect to time
ft	foot
J	joule or N·m
kg	kilogram
lb	pound
ln	natural logarithm
m	meter
N	newton
Pa	pascal or N/m^2
rad	radian
s, sec	second
sin	sine
tan	tangent
$(\dot{\ })$	first derivative with respect to time
$(\ddot{\ })$	second derivative with respect to time
Δa	small change in variable *a*

Basic Concepts 1

Mechanics is the science that deals with stationary and moving bodies under the action of forces. Theoretical mechanics is generally the concern of physicists and applied mathematicians; engineering mechanics is primarily of interest to engineers. The study of mechanics may be categorized into a study of the mechanics of fluids, the mechanics of bodies that deform, and the mechanics of rigid bodies. Here, we are concerned primarily with the mechanics of rigid bodies.

The mechanics of rigid bodies can be considered to consist of two parts, statics and dynamics. In general, statics treats the equilibrium of stationary bodies, as well as bodies moving at constant velocity, under the influence of various kinds of forces. In this volume we restrict our treatment of statics to include only the equilibrium of stationary bodies. The study of the motion of bodies and the forces that cause the motion is the subject of a separate volume entitled *Dynamics*.

Mechanics is among the oldest of the physical sciences. Interest in mechanical problems goes back to the time of Aristotle (384–322 B.C.). The lever fulcrum, as well as the theory of buoyancy, was first clearly explained by Archimedes (287–212 B.C.). After that, there were very few developments until, in the fifteenth century A.D., Leonardo da Vinci (1452–1519) continued Archimedes' work on levers and deduced the concept of moments as they apply to the equilibrium of bodies. Next, the laws of equilibrium and the parallelogram law were developed by Stevinus (1548–1620), and the relationship between the moment of a force and its components was established by Varignon (1654–1722). The idea of virtual work, which is an important concept in the formulation of advanced techniques used in mechanics, was established by Descartes

(1596–1650), whereas the concept of virtual displacements was used by Pascal (1623–1662) to establish analytically the direction of the propagation of stresses.

Despite the many important contributions made to mechanics before him, Newton's (1642–1727) three famous laws and his law of universal gravitation were perhaps the most important steps in the progress of mechanics. Newton's work on particles, based on geometry, was extended to rigid-body systems by Euler (1707–1783), whose work was based on calculus. At about this time, much of the work in mechanics was reformulated by Lagrange (1736–1813), whose analytical approach used the concepts of energy. The contributions of D'Alembert, Hamilton, Routh, and others, coupled with the fundamental work described earlier, has helped establish what we know as engineering mechanics, statics and dynamics.

We begin our study of statics by first reviewing some basic concepts. These include a few fundamental ideas pertaining to newtonian mechanics, dimensions and units, and finally a review of vectors.

1.1 Fundamental Concepts

The concepts fundamental to a study of mechanics are those of space, time, inertia, and force. The following is a brief description of these and other important concepts:

Space. By space we mean a geometric region in which physical events occur. It can have one, two, or three dimensions. Indeed, more than three dimensions can be conceptualized. Here we shall be concerned with, at most, three-dimensional space. Position in space is established relative to some reference system. The basic reference system necessary for newtonian mechanics is one that is considered fixed in space.* Measurements relative to this system are called absolute.

Time. Essentially a measure of the orderly succession of events occurring in space, time is considered as an absolute quantity. The unit of time is a second, which was originally related directly to the earth's rate of spin. Today, the standard of time is established by the frequency of oscillation of a cesium atom.

* In dynamics we show that a nonaccelerating reference frame is also an acceptable reference frame for newtonian mechanics.

Inertia. The ability of a body to resist a change in motion is called inertia.

Mass. The mass of a body is a quantitative measure of its inertia.

Force. A force is the action of one body on another. This action may exist because of contact between the bodies, which is called the push-pull effect; or it may exist with the bodies apart, which is called the field-of-force effect.

Particle. If the dimensions of a body are treated as negligible, the body is said to be a particle. The mass of a particle is assumed to be concentrated at a point, and a particle is sometimes called a mass point.

Rigid Body. A rigid body is characterized by the condition that any two points of the body remain at a fixed distance relative to each other for all time.

The study of mechanics derives from several fundamental laws, the most important of which were formulated by Sir Isaac Newton in 1687.

Newton's Laws

I. Every particle remains at rest or continues to move in a straight line with a uniform motion if there is no unbalanced force acting upon it.

II. The time rate of change of the linear momentum of a particle is proportional to the unbalanced force acting upon it and occurs in the direction in which the force acts. It can be shown that when the mass is constant the time rate of change of the linear momentum is equivalent to the product of the mass and its acceleration. Hence, we often use the more familiar relationship that says force equals mass times acceleration.

III. To every action there is an equal and opposite reaction. The mutual forces of two bodies acting upon each other are equal in magnitude, collinear, and opposite in direction.

In addition to these three basic laws, Newton also formulated the **law of gravitation**, which governs the mutual attraction between two isolated bodies. This law is expressed mathematically as

$$F = G\frac{m_1 m_2}{r^2}$$

where F = magnitude of the mutual force of attraction between the two bodies
G = universal gravitational constant
m_1, m_2 = masses of the bodies
r = distance between the centers of mass of the bodies

The mutual forces between the two bodies obey Newton's third law so that they are equal in magnitude, opposite in direction, and lie along the line joining the centers of the two bodies.

These laws have been verified experimentally over the years. It is important to realize that the first two laws hold for measurements made with respect to a nonaccelerating reference frame. The problems of statics are primarily concerned with the first and third laws, while the second law is important in the study of dynamics. The law of gravitation is of fundamental importance in studying the motion of planets or artificial satellites and will also be treated in dynamics.

1.2 Dimensions and Units

When we measure the length of a string, we may choose to record its length in centimeters, inches, or feet. The concept of length is called a **dimension**, while the terms centimeters, inches, and feet are called **units**. Similarly, the concept of time is a dimension and the units with which we record it may be seconds, minutes, hours, etc.

Three dimensions are essential to newtonian mechanics. It has been agreed that length and time are fundamental to all systems and that mass or force may be selected as convenient. In this text we shall use both the International System (SI) of units as well as the English gravitational system. In the SI system of units the mass is in kilograms (kg), the length is in meters (m), and the time is in seconds (s). To obtain the units for force in the SI system, consider Newton's second law of motion for a particle in mathematical form, i.e.,

$$\text{Force} = \text{Mass} \times \text{Acceleration}$$

In the SI system we assign the newton (N) unit for the force, in which one newton causes a mass of one kilogram to accelerate at one meter per second per second (m/s^2). Table 1.1 lists several quantities in the SI units that we shall be using in statics.

An important comment must be made concerning the units of the weight of a body. Note that the weight of a body is a force and should be expressed in newtons, which is consistent with Newton's

TABLE 1.1 ***International Standard of Units***

Quantity	*SI units*	*Abbreviation*
Length	meters	m
Time	second	s
Mass	kilogram	kg
Force	newton	N
Weight	newton	N
Density	kilogram/meter^3	kg/m^3
Metric ton	1000 kilograms	t
Work	newton-meter = joule	N·m = J
Energy	newton-meter = joule	N·m = J
Power	newton-meter/second = watt	N·m/s = W
Pressure	newton/meter^2 = pascal	N/m^2 = Pa
Moment	meter-newton	m·N
Area moment of inertia	meter^4	m^4
Mass moment of inertia	kilogram-meter^2	kg·m^2
Plane angle	radian	rad

law of gravitation. However, it should be realized that considerable confusion exists in the use of the metric system in which weight and mass are both expressed in kilograms. The correct relationship between the weight and mass is

$$\text{Weight} = \text{Mass} \times g \tag{1.1}$$

where $g = 9.81\ \text{m/s}^2$ is the acceleration of gravity on the earth's surface. For example, a mass of 10 kg weighs 98.1 N. To underscore this concept the term gravity force is sometimes used to refer to the weight of the body.

In the gravitational system the fundamental dimensions are force, length, and time. The unit of force is the pound (lb), the length is the foot (ft), and the time is the second (sec). To obtain the unit of mass, we again use Newton's second law of motion,

$$\text{Force} = \text{Mass} \times \text{Acceleration}$$

If the force is in lb and acceleration in ft/sec^2, then the derived units of mass from Newton's second law become lb·sec^2/ft. This unit for mass is called a slug. The units and dimensions of several quantities in the gravitational system are given in Table 1.2. Table 1.3 gives some useful conversion factors in converting from one system to the other.

TABLE 1.2 ***English Gravitational Units***

Quantity	*Gravitational units*	*Abbreviation*
Length	foot	ft
Time	second	sec
Force	pound	lb
Mass	pound-second^2/foot = slug	lb·sec^2/ft
Weight	pound	lb
Density	pound-second^2/foot^4 = slug/foot^3	(lb·sec^2)/ft^4
Work	pound-foot	lb·ft
Energy	pound-foot	lb·ft
Power	pound-foot/second	lb·ft/sec
Pressure	pound/foot^2	lb/ft^2
Moment	foot-pound	ft·lb
Area moment of inertia	foot^4	ft^4
Mass moment of inertia	pound-second^2-foot = slug-foot^2	lb·sec^2·ft
Plane angle	radian	rad

TABLE 1.3 ***Conversion from the English Gravitational Units to the International Standard of Units***

To convert	*to*	*Multiply by*
Foot	Meter	0.3048
Pound-foot	Joule	1.356
Mile	Meter	1,609
Pound	Newton	4.448
Slug	Kilogram	14.59
Kip (1000 lb)	Newton	4,448

1.3 Vectors

Historically, much of the early work in mechanics relied on geometrical concepts. Later, many of the laws and equations were derived using calculus. In a modern study of statics it is desirable to employ a notation that is concise, simple, and complete. Such a notation is to be found in vectors and vector calculus. Also, the use of vector notation allows us to express a physical law or mathematical operation independently of any coordinate system.

A **vector** is a variable that has two properties, magnitude and direction. These properties are independent of any particular coordinate system. If a variable has magnitude but no direction, such as temperature or density, it is called a **scalar**. A vector shall be denoted by **V**, **A**, etc., having magnitude V, A, etc. Some sample vectors are shown in Fig. 1.1. Here the vector **V** has a magnitude of 3 while the vector **A** has a magnitude of 5. The velocity of an automobile is a physical example of a vector in that we must know its speed (magnitude) as well as its direction of travel. Another physical example of a vector is the force applied to the block shown in Fig. 1.2. Here the direction of the force is given by the angle θ measured from the horizontal plane.

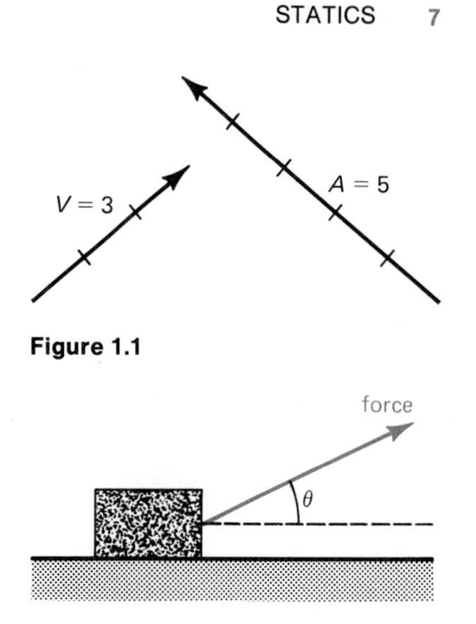

Figure 1.1

Figure 1.2

A vector **A** can be multiplied by a scalar without affecting its direction. If the vector **A** shown in Fig. 1.3(a) is multiplied by 2, we obtain the vector **B** shown in Fig. 1.3(b) whose magnitude is twice that of vector **A** and whose direction is the same as **A**. Thus we express a vector as consisting of a magnitude and a direction. This is most conveniently done by introducing a **unit vector** whose magnitude is unity and whose direction lies along the direction of the represented vector.

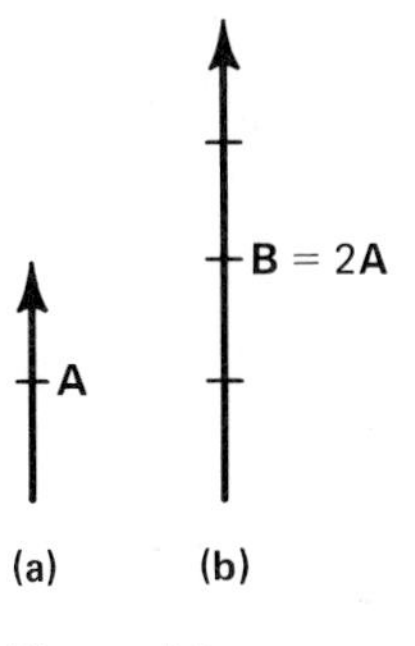

Figure 1.3

$$\mathbf{A} = A\mathbf{e}_A \tag{1.2}$$

is a vector of magnitude A in the direction of the unit vector $\mathbf{e}_A$. Dividing both sides of Eq. (1.2) by the magnitude A, the unit vector becomes

$$\mathbf{e}_A = \frac{\mathbf{A}}{A}$$

The symbol **e** will be used for unit vectors along with an appropriate subscript.

Since the cartesian coordinate system X, Y, Z is used so often, the unit vectors associated with this coordinate system are given the special symbols **i**, **j**, **k** as shown in Fig. 1.4. A vector having a magnitude of 5 in the X direction is written as 5**i**. A vector having a magnitude of 6 in the Y direction is written as 6**j**.

Two vectors are defined to be equal if they have the same magnitude and the same direction.

The sum of two vectors is defined by the **parallelogram rule** which states that two vectors can be replaced by a resultant vector which is the diagonal of a parallelogram that has sides equal to the two given vectors. This is shown in Fig. 1.5 where vector **A** and vector **B** are drawn from a common point. A dashed line is then constructed from the tip of vector **B** parallel with vector **A**. Likewise,

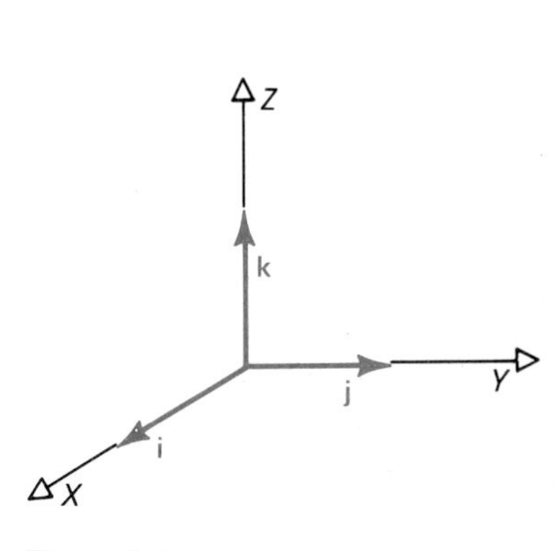

Figure 1.4

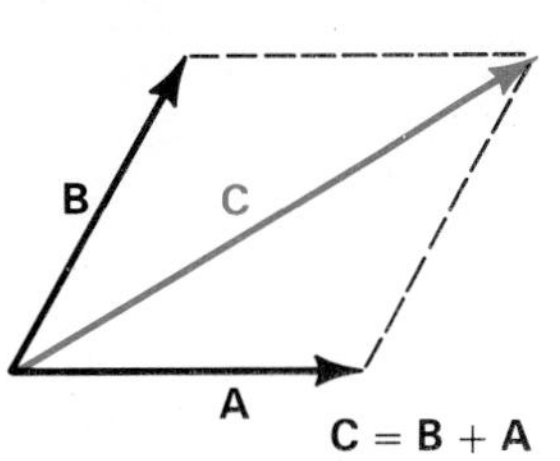

Figure 1.5

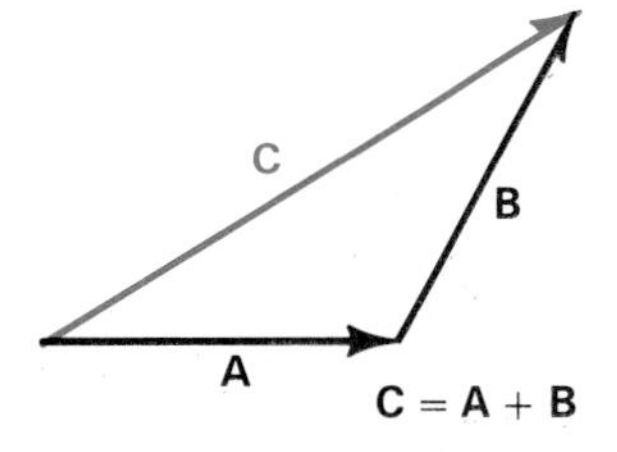

Figure 1.6

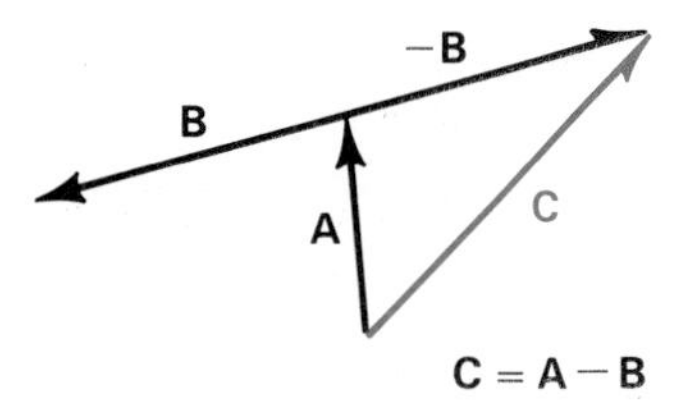

Figure 1.7

a second dashed line is drawn from the tip of vector **A** parallel with vector **B**. The vector **C** is drawn from the common point of **A** and **B** to the point of intersection of the two dashed lines. Vector **C** is defined as the summation of vectors **A** and **B**.

The **triangle rule** is another graphical method for adding vectors. This rule states that two vectors can be replaced by a resultant vector which is the third side of a triangle formed from the two given vectors. Figure 1.6 illustrates the use of the triangle rule. Here the sum of vectors **A** and **B** is obtained by first constructing vector **A**. Then the tail of vector **B** is placed at the tip of vector **A**. Finally, the resultant vector **C** is obtained by constructing a line from the tail of vector **A** to the tip of vector **B**.

The difference between two vectors can also be obtained by using the triangle rule, as shown in Fig. 1.7. Here we wish to subtract vector **B** from vector **A**. Vector **A** and vector **B** are drawn as if we were to add them. The negative of **B** is drawn in the opposite direction of **B**. Using the triangle rule, we obtain the vector $\mathbf{C} = \mathbf{A} - \mathbf{B}$.

In adding or subtracting vectors the following rules should be remembered,

a. Vector addition is commutative since the result is independent of the order of vector addition.
b. Vector addition and subtraction are associative since vectors may be added or subtracted in any order.
c. If k is a scalar, then

$$k(\mathbf{A} + \mathbf{B}) = k\mathbf{A} + k\mathbf{B}$$

and, therefore, multiplication by a scalar is distributive.

A vector can be broken down into several components. The particular components selected for solution of a problem are a matter of convenience. Consider the vector **V** shown in Fig. 1.8(a). When we

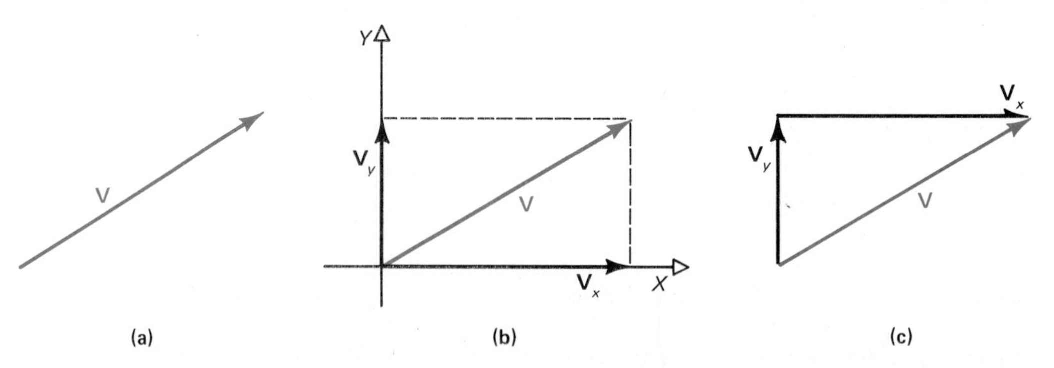

Figure 1.8

construct a cartesian coordinate system whose origin coincides with the tail of this vector, the projections of **V** on the X and Y axes are shown in Fig. 1.8(b). Conversely, if $\mathbf{V}_x$ and $\mathbf{V}_y$ are two component vectors, they can be added to yield **V** as shown in Fig. 1.8(c). Here we construct $\mathbf{V}_y$ and then $\mathbf{V}_x$ to obtain **V** by the triangle rule.

To add or subtract vectors algebraically, it is convenient to use the unit vector representation with a suitable coordinate system. For example, if

$$\mathbf{A} = A_x\mathbf{i} + A_y\mathbf{j} \qquad \text{and} \qquad \mathbf{B} = B_x\mathbf{i} + B_y\mathbf{j}$$

then

$$\mathbf{A} + \mathbf{B} = (A_x + B_x)\mathbf{i} + (A_y + B_y)\mathbf{j}$$

and

$$\mathbf{A} - \mathbf{B} = (A_x - B_x)\mathbf{i} + (A_y - B_y)\mathbf{j}$$

Example 1.1

Vector $\mathbf{A} = 2\mathbf{i} + 3\mathbf{j}$, and $\mathbf{B} = 3\mathbf{i} + \mathbf{j}$. Obtain the sum and difference of these two vectors (a) algebraically, and (b) using the parallelogram rule.

(a)

$$\mathbf{A} + \mathbf{B} = (2\mathbf{i} + 3\mathbf{j}) + (3\mathbf{i} + \mathbf{j})$$
$$= 5\mathbf{i} + 4\mathbf{j} \qquad \textit{Answer}$$

$$\mathbf{A} - \mathbf{B} = (2\mathbf{i} + 3\mathbf{j}) - (3\mathbf{i} + \mathbf{j})$$
$$= -\mathbf{i} + 2\mathbf{j} \qquad \textit{Answer}$$

(b) The graphical solution using the parallelogram rule is shown in Fig. 1.9. The addition of the two vectors is represented by **R** as shown in Fig. 1.9(a) while the difference is shown in Fig. 1.9(b). Note once again that the $-\mathbf{B}$ vector is drawn in the opposite direction of **B**.

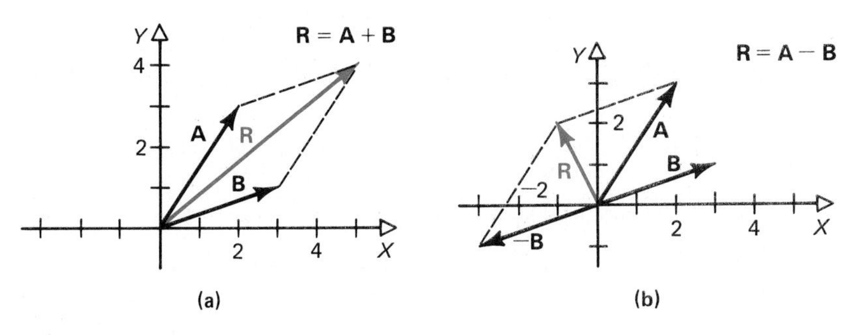

Figure 1.9

Example 1.2

Vector $\mathbf{A} = 3\mathbf{i} + 2\mathbf{j}$, and $\mathbf{B} = \mathbf{i} - 4\mathbf{j}$. Obtain the sum and difference of these two vectors (a) algebraically and (b) using the triangle rule.

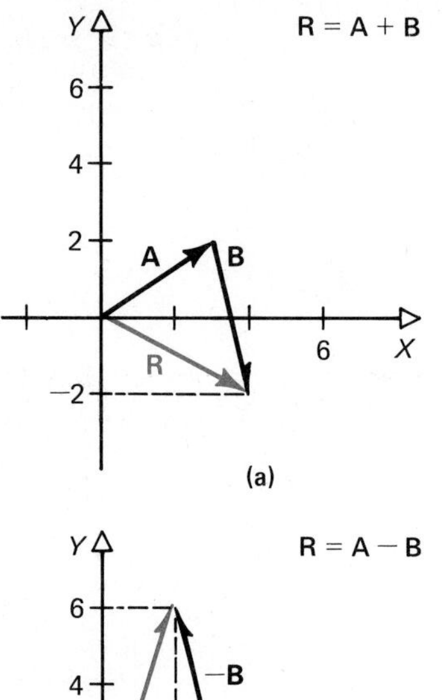

Figure 1.10

(a)
$$\mathbf{A} + \mathbf{B} = (3\mathbf{i} + 2\mathbf{j}) + (\mathbf{i} - 4\mathbf{j})$$
$$= 4\mathbf{i} - 2\mathbf{j} \qquad \textit{Answer}$$
$$\mathbf{A} - \mathbf{B} = (3\mathbf{i} + 2\mathbf{j}) - (\mathbf{i} - 4\mathbf{j})$$
$$= 2\mathbf{i} + 6\mathbf{j} \qquad \textit{Answer}$$

(b) The graphical solution is shown in Fig. 1.10. In Fig. 1.10(a) the resultant **R** gives the sum of the two vectors, whereas the difference is shown in Fig. 1.10(b).

Example 1.3

A vector **A** is given by $\mathbf{A} = 4\mathbf{i} + 3\mathbf{j}$. Express this vector in terms of the magnitude A and the unit vector $\mathbf{e}_A$. Also show the relationship between $\mathbf{e}_A$ and the unit vectors **i** and **j**.

The vector **A** can be expressed as $\mathbf{A} = A\mathbf{e}_A$ where its magnitude is

$$A = \sqrt{(4)^2 + (3)^2} = 5$$

Since
$$\mathbf{A} = 4\mathbf{i} + 3\mathbf{j} = 5\mathbf{e}_A \qquad \textit{Answer}$$

Solving for $\mathbf{e}_A$ in terms of **i** and **j**,

$$\mathbf{e}_A = \frac{4\mathbf{i} + 3\mathbf{j}}{5} = 0.8\mathbf{i} + 0.6\mathbf{j} \qquad \textit{Answer}$$

PROBLEMS

1.1 Vector **A** in the XY plane has a magnitude of 8 and makes a 60° angle counterclockwise with the positive X axis. Express this vector in terms of its X and Y components.

1.2 If $\mathbf{A} = 5\mathbf{i} + 8\mathbf{j}$ and $\mathbf{B} = -4\mathbf{i} + 6\mathbf{j}$, find $\mathbf{A} + \mathbf{B}$ and $\mathbf{A} - \mathbf{B}$ by the parallelogram rule.

1.3 Obtain $\mathbf{A} + \mathbf{B}$ and $\mathbf{A} - \mathbf{B}$ for the vectors $\mathbf{A} = -2\mathbf{i} + 3\mathbf{j}$ and $\mathbf{B} = 5\mathbf{i} - 4\mathbf{j}$ using the parallelogram rule.

1.4 Solve Problem 1.3 by the triangle rule.

1.5 If $\mathbf{A} = 6\mathbf{i} - 6\mathbf{j}$ and $\mathbf{B} = -4\mathbf{i} + 2\mathbf{j}$, find $\mathbf{A} + \mathbf{B}$ and $\mathbf{A} - \mathbf{B}$ by the triangle rule.

1.6 Solve Problem 1.5 by the parallelogram rule.

1.7 Obtain $2\mathbf{A} - 4\mathbf{B}$ for the vectors $\mathbf{A} = 4\mathbf{i}$ and $\mathbf{B} = 2\mathbf{i} - 2\mathbf{j}$ using the triangle rule.

1.8 Solve Problem 1.7 by the parallelogram rule.

1.9 Consider vector $\mathbf{A} = 7\mathbf{i} - 4\mathbf{j}$. Express this vector in terms of the unit vector $\mathbf{e}_A$ and its magnitude A. Show the relationship between $\mathbf{e}_A$, **i** and **j**.

1.10 If $\mathbf{A} + 2\mathbf{B} = 6\mathbf{i} - 2\mathbf{j}$, find $\mathbf{B}$ if $\mathbf{A} = 4\mathbf{i}$.

1.11 An airplane is flying at a velocity $\mathbf{V}$ as shown in Fig. P1.11. Express this vector in terms of the unit vectors $\mathbf{e}_1$ and $\mathbf{e}_2$.

1.12 Obtain the unit vector that gives the direction of the tip of $\mathbf{B}$ relative to $\mathbf{A}$, if $\mathbf{A} = 3\mathbf{i} + \mathbf{j}$ and $\mathbf{B} = \mathbf{i} - 2\mathbf{j}$. (*Hint*: Form $\mathbf{B} - \mathbf{A}$)

1.13 Two cars in radio contact are shown in Fig. P1.13. Their position is measured from a fixed point O and is given by

$$\mathbf{r}_B = 10(\cos \pi t\mathbf{i} + \sin \pi t\mathbf{j})$$

$$\mathbf{r}_A = 10 \cos 3\pi t\mathbf{i} + 10(2 + \sin 3\pi t)\mathbf{j}$$

What is the location of car A relative to car B at $t = 1$ s?

1.14 The velocity of the aircraft shown in Fig. P1.14 consists of two parts, $\mathbf{V}_r$ and $\mathbf{V}_\theta$. Express the resultant velocity $\mathbf{V}$ in cartesian components.

1.15 A belt drive used to transport luggage is shown in Fig. P1.15. Express the position of the point A as a vector in the XY coordinate system at the instant shown.

1.16 Obtain the sum of the vectors **A**, **B**, and **C** shown in Fig. P1.16 and express the answer in terms of its X and Y components.

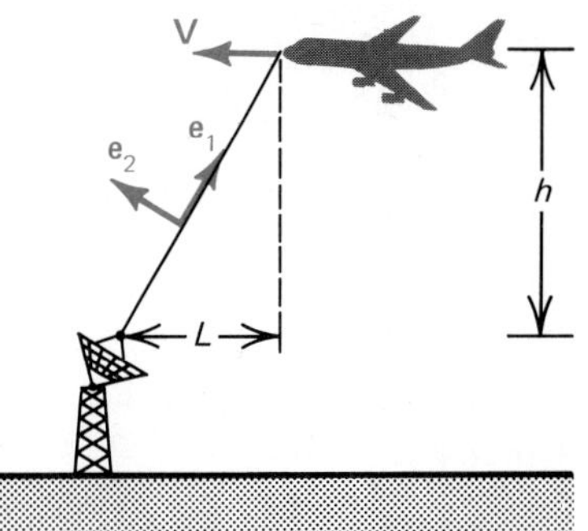

Figure P1.11

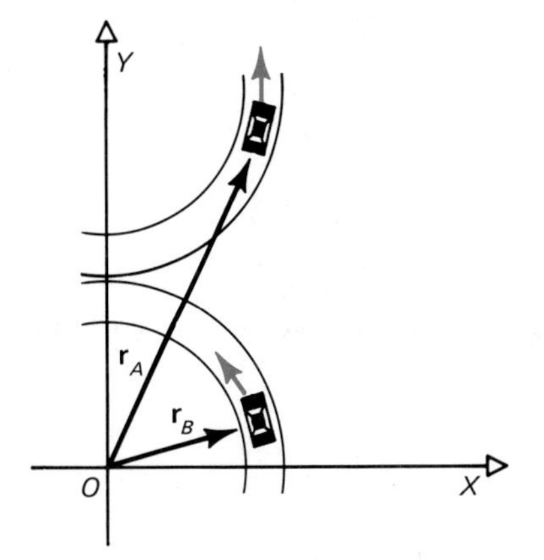

Figure P1.13

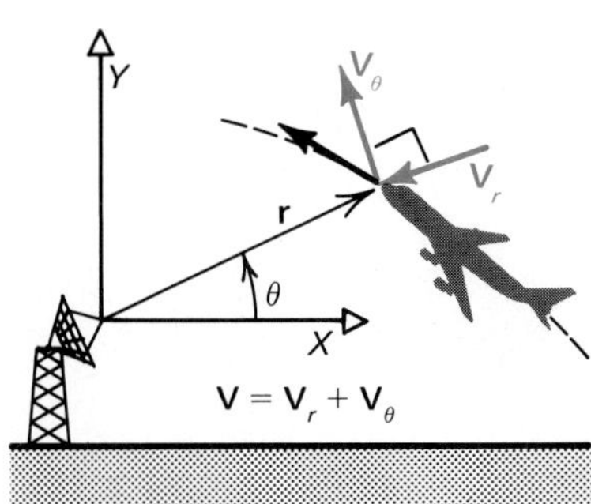

Figure P1.14

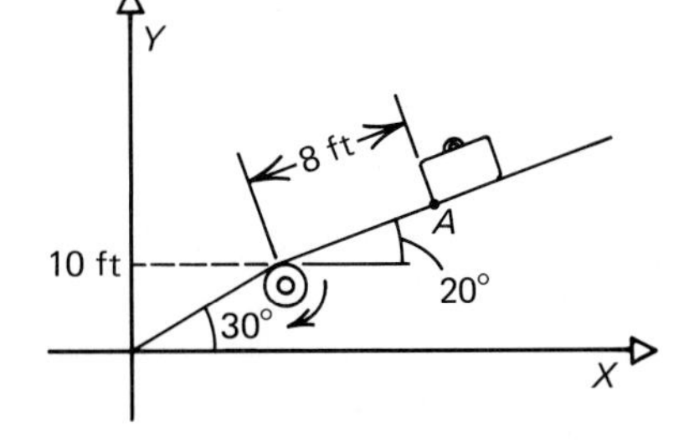

Figure P1.15

Figure P1.16

1.4 Scalar Product of Vectors

There are two types of multiplication involving vector quantities: a scalar or dot product and a vector or cross product. The scalar product of **A** and **B** is written as $\mathbf{A} \cdot \mathbf{B}$, and is defined as

$$\mathbf{A} \cdot \mathbf{B} = AB \cos \theta \qquad (1.3)$$

where θ is the angle between the vectors **A** and **B**. Figure 1.11 shows

Figure 1.11

A and **B** lying in a plane. Note that θ can vary between zero and 360°. If

$$\mathbf{A} \cdot \mathbf{B} = 0$$

and $A \neq 0$, $B \neq 0$, then **A** and **B** are orthogonal, i.e., the vectors are perpendicular to each other. If

$$\mathbf{A} \cdot \mathbf{B} = AB$$

then **A** and **B** are collinear and lie in the same direction. Should

$$\mathbf{A} \cdot \mathbf{B} = -AB$$

then **A** and **B** are collinear but lie in opposite directions ($\theta = 180°$).

The scalar product can be used for obtaining the magnitude of the projection of a vector in a given direction. The magnitude of the projection of the vector **A** on the X axis shown in Fig. 1.12 can be obtained by forming the scalar product of **A** and **i**,

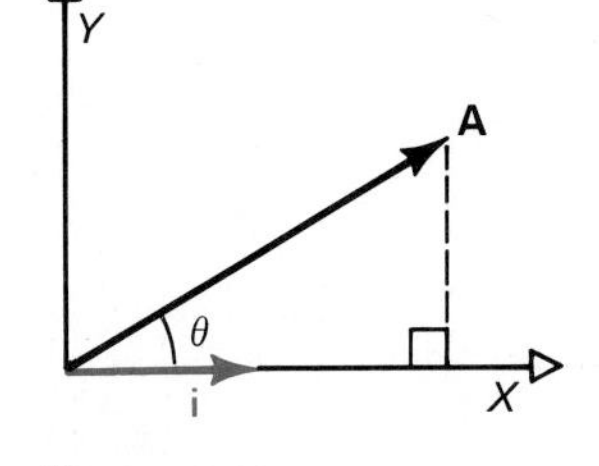

Figure 1.12

$$\mathbf{A} \cdot \mathbf{i} = A \cos \theta$$

Similarly, the magnitude of the projection of vector **A** along the same direction as vector **B** in Fig. 1.13 can be obtained by forming the scalar product of **A** and the unit vector $\mathbf{e}_B$,

$$\mathbf{A} \cdot \mathbf{e}_B = A \cos \beta$$

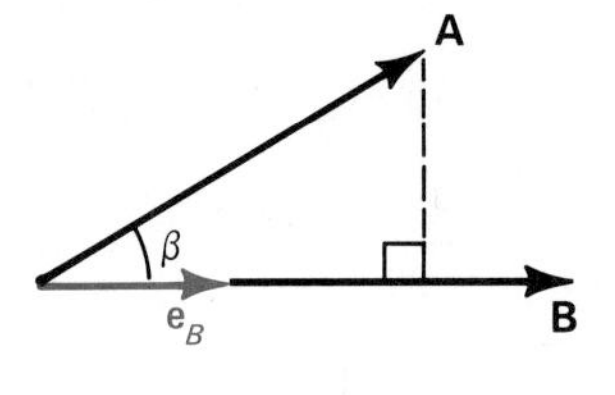

Figure 1.13

where the angle β is measured between **A** and **B**. But the unit vector $\mathbf{e}_B$ is

$$\mathbf{e}_B = \frac{\mathbf{B}}{B}$$

so that the magnitude of the projection of vector **A** on vector **B** may be expressed as

$$\frac{\mathbf{A} \cdot \mathbf{B}}{B}$$

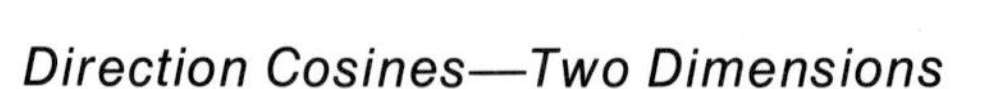

Direction Cosines—Two Dimensions

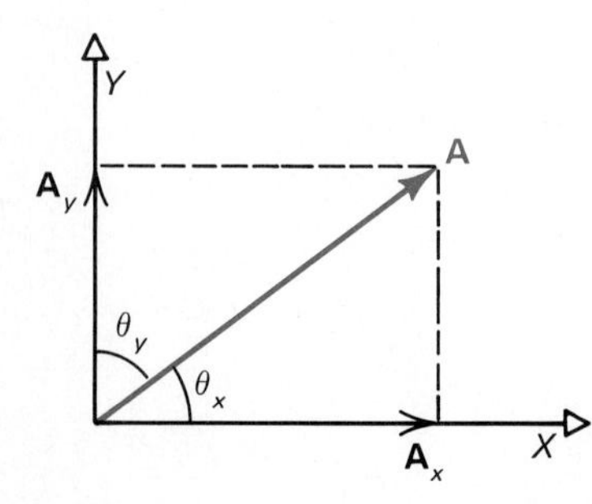

Figure 1.14

The scalar product is also used to define the direction cosines of a vector. Consider the vector **A** lying in the XY plane as shown in Fig. 1.14.

$$\mathbf{A} = A_x\mathbf{i} + A_y\mathbf{j} \tag{1.4}$$

where **i** and **j** are unit vectors along the X and Y axes, respectively, and A_x and A_y are the components along the X and Y axes. Forming the scalar product of both sides of Eq. (1.4) with **i** yields

$$\mathbf{A}\cdot\mathbf{i} = A_x\mathbf{i}\cdot\mathbf{i} + A_y\mathbf{j}\cdot\mathbf{i}$$
$$= A_x$$

since $\mathbf{i}\cdot\mathbf{i} = 1$ and $\mathbf{j}\cdot\mathbf{i} = 0$. Therefore, A_x is the scalar product of **A** and **i**. But from the definition of a scalar product,

$$\mathbf{A}\cdot\mathbf{i} = A_x = A\cos\theta_x \tag{1.5a}$$

where θ_x is the angle between **A** and **i**. Likewise,

$$\mathbf{A}\cdot\mathbf{j} = A_y = A\cos\theta_y \tag{1.5b}$$

Substituting Eqs. (1.5) into Eq. (1.4), we obtain

$$\mathbf{A} = A\cos\theta_x\mathbf{i} + A\cos\theta_y\mathbf{j} \tag{1.6}$$

Recall that we established that a vector may be expressed by its magnitude and unit vector, so that

$$\mathbf{A} = A\mathbf{e}_A$$

Substituting this into the left side of Eq. (1.6),

$$A\mathbf{e}_A = A\cos\theta_x\mathbf{i} + A\cos\theta_y\mathbf{j}$$

or

$$\mathbf{e}_A = \cos\theta_x\mathbf{i} + \cos\theta_y\mathbf{j} \tag{1.7}$$

The unit vector $\mathbf{e}_A$ gives the direction of the vector **A** in terms of the angles θ_x and θ_y as well as the unit vectors **i** and **j**. The two cosines in Eq. (1.7) are called the direction cosines, which are summarized from Eqs. (1.5) as

$$\cos\theta_x = \frac{A_x}{A} \tag{1.8a}$$

$$\cos\theta_y = \frac{A_y}{A} \tag{1.8b}$$

The magnitude A can be obtained from the scalar product of **A** with itself.

$$\mathbf{A}\cdot\mathbf{A} = A^2 = A_x^2 + A_y^2$$

and the magnitude A becomes

$$A = \sqrt{A_x^2 + A_y^2} \tag{1.9}$$

Direction Cosines—Three Dimensions

We can readily extend the definition of the direction cosines for a vector represented in three dimensions. Consider the vector **A** as shown in Fig. 1.15. Using the scalar products $\mathbf{i}\cdot\mathbf{j} = \mathbf{j}\cdot\mathbf{k} = \mathbf{k}\cdot\mathbf{i} = 0$, and following similar steps as taken in the two-dimensional

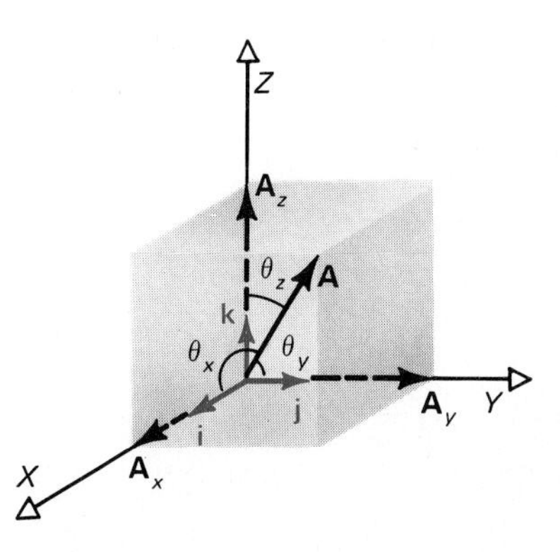

Figure 1.15

case, we can establish the following relationships for the three-dimensional case:

$$\mathbf{e}_A = \cos\theta_x\mathbf{i} + \cos\theta_y\mathbf{j} + \cos\theta_z\mathbf{k} \tag{1.10}$$

$$\cos\theta_x = \frac{A_x}{A} \tag{1.11a}$$

$$\cos\theta_y = \frac{A_y}{A} \tag{1.11b}$$

$$\cos\theta_z = \frac{A_z}{A} \tag{1.11c}$$

where θ_x, θ_y, and θ_z are the angles between **A** and the X, Y, and and Z axes, respectively. The magnitude A is given by

$$A = \sqrt{A_x^2 + A_y^2 + A_z^2} \tag{1.12}$$

The scalar product of two vectors **A** and **B** was written as

$$\mathbf{A} \cdot \mathbf{B} = AB\cos\theta \tag{1.3}$$

However, if the vectors are written in cartesian coordinates,

$$\mathbf{A} = A_x\mathbf{i} + A_y\mathbf{j} + A_z\mathbf{k} \qquad \text{and} \qquad \mathbf{B} = B_x\mathbf{i} + B_y\mathbf{j} + B_z\mathbf{k}$$

then the scalar product becomes

$$\begin{aligned}\mathbf{A} \cdot \mathbf{B} &= (A_x\mathbf{i} + A_y\mathbf{j} + A_z\mathbf{k}) \cdot (B_x\mathbf{i} + B_y\mathbf{j} + B_z\mathbf{k}) \\ &= A_xB_x + A_yB_y + A_zB_z\end{aligned}$$

The angle between the vectors **A** and **B** can be obtained by equating this to Eq. (1.3) and solving for $\cos\theta$

$$\cos\theta = \frac{A_xB_x + A_yB_y + A_zB_z}{AB}$$

where A is the magnitude of **A** and B is the magnitude of **B**.

Example 1.4

A vector **A** is given by $\mathbf{A} = 5\mathbf{i} + \mathbf{j} + 2\mathbf{k}$. (a) What is the magnitude of **A**? (b) What are the direction cosines of **A**? (c) What is the unit vector $\mathbf{e}_A$?

(a) From Eq. (1.12) we find that

$$A = \sqrt{(5)^2 + (1)^2 + (2)^2} = 5.48 \qquad \textit{Answer}$$

(b) Using Eqs. (1.11), the direction cosines are

$$\cos\theta_x = \frac{A_x}{A} = \frac{5}{5.48} = 0.912 \qquad \textit{Answer}$$

$$\cos\theta_y = \frac{A_y}{A} = \frac{1}{5.48} = 0.183 \qquad \textit{Answer}$$

$$\cos\theta_z = \frac{A_z}{A} = \frac{2}{5.48} = 0.365 \qquad \textit{Answer}$$

(c) The unit vector $\mathbf{e}_A$ is obtained by substituting the direction cosines into Eq. (1.10).

$$\begin{aligned}\mathbf{e}_A &= \cos\theta_x\mathbf{i} + \cos\theta_y\mathbf{j} + \cos\theta_z\mathbf{k}\\ &= 0.912\mathbf{i} + 0.183\mathbf{j} + 0.365\mathbf{k} \qquad \textit{Answer}\end{aligned}$$

A more direct way to obtain $\mathbf{e}_A$ is as follows:

$$\mathbf{A} = 5.48\mathbf{e}_A = 5\mathbf{i} + \mathbf{j} + 2\mathbf{k}$$

Therefore

$$\mathbf{e}_A = \frac{5\mathbf{i} + \mathbf{j} + 2\mathbf{k}}{5.48} = 0.912\mathbf{i} + 0.183\mathbf{j} + 0.365\mathbf{k} \qquad \textit{Answer}$$

Example 1.5

Obtain the magnitude of the projection of **A** on **B**, if

$$\mathbf{A} = 2\mathbf{i} + 3\mathbf{j} \qquad \text{and} \qquad \mathbf{B} = 5\mathbf{i} + 2\mathbf{j} + \mathbf{k}$$

The magnitude of the projection of **A** on **B** is $A\cos\theta$. The scalar product of **A** and **B** is

$$\mathbf{A}\cdot\mathbf{B} = AB\cos\theta$$

Therefore,

$$\begin{aligned}A\cos\theta &= \frac{\mathbf{A}\cdot\mathbf{B}}{B} = \frac{(2\mathbf{i} + 3\mathbf{j})\cdot(5\mathbf{i} + 2\mathbf{j} + \mathbf{k})}{\sqrt{(5)^2 + (2)^2 + (1)^2}}\\ &= \frac{16}{\sqrt{30}} = 2.92 \qquad \textit{Answer}\end{aligned}$$

Example 1.6

Given $\mathbf{C} = 2\mathbf{i} - 4\mathbf{j}$ and $\mathbf{D} = 3\mathbf{i} + 4\mathbf{j} - 5\mathbf{k}$, find the angle between these vectors.

Rewrite Eq. (1.3) as

$$\mathbf{C}\cdot\mathbf{D} = CD\cos\theta$$

or

$$\cos\theta = \frac{\mathbf{C}\cdot\mathbf{D}}{CD}$$

Now

$$C = \sqrt{(2)^2 + (-4)^2} = 4.47$$

$$D = \sqrt{(3)^2 + (4)^2 + (-5)^2} = 7.07$$

Therefore,

$$\cos\theta = \frac{(2\mathbf{i} - 4\mathbf{j})\cdot(3\mathbf{i} + 4\mathbf{j} - 5\mathbf{k})}{(4.47)(7.07)}$$

$$= \frac{6 - 16}{31.6} = -0.316$$

so that

$$\theta = \text{arc}\cos(-0.316) = 108^\circ$$ *Answer*

PROBLEMS

1.17 Show that $\mathbf{A}\cdot\mathbf{B} = \mathbf{B}\cdot\mathbf{A}$ assuming $\mathbf{A} = A_x\mathbf{i} + A_y\mathbf{j}$ and $\mathbf{B} = B_x\mathbf{i} + B_y\mathbf{j}$.

1.18 A vector is given by $\mathbf{C} = 5.2\mathbf{i} - 2.6\mathbf{j}$. What are its direction cosines?

1.19 A vector is given by $\mathbf{B} = -4.1\mathbf{i} + 2\mathbf{k}$. What are its direction cosines?

1.20 If $\mathbf{C} = 10\mathbf{i} + 4\mathbf{j}$ and $\mathbf{D} = 4\mathbf{i} - 10\mathbf{j}$, find $\mathbf{C}\cdot\mathbf{D}$.

1.21 If $\mathbf{V}_1 = 2\mathbf{i} + 3\mathbf{k}$ and $\mathbf{V}_2 = -4\mathbf{k}$, find $\mathbf{V}_1\cdot\mathbf{V}_2$.

1.22 A missile is to be fired with velocity $\mathbf{V}_1$. However, it misfires so that its velocity is $\mathbf{V}_2$. Obtain the magnitude and direction of the velocity difference $\mathbf{V}_2 - \mathbf{V}_1$. Let $\mathbf{V}_1 = 5\mathbf{i} + 3\mathbf{j}$ and $\mathbf{V}_2 = 4.9\mathbf{i} + 2.8\mathbf{j} + 0.5\mathbf{k}$.

1.23 Obtain a unit vector that gives the direction of the tip of **B** relative to the tip of **A**, if $\mathbf{A} = 3\mathbf{i} + \mathbf{j}$ and $\mathbf{B} = \mathbf{i} + 2\mathbf{j} + 3\mathbf{k}$.

1.24 If $\mathbf{E} = 6\mathbf{i} + 3\mathbf{j} - 4\mathbf{k}$, find its direction cosines.

1.25 If $\mathbf{V} = 8.3\mathbf{i} - 2.5\mathbf{j} + 5\mathbf{k}$, find its direction cosines.

1.26 What is the magnitude of the projection of vector $\mathbf{V}_1 = 5\mathbf{i} + 3.3\mathbf{k}$ on the vector $\mathbf{V}_2 = 3.3\mathbf{i} + 5\mathbf{k}$?

1.27 What is the magnitude of the projection of $\mathbf{B} = -4\mathbf{i} + 2\mathbf{j}$ on the vector $\mathbf{A} = 6\mathbf{i} + 4\mathbf{j}$?

1.28 Obtain the angle between the vector $\mathbf{C} = 2\mathbf{i} - 4\mathbf{j}$ and $\mathbf{D} = 3\mathbf{i} + \mathbf{j}$.

1.29 Obtain the angle between the vector $\mathbf{A} = \mathbf{i} + 3\mathbf{j} + 2\mathbf{k}$ and the vector $\mathbf{B} = 3\mathbf{j} + \mathbf{k}$.

1.30 Using the scalar product show that if $\mathbf{C} = \mathbf{A} - \mathbf{B}$, then

$$C^2 = A^2 + B^2 - 2AB\cos\theta$$

where θ is the angle between **A** and **B**.

1.31 Three vectors are given by $\mathbf{A}_1 = 5\mathbf{i} + 2\mathbf{j}$, $\mathbf{A}_2 = 3\mathbf{i} + \mathbf{k}$, and $\mathbf{A}_3 = 8\mathbf{j}$. What are (a) the magnitude and (b) the direction cosines of the sum of these three vectors?

1.32 A line passes through the points $A(1, 8, 3)$ and $B(2, 1, 5)$ in a three-dimensional cartesian coordinate system. What are the direction cosines of the line from A to B?

1.33 The position of a satellite in Fig. P1.33 is given by

$$\mathbf{r} = (5\mathbf{i} + 3\mathbf{j} + 4.55\mathbf{k})R$$

where the earth's radius R is approximately 6,370 km. What is the magnitude of $\mathbf{r}$? Obtain a unit vector $\mathbf{e}_r$ coincident with $\mathbf{r}$.

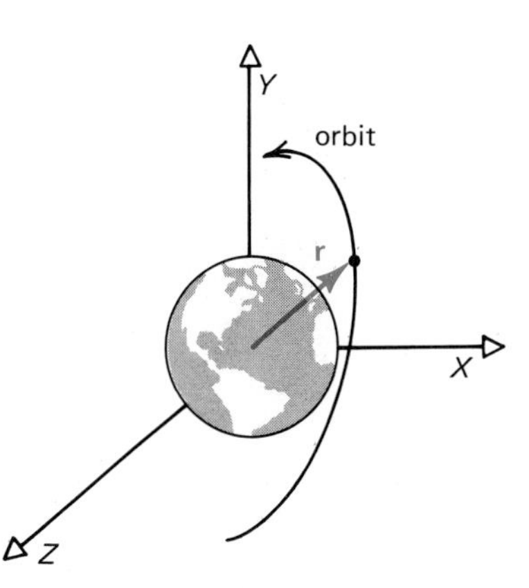

Figure P1.33

1.5 Cross Product of Vectors

The cross product $\mathbf{C}$, also called the vector product, of two vectors $\mathbf{A}$ and $\mathbf{B}$ is defined as

$$\mathbf{C} = \mathbf{A} \times \mathbf{B} = AB \sin \theta \mathbf{e}_c \tag{1.13}$$

where $0 \leq \theta \leq 180°$ and $\mathbf{e}_c$ is a unit vector perpendicular to the plane containing the vectors $\mathbf{A}$ and $\mathbf{B}$ as shown in Fig. 1.16. The direction of $\mathbf{C}$ is obtained by the right-hand rule, i.e., place the right-hand fingers on vector $\mathbf{A}$ and rotate them by the angle θ that brings the vector $\mathbf{A}$ into coincidence with $\mathbf{B}$ so that the thumb points in the direction of vector $\mathbf{C}$. The advancing right-hand screw shown in Fig. 1.17 is another model for the cross product in which the screw advances as vector $\mathbf{A}$ is rotated into vector $\mathbf{B}$.

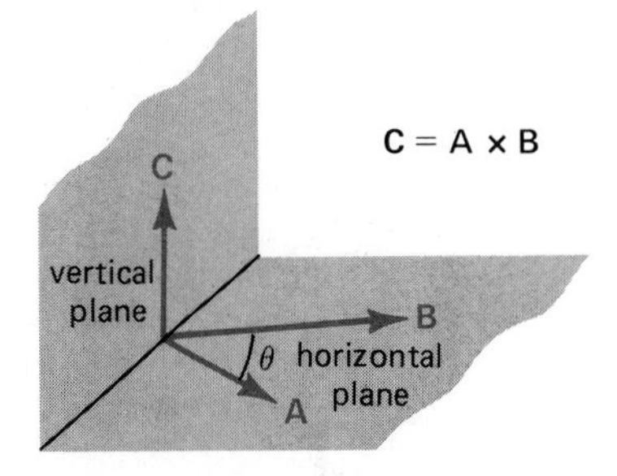

Figure 1.16

From the definition of the cross product, if

$$\mathbf{A} \times \mathbf{B} = 0$$

and $A \neq 0$, $B \neq 0$, then $\mathbf{A}$ and $\mathbf{B}$ are collinear. If $\mathbf{A}$ and $\mathbf{B}$ are orthogonal, then

$$\mathbf{A} \times \mathbf{B} = AB\mathbf{e}_c$$

The cross product is not commutative. That is, in general $\mathbf{A} \times \mathbf{B} \neq \mathbf{B} \times \mathbf{A}$ for, in accordance with the right-hand rule, $\mathbf{B} \times \mathbf{A} = -\mathbf{A} \times \mathbf{B}$.

Let us now consider $\mathbf{A}$ and $\mathbf{B}$ in a right-handed cartesian coordinate system, i.e., a system in which $\mathbf{i} \times \mathbf{j} = \mathbf{k}$. Let

$$\mathbf{A} = A_x\mathbf{i} + A_y\mathbf{j}$$

$$\mathbf{B} = B_x\mathbf{i} + B_y\mathbf{j}$$

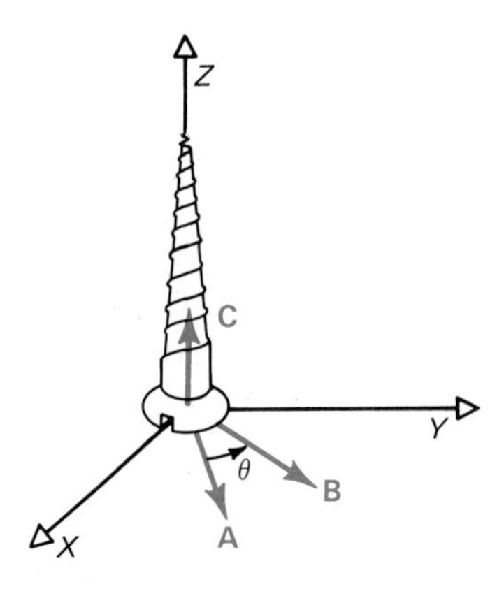

Figure 1.17

The cross product of $\mathbf{A}$ and $\mathbf{B}$ may be expanded by the distributive law as follows:

$$\begin{aligned}\mathbf{A} \times \mathbf{B} &= (A_x\mathbf{i} + A_y\mathbf{j}) \times (B_x\mathbf{i} + B_y\mathbf{j}) \\ &= A_xB_x(\mathbf{i} \times \mathbf{i}) + A_xB_y(\mathbf{i} \times \mathbf{j}) + A_yB_x(\mathbf{j} \times \mathbf{i}) + A_yB_y(\mathbf{j} \times \mathbf{j})\end{aligned}$$

Since
$$\mathbf{i} \times \mathbf{i} = \mathbf{j} \times \mathbf{j} = 0$$
$$\mathbf{i} \times \mathbf{j} = \mathbf{k} = -\mathbf{j} \times \mathbf{i}$$
then
$$\mathbf{A} \times \mathbf{B} = A_x B_y \mathbf{k} - A_y B_x \mathbf{k}$$
$$= (A_x B_y - A_y B_x)\mathbf{k}$$

We observe that the cross product of two vectors lying in the XY plane leads to a vector that always lies in the Z direction.

For the three-dimensional case, in which
$$\mathbf{A} = A_x\mathbf{i} + A_y\mathbf{j} + A_z\mathbf{k}$$
$$\mathbf{B} = B_x\mathbf{i} + B_y\mathbf{j} + B_z\mathbf{k}$$
it can be shown that
$$\mathbf{A} \times \mathbf{B} = (A_yB_z - A_zB_y)\mathbf{i} + (A_zB_x - A_xB_z)\mathbf{j} + (A_xB_y - A_yB_x)\mathbf{k} \tag{1.14}$$

In arriving at this result, it is convenient to summarize the cross products of the unit vectors as shown in Fig. 1.18. Labeling the counterclockwise direction as positive, we observe that as we proceed in either direction, we obtain the following results:

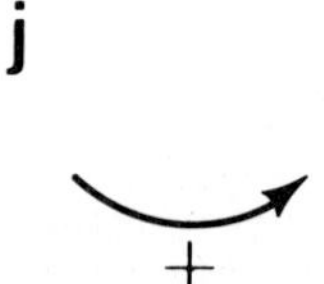

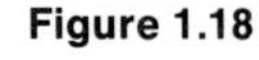

Figure 1.18

$$\mathbf{i} \times \mathbf{j} = \mathbf{k} = -\mathbf{j} \times \mathbf{i}$$
$$\mathbf{j} \times \mathbf{k} = \mathbf{i} = -\mathbf{k} \times \mathbf{j}$$
$$\mathbf{k} \times \mathbf{i} = \mathbf{j} = -\mathbf{i} \times \mathbf{k}$$

Sometimes it is convenient to obtain the results of Eq. (1.14) by the following determinant:

$$\mathbf{A} \times \mathbf{B} = \begin{vmatrix} \mathbf{i} & \mathbf{j} & \mathbf{k} \\ A_x & A_y & A_z \\ B_x & B_y & B_z \end{vmatrix} = \mathbf{i}\begin{vmatrix} A_y & A_z \\ B_y & B_z \end{vmatrix} - \mathbf{j}\begin{vmatrix} A_x & A_z \\ B_x & B_z \end{vmatrix} + \mathbf{k}\begin{vmatrix} A_x & A_y \\ B_x & B_y \end{vmatrix}$$

Note that the unit vectors in the determinant are entered in the first row, the components of **A** in the second row, and the components of **B** in the third row. The expansion of a determinant may be obtained by starting with the element of row 1, column 1, **i**, and multiplying by the minor $(A_yB_z - A_zB_y)$. Then add to this result the product of the negative of the element in row 1, column 2, $(-\mathbf{j})$, and the minor $(A_xB_z - A_zB_x)$. Finally, add the product of the element in row 1, column 3, **k**, by $(A_xB_y - A_yB_x)$ to the sum. This gives us the earlier result shown in Eq. (1.14).

Example 1.7

If $\mathbf{A} = 2\mathbf{i} + 4\mathbf{j}$ and $\mathbf{B} = -\mathbf{i} + 3\mathbf{j}$, find (a) $\mathbf{A} \times \mathbf{B}$, and (b) the angle θ through which **A** is rotated in obtaining the cross product.

(a)
$$\mathbf{A} \times \mathbf{B} = (2\mathbf{i} + 4\mathbf{j}) \times (-\mathbf{i} + 3\mathbf{j})$$
$$= 6\mathbf{k} + 4\mathbf{k} = 10\mathbf{k}$$ *Answer*

(b) Since $\mathbf{A} \times \mathbf{B} = AB \sin \theta \mathbf{k}$

$$A = \sqrt{(2)^2 + (4)^2} = \sqrt{20} = 4.47$$
$$B = \sqrt{(-1)^2 + (3)^2} = \sqrt{10} = 3.16$$

Therefore,
$$\sin \theta \mathbf{k} = \frac{10\mathbf{k}}{(4.47)(3.16)}$$

or
$$\sin \theta = 0.707$$
$$\theta = \text{arc sin } (0.707) = 45.0°$$ *Answer*

Example 1.8

If $\mathbf{w} = 3\mathbf{k}$ and $\mathbf{r} = 2\mathbf{i} + \mathbf{j}$, find $\mathbf{w} \times \mathbf{r}$ using the determinant form.

$$\mathbf{w} \times \mathbf{r} = \begin{vmatrix} \mathbf{i} & \mathbf{j} & \mathbf{k} \\ 0 & 0 & 3 \\ 2 & 1 & 0 \end{vmatrix} = \mathbf{i}\begin{vmatrix} 0 & 3 \\ 1 & 0 \end{vmatrix} - \mathbf{j}\begin{vmatrix} 0 & 3 \\ 2 & 0 \end{vmatrix} + \mathbf{k}\begin{vmatrix} 0 & 0 \\ 2 & 1 \end{vmatrix}$$
$$= \mathbf{i}(0 - 3) - \mathbf{j}(0 - 6) + \mathbf{k}(0 - 0)$$
$$= -3\mathbf{i} + 6\mathbf{j}$$ *Answer*

Example 1.9

If $\mathbf{C} = 4\mathbf{i} + 3\mathbf{j} - 2\mathbf{k}$ and $\mathbf{D} = -\mathbf{i} + 2\mathbf{j} + 2\mathbf{k}$, find $\mathbf{C} \times \mathbf{D}$ and $\mathbf{D} \times \mathbf{C}$.

$$\mathbf{C} \times \mathbf{D} = \begin{vmatrix} \mathbf{i} & \mathbf{j} & \mathbf{k} \\ 4 & 3 & -2 \\ -1 & 2 & 2 \end{vmatrix} = \mathbf{i}\begin{vmatrix} 3 & -2 \\ 2 & 2 \end{vmatrix} - \mathbf{j}\begin{vmatrix} 4 & -2 \\ -1 & 2 \end{vmatrix} + \mathbf{k}\begin{vmatrix} 4 & 3 \\ -1 & 2 \end{vmatrix}$$
$$= \mathbf{i}(6 + 4) - \mathbf{j}(8 - 2) + \mathbf{k}(8 + 3)$$
$$= 10\mathbf{i} - 6\mathbf{j} + 11\mathbf{k}$$ *Answer*

$$\mathbf{D} \times \mathbf{C} = \begin{vmatrix} \mathbf{i} & \mathbf{j} & \mathbf{k} \\ -1 & 2 & 2 \\ 4 & 3 & -2 \end{vmatrix} = \mathbf{i}\begin{vmatrix} 2 & 2 \\ 3 & -2 \end{vmatrix} - \mathbf{j}\begin{vmatrix} -1 & 2 \\ 4 & -2 \end{vmatrix} + \mathbf{k}\begin{vmatrix} -1 & 2 \\ 4 & 3 \end{vmatrix}$$
$$= \mathbf{i}(-4 - 6) - \mathbf{j}(2 - 8) + \mathbf{k}(-3 - 8)$$
$$= -10\mathbf{i} + 6\mathbf{j} - 11\mathbf{k}$$ *Answer*

PROBLEMS

1.34 Using $\mathbf{A} = A_1\mathbf{i} + A_2\mathbf{j}$ and $\mathbf{B} = B_1\mathbf{i} + B_2\mathbf{j}$, show that $\mathbf{A} \times \mathbf{B} = -\mathbf{B} \times \mathbf{A}$.

1.35 If $\mathbf{A} = -6\mathbf{j}$ and $\mathbf{B} = 2\mathbf{i} - \mathbf{j}$, find $\mathbf{A} \times \mathbf{B}$.

1.36 If $\mathbf{A} = 2\mathbf{i}$ and $\mathbf{B} = 3\mathbf{i} - 2\mathbf{j}$, (a) find $\mathbf{A} \times \mathbf{B}$ and (b) the angle θ through which **A** rotates in forming the cross product.

1.37 If $\mathbf{C} = -2\mathbf{k}$ and $\mathbf{D} = 5\mathbf{j} + \mathbf{k}$, (a) find $\mathbf{C} \times \mathbf{D}$ and (b) the angle θ through which **C** rotates in forming the cross product.

1.38 If $\mathbf{E} = 2\mathbf{i} - \mathbf{j}$ and $\mathbf{F} = -3\mathbf{i} + 2\mathbf{k}$, find $\mathbf{E} \times \mathbf{F}$.

1.39 What is the magnitude of the vector obtained by crossing $\mathbf{A} = \mathbf{i} + 2\mathbf{j}$ with $\mathbf{B} = -3\mathbf{k}$?

1.40 If $\mathbf{A} = 5\mathbf{i} + 5\mathbf{j} + \mathbf{k}$ and $\mathbf{B} = 2\mathbf{i} + 4\mathbf{k}$, find $\mathbf{A} \times \mathbf{B}$.

1.41 If $\mathbf{V}_1 = 2\mathbf{i} + \mathbf{j} + \mathbf{k}$ and $\mathbf{V}_2 = \mathbf{j} + 2\mathbf{k}$, find $\mathbf{V}_1 \times \mathbf{V}_2$.

1.42 If $\mathbf{w} = \mathbf{j} - \mathbf{k}$ and $\mathbf{r} = \mathbf{i} + 2\mathbf{j}$, find $\mathbf{w} \times \mathbf{r}$.

1.43 What is the magnitude of the vector obtained by crossing $\mathbf{r} = \mathbf{i} + 2\mathbf{j} + 3\mathbf{k}$ with $\mathbf{w} = 6\mathbf{k}$?

1.44 Using $\mathbf{A} = A_1\mathbf{i} + A_2\mathbf{j} + A_3\mathbf{k}$ and $\mathbf{B} = B_1\mathbf{i} + B_2\mathbf{j} + B_3\mathbf{k}$, show that $\mathbf{A} \cdot (\mathbf{B} \times \mathbf{A})$ is zero.

1.45 A vector $\mathbf{A} = 5\mathbf{i} + \mathbf{j} + Z\mathbf{k}$ is crossed into $\mathbf{B} = 2\mathbf{i} + 3\mathbf{j}$ to obtain $-6\mathbf{i} + 4\mathbf{j} + 13\mathbf{k}$. What must be the value of Z?

1.46 A triangle ABC is defined by the points $A(3, 0, 0)$, $B(0, 6, 0)$, and $C(0, 0, 9)$. Obtain a unit vector perpendicular to the triangle surface ABC.

1.47 Using the definition of a cross product show that

$$\frac{\sin \theta_1}{B} = \frac{\sin \theta_2}{C}$$

where θ_1 is the angle between **A** and **C**, and θ_2 is the angle between **A** and **B**. Let **A**, **B**, and **C** be the sides of a triangle where $\mathbf{C} = \mathbf{A} + \mathbf{B}$.

1.6 Problem Solving

In solving problems in statics, it is necessary to obtain a quantitative description of the forces present. The relationships between these forces are established from the requirements of equilibrium and can be stated in mathematical form. The mathematical equations can be solved assuming that the number of unknowns and the number of independent equations are the same. In general, the difficulty is not in the mathematics but in the thought process entailed in the quantitative description of the forces acting on a body, i.e., in obtaining the model.

Obtaining the model is an essential part of formulating the problem. It requires thinking in terms of a physical situation and then translating the situation into a mathematical expression. A neat sketch of the body and all forces acting on it must be drawn. This sketch is called the **free-body diagram**. We shall treat free-body diagrams in detail throughout the text.

Often it is necessary that approximations be made in formulating the problem. For example, small distances, angles, or forces may be neglected in favor of large distances, angles, or forces. If a force is acting on a very small area, it may be approximated as a force acting at a point. The formulation of a problem includes stating what data are given and what results are desired.

Once the problem has been formulated, the fundamental laws are used to derive the relationships between the various unknowns. In statics, application of these laws yields the equations of equilibrium. These equations are solved and the answers are checked for units and computational error. This can be done by substituting the answers back into the equations to see if the equations are satisfied.

We can now summarize the following musts in solving a problem.

1. Formulate the problem. This includes the
   - Free-body diagram
   - Necessary approximations
   - Given data
   - Results desired
2. Write the mathematical equations and solve for the unknowns. A graphical approach may be preferred for obtaining the unknowns.
3. Check the answers for units and errors made in computations.
4. Check if the results appear to be reasonable.

The necessity for adherence to this procedure cannot be overemphasized. Such an orderly approach toward problem solving not only yields correct and quick results, it is what makes engineers unique as problem solvers.

Computations for problems in this text are made with an electronic calculator. Normally, final answers are rounded off to three or four significant figures, although intermediate steps may contain additional significant figures. For example, in obtaining the product of $\sqrt{2}$ and π, we enter these quantities in the calculator (which may express them in as many as ten significant figures), perform the operation, so that we obtain 4.44 as our answer. When long calculations are performed sequentially, the rounding should be generally done at the end. For convenience, however, when the calculator has limited capability it may be necessary to round off at intermediate steps.

As a final comment on computations, let us consider the role of significant digits. When we have performed the calculations in a problem, we should round off the answer based on the least significant data. For example, if the data shows one quantity $A = 10.52$ and $B = 13.7$, then the product $AB = 144.127$. Since B contains one number beyond the decimal point we may round off the answer to 144.1. We will generally adhere to this concept when the number of significant figures is important in a problem.

1.7 Summary

Some basic concepts and laws fundamental to our study of statics were presented, along with a review of vectors and vector algebra. The importance of units and dimensions was emphasized, as well as some recommended procedures for solving problems in engineering mechanics.

A. The important *ideas*:

1. Our study of statics in this text is concerned with the equilibrium of rigid bodies under the influence of various kinds of forces.
2. The study of statics is based on newtonian mechanics and is conducted via a vector approach.
3. The weight of a body is a force and is sometimes called the gravity force.
4. A vector has magnitude and direction.
5. An orderly procedure is necessary in solving problems efficiently and correctly in statics.

B. The important *equations*:

$$\text{Weight} = \text{Mass} \times g \tag{1.1}$$

Vector:
$$\mathbf{A} = A\mathbf{e}_A \tag{1.2}$$

Scalar product:
$$\mathbf{A} \cdot \mathbf{B} = AB \cos \theta \tag{1.3}$$

Direction cosines:
$$\mathbf{e}_A = \cos \theta_x \mathbf{i} + \cos \theta_y \mathbf{j} + \cos \theta_z \mathbf{k} \tag{1.10}$$

Vector product:
$$\mathbf{C} = \mathbf{A} \times \mathbf{B} = AB \sin \theta \mathbf{e}_c \tag{1.13}$$

2 Forces and Moments

An engineering structure may be subjected to many forces that it must withstand during its lifetime. An analysis of these forces is an important part in the study of statics. In this chapter we are concerned with these forces and how they can be reduced into a resultant force and a resultant moment. The material here and in the next chapter is fundamental to obtaining the understanding necessary for problem solving not only in statics but in dynamics as well. It is therefore essential that this material be mastered.

2.1 Resolution of Forces

Graphical Solution

A force has been characterized as the action of one body on another. A force is a **vector** quantity, i.e., it has **magnitude** and **direction**. As it is applied to a body or a structure it is also characterized by its **point of application**. The magnitude of a force is given by a certain number of units. In this text, the unit of force is either the newton (N) or the pound (lb). We shall also use the kilonewton (kN) which equals 1000 N and the kilopound (kip) which equals 1000 lbs. The direction of a force is defined by its line of action and its sense, which is given by an arrow, as shown in Fig. 2.1(a) on page 24. Note that the vector representing the 50-N force in Fig. 2.1(a) has either its tip or its tail at the point of application. Both representations are acceptable. Fig. 2.1(b) shows forces $\mathbf{F}_1$ and $\mathbf{F}_2$ that act at a common point of application. When two or more forces act at a common point of application, the forces are called **concurrent forces**.

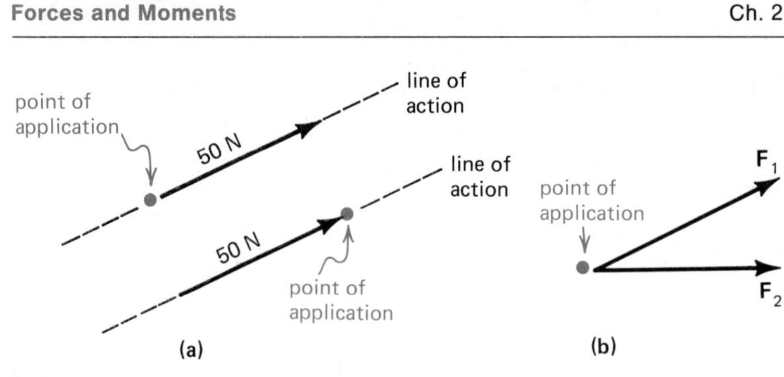

Figure 2.1

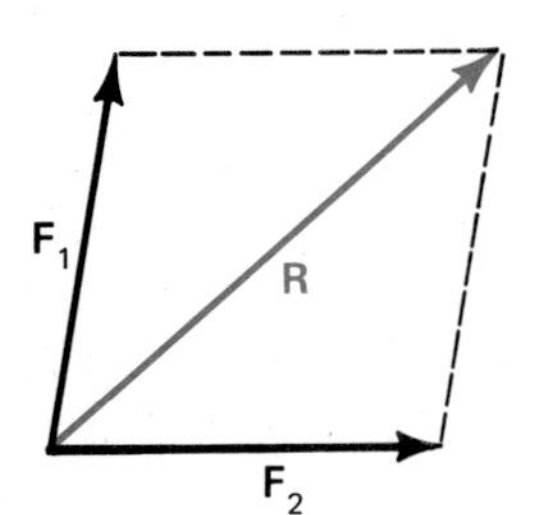

(a)

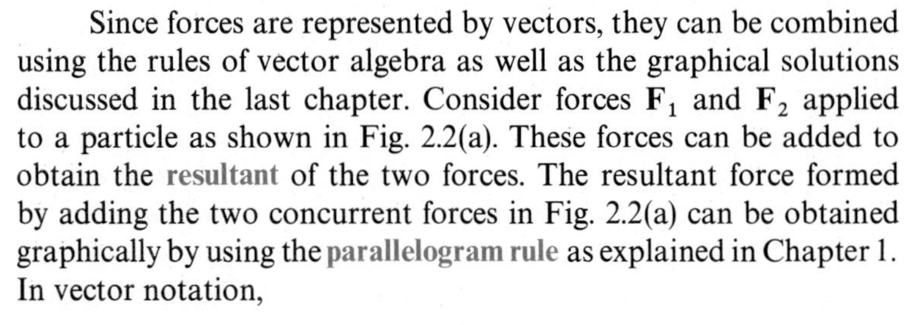

Since forces are represented by vectors, they can be combined using the rules of vector algebra as well as the graphical solutions discussed in the last chapter. Consider forces $\mathbf{F}_1$ and $\mathbf{F}_2$ applied to a particle as shown in Fig. 2.2(a). These forces can be added to obtain the **resultant** of the two forces. The resultant force formed by adding the two concurrent forces in Fig. 2.2(a) can be obtained graphically by using the **parallelogram rule** as explained in Chapter 1. In vector notation,

$$\mathbf{R} = \mathbf{F}_1 + \mathbf{F}_2$$

Here forces $\mathbf{F}_1$ and $\mathbf{F}_2$ form the two sides of a parallelogram whose diagonal represents the resultant force. The resultant force formed by subtracting $\mathbf{F}_2$ from $\mathbf{F}_1$ is shown graphically in Fig. 2.2(b), and is given by

$$\mathbf{R} = \mathbf{F}_1 - \mathbf{F}_2$$

Here again the forces $\mathbf{F}_1$ and $-\mathbf{F}_2$ form the two sides of a parallelogram whose diagonal represents the resultant force. Concurrent forces $\mathbf{F}_1$ and $\mathbf{F}_2$ may also be added by the **triangle rule**, as shown in Fig. 2.3(a), where the resultant is obtained by arranging the vectors in a **tip-to-tail** fashion, and then joining the tail of $\mathbf{F}_1$ to the tip of $\mathbf{F}_2$. The order of combining is not important, as seen in Fig. 2.3(b).

R

F_1

$-F_2$

(b)

Figure 2.2

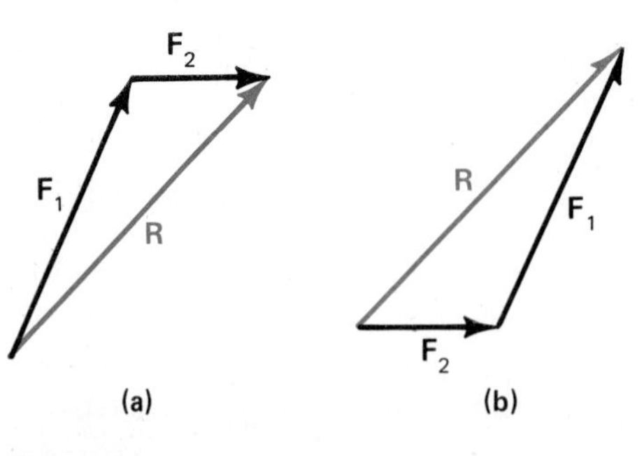

Figure 2.3

Here $\mathbf{F}_2$ is first constructed, then $\mathbf{F}_1$ is drawn to obtain the resultant $\mathbf{R}$.

To obtain the difference of two concurrent forces by the triangle rule, consider Fig. 2.4. In this case we construct $\mathbf{F}_1$ followed by $-\mathbf{F}_2$ to obtain $\mathbf{R}$ in the tip-to-tail fashion as explained in Chapter 1.

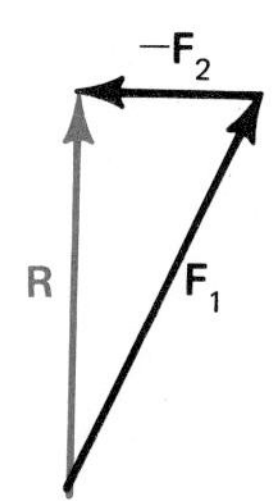

Figure 2.4

By following the tip-to-tail method, a resultant force can be obtained for the addition of several concurrent forces acting in the same plane. For example, consider forces $\mathbf{F}_1$, $\mathbf{F}_2$, and $\mathbf{F}_3$ acting at a point on the column shown in Fig. 2.5(a). The addition of these three forces is shown in Fig. 2.5(b), from which we obtain the resultant force $\mathbf{R}$ that acts on the column. The method of combining several vectors in a tip-to-tail fashion to obtain a resultant vector is called the polygon rule, which is an extension and generalization of the triangle rule. Notice that these graphical procedures can be easily carried out if all the concurrent forces are in the same plane.

The concept of the resultant force is important because the resultant force is equivalent to the original forces acting at the point. We shall have more to say about equivalent forces later in this chapter. For now, we can state that the resultant force $\mathbf{R}$ is the sum of n concurrent forces,

$$\mathbf{R} = \mathbf{F}_1 + \mathbf{F}_2 + \cdots + \mathbf{F}_n$$

This addition can be performed graphically, as discussed earlier, or algebraically, as will be discussed in the next section. The equation for the resultant force $\mathbf{R}$ can be expressed in the following abbreviated form:

$$\mathbf{R} = \sum_{i=1}^{n} \mathbf{F}_i \tag{2.1}$$

If $n = 2$, we can use the parallelogram or triangle rule. If n is larger than 2, then we make use of the polygon rule. We emphasize that the sum in Eq. (2.1) is a vector sum. A vector sum differs from the algebraic sum of scalar quantities, which involves only magnitudes.

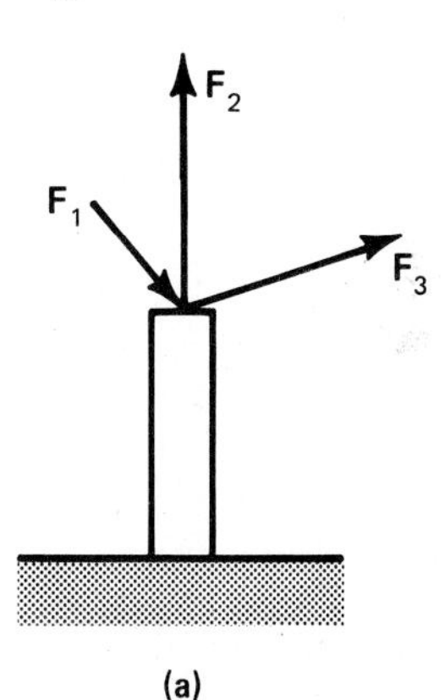

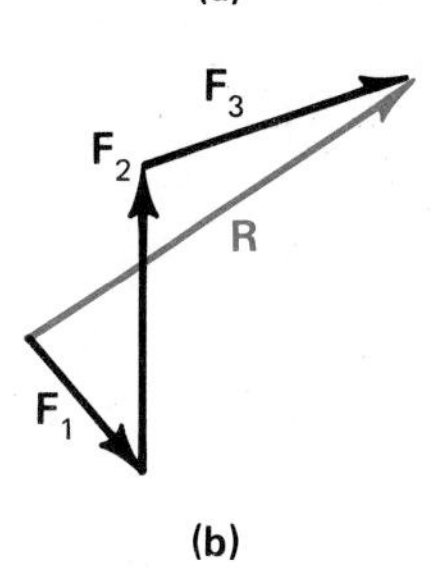

Figure 2.5

Example 2.1

A steamer is pulled by two tugboats as shown in Fig. 2.6(a). Obtain the resultant force applied to the steamer: (a) using the parallelogram rule and (b) using the triangle rule. Assume that $F_1 = 1.5$ kN and $F_2 = 2$ kN.

(a) Figure 2.6(b) shows the construction for obtaining the resultant force by the parallelogram rule. Note that we should use a protractor and a convenient scale for this construction. From

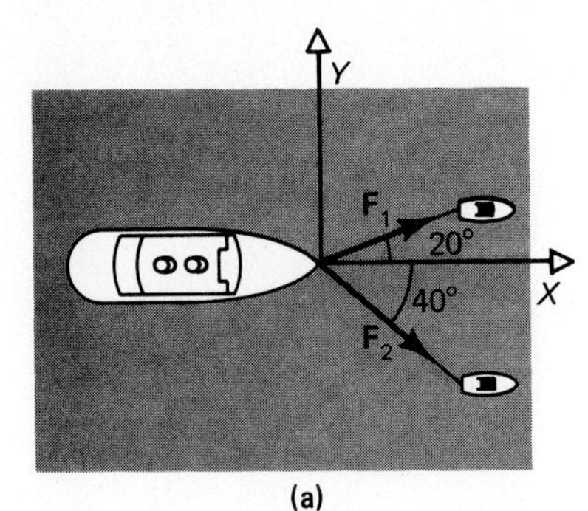

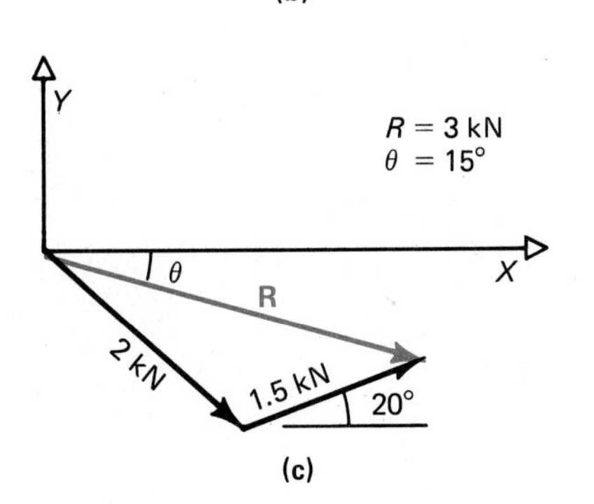

Figure 2.6

the graphical result we obtain a resultant force whose magnitude equals 3 kN and whose direction is 15° measured clockwise from the X axis.

(b) The graphical solution by the triangle rule is shown in Fig. 2.6(c). In this case the 2-kN force is constructed first, followed by the 1.5-kN force. The results are identical with those found in part (a).

Example 2.2

Given $\mathbf{F}_1 = (6\mathbf{i} - 4\mathbf{j})$, $\mathbf{F}_2 = (-3\mathbf{i} - 5\mathbf{j})$, and $\mathbf{F}_3 = -6\mathbf{i}$, solve for the resultant force $\mathbf{R}$ using the polygon rule. Each force is expressed in pounds.

Figure 2.7 shows the construction of the polygon starting with $\mathbf{F}_1$. We note that $\mathbf{F}_1$, $\mathbf{F}_2$, and $\mathbf{F}_3$ are arranged in a tip-to-tail fashion starting from the fixed point at 0. The resultant vector $\mathbf{R}$ goes from 0 to the tip of the last vector, which is $\mathbf{F}_3$. The magnitude of the resultant force equals 9.5 lb and the direction is given by $\theta = 71.5°$ measured counterclockwise from the negative X axis or 251.5° measured counterclockwise from the positive X axis.

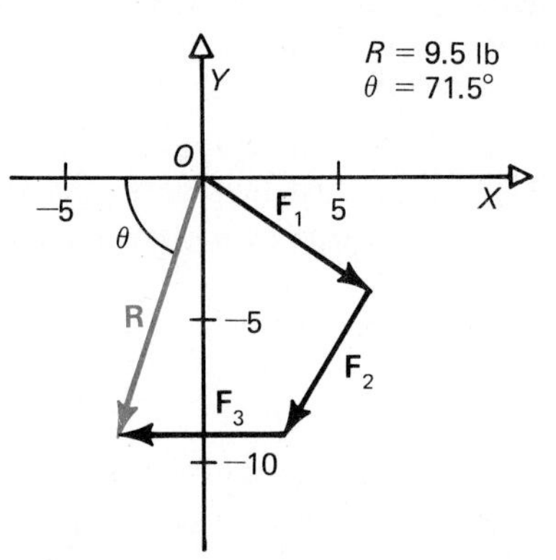

Figure 2.7

Example 2.3

A truck is towed using three ropes as shown in Fig. 2.8(a). The magnitudes of these forces are $F_1 = 1.5$ kN, $F_2 = 1$ kN, and $F_3 = 2$ kN. Using the polygon rule, obtain the resultant of the three forces.

Figure 2.8(b) shows the graphical construction, beginning with $\mathbf{F}_1$ and proceeding to $\mathbf{F}_2$ and $\mathbf{F}_3$. The magnitude of the resultant force $\mathbf{R}$ equals 4.16 kN and its direction $\theta = 6.5°$ measured clockwise from the X axis.

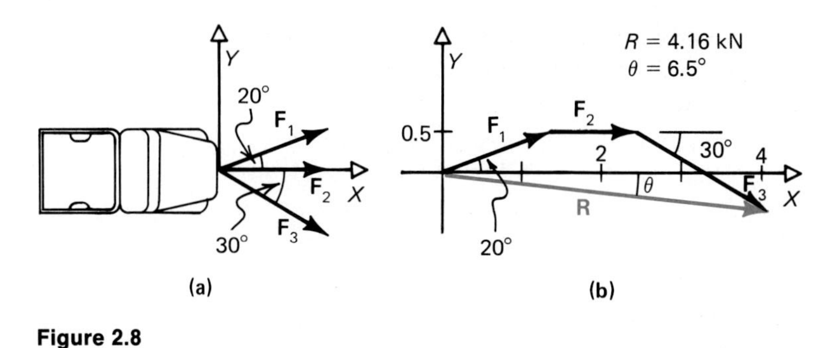

Figure 2.8

Algebraic Solution

A resultant force is the summation of the forces acting at a common point of application. Conversely, a force may be resolved into its components. The force **F** is resolved into two different sets of **A** and **B** components in Figs. 2.9(a) and 2.9(b). Clearly, the choice of the particular components is limitless. In engineering, a force is often resolved into rectangular or cartesian components. When a force expressed in rectangular coordinates has only two components rather than three, we speak of the force as being planar.

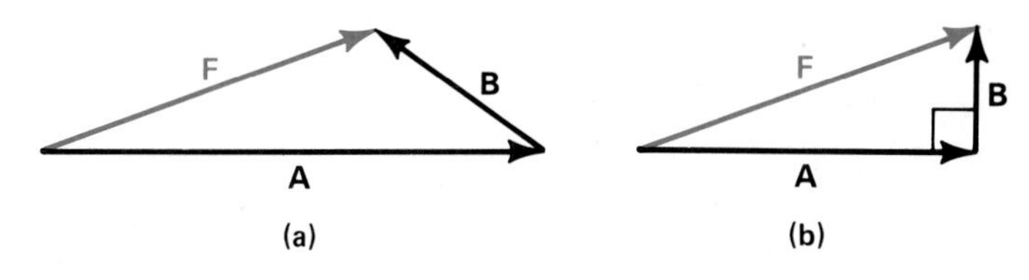

Figure 2.9

A planar force represented in the cartesian system is shown in Fig. 2.10. The force **F** has two components. $\mathbf{F}_x$ is the projection of **F** on the X axis and $\mathbf{F}_y$ is the projection of **F** on the Y axis, so that

$$\mathbf{F} = \mathbf{F}_x + \mathbf{F}_y \tag{2.2}$$

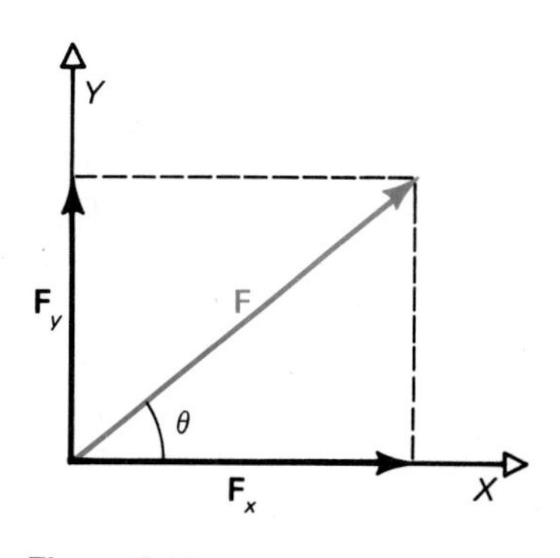

Figure 2.10

When a force is expressed in this way, it is convenient to use the unit vector notation,

$$\mathbf{F}_x = F_x\mathbf{i} \qquad \mathbf{F}_y = F_y\mathbf{j}$$

We call F_x and F_y the components of the force on the X axis and Y axis, respectively. The vectors **i** and **j** are unit vectors coincident with the X and Y axes as shown in Fig. 2.11. The force **F** can now be written as

$$\mathbf{F} = F_x\mathbf{i} + F_y\mathbf{j} \tag{2.3}$$

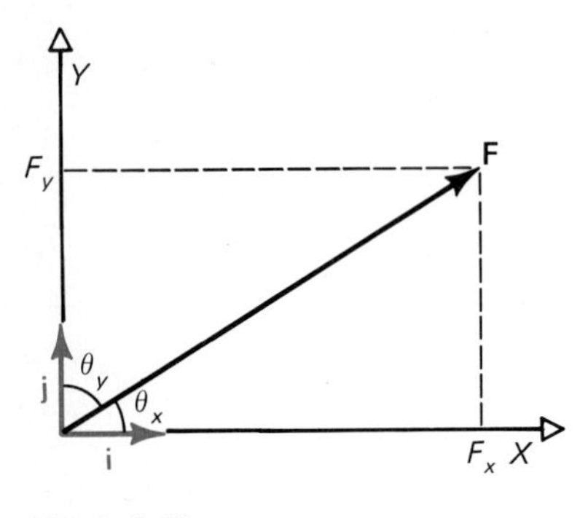

Figure 2.11

We also notice that since F_x and F_y are the components of **F** on the X and Y axes respectively, they can be expressed as

$$F_x = \mathbf{F} \cdot \mathbf{i} = F \cos \theta_x$$

$$F_y = \mathbf{F} \cdot \mathbf{j} = F \cos \theta_y = F \sin \theta_x$$

We notice that since $\theta_y = \pi/2 - \theta_x$, $\cos \theta_y$ equals $\sin \theta_x$, so that only one angle needs to be defined.

If the components F_x and F_y are known, we may compute the magnitude and direction of the resultant force **F** using the following notation:

$$F = \sqrt{F_x^2 + F_y^2} \tag{2.4a}$$

$$\theta_x = \tan^{-1} \frac{F_y}{F_x} \tag{2.4b}$$

Note that a positive θ_x is counterclockwise measured from the X axis.

We have shown how several concurrent forces can be added graphically by the parallelogram rule, the triangle rule, or the polygon rule to obtain the resultant force. It is, of course, possible to add several forces by combining their components. For example, let us add **A** and **B** by adding their components. We have

$$\mathbf{A} = A_x\mathbf{i} + A_y\mathbf{j} \qquad \mathbf{B} = B_x\mathbf{i} + B_y\mathbf{j}$$

and, since the resultant is vector **R**,

$$\mathbf{R} = \mathbf{A} + \mathbf{B}$$

We substitute for **A** and **B** to obtain

$$\mathbf{R} = (A_x + B_x)\mathbf{i} + (A_y + B_y)\mathbf{j}$$

We may now think of the resultant as having two components, as shown in Fig. 2.12.

$$\mathbf{R} = R_x\mathbf{i} + R_y\mathbf{j}$$

where $\qquad R_x = A_x + B_x \qquad R_y = A_y + B_y$

The magnitude and direction of **R** can be written as

$$R = \sqrt{R_x^2 + R_y^2}$$

$$\theta_x = \tan^{-1} \frac{R_y}{R_x}$$

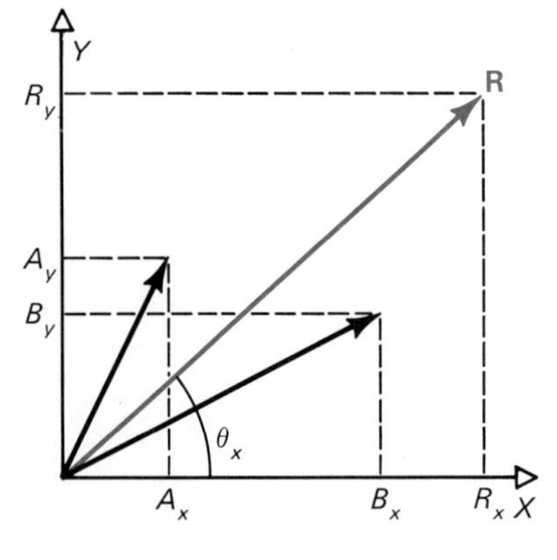

Figure 2.12

The components of the resultant of several forces are, therefore, obtained by algebraically adding the corresponding components of the individual forces. We can write the components of a resultant force as

$$R_x = \sum F_x \qquad R_y = \sum F_y$$

A force can also be characterized by specifying its magnitude and two of the points through which it passes. Consider a force **F** whose line of action passes through two points, $P(x_1, y_1)$ and $Q(x_2, y_2)$ as shown in Fig. 2.13. The force **F** can be written as

$$\mathbf{F} = F\mathbf{e}_F \tag{2.5}$$

where $\mathbf{e}_F$ is the unit vector shown in Fig. 2.13. The vector **PQ** has the same direction as the force **F**, so that

$$\mathbf{PQ} = D\mathbf{e}_F$$

where D is the distance between P and Q. From Fig. 2.13 we can write

$$\mathbf{PQ} = (x_2 - x_1)\mathbf{i} + (y_2 - y_1)\mathbf{j} = D\mathbf{e}_F$$

Since $\qquad D_x = x_2 - x_1 \qquad$ and $\qquad D_y = y_2 - y_1$

then $$\mathbf{PQ} = D_x\mathbf{i} + D_y\mathbf{j} = D\mathbf{e}_F$$

where $$D = \sqrt{D_x^2 + D_y^2}$$

We now obtain the unit vector

$$\mathbf{e}_F = \frac{D_x}{D}\mathbf{i} + \frac{D_y}{D}\mathbf{j} \tag{2.6}$$

The terms D_x/D and D_y/D are the direction cosines, i.e.,

$$\cos\theta_x = \frac{D_x}{D} \qquad \cos\theta_y = \frac{D_y}{D} \tag{2.7}$$

where θ_x is the angle between the vector **F** and the X axis, and θ_y is the angle between the vector **F** and the Y axis.

Substituting Eq. (2.6) into Eq. (2.5), we have

$$\mathbf{F} = F\left(\frac{D_x}{D}\mathbf{i} + \frac{D_y}{D}\mathbf{j}\right)$$

However, the vector **F** can be written as

$$\mathbf{F} = F_x\mathbf{i} + F_y\mathbf{j}$$

so that by comparison,

$$F_x = F\frac{D_x}{D} = F\cos\theta_x$$

$$F_y = F\frac{D_y}{D} = F\cos\theta_y$$

Consequently,

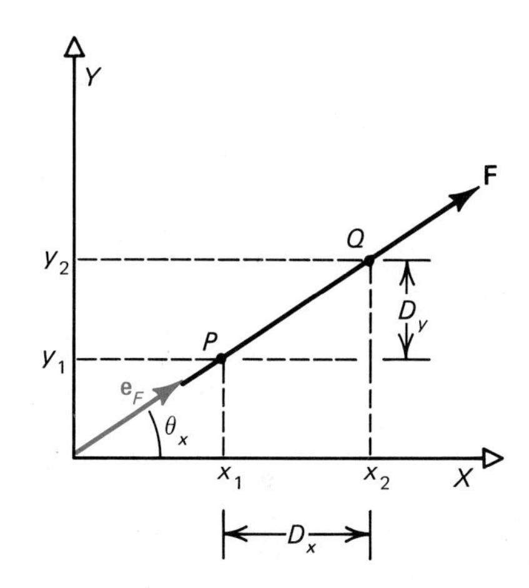

Figure 2.13

$$\mathbf{F} = F(\cos\theta_x\mathbf{i} + \cos\theta_y\mathbf{j}) \tag{2.8}$$

In computing $\cos\theta_x$ and $\cos\theta_y$ we must remember that D_x and D_y are the distances along the X and Y axes between P and Q, whereas D is the straight-line distance between P and Q.

Example 2.4

A 1-ton weight is lifted by a crane as shown in Fig. 2.14(a). Express the weight of this object as components along the X and Y axes.

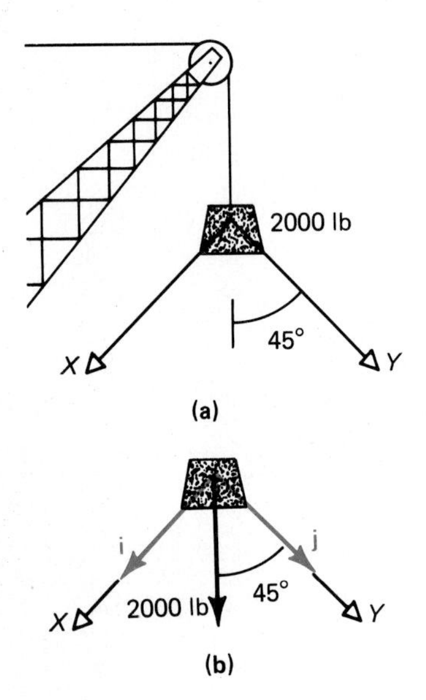

Figure 2.14

Note the arrangement of the X and Y axes. While we normally place the X axis horizontal to the right and the Y axis vertical and upward, we are actually free to arrange the coordinate system for our convenience. Once we have done so, then our answers must conform to the chosen arrangement of the coordinate system. Figure 2.14(b) shows the 2000-lb force acting vertically downward. Now

$$\mathbf{w} = w_x\mathbf{i} + w_y\mathbf{j}$$

$$w_x = w\sin 45^\circ = 2{,}000\sin 45^\circ = 1{,}414\text{ lb} \qquad \textit{Answer}$$

$$w_y = w\cos 45^\circ = 2{,}000\cos 45^\circ = 1{,}414\text{ lb} \qquad \textit{Answer}$$

Example 2.5

Obtain algebraically the resultant force acting on the truck being towed by the three ropes in Example 2.3.

Equation (2.1) is written for the three concurrent forces.

$$\mathbf{R} = \mathbf{F}_1 + \mathbf{F}_2 + \mathbf{F}_3$$

where
$$\mathbf{F}_1 = 1.5(\cos 20^\circ\mathbf{i} + \sin 20^\circ\mathbf{j})$$

$$\mathbf{F}_2 = 1\mathbf{i}$$

$$\mathbf{F}_3 = 2(\cos 30^\circ\mathbf{i} - \sin 30^\circ\mathbf{j})$$

Substituting, we obtain

$$\begin{aligned}\mathbf{R} &= (1.5\cos 20^\circ + 1 + 2\cos 30^\circ)\mathbf{i} + (1.5\sin 20^\circ - 2\sin 30^\circ)\mathbf{j}\\ &= 4.14\mathbf{i} - 0.487\mathbf{j}\end{aligned}$$

The magnitude of the resultant force is

$$R = \sqrt{(4.14)^2 + (-0.487)^2} = 4.17\text{ kN} \qquad \textit{Answer}$$

and the direction of the resultant force measured from the X axis is

$$\theta = \tan^{-1}\left(-\frac{0.487}{4.14}\right) = -6.71^\circ \qquad \textit{Answer}$$

These results agree favorably with the graphical solution obtain in Example 2.3.

Example 2.6

The tension T in the rope shown in Fig. 2.15 supports the weight W. The rope is known to pass through $P(-5, -5)$ and $Q(-1, -2)$. If the tension is directed along the rope, express the tensile force $\mathbf{T}$ as a vector quantity.

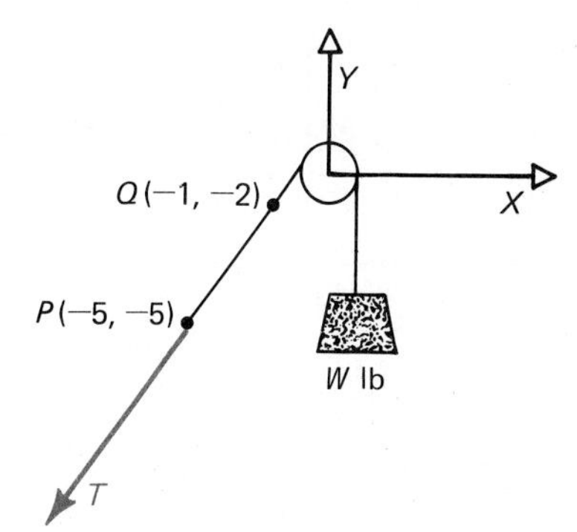

Figure 2.15

Let the unit vector and the force vector be written as

$$\mathbf{e}_T = \cos\theta_x \mathbf{i} + \cos\theta_y \mathbf{j} \qquad \mathbf{T} = T\mathbf{e}_T$$

The distances of the vector passing from point Q to point P are

$$D_x = -5 + 1 = -4$$

$$D_y = -5 + 2 = -3$$

$$D = \sqrt{4^2 + 3^2} = 5$$

Therefore, from Eq. (2.7),

$$\cos\theta_x = \frac{D_x}{D} \qquad \cos\theta_y = \frac{D_y}{D}$$

$$\cos\theta_x = -\frac{4}{5} = -0.8 \qquad \cos\theta_y = -\frac{3}{5} = -0.6$$

so that the unit vector and force vector become

$$\mathbf{e}_T = -0.8\mathbf{i} - 0.6\mathbf{j} \qquad \text{and} \qquad \mathbf{T} = T(-0.8\mathbf{i} - 0.6\mathbf{j})$$ *Answer*

PROBLEMS

2.1 Determine the resultant force on the hook shown in Fig. P2.1 (a) by the parallelogram rule and (b) by the triangle rule.

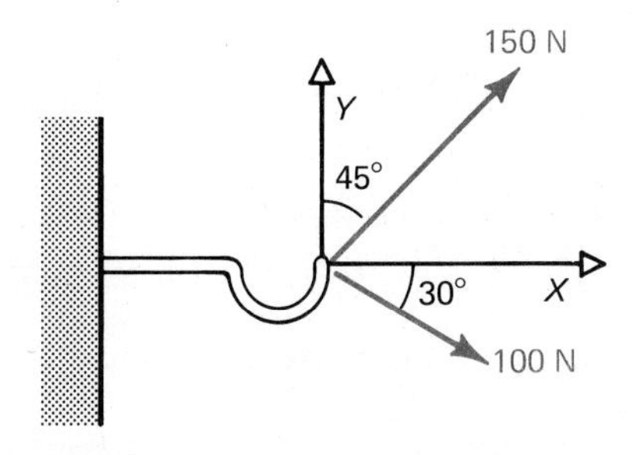

Figure P2.1

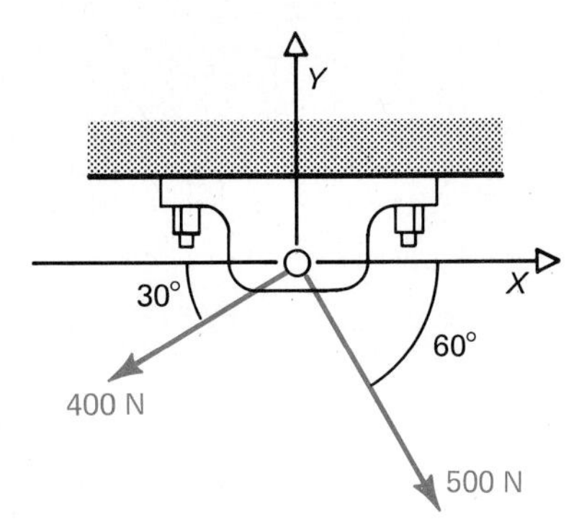

Figure P2.2

2.2 Obtain the resultant force on the structure of Fig. P2.2 by a graphical method.

2.3 Given $\mathbf{F}_1 = (2\mathbf{i} + 2\mathbf{j})$, $\mathbf{F}_2 = (-4\mathbf{j})$, and $\mathbf{F}_3 = (-2\mathbf{i})$, find the resultant force using the polygon rule. Assume each force is expressed in newtons.

2.4 Given $\mathbf{F}_1 = (-3\mathbf{i} + 2\mathbf{j})$, $\mathbf{F}_2 = (-3\mathbf{i})$, $\mathbf{F}_3 = (3\mathbf{i} + 4\mathbf{j})$, and $\mathbf{F}_4 = (-6\mathbf{j})$, find the resultant force using the polygon rule. Assume each force is expressed in newtons.

2.5 Obtain graphically the resultant force on the rod AB shown in Fig. P2.5.

2.6 The hook in Fig. P2.6 is connected to three cables whose forces are shown. Obtain the resultant force on the hook by the polygon rule.

2.7 A barge is pulled by two ropes that develop the forces shown in Fig. P2.7. Graphically obtain the resultant force applied on the barge.

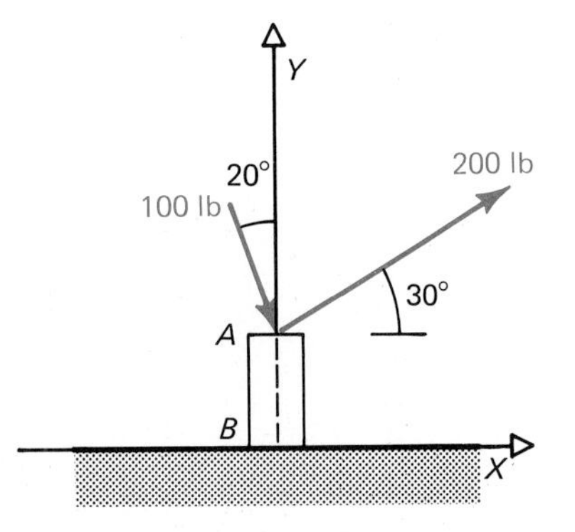

Figure P2.5

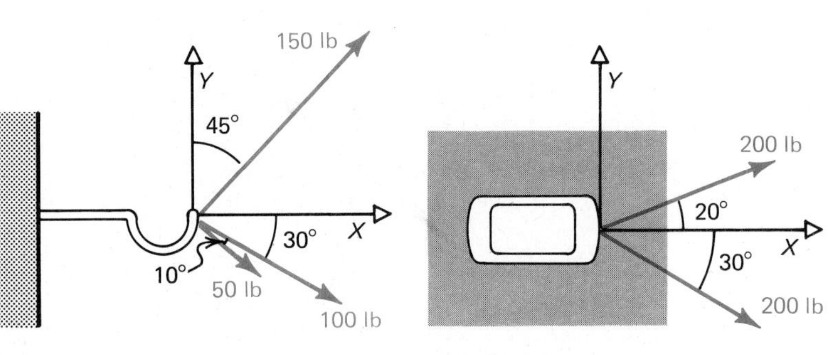

Figure P2.6 **Figure P2.7**

2.8 Using Fig. P2.1, obtain the magnitude and direction of the resultant force by vector algebra.

2.9 Using Fig. P2.2, obtain the magnitude and direction of the resultant force by vector algebra.

2.10 Solve Problem 2.3 by vector algebra.

2.11 Solve Problem 2.4 by vector algebra.

2.12 Using Fig. P2.5, obtain the cartesian components of the resultant force by vector algebra.

2.13 Using Fig. P2.6, obtain the cartesian components of the resultant force by vector algebra.

2.14 Obtain the angle α and the magnitude of $\mathbf{F}_1$ in Fig. P2.14 if $|\mathbf{F}_2| =$ 100 lb and $\mathbf{R} = \mathbf{F}_1 + \mathbf{F}_2$. Use (a) the parallelogram rule and (b) the triangle rule.

2.15 Solve Problem 2.14 by vector algebra.

2.16 The tension in a rope used to support a weight is $T = 1500$ N and is

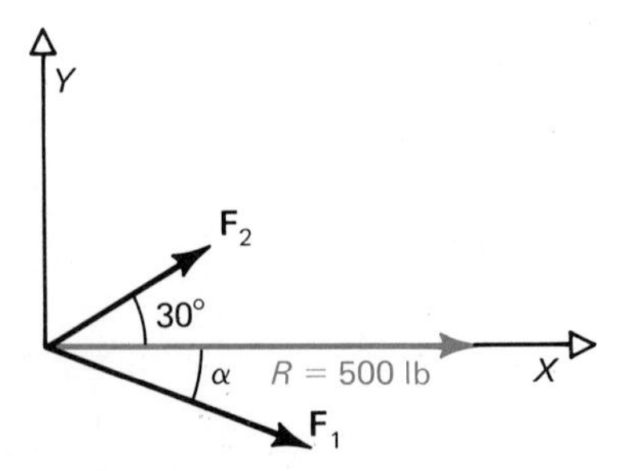

Figure P2.14

directed along the rope as shown in Fig. P2.16. If the rope passes through $A(5, 2)$ and $B(8, 3)$, express the tension force as a vector quantity.

2.17 A force $\mathbf{F}_1$ is given by $\mathbf{F}_1 = (8\mathbf{i} + 9\mathbf{j})$ lb. What is the projection of this force on a line whose direction is given by the unit vector $\boldsymbol{\lambda} = 0.8\mathbf{i} + 0.6\mathbf{j}$?

2.18 Obtain the resultant of the concurrent forces shown in Fig. P2.18 using the polygon rule.

2.19 Obtain the magnitude and unit vector of the resultant force of Problem 2.18 by summing the components of the four concurrent forces.

2.20 A 250-lb force passes through point $A(-3, 3)$ and point $B(-8, 9)$. Express the force as a vector passing (a) from A to B, and (b) from B to A.

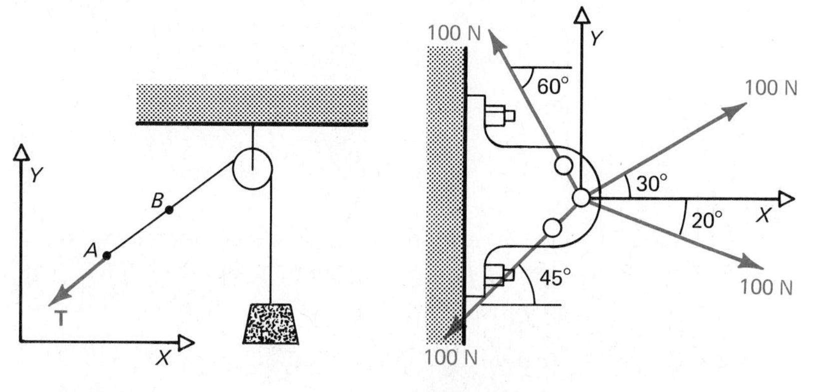

Figure P2.16

Figure P2.18

2.2 Three-Dimensional Forces

So far we have considered concurrent forces that act in a plane. Now we address the problem of finding the resultant force when the concurrent forces no longer lie in a plane. We call these forces three-dimensional forces. While graphical rules can be used in principle, they are difficult to implement since three-dimensional drawings are required. It is customary to represent a three-dimensional force by its components in three-dimensional space and then combine algebraically the components of each force to obtain the resultant force.

A force $\mathbf{F}$ in space can be represented in three dimensions by its projections on the X, Y, and Z axes as

$$\mathbf{F} = F_x\mathbf{i} + F_y\mathbf{j} + F_z\mathbf{k}$$

where $\mathbf{i}$, $\mathbf{j}$, $\mathbf{k}$ are the unit vectors as shown in Fig. 2.16. Note that the components F_x and F_z may be combined by the parallelogram rule to obtain a resultant in the XZ plane. This resultant may be

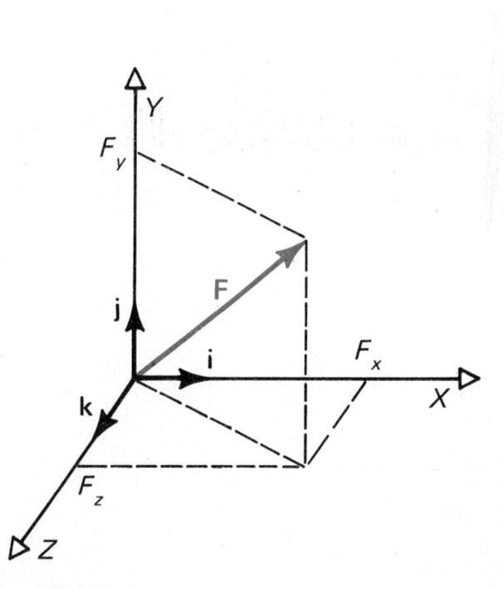

Figure 2.16

combined with the component F_y by the parallelogram rule to obtain the force **F**.

The three components can be written by forming the following scalar products:

$$F_x = \mathbf{F} \cdot \mathbf{i} \qquad F_y = \mathbf{F} \cdot \mathbf{j} \qquad F_z = \mathbf{F} \cdot \mathbf{k}$$

Evaluating these scalar products, we obtain

$$F_x = F \cos \theta_x \qquad F_y = F \cos \theta_y \qquad F_z = F \cos \theta_z$$

Here, $\cos \theta_x$, $\cos \theta_y$, and $\cos \theta_z$ are the direction cosines of the force **F** relative to the X, Y, and Z axes, respectively.

If we have the components of a vector, then its magnitude is given by

$$F = \sqrt{F_x^2 + F_y^2 + F_z^2}$$

and the direction cosines are

$$\cos \theta_x = \frac{F_x}{F} \qquad \cos \theta_y = \frac{F_y}{F} \qquad \cos \theta_z = \frac{F_z}{F}$$

If we have n forces represented in three-dimensional space, the resultant force **R** is obtained by summing all the forces:

$$\mathbf{R} = \mathbf{F}_1 + \mathbf{F}_2 + \cdots + \mathbf{F}_n$$

where each force is represented by its components as

$$\mathbf{F}_1 = F_{1x}\mathbf{i} + F_{1y}\mathbf{j} + F_{1z}\mathbf{k}$$
$$\vdots$$
$$\mathbf{F}_n = F_{nx}\mathbf{i} + F_{ny}\mathbf{j} + F_{nz}\mathbf{k}$$

If we substitute this in the right side of the equation for the resultant force, we obtain

$$\mathbf{R} = R_x\mathbf{i} + R_y\mathbf{j} + R_z\mathbf{k}$$

where $$R_x = F_{1x} + F_{2x} + \cdots + F_{nx} = \sum_{i=1}^{n} F_{ix} = \sum F_x$$

$$R_y = F_{1y} + F_{2y} + \cdots + F_{ny} = \sum_{i=1}^{n} F_{iy} = \sum F_y$$

$$R_z = F_{1z} + F_{2z} + \cdots + F_{nz} = \sum_{i=1}^{n} F_{iz} = \sum F_z$$

Note that we have abbreviated the summation notation for convenience.

As in the two-dimensional case, a force can also be characterized by specifying its magnitude and two points in space through which the line of action must pass. Consider the force **F** whose line of action passes through P and Q as shown in Fig. 2.17. Now,

$$\mathbf{F} = F\mathbf{e}_F \tag{2.9}$$

where $\mathbf{e}_F$ is the unit vector shown in Fig. 2.17. The vector **PQ** has the same direction as force **F**, so that

$$\mathbf{PQ} = D\mathbf{e}_F$$

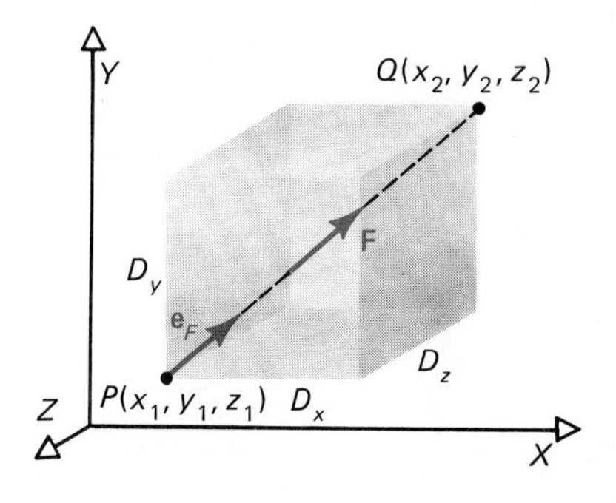

Figure 2.17

Following the development of the two-dimensional force components in the previous section, we can write

$$\mathbf{e}_F = \frac{D_x}{D}\mathbf{i} + \frac{D_y}{D}\mathbf{j} + \frac{D_z}{D}\mathbf{k}$$

where $D_x = x_2 - x_1 \qquad D_y = y_2 - y_1 \qquad D_z = z_2 - z_1$

and $$D = \sqrt{D_x^2 + D_y^2 + D_z^2}$$

The direction cosines are defined as in Eq. (2.7), so that

$$\cos\theta_x = \frac{D_x}{D} \qquad \cos\theta_y = \frac{D_y}{D} \qquad \cos\theta_z = \frac{D_z}{D}$$

Consequently, the unit vector can be written as

$$\mathbf{e}_F = \cos\theta_x\mathbf{i} + \cos\theta_y\mathbf{j} + \cos\theta_z\mathbf{k} \tag{2.10}$$

The force passing through points P and Q can now be written

$$\mathbf{F} = F(\cos\theta_x\mathbf{i} + \cos\theta_y\mathbf{j} + \cos\theta_z\mathbf{k})$$

Example 2.7

Two forces $\mathbf{T}_A$ and $\mathbf{T}_B$ act at a point where $\mathbf{T}_A = (3\mathbf{i} - 4\mathbf{j} + 2\mathbf{k})$ N and $\mathbf{T}_B = (-7\mathbf{i} - 2\mathbf{j} + 3\mathbf{k})$ N. What is the magnitude R and unit vector $\mathbf{e}_R$ of the resultant of the two forces?

The resultant force is

$$\begin{aligned}\mathbf{R} &= \mathbf{T}_A + \mathbf{T}_B \\ &= (3\mathbf{i} - 4\mathbf{j} + 2\mathbf{k}) + (-7\mathbf{i} - 2\mathbf{j} + 3\mathbf{k}) \\ &= (3 - 7)\mathbf{i} + (-4 - 2)\mathbf{j} + (2 + 3)\mathbf{k} \\ &= -4\mathbf{i} - 6\mathbf{j} + 5\mathbf{k}\end{aligned}$$

The magnitude of the resultant force is

$$\begin{aligned}R &= \sqrt{(-4)^2 + (-6)^2 + (5)^2} \\ &= 8.77 \text{ N}\end{aligned}$$ *Answer*

The unit vector is

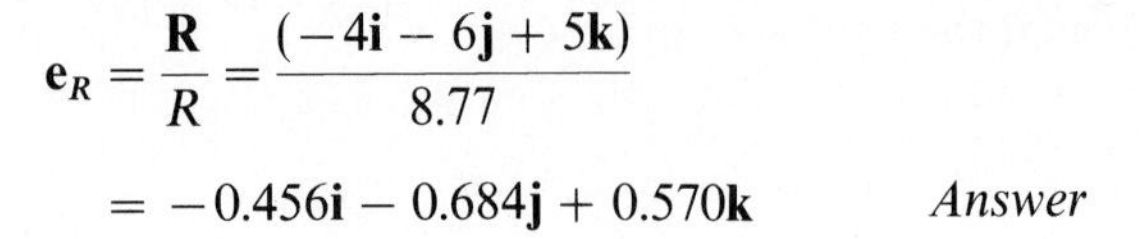

$$\mathbf{e}_R = \frac{\mathbf{R}}{R} = \frac{(-4\mathbf{i} - 6\mathbf{j} + 5\mathbf{k})}{8.77}$$

$$= -0.456\mathbf{i} - 0.684\mathbf{j} + 0.570\mathbf{k} \qquad \textit{Answer}$$

Example 2.8

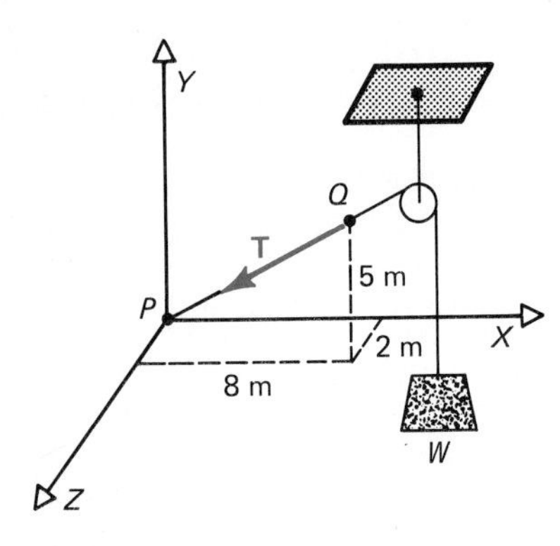

Figure 2.18

A rope tied to an anchor at point P supports a weight as shown in Fig. 2.18. Find the components of the tension force $\mathbf{T}$ of the rope along the X, Y, Z axes if the rope passes through the points $P(0, 0, 0)$ and $Q(8, 5, 2)$ and $T = 1000$ lb.

The tension in the rope is

$$\mathbf{T} = T\mathbf{e}_T = 1000\mathbf{e}_T$$

where

$$\mathbf{e}_T = \cos\theta_x\mathbf{i} + \cos\theta_y\mathbf{j} + \cos\theta_z\mathbf{k}$$

From Figure 2.18

$$D_x = -8 \qquad D_y = -5 \qquad D_z = -2$$

Now we obtain the distance between P and Q,

$$D = \sqrt{(-8)^2 + (-5)^2 + (-2)^2} = 9.64$$

The direction cosines become

$$\cos\theta_x = -\frac{8}{9.64} = -0.830$$

$$\cos\theta_y = -\frac{5}{9.64} = -0.519$$

$$\cos\theta_z = -\frac{2}{9.64} = -0.207$$

$$\mathbf{T} = 1000(-0.830\mathbf{i} - 0.519\mathbf{j} - 0.207\mathbf{k})$$
$$= -830\mathbf{i} - 519\mathbf{j} - 207\mathbf{k}$$

Therefore,

$$T_x = -830 \text{ lb} \qquad T_y = -519 \text{ lb} \qquad T_z = -207 \text{ lb} \qquad \textit{Answer}$$

Example 2.9

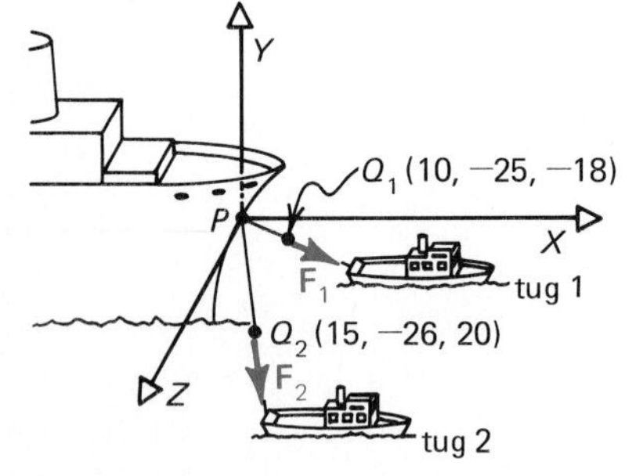

Figure 2.19

A steamer is being pulled by two tugs as shown in Fig. 2.19. The origin of the coordinate system is at point P on the steamer. Tug 1 pulls with a force of $\mathbf{F}_1$ and tug 2 with a force of $\mathbf{F}_2$ as shown. What is the resultant force on the steamer? The magnitude of both $\mathbf{F}_1$ and $\mathbf{F}_2$ is 10 kN.

Force $\mathbf{F}_1$ is

$$\mathbf{F}_1 = F_1\mathbf{e}_1$$

where
$$\mathbf{e}_1 = \cos\theta_{x1}\mathbf{i} + \cos\theta_{y1}\mathbf{j} + \cos\theta_{z1}\mathbf{k}$$

where $\cos\theta_{x1}$, $\cos\theta_{y1}$, and $\cos\theta_{z1}$ are direction cosines of $\mathbf{PQ}_1$. The distances are

$$D_{x1} = 10 \qquad D_{y1} = -25 \qquad D_{z1} = -18$$

and
$$D_1 = \sqrt{(10)^2 + (-25)^2 + (-18)^2} = 32.4$$

Therefore,

$$\cos\theta_{x1} = 0.309 \qquad \cos\theta_{y1} = -0.772 \qquad \cos\theta_{z1} = -0.556$$

so that
$$\mathbf{F}_1 = 10(0.309\mathbf{i} - 0.772\mathbf{j} - 0.556\mathbf{k})$$
$$= 3.09\mathbf{i} - 7.72\mathbf{j} - 5.56\mathbf{k}$$

Similarly,
$$\mathbf{F}_2 = F_2\mathbf{e}_2$$
$$\mathbf{e}_2 = \cos\theta_{x2}\mathbf{i} + \cos\theta_{y2}\mathbf{j} + \cos\theta_{z2}\mathbf{k}$$

where $\cos\theta_{x2}$, $\cos\theta_{y2}$, and $\cos\theta_{z2}$ are direction cosines of $\mathbf{PQ}_2$. The distances are

$$D_{x2} = 15 \qquad D_{y2} = -26 \qquad D_{z2} = 20$$

and
$$D_2 = \sqrt{(15)^2 - (-26)^2 + (20)^2} = 36.1$$

so that the direction cosines become

$$\cos\theta_{x2} = 0.416 \qquad \cos\theta_{y2} = -0.720 \qquad \cos\theta_{z2} = 0.554$$

and the force $\mathbf{F}_2$ can be written as

$$\mathbf{F}_2 = 10(0.416\mathbf{i} - 0.720\mathbf{j} + 0.554\mathbf{k})$$
$$= 4.16\mathbf{i} - 7.20\mathbf{j} + 5.54\mathbf{k}$$

The resultant force $\mathbf{R}$ is obtained from

$$\mathbf{R} = \mathbf{F}_1 + \mathbf{F}_2 = R_x\mathbf{i} + R_y\mathbf{j} + R_z\mathbf{k}$$

where
$$R_x = \sum F_x = 3.09 + 4.16 = 7.25$$
$$R_y = \sum F_y = -7.72 - 7.20 = -14.92$$
$$R_z = \sum F_z = -5.56 + 5.54 = -0.02$$

Substituting, we obtain

$$\mathbf{R} = (7.25\mathbf{i} - 14.9\mathbf{j} - 0.02\mathbf{k})\ \text{kN} \qquad \textit{Answer}$$

PROBLEMS

2.21 What are the direction cosines of the force $\mathbf{F}$ given by $\mathbf{F} = 1000\mathbf{i} + 600\mathbf{j} + 565\mathbf{k}$?

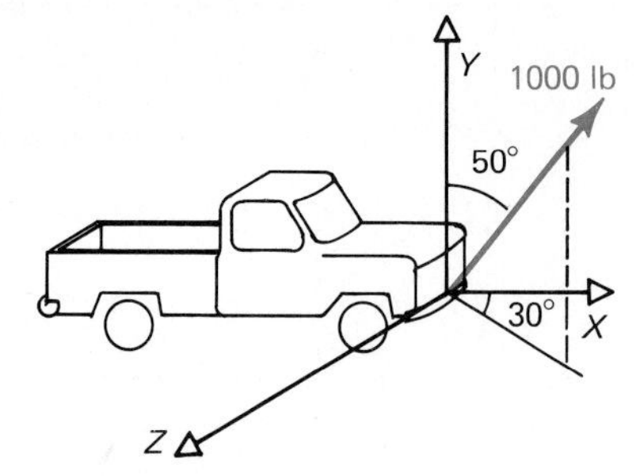

Figure P2.22

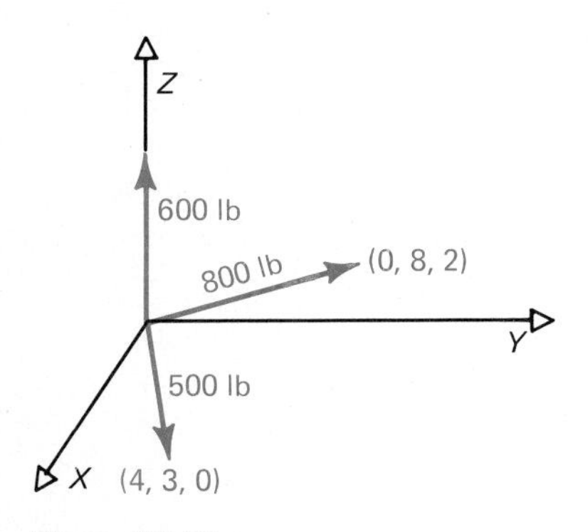

Figure P2.26

2.22 A 1000-lb force is applied to a truck as shown in Fig. P2.22. Resolve this force into components in the XYZ directions.

2.23 Two forces $\mathbf{T}_1 = (4\mathbf{i} - 5\mathbf{k})$ N and $\mathbf{T}_2 = (2\mathbf{i} + 3\mathbf{j} - 2\mathbf{k})$ N, act at a point on a beam. Find the resultant force $\mathbf{R}$ acting on the beam in terms of its magnitude R and unit vector $\mathbf{e}_R$.

2.24 The resultant force $\mathbf{R} = 20(0.519\mathbf{i} + 0.830\mathbf{j} + 0.207\mathbf{k})$ N is obtained by combining forces $\mathbf{F}_1$ and $\mathbf{F}_2$. If $\mathbf{F}_1 = (-8\mathbf{i} + 2\mathbf{j} + 3\mathbf{k})$ N, what is $\mathbf{F}_2$?

2.25 If $\mathbf{T}_1 = (12\mathbf{i} + 12\mathbf{j} - 10\mathbf{k})$ lb, $\mathbf{T}_2 = (-4\mathbf{i} + 2\mathbf{j} - 3\mathbf{k})$ lb, and $\mathbf{T}_3 = (3\mathbf{i} + 4\mathbf{k})$ lb, find the resultant force in terms of its magnitude R and unit vector $\mathbf{e}_R$.

2.26 Find the magnitude and direction cosines of the resultant force obtained by adding the three forces shown in Fig. P2.26.

2.27 A force whose magnitude is 500 N goes through points $A(5, 6, 2)$ and $B(2, 1, 3)$ in a cartesian coordinate system. Represent the force as a vector passing from A to B.

2.28 Find the magnitude and direction cosines of the resultant force obtained by the expression

$$\mathbf{R} = \mathbf{F}_1 + \mathbf{F}_2 + \mathbf{F}_3$$

where
$$\mathbf{F}_1 = (100\mathbf{i} + 200\mathbf{j} + 50\mathbf{k}) \text{ N}$$
$$\mathbf{F}_2 = (50\mathbf{i} - 10\mathbf{j} + 10\mathbf{k}) \text{ N}$$
$$\mathbf{F}_3 = (20\mathbf{i} + 60\mathbf{j} - 400\mathbf{k}) \text{ N}$$

2.29 A force of 25 lb goes through points $P(-2, 6, 1)$ and $Q(-5, -2, -4)$. Express this force as a vector quantity passing from P to Q.

2.30 Force $\mathbf{F}_1$ passes from point $P(1, 1, 1)$ to $Q_1(8, 6, 4)$ and force $\mathbf{F}_2$ passes from point $P(1, 1, 1)$ to $Q_2(6, -4, -3)$. The magnitude of $\mathbf{F}_1$ equals 16 lb and the magnitude of $\mathbf{F}_2$ equals 12 lb. What is the resultant force acting at point P?

2.3 Equilibrium of a Particle

From Newton's first law we know that a particle will remain at rest if the resultant force acting on the particle is zero. This condition of no motion and, therefore, zero resultant force, is called a **condition of equilibrium**.

If a particle is subjected to forces $\mathbf{F}_1$, $\mathbf{F}_2$, and $\mathbf{F}_3$, the sum of these forces must be zero for equilibrium. For simplicity, let these three forces lie in the XY plane as shown in Fig. 2.20(a). Using the polygon rule in Fig. 2.20(b), we observe that the three forces must complete the polygon since the resultant force is zero.

If the resultant force is zero, all of the components of the resultant force must also be zero. Therefore, for equilibrium

$$R_x = \sum F_x = 0 \qquad R_y = \sum F_y = 0$$

provided the set of forces acting on the particle lie in the XY plane. These equations obtained by setting the resultant force equal to zero are called the **equations of equilibrium**. In this two-dimensional case, we note that there are only two equations of equilibrium, and therefore at most two unknowns can be determined from these equations.

If the forces acting on the particle are three-dimensional, i.e., there are components along the X, Y, and Z axes, the resultant force $\mathbf{R}$ is also set equal to zero for the condition of equilibrium.

$$\mathbf{R} = \sum \mathbf{F} = 0$$

or

$$(\sum F_x)\mathbf{i} + (\sum F_y)\mathbf{j} + (\sum F_z)\mathbf{k} = 0$$

This vector equation is actually equivalent to three scalar equations of equilibrium that we obtain by setting the coefficient of each independent unit vector equation equal to zero, i.e.,

$$\sum F_x = 0$$
$$\sum F_y = 0 \qquad (2.11)$$
$$\sum F_z = 0$$

Since there are three equations of equilibrium for a particle in equilibrium for the three-dimensional case, there can be at most three unknowns that can be determined from these equations.

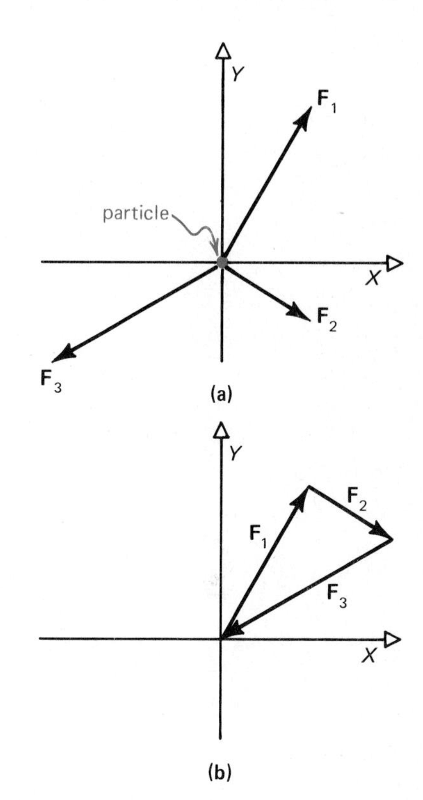

Figure 2.20

Free-Body Diagrams

In examining the conditions for the equilibrium of a particle, we introduce the free-body diagram. This is a sketch in which we show all forces acting on the particle. These forces act concurrently at the point that represents the particle in equilibrium. By way of example, let us examine a two-dimensional case in which a 10-kg traffic light is suspended as shown in Fig. 2.21(a). Its weight $w = 10(9.81) = 98.1$ N. Three forces intersect at point P, which represents the particle in equilibrium. These forces include the force in cable AP, the force in cable PB, and the weight of the traffic light. Recognizing this, we construct the free-body diagram of these three forces acting concurrently on the particle as shown in Fig. 2.21(b). Since point P is in equilibrium for this two-dimensional case, we should be able to obtain the magnitudes of the two forces $\mathbf{T}_A$ and $\mathbf{T}_B$ to ensure this equilibrium. These unknowns can be determined either graphically or algebraically. Note that we do know the direction of these forces.

For the graphical solution, the forces are combined to form the polygon shown in Fig. 2.21(c). To obtain this result, we draw

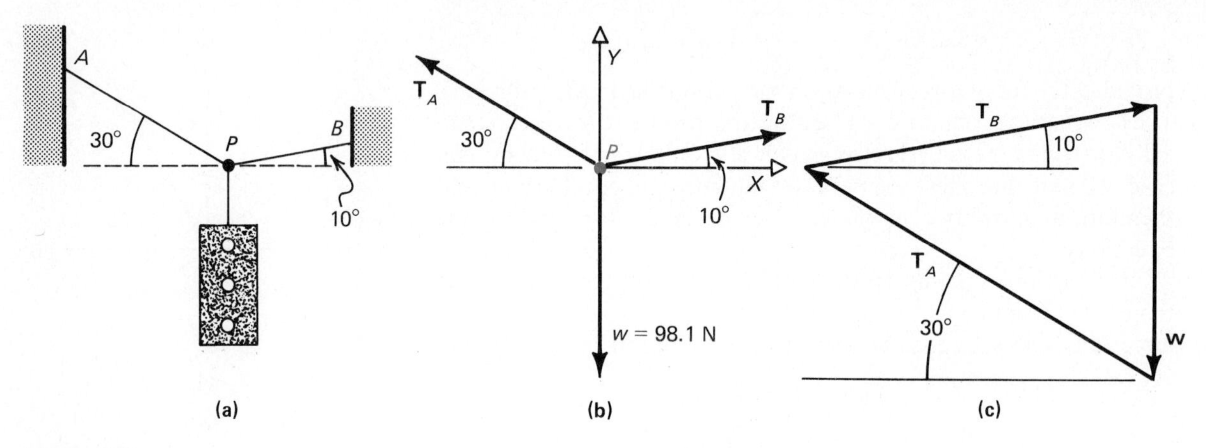

Figure 2.21

the vector **w** to some convenient scale. Next, we construct a line through the tip of **w** and at 30° off the horizontal to represent $\mathbf{T}_A$. Finally, we construct a line through the tail of **w** at 10° off the horizontal to represent $\mathbf{T}_B$. The intersection of these two lines closes the polygon, and the unknown magnitudes are measured to scale.

For the algebraic solution of the traffic light example, we use the two equations of equilibrium for the two-dimensional case,

$$\sum F_x = 0 \qquad \sum F_y = 0$$

Referring to the free-body diagram in Fig. 2.21(b),

$$\sum F_x = -T_A \cos 30^\circ + T_B \cos 10^\circ = 0$$

$$\sum F_y = T_A \sin 30^\circ + T_B \sin 10^\circ - 98.1 = 0$$

Rearranging the first equation,

$$T_B = \frac{T_A \cos 30^\circ}{\cos 10^\circ} = 0.879 T_A$$

Substituting this result into the second equation, we obtain

$$T_A \sin 30^\circ + (0.879 T_A) \sin 10^\circ = 98.1$$

$$T_A = 150 \text{ N}$$

Therefore $\qquad T_B = 0.879 T_A = 132 \text{ N}$

The following examples demonstrate the important role that free-body diagrams play in analyzing the equilibrium of forces acting on a particle. Drawing a free-body diagram is an essential step and one that we shall use extensively throughout our study of statics.

Example 2.10

A 220-lb man is walking a tightrope as shown in Fig. 2.22(a). What is the tension in the rope between F and P and between P and S to ensure equilibrium?

First we construct a free-body diagram of a particle at point P where the man is standing. The magnitudes of the three forces acting at point P are the weight $w = 220$ lb, a tensile force T_S in PS, and a tensile force T_F in PF as shown in Fig. 2.22(b). These tensions may be obtained by the polygon rule as shown in Fig. 2.22(c). The results are

$$T_S = 414 \text{ lb} \qquad \textit{Answer}$$

$$T_F = 434 \text{ lb} \qquad \textit{Answer}$$

For the algebraic solution, refer to the free-body diagram of Fig. 2.22(b). Summing the forces in the X and Y direction yields the two equations of equilibrium,

$$\sum F_x = T_S \cos 10^\circ - T_F \cos 20^\circ = 0$$

$$\sum F_y = T_S \sin 10^\circ + T_F \sin 20^\circ - 220 = 0$$

From the first equation,

$$T_F = 1.05 T_S$$

Substituting this result into the second equation gives

$$T_S \sin 10^\circ + (1.05 T_S) \sin 20^\circ = 220$$

$$T_S = 413 \text{ lb} \qquad \textit{Answer}$$

Consequently, $$T_F = 434 \text{ lb} \qquad \textit{Answer}$$

These results agree favorably with the graphical results shown in Fig. 2.22(c).

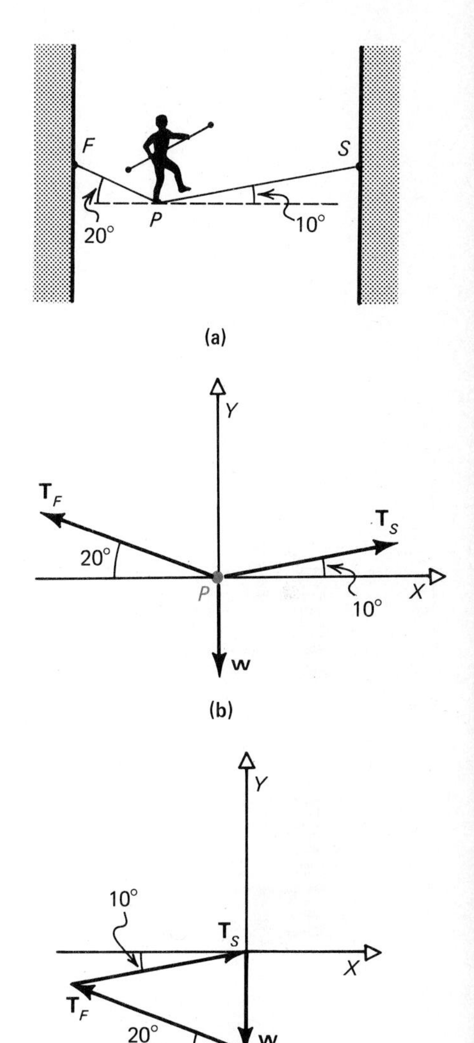

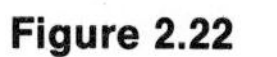
Figure 2.22

Example 2.11

A 52-kg block is suspended by three ropes lying in the XY plane as shown in Fig. 2.23(a). It is known that the tension in rope C is 150 N. What is the tension in ropes A and B for equilibrium?

The weight of the block is first established.

$$w = 9.81(52) = 510 \text{ N}$$

The free-body diagram of the forces acting at point P, which represents the particle in equilibrium, is shown in Fig. 2.23(b). The equilibrium equations are:

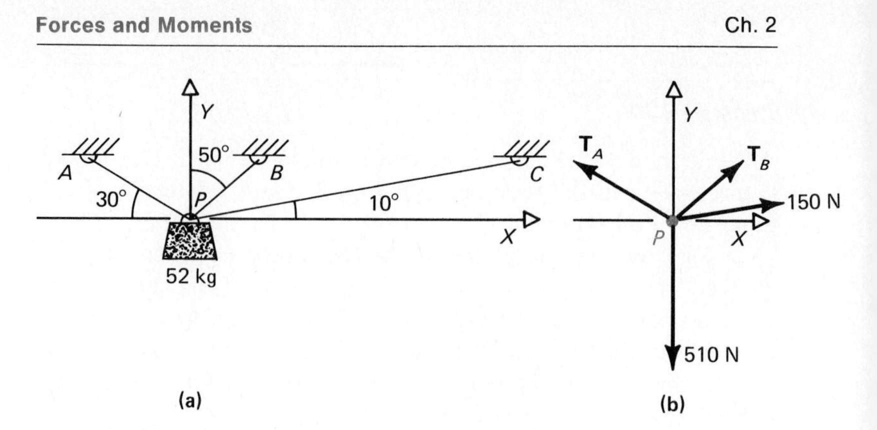

Figure 2.23

$$\sum F_x = -T_A \cos 30^\circ + T_B \sin 50^\circ + 150 \cos 10^\circ = 0$$

$$\sum F_y = T_A \sin 30^\circ + T_B \cos 50^\circ + 150 \sin 10^\circ - 510 = 0$$

From the first equation,

$$T_A = \frac{0.766T_B + 148}{0.866} = 0.885T_B + 171$$

Substituting into the second equilibrium equation,

$$0.5(0.885T_B + 171) + 0.643T_B + 26.0 - 510 = 0$$

$$(0.443 + 0.643)T_B = 510 - 26.0 - 85.5$$

$$T_B = \frac{398.5}{1.086} = 367 \text{ N} \qquad \textit{Answer}$$

Consequently, $T_A = 496$ N *Answer*

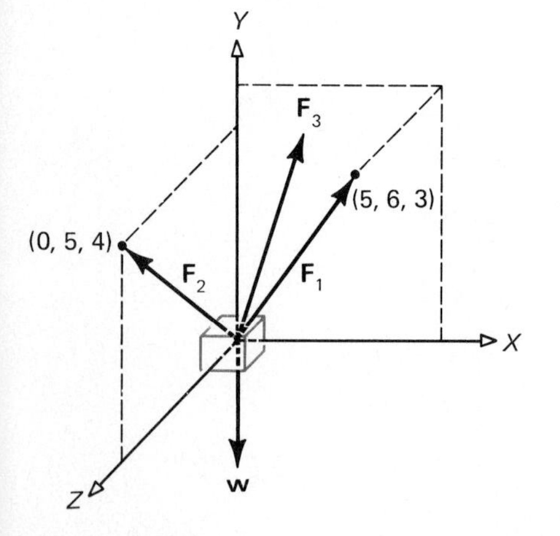

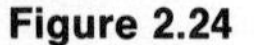

Figure 2.24

Example 2.12

Three forces $\mathbf{F}_1$, $\mathbf{F}_2$, and $\mathbf{F}_3$ support a 200-lb block as shown in the free-body diagram in Fig. 2.24. The weight of the block is represented by the force **w**. If the magnitudes of $\mathbf{F}_1$ and $\mathbf{F}_2$ are 60 lb and 50 lb, respectively, find the magnitude and direction of $\mathbf{F}_3$ for equilibrium.

In solving equilibrium problems in three dimensions, we shall first represent each force in vector form before writing the equations of equilibrium.

$$\mathbf{w} = -200\mathbf{j}$$

$$\mathbf{F}_1 = F_1\mathbf{e}_1 \qquad \mathbf{F}_2 = F_2\mathbf{e}_2 \qquad \mathbf{F}_3 = F_3\mathbf{e}_3$$

The coordinates through which $\mathbf{F}_1$ and $\mathbf{F}_2$ pass are given in Fig. 2.24.

$$\mathbf{e}_1 = \frac{5\mathbf{i} + 6\mathbf{j} + 3\mathbf{k}}{D_1} \qquad D_1 = \sqrt{5^2 + 6^2 + 3^2} = 8.37$$

$$\mathbf{e}_2 = \frac{5\mathbf{j} + 4\mathbf{k}}{D_2} \qquad D_2 = \sqrt{5^2 + 4^2} = 6.40$$

so that,

$$\mathbf{e}_1 = 0.597\mathbf{i} + 0.717\mathbf{j} + 0.358\mathbf{k}$$

$$\mathbf{e}_2 = 0.781\mathbf{j} + 0.625\mathbf{k}$$

The condition for equilibrium is

$$\mathbf{F}_1 + \mathbf{F}_2 + \mathbf{F}_3 + \mathbf{w} = 0$$

Substituting into this equation for each term, we obtain

$$60(0.597\mathbf{i} + 0.717\mathbf{j} + 0.358\mathbf{k}) + 50(0.781\mathbf{j} + 0.625\mathbf{k}) + \mathbf{F}_3 - 200\mathbf{j} = 0$$

$$35.8\mathbf{i} - 117.9\mathbf{j} + 52.7\mathbf{k} + \mathbf{F}_3 = 0$$

$$\mathbf{F}_3 = -35.8\mathbf{i} + 117.9\mathbf{j} - 52.7\mathbf{k}$$

The magnitude of the force is

$$F_3 = \sqrt{(-35.8)^2 + (117.9)^2 + (-52.7)^2} = 134.0 \text{ lb} \qquad \textit{Answer}$$

The direction of the force is given by its unit vector

$$\mathbf{e}_3 = \frac{\mathbf{F}_3}{F_3} = -0.267\mathbf{i} + 0.880\mathbf{j} - 0.393\mathbf{k} \qquad \textit{Answer}$$

Example 2.13

A 100-kg chandelier is suspended by three chains as shown in Fig. 2.25(a). Find the magnitude of the force in each chain for equilibrium.

The weight of the chandelier is

$$w = 9.81(100) = 981 \text{ N}$$

The free-body diagram is shown in Fig. 2.25(b). Here we have three unknowns, namely the magnitude of the force in each chain. Before proceeding to the equations of equilibrium, each force acting on the particle P is first written as a vector quantity.

$$\mathbf{w} = -981\mathbf{j}$$

and

$$\mathbf{T}_A = T_A\mathbf{e}_A \qquad \mathbf{T}_B = T_B\mathbf{e}_B \qquad \mathbf{T}_C = T_C\mathbf{e}_C$$

where

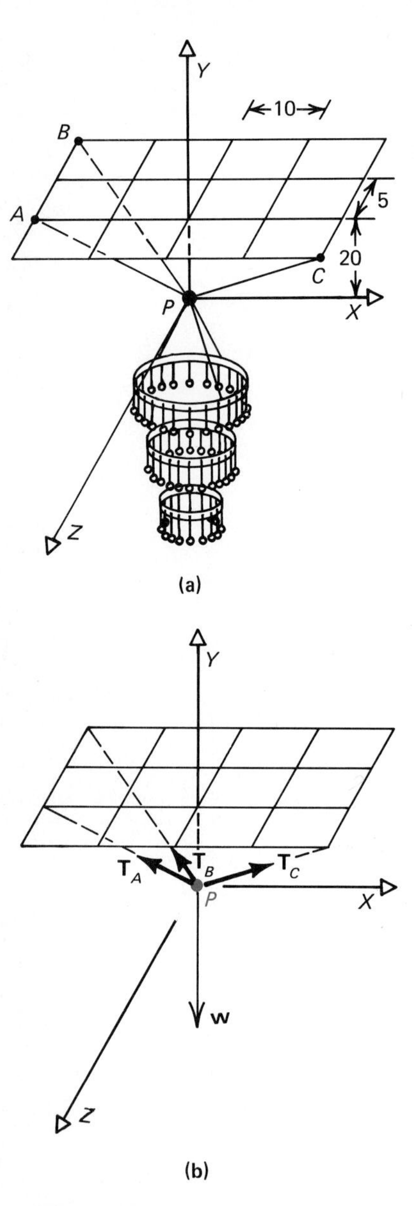

Figure 2.25

$$\mathbf{e}_A = \frac{-20\mathbf{i} + 20\mathbf{j}}{D_A} \qquad D_A = \sqrt{(-20)^2 + (20)^2} = 28.3$$

$$\mathbf{e}_B = \frac{-20\mathbf{i} + 20\mathbf{j} - 10\mathbf{k}}{D_B} \qquad D_B = \sqrt{(-20)^2 + (20)^2 + (-10)^2} = 30.0$$

$$\mathbf{e}_C = \frac{20\mathbf{i} + 20\mathbf{j} + 5\mathbf{k}}{D_C} \qquad D_C = \sqrt{(20)^2 + (20)^2 + (5)^2} = 28.7$$

Thus,

$$\mathbf{e}_A = -0.707\mathbf{i} + 0.707\mathbf{j}$$

$$\mathbf{e}_B = -0.667\mathbf{i} + 0.667\mathbf{j} - 0.333\mathbf{k}$$

$$\mathbf{e}_C = 0.697\mathbf{i} + 0.697\mathbf{j} + 0.174\mathbf{k}$$

The condition for equilibrium is

$$\mathbf{w} + \mathbf{T}_A + \mathbf{T}_B + \mathbf{T}_C = 0$$

where the magnitudes T_A, T_B, and T_C are unknown. Substituting into the equilibrium equation we obtain

$$-981\mathbf{j} + T_A(-0.707\mathbf{i} + 0.707\mathbf{j}) + T_B(-0.667\mathbf{i} + 0.667\mathbf{j} - 0.333\mathbf{k}) + T_C(0.697\mathbf{i} + 0.697\mathbf{j} + 0.174\mathbf{k}) = 0$$

Now write the separate scalar equations for equilibrium:

for **i**: $$-0.707T_A - 0.667T_B + 0.697T_C = 0$$

for **j**: $$0.707T_A + 0.667T_B + 0.697T_C = 981$$

for **k**: $$-0.333T_B + 0.174T_C = 0$$

Adding the first two equations,

$$1.394T_C = 981$$

$$T_C = 704 \text{ N} \qquad \textit{Answer}$$

From the third equilibrium equation,

$$T_B = 0.523T_C = 368 \text{ N} \qquad \textit{Answer}$$

From the first equilibrium equation,

$$T_A = \frac{-0.667(368) + 0.697(704)}{0.707} = 347 \text{ N} \qquad \textit{Answer}$$

PROBLEMS

2.31 Graphically obtain T_1 and T_2 so that the particle P in Fig. P2.31 is in equilibrium.

2.32 Graphically obtain T_1 and T_2 so that the particle P in Fig. P2.32 is in equilibrium.

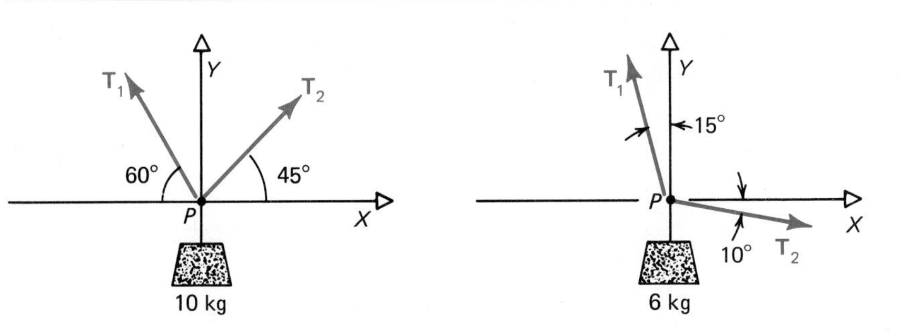

Figure P2.31 **Figure P2.32**

2.33 Considering the equilibrium of the particle at C, obtain graphically the tension in cables AC and BC in Fig. P2.33.

2.34 Considering the equilibrium of the particle at C, obtain graphically the tension in cables AC and BC in Fig. P2.34

2.35 Graphically obtain T and α so that the particle P in Fig. P2.35 is in equilibrium.

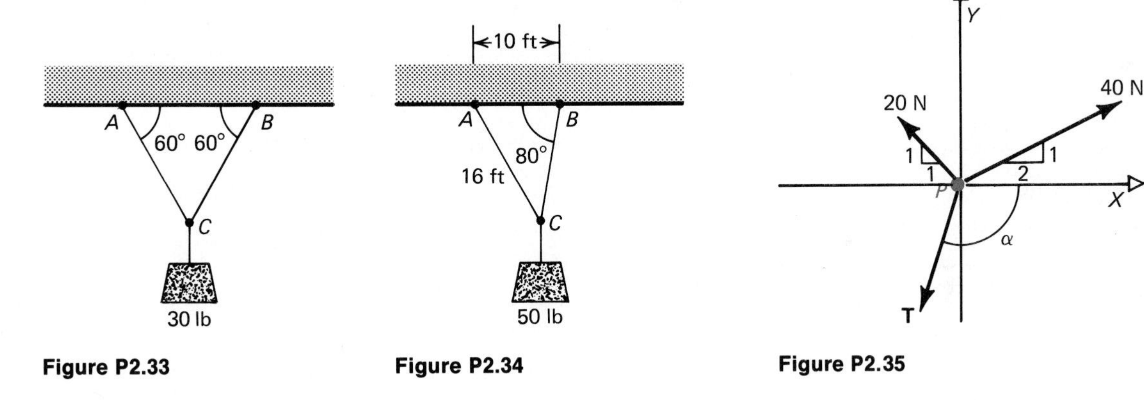

Figure P2.33 **Figure P2.34** **Figure P2.35**

2.36 Graphically obtain the magnitude of the force **T** and α so that the particle P in Fig. P2.36 is in equilibrium.

2.37 Solve Problem 2.31 algebraically.

2.38 Solve Problem 2.32 algebraically.

2.39 Solve Problem 2.33 algebraically.

2.40 Solve Problem 2.34 algebraically.

2.41 Solve Problem 2.35 algebraically.

2.42 Solve Problem 2.36 algebraically.

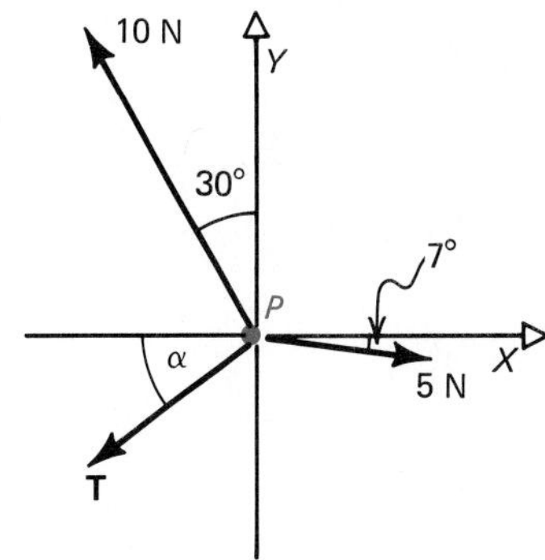

Figure P2.36

2.43 A particle is subjected to two forces given by $\mathbf{F}_1 = 8\mathbf{i} + 6\mathbf{j}$ and $\mathbf{F}_2 = 9\mathbf{i} + 15\mathbf{j}$. What additional force must be applied in order to have the particle at equilibrium? Assume all forces are in pounds.

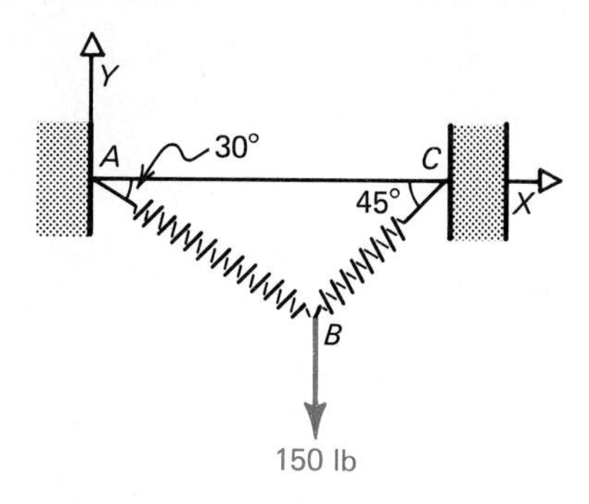

Figure P2.44

2.44 A spring is stretched by a 150-lb force as shown in Fig. P2.44. What is the tension in AB and BC?

2.45 Write the equations of equilibrium for the particle at P in terms of a and h as shown in Fig. P2.45.

2.46 A particle is subjected to four planar forces as shown in Fig. P2.46. Determine the magnitude of the forces $\mathbf{F}_1$ and $\mathbf{F}_2$ necessary for the particle to be in equilibrium.

2.47 The maximum tension in ropes AB and BC shown in Fig. P2.47 can be 500 lb. What is the maximum value of w to ensure equilibrium? Assume the tension in the cable from B to D equals the magnitude of the weight.

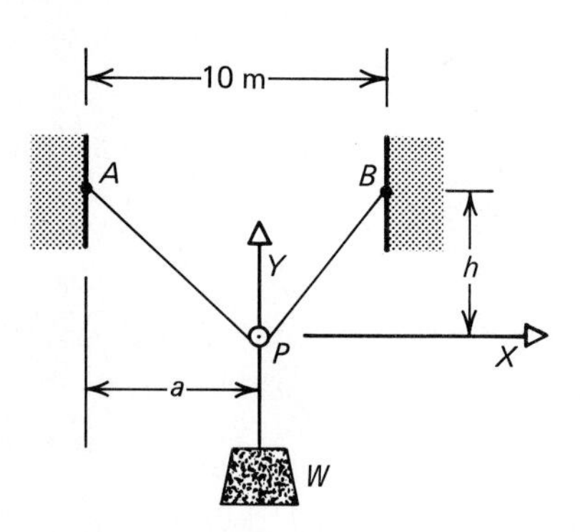

Figure P2.45

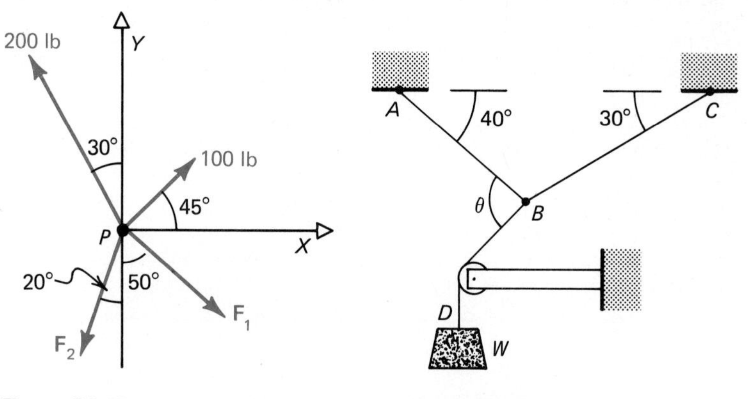

Figure P2.46

Figure P2.47

Figure P2.48

2.48 A crane is used to pick up a bundle of logs weighing 1 ton, as shown in Fig. P2.48. A person pulls to prevent any swaying. For the instant shown, what is the tension in the two cables AB and BC?

2.49 A particle is subjected to two forces $\mathbf{F}_1$ and $\mathbf{F}_2$ given by $\mathbf{F}_1 = 5\mathbf{i} + 6\mathbf{j} + 2\mathbf{k}$ and $\mathbf{F}_2 = \mathbf{i} + 3\mathbf{j} - \mathbf{k}$, where the magnitude is in newtons. (a) What is the magnitude and (b) what are the direction cosines of the third force necessary for equilibrium?

2.50 Find the force $\mathbf{F}_3$ for equilibrium of particle P in Fig. P2.50 if $F_1 = 10$ N and $F_2 = 6$ N.

2.51 Find the direction cosines of the force $\mathbf{F}_3$ in Fig. P2.50 if $F_1 = 8$ lb and $F_2 = 10$ lb.

2.52 Find the force $\mathbf{F}_3$ for equilibrium of particle P in Fig. P2.52. Assume w equals 40 lb.

2.53 Solve Problem 2.52 for $w = 60$ lb.

2.54 A 100-kg block is supported by three cables as shown in Fig. P2.54. Determine the tension in each cable.

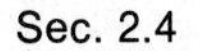

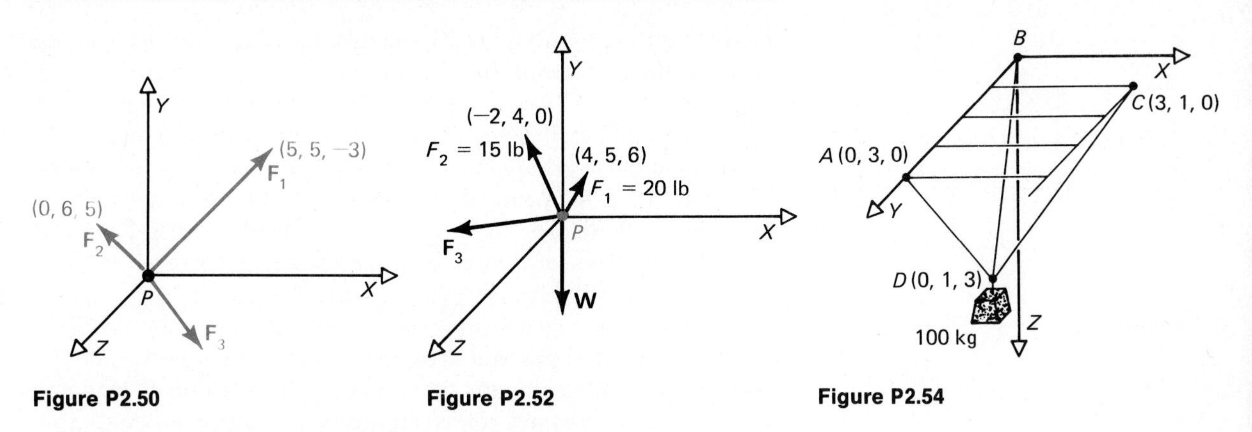

Figure P2.50 **Figure P2.52** **Figure P2.54**

2.55 Determine the forces in the three rigid members PA, PB, and PC shown in Fig. P2.55.

2.56 A traffic light whose mass equals 81.55 kg is suspended as shown in Fig. P2.56. Determine the tension in cables AB, AC, and AD.

2.57 A 240-lb block is suspended from a horizontal circular plate by three wires as shown in Fig. P2.57. The plate is 3 ft in diameter and the point D is 4 ft below the center of the plate. Determine the magnitude of the tension in each wire.

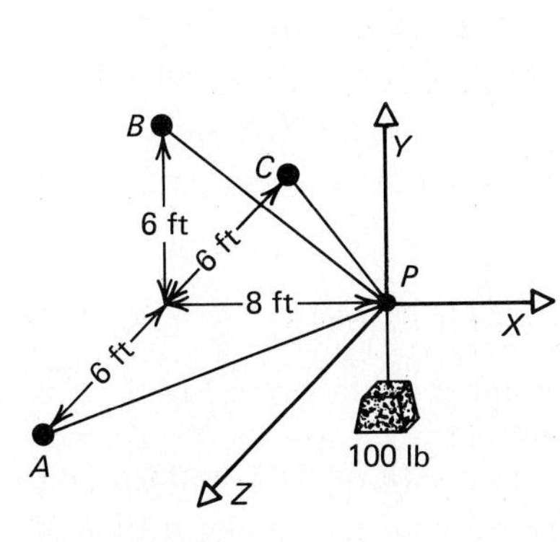

Figure P2.55

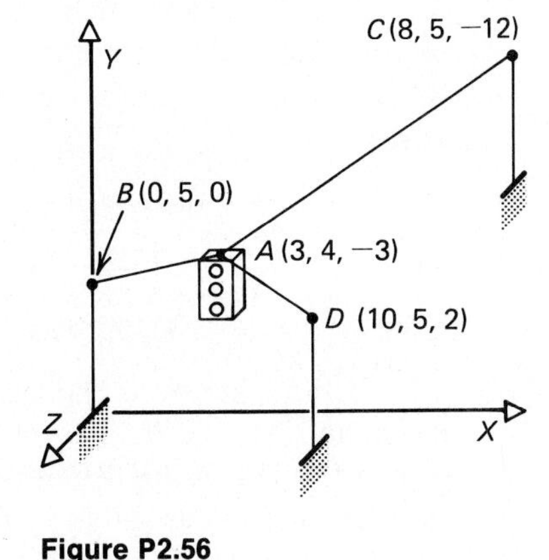

Figure P2.56

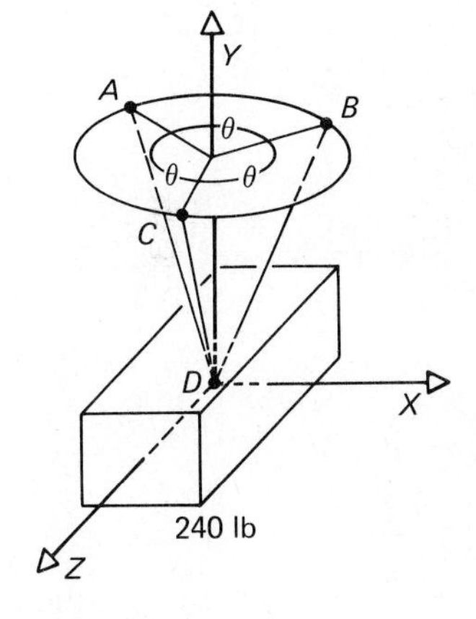

Figure P2.57

2.4 Moment of a Force

A moment applied to a body creates a tendency for the body to rotate. Figure 2.26 shows an example of a moment produced by a force. Here we have a force in the XY plane applied perpendicular to a balanced beam that will rotate about the pivot point O. The

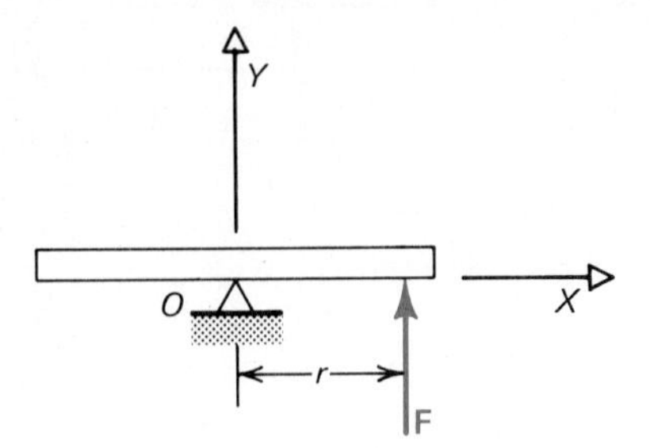

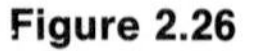
Figure 2.26

moment developed about O, $\mathbf{M}_o$, equals the product of the distance r, called the lever arm, and the force $\mathbf{F}$ acting perpendicular to $\mathbf{r}$. The magnitude of the moment is $M_o = rF$, and its direction of rotation is counterclockwise. Since a moment has both a magnitude and a direction associated with it, we represent moments by vectors. The units of a moment are normally in meter-newton (m·N) or foot-pound (ft·lb). The moment of a force is also called a torque.

The moment of a force is defined in vector form as

$$\mathbf{M} = \mathbf{r} \times \mathbf{F} \tag{2.12}$$

The vector $\mathbf{r}$ is the position vector from the point about which the moment is calculated to any point on the line of action of the force $\mathbf{F}$. Note that this expression is applicable for both the two-dimensional case (forces in the XY plane), and the three-dimensional case. In the example of the balanced beam, $\mathbf{r} = r\mathbf{i}$ and $\mathbf{F} = F\mathbf{j}$, so that the moment developed about O is

$$\mathbf{M}_o = r\mathbf{i} \times (F\mathbf{j}) = rF\mathbf{k}$$

The result provides us with both the magnitude and the direction of the moment. This direction is determined by the right-hand rule for the cross product. By pointing our thumb out of the page to represent the unit vector $\mathbf{k}$, our fingers indicate a rotation in the counterclockwise direction.

As a second example, consider the moment developed about point O due to the force $\mathbf{P}$ in the XY plane as shown in Fig. 2.27 (a). In this case, the position vector from point O to the force applied at point A is

$$\mathbf{r} = a\mathbf{i} + b\mathbf{j}$$

and $$\mathbf{F} = -P\mathbf{i}$$

Now $$\mathbf{M}_o = (a\mathbf{i} + b\mathbf{j}) \times (-P\mathbf{i}) = bP\mathbf{k}$$

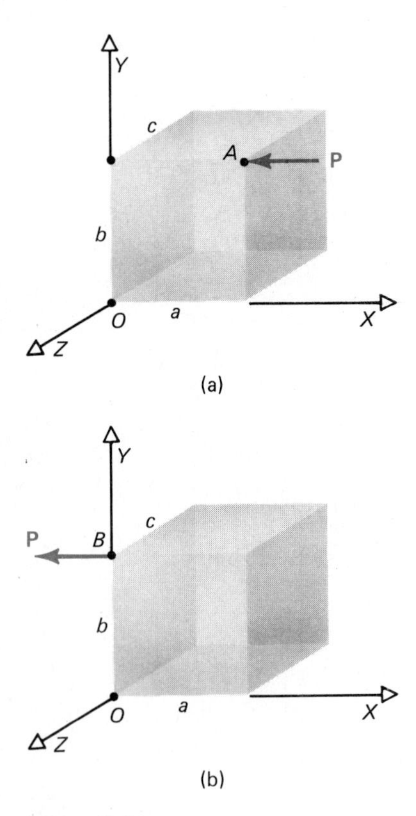

Figure 2.27

If the force were applied at point B as shown in Fig. 2.27(b), $\mathbf{r} = b\mathbf{j}$ and $\mathbf{M}_o = bP\mathbf{k}$. That is, we obtain the same moment about O if the force $\mathbf{P}$ is applied at point A or point B. This is always the case, i.e., as long as the force is applied anywhere along its line of action, the resulting moment about the fixed point O is the same.

Stated more formally, we say that the moment $\mathbf{M}_o$ depends upon $\mathbf{r}$ as well as upon the magnitude and direction of the force $\mathbf{F}$, but does not depend on its point of application. Therefore, if we know only $\mathbf{M}_o$, we obviously can say nothing about the point of application of the force $\mathbf{F}$ that produced it. Also, from the transmissibility principle, we say that the two forces shown in Fig. 2.28 are equivalent provided they have the same magnitude, direction, and line of action. We may now expand this to say that two forces are equivalent if, in addition, they produce the same moment about any given point. If the two forces in Fig. 2.28 have the same magnitude, they will produce the same moment about the point O and are therefore equivalent. We will have more to say about equivalent forces in Section 2.6.

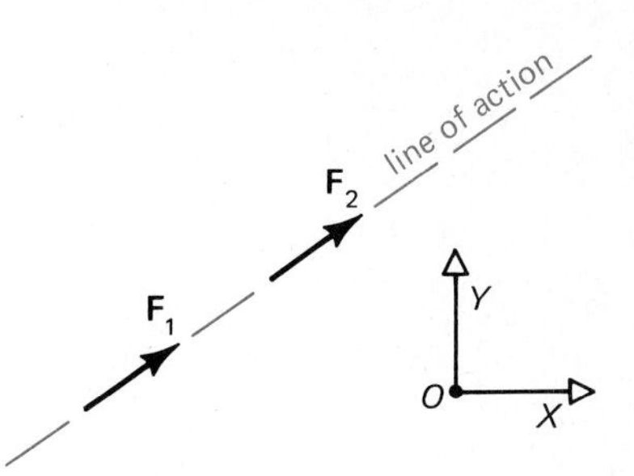

Figure 2.28

Let us now enlarge upon the previous example and consider the same block subjected to the forces **P** and **Q** shown in Fig. 2.29. Now, the moment about O is

$$\mathbf{M}_o = \mathbf{r}_1 \times \mathbf{F}_1 + \mathbf{r}_2 \times \mathbf{F}_2$$

where $\mathbf{r}_1 = (a\mathbf{i} + b\mathbf{j}) \qquad \mathbf{F}_1 = -P\mathbf{i}$

$\mathbf{r}_2 = b\mathbf{j} \qquad \mathbf{F}_2 = Q\mathbf{k}$

Therefore
$$\mathbf{M}_o = (a\mathbf{i} + b\mathbf{j}) \times (-P\mathbf{i}) + (b\mathbf{j}) \times (Q\mathbf{k})$$
$$= bP\mathbf{k} + bQ\mathbf{i}$$

Note that the moment tends to rotate not only about the Z axis, but also the X axis.

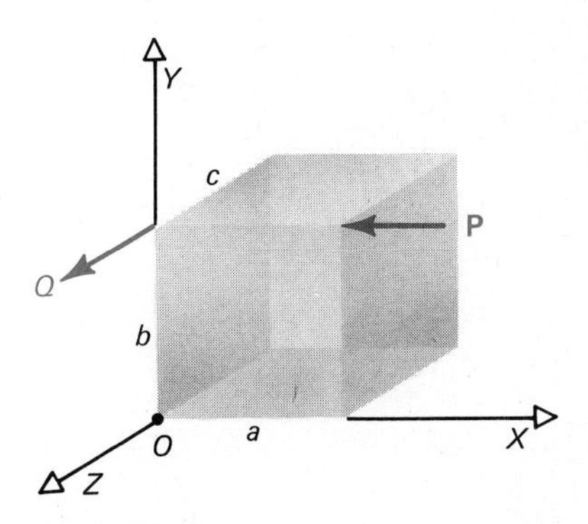

Figure 2.29

Components of a Moment

Consider a force in the XY plane acting at point A, where

$$\mathbf{F} = F_x\mathbf{i} + F_y\mathbf{j}$$

and the position vector to point A from the origin is

$$\mathbf{r} = x\mathbf{i} + y\mathbf{j}$$

as shown in Fig. 2.30. The moment about O due to this force is

$$\mathbf{M}_o = \mathbf{r} \times \mathbf{F} \tag{2.12}$$

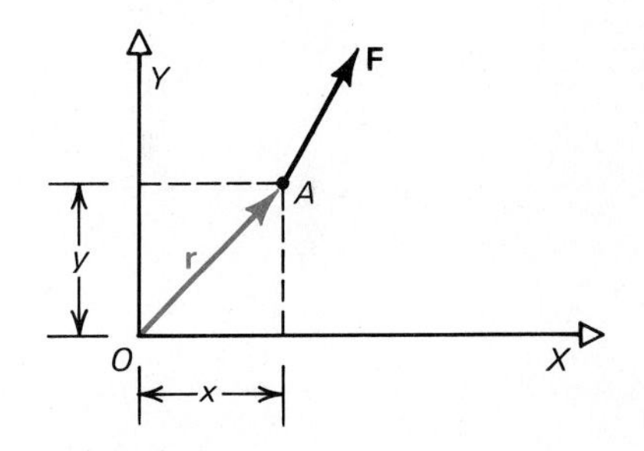

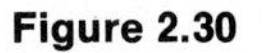
Figure 2.30

Recall from Chapter 1 that a vector product can be written in matrix form, so that

$$\mathbf{M}_o = \mathbf{r} \times \mathbf{F} = \begin{vmatrix} \mathbf{i} & \mathbf{j} & \mathbf{k} \\ x & y & 0 \\ F_x & F_y & 0 \end{vmatrix}$$
$$= (xF_y - yF_x)\mathbf{k}$$
$$= M_z\mathbf{k}$$

where $M_z = (xF_y - yF_x)$

The moment vector due to a force in the XY plane is directed along the Z axis and has a magnitude of M_z. In this case the moment has only one component which is along the Z axis.

For the three-dimensional case,

$$\mathbf{F} = F_x\mathbf{i} + F_y\mathbf{j} + F_z\mathbf{k} \qquad \text{and} \qquad \mathbf{r} = x\mathbf{i} + y\mathbf{j} + z\mathbf{k}$$

then the expression for the moment becomes

$$\mathbf{M}_o = \mathbf{r} \times \mathbf{F} = \begin{vmatrix} \mathbf{i} & \mathbf{j} & \mathbf{k} \\ x & y & z \\ F_x & F_y & F_z \end{vmatrix}$$
$$= (yF_z - zF_y)\mathbf{i} + (zF_x - xF_z)\mathbf{j} + (xF_y - yF_x)\mathbf{k}$$

Here, the moment has three components and may be written as

$$\mathbf{M}_o = M_x\mathbf{i} + M_y\mathbf{j} + M_z\mathbf{k}$$

where

$$M_x = yF_z - zF_y \qquad M_y = zF_x - xF_z \qquad M_z = xF_y - yF_x$$

The magnitude of the moment is given by

$$M_o = \sqrt{M_x^2 + M_y^2 + M_z^2}$$

Finally, as an alternate form we can write the moment as

$$\mathbf{M}_o = M_o\mathbf{e}_M$$

where $\mathbf{e}_M = \cos\theta_x\mathbf{i} + \cos\theta_y\mathbf{j} + \cos\theta_z\mathbf{k}$

and the direction cosines are obtained as introduced earlier, i.e.,

$$\cos\theta_x = \frac{M_x}{M_o} \qquad \cos\theta_y = \frac{M_y}{M_o} \qquad \cos\theta_z = \frac{M_z}{M_o}$$

Moment Due to Several Forces

If n concurrent forces are applied to a point P and $\mathbf{r}$ is the position vector from point O to point P, the moment caused by these forces is

$$\mathbf{M}_o = \mathbf{r} \times (\mathbf{F}_1 + \mathbf{F}_2 + \cdots + \mathbf{F}_n)$$

Since vector products are distributive, this can be written as

$$\mathbf{M}_o = \mathbf{r} \times (\mathbf{F}_1 + \mathbf{F}_2 + \cdots + \mathbf{F}_n)$$
$$= (\mathbf{r} \times \mathbf{F}_1) + (\mathbf{r} \times \mathbf{F}_2) + \cdots + (\mathbf{r} \times \mathbf{F}_n)$$

This equation states that the moment due to several concurrent forces is equal to the sum of the moments caused by the individual forces. This observation is called **Varignon's theorem**, after the French mathematician Varignon (1654–1722) who first deduced it.

Example 2.14

Compute the moment about O due to force $\mathbf{F}$ applied at point A on the structural member shown in Fig. 2.31(a).

We first represent the force as a vector,

$$\mathbf{F} = 10(\cos 30^\circ \mathbf{i} - \sin 30^\circ \mathbf{j})$$
$$= 8.66\mathbf{i} - 5\mathbf{j}$$

The position vector $\mathbf{r}$ from point O to point A is

$$\mathbf{r} = (0.12\mathbf{i} + 0.10\mathbf{j})\,\text{m}$$

The moment about O is

$$\mathbf{M}_o = \mathbf{r} \times \mathbf{F} = \begin{vmatrix} \mathbf{i} & \mathbf{j} & \mathbf{k} \\ 0.12 & 0.10 & 0 \\ 8.66 & -5 & 0 \end{vmatrix}$$

$$= -1.466\mathbf{k}\ \text{m}\cdot\text{N} \qquad \textit{Answer}$$

Note that the negative sign means that the moment acts in the clockwise (CW) direction.

As an alternative solution, we could solve the problem by using the components of the force as shown in Fig. 2.31(b). Multiply each component by its perpendicular lever arm to point O keeping in mind that counterclockwise (CCW) moments are positive.

$$M_o = -8.66(0.10) - 5.0(0.12) = -1.466$$

or $$M_o = 1.466\ \text{m}\cdot\text{N} \curvearrowleft$$

which is the answer obtained by the first method. Notice that $\curvearrowleft$ represents the clockwise direction and is the direction in which the body tends to rotate. From the right-hand rule, a screw tending to rotate this way advances in the $-\mathbf{k}$ direction. Therefore, $\curvearrowleft$ and $-\mathbf{k}$ are equivalent in that they represent the same direction. Similarly, $\curvearrowright$ is equivalent to the $+\mathbf{k}$ direction.

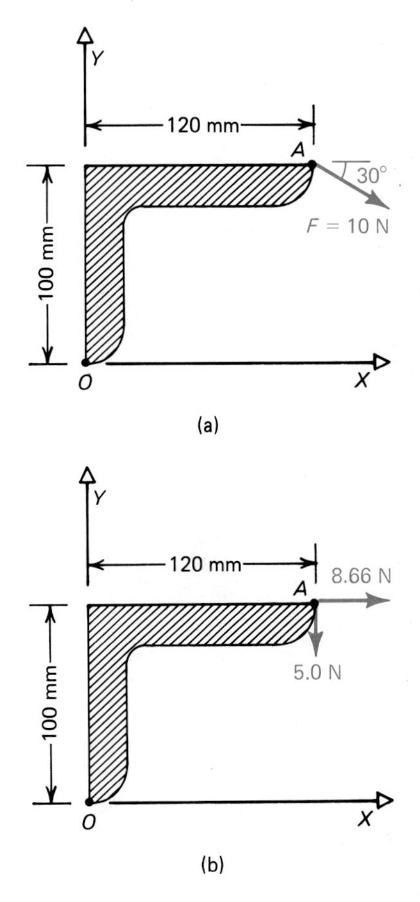

Figure 2.31

Example 2.15

A transmitting antenna is secured by two guy wires lying in the XY plane as shown in Fig. 2.32(a). If the tension in wire AB equals 40 lb and the tension in wire CD equals 30 lb, find the moment due to these two tensions about point O.

The forces in each wire are shown acting on the antenna in Fig. 2.32(b). Let $\mathbf{F}_1$ be the force in AB and $\mathbf{F}_2$ be the force in CD.

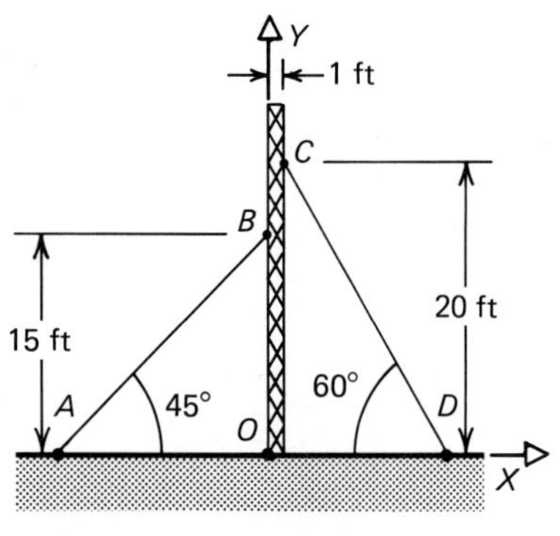

(a)

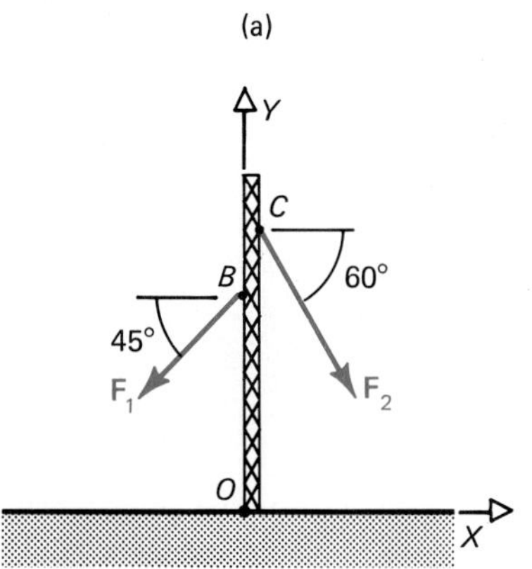

(b)

Figure 2.32

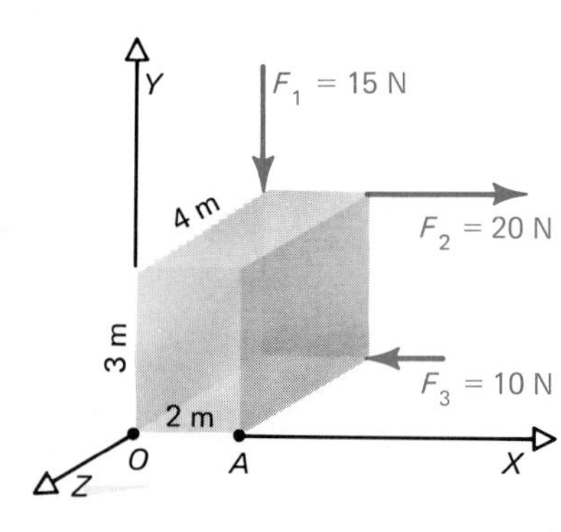

Figure 2.33

The position vectors and corresponding forces are

$$\mathbf{r}_1 = 15\mathbf{j} \qquad \mathbf{F}_1 = 40(-\cos 45°\mathbf{i} - \sin 45°\mathbf{j}) = -28.3\mathbf{i} - 28.3\mathbf{j}$$

$$\mathbf{r}_2 = \mathbf{i} + 20\mathbf{j} \qquad \mathbf{F}_2 = 30(\cos 60°\mathbf{i} - \sin 60°\mathbf{j}) = 15.0\mathbf{i} - 26.0\mathbf{j}$$

The moment developed by each force is as follows:

$$\mathbf{r}_1 \times \mathbf{F}_1 = \begin{vmatrix} \mathbf{i} & \mathbf{j} & \mathbf{k} \\ 0 & 15 & 0 \\ -28.3 & -28.3 & 0 \end{vmatrix} = 425\mathbf{k}$$

$$\mathbf{r}_2 \times \mathbf{F}_2 = \begin{vmatrix} \mathbf{i} & \mathbf{j} & \mathbf{k} \\ 1 & 20 & 0 \\ 15.0 & -26.0 & 0 \end{vmatrix} = -326\mathbf{k}$$

The total moment developed about point O is

$$\begin{aligned} \mathbf{M}_o &= (\mathbf{r}_1 \times \mathbf{F}_1) + (\mathbf{r}_2 \times \mathbf{F}_2) \\ &= 425\mathbf{k} - 326\mathbf{k} = 99\mathbf{k} \text{ ft}\cdot\text{lb} \end{aligned}$$ *Answer*

The positive sign means that the moment acts in the counterclockwise direction.

Example 2.16

Find the moment developed by the forces acting on the block in Fig. 2.33 (a) about point O, and (b) about point A.

(a) The forces and position vectors relative to point O are as follows:

$$\mathbf{r}_1 = 3\mathbf{j} - 4\mathbf{k} \qquad \mathbf{F}_1 = -15\mathbf{j}$$

$$\mathbf{r}_2 = 2\mathbf{i} + 3\mathbf{j} - 4\mathbf{k} \qquad \mathbf{F}_2 = 20\mathbf{i}$$

$$\mathbf{r}_3 = 2\mathbf{i} - 4\mathbf{k} \qquad \mathbf{F}_3 = -10\mathbf{i}$$

The moment about point O is the sum of the moments due to each force.

$$\begin{aligned} \mathbf{M}_o &= (\mathbf{r}_1 \times \mathbf{F}_1) + (\mathbf{r}_2 \times \mathbf{F}_2) + (\mathbf{r}_3 \times \mathbf{F}_3) \\ &= (3\mathbf{j} - 4\mathbf{k}) \times (-15\mathbf{j}) + (2\mathbf{i} + 3\mathbf{j} - 4\mathbf{k}) \\ &\quad \times (20\mathbf{i}) + (2\mathbf{i} - 4\mathbf{k}) \times (-10\mathbf{i}) \\ &= -60\mathbf{i} - 60\mathbf{k} - 80\mathbf{j} + 40\mathbf{j} \\ &= (-60\mathbf{i} - 40\mathbf{j} - 60\mathbf{k}) \text{ m}\cdot\text{N} \end{aligned}$$ *Answer*

(b) The position vectors from point A to the three forces are as follows:

$$\mathbf{r}_1 = -2\mathbf{i} + 3\mathbf{j} - 4\mathbf{k}$$

$$\mathbf{r}_2 = 3\mathbf{j} - 4\mathbf{k}$$

$$\mathbf{r}_3 = -4\mathbf{k}$$

The moment about point A is obtained by cross multiplying each new position vector with its appropriate force and then adding the results.

$$\begin{aligned}\mathbf{M}_A &= (-2\mathbf{i} + 3\mathbf{j} - 4\mathbf{k}) \times (-15\mathbf{j}) + (3\mathbf{j} - 4\mathbf{k}) \\ &\quad \times (20\mathbf{i}) + (-4\mathbf{k}) \times (-10\mathbf{i}) \\ &= 30\mathbf{k} - 60\mathbf{i} - 60\mathbf{k} - 80\mathbf{j} + 40\mathbf{j} \\ &= (-60\mathbf{i} - 40\mathbf{j} - 30\mathbf{k})\ \text{m}\cdot\text{N}\end{aligned}$$ *Answer*

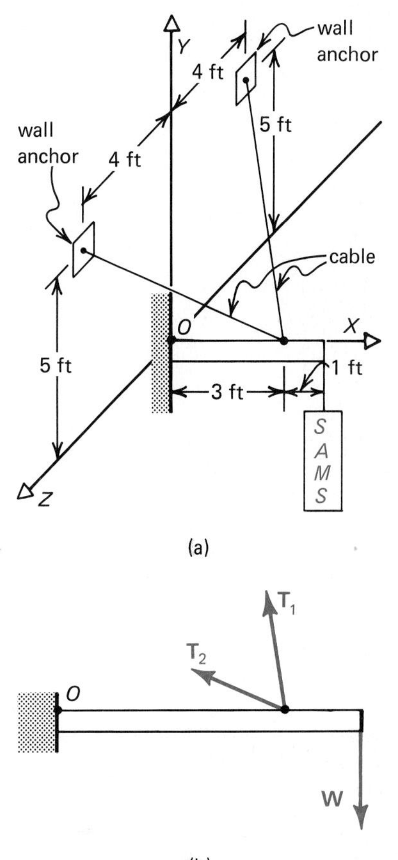

(a)

(b)

Figure 2.34

Example 2.17

The two cables in Fig. 2.34(a) are used to support a thin rod on which a 150-lb sign is hung. The tension in each cable is 100 lb. What is the resultant moment about O due to the forces in the cables and the weight of the sign?

Figure 2.34(b) shows the cable forces and the weight acting on the rod. The forces and position vectors are as follows:

$$\mathbf{r}_w = 4\mathbf{i} \qquad \mathbf{w} = -150\mathbf{j}$$

$$\mathbf{r}_1 = 3\mathbf{i} \qquad \mathbf{T}_1 = 100\left(\frac{-3\mathbf{i} + 5\mathbf{j} - 4\mathbf{k}}{D_1}\right)$$

where

$$D_1 = \sqrt{(-3)^2 + (5)^2 + (-4)^2} = 7.07$$

$$\therefore \quad \mathbf{T}_1 = -42.4\mathbf{i} + 70.7\mathbf{j} - 56.6\mathbf{k}$$

$$\mathbf{r}_2 = 3\mathbf{i} \qquad \mathbf{T}_2 = 100\left(\frac{-3\mathbf{i} + 5\mathbf{j} + 4\mathbf{k}}{D_2}\right)$$

where

$$D_2 = \sqrt{(-3)^2 + (5)^2 + (4)^2} = 7.07$$

$$\therefore \quad \mathbf{T}_2 = -42.4\mathbf{i} + 70.7\mathbf{j} + 56.6\mathbf{k}$$

The moment from each force is now computed separately.

$$\mathbf{r}_w \times \mathbf{w} = 4\mathbf{i} \times (-150\mathbf{j}) = -600\mathbf{k}$$

$$\begin{aligned}\mathbf{r}_1 \times \mathbf{T}_1 &= 3\mathbf{i} \times (-42.4\mathbf{i} + 70.7\mathbf{j} - 56.6\mathbf{k}) \\ &= 212\mathbf{k} + 170\mathbf{j}\end{aligned}$$

$$\begin{aligned}\mathbf{r}_2 \times \mathbf{T}_2 &= 3\mathbf{i} \times (-42.4\mathbf{i} + 70.7\mathbf{j} + 56.6\mathbf{k}) \\ &= 212\mathbf{k} - 170\mathbf{j}\end{aligned}$$

Summing each moment, we obtain the resultant moment about O.

$$\mathbf{M}_o = -600\mathbf{k} + 212\mathbf{k} + 170\mathbf{j} + 212\mathbf{k} - 170\mathbf{j}$$
$$= -176\mathbf{k}\ \text{ft}\cdot\text{lb} \qquad \textit{Answer}$$

PROBLEMS

2.58 Determine the moment about point A and about point B in Fig. P2.58.

2.59 Determine the moment about point A and about point B in Fig. P2.59.

2.60 Determine the moment about point A and about point B in Fig. P2.60.

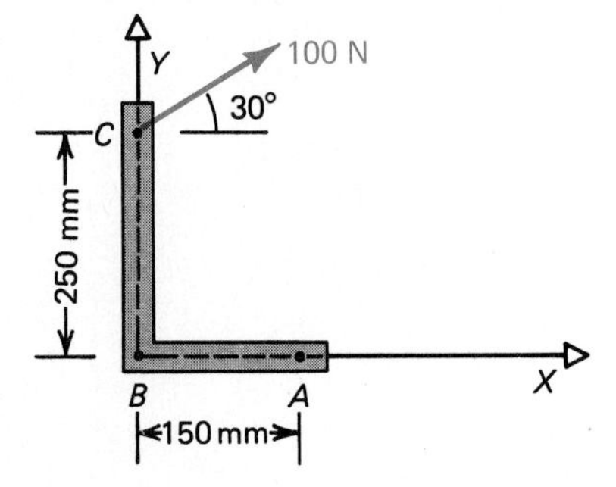

Figure P2.58

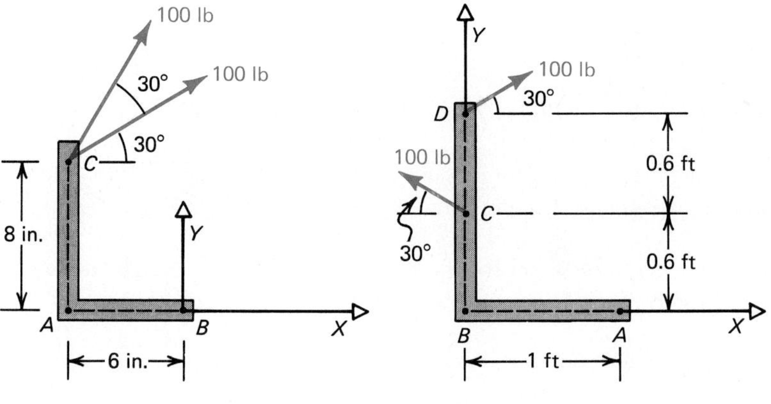

Figure P2.59 **Figure P2.60**

2.61 Determine the moment about point A and about point B in Fig. P2.61.

2.62 Determine the moment about point A and about point B in Fig. P2.62.

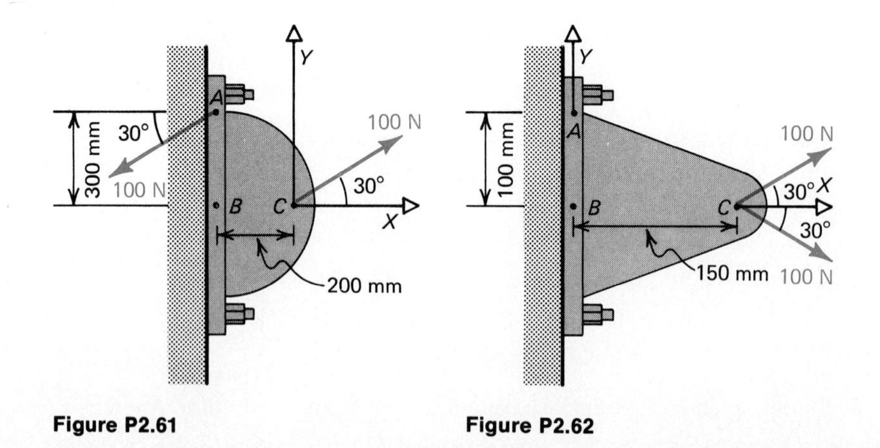

Figure P2.61 **Figure P2.62**

2.63 Determine the moment about point A in Fig. P2.63.

2.64 A force $\mathbf{F} = (50\mathbf{i} + 60\mathbf{j})$ lb is applied at a point whose position vector from O is given by $\mathbf{r} = (\mathbf{i} + 4\mathbf{j})$ ft. What is the resulting moment about the point O?

2.65 Compute the moment about point O for the force acting on the plate shown in Fig. P2.65.

2.66 Determine the moment about point A due to the 250-N force as shown in Fig. P2.66.

2.67 What moment does the 20-lb force create about point A in Fig. P2.67?

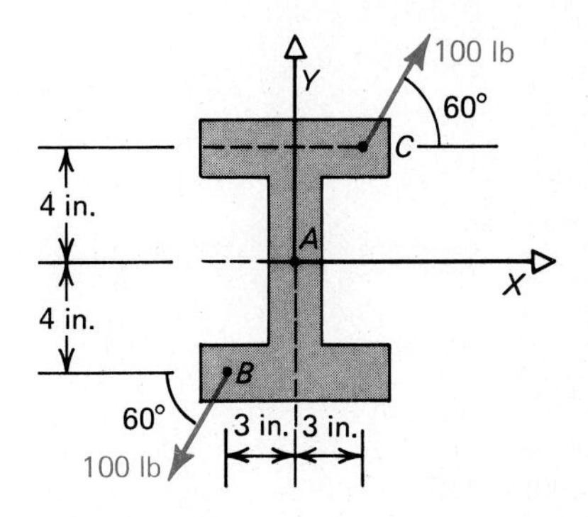

Figure P2.63

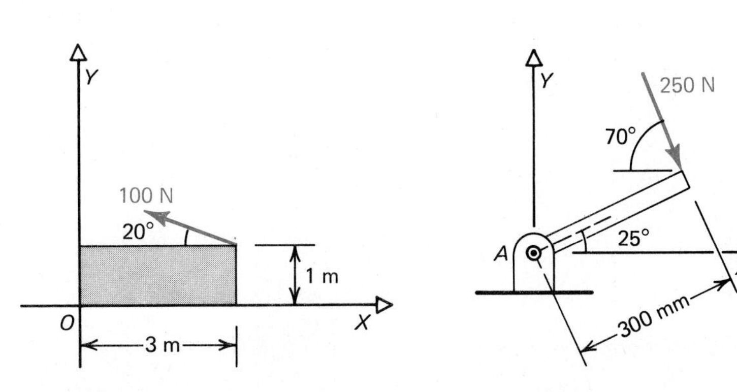

Figure P2.65 **Figure P2.66**

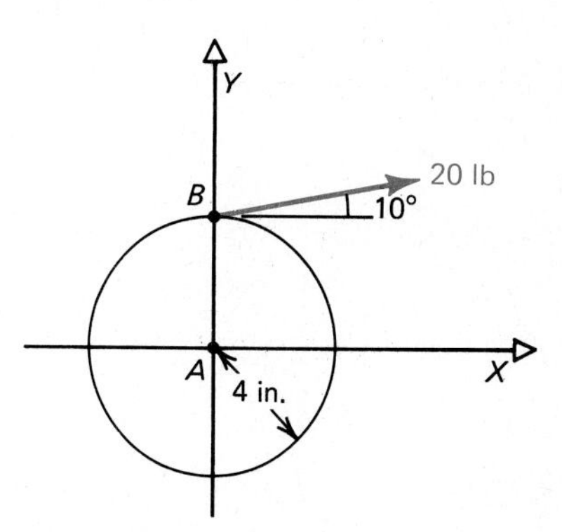

Figure P2.67

2.68 Calculate the moment of the 50-lb force about point P of the plate shown in Fig. P2.68.

2.69 Calculate the moment of the 50-lb force about point Q of the plate shown in Fig. P2.68.

2.70 What is the resultant moment about point P of the plate shown in Fig. P2.70?

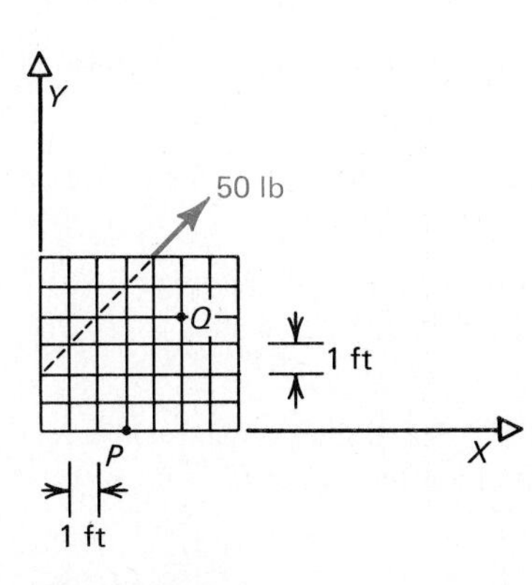

Figure P2.68

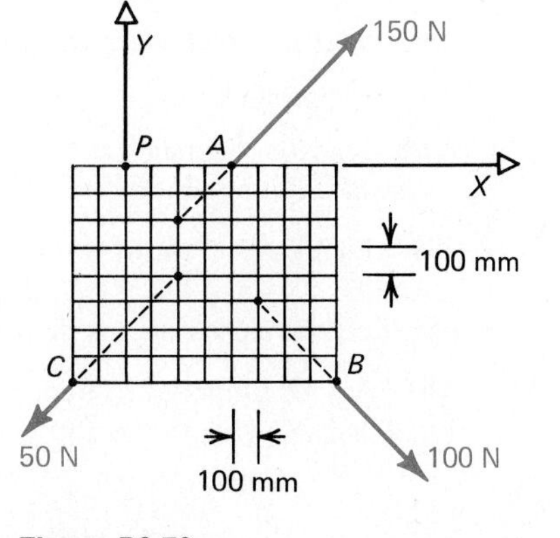

Figure P2.70

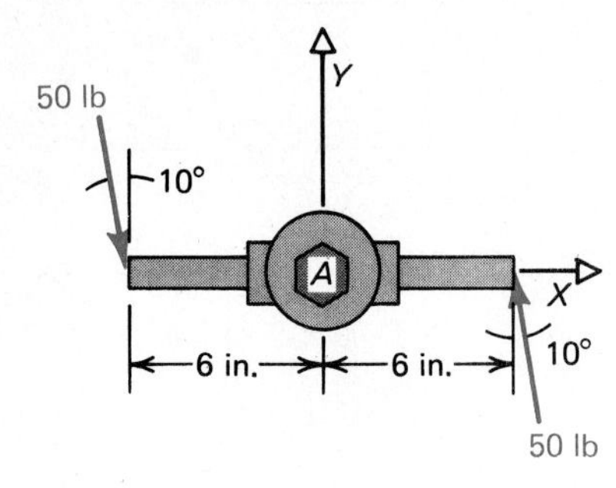

Figure P2.72

2.71 What is the resultant moment about point A of the plate shown in Fig. P2.70?

2.72 A bolt is to be tightened as shown in Fig. P2.72. Determine the moment applied about point A.

2.73 A force $\mathbf{F} = (8\mathbf{i} - 10\mathbf{j} + 4\mathbf{k})$ lb acts at a point whose position vector from O is given by $\mathbf{r} = (2\mathbf{i} + 4\mathbf{k})$ ft. Calculate the magnitude and unit vector of the moment about point O.

2.74 Obtain the magnitude of the moment of the 50-lb force about point A shown in Fig. P2.74.

2.75 Obtain the moment about point A due to the 100-N force shown in Fig. P2.75.

2.76 A force $\mathbf{F} = (8\mathbf{i} + 5\mathbf{j} + 3\mathbf{k})$ N is applied at a point whose position vector from O is given by $\mathbf{r} = 3\mathbf{i} + 2\mathbf{j} + \mathbf{k}$, where distances are measured in meters. What is the resulting moment about O?

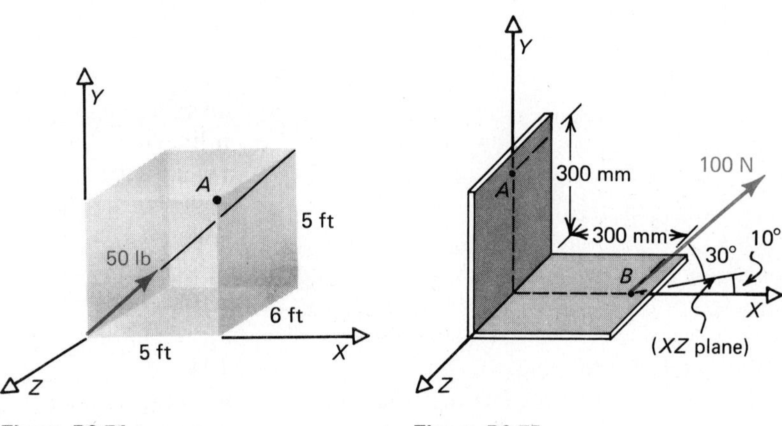

Figure P2.74

Figure P2.75

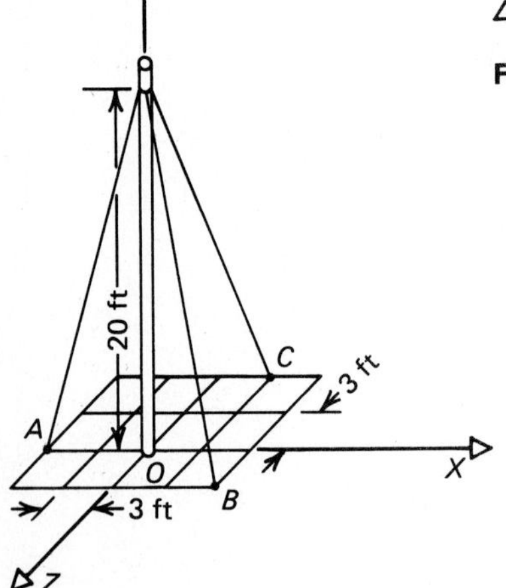

Figure P2.77

2.77 A thin flagpole is supported by three ropes as shown in Fig. P2.77. If the tension in each rope is 50 lb, what is the resulting moment about O?

2.78 Find the moment developed about point O due to the forces acting on the block in Fig. P2.78.

2.79 Find the moment developed about point A due to the forces acting on the block in Fig. P2.78.

2.80 Figure P2.80 shows a thin tube supporting a 10-kg mass at point C. If members AB and BC of the tube lie in the XZ plane, calculate the moment developed by the load about point O.

2.81 A door is opened by applying a 20-N force to the handle as shown in Fig. P2.81. What is the resulting moment about the axis AA?

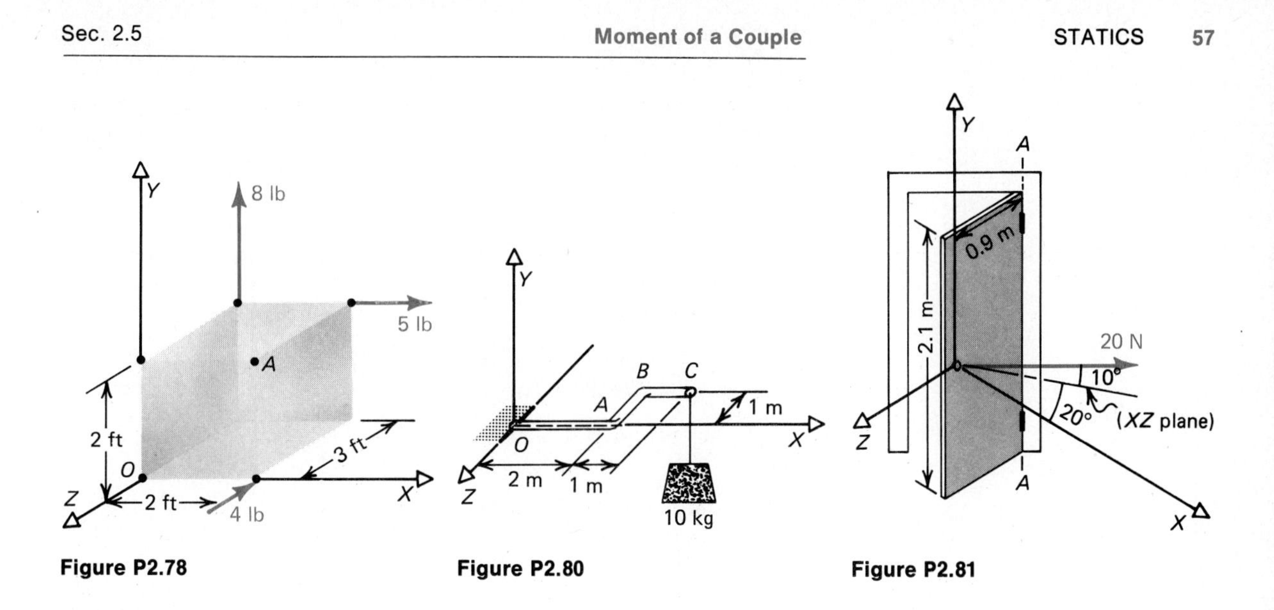

Figure P2.78 Figure P2.80 Figure P2.81

2.5 Moment of a Couple

Consider two forces applied to the flat plate shown in Fig. 2.35. These forces are equal in magnitude, parallel in their lines of action, and opposite in direction. Such a set of forces forms what is called a couple. The magnitude of the moment of the couple equals the product of one force and the distance between the forces, while the direction of the couple is obtained by applying the right-hand rule as before. In the above case, the magnitude of the moment of the couple equals cF and the direction of the couple is counterclockwise. Using vector notation, the moment of the couple is $cF\mathbf{k}$.

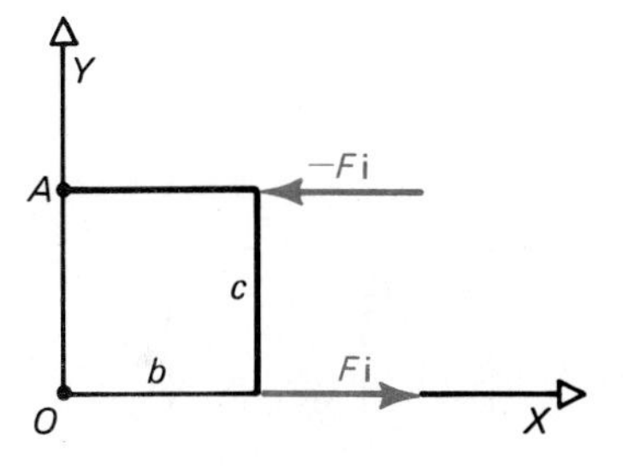

Figure 2.35

It is interesting to note that the moment about any point on the plate in Fig. 2.35 is the same everywhere. For example, consider the moment about point O. Here,

$$\mathbf{r}_1 = b\mathbf{i} \qquad \mathbf{F}_1 = F\mathbf{i}$$

$$\mathbf{r}_2 = b\mathbf{i} + c\mathbf{j} \qquad \mathbf{F}_2 = -F\mathbf{i}$$

so that

$$\begin{aligned} \mathbf{M}_o &= \mathbf{r}_1 \times \mathbf{F}_1 + \mathbf{r}_2 \times \mathbf{F}_2 \\ &= (b\mathbf{i} \times F\mathbf{i}) + (b\mathbf{i} + c\mathbf{j}) \times (-F\mathbf{i}) \\ &= cF\mathbf{k} \end{aligned}$$

This equals the moment of the couple. Likewise, the same result is obtained by calculating the moment about point A or any other point on the body. The moment due to a couple is called a free vector, that is, the vector representing the moment need not pass through any particular point on the body. This is shown pictorially in Fig. 2.36. Here, the original loading is replaced by a vector **M** applied to some point on the right edge of the plate or by **M** applied to point A. Each loading on the plate is equivalent to the other.

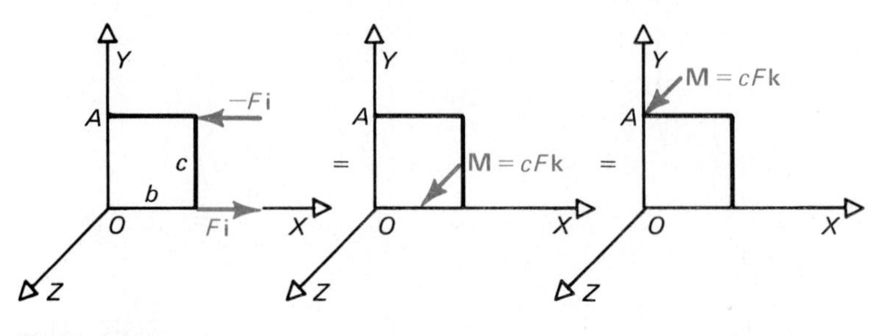

Figure 2.36

The moments of several couples can be combined using the rules of vector algebra. For example, the block in Fig. 2.37(a) has three couples acting on it. Figure 2.37(b) shows the moment of each couple. The two moments acting along the Z axis direction, namely $2cF\mathbf{k}$ and $-3cF\mathbf{k}$ are added so that the moments of the original couples reduce to that shown in Fig. 2.37(c). Finally, we may combine these two moments to obtain a resultant moment.

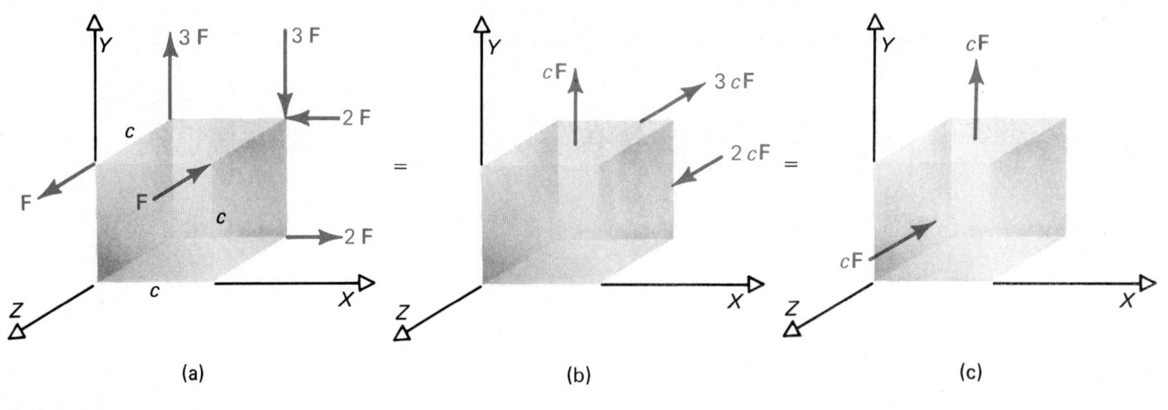

Figure 2.37

Example 2.18

What is the moment of the couple acting on the Z angle shown in Fig. 2.38?

Since we can obtain the moment of the couple by choosing any reference point on the body, select point O for convenience. This means that we include only the force acting at A to calculate the moment since the other force acts through the point O.

$$\begin{aligned}\mathbf{M}_o &= \mathbf{r} \times \mathbf{F} \\ &= (0.2\mathbf{i} + 0.1\mathbf{j}) \times (-50 \cos 45°\mathbf{i} + 50 \sin 45°\mathbf{j}) \\ &= 7.07\mathbf{k} + 3.54\mathbf{k} \\ &= 10.6\mathbf{k} \text{ m} \cdot \text{N} \end{aligned}$$

Answer

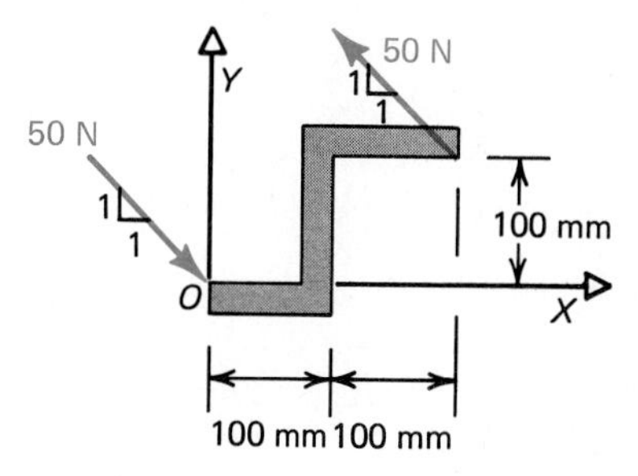

Figure 2.38

Example 2.19

Find the resultant moment of the couples acting on the loaded beam in Fig. 2.39.

By observing the loading in Fig. 2.39, we have the moments of each couple: $-20\mathbf{k}$, $30\mathbf{k}$, and $40\mathbf{j}$. The resultant moment of the couples equals

$$\begin{aligned}\mathbf{M} &= -20\mathbf{k} + 30\mathbf{k} + 40\mathbf{j} \\ &= (40\mathbf{j} + 10\mathbf{k}) \text{ ft} \cdot \text{lb}\end{aligned}$$

Answer

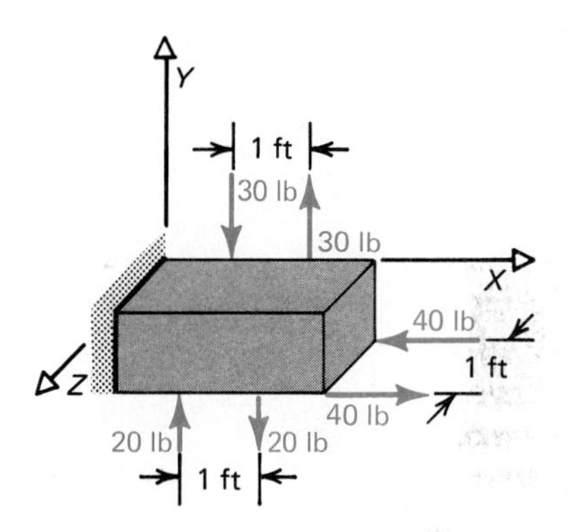

Figure 2.39

PROBLEMS

2.82 Determine the moment of the couple applied to the bracket shown in Fig. P2.61.

2.83 Determine the moment of the couple applied to the I-beam shown in Fig. P2.63.

2.84 Determine the moment of the couple applied to the bolt in Fig. P2.72.

2.85 Obtain the moment of the couple in each of the configurations shown in Fig. P2.85 if $P = 10$ N.

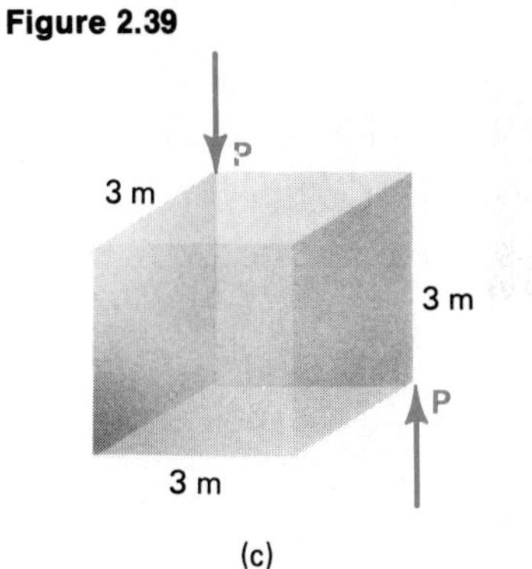

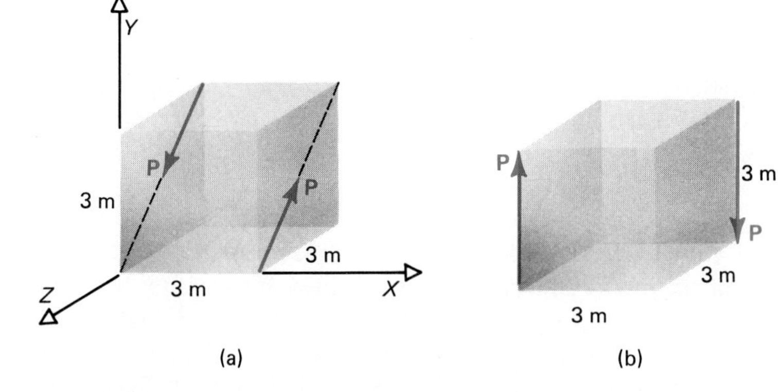

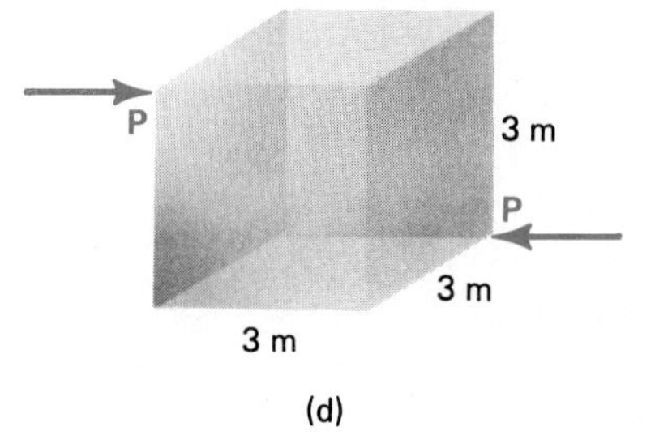

Figure P2.85

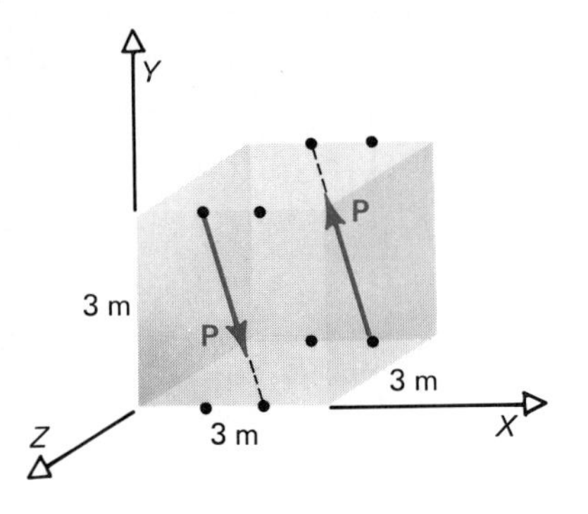

Figure P2.86

2.86 Obtain the moment of the couple for the configuration shown in Fig. P2.86 if $P = 10$ N.

2.87 What is the magnitude of the projection of the couple $994\mathbf{k}$ ft·lb on a line passing through the points $A(1, 2, 3)$ and $B(-1, -2, -3)$?

2.88 Compute the magnitude and direction cosines of the couple produced by $\mathbf{F}$ and $-\mathbf{F}$ where $\mathbf{F} = (5\mathbf{i} + 4\mathbf{j} + \mathbf{k})$ N. The position vector from the point of application of $-\mathbf{F}$ to $\mathbf{F}$ is given by $\mathbf{r} = (3\mathbf{i} + \mathbf{j} + 2\mathbf{k})$ m.

2.89 Calculate the resultant moment of the couples acting on the block in Fig. P2.89.

2.90 Calculate the resultant moment of the couples acting on the block in Fig. P2.90.

2.91 Two forces act on the bracket as shown in Fig. P2.91. What is the moment of the couple acting on the bracket?

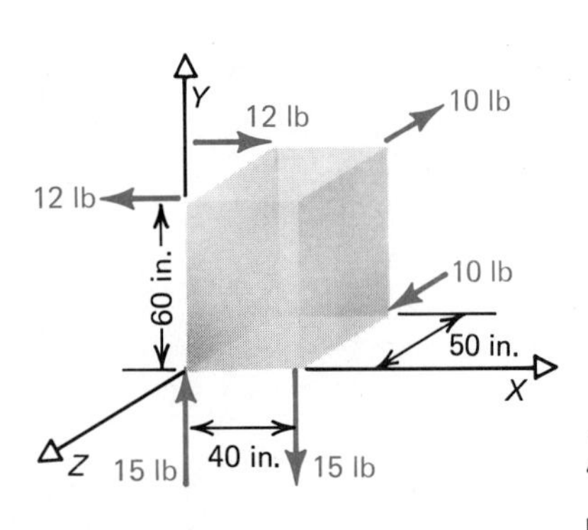

Figure P2.89

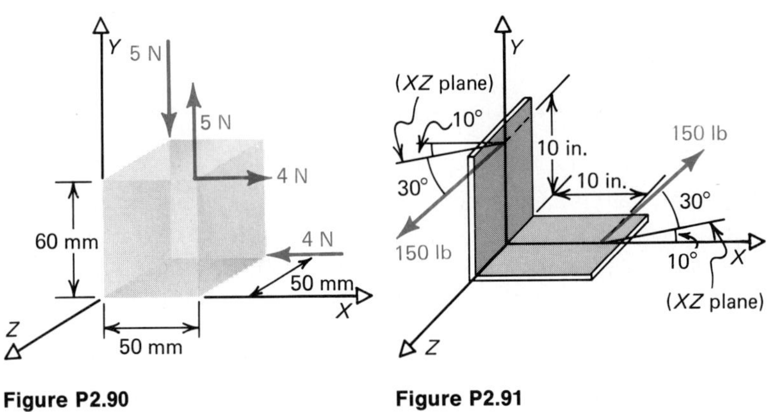

Figure P2.90

Figure P2.91

2.6 Combining Forces and Moments

We have been dealing with forces and moments individually. It often happens that several forces and moments are applied simultaneously to a rigid body. We shall consider combining these forces and moments in a systematic way so as to obtain simpler but equivalent loads acting on the body.

By way of introduction, consider the force $\mathbf{F}$ applied to the block at point A as shown in Fig. 2.40(a). We wish to move the force to a different point but retain an equivalent system. For example, suppose we select a second point B, and apply forces $\mathbf{F}$ and $-\mathbf{F}$ as shown in Figure 2.40(b). Notice that since $\mathbf{F} + (-\mathbf{F}) = 0$, we have not really done anything to produce an unbalance in the system. The original force in Fig. 2.40(a) and the set of forces in Fig. 2.40(b)

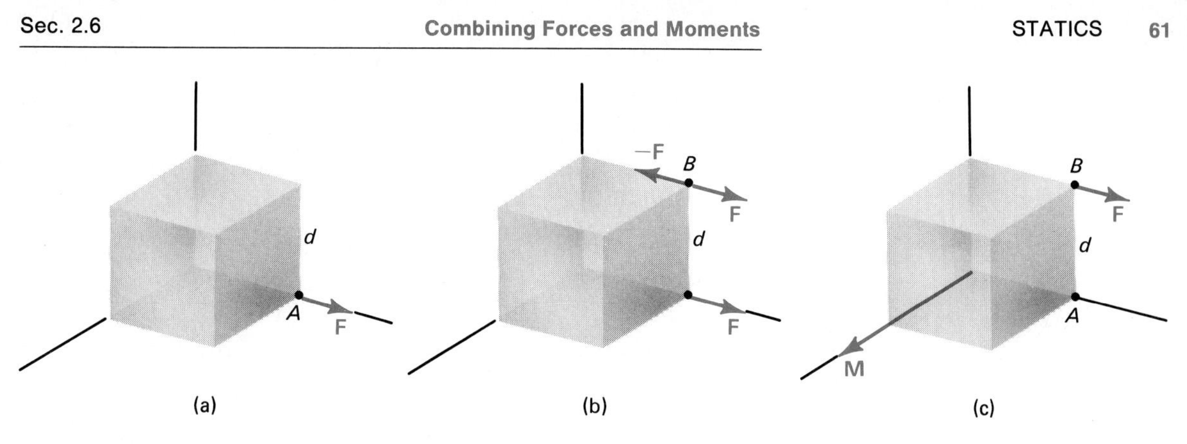

Figure 2.40

form an equivalent system of forces. We also observe that the force $\mathbf{F}$ applied at A and the force $-\mathbf{F}$ applied at B constitute a couple. We therefore remove $\mathbf{F}$ at A and $-\mathbf{F}$ at B and replace them by the moment of the couple to obtain the equivalent system shown in Fig. 2.40(c). Recall that the point of application of the moment of the couple $\mathbf{M}$ is arbitrary since the moment of a couple is a free vector.

As another case in point, consider the system in Fig. 2.41(a), where $\mathbf{F}_1$ and $\mathbf{F}_2$ are in the XZ plane and $\mathbf{M}_3$ is a moment parallel to the Y axis. We first place $\mathbf{F}_1$ and $-\mathbf{F}_1$ at point A as shown in Fig. 2.41(b). Since $\mathbf{F}_1$ and $-\mathbf{F}_1$ constitute a couple whose moment equals

$$\mathbf{M}_1 = \mathbf{r}_1 \times \mathbf{F}_1$$

then Fig. 2.41(c) is an equivalent system. Next, we place $\mathbf{F}_2$ and $-\mathbf{F}_2$ at A as shown in Fig. 2.41(d), and define a couple whose moment equals

$$\mathbf{M}_2 = \mathbf{r}_2 \times \mathbf{F}_2$$

This leads to the system shown in Fig. 2.41(e). Finally, by adding the moments and forces, i.e.,

$$\mathbf{M} = \mathbf{M}_1 + \mathbf{M}_2 + \mathbf{M}_3$$

$$\mathbf{R} = \mathbf{F}_1 + \mathbf{F}_2$$

we obtain the equivalent system shown in Fig. 2.41(f). We have reduced the original system to one resultant force and one resultant moment. It is important to recognize that in this general situation, the resulting magnitude of the moment depends upon the location of point A.

We now examine three common cases in which we are able to reduce a system of forces to an equivalent system. The first case is

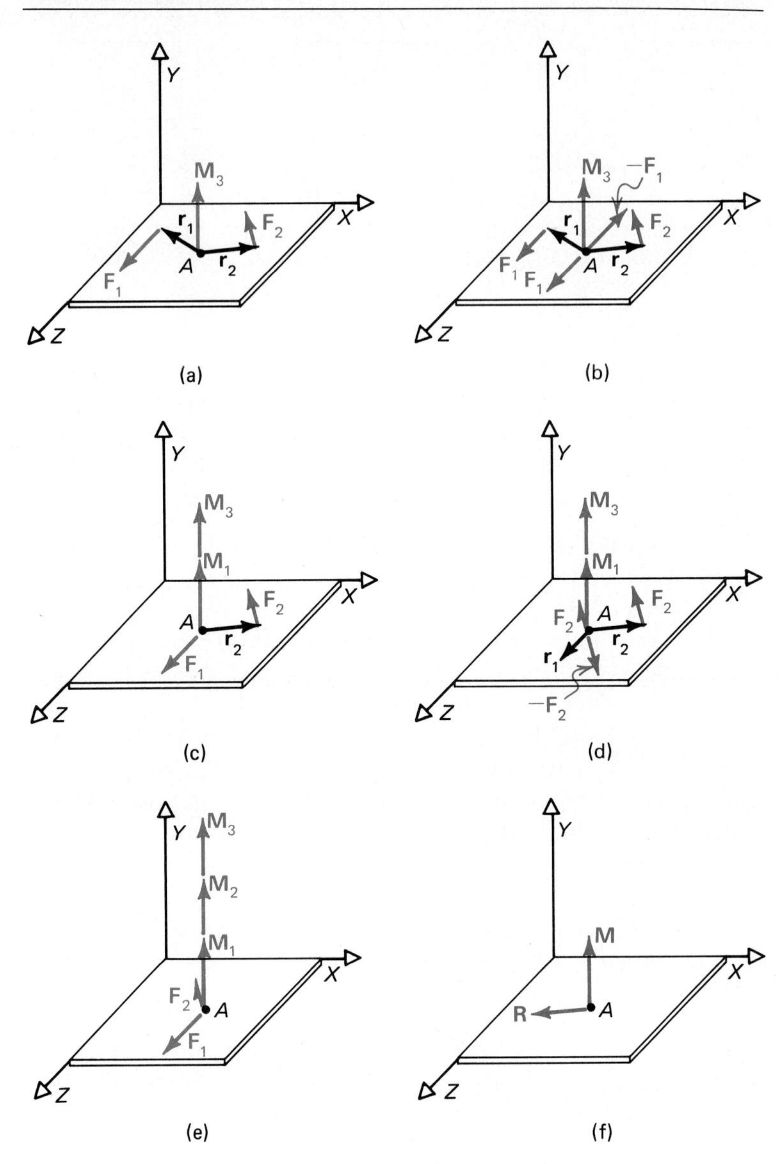

Figure 2.41

for concurrent forces applied to a body as shown in Fig. 2.42(a). We may obtain the resultant force **R** by vector addition as described earlier in this chapter.

$$\mathbf{R} = \sum \mathbf{F}$$

where **R** is applied at O. We now have an equivalent system in Fig. 2.42(b) with one resultant force.

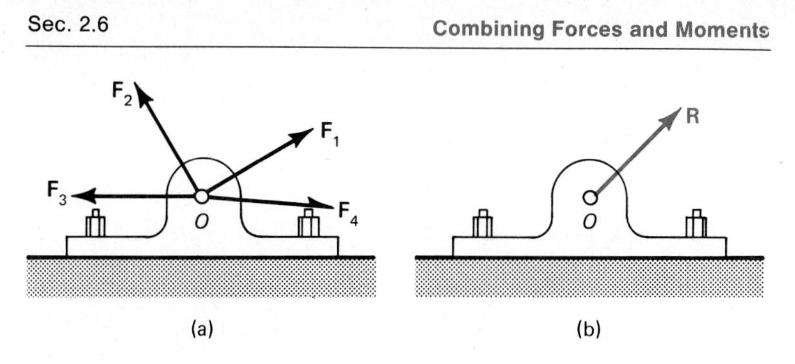

Figure 2.42

Next we consider the case of coplanar forces as shown in Fig. 2.43(a). These forces can be replaced by one moment $\mathbf{M}_o$ and one force $\mathbf{R}$ acting at O as shown in Fig. 2.43(b). By shifting the force $\mathbf{R}$ to a new point A we can reduce the equivalent system to one force only as shown in Fig. 2.43(c). The distance from O to A is

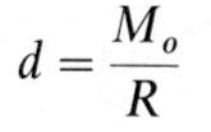

$$d = \frac{M_o}{R}$$

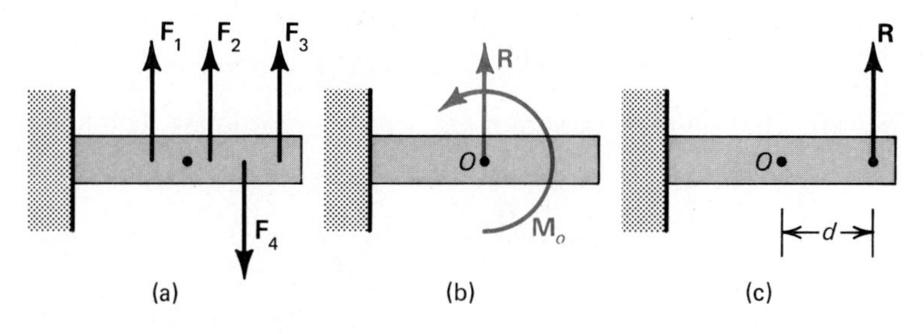

Figure 2.43

For the more general case consider the resultant force $\mathbf{R}$ lying in the XY plane and a moment $\mathbf{M}_o$ applied to the plate in Fig. 2.44(a). The equivalent system is shown in Fig. 2.44(b), where the resultant force $\mathbf{R}$ is introduced at an arbitrary point P defined by

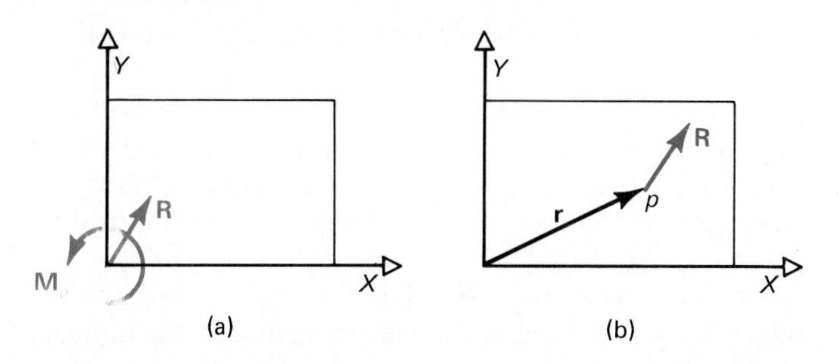

Figure 2.44

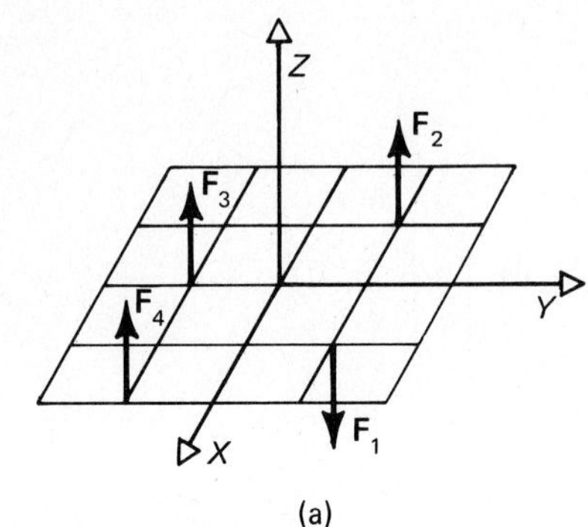

(a)

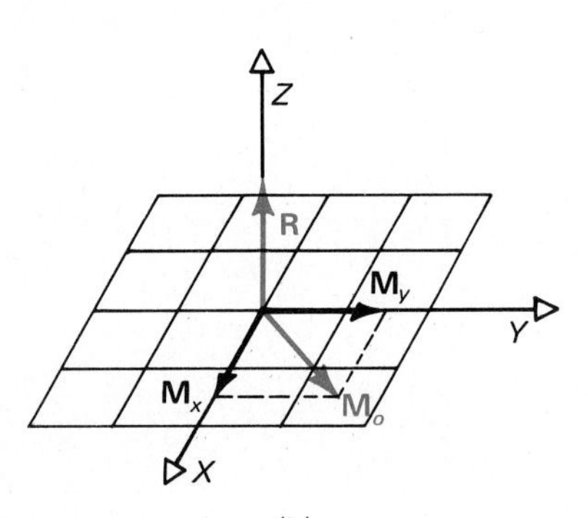

(b)

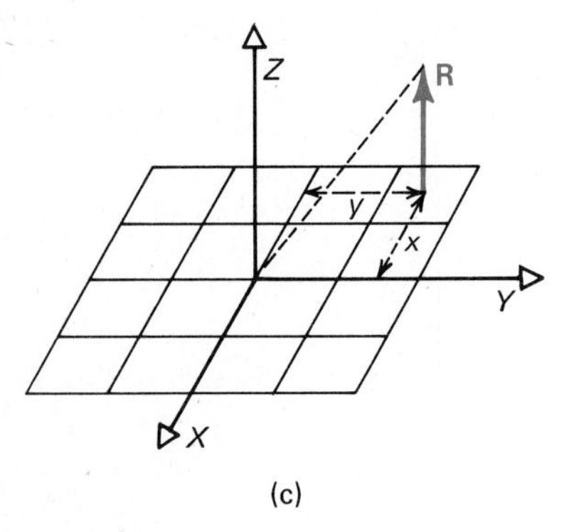

(c)

Figure 2.45

$$\mathbf{r} = x\mathbf{i} + y\mathbf{j} \qquad \mathbf{R} = R_x\mathbf{i} + R_y\mathbf{j}$$

so that

$$\mathbf{M}_o = \mathbf{r} \times \mathbf{R}$$

$$M_o\mathbf{k} = (x\mathbf{i} + y\mathbf{j}) \times (R_x\mathbf{i} + R_y\mathbf{j}) = (xR_y - yR_x)\mathbf{k}$$

$$M_o = xR_y - yR_x$$

$$y = \frac{xR_y}{R_x} - \frac{M_o}{R_x}$$

Since R_x, R_y, and M_o are known, this establishes a relationship between x and y, i.e., we have obtained the line of action along which **R** passes.

Finally, we consider the third case in which parallel forces act on a body as shown in Fig. 2.45(a). The forces, since they are parallel with the Z axis, give rise to moments about the X and Y axes. The resultant moment $\mathbf{M}_o$ will be in the XY plane, as shown in Fig. 2.45(b). The final equivalent system is shown in Fig. 2.45(c). The resultant force **R** is located at the coordinate point (x, y) so as to produce the same moments about the X and Y axes. Since

$$\mathbf{M}_o = \mathbf{r} \times \mathbf{R} \qquad \text{or} \qquad M_x\mathbf{i} + M_y\mathbf{j} = yR\mathbf{i} - xR\mathbf{j}$$

we can establish the location of the x and y coordinate. It follows that

$$x = -\frac{M_y}{R} \qquad y = \frac{M_x}{R}$$

The system shown in Fig. 2.45(c), where only one resultant force **R** is applied, is therefore equivalent to that shown in Fig. 2.45(a).

The condition in which a resultant force **R** and a moment **M** have the same line of action is known as a **wrench**. When the directions of **R** and **M** are the same the wrench is positive, and if they are opposite the wrench is negative, as shown in Fig. 2.46. The screwdriver is a common example of a positive wrench in that a thrust R and a moment M are both exerted on the screw in the direction of its axis.

Any general force-moment system may be represented by a wrench located along a unique line of action. The line of action is located in a plane perpendicular to the plane containing the resultant force **R** and moment **M**. Example 2.24 demonstrates this concept.

For the general case we can also calculate the pitch of the wrench (PW) which is defined as the projection of the moment **M** on the force **R** as shown in Fig. 2.47. Recalling the discussion on scalar products,

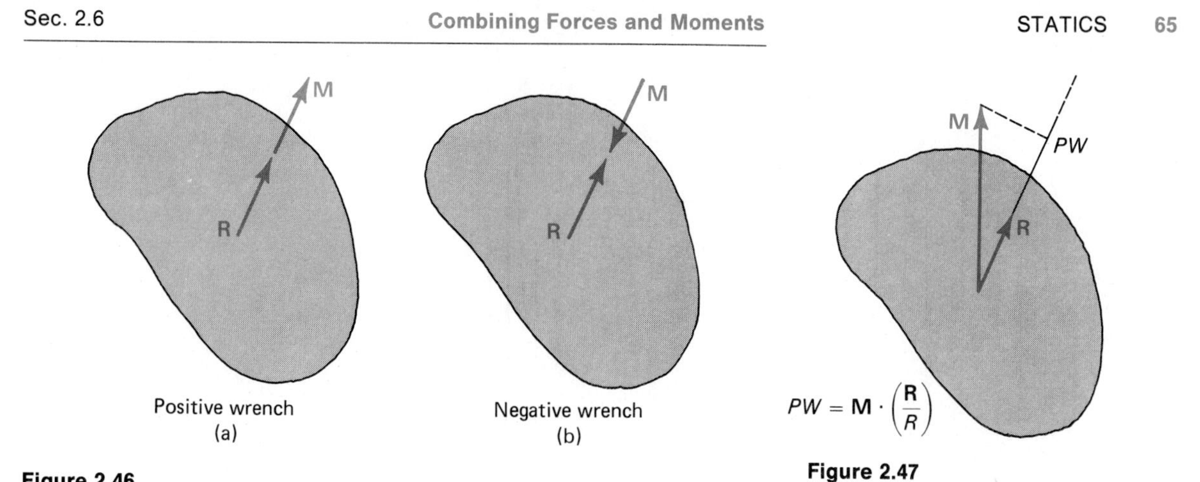

Figure 2.46

Figure 2.47

$$PW = \mathbf{M} \cdot \left(\frac{\mathbf{R}}{R}\right)$$

in which $\mathbf{R}/R$ is the unit vector along $\mathbf{R}$.

Example 2.20

A beam is subjected to two forces as shown in Fig. 2.48(a). Reduce the system to only one force and one moment.

The 150-N force is moved to A so that a moment $\mathbf{M}$ is added as shown in Fig. 2.48(b), where the magnitude of $\mathbf{M}$ is

$$M = 2(150) = 300 \text{ m} \cdot \text{N}$$

In vector notation this is

$$\mathbf{M} = 300\mathbf{k} \text{ m} \cdot \text{N} \qquad \textit{Answer}$$

Also, summing the two forces in Fig. 2.48(b) we obtain the resultant force

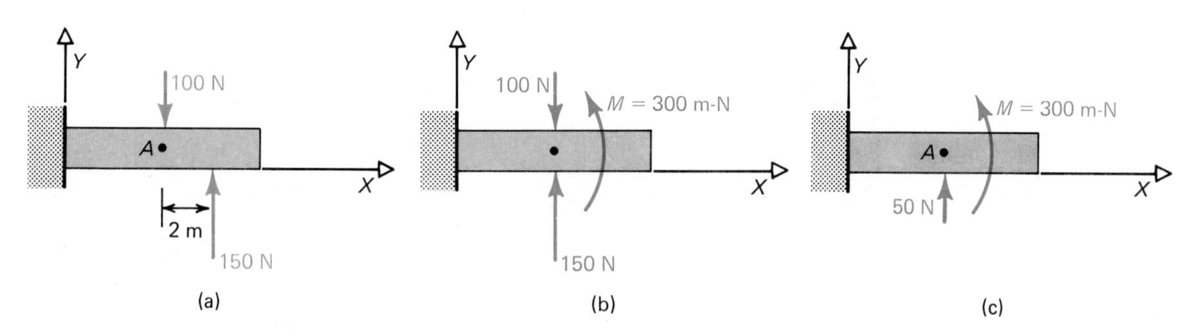

Figure 2.48

$$\mathbf{R} = 150\mathbf{j} - 100\mathbf{j} = 50\mathbf{j}\ \text{N} \qquad \textit{Answer}$$

The resultant force and resultant moment are shown in Fig. 2.48(c).

Example 2.21

Reduce the loading shown in Fig. 2.49(a) to a system of one force only.

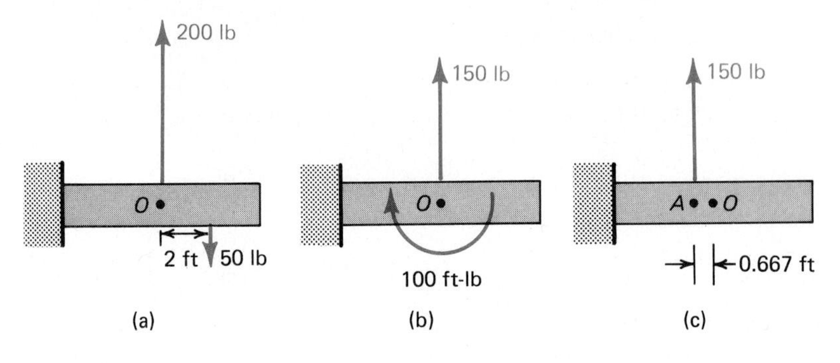

Figure 2.49

We move the 50-lb force to point O and introduce a moment of 100↺ ft·lb as shown in Fig. 2.49(b). This leaves us with the 150-lb force that we wish to locate so that it produces the 100 ft·lb moment on the body. Assume that point A is this point. We move the 150-lb force to A as shown in Fig. 2.49(c). For the loading in Fig. 2.49(c) to be equivalent to Fig. 2.49(b), locate the 150-lb force at a distance to the left of point A such that

$$150d = 100$$

$$d = 0.667\ \text{ft} \qquad \textit{Answer}$$

Example 2.22

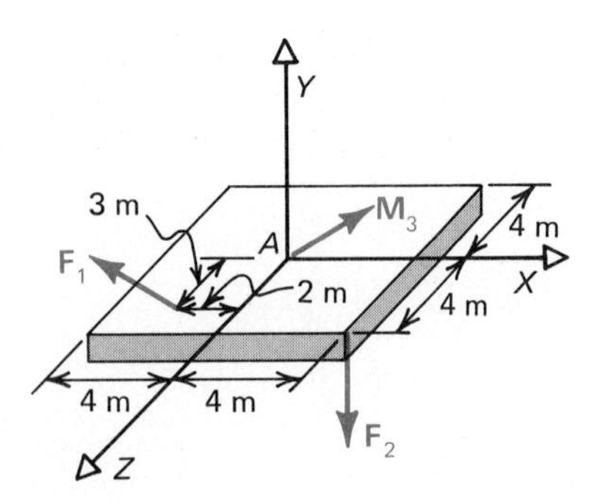

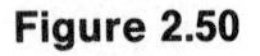

Figure 2.50

Reduce the loading on the plate in Fig. 2.50 to one force acting on point A and one resultant moment. Let $\mathbf{M}_3 = (100\mathbf{i} + 400\mathbf{j} + 300\mathbf{k})$ m·N,

$$\mathbf{F}_1 = (-100\mathbf{i} + 100\mathbf{j} - 50\mathbf{k})\ \text{N} \qquad \text{and} \qquad \mathbf{F}_2 = -100\mathbf{j}\ \text{N}$$

Using point A as the reference point, we apply $\mathbf{F}_1$ and $-\mathbf{F}_1$ at point A as well as $\mathbf{F}_2$ and $-\mathbf{F}_2$. We now have a set of couples acting on the plate. The moment of each couple is obtained as follows:

$$\text{For } \mathbf{F}_1\text{:} \quad \mathbf{M}_1 = \mathbf{r}_1 \times \mathbf{F}_1$$
$$= (-2\mathbf{i} + 3\mathbf{k}) \times (-100\mathbf{i} + 100\mathbf{j} - 50\mathbf{k})$$
$$= -300\mathbf{i} - 400\mathbf{j} - 200\mathbf{k}$$

For $\mathbf{F}_2$: $$\mathbf{M}_2 = \mathbf{r}_2 \times \mathbf{F}_2$$
$$= (4\mathbf{i} + 4\mathbf{k}) \times (-100\mathbf{j})$$
$$= 400\mathbf{i} - 400\mathbf{k}$$

The resultant moment therefore becomes

$$\mathbf{M} = (100\mathbf{i} + 400\mathbf{j} + 300\mathbf{k}) + (-300\mathbf{i} - 400\mathbf{j} - 200\mathbf{k})$$
$$+ (400\mathbf{i} - 400\mathbf{k})$$
$$= (200\mathbf{i} - 300\mathbf{k})\ \text{m} \cdot \text{N} \qquad \textit{Answer}$$

The remaining forces $\mathbf{F}_1$ and $\mathbf{F}_2$ acting at A are added to obtain the resultant force $\mathbf{R}$.

$$\mathbf{R} = (-100\mathbf{i} + 100\mathbf{j} - 50\mathbf{k}) + (-100\mathbf{j})$$
$$= (-100\mathbf{i} - 50\mathbf{k})\ \text{N} \qquad \textit{Answer}$$

Example 2.23

Reduce the force and moment acting on the plate in Fig. 2.51(a) to an equivalent system of one force.

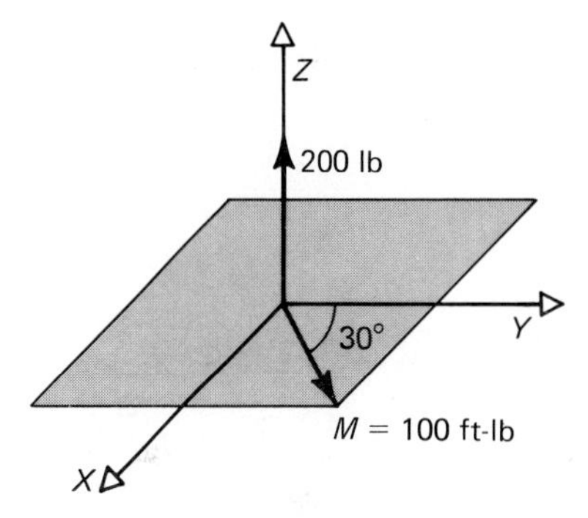

The moment is written in vector notation as

$$\mathbf{M} = 100(\sin 30^\circ \mathbf{i} + \cos 30^\circ \mathbf{j})$$
$$= 50\mathbf{i} + 86.6\mathbf{j}$$

Move the 200-lb force to a point whose coordinates are (x, y) such that

$$\mathbf{r} \times \mathbf{F} = \mathbf{M}$$

Letting $\mathbf{r} = x\mathbf{i} + y\mathbf{j}$ and since $\mathbf{F} = 200\mathbf{k}$,

$$(x\mathbf{i} + y\mathbf{j}) \times 200\mathbf{k} = 50\mathbf{i} + 86.6\mathbf{j}$$

$$-200x\mathbf{j} + 200y\mathbf{i} = 50\mathbf{i} + 86.6\mathbf{j}$$

Solving for the coordinates x and y, we obtain

$$x = -\frac{86.6}{200} \qquad y = \frac{50}{200}$$

$$= -0.433\ \text{ft} \qquad = 0.25\ \text{ft} \qquad \textit{Answer}$$

Figure 2.51(b) shows the equivalent system of one force.

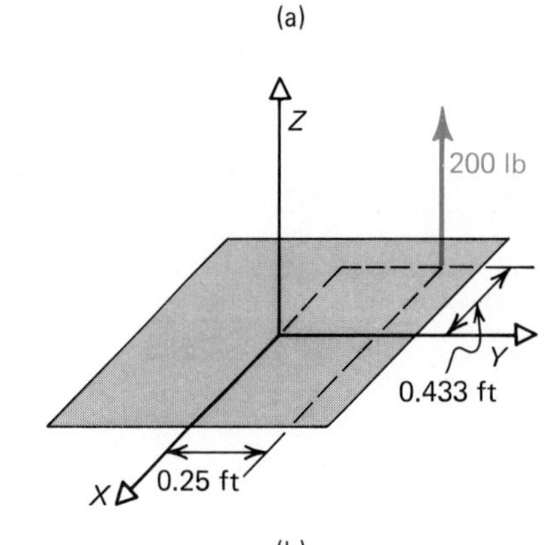

Figure 2.51

Example 2.24

Locate the wrench for the loaded plate in Fig. 2.52(a).

Figure 2.52(b) shows the component M_x along the X axis, the component M_y along the Y axis, and the resultant force $\mathbf{R}$.

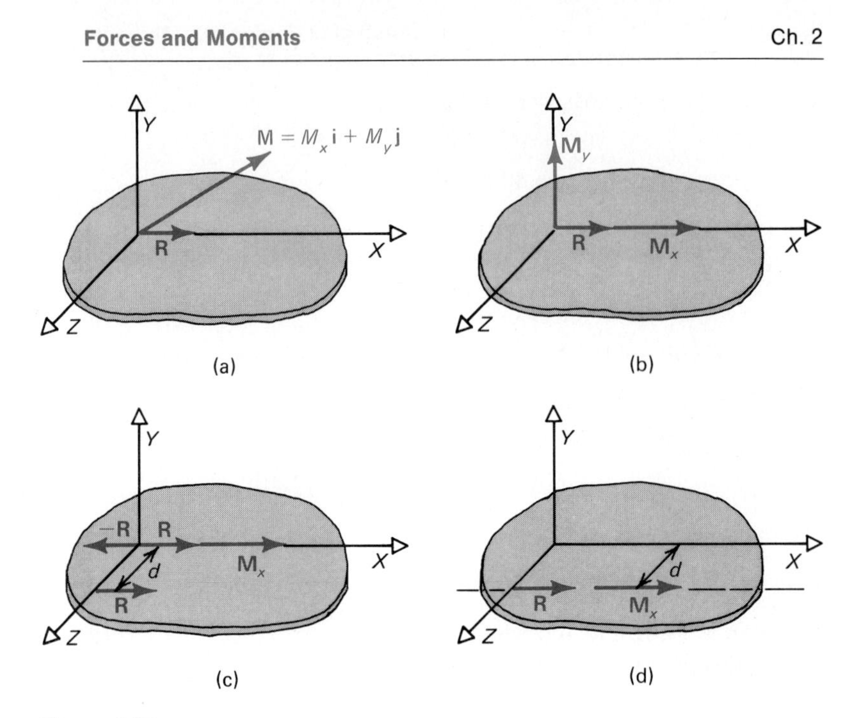

Figure 2.52

In Fig. 2.52(c), M_y is replaced by a couple such that $d = M_y/R$. Finally, since M_x is a free vector, it is located in Fig. 2.52(d) along the line of action of **R** which forms a positive wrench.

PROBLEMS

2.92 Reduce the force applied to the bracket shown in Fig. P2.58 to an equivalent system of a force and moment applied at *A*.

2.93 Reduce the forces applied to the bracket shown in Fig. P2.59 to an equivalent system of a force and moment applied at *A*.

2.94 Reduce the forces applied to the bracket shown in Fig. P2.60 to an equivalent system of a force and moment applied at *C*.

2.95 Reduce the forces applied to the bracket shown in Fig. P2.61 to an equivalent system of a force and moment applied at *A*.

2.96 Reduce the forces applied to the bracket shown in Fig. P2.62 to an equivalent system of a force and moment applied at *B*.

2.97 Reduce the forces applied to the member shown in Fig. P2.63 to an equivalent system of a force and moment applied at *A*.

2.98 Replace the force applied to the plate in Fig. P2.68 by an equivalent system of a force and moment applied at *Q*.

2.99 Replace the forces applied to the plate in Fig. P2.70 by an equivalent system of a force and moment applied at *P*.

2.100 Replace the force applied to the block in Fig. P2.74 by an equivalent system of a force and moment applied at *A*.

2.101 The four forces shown in Fig. P2.101 are to be replaced by one force and a moment about *A*. What is the force and the moment?

2.102 Determine the resultant force and moment applied at *A* to replace the forces and moment shown in Fig. P2.102. Assume the moment $\mathbf{M} = 5.0\mathbf{k}$ m·N.

2.103 Reduce the system of Problem 2.102 to one resultant force. Locate the point of application along the *X* axis.

2.104 Obtain the equivalent force-moment system at *A* for the pulley shown in Fig. P2.104. (*Note*: Each force is tangent to the pulley.)

2.105 Obtain the equivalent force-moment system at *A* for the gear shown in Fig. P2.105.

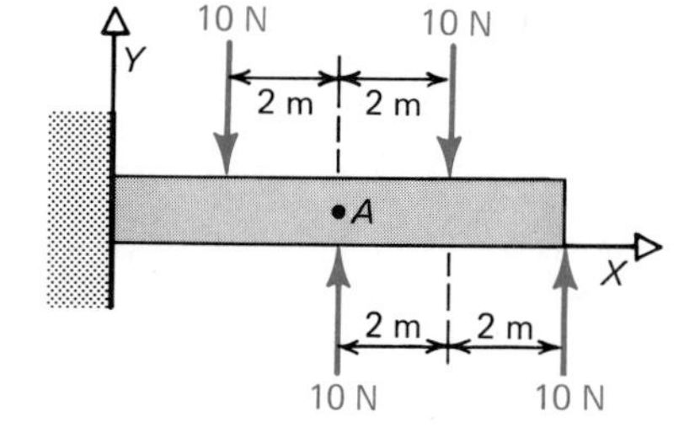

Figure P2.101

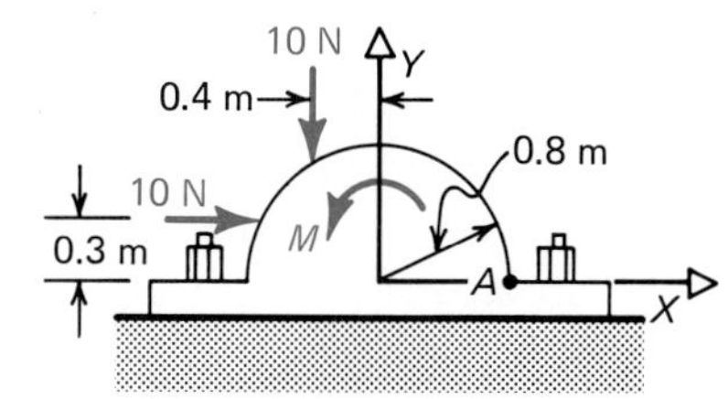

Figure P2.102

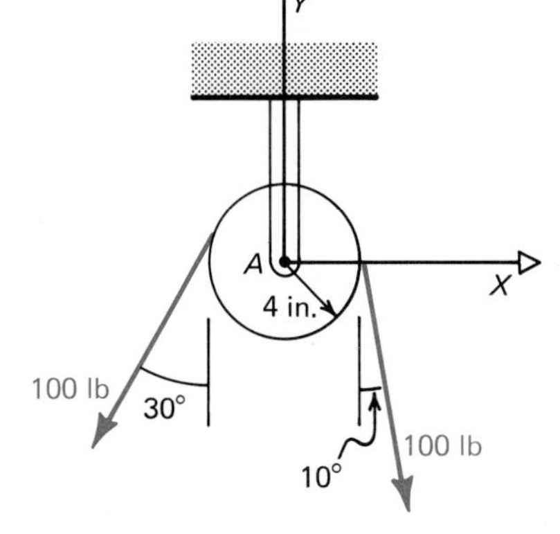

Figure P2.104

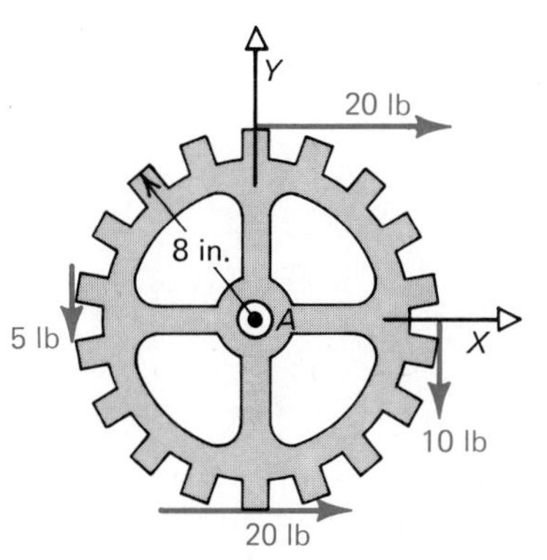

Figure P2.105

2.106 The truss of a roof is loaded as shown in Fig. P2.106. What is the equivalent force-moment system at *A*?

2.107 Obtain the equivalent force-moment system at *A* for the block shown in Fig. P2.107.

2.108 Obtain the equivalent force-moment system at *B* for the block shown in Fig. P2.107.

2.109 Obtain the equivalent force-moment system at *A* for the block shown in Fig. P2.109.

2.110 Obtain the equivalent force-moment system at *B* for the block shown in Fig. P2.109.

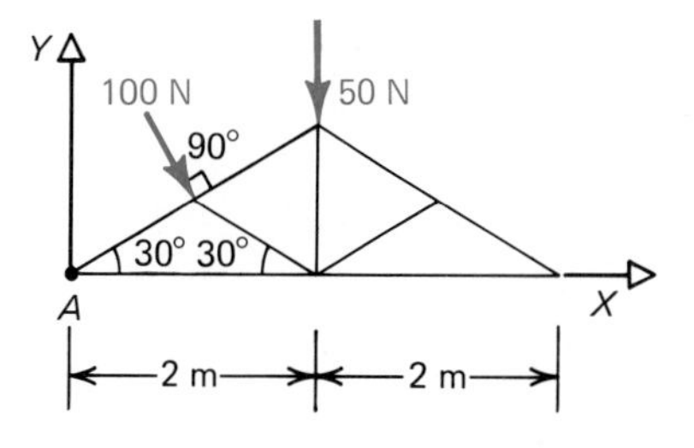

Figure P2.106

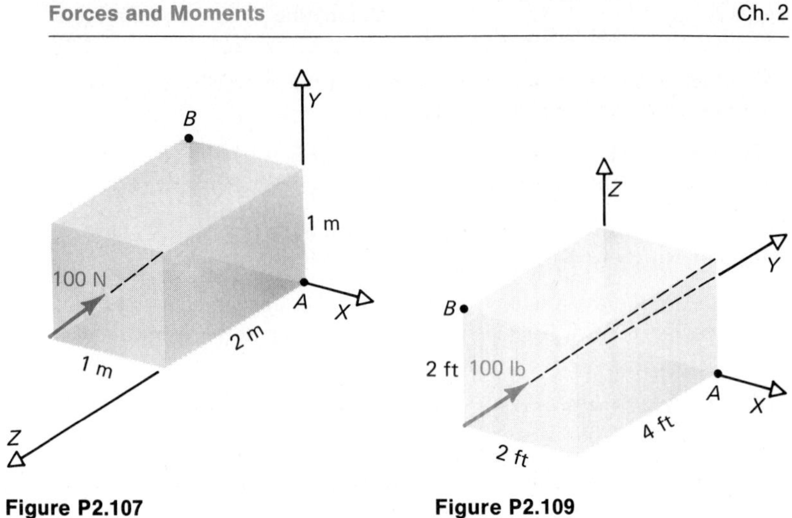

Figure P2.107

Figure P2.109

2.111 Obtain the equivalent force-moment system at A for the block shown in Fig. P2.111. Assume $\mathbf{F}_1 = 100(-0.577\mathbf{i} + 0.577\mathbf{j} - 0.577\mathbf{k})$ N, $\mathbf{F}_2 = -50\mathbf{j}$ N, and $\mathbf{F}_3 = 40(0.707\mathbf{i} + 0.707\mathbf{j})$ N.

2.112 Repeat Problem 2.111 for point B.

2.113 What is the resultant force and moment at A for the system shown in Fig. P2.113?

2.114 Determine the equivalent resultant force and point of application for the loaded plate in Fig. P2.114.

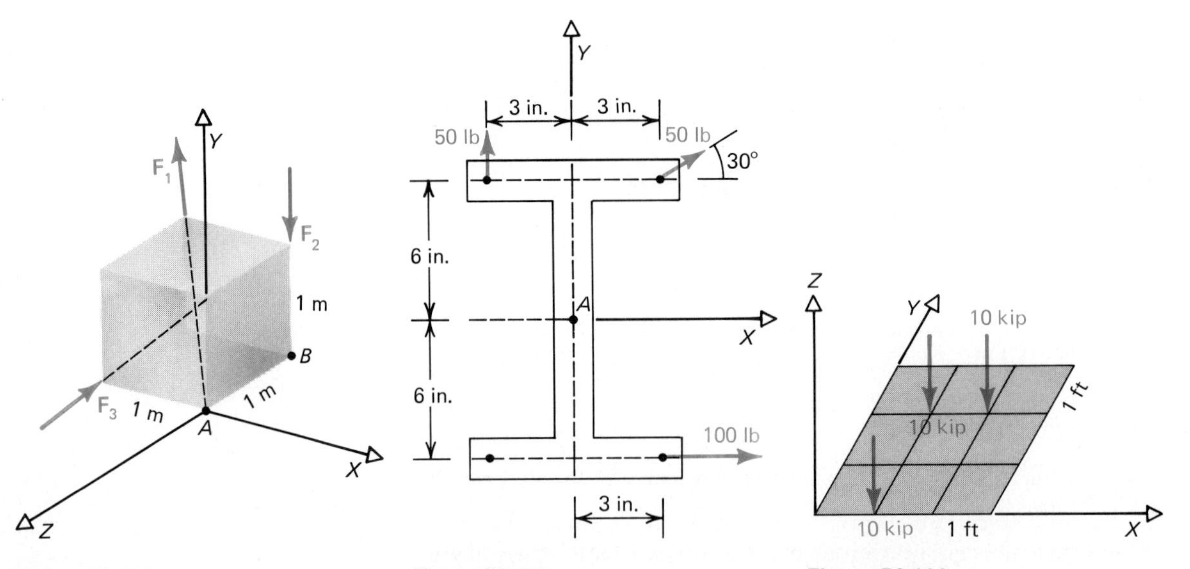

Figure P2.111

Figure P2.113

Figure P2.114

2.115 Determine the equivalent resultant force and point of application for the loaded plate in Fig. P2.115.

2.116 The wrench shown in Fig. P2.116 must be replaced by two forces. One force must go through A and the other must be in the XY plane. Obtain these forces.

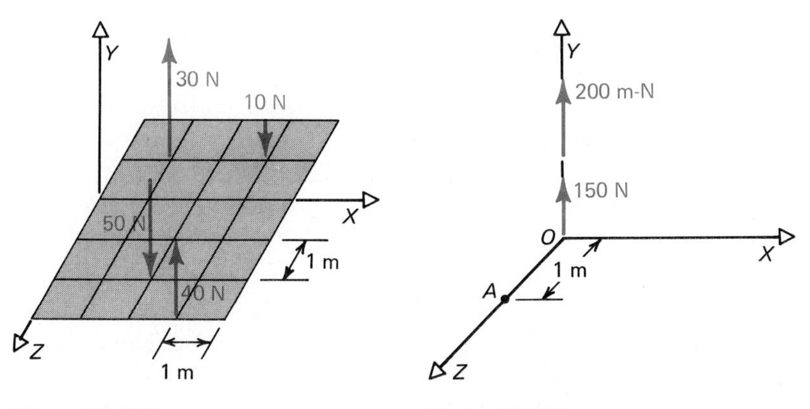

Figure P2.115

Figure P2.116

2.117 Locate the wrench for the loaded plate in Fig. P2.117.

2.118 Locate the wrench for the loaded body in Fig. P2.118.

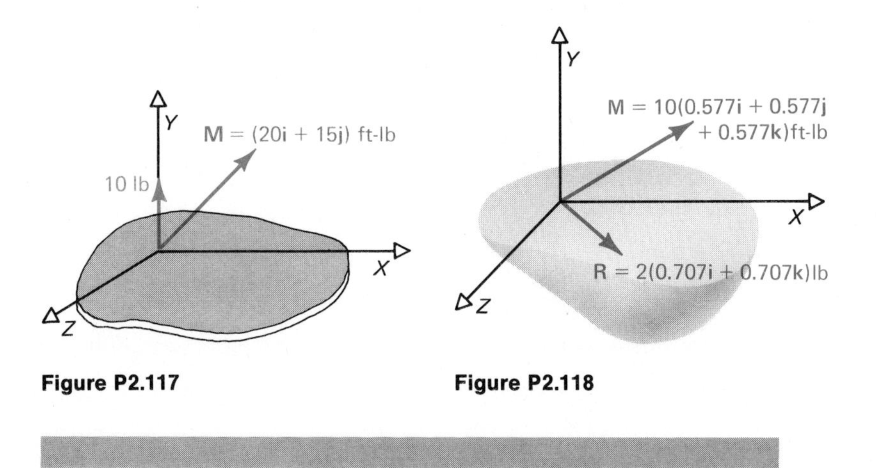

Figure P2.117

Figure P2.118

2.7 Summary

A force was defined and represented as a vector quantity. We have shown how a force is resolved into components and how several forces were combined to form a resultant force. The resultant of many forces applied to a particle must be zero for equilibrium of the particle. The moment of a force and the concept of a couple were introduced. It was shown how one resultant force and one resultant moment may replace several forces and couples applied to a body.

A. The important *ideas*:

1. A force is a vector quantity. It has magnitude, direction, and point of application.
2. For a particle to be in equilibrium, the resultant of all concurrent forces applied to the particle must be zero.
3. Whereas a force creates a tendency for a body to translate, the moment of a force creates a tendency for the body to rotate about the line of action of the moment.
4. Two forces that are equal in magnitude, parallel in their lines of action, and opposite in direction form a couple.
5. Forces and moments acting on a rigid body may be combined to form an equivalent force and moment acting on the body, such that the effect of the original loads on the body and the equivalent loads on the body are the same.

B. The important *equations*:

Resultant force:
$$\mathbf{R} = \sum_{i=1}^{n} \mathbf{F}_i \tag{2.1}$$

Planar force components:
$$\mathbf{F} = F_x\mathbf{i} + F_y\mathbf{j} \tag{2.3}$$
$$\mathbf{F} = F(\cos\theta_x\mathbf{i} + \cos\theta_y\mathbf{j}) \tag{2.8}$$

Three-dimensional force:
$$\mathbf{F} = F\mathbf{e}_F \tag{2.9}$$
$$\mathbf{e}_F = (\cos\theta_x\mathbf{i} + \cos\theta_y\mathbf{j} + \cos\theta_z\mathbf{k}) \tag{2.10}$$

Equilibrium of a particle:
$$\sum F_x = 0 \qquad \sum F_y = 0 \qquad \sum F_z = 0 \tag{2.11}$$

Moment of a force:
$$\mathbf{M} = \mathbf{r} \times \mathbf{F} \tag{2.12}$$

3 Rigid Body Equilibrium

A particle is in equilibrium if the resultant force acting on it is zero. Similarly, a rigid body is in equilibrium if (1) the resultant force is zero and (2) the resultant moment is zero.

Forces and moments can be **external** or **internal**. We shall have more to say about internal forces in the next chapter. The forces that are applied to a rigid body by another body or by the earth are external forces. Snow on the roof produces an external force on the building. The weight of a body is another case of an external force. Since we shall consider the equilibrium of the entire rigid body, our concern in this chapter is with external forces and the moments these external forces produce.

External forces can be further divided into **reaction** forces and **applied** forces. Reaction forces, also called reactive forces, are those forces that are contributed by a body's supports or connections. Applied forces are those forces that are directly applied to a body from some external source, such as the snow load on a roof. The resultant moment and force acting on a rigid body must be obtained from a free-body diagram of the rigid body as was done for the equilibrium of a particle. This free-body diagram must show all the applied forces as well as all the reaction forces.

3.1 Free-Body Diagrams

A free-body diagram of a body is a sketch of the body such that its environment is replaced by equivalent forces. This diagram shows all reaction forces, applied forces, and moments acting on the body. Having the free-body diagram, we are able to examine the body for equilibrium.

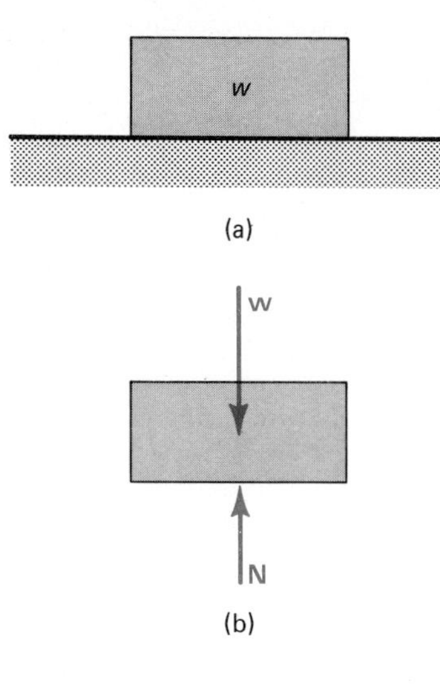

Figure 3.1

Consider the block of weight w at rest as shown in Fig. 3.1(a). The free-body diagram of the block is shown in Fig. 3.1(b). Essentially, we have removed the ground and replaced it with a reaction **N**. We also include the gravity force or weight w. Clearly, for equilibrium, we have $N = w$. As another example, consider the block of weight w on a smooth incline shown in Fig. 3.2(a). Here, in order to draw the free-body diagram, we replace the inclined plane and wall with two reaction forces $\mathbf{N}_1$ and $\mathbf{N}_2$. The free-body diagram in Fig. 3.2(b) now shows four forces applied to the block, **w, P,** $\mathbf{N}_1$, and $\mathbf{N}_2$. We note that $\mathbf{N}_1$ and $\mathbf{N}_2$ are unknowns. We shall see that in this case we can write two independent scalar equations of equilibrium, so that we can solve for the two unknown forces.

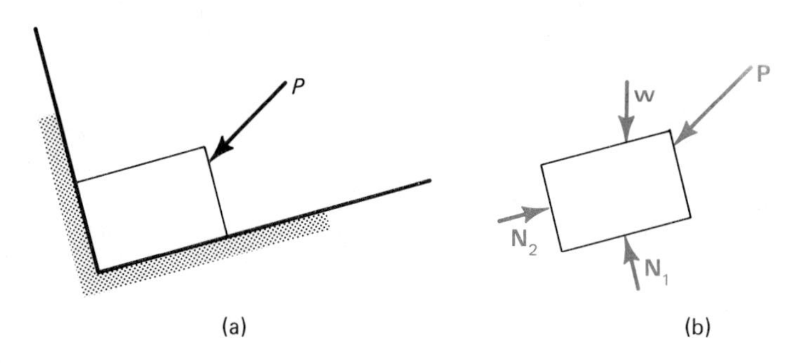

Figure 3.2

In drawing the free-body diagram of a body, the joints or connections must be replaced by the equivalent forces to ensure the equilibrium of the body. Consider the body of weight w_1 shown in Fig. 3.3(a). Here the connection at A is a frictionless pin and at B it is a roller. It is assumed that the wall is frictionless so that when the roller is removed it is replaced by one reaction force $\mathbf{B}_x$ which acts in the X direction. The pin is removed and replaced by two reactions $\mathbf{A}_x$ and $\mathbf{A}_y$. The complete free-body diagram including all the reaction forces, the body weight w_1, and the applied weight w_2 is shown in Fig. 3.3(b).

In many situations, the removal of connections and supports gives rise not only to reaction forces but reaction moments as well. Consider, for example, the cantilever beam shown in Fig. 3.4(a), where a force **P** is applied as shown. Now, when we remove the connection at the wall we have not only the reaction forces $\mathbf{R}_x$ and $\mathbf{R}_y$ but a moment **M** as well, as seen in Fig. 3.4(b). Clearly, if no moment existed, the beam would rotate, which would violate conditions of static equilibrium. Since the magnitudes of the reactions are unknown, we see that there are three unknowns, R_x, R_y, and M. We can write three independent equilibrium equations to obtain

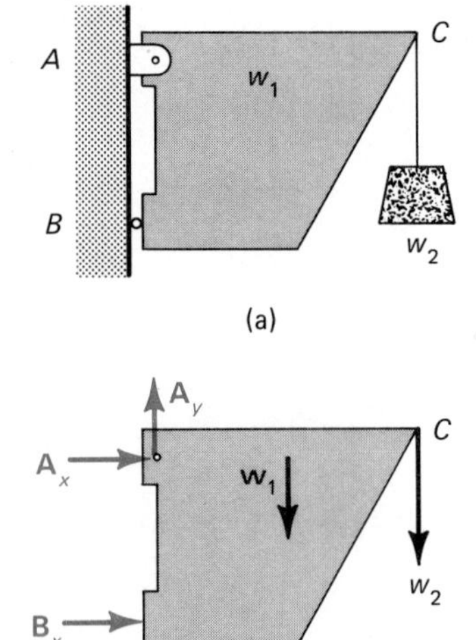

Figure 3.3

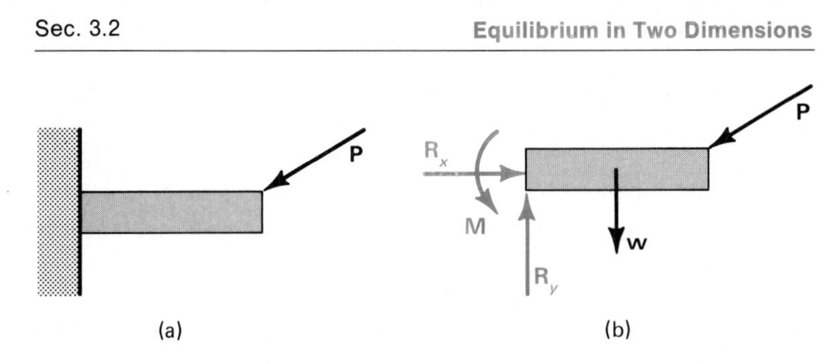

Figure 3.4

the three unknowns; two scalar equations for the forces and one scalar equation for the moments.

A free-body diagram showing all the applied and reaction forces is necessary before we can write the equations of equilibrium. Also, the drawing of free-body diagrams requires an accurate representation of the reaction forces. In the next two sections we will treat the equilibrium of rigid bodies in two and three dimensions, including the reaction forces that generally occur in two- and three-dimensional equilibrium. In each case in which we can solve for all the unknowns, we will find that the free-body diagram allows us to write as many independent equations as there are unknowns shown on the free-body diagram.

3.2 Equilibrium in Two Dimensions

If we restrict all forces acting on a rigid body to the XY plane and all external moments directed parallel to the Z axis, the problem of equilibrium is restricted to two dimensions. We have already established that when a force has components along the X and Y axes, the resulting moment about a point on the body is directed along the Z axis, and there are no components of moments along the X axis or the Y axis. Therefore, for equilibrium in two dimensions, we must satisfy the following equilibrium equations:

$$\sum F_x = 0 \qquad \sum F_y = 0 \qquad \sum M_z = 0 \tag{3.1}$$

Since all forces are restricted to a plane, the reaction forces of a two-dimensional structure must also be confined to a plane. These reaction forces, as well as the reaction moments, must be included in the free-body diagram. In general, the reaction forces of supports and connections can be classified as shown in Table 3.1. A structure or body is often composed of several of these connections, which impose forces and moments on the constrained body as soon as external forces or moments are applied to the body. A reaction force occurs when a rigid body is restricted to a stationary position or

TABLE 3.1

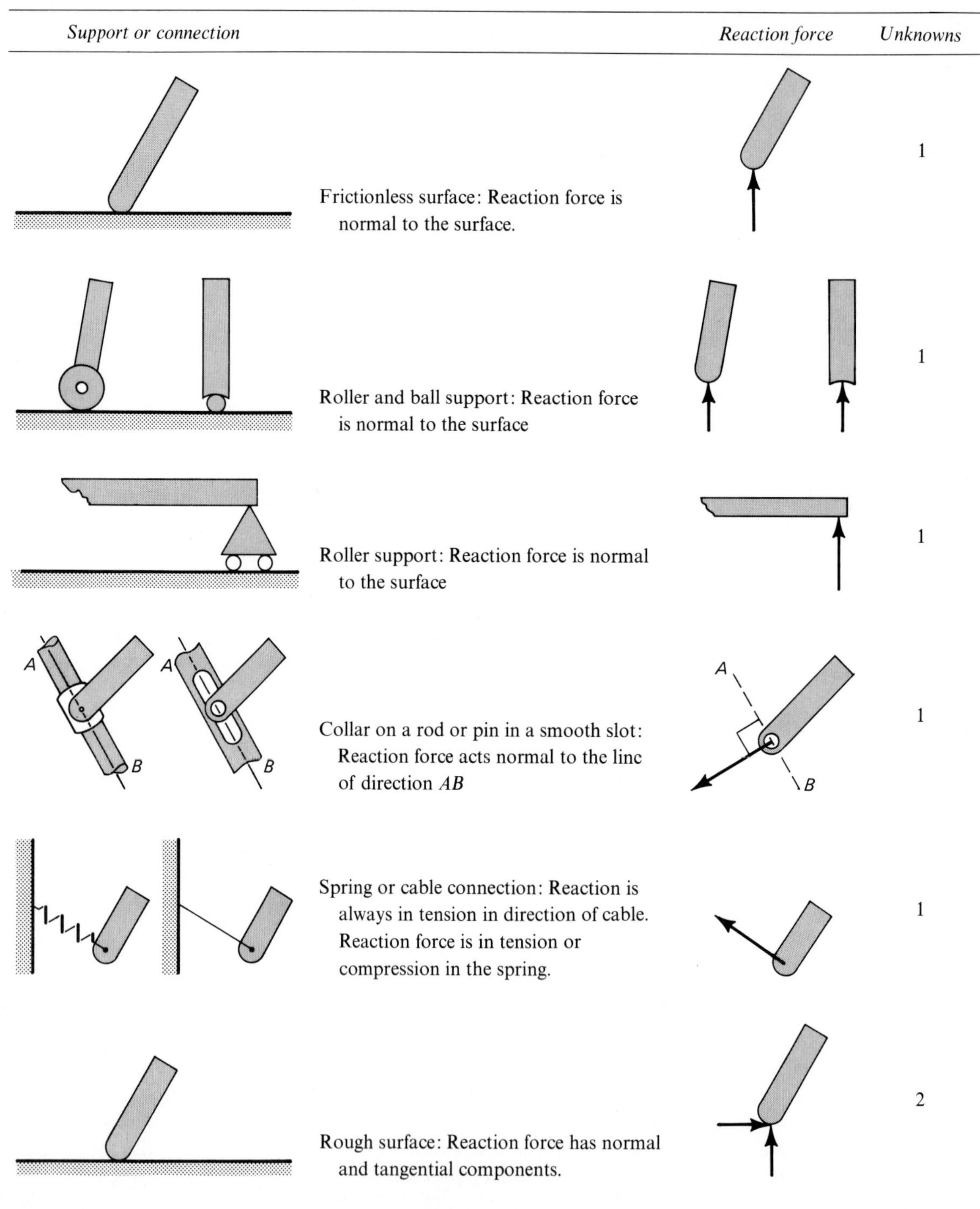

Support or connection		*Reaction force*	*Unknowns*
	Frictionless surface: Reaction force is normal to the surface.		1
	Roller and ball support: Reaction force is normal to the surface		1
	Roller support: Reaction force is normal to the surface		1
	Collar on a rod or pin in a smooth slot: Reaction force acts normal to the linc of direction *AB*		1
	Spring or cable connection: Reaction is always in tension in direction of cable. Reaction force is in tension or compression in the spring.		1
	Rough surface: Reaction force has normal and tangential components.		2

TABLE 3.1—*Cont.*

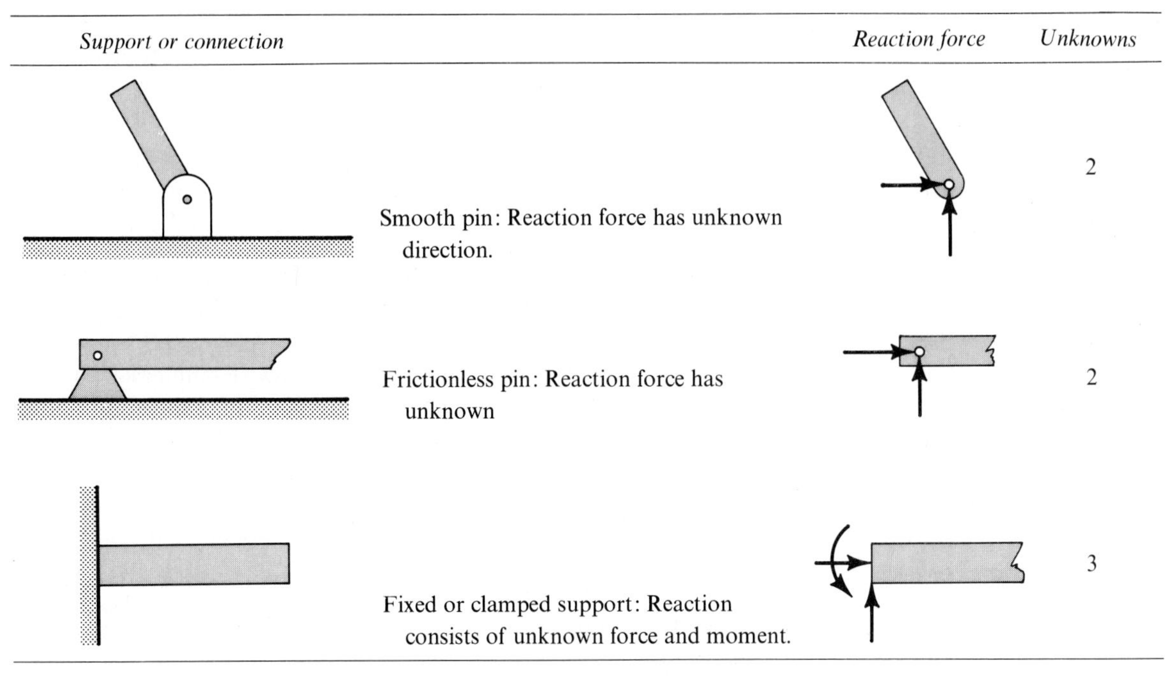

Support or connection		*Reaction force*	*Unknowns*
	Smooth pin: Reaction force has unknown direction.		2
	Frictionless pin: Reaction force has unknown		2
	Fixed or clamped support: Reaction consists of unknown force and moment.		3

to move in a certain direction. For example, a roller support on a frictionless surface shown in Table 3.1 can move horizontally, but not vertically downward owing to the support of the floor. Hence a vertical reaction force acts on the body.

One Reaction Force. Table 3.1 shows five examples of supports and connections that provide a single reaction force. In the first three examples, the structural member is free to move in the horizontal direction. In the fourth example, the collar and pin are free to slide along the member AB. However, a contact force is developed that is perpendicular to the direction of this motion. In the fifth example, we assume that the spring develops either a tensile force or a compressive force. In the case of the cable, only a tensile force may occur.

Two Reaction Forces. This group shown in Table 3.1 includes rough surfaces—where both a normal force and a tangential or frictional force exist—and the frictionless pin connection which allows rotational motion but does not allow any translational motion. The reaction forces of this group are normally represented by their rectangular components in the vertical and horizontal directions.

Three Reaction Forces. This group is exemplified by the fixed or clamped support shown in Table 3.1. Such a support is capable of providing two restraining forces and one restraining moment.

Once the reaction forces and moments present at the supports are identified, the free-body diagram is drawn. A body, or combination of elements considered as a single body, that possesses more external supports or constraints than are necessary to maintain an equilibrium position is said to be **statically indeterminate**. This means that the equations of equilibrium studied in this text are inadequate to solve for the reactions. However, when there are as many equations of equilibrium, i.e., equations of statics, as there are unknowns, the reaction forces are said to be **statically determinate**, and the structure is completely constrained. If there are fewer unknowns than there are available equations of equilibrium, the structure is partially constrained.

Consider again the body supporting the weight w as shown in Fig. 3.3(a). The upper support employs a frictionless pin connection while the lower support is a roller. The free-body diagram is drawn in Fig. 3.3(b). We notice that there are three unknowns, $\mathbf{A}_x$, $\mathbf{A}_y$, and $\mathbf{B}_x$. Here the directions are known but the magnitudes are unknown. Since we can write three equations, the system is statically determinate and the structure is completely constrained. The three unknowns may be found by writing the equilibrium equations,

$$\sum F_x = 0 \qquad \sum F_y = 0 \qquad \sum M_z = 0$$

The moment can be taken about any point since the entire rigid body is in equilibrium. In writing the equations of equilibrium it is important to understand that there are *three independent equations* of equilibrium for two-dimensional equilibrium. These three equations may be obtained by using any one of the following set of equations,

$$\sum M_A = 0 \qquad \sum M_B = 0 \qquad \sum M_C = 0$$

$$\text{or} \qquad \sum M_A = 0 \qquad \sum M_B = 0 \qquad \sum F_y = 0$$

$$\text{or} \qquad \sum M_A = 0 \qquad \sum F_x = 0 \qquad \sum F_y = 0$$

Returning to the free-body diagram shown in Fig. 3.3(b) we see that summing the moments about A will yield B_x and summing moments about B yields A_x. Finally A_y can be obtained by either summing moments about C or by summing the forces in the Y direction. Also, the points A, B, and C could be arbitrarily chosen on the rigid body although they should be chosen carefully so that the calculations are minimized. In this example, A is an excellent choice for summing moments since B_x can be obtained immediately.

As another example, consider the structure shown in Fig. 3.5(a). Here the inclined beam is clamped at B and restrained by the cable AC. The free-body diagram is shown in Fig. 3.5(b). At the clamped end of the beam there are two reaction forces $\mathbf{B}_x$ and $\mathbf{B}_y$ and a reaction moment $\mathbf{M}$. Also, there is a tensile force $\mathbf{T}$ acting on the beam at point C. Since there are four unknown reactions and three equations of equilibrium, this system is statically indeterminate. If the fixed support were replaced by a frictionless pin connection, the number of unknowns would become three and the system would become determinate.

Next, consider the structure shown in Fig. 3.6(a) that is supported by rollers at A and B. Assume the surface is smooth. The free-body diagram is shown in Fig. 3.6(b). We observe that there are two unknown reactions although we have three equilibrium equations to satisfy. This is an example of a partially constrained structure.

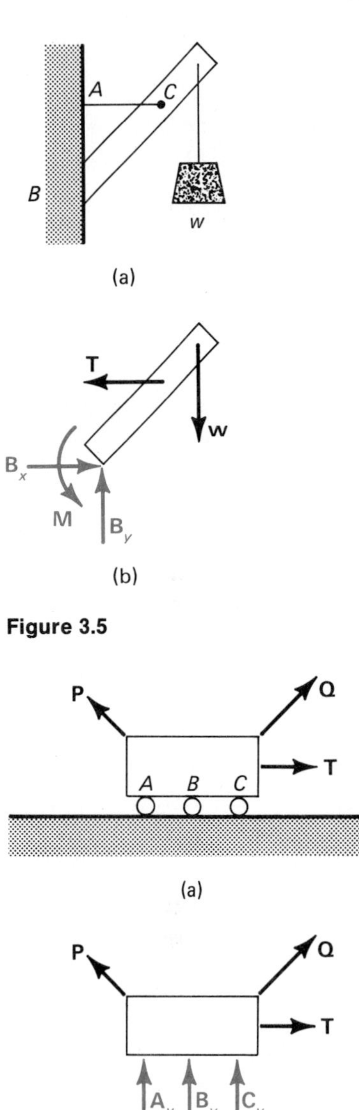

Figure 3.5

Figure 3.7

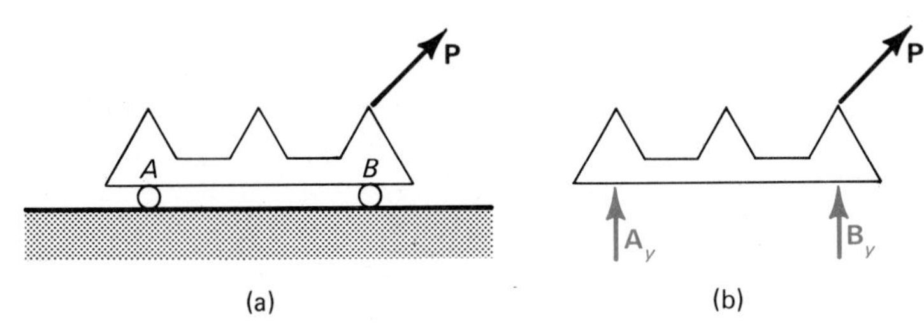

Figure 3.6

It appears that when the number of unknowns and equations are the same the reaction forces are determinate. However, this is not true if a body is improperly constrained. This is demonstrated by the example shown in Fig. 3.7(a). The roller support on the smooth surface gives rise to one reaction at each roller as shown in the free-body diagram of Fig. 3.7(b). When we write the equations of equilibrium, we find that summing the horizontal forces does not include our unknown forces at all. We are, therefore, left with two independent equations that relate three unknowns. Clearly, the three-roller support has led to a statically indeterminate situation.

Example 3.1

A force of 800 N is applied to a beam as shown in Fig. 3.8(a). The beam has a mass of 50 kg. What are the magnitudes of the reactions at A and B?

The free-body diagram is shown in Fig. 3.8(b). Here the roller support at B has been replaced by a reaction of magnitude B_y,

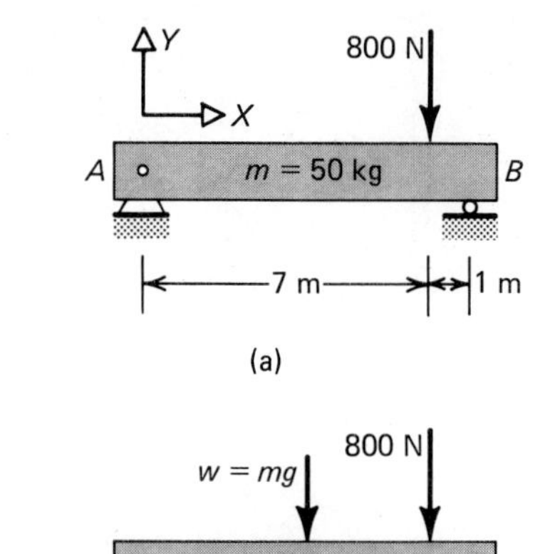

Figure 3.8

whereas the pin connection at A has been replaced by two reaction forces of magnitude A_x and A_y. For equilibrium in the X direction, we sum the forces to obtain

$$\sum F_x = 0$$

Therefore,

$$A_x = 0 \qquad \textit{Answer}$$

The weight of the beam is

$$w = 50(9.81) = 490.5 \text{ N}$$

Now taking the moments about A and setting the sum equal to zero for equilibrium, i.e.,

$$\sum M_A = 0$$

we obtain

$$-4(490.5) - 7(800) + 8B_y = 0$$

$$B_y = 945.3 \text{ N} \qquad \textit{Answer}$$

To find A_y, consider the equilibrium of the forces in the Y direction, i.e.,

$$\sum F_y = 0$$

$$A_y - 490.5 - 800 + 945.3 = 0$$

$$A_y = 345.2 \text{ N} \qquad \textit{Answer}$$

Since the signs of the reaction forces are positive, the assumed directions of these forces as seen in the free-body diagram are correct. As an alternate approach, the third independent equilibrium equation and the magnitude of $\mathbf{A}_y$ could have been obtained by summing the moments about B in place of summing the forces in the Y direction.

$$\sum M_B = 0 = -8A_y + 4(490.5) + 1(800)$$

$$A_y = 345.2 \text{ N}$$

which is the same answer obtained by summing the forces in the Y direction.

Example 3.2

Two forces are applied to a cantilever beam as shown in Fig. 3.9(a). Obtain the magnitudes of the reactions at the support. Neglect the weight of the beam.

The free-body diagram is shown in Fig. 3.9(b). We note that there are two reaction forces $\mathbf{A}_x$, $\mathbf{A}_y$ and one reaction moment $\mathbf{M}$. This moment is sometimes called a **clamping moment**.

The clamping moment may be obtained by summing the moments about A.

$$\sum M_A = 0 = M + 5(100 \sin 45^\circ) - 12(50)$$

therefore, $M = 246 \text{ ft}\cdot\text{lb}$ *Answer*

Summing the forces in the X direction, we find that

$$\sum F_x = 0 = A_x + 100 \cos 45^\circ$$

therefore, $A_x = -70.7 \text{ lb}$ *Answer*

Note that the reaction at A in the X direction is *opposite* to that originally assumed on the free-body diagram since we obtained a negative sign in our solution. That is, $\mathbf{A}_x$ actually acts on the beam to the left.

Finally, we obtain A_y by summing forces in the Y direction.

$$\sum F_y = 0 = A_y + 100 \sin 45^\circ - 50$$

therefore, $A_y = -20.7 \text{ lb}$ *Answer*

The vertical reaction at A is also opposite to the assumed direction in Fig. 3.9(b).

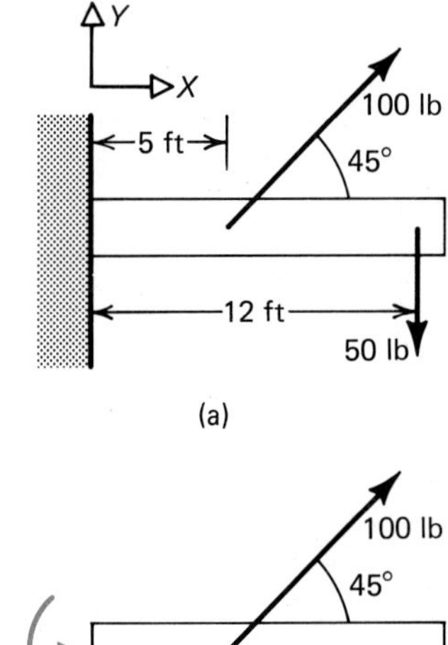

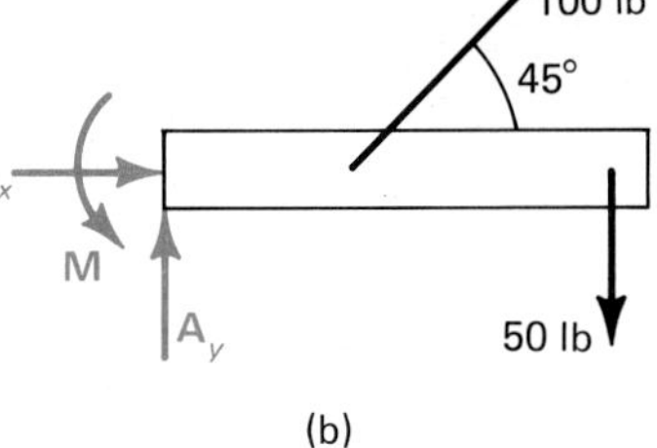

Figure 3.9

Example 3.3

The frame in Fig. 3.10(a) is restrained by a spring while supporting the 100-kg load. What are the reactions at A and the magnitude of the force in the spring?

The weight of the load is

$$w = 9.81(100) = 981 \text{ N}$$

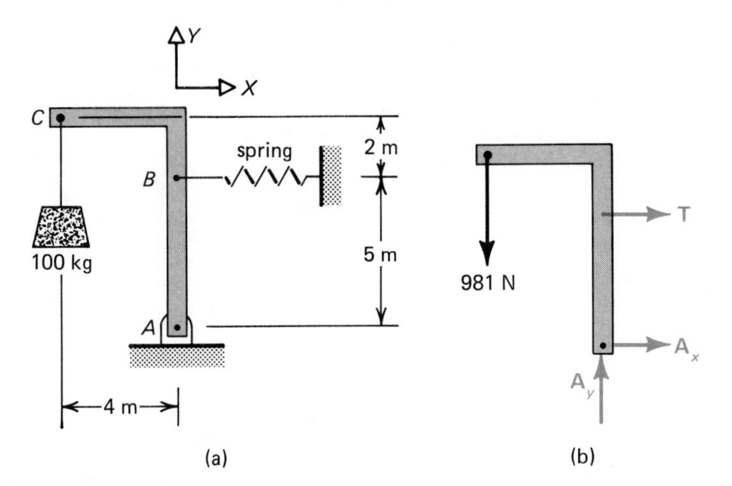

Figure 3.10

Referring to the free-body diagram in Fig. 3.10(b), sum moments about A to find the magnitude of the tensile force **T** in the spring.

$$\sum M_A = 0$$

$$4(981) - 5T = 0$$

$$T = 785 \text{ N} \qquad \textit{Answer}$$

To find $\mathbf{A}_y$,

$$\sum F_y = 0$$

$$A_y - 981 = 0$$

$$\mathbf{A}_y = 981\mathbf{j} \text{ N} \qquad \textit{Answer}$$

To find $\mathbf{A}_x$,

$$\sum F_x = 0$$

$$A_x + T = 0$$

$$A_x = -T = -785 \text{ N}$$

$$\mathbf{A}_x = -785\mathbf{i} \text{ N} \qquad \textit{Answer}$$

Again the negative sign indicates that the reaction at A actually acts to the left.

Example 3.4

Determine the force **T** necessary to keep the pulley in Fig. 3.11(a) at equilibrium. Neglect the weight of the pulley.

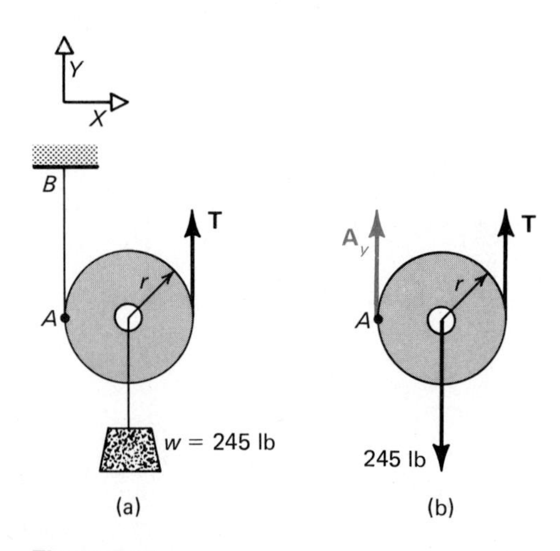

Figure 3.11

Figure 3.11(b) shows the free-body diagram that consists of the supported weight w, the force of magnitude A_y in the rope AB acting on the pulley, and the force of magnitude T. To find T, sum moments about point A on the pulley.

$$\sum M_A = 0$$

$$-r(245) + 2rT = 0$$

$$T = 122.5$$

Therefore, $\mathbf{T} = 122.5\mathbf{j}$ lb *Answer*

Next, summing the forces in the Y direction

$$\sum F_y = A_y - 245 + T = 0$$

$$A_y = 122.5$$

Therefore, $\mathbf{A}_y = 122.5\mathbf{j}$ lb *Answer*

Example 3.5

A car is hoisted by a jack as shown in Fig. 3.12(a). The mass of the car is m kg. (a) Obtain the magnitudes of the reactions at A and B as functions of a, b, c, w, and θ. (b) Obtain the magnitudes of the reactions for $a = 2$ m, $b = 3$ m, $c = 1.5$ m, $m = 2500$ kg, and $\theta = 10°$. Assume that the only reaction at A is that which is normal to the ground while the reaction at B may have a horizontal and vertical component.

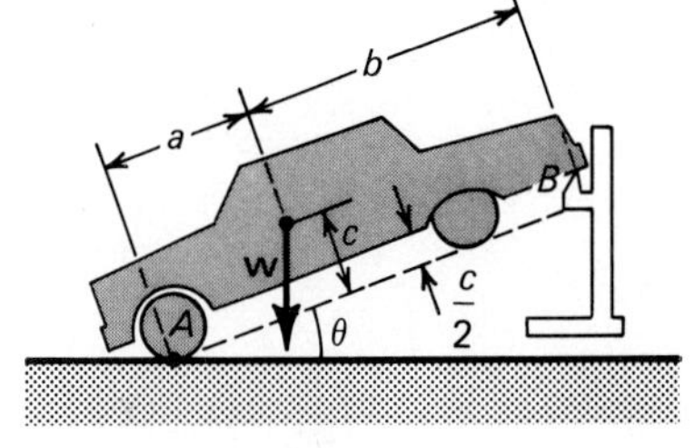

(a)

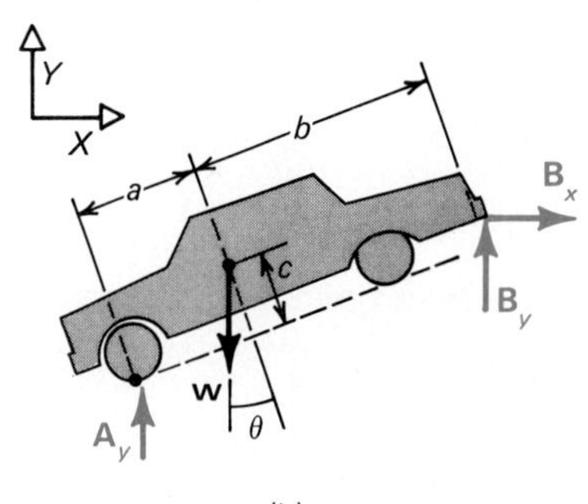

(a) The weight of the car is

$$w = mg$$

Next, we note from the free-body diagram in Fig. 3.12(b) that summing the moments about B yields A_y.

$$\sum M_B = 0 = w\left(b \cos\theta + \frac{c}{2}\sin\theta\right) + A_y\frac{c}{2}\sin\theta - A_y(a+b)\cos\theta$$

therefore, $$A_y = \frac{w(2b\cos\theta + c\sin\theta)}{2(a+b)\cos\theta - c\sin\theta}$$ *Answer*

Summing the forces in the vertical direction, we have

$$\sum F_y = 0 = A_y + B_y - w$$

therefore, $$B_y = w - \frac{(2b\cos\theta + c\sin\theta)w}{2(a+b)\cos\theta - c\sin\theta}$$

$$= \frac{2w(a\cos\theta - c\sin\theta)}{2(a+b)\cos\theta - c\sin\theta}$$ *Answer*

Finally, we obtain B_x by summing horizontal forces.

$$\sum F_x = 0$$

so that

$$B_x = 0 \qquad \textit{Answer}$$

(b) First calculate the weight for $m = 2500$ kg.

$$w = 2500(9.81) = 24\,525 \text{ N}$$

Now substitute the given dimensions into the equations derived in part a.

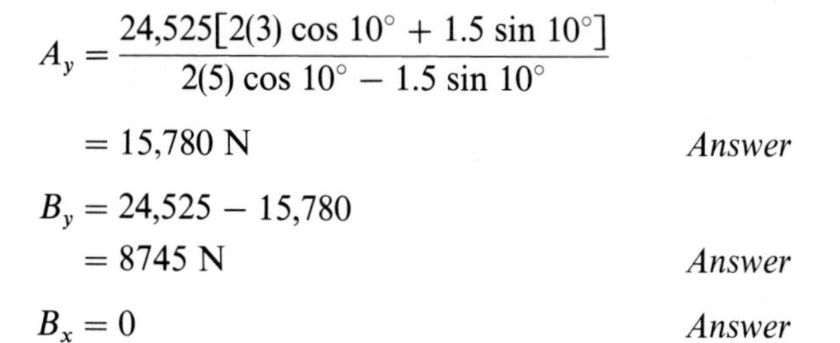

$$A_y = \frac{24{,}525[2(3)\cos 10^\circ + 1.5 \sin 10^\circ]}{2(5)\cos 10^\circ - 1.5 \sin 10^\circ}$$

$$= 15{,}780 \text{ N} \qquad \textit{Answer}$$

$$B_y = 24{,}525 - 15{,}780$$

$$= 8745 \text{ N} \qquad \textit{Answer}$$

$$B_x = 0 \qquad \textit{Answer}$$

From this example we can make a very important observation about the reaction forces and moments. That is, the reaction forces and moments *depend* upon the magnitude and location of the applied forces. In the case of the car hoisted by the jack in the previous example, the magnitude and location of the weight w clearly dictated the magnitude of the reaction forces.

PROBLEMS

3.1 Neglecting the weight of the beam, obtain the reactions for the beam in Fig. P3.1.

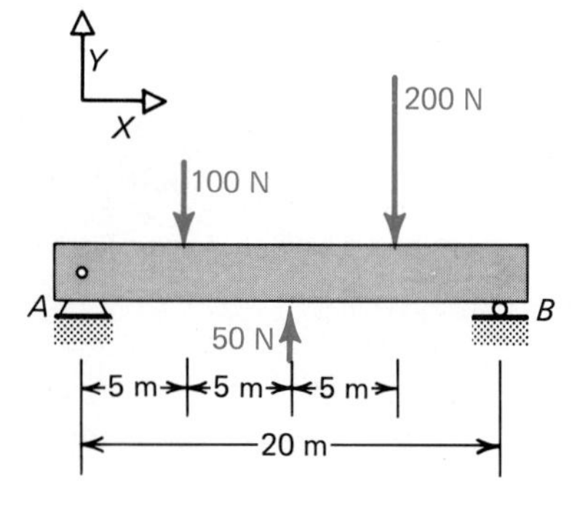

Figure P3.1

3.2 Neglecting the weight of the beam, obtain the reactions for the beam in Fig. P3.2.

3.3 Neglecting the weight of the beam, obtain the reactions for the beam in Fig. P3.3.

3.4 Neglecting the weight of the beam, obtain the reactions for the beam in Fig. P3.4.

3.5 Neglecting the weight of the beam, obtain the reactions for the beam in Fig. P3.5.

3.6 Neglecting the weight of the beam, obtain the reactions for the beam in Fig. P3.6.

3.7 Neglecting the weight of the column, obtain the reactions for the column in Fig. P3.7.

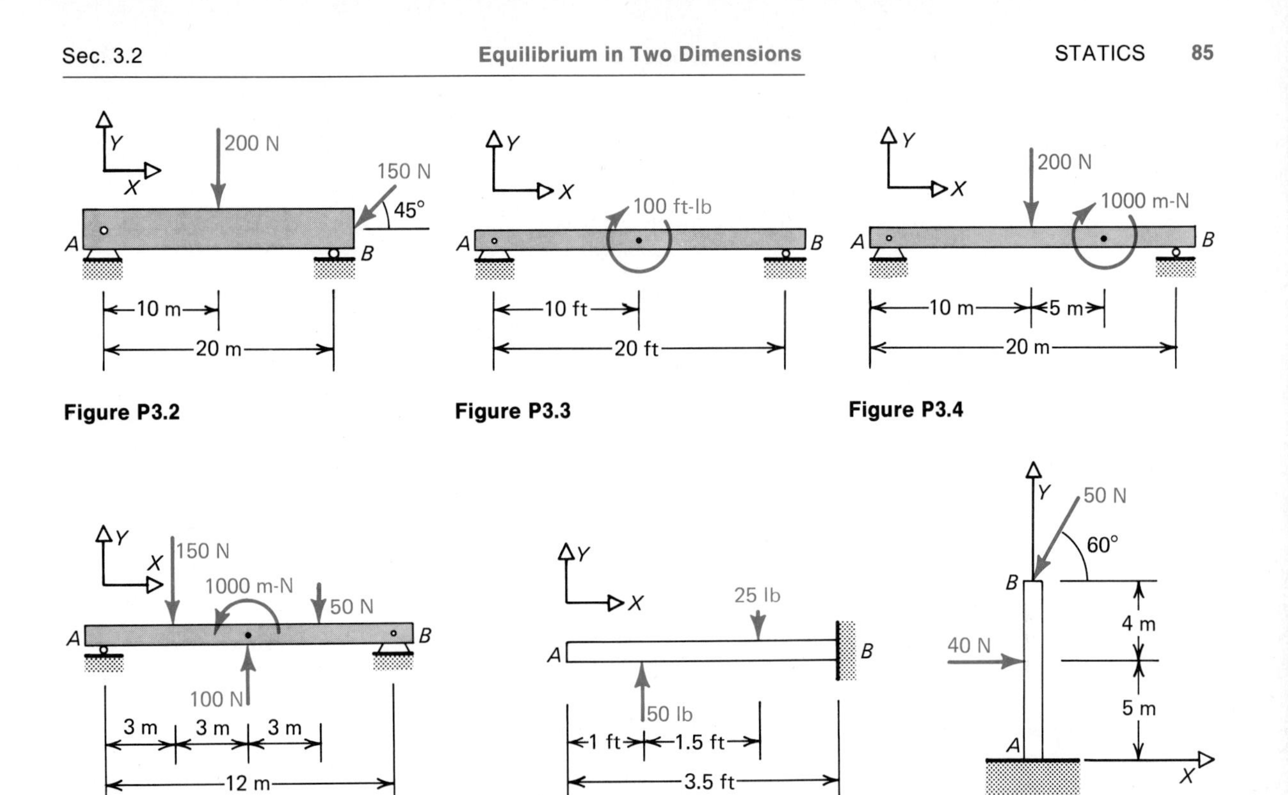

Figure P3.2

Figure P3.3

Figure P3.4

Figure P3.5

Figure P3.6

Figure P3.7

3.8 Neglecting the weight of the overhanging beam, obtain the reactions for the beam in Fig. P3.8.

3.9 Neglecting the weight of the overhanging beam, obtain the reactions for the beam in Fig. P3.9.

3.10 Determine the tension T necessary to hold the two blocks in equilibrium on the inclined plane shown in Fig. P3.10. Assume that the inclined surface is frictionless.

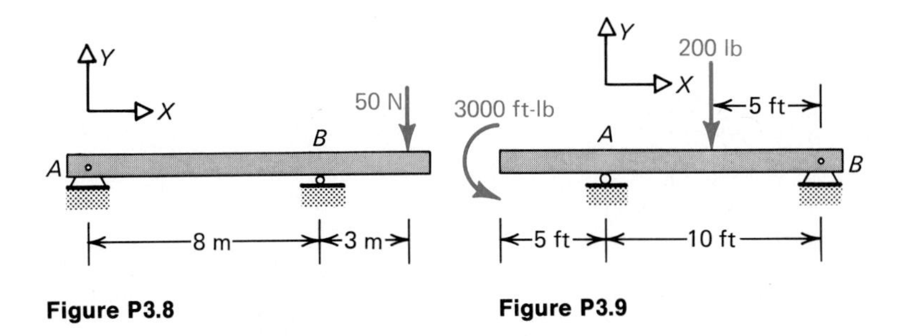

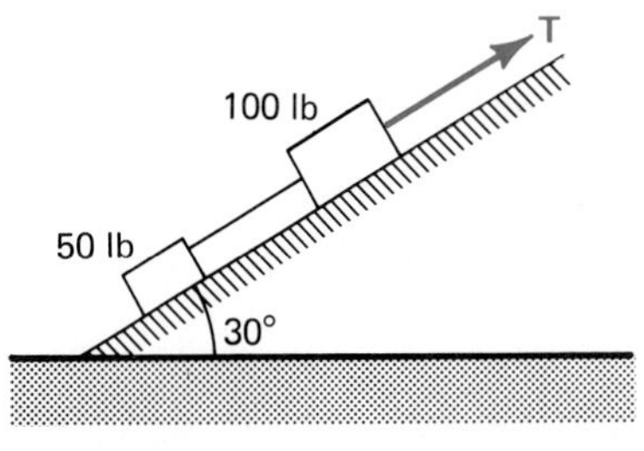

Figure P3.8

Figure P3.9

Figure P3.10

3.11 A car for carrying coal on an inclined plane is shown in Fig. P3.11. The weight of the car and coal is represented by the 2500-N force going through C. (a) Determine the tension T necessary to hold the car stationary on the frictionless surface. (b) What is the reaction at each wheel? Let $\theta = 40°$ and $h = 0.75$ m.

3.12 Determine the tension T necessary to keep the 400-lb block shown in Fig. P3.12 at equilibrium.

3.13 A 100-kg block is supported by three pulleys as shown in Fig. P3.13. What value of T will ensure the equilibrium of the block?

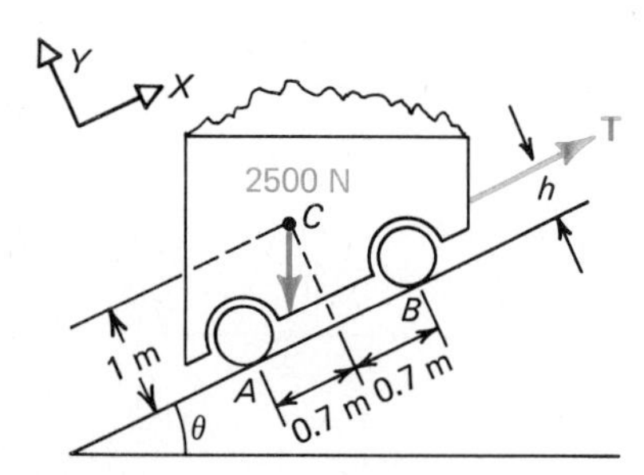

Figure P3.11

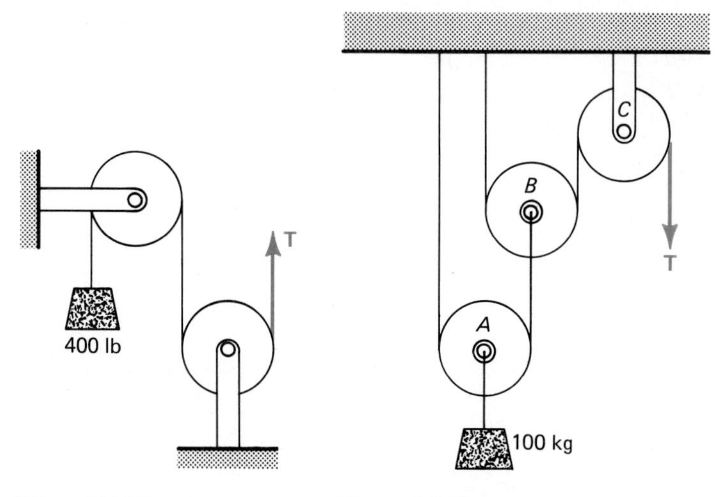

Figure P3.12

Figure P3.13

3.14 Calculate (a) the tension in the cable and (b) the reaction at O for the rig shown in Fig. P3.14 for $\theta = 10°$, $r = 0.5$ ft and $P = 1000$ lb.

3.15 Derive expressions for the reactions at A and B for the loading shown in Fig. P3.15.

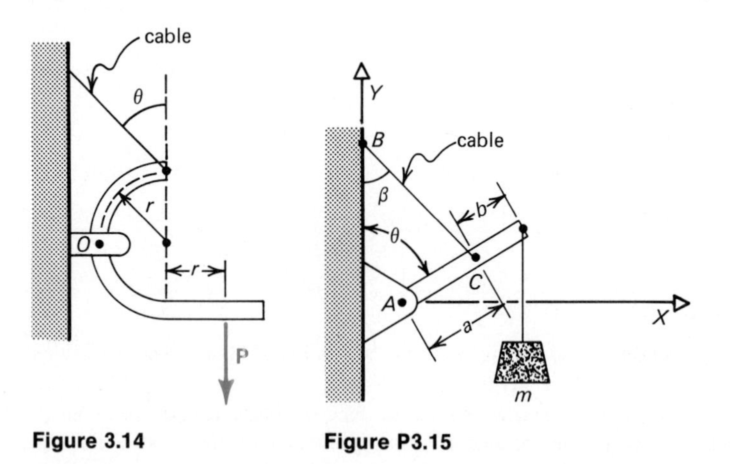

Figure 3.14

Figure P3.15

3.16 Determine (a) the tension in the rope CB and (b) the reactions at A for the system shown in Fig. P3.15. Let $\theta = 30°$, $\beta = 60°$, $a = 1$ m, $b = 1.5$ m, and $m = 40.7$ kg.

3.17 A flagpole weighing 300 lb is supported as shown in Fig. P3.17. It is known that the tension in cable BE is 200 lb and in CD it is 175 lb. Determine the reactions at A assuming a clamped support.

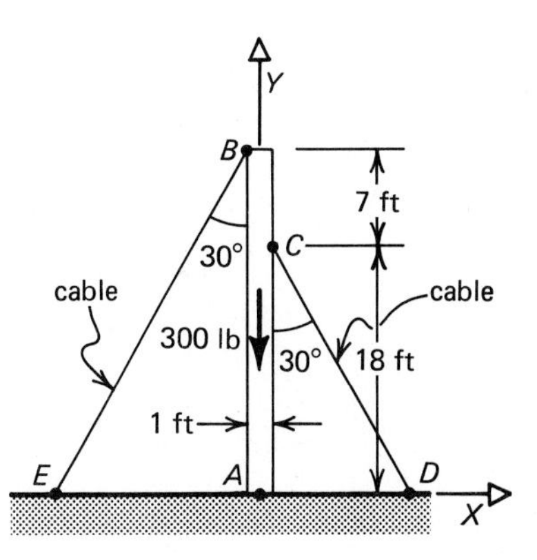

Figure P3.17

3.18 A 100-kg body is supported as shown in Fig. P3.18. What are (a) the reactions at A and (b) the tension in the cable?

3.19 If the connection at A in Problem 3.18 were made into a clamped connection, is it possible to obtain the reactions by the equations of static equilibrium?

3.20 A beam is loaded as shown in Fig. P3.20. Obtain (a) the tension in the cable and (b) the reactions at A.

3.21 Calculate the reactions at A and the magnitude of the force in the spring for the beam in Fig. P3.21. Let $P = 750$ lb, $\theta = 60°$, $AB = 1.5$ ft, and $BC = 1.2$ ft.

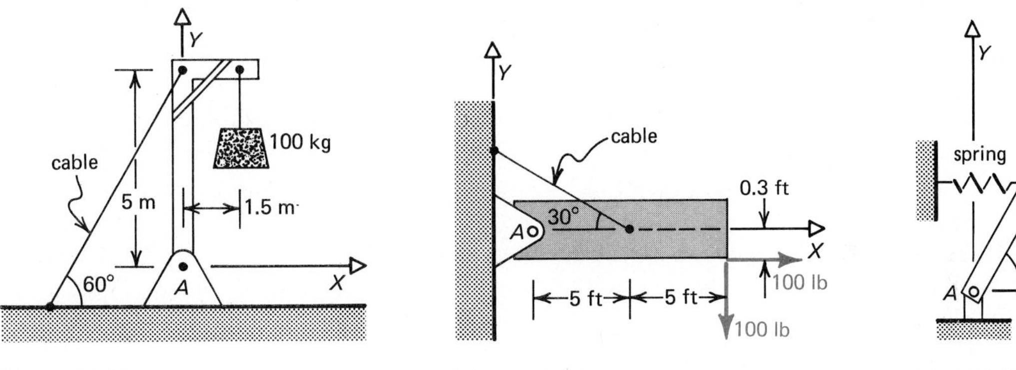

Figure P3.18

Figure P3.20

Figure P3.21

3.22 Calculate the reactions at A and the magnitude of the force in the spring for the beam in Fig. P3.21. Let $P = 1000$ N, $\theta = 45°$, $AB = 2$ m, and $BC = 1.5$ m.

3.23 Compute the magnitude of the spring force for the loaded beam in Fig. P3.23.

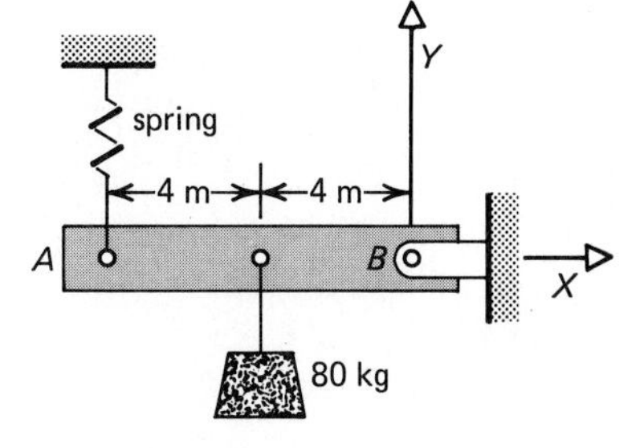

Figure P3.23

3.24 The body in Fig. P3.24 is supported by the springs with reactions $\mathbf{A}_y = 171.5\mathbf{j}$ and $\mathbf{B}_y = 428.5\mathbf{j}$ lb. At what position X should the load P be placed so that the body is oriented as shown in the figure.

3.25 A car-and-trailer system is shown in Fig. P3.25. (a) What is the reaction at A? (b) What are the reactions at the three wheels B, C, and D? Assume that the point at A may be considered as a ball support. (*Hint*: Draw a separate free-body diagram for the car and for the trailer.)

3.26 Obtain the reactions at A and B for the two-bar system lying in the horizontal plane as shown in Fig. P3.26.

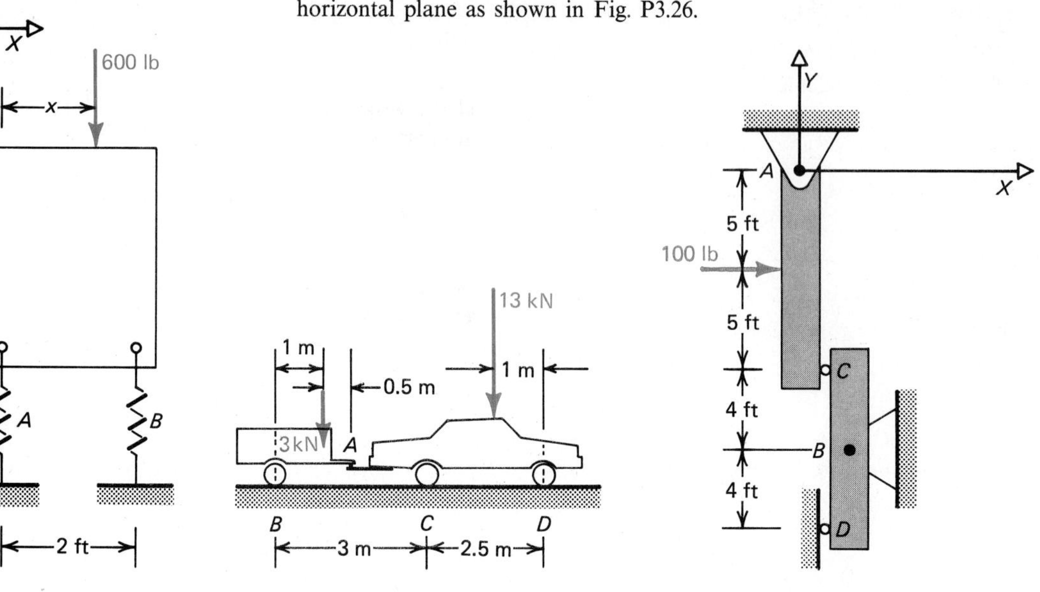

Figure P3.24

Figure P3.25

Figure P3.26

3.27 Determine the reactions at A and B for the rigid mechanism shown in Fig. P3.27. Assume $P = 200$ lb, $\theta = 25°$, $a = 3$ in, $b = 6$ in, and $c = 10$ in.

3.28 A pair of tongs is used to hold a steel ball as shown in Fig. P3.28. If the magnitude of the applied forces is 10 N as shown, what is the magnitude of the compressive forces acting on the steel ball?

3.29 For the structure in Fig. P3.29, obtain the reaction at the smooth pin at D. Assume that $\theta = 60°$, and $w = 100$ lb.

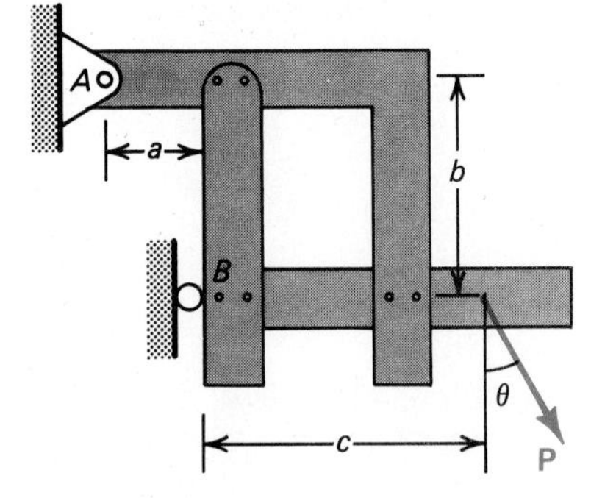

Figure P3.27

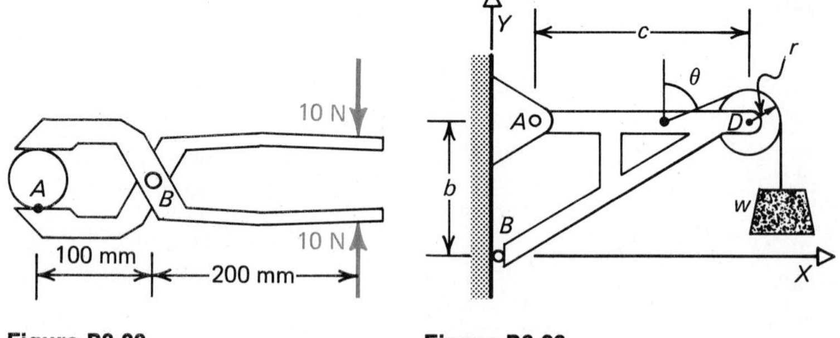

Figure P3.28

Figure P3.29

3.30 Determine the reaction at the clamped support at A for the rigid bar shown in Fig. P3.30. Let $P = 200$ N, $Q = 400$ N, and $\theta = 20°$.

3.31 An antenna 60 ft high, shown in Fig. P3.31, has a horizontal force of magnitude 1000 lb applied due to wind pressure. What is the reaction at the supports A and B? Neglect the weight of the antenna and assume a frictionless pin at A and a roller support at B.

3.32 If the antenna in Problem 3.31 weighs 2000 lb and the weight acts down through C vertically, what is the reaction at the supports with the wind load applied?

3.33 Determine the reaction at the wheels of the mobile crane shown in Fig. P3.33. Assume the weight of the load $w = 3$ kN.

3.34 Derive an expression for the tension in the rope and the reaction at A for the crane shown in Fig. P3.34. Calculate the answer for $w_1 = 4$ kip, $w_2 = 1$ kip, $a = 9$ ft, $b = 1$ ft, and $\theta = 30°$.

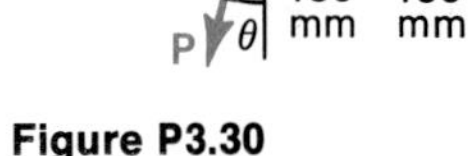

Figure P3.30

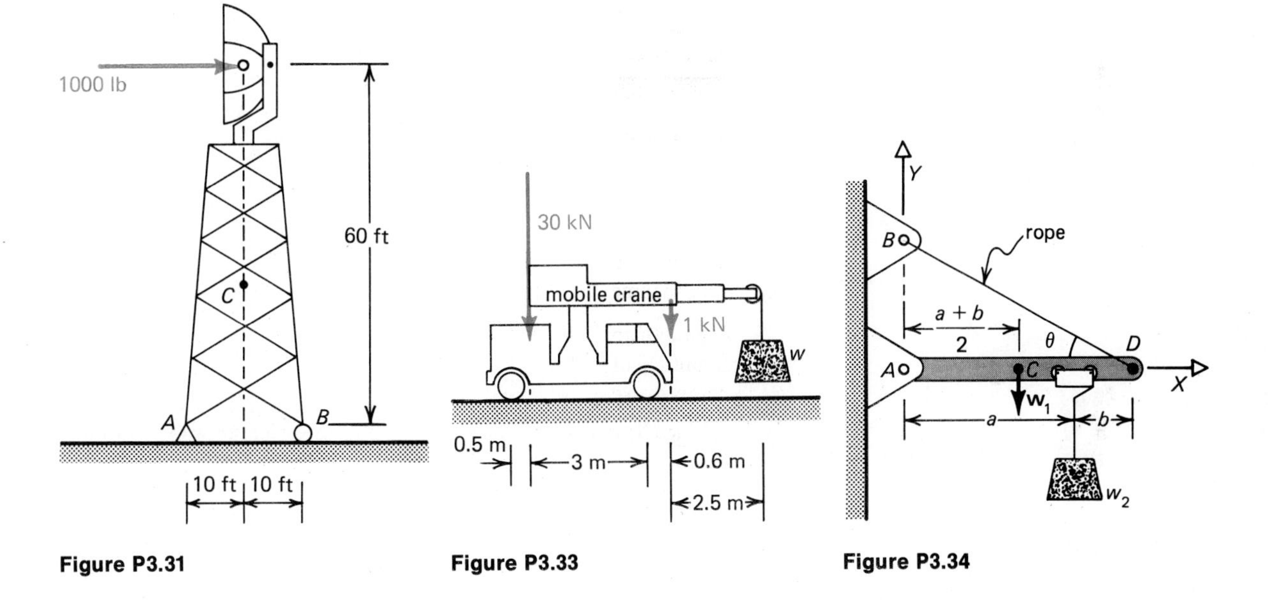

Figure P3.31

Figure P3.33

Figure P3.34

3.35 Repeat Problem 3.34 assuming that $w_1 = 3$ kip, $w_2 = 2$ kip, $a = 6$ ft, $b = 4$ ft, and $\theta = 35°$.

3.36 A 500-kg load is raised by the arrangement shown in Fig. P3.36. Calculate the magnitude of the force in the cable and the reaction at A.

3.37 Repeat Problem 3.36 taking into account that the mass of member AB equals 100 kg. Assume this weight acts vertically downward at the center of member AB.

3.38 A uniform bar is supported in the equilibrium position as shown in Fig. P3.38. Determine the magnitude of the horizontal force **P** needed to ensure equilibrium. The floor and wall are to be considered frictionless and $w = 100$ lb.

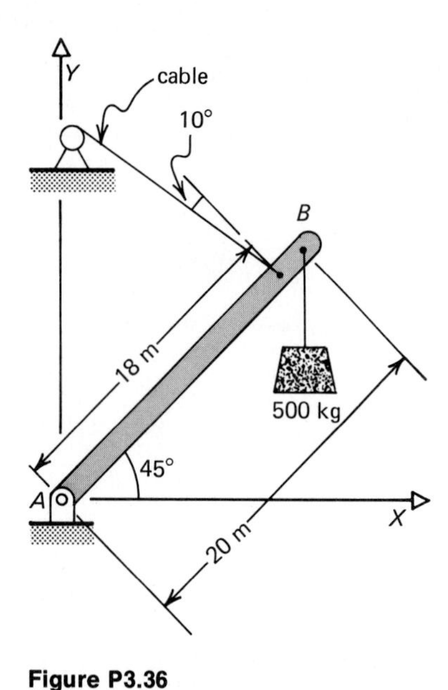

Figure P3.36

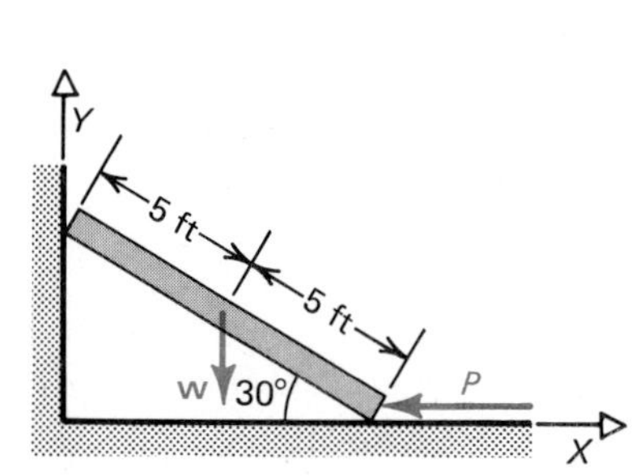

Figure P3.38

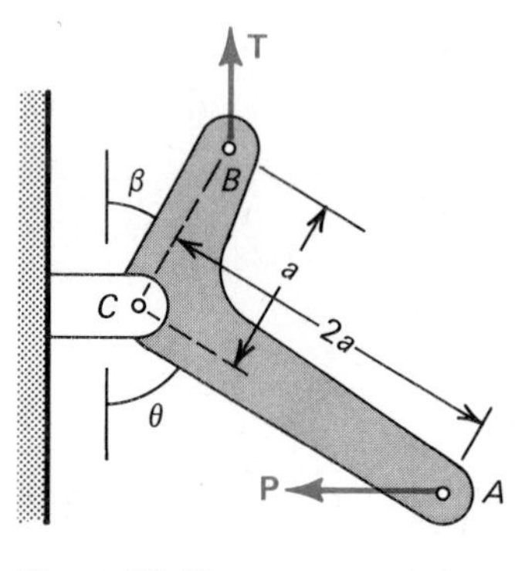

Figure P3.39

3.39 The bell crank shown in Fig. P3.39 has a force **P** applied at A. Determine (a) the tension T and (b) the reactions at C. Assume $\theta = 45°$, $\beta = 45°$, $P = 50$ lb, and $a = 4$ in.

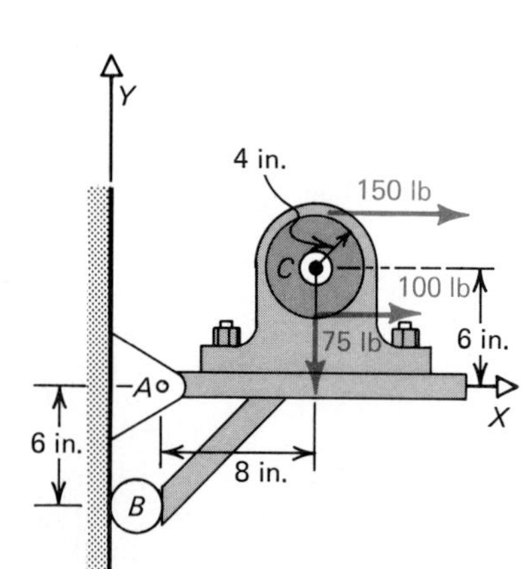

Figure P3.40

3.40 Find the tension T and the reactions at A for the beam in Fig. P3.40.

3.41 A motor is mounted on a bracket as shown in Fig. P3.41. If the tensions in the belt are 100 lb and 150 lb when the motor is running, determine the reactions at A and B.

3.42 Two small control motors are mounted on a cantilever beam as shown in Fig. P3.42. Determine the reaction at the clamped point A for the given loading.

3.43 Obtain the magnitude of the vertical reaction at A in Fig. P3.43 in terms of d, b, and w for the case where the magnitude of the reaction moment at A is zero.

Figure P3.41

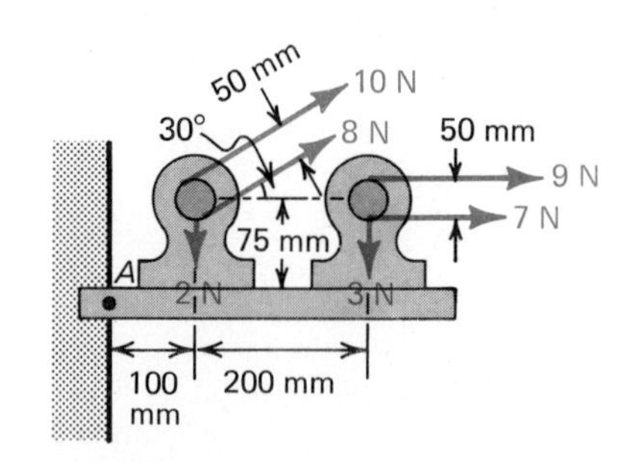

Figure P3.42

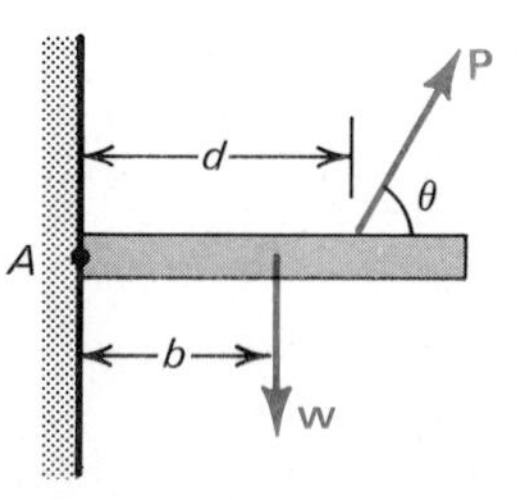

Figure P3.43

3.3 Equilibrium in Three Dimensions

In general we can say that for equilibrium of a rigid body the resultant force as well as the resultant moment must be zero. This is represented in vector form as

$$\sum \mathbf{F} = 0 \qquad \sum \mathbf{M} = 0$$

The equilibrium of a rigid body in all three dimensions requires that the resultant force and the resultant moment about the X, Y, and Z axes be equal to zero. Therefore, from each of the vector equations we obtain three scalar equations from the summation of the forces and three scalar equations from the summation of the moments, namely

$$\sum F_x = 0 \qquad \sum F_y = 0 \qquad \sum F_z = 0 \tag{3.2}$$

$$\sum M_x = 0 \qquad \sum M_y = 0 \qquad \sum M_z = 0 \tag{3.3}$$

There are, therefore, at most six independent equations of equilibrium. These six equations of equilibrium are obtained from the free-body diagram showing all the applied forces and moments as well as all the reaction forces and moments. From these six equations of equilibrium, we can solve for six unknowns. If there are more than six unknowns, the system is statically indeterminate. If we have six unknowns and can write six equations of equilibrium that relate these unknowns, the reactions are statically determinate and the structure is completely constrained.

Supports and connections in three dimensions give rise to as many as six reactions as illustrated in Table 3.2. The number and type of reactions depend upon the motion that the support or connection prevents. Note that we are including reaction moments in counting the number of reaction forces in Table 3.2. Hence, when we talk about reactions we mean both forces and moments developed at the support. For example, if a connection allows five motions while it constrains one motion, then there is only one reaction force. As with two-dimensional equilibrium, supports and connections can be grouped on the basis of the number of reaction forces as shown in Table 3.2.

One Reaction Force. The ball on a frictionless surface, and the cable and spring connections all contribute only one reaction force.

Three Reaction Forces. A roller support has up to three reaction forces, one due to the reaction from the surface, one due to friction developed on the rough surface, and a reaction moment due to twisting of the roller about the vertical axis. Table 3.2 also shows

TABLE 3.2

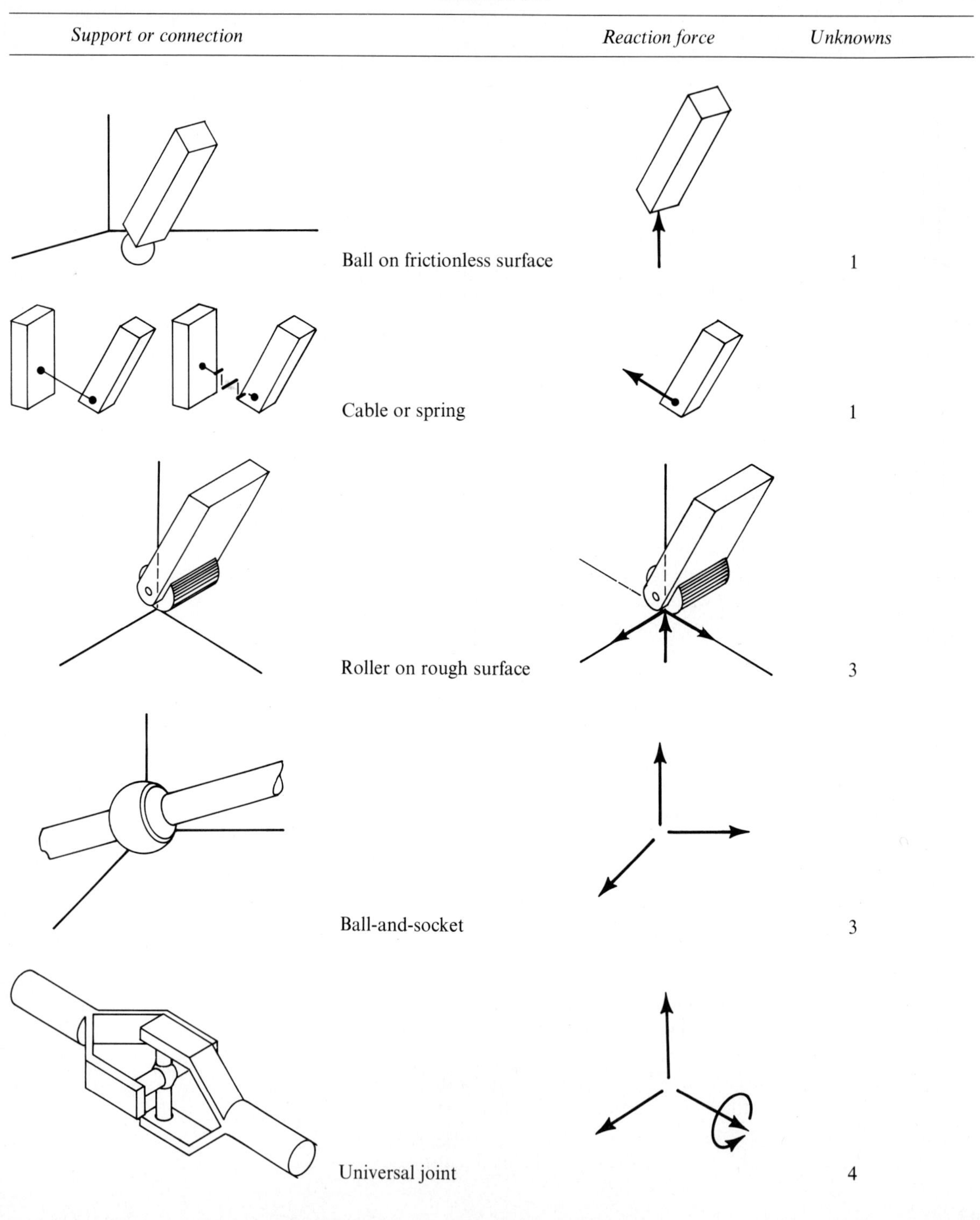

Support or connection		*Reaction force*	*Unknowns*
	Ball on frictionless surface		1
	Cable or spring		1
	Roller on rough surface		3
	Ball-and-socket		3
	Universal joint		4

TABLE 3.2—*Cont.*

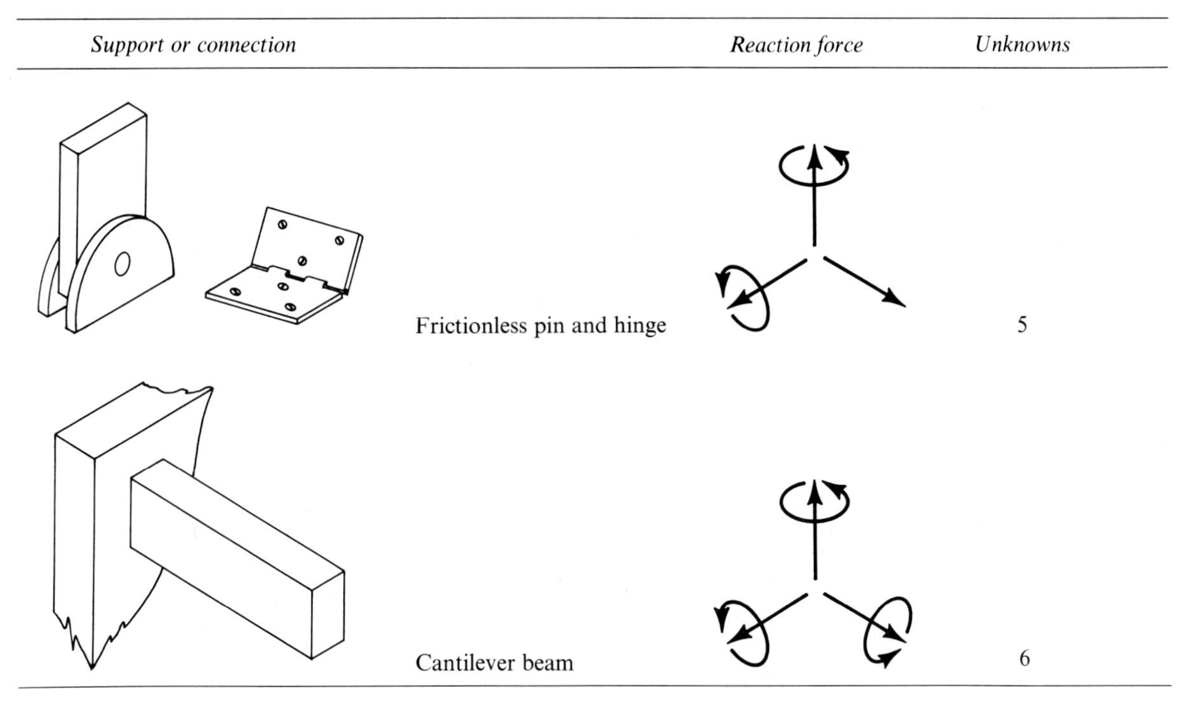

Support or connection	Reaction force	Unknowns
Frictionless pin and hinge		5
Cantilever beam		6

a ball-and-socket joint that allows rotational freedom but constrains translational motion in all three translational directions.

Four Reaction Forces. Here, four possible motions are constrained. The universal joint shown in Table 3.2 satisfies this criterion, since it resists all translational motion and one rotational motion around the longitudinal axis of the joint.

Five Reaction Forces. Both the smooth pin and hinge are examples of devices that allow rotation only about their hinged axis.

Six Reaction Forces. The cantilever beam resists all translational motion and rotational motion.

The equations of equilibrium in three dimensions are obtained as in the two-dimensional case, i.e., we should draw a free-body diagram first. This free-body diagram must show all of the external forces on the rigid body as well as all the reactions at the various supports and connections.

Consider the flagpole shown in Fig. 3.13(a), where the point C is a ball-and-socket joint and the flagpole is constrained by the cables AD and BD. The free-body diagram is shown in Fig. 3.13(b). The reactions at C consist of three forces, $\mathbf{C}_x$, $\mathbf{C}_y$, and $\mathbf{C}_z$. The weight w of the pole acts vertically downward at the center of gravity of the pole. The cables connected at A and B can support tension only, so that they are replaced by two tensile forces $\mathbf{T}_A$ and $\mathbf{T}_B$. Here we have five unknowns, $\mathbf{C}_x$, $\mathbf{C}_y$, $\mathbf{C}_z$, $\mathbf{T}_A$, and $\mathbf{T}_B$. We shall see below, in Example 3.7, that we can indeed obtain these unknowns by obtaining five distinct equations of equilibrium.

(a)

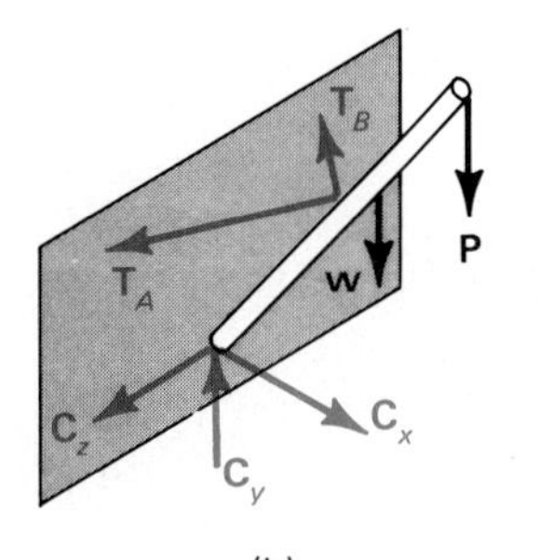

(b)

Figure 3.13

A different connection for the boom is illustrated in Fig. 3.14 where forces $\mathbf{P}$ and $\mathbf{Q}$ are acting on the boom that has a frictionless smooth pin at C and a restraining cable AB. The free-body diagram in Fig. 3.14(b) shows three reaction forces $\mathbf{C}_x$, $\mathbf{C}_y$, and $\mathbf{C}_z$ and two reaction moments $\mathbf{M}_x$ and $\mathbf{M}_y$ developed at C. The weight of the boom acts vertically down through the center of gravity of the boom. The five reactions plus the tension in the cable total six unknowns that can be solved in this three-dimensional case. It is interesting to note that if $\mathbf{Q}$, which acts in the Z direction, is zero, then $\mathbf{C}_z = \mathbf{M}_x = \mathbf{M}_y = 0$ since the remaining forces $\mathbf{T}$, $\mathbf{w}$, and $\mathbf{P}$ lie in the XY plane.

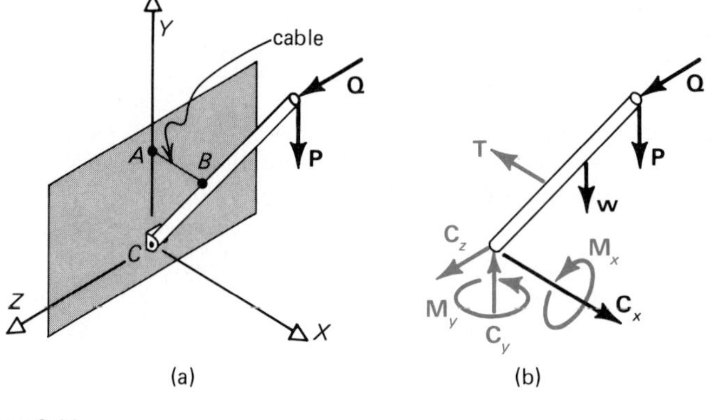

Figure 3.14

Example 3.6

Obtain the reactions at A for the uniform cantilever beam in Fig. 3.15 if $\mathbf{F} = (30\mathbf{i} + 40\mathbf{j} - 50\mathbf{k})$ kip and the beam weighs 10 kip.

Figure 3.15(b) shows the free-body diagram of the cantilever beam. Note that all six reactions at A are assumed to act in the positive direction. The weight of the beam is represented by $\mathbf{w}$ acting through the center of the beam,

$$\mathbf{w} = -10\mathbf{j}$$

For equilibrium we first sum all the forces,

$$\mathbf{A} + \mathbf{w} + \mathbf{F} = 0$$

Substituting for **F** and **w** and expressing **A** in terms of its components we obtain the sum of the forces,

$$A_x\mathbf{i} + A_y\mathbf{j} + A_z\mathbf{k} - 10\mathbf{j} + 30\mathbf{i} + 40\mathbf{j} - 50\mathbf{k} = 0$$

Now consider each scalar equation of equilibrium.

X components: $$A_x + 30 = 0$$
$$A_x = -30$$

Y components: $$A_y - 10 + 40 = 0$$
$$A_y = -30$$

Z components: $$A_z - 50 = 0$$
$$A_z = 50$$

Therefore, the reaction force at *A* is

$$\mathbf{A} = (-30\mathbf{i} - 30\mathbf{j} + 50\mathbf{k}) \text{ kips} \qquad \textit{Answer}$$

Once again the negative signs mean that the assumed directions for $\mathbf{A}_x$ and $\mathbf{A}_y$ in Fig. 3.15(b) are incorrect.

To find the restraining moments at the support, find the moment developed at *A* due to each applied load considered one at a time. For the weight **w**

$$\mathbf{M}_1 = \mathbf{r}_1 \times \mathbf{w}$$
$$= 4\mathbf{i} \times (-10\mathbf{j}) = -40\mathbf{k}$$

For the force **F**,

$$\mathbf{M}_2 = \mathbf{r}_2 \times \mathbf{F}$$
$$= 8\mathbf{i} \times (30\mathbf{i} + 40\mathbf{j} - 50\mathbf{k})$$
$$= 320\mathbf{k} + 400\mathbf{j}$$

The restraining moment **M** at the support is represented as

$$\mathbf{M} = M_x\mathbf{i} + M_y\mathbf{j} + M_z\mathbf{k}$$

For equilibrium, the sum of the moments acting on the beam must equal zero.

$$\mathbf{M} + \mathbf{M}_1 + \mathbf{M}_2 = 0$$

$$M_x\mathbf{i} + M_y\mathbf{j} + M_z\mathbf{k} - 40\mathbf{k} + 320\mathbf{k} + 400\mathbf{j} = 0$$

Once again consider each scalar equation of equilibrium.

X components: $$M_x = 0$$

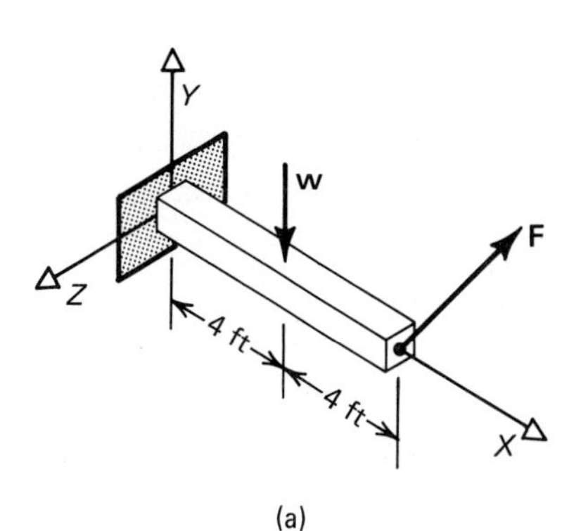

(a)

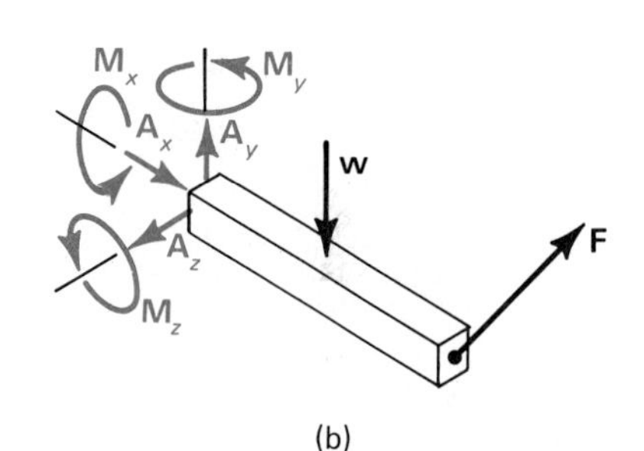

(b)

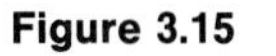

Figure 3.15

Y components: $\quad M_y + 400 = 0$

$$M_y = -400$$

Z components: $\quad M_z - 40 + 320 = 0$

$$M_z = -280$$

Therefore, the restraining moment at the support is

$$\mathbf{M} = (-400\mathbf{j} - 280\mathbf{k}) \times 10^3 \text{ ft} \cdot \text{lb} \qquad \textit{Answer}$$

Example 3.7

Obtain the reactions at the ball-and-socket joint at C and the tension in the cables AE and BE for the boom supporting the block in Fig. 3.16(a). The distance along the boom from C to D is 3 m and from D to E is 1 m. The mass of the block equals 100 kg. Neglect the mass of the boom.

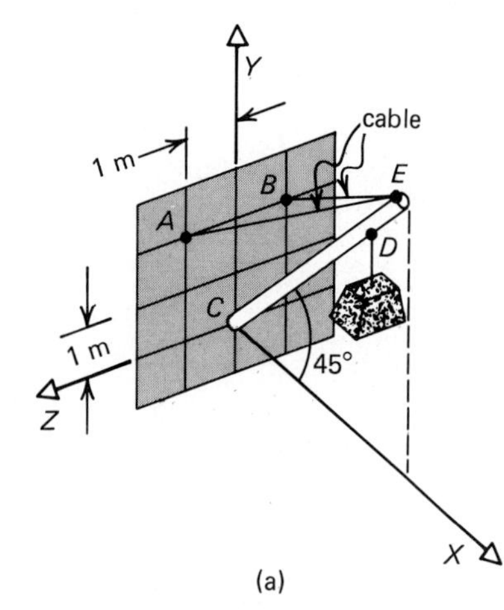

(a)

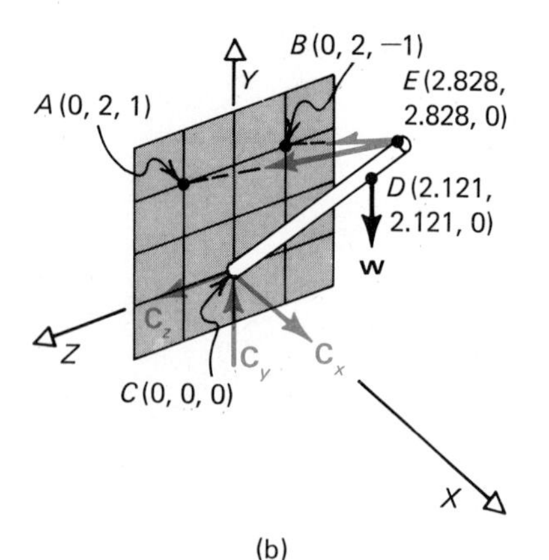

(b)

Figure 3.16

The free-body diagram is shown in Fig. 3.16(b). It is convenient to label the coordinate points in this same figure to establish the vector expressions for the forces. The forces are as follows:

$$\mathbf{T}_A = T_A\left[\frac{(0 - 2.828)\mathbf{i} + (2 - 2.828)\mathbf{j} + (1 - 0)\mathbf{k}}{\sqrt{(-2.828)^2 + (0.828)^2 + (1)^2}}\right]$$

$$= T_A(-0.909\mathbf{i} - 0.266\mathbf{j} + 0.321\mathbf{k})$$

$$\mathbf{T}_B = T_B(-0.909\mathbf{i} - 0.266\mathbf{j} - 0.321\mathbf{k})$$

$$\mathbf{w} = -9.81(100)\mathbf{j} = -981\mathbf{j}$$

The reactions at C are

$$\mathbf{C} = C_x\mathbf{i} + C_y\mathbf{j} + C_z\mathbf{k}$$

For equilibrium,

$$\mathbf{T}_A + \mathbf{T}_B + \mathbf{w} + \mathbf{C} = 0 \qquad \text{(a)}$$

Before substituting into Eq. (a), let us establish the moments about C due to the loading. Again, we calculate the moment of each force one at a time.

$$\mathbf{M}_1 = \mathbf{r}_1 \times \mathbf{T}_A$$

where $\quad \mathbf{r}_1 = 2.828\mathbf{i} + 2.828\mathbf{j}$

$$\mathbf{M}_1 = \begin{vmatrix} \mathbf{i} & \mathbf{j} & \mathbf{k} \\ 2.828 & 2.828 & 0 \\ -0.909T_A & -0.266T_A & 0.321T_A \end{vmatrix}$$

$$\mathbf{M}_1 = (0.908\mathbf{i} - 0.908\mathbf{j} + 1.818\mathbf{k})T_A$$

Likewise, $\mathbf{M}_2 = \mathbf{r}_1 \times \mathbf{T}_B$

$$= (-0.908\mathbf{i} + 0.908\mathbf{j} + 1.818\mathbf{k})T_B$$

Finally, $\mathbf{M}_3 = \mathbf{r}_2 \times \mathbf{w}$

$$= (2.121\mathbf{i} + 2.121\mathbf{j}) \times (-981\mathbf{j}) = -2081\mathbf{k}$$

For equilibrium about point C,

$$\mathbf{M}_1 + \mathbf{M}_2 + \mathbf{M}_3 = 0$$

$$0.908(T_A - T_B)\mathbf{i} + 0.908(-T_A + T_B)\mathbf{j} + (1.818T_A + 1.818T_B - 2081)\mathbf{k} = 0$$

Now consider each scalar equation of equilibrium.

$$0.908(T_A - T_B) = 0$$

$$T_A = T_B$$

Note that we also obtain this result with the Y component equation.

Z components: $1.818(T_A + T_B) - 2081 = 0$

$$T_A = T_B = 572 \text{ N}$$

The tensile forces in the cables are

$$\mathbf{T}_A = (-520\mathbf{i} - 152\mathbf{j} + 184\mathbf{k})\text{ N} \qquad \textit{Answer}$$

$$\mathbf{T}_B = (-520\mathbf{i} - 152\mathbf{j} - 184\mathbf{k})\text{ N} \qquad \textit{Answer}$$

To find the reactions at C, return to Eq. (a).

$$(-520\mathbf{i} - 152\mathbf{j} + 184\mathbf{k}) + (-520\mathbf{i} - 152\mathbf{j} - 184\mathbf{k}) - 981\mathbf{j} + C_x\mathbf{i} + C_y\mathbf{j} + C_z\mathbf{k} = 0$$

X components: $-520 - 520 + C_x = 0$

$$C_x = 1040$$

Y components: $-152 - 152 - 981 + C_y = 0$

$$C_y = 1285$$

Z components: $184 - 184 + C_z = 0$

$$C_z = 0$$

Thus, $\mathbf{C} = (1040\mathbf{i} + 1285\mathbf{j})\text{ N}$ *Answer*

PROBLEMS

3.44 Find the reactions of the cantilever beam at A in Fig. P3.44 if $\mathbf{F}_1 = -200\mathbf{j}$ lb and $\mathbf{F}_2 = (100\mathbf{j} - 50\mathbf{k})$ lb and $L = 6$ ft.

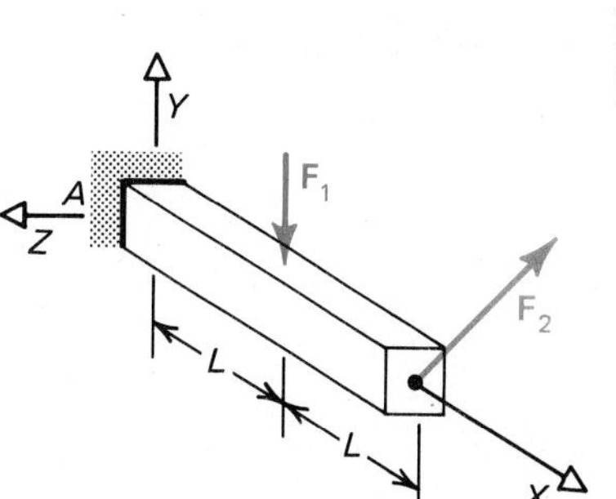

Figure P3.44

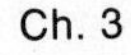

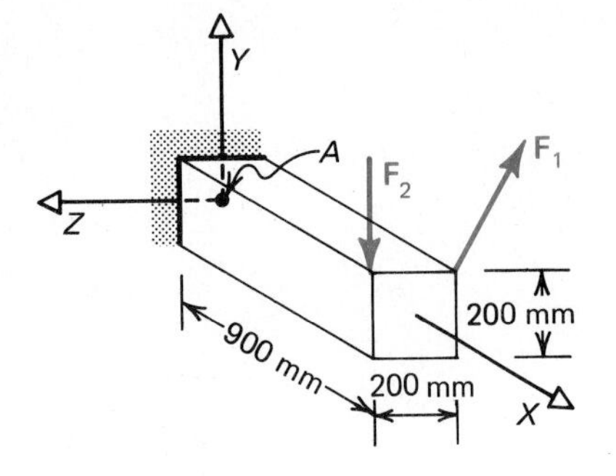

Figure P3.46

3.45 Repeat Problem 3.44 with $\mathbf{F}_1 = -350\mathbf{j}$ N, $\mathbf{F}_2 = (60\mathbf{i} + 80\mathbf{j} - 100\mathbf{k})$ N and $L = 6$ m.

3.46 Find the reactions of the cantilever beam at A in Fig. P3.46 if $\mathbf{F}_1 = (200\mathbf{j} - 100\mathbf{k})$ N and $\mathbf{F}_2 = -150\mathbf{j}$ N.

3.47 The boom in Fig. P3.47 is restrained by a frictionless pin connection at A and a cable BC. Find the reactions at A if $\mathbf{P} = (-600\mathbf{j} + 800\mathbf{k})$ lb.

3.48 Repeat Problem 3.47 if $\mathbf{P} = (200\mathbf{i} - 400\mathbf{j} - 600\mathbf{k})$ lb.

3.49 The boom shown in Fig. P3.49 lies in the XY plane, has a ball-and-socket joint at C, and is supported by two cables. Determine the reactions at C.

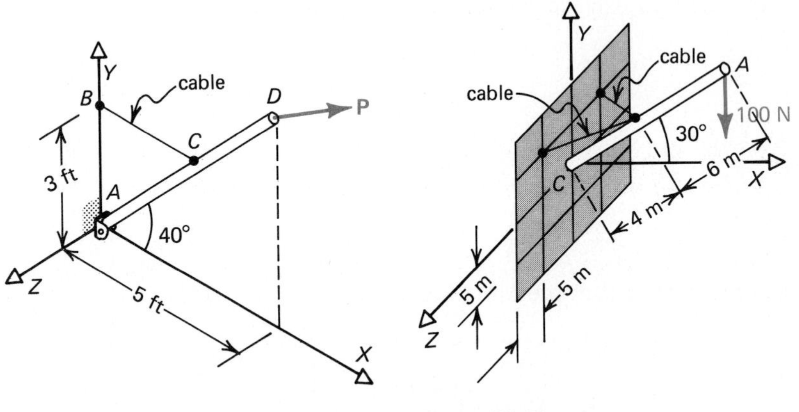

Figure P3.47

Figure P3.49

3.50 Repeat Problem 3.49 for an additional load $\mathbf{P} = (86.7\mathbf{i} + 50\mathbf{j})$ N applied at point A.

3.51 A 20-ft boom is held by a ball-and-socket joint at O and three cables as shown in Fig. P3.51. Determine the tension in each cable if the applied force $\mathbf{P} = (500\mathbf{i} + 100\mathbf{j} - 50\mathbf{k})$ lb. The cable from A to C passes over a common pulley at D. The coordinates are in feet.

3.52 A steel bracket is mounted, as shown in Fig. P3.52, in order to support a motor weighing 25 lb which develops a couple equal to $-50\mathbf{j}$ ft·lb. Assuming

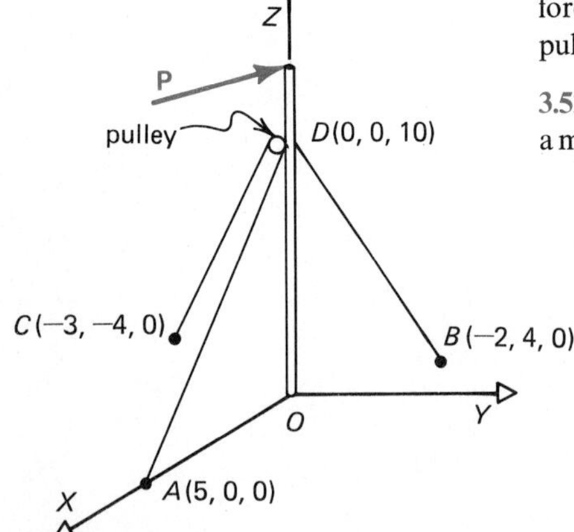

Figure P3.51

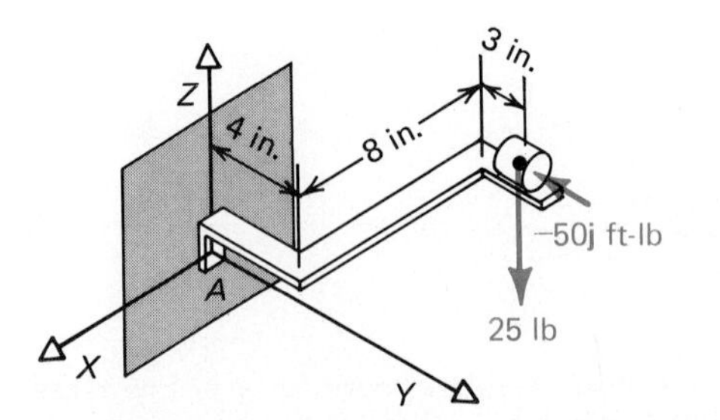

Figure P3.52

that the bracket at A is clamped, determine the reactions at A. Neglect the mass of the bracket.

3.53 A rigid piece of pipe is bent as shown in Fig. P3.53. A 100-kg and a 10-kg mass are hung as shown, and a cable connected at a corner applies a tension $\mathbf{P} = 2\mathbf{j}$ kN. Determine the reactions at A assuming that it is a clamped connection. Neglect the weight of the pipe.

3.54 A steel bar is hinged to the wall and supported by a cable as shown in Fig. P3.54. (a) What is the tension in the cable? (b) What are the reactions at A? Neglect the weight of the bar.

3.55 If the weight of the bar of Problem 3.54 is represented as a 600-N force directed vertically down through the point C, what is the tension of the cable? Express the results in English gravitational units.

3.56 A cantilever beam is bent and subjected to four forces as shown in Fig. P3.56. Neglecting the weight of the beam, determine the reaction at the clamped support at A.

3.57 A 100-kg sign is hung as shown in Fig. P3.57. If the joint at O is a ball-and-socket, determine (a) the reaction at O and (b) the tension in cables AC and BC.

3.58 A plate weighing 50 N is rigidly fixed to the wall as shown in Fig. P3.58. Determine the reactions for the loading shown.

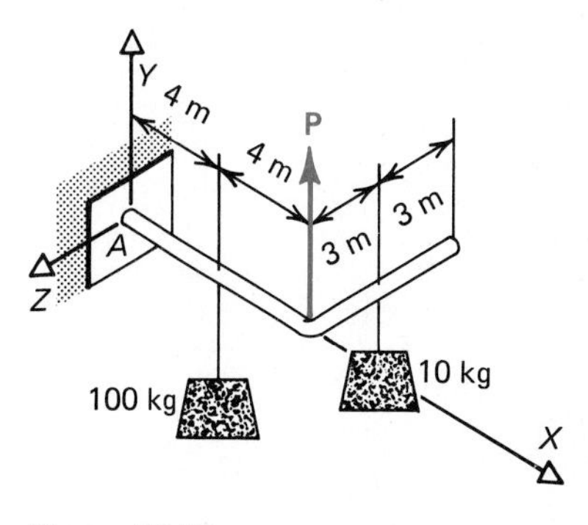

Figure P3.53

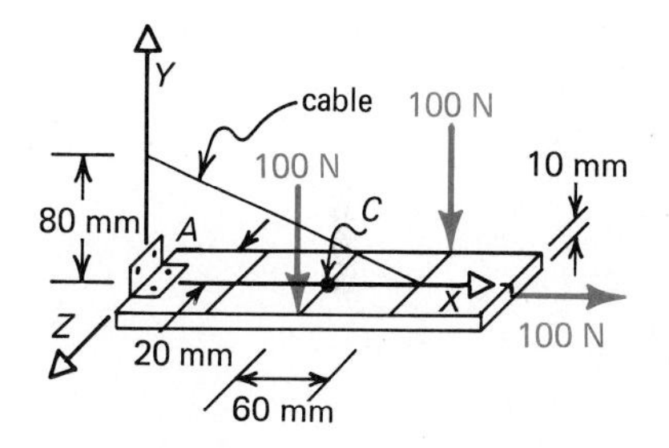

Figure P3.54

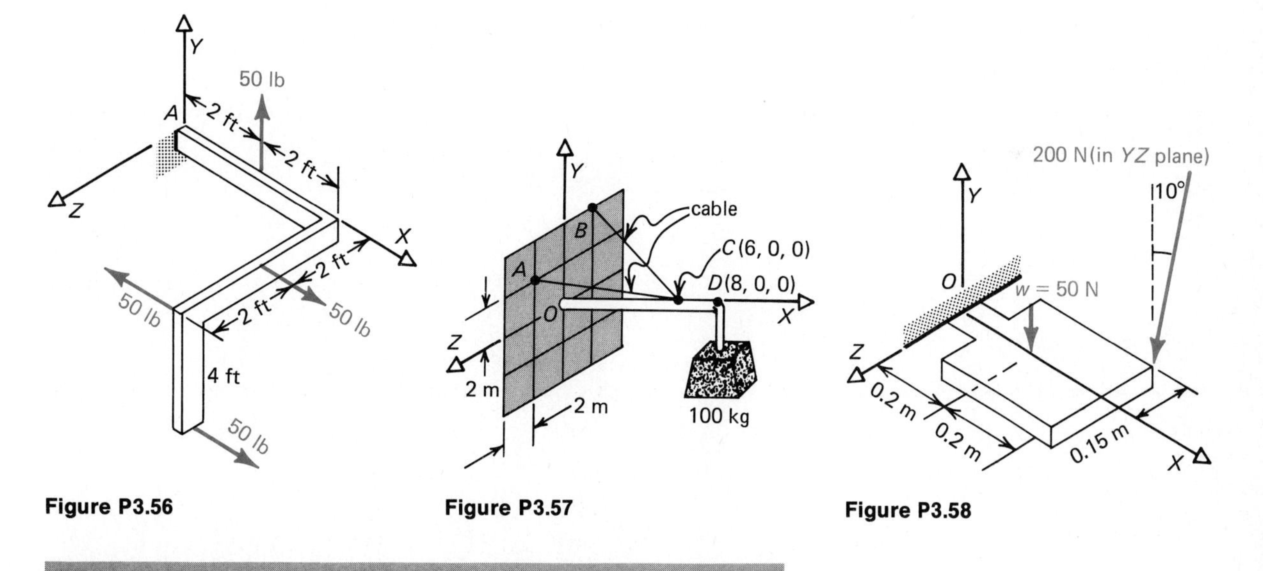

Figure P3.56

Figure P3.57

Figure P3.58

3.4 Summary

The free-body diagram of a rigid body in equilibrium must show the externally applied forces and moments as well as the reactions at the supports. The resultant force and resultant moment obtained from the free-body diagram must be zero for the rigid body to be in equilibrium.

A. The important *ideas*:

1. A rigid body is in equilibrium if the resultant force is zero and the resultant moment is zero.
2. A free-body diagram must include the applied as well as the reaction forces and moments.
3. The term reaction refers to both reaction forces and reaction moments.
4. When the number of unknown reactions is equal to the number of independent equations of equilibrium, a rigid body is completely constrained, and the rigid body is said to be statically determinate. If the number of unknowns are more than the equations of equilibrium, the rigid body is statically indeterminate.

B. The important *equations*:

Equilibrium in two dimensions

$$\sum F_x = 0 \qquad \sum F_y = 0 \qquad \sum M_z = 0 \tag{3.1}$$

Equilibrium in three dimensions

$$\sum F_x = 0 \qquad \sum F_y = 0 \qquad \sum F_z = 0 \tag{3.2}$$

$$\sum M_x = 0 \qquad \sum M_y = 0 \qquad \sum M_z = 0 \tag{3.3}$$

4 Structures

We have been concerned with analysis of externally applied forces and reaction forces acting on an isolated rigid body. Problems in engineering are concerned not only with single rigid bodies but with structures that are formed by a series of connected rigid bodies. Examples of such structures can be observed in our everyday experiences—e.g., cranes, electric transmission and TV towers, bridges, and roof trusses. These structures can be analyzed by means of the basic concepts of equilibrium developed in earlier chapters, whereby a free-body diagram of an isolated rigid body is used to obtain the equilibrium equations of statics in order to find the unknown reactions. However, a structure composed of several rigid bodies introduces a new concept—that of **internal forces**, which we treat in this chapter. When a separate free-body diagram is drawn of each member of the structure, the internal forces that hold the structure together are shown along with any reaction forces that may be present.

In our analysis of structures, we will be concerned with trusses, frames, and machines. While the basic equilibrium equations hold for each of these structures, certain distinctions among them do exist and will be taken into account in what follows.

4.1 Equilibrium of Plane Trusses

A truss consists of a series of straight structural members connected to each other such that no member is continuous through a joint. The structural members can be bars, angles, or beams, and the connection of each member may be accomplished by bolts, rivets, welds, etc. A truss that is designed to carry applied loads in a plane is a two-dimensional structure and is called a **plane** truss. An example of a plane truss is shown in Fig. 4.1. This truss consists of seven distinct members joined at five joints by riveted gusset plates. Although trusses have joints that are riveted, for purposes of simplicity they will not be shown in detail in future figures.

Trusses are used to carry heavy loads and are widely used in a variety of applications. Bridges and roofs of buildings are two common applications of trusses. Figure 4.2(a) shows a sketch of a bridge which consists of two trusses and several other structural members, whereas Fig. 4.2(b) shows a sketch of a building with several roof trusses. Various types of bridge and roof trusses are shown in Figs. 4.3 and 4.4. It is not our intention here to learn how to analyze and design these entire structures, but to indicate that the trusses we will be analyzing are, in reality, part of an overall structure such as a bridge or a building.

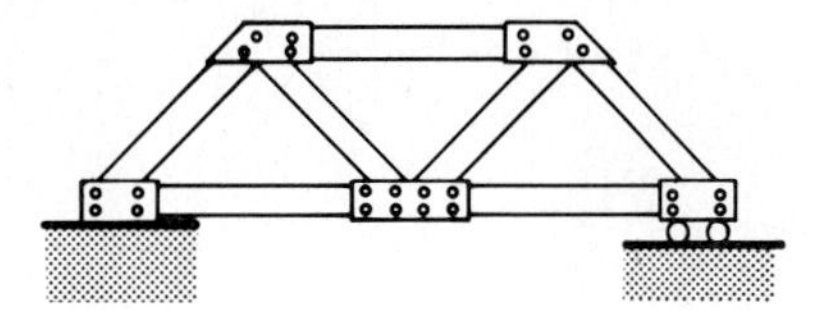

Figure 4.1

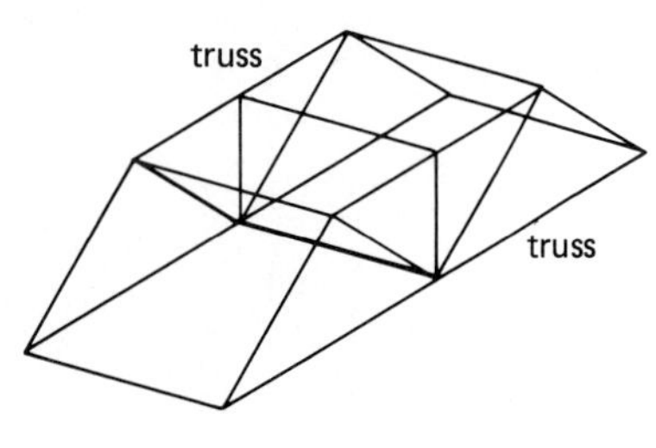

(a)

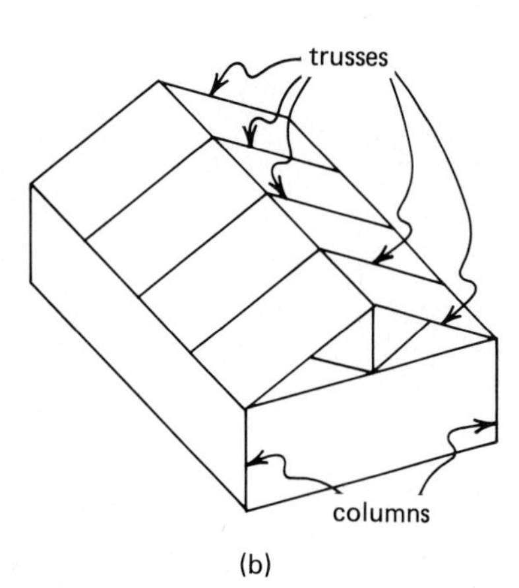

(b)

Figure 4.2

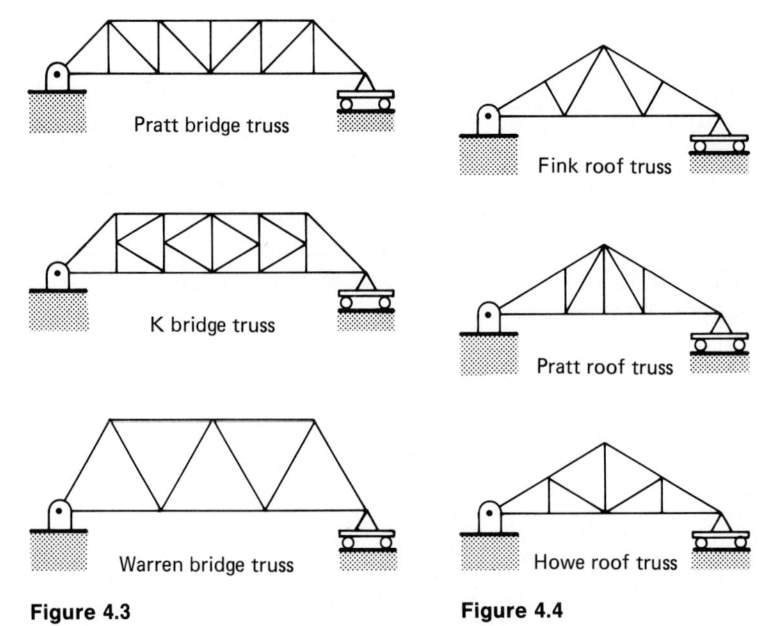

Figure 4.3

Figure 4.4

It can be seen from the last two figures that trusses may be formed by repeating a basic geometric pattern. A triangle is the most common pattern. In general, a simple truss is created by adding two new members to the joints of a triangle and thereby creating a new joint. Thus, in simple trusses, the addition of two members creates one new joint. For example, consider the simple truss shown in Fig. 4.5(a) which consists of three members connected at three joints A, B, and C. If we add two more members, as shown in Fig.

4.5(b), then another joint D is created in the simple truss. The total number of members M in a simple truss is given by

$$M = 2J - 3 \tag{4.1}$$

where J is the number of joints.

In this chapter, we are interested in obtaining the magnitude and direction of the internal forces acting on each member of a truss. It is this information that enables a designer to select the sizes and shapes of the structural members forming a truss. Four assumptions are basic to the following analyses.

1. All trusses considered are plane trusses.
2. Frictionless pins connect the truss members at each joint.
3. All loads and reactions act at the joints.
4. The weight of each structural member is neglected.

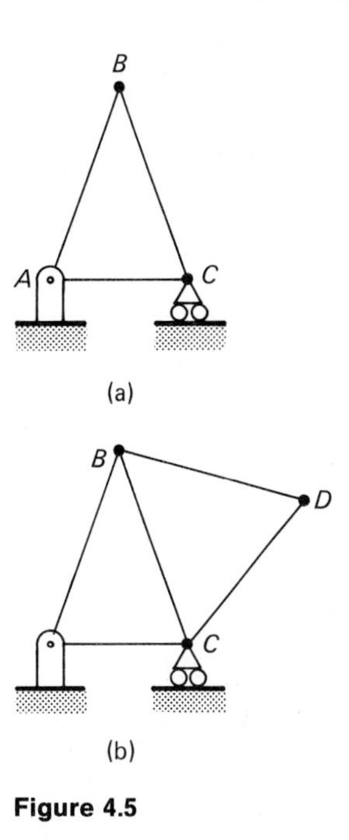

Figure 4.5

Since the members are assumed to be pinned together, the resultant force acting at the end of a member consists only of a force because the frictionless pin joint cannot carry a moment. Furthermore, since each force is applied to each end of a member, it is called a **two-force member**. Therefore

Plane trusses are composed of two-force members connected by frictionless pins.

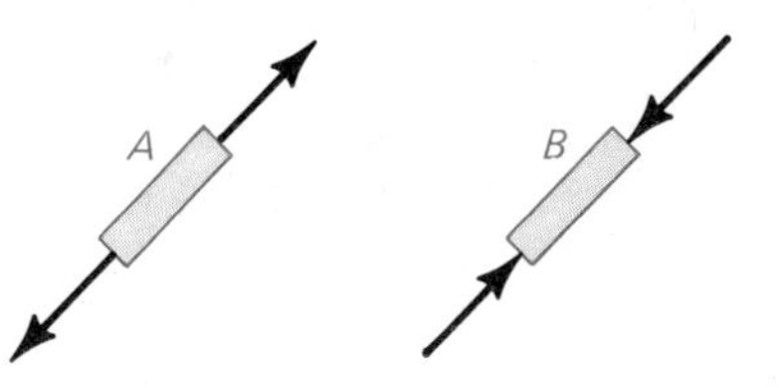

Figure 4.6

In Fig. 4.6, the member A is subjected to tensile forces, while member B is under compressive forces. Note that these forces must be equal in magnitude, opposite in direction, and collinear, in accord with Newton's third law. All external loads and reaction forces are applied to the truss at the joints. While we will neglect the weight of the structural members, in practice they are accounted for by distributing their weight at the joints as additional applied loads.

For purposes of analysis the many pins and members in a truss should be labeled in an orderly fashion. Consider the Warren truss shown in Fig. 4.7(a). The pin joints are labeled A, B, C, D, and E. The member joining pins A and B is referred to as AB. The magnitude of the force in this member is F_{AB}. When the number of pins becomes very large, each pin is assigned a number as shown in Fig. 4.7(b). The member joining pin 2 and 3 is 23 and the force in this member has a magnitude of F_{23}. This latter method of labeling becomes very attractive for computer programming. In this text we shall use the alphabet system shown in Fig. 4.7(a).

With the above ideas in mind we may proceed to draw free-body diagrams for the truss and each of its components and to

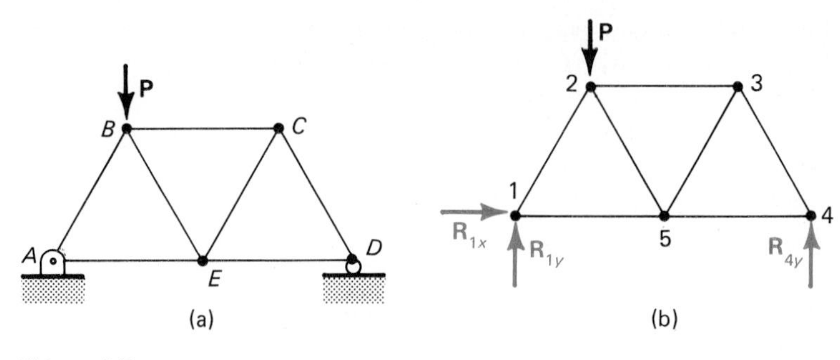

Figure 4.7

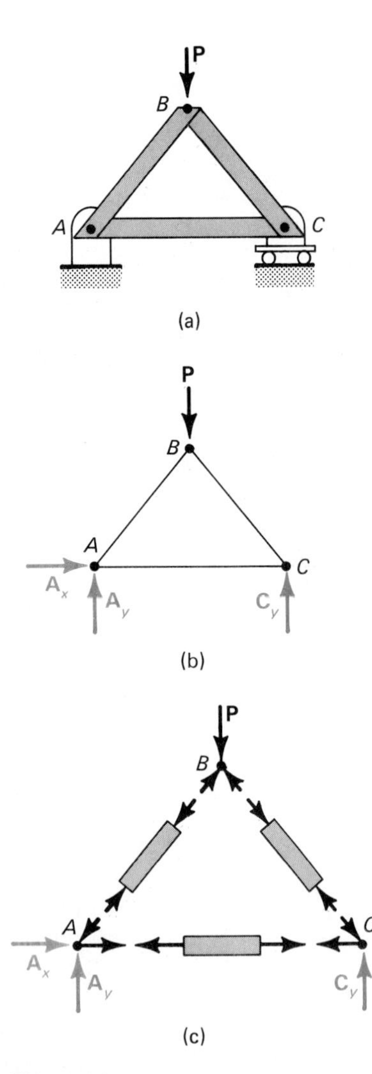

Figure 4.8

write the equilibrium equations. Consider the truss shown in Fig. 4.8(a). The free-body diagram of the entire truss is shown in Fig. 4.8(b) where $\mathbf{A}_x$ and $\mathbf{A}_y$ are reaction forces at joint A, $\mathbf{C}_y$ is the reaction force at joint C, and $\mathbf{P}$ is the applied load at the joint B. The free-body diagrams for each member and each pin are shown in Fig. 4.8(c). We can solve for the forces acting at a pin by writing the sum of the forces in the X and Y directions and setting them equal to zero. This means that each pin is in equilibrium, which is true, since the entire truss is in static equilibrium. Since there are three pins in this example, i.e., $J = 3$ in Eq. (4.1), we have six equilibrium equations for six unknowns—three reaction forces and three internal forces.

It is instructive to examine an end-view section of joint A of Fig. 4.8 shown in Fig. 4.9. Here we see members AC and AB secured by means of the connecting pin and rigid foundation bracket. The forces from the truss members that act at a joint must eventually be transmitted to the foundation bracket via the connecting pin. We shall return to this concept when three or more members are connected to the same joint.

One approach to a solution is to obtain the reaction forces first by writing the equilibrium equations of the entire truss using a free-body diagram as in Fig. 4.8(b). Having these reaction forces, we then proceed to calculate the internal forces. It should be noted that the equilibrium equations of the entire truss are not independent, but rather they are a subset of the $2J$ equations obtained by analyzing the pins. Another important point to remember is that

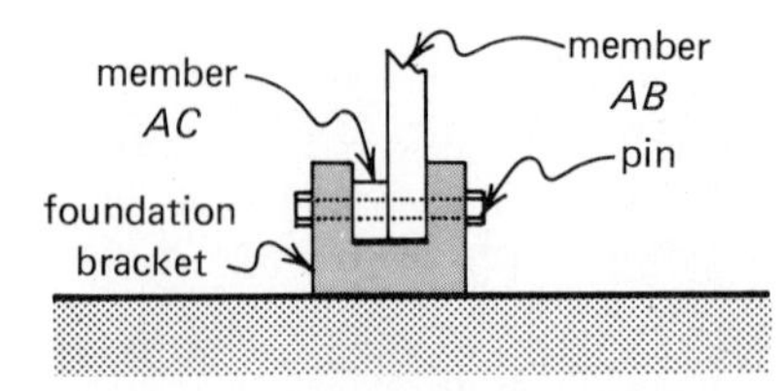

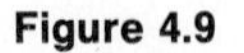

Figure 4.9

the reaction and internal forces are dependent upon the applied loads. As the loads shift from pin to pin or change in magnitude, the reaction and internal forces change accordingly.

The following sections introduce three common methods employed in analyzing trusses: (1) the method of joints, (2) the graphical method, and (3) the method of sections.

4.2 Method of Joints

The method of joints examines the conditions of equilibrium of forces at each pin. Since each member of the truss transmits a single force to its joints, the forces are concurrent at each pin, i.e., all the forces pass through the point represented by the pin. For static equilibrium of the pin, it follows that

$$\sum F_x = 0 \qquad \sum F_y = 0$$

We therefore apply these equilibrium equations at each joint of the truss.

Before we consider examples, we need a method of ascertaining the direction of the internal force. When a force acts toward the pin as shown in Fig. 4.10(a) we observe that the same force acts to compress the truss member. We say that the force is a compressive force. Likewise, if the force acts away from the pin as shown in Fig. 4.10(b), the same force acts to stretch the truss member. In this case, the force is a tensile force. As a matter of convention, a force F_{AB} is written as F_{AB}C for compression and F_{AB}T for tension.

The steps in the procedure of using the method of joints can be summarized as follows:

1. Label all pin joints as shown in Fig. 4.11(a).
2. Draw and label a free-body diagram of the entire truss as shown in Fig. 4.11(b).

truss member

pin

(a)

truss member

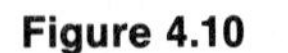

pin

(b)

Figure 4.10

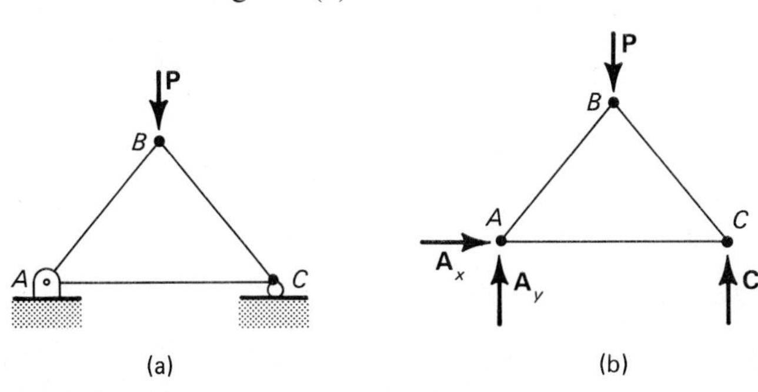

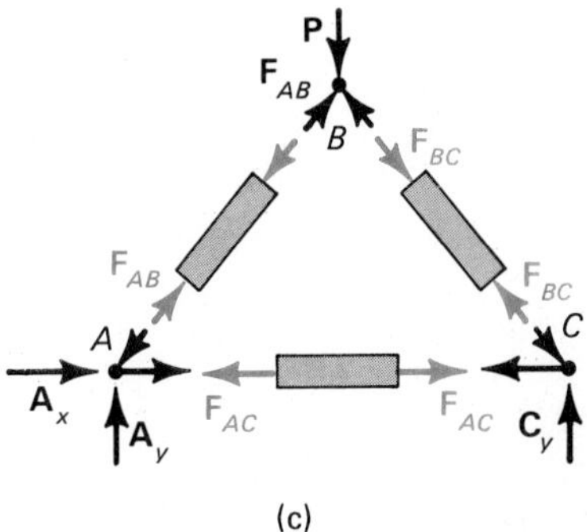

Figure 4.11

3. Calculate the reaction forces using three equations of statics applied to the whole truss.
4. Draw a free-body diagram for each member and each pin as shown in Fig. 4.11(c). While it is often possible to predict the correct direction of the internal forces, there are many cases where it is necessary to assume their direction.
5. Apply the two equations of statics at each joint, namely,

$$\sum F_x = 0 \qquad \sum F_y = 0$$

If a calculated internal force is negative, it means that the assumed direction of the force is incorrect. Note that it is not necessary to change the direction of the internal force. Continue to apply the equilibrium equations from joint to joint using the original free-body diagram and all calculated forces including their algebraic sign.

Before we proceed with examples of the method of joints, let us consider the free-body diagram of the truss joint E shown in Fig. 4.12(a). Having found the reactive forces, consider the free-body diagram of joint E shown in Fig. 4.12(b). Summing forces in the vertical direction yields $F_{BE} = 0$. Member BE does not carry a force and is called a **zero-force member**. If we consider joint B as shown in Fig. 4.12(c), we see that $F_{BD} = 0$, so that member BD is also a zero-force member. Recognizing that certain members of a truss are zero-force members for certain loading conditions facilitates the analysis by the method of joints.

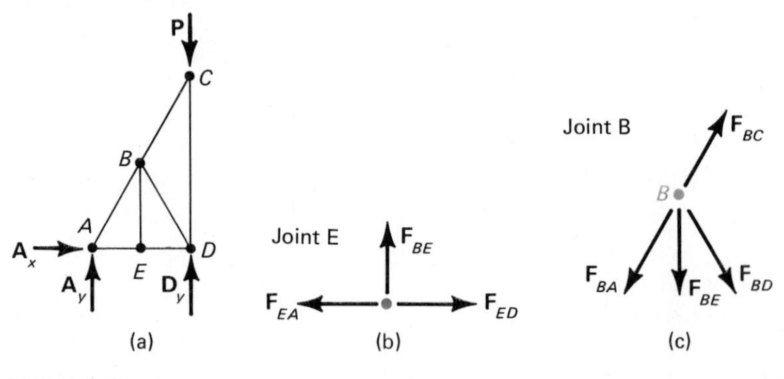

Figure 4.12

Example 4.1

Obtain the magnitudes of the reactions and internal forces in each member of the truss shown in Fig. 4.13(a). State if the member is in tension or compression.

Figure 4.13(b) shows the free-body diagram of the entire truss. The magnitudes of the reaction forces are labeled A_x, A_y, and C_y.

Equilibrium of the entire truss:

$$\sum M_A = 0 \qquad -2.5(100) + 5(C_y) = 0$$
$$C_y = 50 \text{ N} \qquad \textit{Answer}$$
$$\sum F_x = 0 \qquad A_x = 0 \qquad \textit{Answer}$$
$$\sum F_y = 0 \qquad -100 + A_y + 50 = 0$$
$$A_y = 50 \text{ N} \qquad \textit{Answer}$$

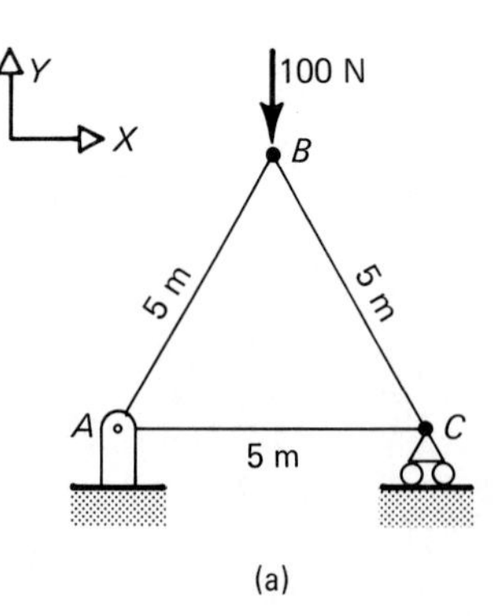

(a)

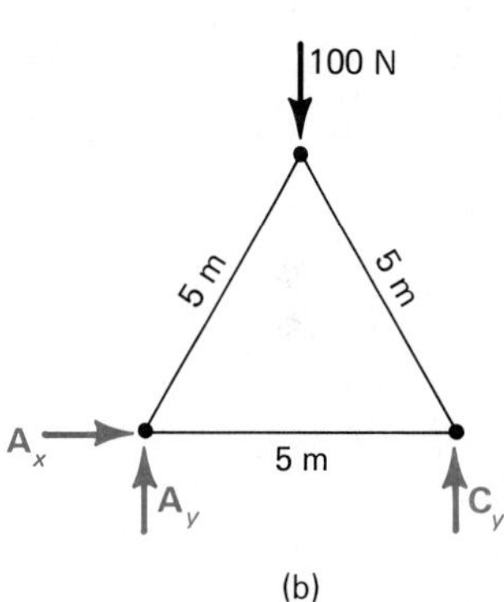

(b)

The free-body diagrams for the three joints are shown in Figs. 4.13(c), (d), and (e). It has been assumed that all internal forces are tensile forces. The equilibrium of the joints, beginning with joint A, is examined.

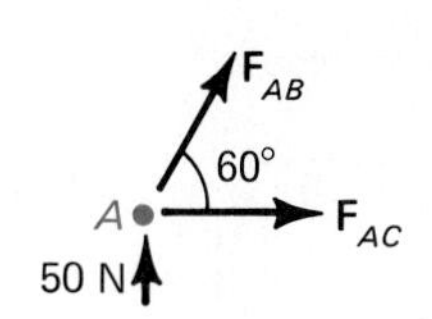

(c)

Joint A:

$$\sum F_y = 0 \qquad F_{AB} \sin 60° + 50 = 0$$
$$F_{AB} = -57.74$$
$$F_{AB} = 57.74 \text{ N C} \qquad \textit{Answer}$$
$$\sum F_x = 0 \qquad F_{AB} \cos 60° + F_{AC} = 0$$
$$(-57.74) \cos 60° + F_{AC} = 0$$
$$F_{AC} = 28.87$$
$$\therefore F_{AC} = 28.87 \text{ N T} \qquad \textit{Answer}$$

Joint B:

$$\sum F_x = 0 \qquad -F_{AB} \cos 60° + F_{CB} \cos 60° = 0$$
$$F_{CB} = F_{AB} = -57.74$$
$$\therefore F_{CB} = 57.74 \text{ N C} \qquad \textit{Answer}$$

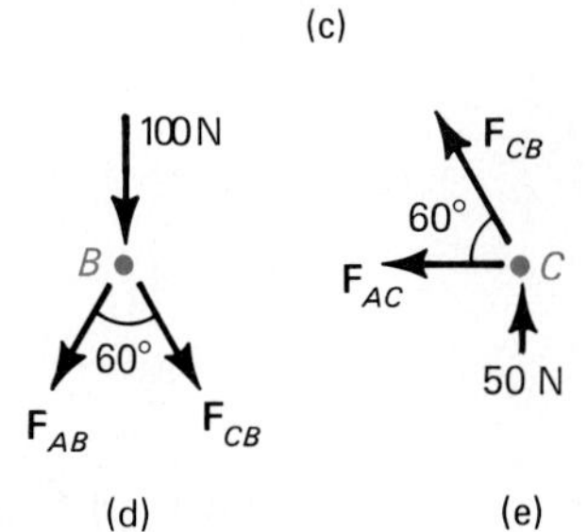

(d) (e)

Figure 4.13

We observe that we have obtained the three internal forces using two joints. The final joint C may be used to check our calculations.

Joint C:

$$\sum F_x = 0 \qquad -F_{CB} \cos 60° - F_{AC} = 0$$
$$(57.74) \cos 60° - 28.87 = 0$$
$$28.87 - 28.87 = 0 \qquad \text{(Checks)}$$
$$\sum F_y = 0 \qquad F_{CB} \sin 60° + 50 = 0$$
$$-57.74 \sin 60° + 50 = 0$$
$$-50 + 50 = 0 \qquad \text{(Checks)}$$

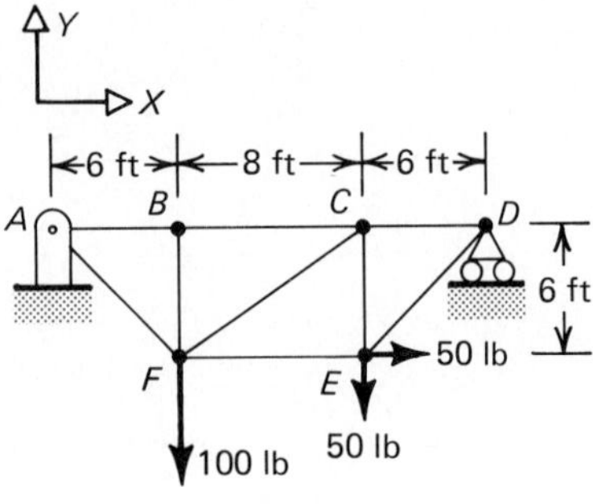

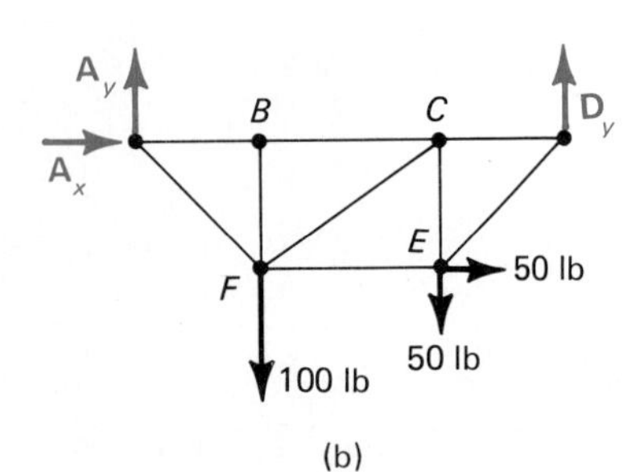

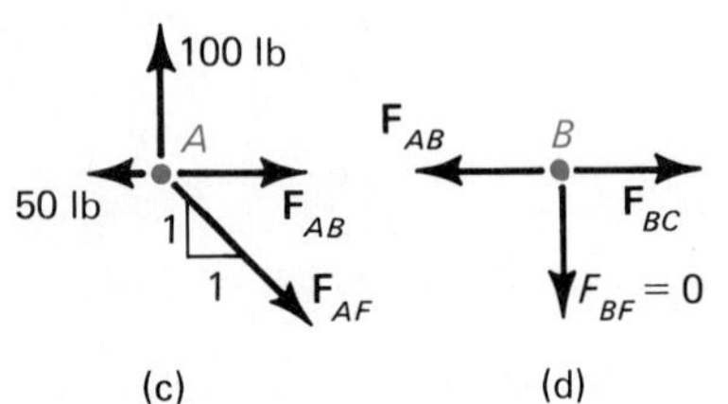

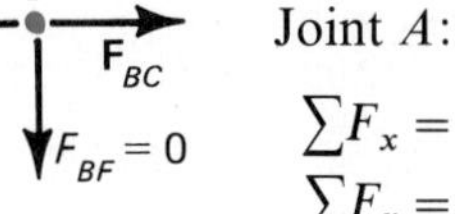

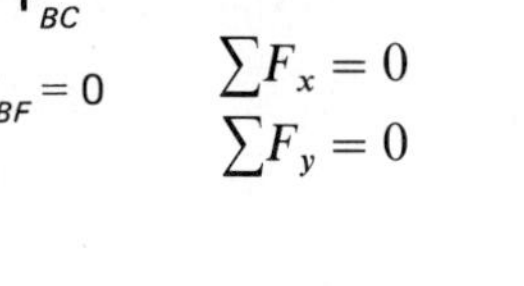

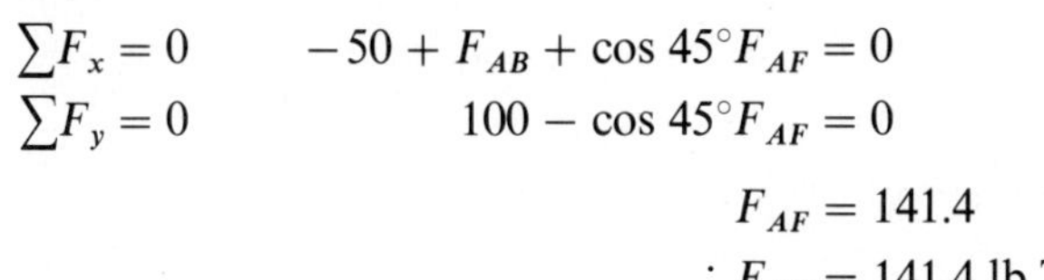

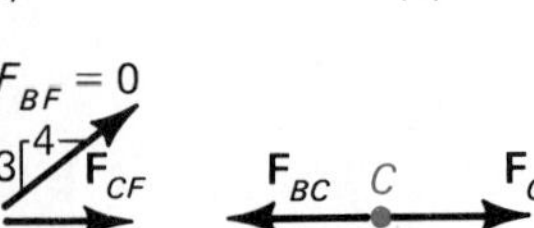

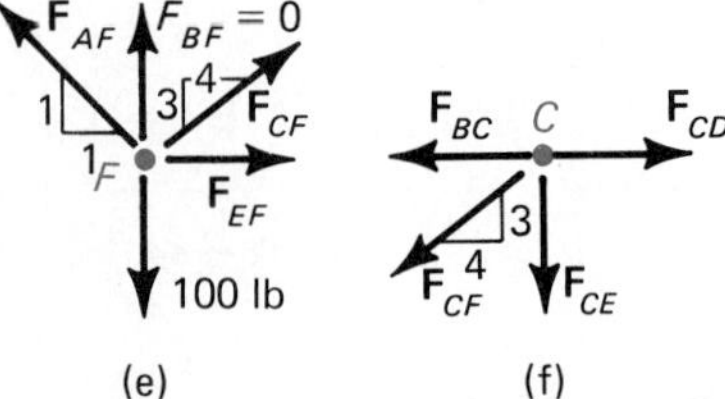

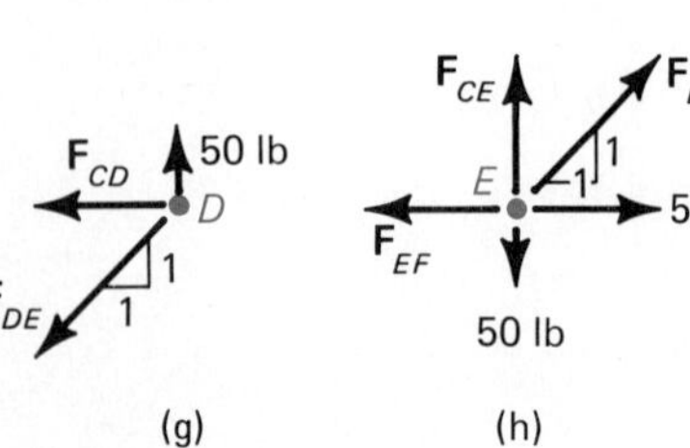

Figure 4.14

Example 4.2

Using the method of joints, find the magnitude of the force in each member for the truss shown in Fig. 4.14(a). Also establish if the members are in tension or compression.

Figure 4.14(b) shows the free-body diagram of the entire truss. Three equations of statics are applied to the entire structure to find the magnitude of the reaction forces.

Equilibrium of the entire truss:

$$\sum M_A = 0 \qquad 20D_y + 50(6) - 50(14) - 100(6) = 0$$
$$D_y = 50 \text{ lb}$$

$$\sum F_x = 0 \qquad A_x + 50 = 0$$
$$A_x = -50 \text{ lb}$$

$$\sum F_y = 0 \qquad A_y + 50 - 50 - 100 = 0$$
$$A_y = 100 \text{ lb}$$

The free-body diagrams for the joints are shown in Fig. 4.14. The reaction forces just calculated are shown in the appropriate diagrams and the internal forces are all assumed tensile. Zero-force member *BF* is also labeled. The equilibrium of the joints, starting with joint *A*, is examined.

Joint *A*:

$$\sum F_x = 0 \qquad -50 + F_{AB} + \cos 45^\circ F_{AF} = 0$$
$$\sum F_y = 0 \qquad 100 - \cos 45^\circ F_{AF} = 0$$
$$F_{AF} = 141.4$$
$$\therefore F_{AF} = 141.4 \text{ lb T} \qquad \textit{Answer}$$

Substituting this result into the first equation gives us

$$F_{AB} = -50$$
$$F_{AB} = 50 \text{ lb C} \qquad \textit{Answer}$$

Joint *B*:

$$\sum F_x = 0 \qquad -F_{AB} + F_{BC} = -(-50) + F_{BC} = 0$$
$$F_{BC} = -50$$
$$\therefore F_{BC} = 50 \text{ lb C} \qquad \textit{Answer}$$

Joint *F*:

$$\sum F_y = 0 \qquad 0.6F_{CF} + \cos 45^\circ F_{AF} - 100 = 0$$
$$0.6F_{CF} + \cos 45^\circ(141.4) - 100 = 0$$
$$F_{CF} = 0 \qquad \textit{Answer}$$

Joint C:

$$\sum F_y = 0 \qquad F_{CE} = 0$$

$$\sum F_x = 0 \qquad -F_{BC} + F_{CD} = -(-50) + F_{CD} = 0$$

$$F_{CD} = -50$$

$$\therefore F_{CD} = 50 \text{ lb C} \qquad \textit{Answer}$$

Joint D:

$$\sum F_x = 0 \qquad -F_{CD} - \cos 45^\circ F_{DE} = 50 - \cos 45^\circ F_{DE} = 0$$

$$F_{DE} = 70.7$$

$$\therefore F_{DE} = 70.7 \text{ lb T} \qquad \textit{Answer}$$

As a check on our computations we sum forces in the Y direction at joint D.

$$\sum F_y = 0 \qquad 50 - \cos 45^\circ F_{DE} = 0$$

$$50 - 50 = 0 \qquad \text{(Checks)}$$

Joint E:

$$\sum F_x = 0 \qquad -F_{EF} + \cos 45^\circ F_{DE} + 50 = 0$$

$$F_{EF} = 100$$

$$\therefore F_{EF} = 100 \text{ lb T} \qquad \textit{Answer}$$

As a check we sum forces in the Y direction at joint E.

$$\sum F_y = 0 \qquad \cos 45^\circ F_{DE} - 50 = 0$$

$$50 - 50 = 0 \qquad \text{(Checks)}$$

We note that although members CF and CE are zero-force members, they could not have been predicted easily as we did for member BF.

Example 4.3

Using the method of joints, find the magnitude of the force in each member for the truss shown in Fig. 4.15(a). State if the members are in tension or compression.

The free-body diagram of the entire truss is shown in Fig. 4.15(b). Equilibrium of the entire truss yields:

$$\sum M_A = 0$$

$$-400\left(\frac{5}{\cos 45^\circ}\right) - 200\left(\frac{10}{\cos 45^\circ}\right) + 20E_y - 1000(10) = 0$$

$$E_y = 782.8 \text{ N}$$

$$\sum F_x = 0$$

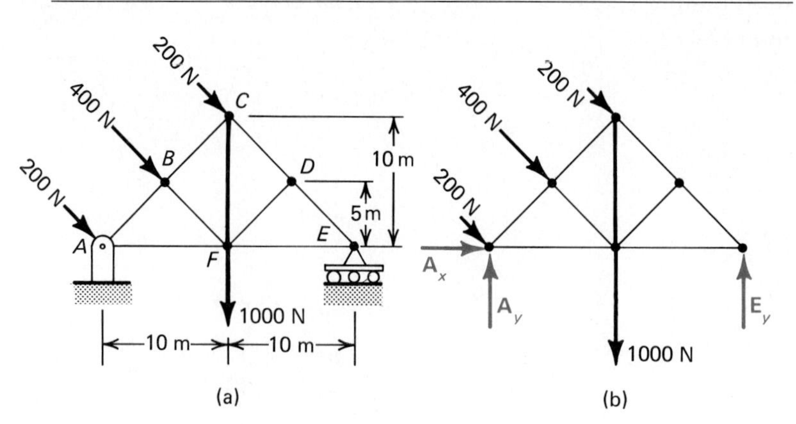

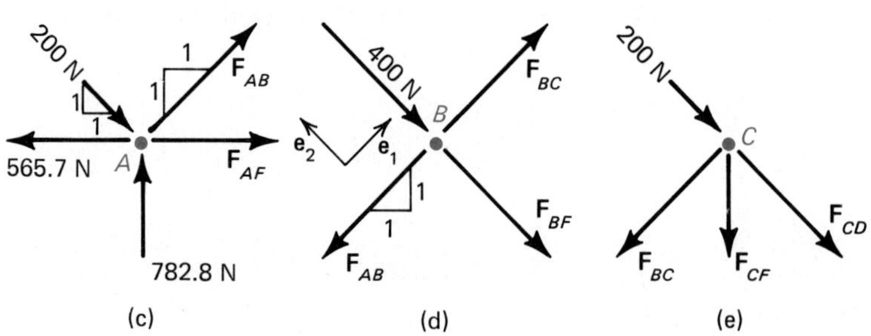

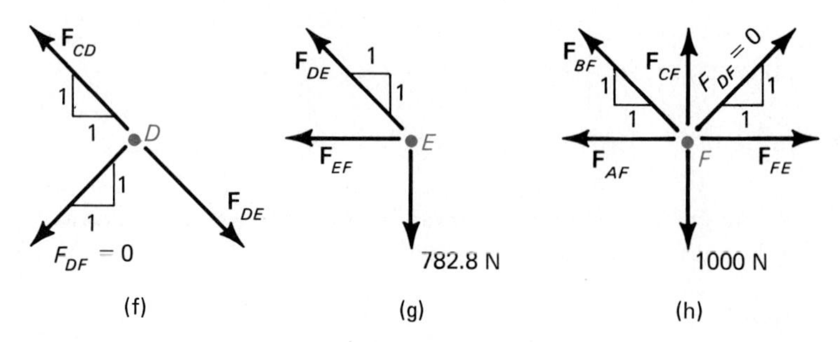

Figure 4.15

$$A_x + (200 + 400 + 200) \cos 45^\circ = 0$$

$$A_x = -565.7 \text{ N}$$

$\sum F_y = 0$

$$A_y - (200 + 400 + 200) \cos 45^\circ + 782.8 - 1000 = 0$$

$$A_y = 782.8 \text{ N}$$

As a check of these calculations for the reaction forces, let us sum moments about some arbitrary joint, say joint F.

$\sum M_F = 0$

$$-10A_y + 200(7.07) - 200(7.07) + 10E_y = 0$$

$$A_y = E_y \qquad \text{(Checks)}$$

The free-body diagrams for the joints shown in Fig. 4.15 include the reaction forces just calculated. Note that the direction of A_x actually acts to the left in Fig. 4.15(c). Zero-force member DF is also labeled.

Joint A:

$$\sum F_y = 0$$

$$782.8 - 200 \cos 45^\circ + F_{AB} \cos 45^\circ = 0$$

$$F_{AB} = -907.1$$

$$\therefore F_{AB} = 907.1 \text{ N C} \qquad \textit{Answer}$$

$$\sum F_x = 0$$

$$-565.7 + 200 \cos 45^\circ + F_{AB} \cos 45^\circ + F_{AF} = 0$$

$$F_{AF} = 1066$$

$$\therefore F_{AF} = 1066 \text{ N T} \qquad \textit{Answer}$$

Joint B: For convenience, vectors $\mathbf{e}_1$ and $\mathbf{e}_2$ as shown in Fig. 4.15(d) are selected rather than the original XY coordinates. This convention simplifies the algebra.

$$\sum F_{e1} = 0$$

$$-F_{AB} + F_{BC} = 0 \qquad F_{BC} = F_{AB} = 907.1 \text{ N C}$$

$$\sum F_{e2} = 0$$

$$-400 - F_{BF} = 0 \qquad F_{BF} = 400 \text{ N C} \qquad \textit{Answer}$$

Joint C:

$$\sum F_x = 0$$

$$-F_{BC} \cos 45^\circ + 200 \cos 45^\circ + F_{CD} \cos 45^\circ = 0$$

$$F_{CD} = -1107 = 1107 \text{ N C} \qquad \textit{Answer}$$

$$\sum F_y = 0$$

$$-F_{CF} - 200 \sin 45^\circ - F_{CD} \cos 45^\circ - F_{BC} \cos 45^\circ = 0$$

$$F_{CF} = -200 \sin 45^\circ - (-1107)$$

$$\times \cos 45^\circ - (-907.1) \cos 45^\circ$$

$$F_{CF} = 1282 \text{ N T} \qquad \textit{Answer}$$

Joint D: $F_{DE} = F_{CD} = 1107 \text{ N C}$ *Answer*

Joint E:

$$\sum F_x = 0 \qquad -F_{EF} - F_{DE} \cos 45^\circ = 0$$

$$F_{EF} = 782.8 \text{ N T} \qquad \textit{Answer}$$

Summing forces in the Y direction adds nothing new in our calculations for internal forces. However, it does provide a valuable check on our calculations.

Joint F: The equilibrium equations at joint F also provide a check on our calculations.

$$\sum F_x = 0 \qquad -F_{AF} - \cos 45^\circ F_{BF} + F_{EF} = 0$$
$$-1066 + 400 \cos 45^\circ + 782.8 = 0$$
$$-1066 + 282.8 + 782.8 \cong 0 \qquad \text{(Checks)}$$

$$\sum F_y = 0 \qquad F_{BF} \cos 45^\circ - 1000 + 1282 = 0$$
$$-282.8 - 1000 + 1282 \cong 0 \qquad \text{(Checks)}$$

PROBLEMS

Note: When obtaining the magnitude of the force of a member, state if the member is in tension or compression in all the problems in this chapter.

4.1 Obtain the magnitude of the forces at joint A for the truss of Fig. P4.1.

4.2 Calculate the magnitude of the forces at all the joints shown in Fig. P4.2.

4.3 Obtain the magnitude of the forces at joints A and B for the Warren truss of Fig. P4.3.

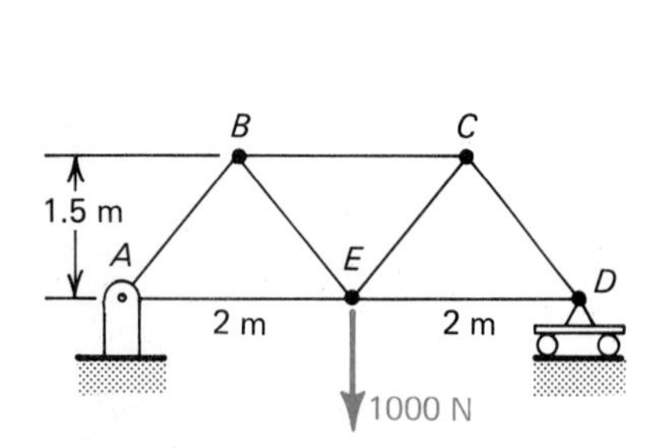

Figure P4.1

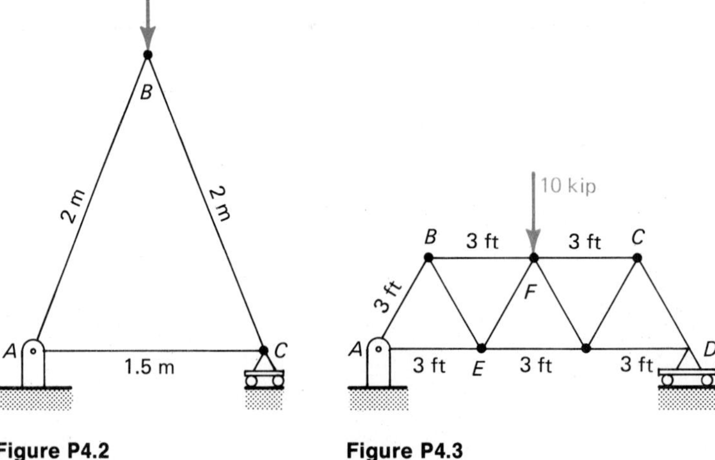

Figure P4.2

Figure P4.3

4.4 Obtain the magnitude of the forces at joints A and B for the Pratt truss shown in Fig. P4.4.

4.5 Obtain the magnitude of the forces at joints E and D for the Pratt truss in Problem 4.4.

4.6 By the method of joints obtain the magnitude of the forces in members *BF* and *BC* of the truss shown in Fig. P4.6.

4.7 Use the method of joints to obtain the magnitude of the forces in members *FE* and *ED* for the truss in Problem 4.6.

4.8 Obtain the magnitude of the forces at joints *D* and *E* for the truss shown in Fig. P4.8.

4.9 Obtain the magnitude of the forces at joints *B* and *C* for the truss in Problem 4.8.

4.10 Use the method of joints to obtain the magnitude of the forces in members *AE*, *AB*, *BE*, and *BC* of the truss shown in Fig. P4.10.

4.11 Obtain the magnitude of the forces in members *ED*, *EC*, and *CD* of the truss in Problem 4.10.

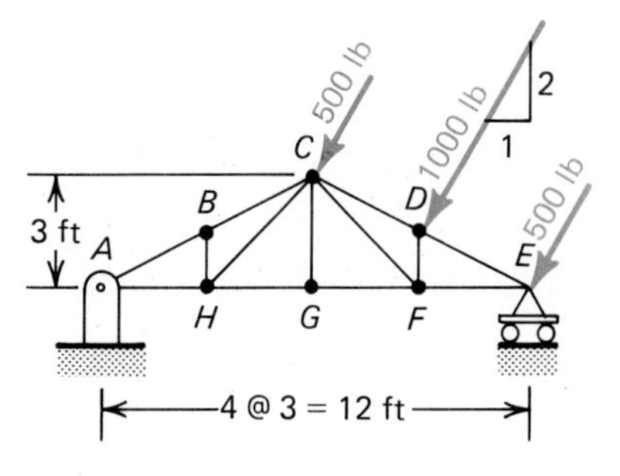

Figure P4.4

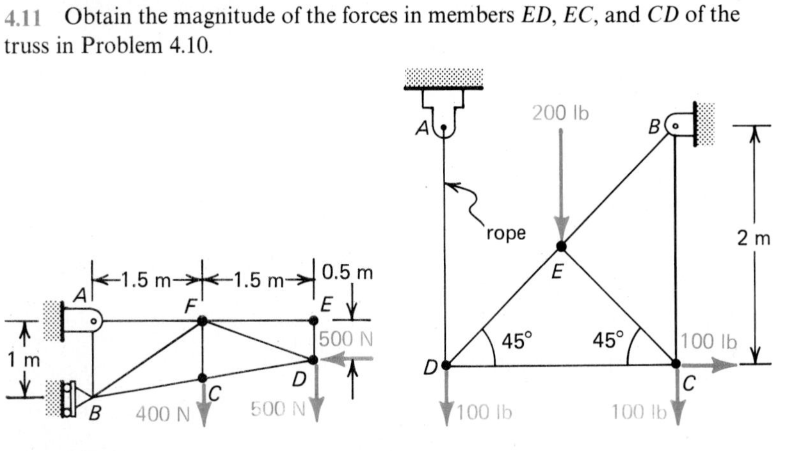

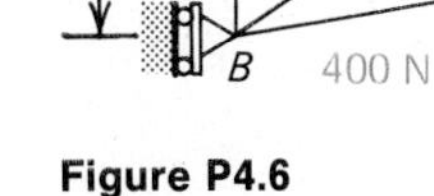

Figure P4.6

Figure P4.8

Figure P4.10

4.12 Calculate the magnitude of the forces at joints *A* and *B* of the truss shown in Fig. P4.12.

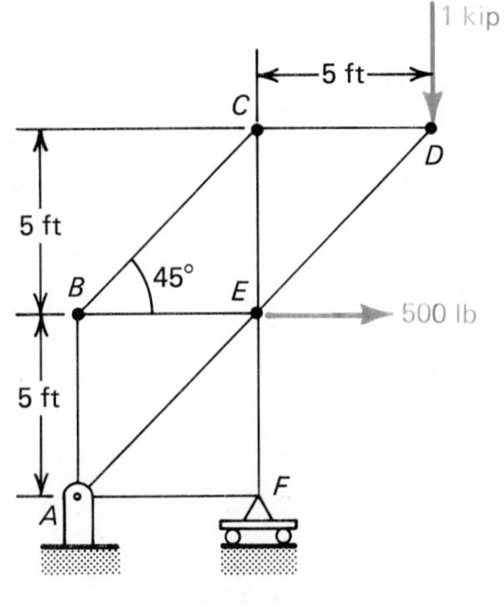

Figure P4.12

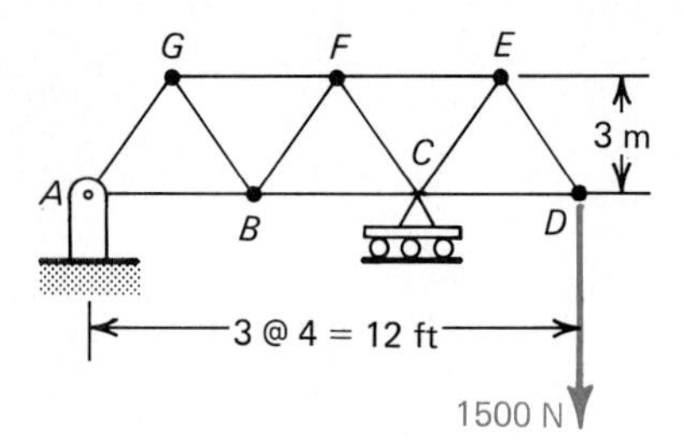

Figure P4.14

4.13 Calculate the magnitude of the forces at joints C and D of the truss in Problem 4.12.

4.14 Obtain the magnitude of the forces at joints A and G of the truss shown in Fig. P4.14.

4.15 Calculate the magnitude of the forces at joints D and E for the truss in Problem 4.14.

4.16 Find the magnitude of the force in member BC for the truss shown in Fig. P4.16.

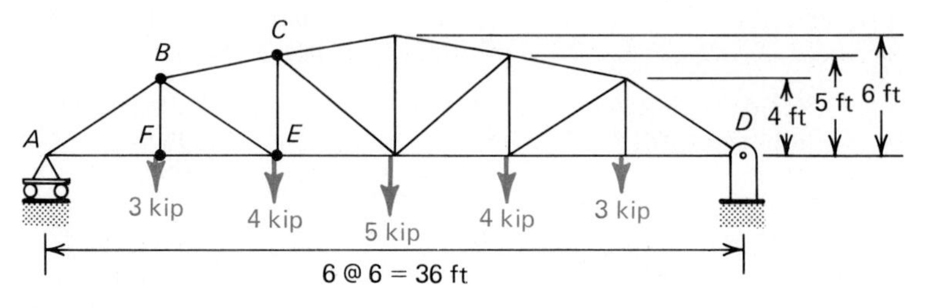

Figure P4.16

4.17 Find the magnitude of the forces in members FB, FA, and FE for the truss in Problem 4.16.

4.18 Obtain the magnitude of the forces in members AE, AD, DB, and CB in the truss shown in Fig. P4.18.

4.19 Find the magnitude of the force in member CE for the truss shown in Fig. P4.19.

4.20 Find the magnitude of the forces in all the members of the truss shown in Fig. P4.20.

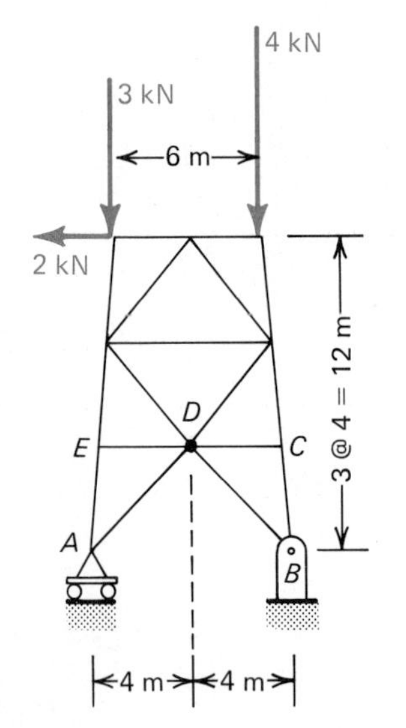

Figure P4.18

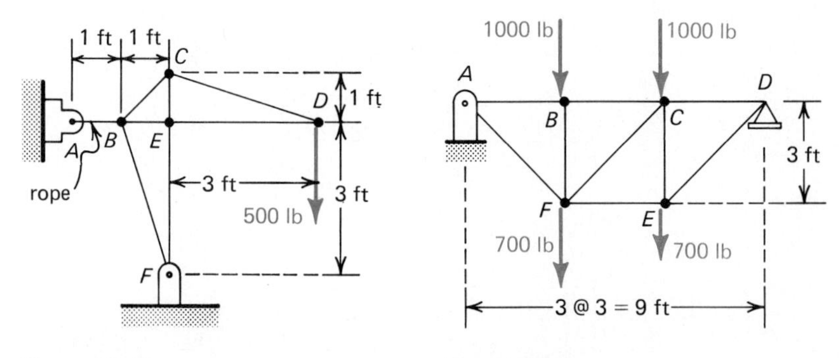

Figure P4.19

Figure P4.20

4.21 Find the magnitude of the forces in the cross bars CG and BF for the truss shown in Fig. P4.21.

4.22 Obtain the magnitude of the forces in all the members of the truss shown in Fig. P4.22.

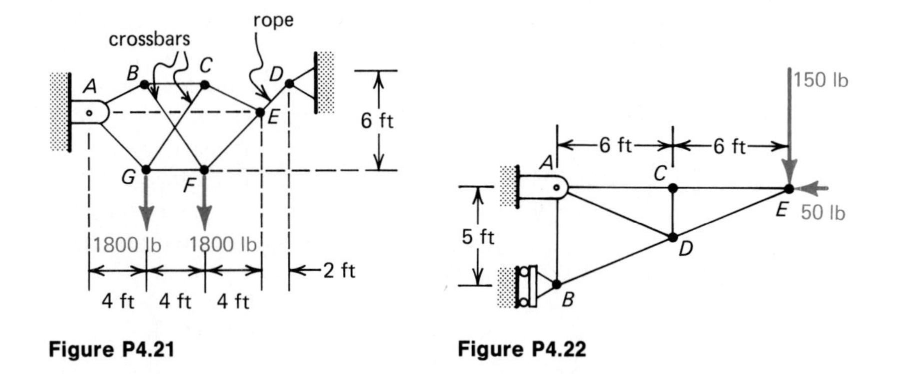

Figure P4.21 **Figure P4.22**

4.23 Find the magnitude of the forces in members *AB*, *AC*, and *AD* for the structure shown in Fig. P4.23. (*Hint*: Draw a separate free-body diagram for the left truss and right truss.)

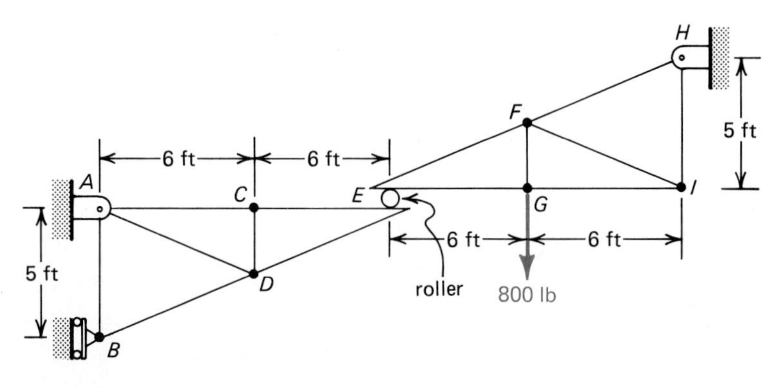

Figure P4.23

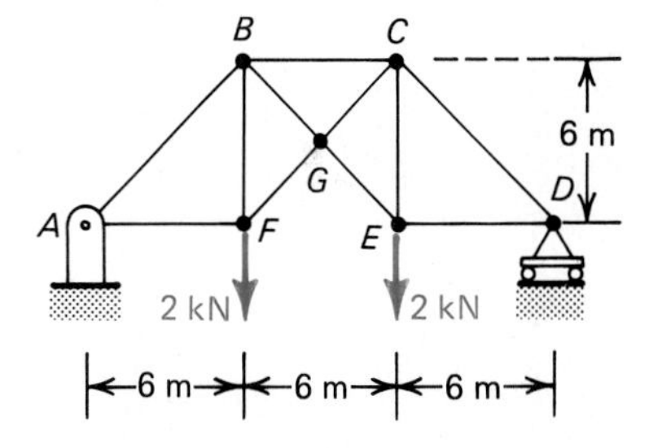

Figure P4.24

4.24 Obtain the magnitude of the forces in the members *AF*, *BF*, and *FG* for the truss shown in Fig. P4.24.

4.25 Find the magnitude of the forces in members *GB*, *GC*, and *GE* for the truss in Problem 4.24.

4.3 Graphical Method

The following graphical method for analyzing trusses is the graphical equivalent of the method of joints. Since the equations of equilibrium are satisfied at each joint and the forces are concurrent at each joint, a force polygon can be drawn at each joint. We begin the solution by finding the reaction forces using the equilibrium equations for the entire structure and then begin the graphical solution starting with a joint which has no more than two unknown internal forces. When the force polygons for each joint have been obtained, they can be combined into a single diagram called the **Maxwell diagram**.

The graphical method affords a quick procedure for obtaining the internal forces in the truss members and provides an independent check on other solutions used. However, it has the disadvantage of being cumbersome when a large-scale drawing is required and is limited to two-dimensional problems. The following two examples illustrate how a force polygon is constructed for each joint and then combined to obtain a Maxwell diagram.

Example 4.4

Use the graphical method to find the magnitude of all internal forces for the truss shown in Fig. 4.13(a).

The reaction forces were computed in Example 4.1 and are shown in Fig. 4.16(a) along with the internal forces. We start the graphical procedure by selecting a joint that has only two unknowns.

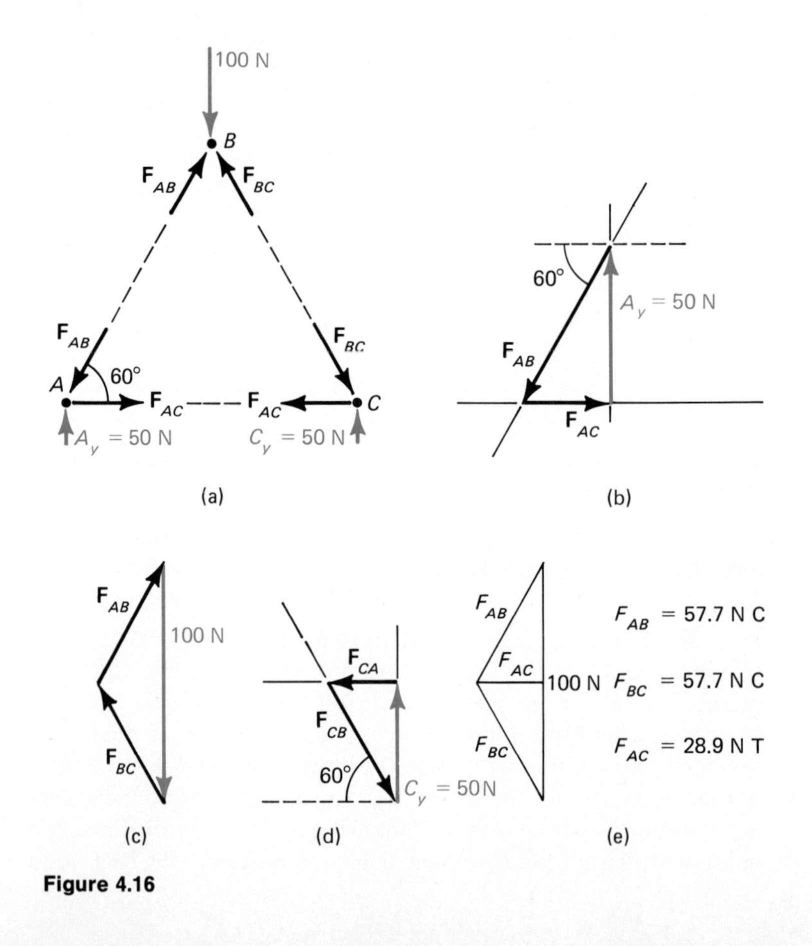

Figure 4.16

In this case any joint will suffice. We therefore begin with joint A. At this joint we know A_y and the assumed directions of the internal forces. As shown in Fig. 4.16(b) we lay off A_y to scale. From the tip of $\mathbf{A}_y$ we lay off a line at an angle of 60° to the horizontal along which $\mathbf{F}_{AB}$ lies. At the tail of $\mathbf{A}_y$, lay off a horizontal line along which $\mathbf{F}_{AC}$ lies. The two lines intersect to close the force polygon. For equilibrium, these three forces must add up to zero. Therefore, the directions of $\mathbf{F}_{AB}$ and $\mathbf{F}_{AC}$ must be as shown in Fig. 4.16(b). The force $\mathbf{F}_{AC}$ acts away from joint A while $\mathbf{F}_{AB}$ acts toward A. Thus member AC is in tension and AB in compression. Figure 4.16(c) and Fig. 4.16(d) show the force polygon for joints B and C, respectively. We note that since $\mathbf{F}_{AB}$ is known, only $\mathbf{F}_{BC}$ can be determined from joint B. The force polygon for joint C simply checks the forces already obtained. The Maxwell diagram is shown in Fig. 4.16(e) with a tabulation of forces. It is seen that the diagram is a combination of the force polygons of all the joints.

Example 4.5

Obtain the Maxwell diagram for the truss shown in Fig. 4.17.

First, from the free-body diagram of the truss shown in Fig. 4.17(b), determine the reaction forces from the equilibrium of the entire truss.

$$\sum M_A = 0 \qquad -2000(4) - 1000(8) + E_x(8) = 0$$

$$E_x = 2000 \text{ lb}$$

$$\sum F_x = 0 \qquad -A_x + E_x = 0$$

$$A_x = E_x = 2000 \text{ lb}$$

$$\sum F_y = 0 \qquad A_y - 2000 - 1000 = 0$$

$$A_y = 3000 \text{ lb}$$

We begin the graphical analysis by selecting a joint that has only two unknowns. In this example we have a choice between joints C and E. Let us begin with joint E. At joint E there are three forces in equilibrium; of these, we know the magnitude and direction of the force $\mathbf{E}_x$ and the lines of action for the other two forces $\mathbf{F}_{EA}$ and $\mathbf{F}_{ED}$. Referring to Fig. 4.17(c), lay off force $\mathbf{E}_x$ to scale. From the tip of $\mathbf{E}_x$, lay off a vertical line along which $\mathbf{F}_{EA}$ lies. At the tail of $\mathbf{E}_x$, lay off a line 45° off the vertical; $\mathbf{F}_{ED}$ must lie along this line. The two lines intersect to close the force polygon. The force $\mathbf{F}_{EA}$ acts away from the joint E while $\mathbf{F}_{ED}$ acts toward the joint E as shown in Fig. 4.17(c). Thus, member EA is in tension while member ED is in compression. Having the directions of the forces acting at joint E, it is useful to show these directions on Fig. 4.17(b) for future use.

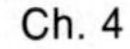

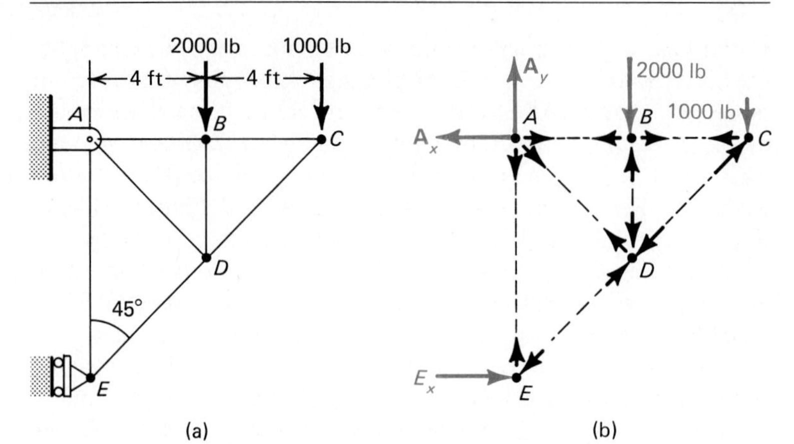

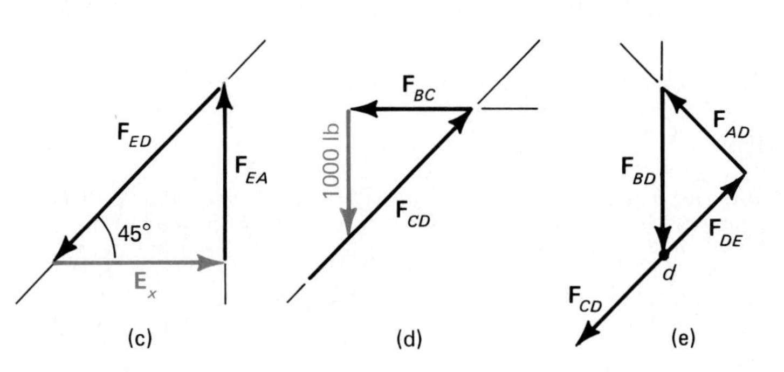

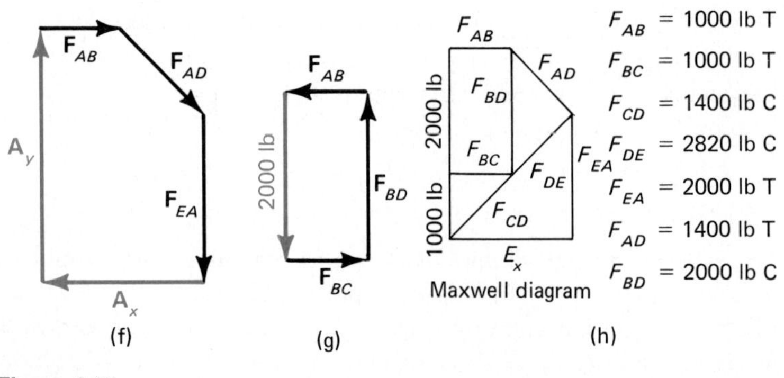

Figure 4.17

Figure 4.17(d) shows the force polygon for joint *C*. We begin by laying off the applied force, which is 1000 lb downward. At the tip of this force, lay off a 45° line with the vertical and a horizontal line through its tail. The intersection of these two lines establishes the force polygon as labeled in the figure. In this case force $\mathbf{F}_{CD}$ acts toward joint *C* (compression) and force $\mathbf{F}_{BC}$ acts away from the joint (tension).

In drawing the force polygon for joint *D* shown in Fig. 4.17(e), lay off force $\mathbf{F}_{CD}$, then from the tip of this vector lay off $\mathbf{F}_{DE}$. From the tip of $\mathbf{F}_{DE}$ lay off a line parallel to member AD and from the tail of $\mathbf{F}_{CD}$ identified by point d lay off a vertical line. These latter two lines intersect to form the force polygon which starts at point d then to the tip of $\mathbf{F}_{CD}$ from where it follows the same line of action to the tip of $\mathbf{F}_{DE}$. From that point on, the polygon follows $\mathbf{F}_{AD}$ and $\mathbf{F}_{BD}$ to close again at point d.

Figure 4.17(f) shows the force polygon for joint A, which established the force $\mathbf{F}_{AB}$. Figure 4.17(g) represents the force polygon for joint B and serves as a check on the forces $\mathbf{F}_{AB}$, $\mathbf{F}_{BC}$, and $\mathbf{F}_{BD}$.

The Maxwell diagram is shown in Fig. 4.17(h). This diagram is constructed as we develop the force polygon at each joint. Also, it is advantageous to tabulate the forces in each member as we proceed from joint to joint. The tabulated forces are shown in Fig. 4.17(h).

PROBLEMS

4.26 Construct a force polygon to scale for the joints A, E, and D in Fig. P4.1.

4.27 Construct a force polygon to scale for the forces acting at joints D and E in the truss in Fig. P4.4.

4.28 Draw the force polygon for the joints A, B, and C of the truss shown in Fig. P4.6.

4.29 Construct a force polygon to scale for joints B and D in Fig. P4.8.

4.30 Construct a force polygon for joints A and D of the truss shown in Fig. P4.10.

4.31 Construct a force polygon for joints C and E and sketch the Maxwell's diagram for the truss in Fig. P4.8.

4.32 Construct a force polygon for joint A for the truss in Fig. P4.12.

4.33 Obtain the magnitude of the force in member GF of the truss shown in Fig. P4.14 using the graphical method.

4.34 Construct force polygons for all the joints and then sketch the Maxwell's diagram for the truss in Fig. P4.20.

4.4 Method of Sections

The method of joints and the equivalent graphical method provide the internal forces of each member of a plane truss, after the reaction forces are evaluated, by employing the conditions of equilibrium at each joint, i.e., $\sum F_x = 0$, $\sum F_y = 0$. This is possible because the forces acting at the joint are concurrent. However, it is possible to construct a free-body diagram of a portion of the truss

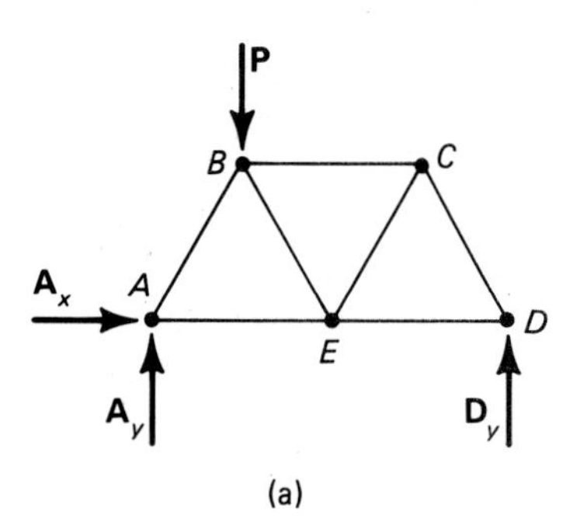

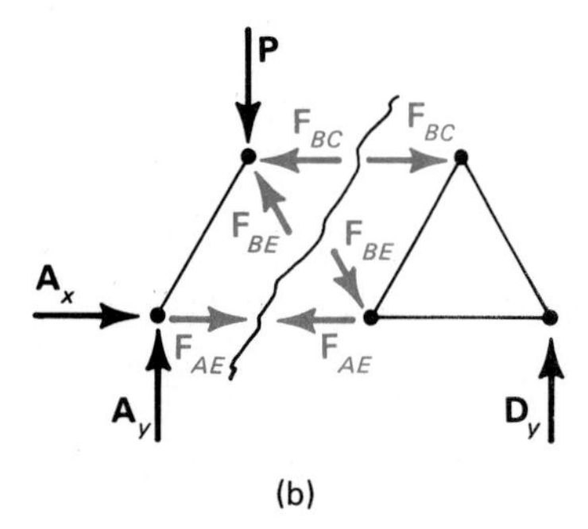

Figure 4.18

containing at least two joints of the truss and employ the third condition of equilibrium as well, namely, $\sum M = 0$. The free-body diagram of a part of a truss will generally contain nonconcurrent forces. This approach is called the method of sections. With this method we can find the internal forces of selected members of a truss more rapidly than with an analysis that proceeds from one joint to the next.

Consider the free-body diagram of a truss shown in Fig. 4.18(a) that shows reaction forces $\mathbf{A}_x$, $\mathbf{A}_y$, and $\mathbf{D}_y$ and an applied load $\mathbf{P}$. Let us find the forces in members BC, BE, and AE. First we cut the truss across these members and label the assumed directions of the internal forces as shown in Fig. 4.18(b). Now, since the section of the truss to the left of the cut and the section of the truss to the right of the cut are each in equilibrium, we can write for either section of the truss,

$$\sum F_x = 0 \qquad \sum F_y = 0 \qquad \sum M = 0$$

where the moments are taken about any convenient reference point, i.e., the reference point may be a point on the truss or off the truss. These three equations allow us to solve for the three unknown forces. Care must be exercised to ensure that the part of the truss under consideration has no more than three cut members with unknown internal forces since only three unknown forces can be found from three equations of equilibrium.

Example 4.6

Obtain the magnitude of the forces in members AC and BC using the method of sections for the truss shown in Fig. 4.19(a).

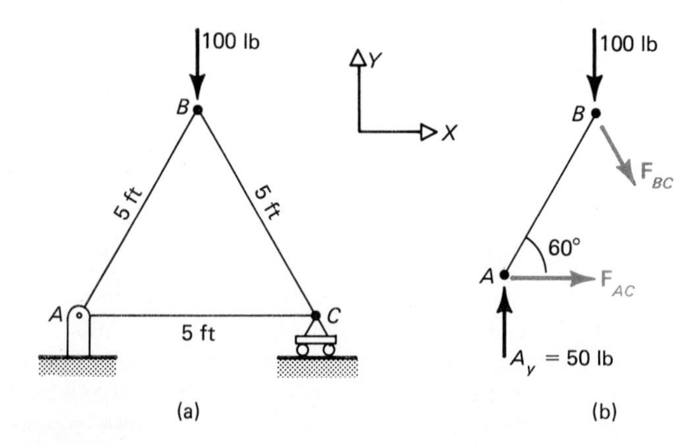

Figure 4.19

The left portion of the truss is shown in Fig. 4.19(b). The 50-lb reaction force is also shown. Summing moments about B yields,

$$\sum M_B = 0 \qquad -50(5\cos 60°) + F_{AC}(5\sin 60°) = 0$$

$$F_{AC} = 28.87$$

$$\therefore F_{AC} = 28.87 \text{ lb T}$$

Answer

Now sum forces in the X direction to find F_{BC}.

$$\sum F_x = 0 \qquad F_{BC}\cos 60° + F_{AC} = 0$$

$$F_{BC} = -57.74$$

$$\therefore F_{BC} = 57.74 \text{ lb C} \qquad \textit{Answer}$$

Example 4.7

Obtain the magnitude of the forces in members FE, FC, and BC of the truss shown in Fig. 4.20(a) by the method of sections.

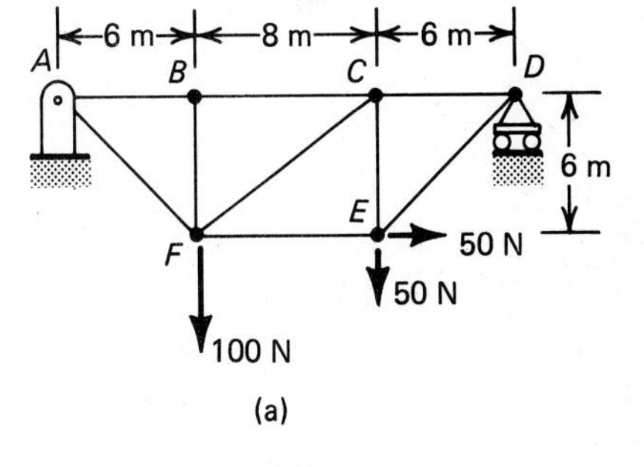

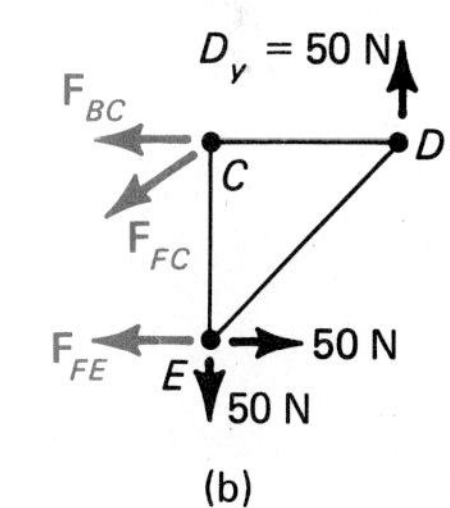

Figure 4.20

The truss is cut across members FE, FC, and BC and the right part of the truss is shown in Fig. 4.20(b). Also shown is the reaction force of magnitude 50 N acting at D. Referring to Fig. 4.20(b), take moments about point C and then sum forces in the X and Y direction, respectively.

$$\sum M_C = 0 \qquad -6F_{FE} + 6(50) + 6(50) = 0$$

$$F_{FE} = 100$$

$$\therefore F_{FE} = 100 \text{ N T} \qquad \textit{Answer}$$

$$\sum F_y = 0 \qquad -0.6F_{FC} - 50 + 50 = 0$$

$$F_{FC} = 0 \qquad \textit{Answer}$$

$$\sum F_x = 0 \qquad -F_{BC} - 100 + 50 = 0$$

$$F_{BC} = -50$$

$$\therefore F_{BC} = 50 \text{ N C} \qquad \textit{Answer}$$

Example 4.8

Use the method of sections to find the magnitude of the forces in members KJ, DE, and EI for the truss shown in Fig. 4.21(a).

In Fig. 4.21(b), the left portion of the truss cut across members CD, CJ, and KJ is shown with F_{KJ} as the desired unknown. For convenience, sum moments about point C since internal forces $\mathbf{F}_{CD}$ and $\mathbf{F}_{CJ}$ pass through this point.

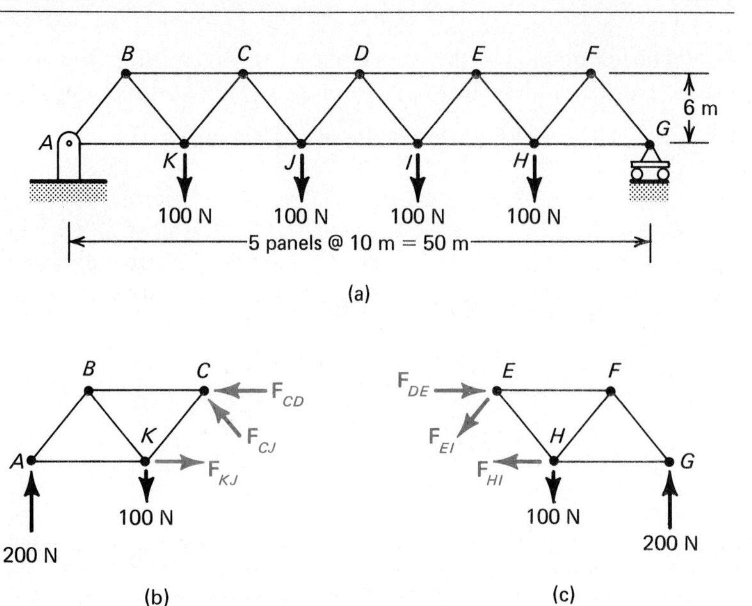

Figure 4.21

$$\sum M_C = 0 \qquad 6F_{KJ} + 100(5) - 200(15) = 0$$

$$F_{KJ} = 416.7$$

$$\therefore F_{KJ} = 416.7 \text{ N T} \qquad \textit{Answer}$$

Figure 4.21(c) shows the right portion of the truss cut across members *DE*, *EI*, and *IH*. We note that member *DE* is under compression. From this free-body diagram we have the following:

$$\sum F_y = 0 \qquad -\frac{6}{7.81} F_{EI} + 200 - 100 = 0$$

$$F_{EI} = 130.2$$

$$\therefore F_{EI} = 130.2 \text{ N T} \qquad \textit{Answer}$$

$$\sum M_H = 0 \qquad -6F_{DE} + 5\left(\frac{6}{7.81}\right)F_{EI}$$

$$+ 6\left(\frac{5}{7.81}\right)F_{EI} + 200(10) = 0$$

$$-6F_{DE} + 500 + 500 + 2000 = 0$$

$$F_{DE} = 500$$

$$F_{DE} = 500 \text{ N C}$$

Answer

Example 4.9

Use the method of sections to find the magnitude of the force in member BG for the truss shown in Fig. 4.22(a).

The left portion of the truss, which is cut across members BC, BG, and HG, is shown in Fig. 4.22(b). The most convenient approach to finding F_{BG} is to extend the line BC to where it intersects with the extended line AH at point I, and then sum moments about point I:

$$\sum M_I = 0 \qquad 4000(90) - 2000(120)$$

$$-\frac{3}{5}F_{BG}(40) - \frac{4}{5}F_{BG}(120) = 0$$

$$F_{BG} = 1000$$

$$\therefore F_{BG} = 1000 \text{ lb T}$$

Answer

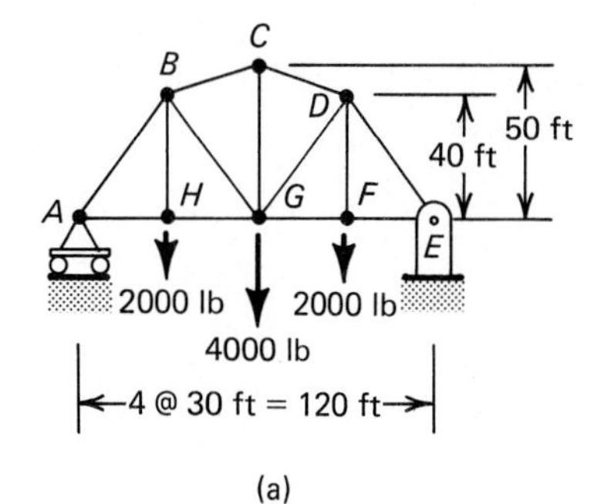

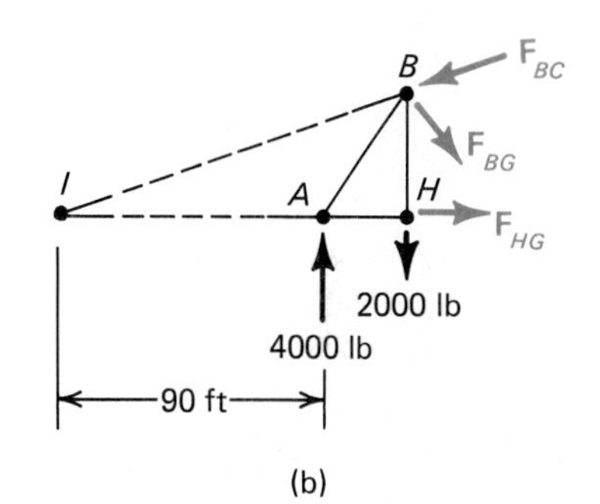

Figure 4.22

PROBLEMS

4.35 Find the magnitude of the forces in members BC, BF, and AF of the truss shown in Fig. P4.35 using the method of sections.

4.36 Obtain the magnitude of the force in member FC of the truss in Problem 4.35.

4.37 Use the method of sections to find the magnitude of the forces in members BF and BC in Fig. P4.20.

4.38 Use the method of sections to find the magnitude of the force in member EG in Fig. P4.23.

4.39 Find the magnitude of the forces in members BG and FG in Fig. P4.24 by the method of sections.

4.40 Find the magnitude of the forces in members BC, CL, and LK in Fig. P4.40 by the method of sections.

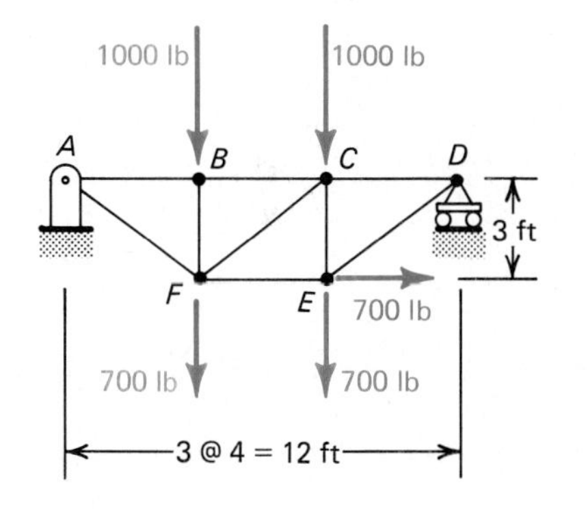

Figure P4.35

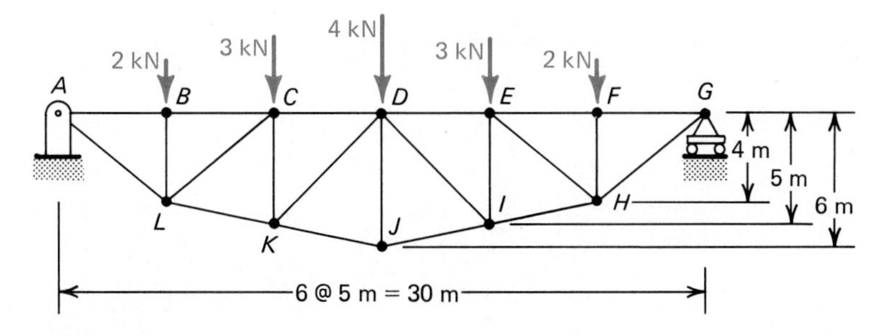

Figure P4.40

4.41 Obtain the magnitude of the forces in members BL and AL for the truss in Problem 4.40.

4.42 Using the method of sections calculate the magnitude of the forces in members DC and DE for the truss in Problem 4.40.

4.43 Calculate the magnitude of the forces in the crossbar BF of the truss shown in Fig. P4.21 by the method of sections.

4.44 Use the method of sections to find the magnitude of the forces in members GI and GH of the truss shown in Fig. P4.44.

4.45 Use the method of sections to find the magnitude of the forces in members CE, ED, and FD in Fig. P4.44.

4.46 Find the magnitude of the forces in members HJ and IL in Fig. P4.46 by the method of sections. (Use suggested cut section.)

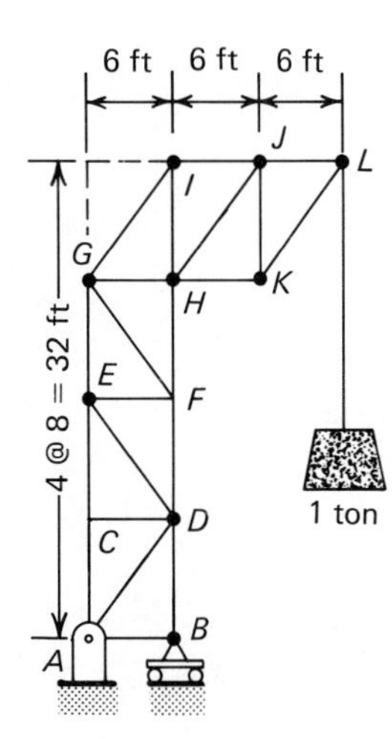

Figure P4.44

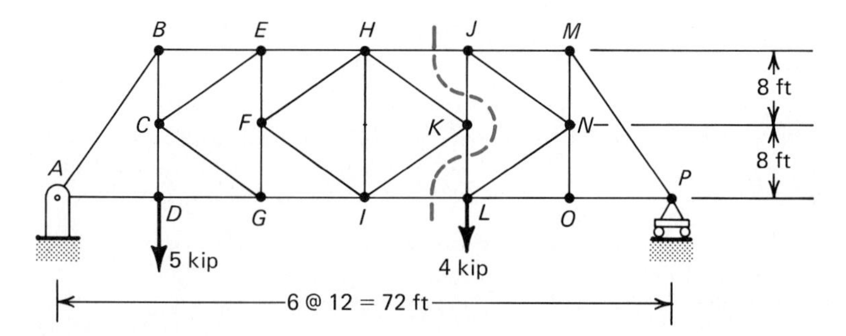

Figure P4.46

4.47 Find the magnitude of the force in member JM in Fig. P4.46.

4.48 Obtain the magnitude of the forces in members FD, ED, and EC of the truss shown in Fig. P4.48.

4.49 Obtain the magnitude of the forces in members DB and CA of the truss shown in Fig. P4.48.

4.50 Calculate the magnitude of the forces in members BC and GE of the truss shown in Fig. P4.24.

4.51 Calculate the magnitude of the forces in members AC, DE, and CD of the truss shown in Fig. P4.22.

4.52 Find the magnitude of the force in member BD of the truss shown in Fig. P4.52.

4.53 Find the magnitude of the force in members CB and CD of the truss shown in Fig. P4.52.

4.54 Find the magnitude of the force in member HG in Fig. P4.54.

4.55 Find all the zero-force members in Fig. P4.54 by observation.

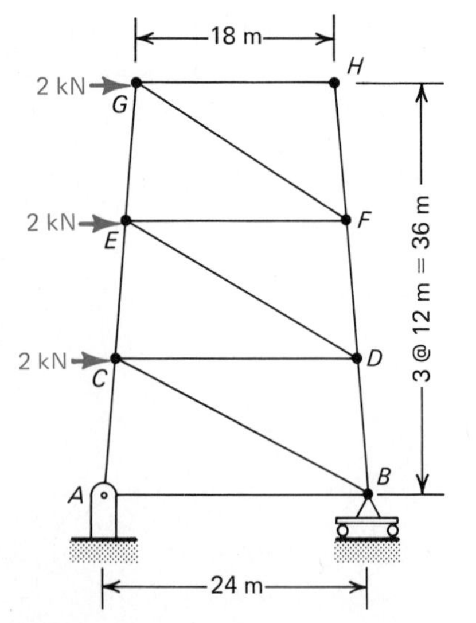

Figure P4.48

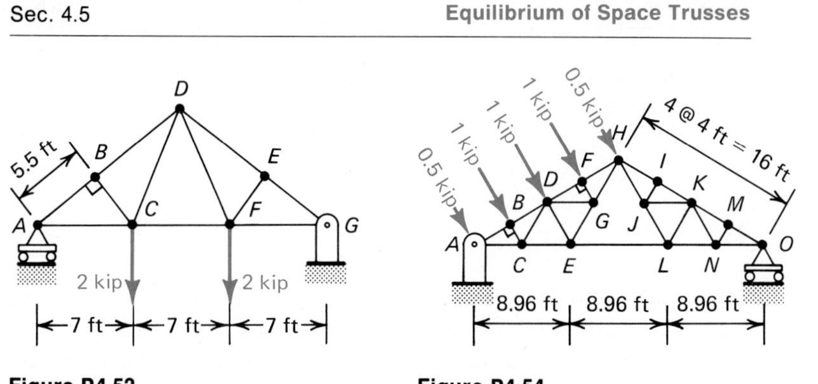

Figure P4.52 **Figure P4.54**

4.56 Add a 3-kip vertical load at L to the truss in Fig. P4.54 and find the magnitude of the force in member EL.

4.5 Equilibrium of Space Trusses

A three-dimensional truss is called a space truss. An elementary space truss consists of six straight members joined together to form a tetrahedron as shown in Fig. 4.23(a). Notice that we can add three new members, not in the same plane, to create another joint as shown in Fig. 4.23(b). The space truss in Fig. 4.23(a) started with six members and four joints, and was expanded to nine members and five joints in Fig. 4.23(b). In general, the number of joints J and members M of a simple space truss can be shown to be related by the simple formula,

$$M = 3J - 6 \tag{4.2}$$

For a space truss to be in equilibrium, all of the six equilibrium equations must be satisfied, namely,

$$\sum F_x = 0 \qquad \sum F_y = 0 \qquad \sum F_z = 0$$
$$\sum M_x = 0 \qquad \sum M_y = 0 \qquad \sum M_z = 0$$

In order for the space truss to be statically determinate there must be no more than six unknown reactions.

Consequently, for our purposes the supports of the truss must consist of rollers, ball and sockets, or balls to ensure no more than six unknown reaction forces. Although the supports in a space truss are not illustrated, the number of unknowns at each support is generally provided or a description of the support is given so that there is no confusion as to the total number of unknown reaction forces.

Likewise, we assume that the members of a space truss are connected by ball-and-socket joints so that we are ignoring bending effects. This means that we are not applying couples to the members.

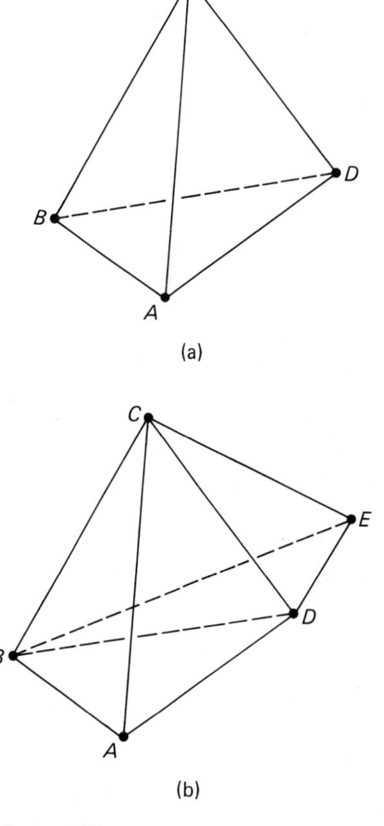

Figure 4.23

Consequently, each member of the truss is a two-force member either in tension or compression.

Once the free-body diagram of each joint is obtained, we can write the equations of equilibrium at each joint. Since we can write three equations at each joint, namely,

$$\sum F_x = 0 \qquad \sum F_y = 0 \qquad \sum F_z = 0$$

and there are J joints in the space truss, we have $3J$ equations. And, since we know that $3J = M + 6$, we can solve for the internal force in each member as well as for the six reaction forces. Often, however, it is easier to solve for the reactions first by considering the equilibrium of the entire space truss and then treat the equilibrium at each joint throughout the truss.

The method of sections as well as the method of joints can be applied to the space truss. A judicious selection of the order of analysis helps to avoid excessive algebraic equations. In today's computerized age there exist numerous computer programs capable of solving the structural problems of large buildings, transmission towers, missiles, etc. The graphical method is, however, difficult to extend to the three-dimensional case.

Example 4.10

Obtain the magnitude of the force in member AB from the free-body diagram of a truss shown in Fig. 4.24(a). The applied load is given by $\mathbf{P} = (8\mathbf{i} + 6\mathbf{j} - 4\mathbf{k})$ N. The reactions at A, B, and C are shown in the figure.

The reactions at A, B, and C are computed by considering the equilibrium of the entire truss. Referring to the free-body diagram in

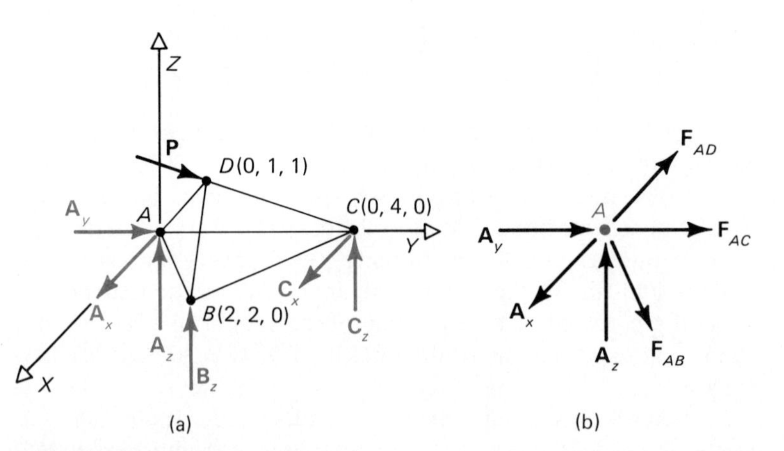

Figure 4.24

Fig. 4.24(a), let us first calculate the moment due to the applied load.

$$\mathbf{M} = \mathbf{r} \times \mathbf{P}$$

where $\mathbf{r} = \mathbf{j} + \mathbf{k}$

$$\mathbf{M} = \begin{vmatrix} \mathbf{i} & \mathbf{j} & \mathbf{k} \\ 0 & 1 & 1 \\ 8 & 6 & -4 \end{vmatrix}$$

$$= -10\mathbf{i} + 8\mathbf{j} - 8\mathbf{k}$$

Having this moment, we now apply the equilibrium equations.

$$\sum M_y = 0 \qquad -2B_z + 8 = 0$$
$$B_z = 4 \text{ N}$$

$$\sum M_x = 0 \qquad 2B_z + 4C_z - 10 = 0$$
$$C_z = 0.5 \text{ N}$$

$$\sum M_z = 0 \qquad -4C_x - 8 = 0$$
$$C_x = -2 \text{ N}$$

$$\sum F_x = 0 \qquad A_x + C_x + 8 = 0$$
$$A_x = -6 \text{ N}$$

$$\sum F_y = 0 \qquad A_y + 6 = 0$$
$$A_y = -6 \text{ N}$$

$$\sum F_z = 0 \qquad A_z - 4 + B_z + C_z = 0$$
$$A_z = -0.5 \text{ N}$$

We now consider the equilibrium of the joint A by summing the forces shown in Fig. 4.24(b). We note that

$$\mathbf{F}_{AB} = F_{AB}(0.707\mathbf{i} + 0.707\mathbf{j})$$

$$\mathbf{F}_{AC} = F_{AC}\mathbf{j}$$

$$\mathbf{F}_{AD} = F_{AD}(0.707\mathbf{j} + 0.707\mathbf{k})$$

The forces in the X direction become

$$\sum F_x = 0 \qquad 0.707F_{AB} + A_x = 0$$
$$0.707F_{AB} - 6 = 0$$
$$F_{AB} = 8.49$$
$$\therefore F_{AB} = 8.49 \text{ N T} \qquad \textit{Answer}$$

Example 4.11

Find the reaction forces at A, B, and D and the magnitude of the internal forces for the space truss shown in Fig. 4.25(a). The reactions

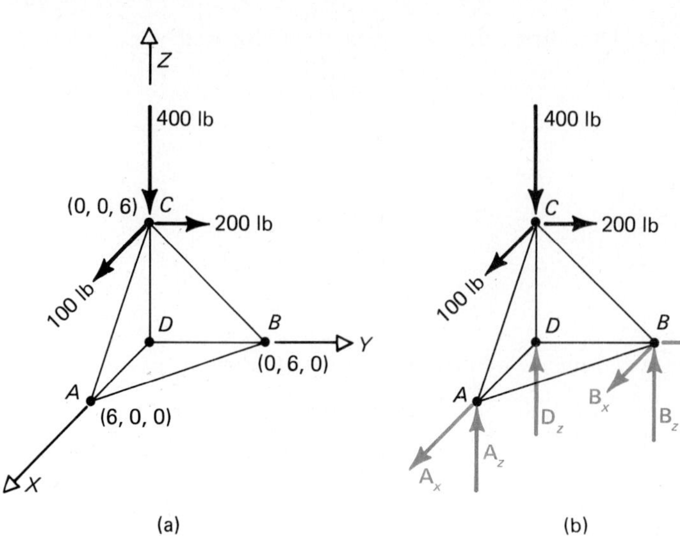

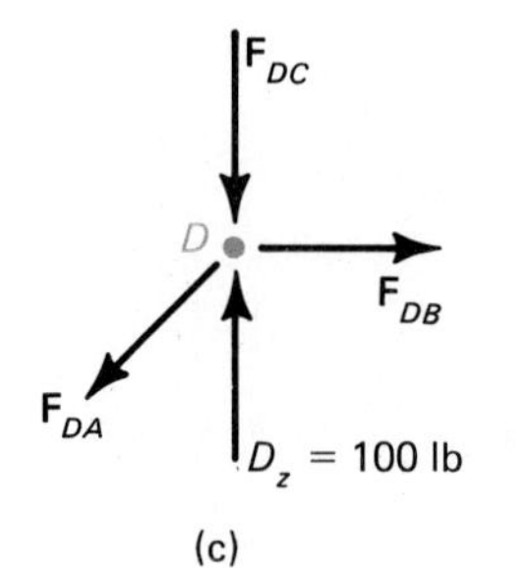

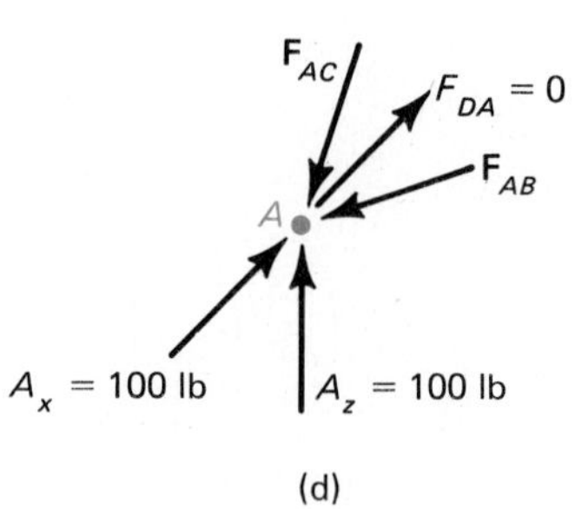

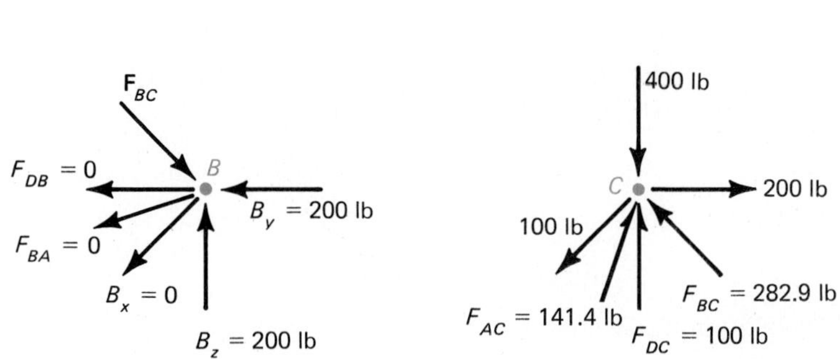

Figure 4.25

due to the supports are shown in the free-body diagram of Fig. 4.25(b).

Referring to Fig. 4.25(b), the equilibrium of the entire truss is treated first. Let us calculate the moment due to the applied load.

$$\mathbf{M} = \mathbf{r} \times \mathbf{P}$$

where $\mathbf{r} = 6\mathbf{k}$

$$\mathbf{M} = \begin{vmatrix} \mathbf{i} & \mathbf{j} & \mathbf{k} \\ 0 & 0 & 6 \\ 100 & 200 & -400 \end{vmatrix}$$

$$= -1200\mathbf{i} + 600\mathbf{j}$$

Now apply the equilibrium equations.

$$\sum M_x = 0 \qquad 6B_z - 1200 = 0$$
$$B_z = 200 \text{ lb}$$

$$\sum M_y = 0 \qquad -6A_z + 600 = 0$$
$$A_z = 100 \text{ lb}$$

$$\sum M_z = 0 \qquad 6B_x = 0$$
$$B_x = 0$$

$$\sum F_x = 0 \qquad A_x + B_x + 100 = 0$$
$$A_x = -100 \text{ lb}$$

$$\sum F_y = 0 \qquad B_y + 200 = 0$$
$$B_y = -200 \text{ lb}$$

$$\sum F_z = 0 \qquad A_z + B_z + D_z - 400 = 0$$
$$D_z = -100 - 200 + 400$$
$$D_z = 100 \text{ lb}$$

The reactions are

$$\mathbf{A} = (-100\mathbf{i} + 100\mathbf{k}) \text{ lb}$$
$$\mathbf{B} = (-200\mathbf{j} + 200\mathbf{k}) \text{ lb} \qquad \textit{Answer}$$
$$\mathbf{D} = 100\mathbf{k} \text{ lb}$$

We now consider the equilibrium of the joints. For convenience, consider starting with joint D as shown in Fig. 4.25(c).

Joint D:

$$\sum F_x = 0 \qquad F_{DA} = 0$$
$$\sum F_y = 0 \qquad F_{DB} = 0$$
$$\sum F_z = 0 \qquad -F_{DC} + 100 = 0$$
$$F_{DC} = 100 \text{ lb C} \qquad \textit{Answer}$$

Joint A:

$$\sum F_z = 0 \qquad 100 - 0.707F_{AC} = 0$$
$$F_{AC} = 141.4 \text{ lb C} \qquad \textit{Answer}$$

$$\sum F_x = 0 \qquad -100 + 0.707F_{AC} + 0.707F_{AB} = 0$$
$$F_{AB} = 0 \qquad \textit{Answer}$$

Joint B:

$$\sum F_z = 0 \qquad 200 - 0.707F_{BC} = 0$$
$$F_{BC} = 282.9 \text{ lb C} \qquad \textit{Answer}$$

As a check, we can use

$$\sum F_y = 0 \qquad -200 + 0.707F_{BC} = 0$$
$$F_{BC} = 282.9 \text{ lb C} \qquad \text{(Checks)}$$

Joint C: The following calculations are a check on the results already obtained.

$$\sum F_x = 0 \qquad 100 - 0.707F_{AC} = 100 - 0.707(141.4) = 0$$
$$100 - 100 = 0 \qquad \text{(Checks)}$$

$$\sum F_y = 0 \qquad 200 - 0.707F_{BC} = 200 - 0.707(282.9) = 0$$
$$200 - 200 = 0 \qquad \text{(Checks)}$$

$$\sum F_z = 0 \qquad -400 + 0.707F_{AC} + F_{DC} + 0.707F_{BC} = 0$$
$$-400 + 100 + 100 + 200 = 0 \qquad \text{(Checks)}$$

PROBLEMS

4.57 Calculate the magnitude of the forces in members AC and AD from the free-body diagram of the space truss shown in Fig. 4.24 for $\mathbf{P} = (8\mathbf{i} + 6\mathbf{j} - 4\mathbf{k})$ N.

4.58 Using the method of joints obtain the magnitude of the forces acting at C for the free-body diagram of the space truss shown in Fig. 4.24. Use the same value of $\mathbf{P}$ as given in Problem 4.57.

4.59 Rework Example 4.10 for $\mathbf{P} = (10\mathbf{i} + 4\mathbf{j} + 9\mathbf{k})$ N.

4.60 Obtain the magnitude of the forces in members AC, AB, and AD of the space truss shown in Fig. 4.25 assuming that the 200-lb applied force at point C in the Y direction is eliminated.

4.61 Find all the reaction forces in the free-body diagram of the space truss shown in Fig. P4.61.

4.62 Obtain the magnitude of the force in member CB of the structure in Problem 4.61.

4.63 Find the magnitude of the forces in members AC, AD, and AB of the structure in Problem 4.61.

4.64 Find the reactions for the truss shown in Fig. P4.64.

4.65 Obtain the magnitude of the force acting in the member AC for the space truss in Problem 4.64.

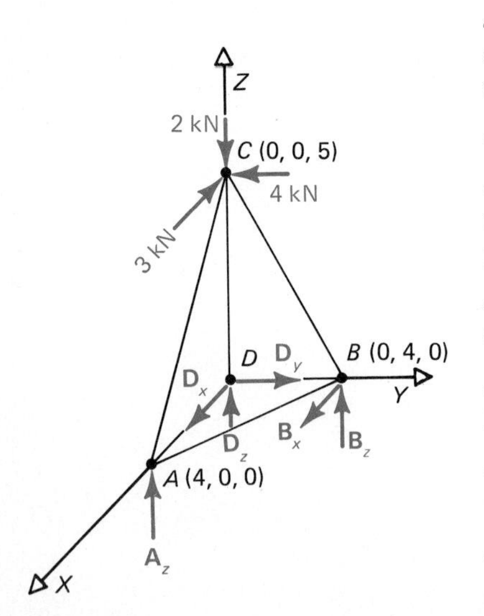

Figure P4.61

4.66 Find the reactions and the magnitude of the force in member AE for the truss shown in Fig. P4.66.

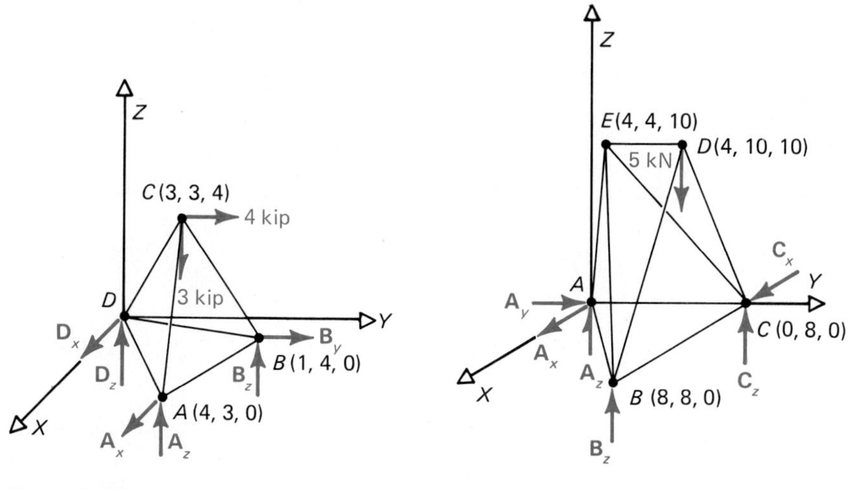

Figure P4.64

Figure P4.66

4.6 Frames and Machines

So far, we have analyzed trusses that are composed of two-force members. In this type of structure, each truss member begins and ends at joints and the external loads are applied at the joints only. When three or more forces are applied to a member, the member is called a multiforce member and the structure is called a frame or machine. As an example consider the structure shown in Figure 4.26(a). Here we have a frame consisting of the following members:

Member EC—a two-force member since forces occur only at two points, C and E

Member EB—a multiforce member since the external load P is applied to it at a point other than one of its joints

Member AD—a multiforce member since it is a continuous member that is carrying not only the reaction forces at A and D, but also the loads transmitted at joints B and C

In carrying out an analysis of the frame in Fig. 4.26(a), the free-body diagram of the entire structure should be treated first, as shown in Fig. 4.26(b). Since this is a plane rigid body, the following three

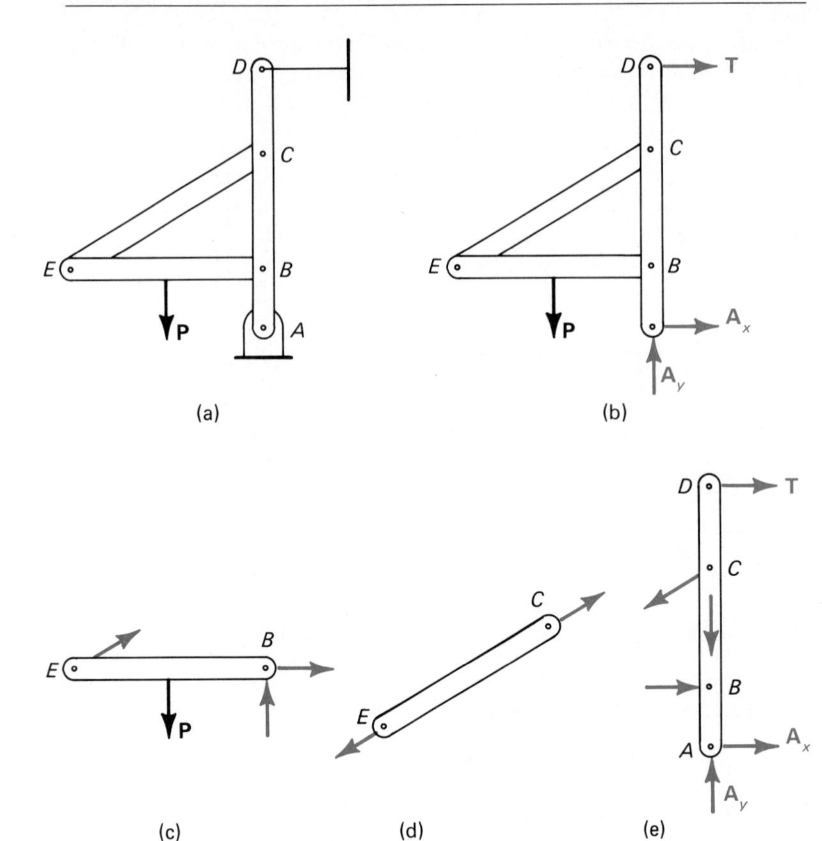

Figure 4.26

equations of equilibrium enable us to find the three reaction forces A_x, A_y, and T:

$$\sum F_x = 0 \qquad \sum F_y = 0 \qquad \sum M = 0$$

Having found the reaction forces, we proceed to the free-body diagram of member EB shown in Fig. 4.26(c). Applying the three equations of equilibrium to this member, we can obtain the internal forces at B and E. From the free-body diagram of the two-force member EC shown in Fig. 4.26(d), the force at joint C is immediately known. Note that these internal forces occur in equal and opposite pairs at each joint of the structure. This is a consequence of Newton's third law which states that every action has an equal and opposite reaction. Finally, as a check on these calculations, the equations of equilibrium should now be satisfied for the free-body diagram of Fig. 4.26(e). If these do not check, we should review our previous calculations for errors.

Frames

A frame is usually a stationary structure consisting of one or more multiforce members. Frames can be further classified into two categories: (1) rigid frame, wherein the shape of the frame does not change; and (2) nonrigid frame, wherein the removal or alteration of the supports of a frame causes the frame to change its shape. Figure 4.27(a) shows a four-link mechanism which is an example of a nonrigid frame. Notice that the ground itself forms the fourth link, and that if either support were removed, or replaced by a roller, this frame would alter its shape.

Rigid frames are analyzed by first drawing the free-body diagram of the entire structure so that the reaction forces may be determined. A free-body diagram of each member is then drawn and the equilibrium conditions are applied to each member in the most convenient order apparent to us in order to find the internal forces.

Nonrigid frames are analyzed in a similar way. However, not all the reaction forces can be obtained from the equilibrium of the entire nonrigid frame, as is apparent from Fig. 4.27(b). Here we have four unknown reaction forces and three equations of equilibrium. But, referring to Fig. 4.27(c), we see that the free-body diagrams of the members show eight unknown forces, the four

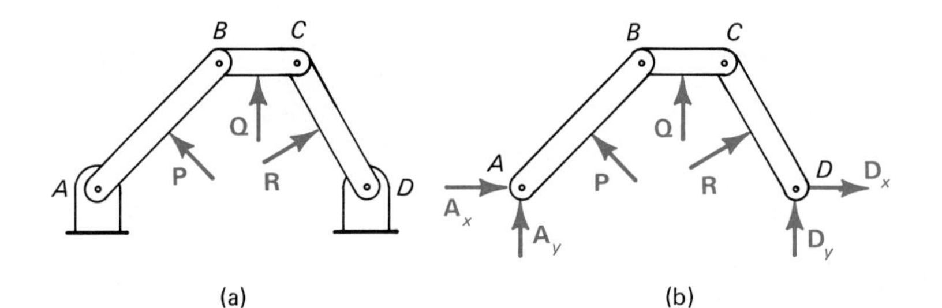

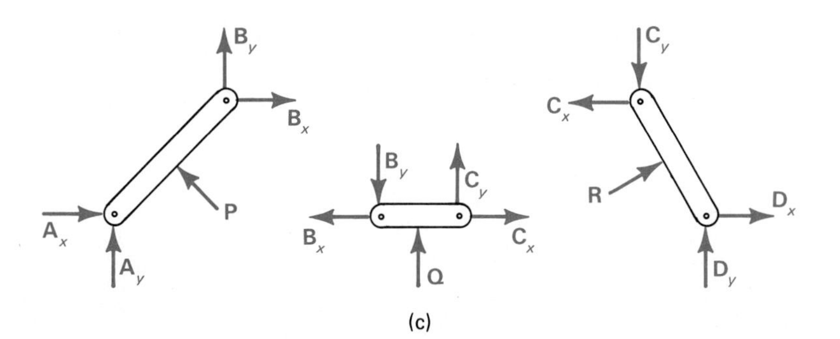

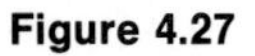
Figure 4.27

reaction forces of magnitude A_x, A_y, D_x, D_y, and four internal forces of magnitude B_x, B_y, C_x, and C_y. Since there are eight independent equilibrium equations, the structure is statically determinate. It may require, as in this case, that we work back and forth with the free-body diagrams to find these forces. For example, by summing moments about joint C for member BC in Fig. 4.27(c) we obtain B_y. Having B_y and summing moments about joint A for member AB, we obtain B_x. By this process of working with each member in a judicious way, we can obtain the external and internal forces for the frame.

Figure 4.28

Machines

A machine has moving parts and is usually not considered a rigid structure. Machines are designed to transmit loads rather than to support them. For example, the pair of tongs shown in Fig. 4.28(a) has a force **P** applied to each tong that transmits the gripping force **Q**. The gripping force as well as the reactions at the pin can be obtained as a function of the applied force **P** by drawing a free-body diagram of a part of the machine as shown in Fig. 4.28(b). We can now apply the equations of equilibrium to compute Q as well as the reactions. Other forms of hand tools are further examples of familiar machines.

Example 4.12

Determine the components of all the forces acting on each member of the frame shown in Fig. 4.29(a). Note that point A and point E are at the same level.

The free-body diagram of the frame is shown in Fig. 4.29(b), while the members are shown in Figs. 4.29(c) and 4.29(d). Note that member BE is a two-force member since no external load acts on it. Also note that triangle BCE is an equilateral triangle.

Equilibrium of the entire frame:

$$\sum M_F = 0$$

$$0.5A_x - 500\frac{\sqrt{3}}{2}(0.5)\left(\frac{\sqrt{3}}{2}\right) - 500\left(\frac{1}{2}\right)\left[1 + 0.5\left(\frac{1}{2}\right)\right] = 0$$

$$A_x = 1000 \text{ N} \qquad \textit{Answer}$$

$$\sum F_x = 0 \qquad -1000 + F_x + 500\left(\frac{1}{2}\right) = 0$$

$$F_x = 750 \text{ N} \qquad \textit{Answer}$$

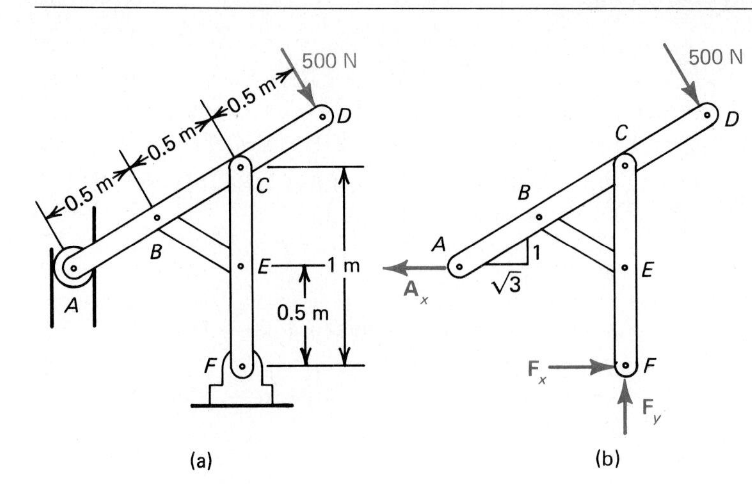

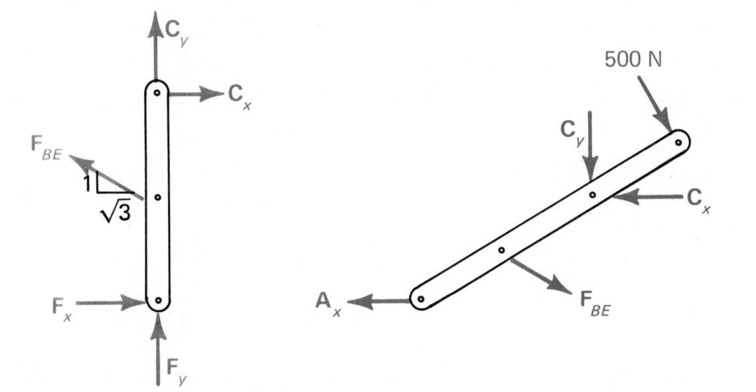

Figure 4.29

$$\sum F_y = 0 \qquad F_y - 500\frac{\sqrt{3}}{2} = 0$$

$$F_y = 433 \text{ N} \qquad \textit{Answer}$$

Member CEF:

$$\sum M_c = 0 \qquad -\frac{\sqrt{3}}{2} F_{BE}(0.5) + F_x(1.0) = 0$$

$$F_{BE} = 1732 \text{ N} \qquad \textit{Answer}$$

$$\sum F_x = 0 \qquad C_x - \frac{\sqrt{3}}{2} F_{BE} + F_x = 0$$

$$C_x = 1500 - 750 = 750 \text{ N} \qquad \textit{Answer}$$

$\sum F_y = 0 \qquad C_y + \frac{1}{2} F_{BE} + F_y = 0$

$C_y = -866 - 433 = -1299$ N *Answer*

The equilibrium equations for member $ABCD$ shown in Fig. 4.29(d) provide a check on our calculations.

$\sum F_x = 0$

$$-A_x + \frac{\sqrt{3}}{2} F_{BE} - C_x + \frac{1}{2}(500) = -1000 + 1500 - 750 + 250 = 0$$

$$-1750 + 1750 = 0 \qquad \text{(Checks)}$$

$\sum F_y = 0$

$$-\frac{1}{2} F_{BE} - C_y - \frac{\sqrt{3}}{2}(500) = -866 + 1299 - 433 = 0$$

$$-1299 + 1299 = 0 \qquad \text{(Checks)}$$

Example 4.13

Determine the components of all the forces acting on each member of the frame shown in Fig. 4.30(a).

The free-body diagram of the entire frame is shown in Fig. 4.30(b).

Equilibrium of the entire frame:

$\sum M_A = 0 \qquad -5E_x - \frac{4}{5}(100)(2) + \frac{3}{5}(100)(3) = 0$

$E_x = 4$ lb *Answer*

$\sum F_x = 0 \qquad E_x + R_x - \frac{3}{5}(100) = 0$

$R_x = 56$ lb *Answer*

$\sum F_y = 0 \qquad R_y - \frac{4}{5}(100) = 0$

$R_y = 80$ lb *Answer*

Note that $\mathbf{R}_x$ and $\mathbf{R}_y$ are the resultant forces transmitted to the foundation bracket.

The free-body diagrams of the various members and pin are shown in Figs. 4.30(c) through 4.30(f).

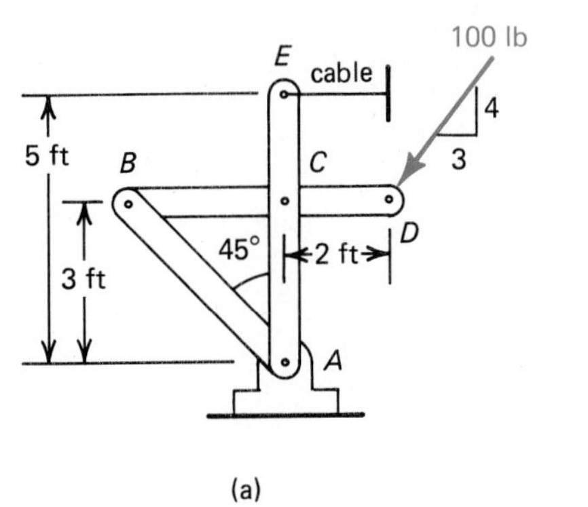

(a)

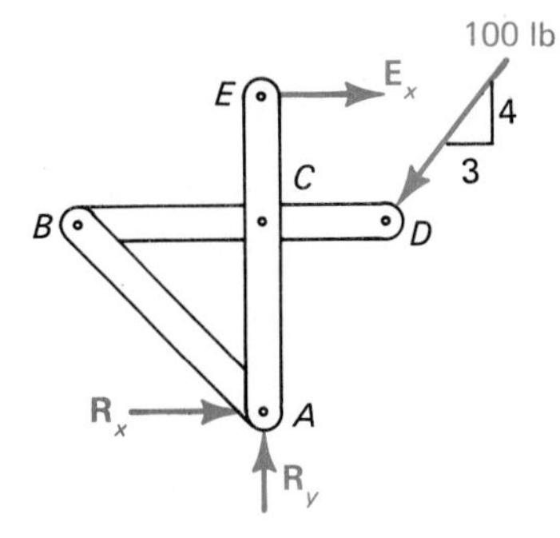

(b)

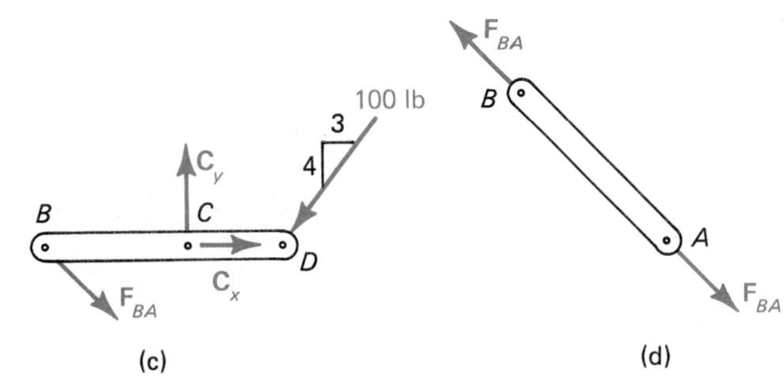

(c) (d)

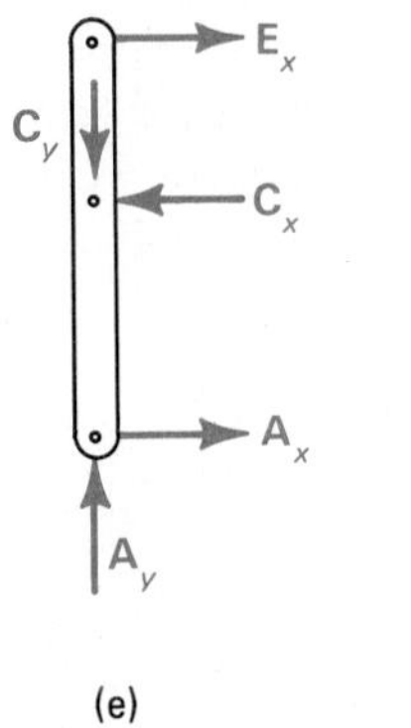

(e)

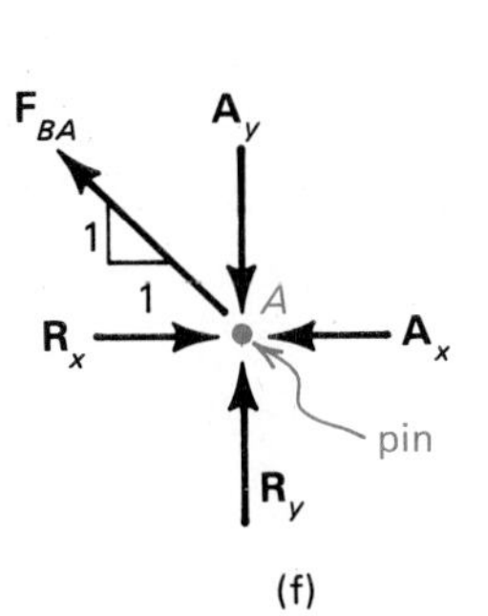

(f)

Figure 4.30

Member BCD:

$$\sum M_c = 0 \qquad 3F_{BA}\sin 45^\circ - 2(100)\frac{4}{5} = 0$$

$$F_{BA} = 75.4 \text{ lb} \qquad \textit{Answer}$$

$$\sum F_x = 0 \qquad F_{BA}\cos 45^\circ + C_x - \frac{3}{5}(100) = 0$$

$$C_x = 6.67 \text{ lb} \qquad \textit{Answer}$$

$$\sum F_y = 0 \qquad -F_{BA} \sin 45^\circ + C_y - \frac{4}{5}(100) = 0$$

$$C_y = 133.3 \text{ lb} \qquad \textit{Answer}$$

Member ACE:

$$\sum F_x = 0 \qquad A_x - C_x + E_x = 0$$

$$A_x = 2.67 \text{ lb} \qquad \textit{Answer}$$

$$\sum F_y = 0 \qquad A_y - C_y = 0$$

$$A_y = 133.3 \text{ lb} \qquad \textit{Answer}$$

The moment about point C can also be used as a check. Finally we use the pin at joint A to check our work. From Fig. 4.30(f) we have

$$\sum F_x = 0 \qquad R_x - A_x - F_{BA} \cos 45^\circ = 0$$

$$56 - 2.67 - 53.3 = 0 \qquad \text{(Checks)}$$

$$\sum F_y = 0 \qquad R_y - A_y + F_{BA} \sin 45^\circ = 0$$

$$80 - 133.3 + 53.3 = 0 \qquad \text{(Checks)}$$

Example 4.14

Find the reaction forces for the nonrigid frame shown in Fig. 4.31(a).

The free-body diagram of the nonrigid frame shows four unknown reaction forces, as seen in Fig. 4.31(b). Since we have only three equilibrium equations, we cannot solve directly for the reaction forces. However, if we proceed to Figs. 4.31(c) and (d), we have six unknowns and six equilibrium equations.

Member AC:

$$\sum M_A = 0 \qquad 4C_y - 2(600) = 0$$

$$C_y = 300 \text{ N}$$

$$\sum F_y = 0 \qquad A_y - 600 + C_y = 0$$

$$\mathbf{A}_y = 300\mathbf{j} \text{ N} \qquad \textit{Answer}$$

Member BCD:

$$\sum M_B = 0 \qquad -3C_x + \frac{4}{5}(1000)(4) = 0$$

$$C_x = 1067 \text{ N}$$

$$\sum F_x = 0 \qquad B_x - C_x + \frac{4}{5}(1000) = 0$$

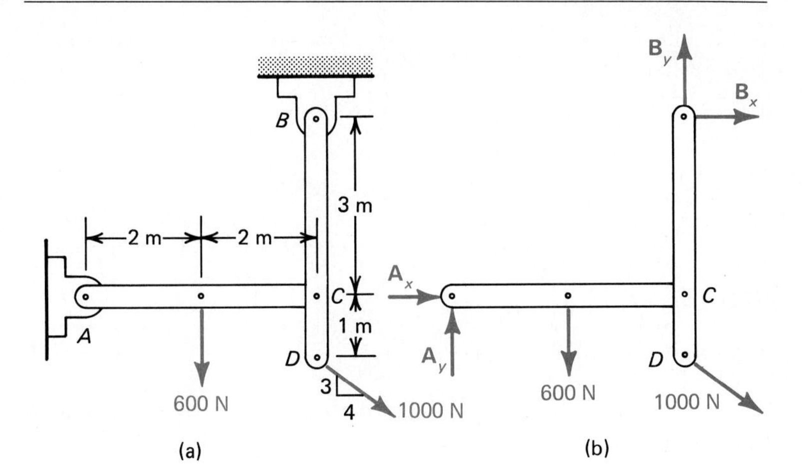

Figure 4.31

$$B_x = 1067 - 800 = 267 \text{ N}$$

$$\mathbf{B}_x = 267\mathbf{i} \text{ N} \qquad \textit{Answer}$$

$$\sum F_y = 0 \qquad B_y - C_y - \frac{3}{5}(1000) = 0$$

$$B_y = 900 \text{ N}$$

$$\mathbf{B}_y = 900\mathbf{j} \text{ N} \qquad \textit{Answer}$$

Member AC:

$$\sum F_x = 0 \qquad A_x + C_x = 0$$

$$\mathbf{A}_x = -1067\mathbf{i} \text{ N} \qquad \textit{Answer}$$

As a check on the calculations, consider the equilibrium of the entire frame shown in Fig. 4.31(b).

$$\sum F_x = 0 \qquad A_x + \frac{4}{5}(1000) + B_x = 0$$

$$-1067 + 800 + 267 = 0 \qquad \text{(Checks)}$$

$$\sum F_y = 0 \qquad A_y + B_y - 600 - \frac{3}{5}(1000) = 0$$

$$300 + 900 - 600 - 600 = 0 \qquad \text{(Checks)}$$

Example 4.15

Calculate Q and the reactions at A for the pliers shown in Fig. 4.32(a).

Consider a free-body diagram of a section of the pliers as shown in Fig. 4.32(b).

$$\sum M_A = 0 \qquad -2Q + 6(100) = 0$$

$$Q = 300 \text{ lb} \qquad \textit{Answer}$$

$$\sum F_x = 0 \qquad A_x = 0$$

$$\sum F_y = 0 \qquad Q + 100 + A_y = 0$$

$$A_y = -400$$

$$\therefore \mathbf{A}_y = -400\mathbf{j} \text{ lb} \qquad \textit{Answer}$$

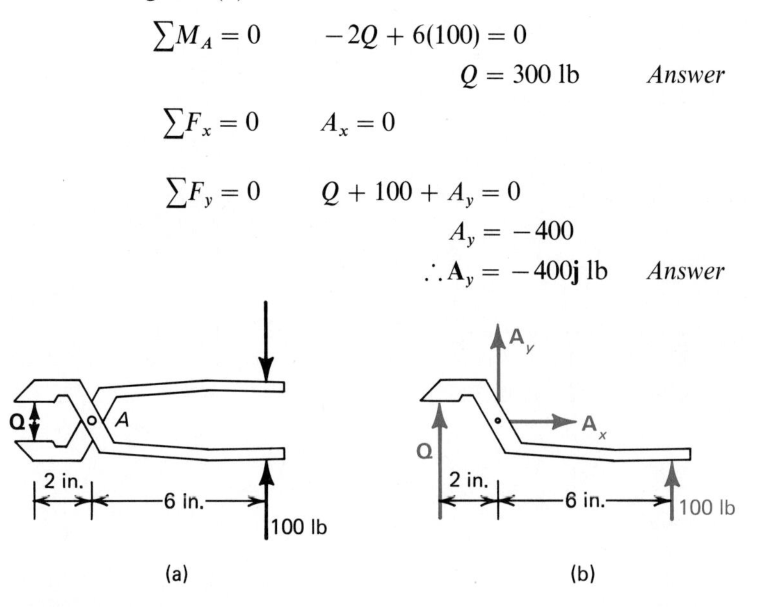

Figure 4.32

Example 4.16

Find the value of P needed to keep the crank-shape mechanism in equilibrium for the applied moment $\mathbf{M}$ shown in Fig. 4.33(a).

The free-body diagram for each member is shown in Figs. 4.33(b) and 4.33(c). Note that for this machine F_{CA} acts perpendicular to member BCD, since the pin at C slides along the grooved slot in the member BCD.

Member AC:

$$\sum M_A = 0$$

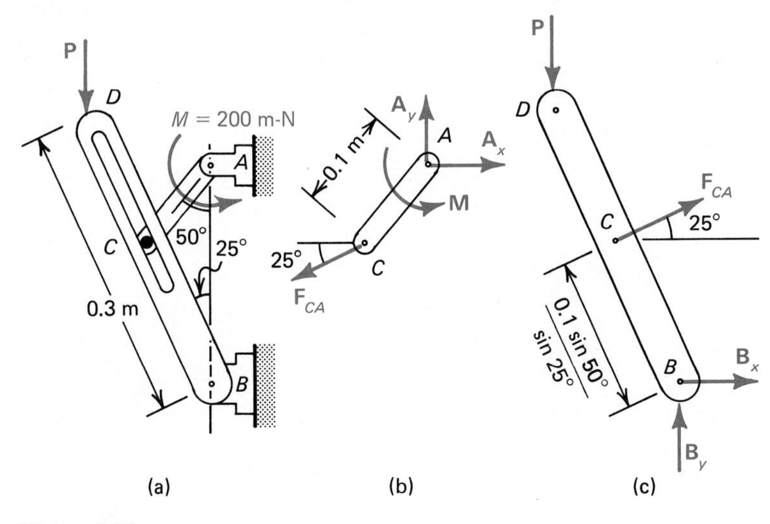

Figure 4.33

$$-(F_{CA} \cos 25°)(0.1 \cos 50°)$$
$$+ (F_{CA} \sin 25°)(0.1 \sin 50°) + M = 0$$
$$-0.1F_{CA} \cos 75° + M = 0$$
$$F_{CA} = 7727 \text{ N}$$

Member BCD:

$$\sum M_B = 0$$

$$P(0.3 \sin 25°) - \frac{0.1 \sin 50°}{\sin 25°} F_{CA} = 0$$

$$P = 11{,}050 \text{ N} \quad \textit{Answer}$$

PROBLEMS

4.67 Obtain the reactions at A and the tension in the cable FC for the frame of Fig. P4.67.

4.68 Obtain the magnitude of the forces in members DE and BE of the frame in Problem 4.67. Use the forces acting on member AD to check your answer.

4.69 Find the magnitude of the forces acting on member AB in Fig. P4.69.

4.70 Find the magnitude of the forces acting on member AD in Fig. P4.70 if $\theta = 45°$.

4.71 Find the magnitude of the force in member AC for the frame in Problem 4.69.

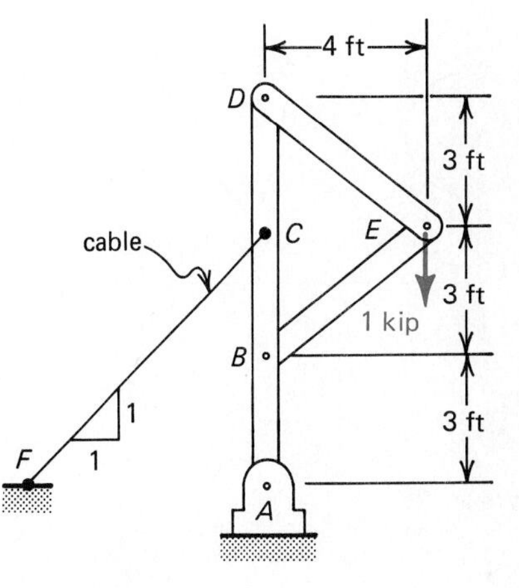

Figure P4.67

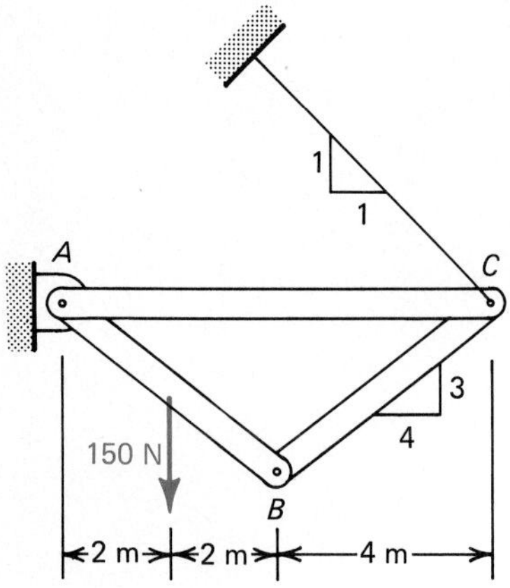

Figure P4.69

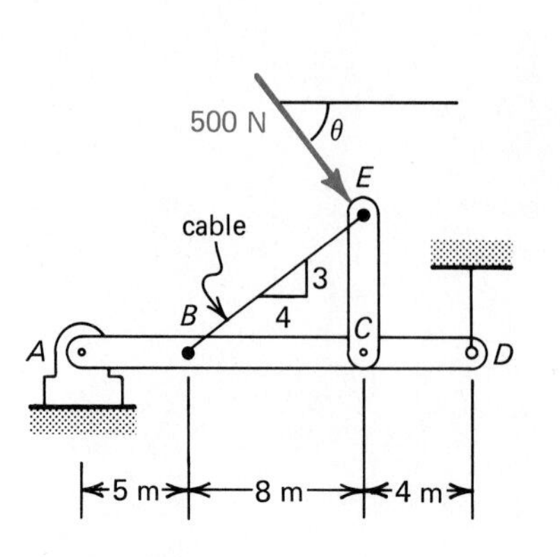

Figure P4.70

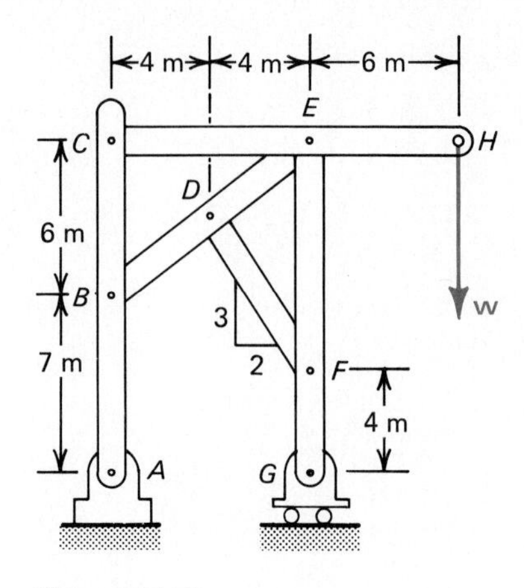

Figure P4.72

4.72 For the structure in Fig. P4.72, what is the maximum allowable load w if member EG can support a maximum compressive force of 280 N?

4.73 If $w = 4$ kN in Problem 4.72 and neglecting the restriction on member EG, find the magnitude of the forces acting on member EG.

4.74 Find the magnitude of the forces acting on member EC in Problem 4.70 if $\theta = 60°$.

4.75 Calculate the magnitude of the forces acting on member GE in Fig. P4.75 if $w = 1$ kip.

4.76 What is the magnitude of the forces acting on member ACD of the frame shown in Fig. P4.76?

4.77 What is the magnitude of the forces acting on member FC of the frame shown in Fig. P4.77?

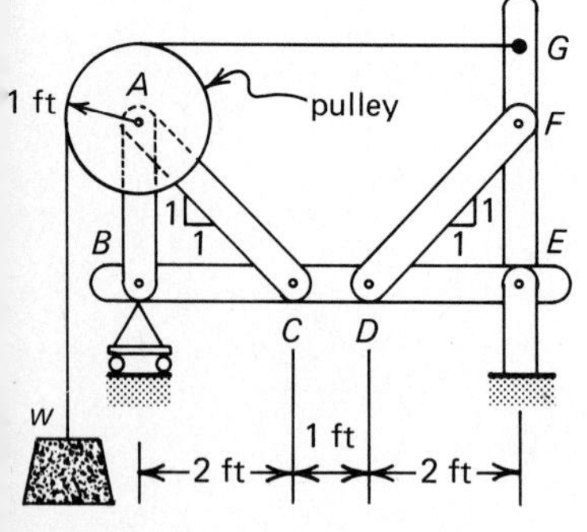

Figure P4.75

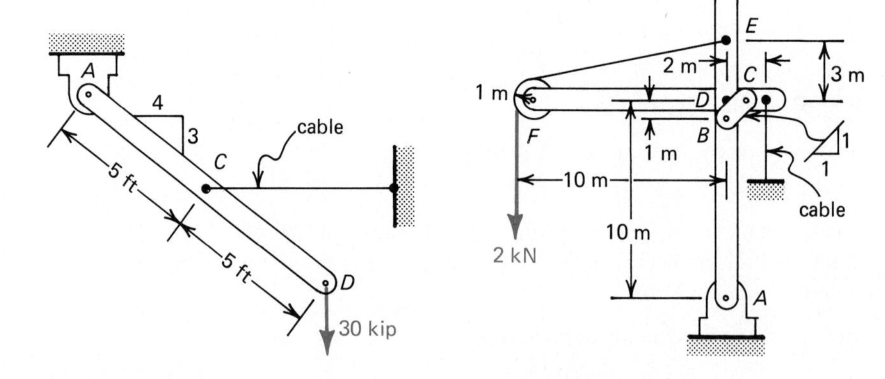

Figure P4.76

Figure P4.77

4.78 What is the magnitude of the forces acting on the member AE of the frame in Problem 4.77?

4.79 Obtain the reactions at A and B for the frame shown in Fig. P4.79. Assume $\theta = 30°$.

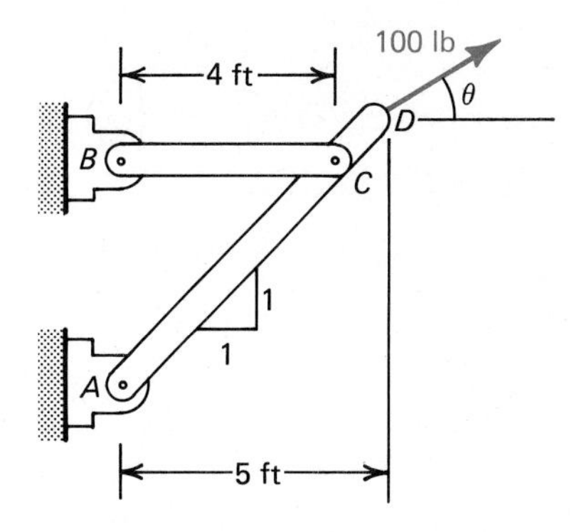

Figure P4.79

4.80 If $\theta = 45°$, calculate the reactions at A and B for the frame in Problem 4.79.

4.81 Calculate the magnitude of all the forces on member $ABCD$ for the frame shown in Fig. P4.81.

4.82 Obtain the magnitude of the forces in member BGE for the frame in Problem 4.81.

4.83 Find the reactions at A and B of the frame shown in Fig. P4.83.

4.84 If the wire in Fig. P4.84 breaks under a tension of 150 N, what is the maximum load P that the frame can carry?

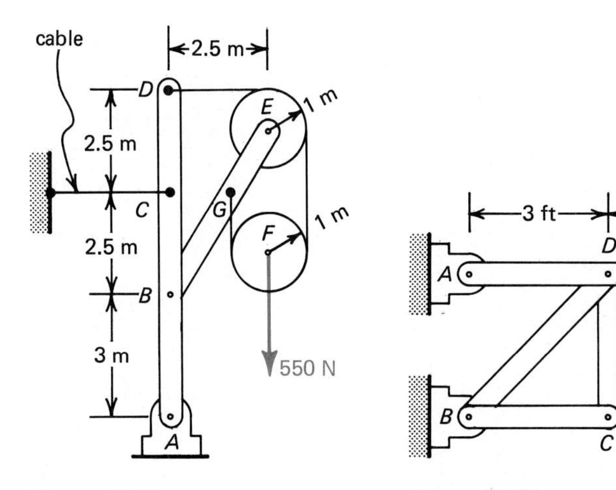

Figure P4.81

Figure P4.83

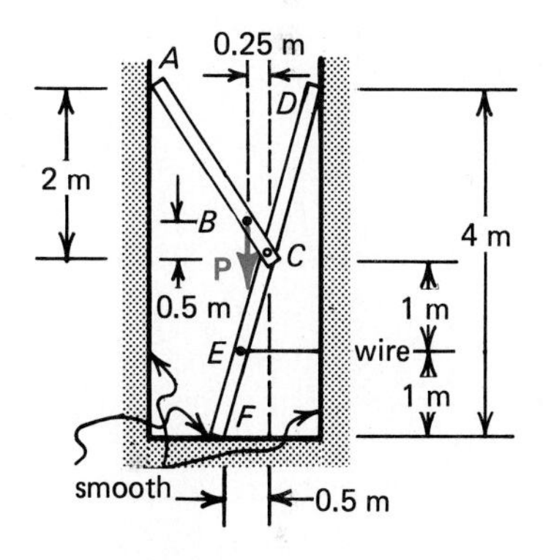

Figure P4.84

4.85 If the load P in Problem 4.84 is 100 N, what is the tension in the wire?

4.86 Calculate the reactions at A and B for the frame shown in Fig. P4.86.

4.87 Compute the magnitude of all the forces that are acting on the member BCD of the frame in Problem 4.86.

4.88 If the wire in Problem 4.84 breaks under a tension of 80 N, what is the maximum load P?

4.89 What forces are exerted on the bolt at A by the tongs in Fig. P4.89?

4.90 The wire cutter in Fig. P4.90 will cut the wire if a 150-N force is applied to the wire. What is P if the wire is at A?

4.91 Redo Problem 4.90 if the wire is at B.

4.92 If $P = 1$ kip in Fig. P4.92, what is the magnitude of the moment M on the crank arm for equilibrium?

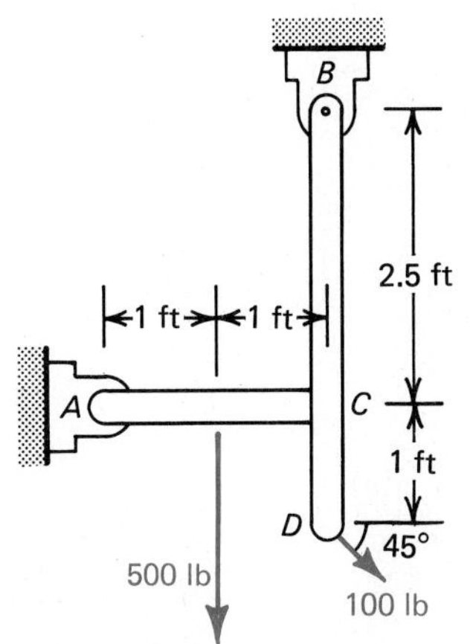

Figure P4.86

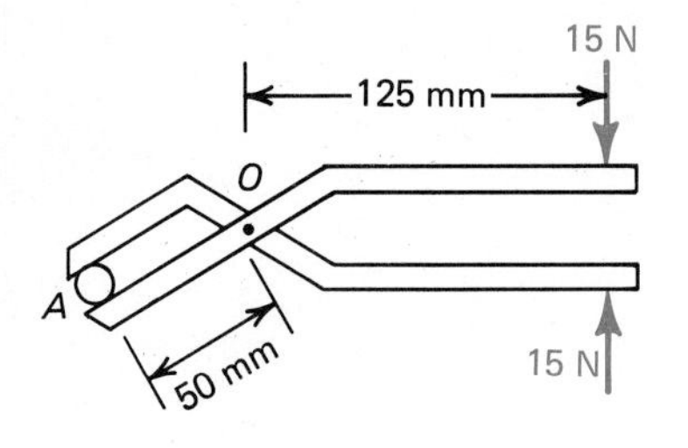

Figure P4.89

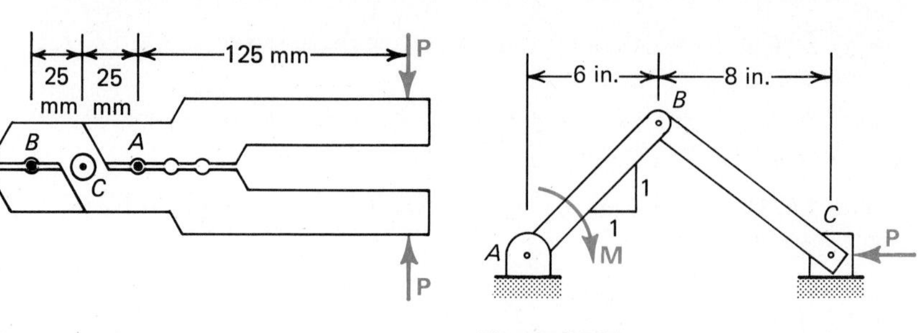

Figure P4.90

Figure P4.92

4.93 If $M_B = 80$ m·N in Fig. P4.93, what is the value of M_A for equilibrium?

4.94 For the gear arrangement in Fig. P4.94, $\mathbf{M} = 50\mathbf{i}$ m·N. What is the moment $\mathbf{M}_R$ for equilibrium?

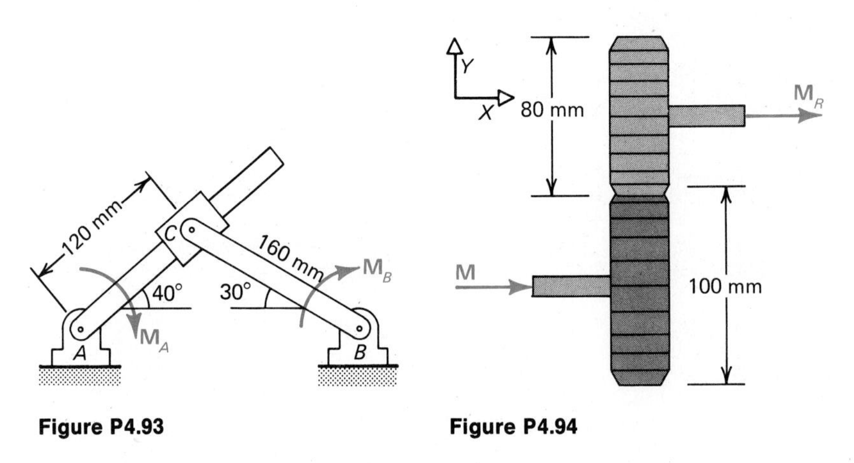

Figure P4.93

Figure P4.94

4.7 Summary

A structure consists of a series of connected rigid bodies. The equilibrium of a structure requires that the resultant forces and moments acting on the entire structure, as well as any portion of the structure, be zero. Several methods were discussed for analyzing the equilibrium of different types of structures and for obtaining the internal forces acting on structural members.

A. The important *ideas*:

1. Plane trusses and space trusses are formed by two-force members, that is, each member is acted on by a pair of forces in tension or compression. These forces are equal in magnitude, collinear, and act in opposite directions.

2. An internal force that acts on a truss member also acts on the joint but in the opposite direction.
3. The method of joints provides a step-by-step approach for finding the internal forces of a truss.
4. The method of sections is particularly useful for finding internal forces of selected members of a truss.
5. Frames and machines include multiforce members that are characterized by the action of three or more forces.

B. The important *equations*:

Plane Truss

Method of joints: $\sum F_x = 0 \qquad \sum F_y = 0$

Method of sections: $\sum F_x = 0 \qquad \sum F_y = 0 \qquad \sum M = 0$

Space truss: $\sum F_x = 0 \qquad \sum F_y = 0 \qquad \sum F_z = 0$

$\sum M_x = 0 \qquad \sum M_y = 0 \qquad \sum M_z = 0$

Frames and machines: $\sum F_x = 0 \qquad \sum F_y = 0 \qquad \sum M = 0$

5 Distributed Forces

We have represented a force as a vector that has magnitude, direction, and a point of application, i.e., a force appears as a load applied at a point. In reality, most forces are applied over a specified area rather than at a single point. The assumption of a point loading is valid in those cases where the area over which the force is applied is small in comparison with the size of the body. If, however, the area of loading is large relative to the body, the forces are called distributed forces and their distribution must be included in the analysis.

There is a distinction between push-pull forces and field forces such as the gravitational force. The quantity that is called the "weight" of a body is a gravitational force. Since gravity applies a force to each particle of a body, its weight should be represented by many small forces distributed over the entire body. It will be seen, however, that these distributed forces may be reduced to a single force, that is, the weight acts through a unique point in the body that is called the center of gravity.

In this chapter we introduce the center of gravity, centroids, and some theorems useful for the treatment of distributed forces. Applications of distributed forces include fluid pressures, buoyancy, and cables.

5.1 Center of Gravity

Consider a thin, flat plate of small thickness t as shown in Fig. 5.1(a). Let the plate be divided into n small elements so that the weight of the ith element is Δw_i located at the coordinate point (x_i, y_i) as shown in the figure. The total weight of the plate is

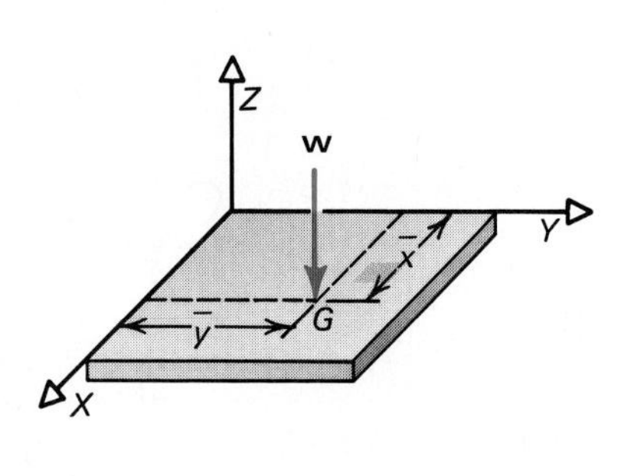

(b)

Figure 5.1

$$w = \sum_{i=1}^{n} \Delta w_i$$

The magnitude of the moment about the X axis produced by all the elements of the weight is

$$M_x = y_1 \Delta w_1 + y_2 \Delta w_2 + \cdots + y_n \Delta w_n$$
$$= \sum_{i=1}^{n} y_i \Delta w_i$$

If we let the size of the largest element approach zero and the number of elements approach infinity, we have the magnitude of the moment in integral form, i.e.,

$$M_x = \int_v y\,dw \tag{5.1a}$$

where the v means that the integration is taken over the entire volume of the plate. This is an example of a multiple integral from calculus. Note that Eq. (5.1a) can be expressed in differential form as

$$dM_x = y\,dw$$

In a similar manner we can represent the magnitude of the moment about the Y axis due to the weight of the plate as

$$M_y = \int_v x\,dw \tag{5.1b}$$

Let us select some point G with coordinates $\bar{x}$ and $\bar{y}$ and represent the total weight w of the plate at G by a force applied perpendicular to the plate, as seen in Fig. 5.1(b). Now, the total weight of the plate in integral form is

$$w = \int_v dw = \int_v \gamma\,dV \tag{5.2}$$

where γ is the specific weight of the body in N/m^3 or lb/ft^3. We select point G so that the magnitudes of the moments caused by w about the X and Y axes, namely

$$M_x = w\bar{y} \qquad M_y = w\bar{x} \tag{5.3}$$

are identical to the moments given by Eqs. (5.1). Equating the right sides of Eqs. (5.1) and (5.3) and rearranging, we obtain

$$M_x = w\bar{y} = \int_v y\,dw \tag{5.4a}$$

$$M_y = w\bar{x} = \int_v x\,dw \tag{5.4b}$$

which yields

$$\bar{x} = \frac{1}{W}\int_v x\,dw \tag{5.5a}$$

$$\bar{y} = \frac{1}{W}\int_v y\,dw \tag{5.5b}$$

The point G is called the **center of gravity** of the body and its coordinates are obtained from Eqs. (5.5a) and (5.5b). We have assumed that the plate thickness t was small. If, however, the body is not uniformly thin, it can be shown that the center of gravity along the Z coordinate is given by

$$\bar{z} = \frac{1}{W}\int_v z\,dw \tag{5.5c}$$

5.2 Centroids

Equations (5.5) are simplified if the body is homogeneous so that γ is constant throughout. For example, the $\bar{x}$ coordinate becomes

$$\bar{x} = \frac{1}{\gamma V}\int_v x\gamma\,dV$$

so that γ cancels in the numerator and denominator, giving

$$\bar{x} = \frac{1}{V}\int_v x\,dV$$

Thus, the location of the center of gravity for a homogeneous body is given by

$$\bar{x} = \frac{1}{V}\int_v x\,dV \qquad \bar{y} = \frac{1}{V}\int_v y\,dV \qquad \bar{z} = \frac{1}{V}\int_v z\,dV \tag{5.6}$$

The coordinates $\bar{x}, \bar{y}, \bar{z}$ that are dependent only on the geometry of the body locate the **centroid** of the body. Note that the centroid of a body is in the same position as the center of gravity if the body is homogeneous. The integrals in Eq. (5.6) are also called the **first moments** about the X, Y, Z axes.

Centroid of Lines

Let us assume that we have a homogeneous slender rod or curved wire of length L in the XY plane as shown in Fig. 5.2. Let A be the uniform cross-sectional area of the wire and dL be the differential length, so that the differential volume is given by

$$dV = A\,dL \tag{5.7}$$

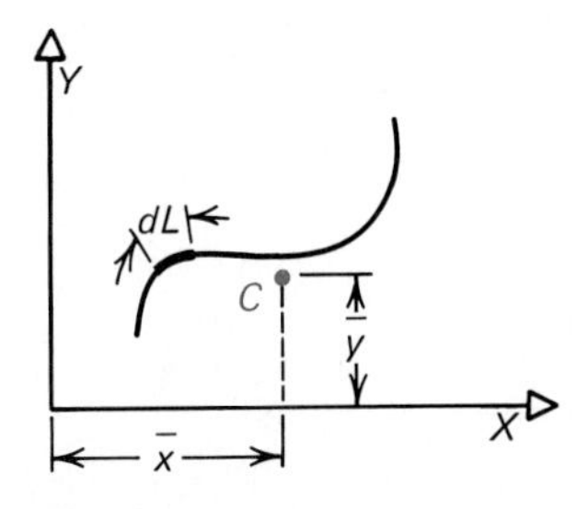

Figure 5.2

Consequently, the total volume is

$$V = AL \tag{5.8}$$

Now substitute Eqs. (5.7) and (5.8) into the first two expressions in Eq. (5.6) to obtain the centroid of a line in a plane.

$$\bar{x} = \frac{1}{L}\int x\,dL \qquad \text{and} \qquad \bar{y} = \frac{1}{L}\int y\,dL \tag{5.9}$$

Depending upon the shape of the wire, the centroid may or may not be on the wire itself. Since the wire is homogeneous, the center of gravity C coincides with the centroid. If the origin of the XY axes is selected or moved to coincide with the centroid C as shown in Fig. 5.3, the axes are called **centroidal axes.** Note that in this case, if we compute the centroids using the centroidal axes as in Fig. 5.3, we obtain a value of zero for both $\bar{x}$ and $\bar{y}$. Table 5.1 lists the centroids for common geometric shapes of lines in a plane.

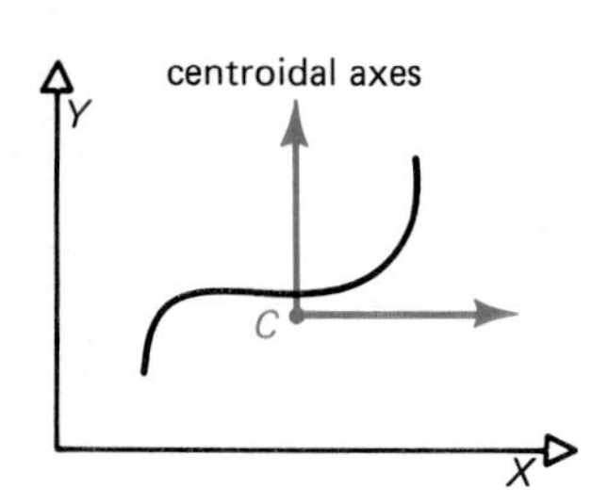

Figure 5.3

Centroid of Areas

We will now consider a homogeneous thin plate of uniform thickness t. Let dA be the differential area of the plate in the XY plane so that the differential element of volume is

$$dV = t\,dA \tag{5.10}$$

The total volume is

$$V = tA \tag{5.11}$$

where A is the area of the plate shown in Fig. 5.4. We can now substitute Eqs. (5.10) and (5.11) into the first two expressions in Eq. (5.6) to obtain

$$\bar{x} = \frac{1}{A}\int x\,dA \qquad \text{and} \qquad \bar{y} = \frac{1}{A}\int y\,dA \tag{5.12}$$

Here, $\bar{x}$ and $\bar{y}$ are the coordinates of the centroid C of the area as shown in Fig. 5.4. The integrals in Eq. (5.12) are called **first moments of area** about the X and Y axes.

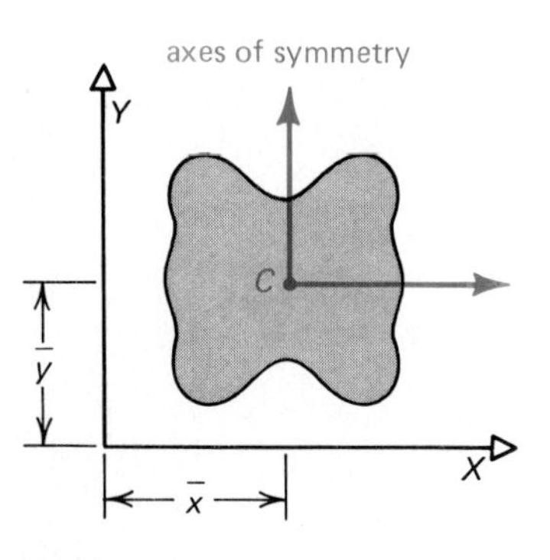

Figure 5.4

We note that in Fig. 5.4 we have identified the centroidal axes as the axes of symmetry. For the centroidal axes to qualify as axes of symmetry, the area of the thin plate must be equally distributed on both sides of each axis. If two axes of symmetry are perpendicular to each other, they locate the centroid of an area at the point of their intersection. This observation enables us to locate centroids of regular shapes such as rectangles, circles, ellipses, etc., in a straightforward manner.

Since we are considering homogeneous cases, the weight on

TABLE 5.1 ***Centroids for Lines in a Plane***

		Centroid	Length
Straight line	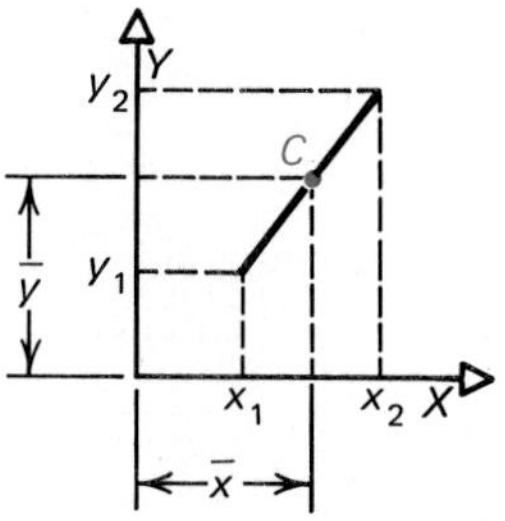	$\bar{x} = \dfrac{x_1 + x_2}{2}$ $\bar{y} = \dfrac{y_1 + y_2}{2}$	$\sqrt{(x_2 - x_1)^2 + (y_2 - y_1)^2}$
Quarter-circular arc	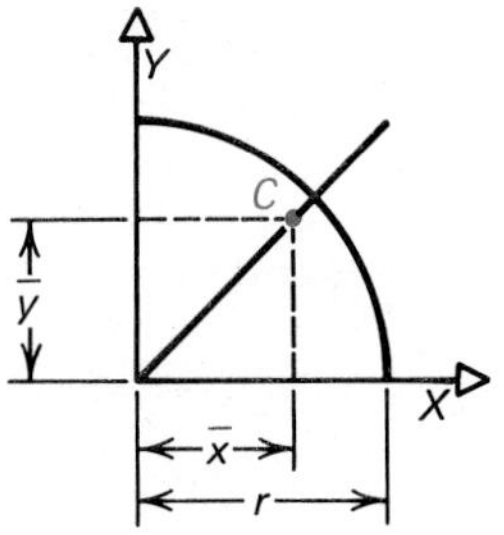	$\bar{x} = \dfrac{2r}{\pi}$ $\quad \bar{y} = \dfrac{2r}{\pi}$	$\dfrac{\pi r}{2}$
Semicircular arc	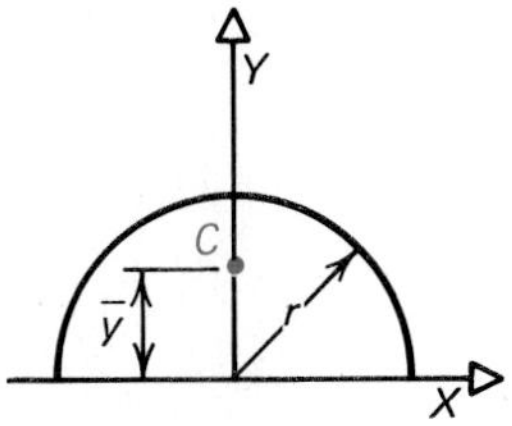	$\bar{x} = 0$ $\quad \bar{y} = \dfrac{2r}{\pi}$	πr
Arc of circle	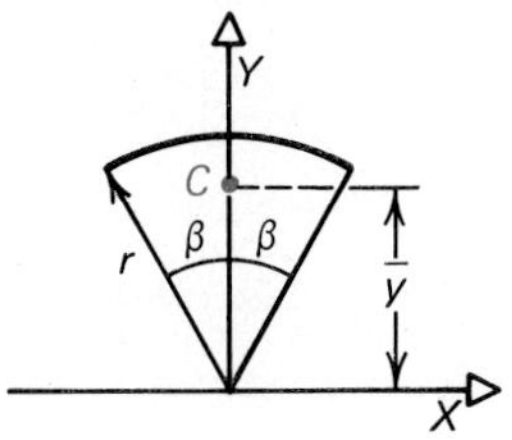	$\bar{x} = 0$ $\quad \bar{y} = \dfrac{r \sin \beta}{\beta}$	$2\beta r$

one side of an axis of symmetry equals that on the other side. This is why the centroid and the center of gravity coincide. Clearly, if we do not have homogeneity, the center of gravity will be different from the centroid. The centroids for a variety of areas are shown in Table 5.2.

Centroid of Volumes

The location of the centroid of a volume is given by Eq. (5.6) and is repeated here for convenience.

$$\bar{x} = \frac{1}{V}\int_v x\,dV \qquad \bar{y} = \frac{1}{V}\int_v y\,dV \qquad \bar{z} = \frac{1}{V}\int_v z\,dV \tag{5.6}$$

Table 5.3 lists the centroids for some common shapes.

Example 5.1

Find the centroid of the line in Fig. 5.5 from point A to point B.

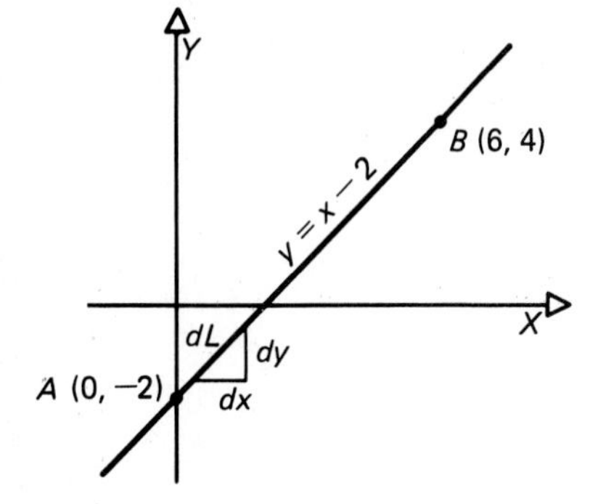

Figure 5.5

The differential element dL shown in Fig. 5.5 is

$$dL = \sqrt{dx^2 + dy^2} = dx\sqrt{1 + (dy/dx)^2}$$

Since the equation of the line is $y = x - 2$,

$$\frac{dy}{dx} = 1$$

and therefore

$$dL = \sqrt{2}\,dx$$

Integrating from point A to point B,

$$L = \sqrt{2}\int_0^6 dx = 6\sqrt{2}$$

From Eqs. (5.9) we obtain

$$\bar{x} = \frac{1}{L}\int x\,dL = \frac{1}{L}\int_0^6 x(\sqrt{2}\,dx)$$

$$= \frac{1}{6}\left[\frac{x^2}{2}\right]_0^6 = 3 \qquad \textit{Answer}$$

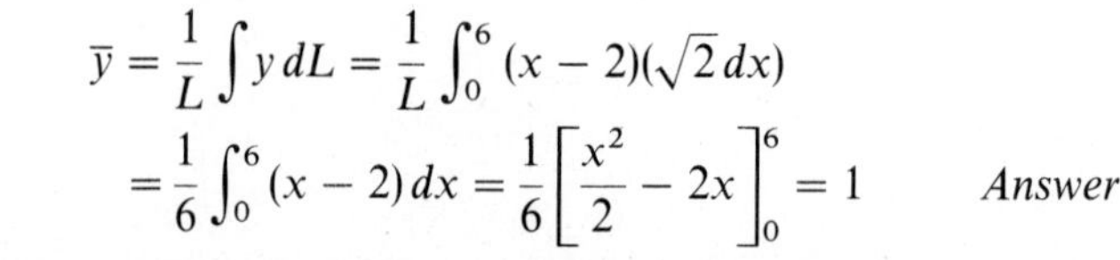

and $$\bar{y} = \frac{1}{L}\int y\,dL = \frac{1}{L}\int_0^6 (x-2)(\sqrt{2}\,dx)$$

$$= \frac{1}{6}\int_0^6 (x-2)\,dx = \frac{1}{6}\left[\frac{x^2}{2} - 2x\right]_0^6 = 1 \qquad \textit{Answer}$$

We note that the centroid for a straight line is coincident with the midpoint of the X and Y projections as shown in Table 5.1. As an

TABLE 5.2 *Centroid for Areas*

	Centroid		Area
Rectangle	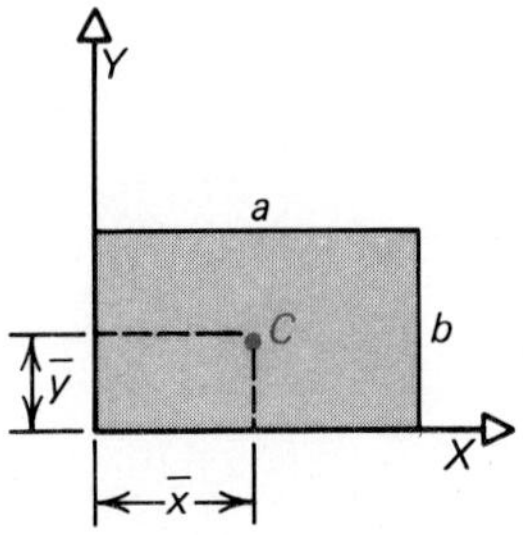 $\bar{x} = \dfrac{a}{2}$	$\bar{y} = \dfrac{b}{2}$	ab
Triangle	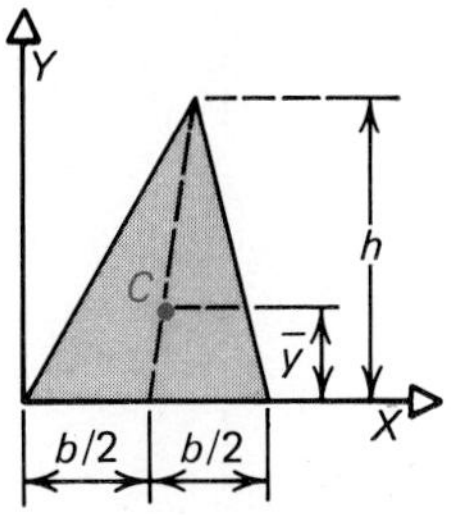 $\bar{y} = \dfrac{h}{3}$		$\dfrac{bh}{2}$
Circular sector	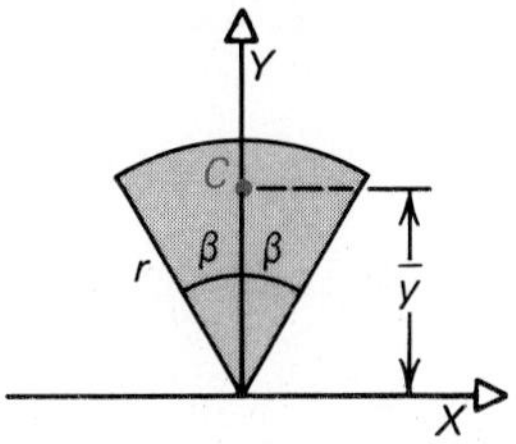 $\bar{x} = 0$	$\bar{y} = \dfrac{2r\sin\beta}{3\beta}$	βr^2
Semicircle	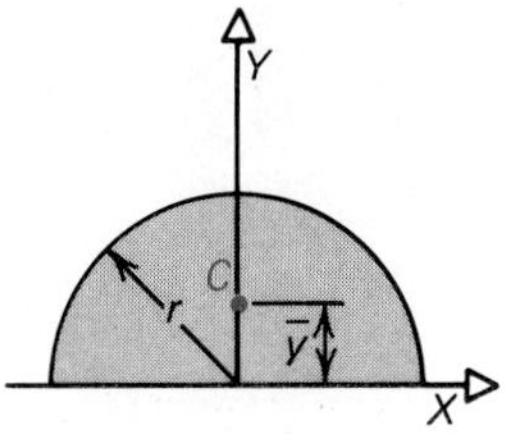 $\bar{x} = 0$	$\bar{y} = \dfrac{4r}{3\pi}$	$\dfrac{\pi r^2}{2}$
Quarter-circular arc	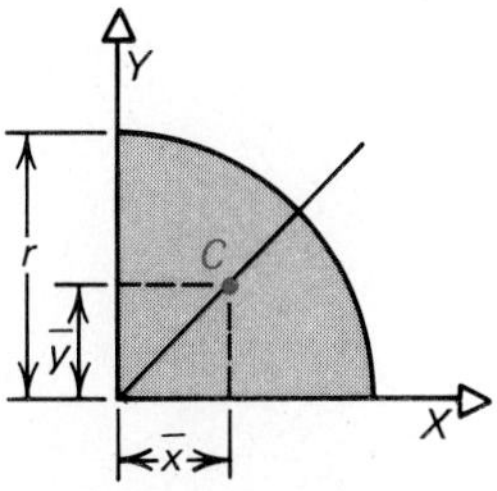 $\bar{x} = \dfrac{4r}{3\pi}$	$\bar{y} = \dfrac{4r}{3\pi}$	$\dfrac{\pi r^2}{4}$

TABLE 5.3 *Centroid for Volumes*

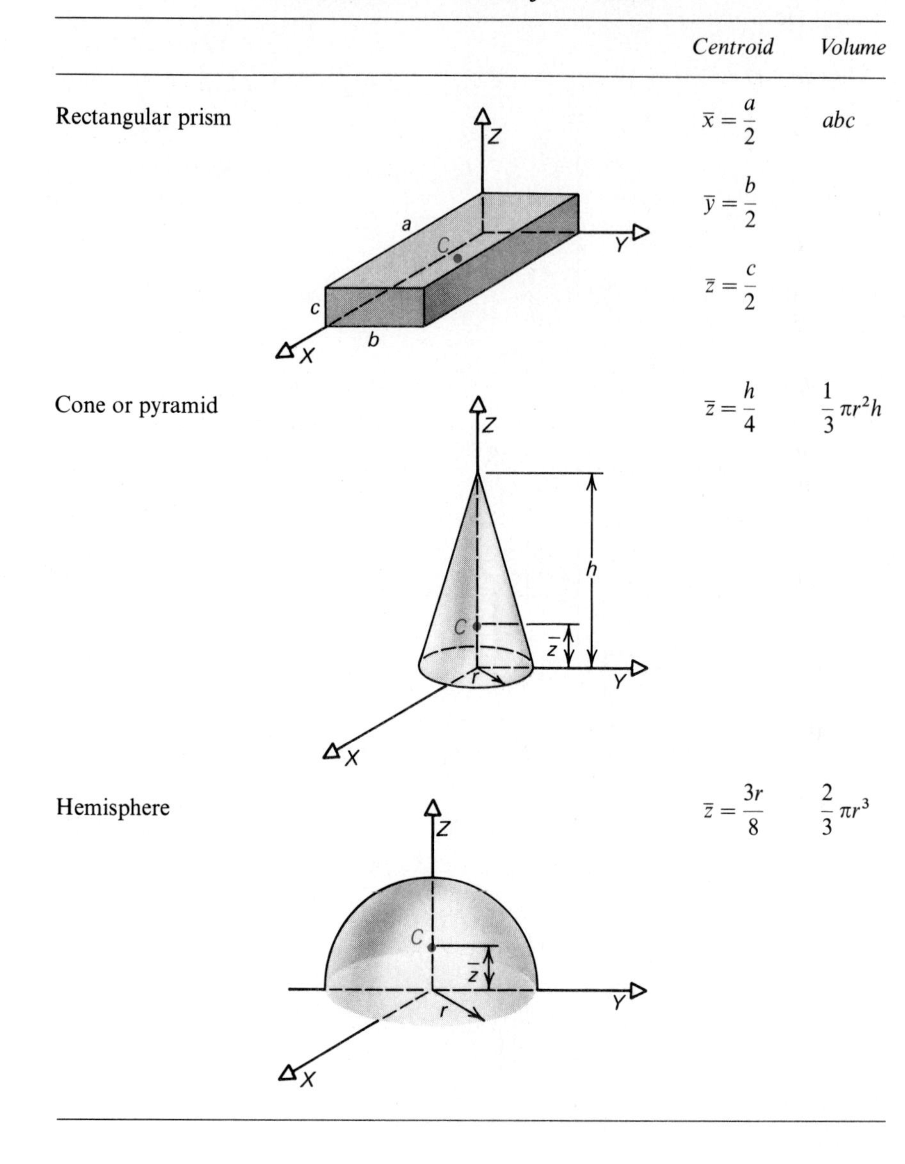

	Centroid	Volume
Rectangular prism	$\bar{x} = \frac{a}{2}$ $\bar{y} = \frac{b}{2}$ $\bar{z} = \frac{c}{2}$	abc
Cone or pyramid	$\bar{z} = \frac{h}{4}$	$\frac{1}{3}\pi r^2 h$
Hemisphere	$\bar{z} = \frac{3r}{8}$	$\frac{2}{3}\pi r^3$

alternate solution, we have $x_1 = 0$, $y_1 = -2$, $x_2 = 6$, and $y_2 = 4$. Using the equations in Table 5.1,

$$\bar{x} = \frac{x_1 + x_2}{2} = \frac{6}{2} = 3 \qquad \textit{Answer}$$

$$\bar{y} = \frac{y_1 + y_2}{2} = \frac{-2 + 4}{2} = 1 \qquad \textit{Answer}$$

Example 5.2

Locate the centroid for the triangle of base b and height h as shown in Fig. 5.6.

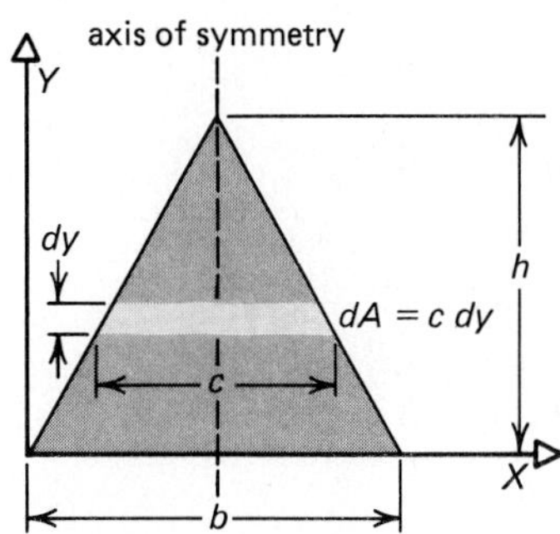

Figure 5.6

We notice that the triangle is symmetric about an axis parallel to the Y axis as shown in Fig. 5.6. This means that the centroid lies along this axis of symmetry and $\bar{x} = b/2$. To find $\bar{y}$, select the element of area

$$dA = c\,dy$$

where, by similar triangles,

$$\frac{c}{b} = \frac{h - y}{h}$$

so that

$$c = \frac{b}{h}(h - y)$$

Then, from Eq. (5.12), we find

$$\bar{y} = \frac{1}{A}\int y\,dA = \frac{2}{bh}\int yc\,dy$$

$$= \frac{2}{bh}\int_0^h \frac{b}{h}(h - y)y\,dy$$

$$= \frac{2}{h^2}\int_0^h (hy - y^2)\,dy$$

$$= \frac{2}{h^2}\left[\frac{hy^2}{2} - \frac{y^3}{3}\right]_0^h = \frac{h}{3} \qquad \textit{Answer}$$

Example 5.3

Find the centroid of an area bounded by the curve $y = 4 - x^2$, the Y axis, and the X axis as shown in Fig. 5.7(a).

Let the element of the area be as shown in Fig. 5.7(a), where

$$dA = y\,dx = (4 - x^2)\,dx$$

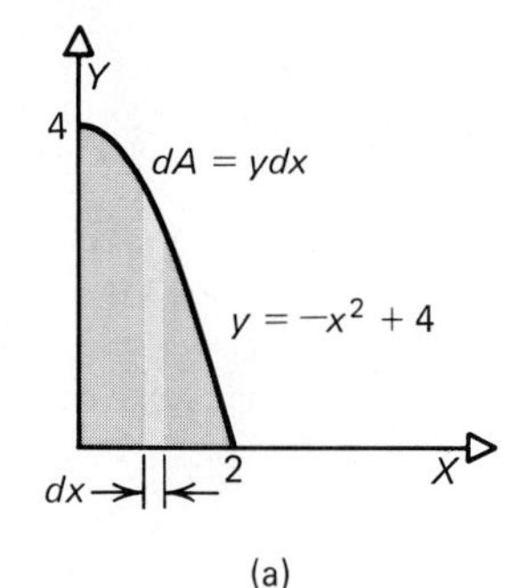

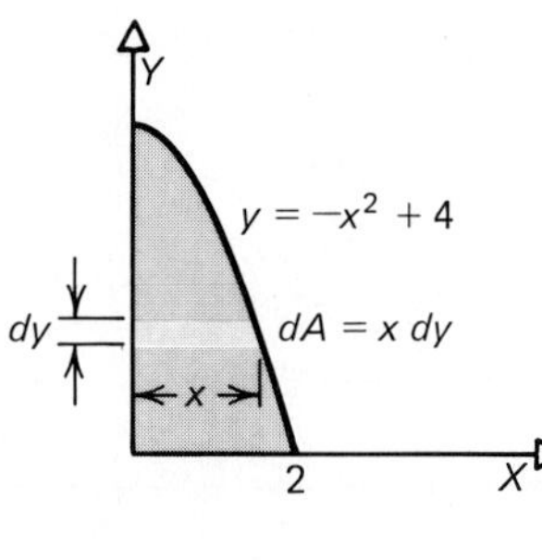
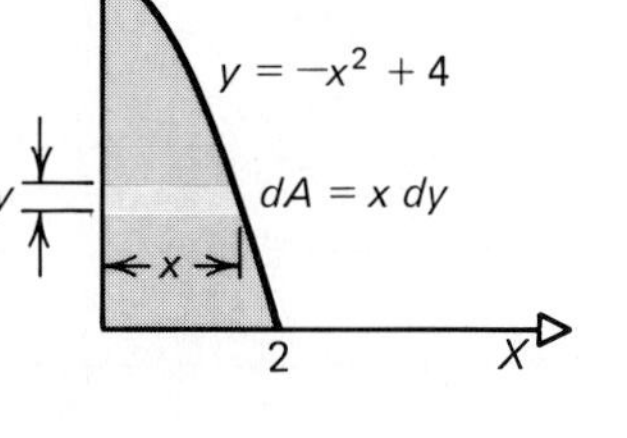

Figure 5.7

$$A = \int_0^2 (4 - x^2)\,dx = \left[4x - \frac{x^3}{3}\right]_0^2 = 5.33$$

The x centroid can now be computed from the first expression in Eq. (5.12).

$$\bar{x} = \frac{1}{A}\int x\,dA = \frac{1}{5.33}\int_0^2 (4x - x^3)\,dx$$

$$= \frac{1}{5.33}\left[2x^2 - \frac{x^4}{4}\right]_0^2 = 0.75 \qquad \textit{Answer}$$

In computing the y centroid we note that the distance from the X axis in Fig. 5.7(a) to the centroid of the differential element dA is $y/2$, so that the second expression in Eq. (5.12) is modified as

$$\bar{y} = \frac{1}{A}\int \frac{y}{2}\,dA = \frac{1}{5.33}\int_0^2 \frac{1}{2}(4 - x^2)^2\,dx$$

$$= \frac{1}{5.33}\left[8x + \frac{x^5}{10} - \frac{4x^3}{3}\right]_0^2 = 1.60$$

As an alternate approach the y centroid can also be obtained by the aid of Fig. 5.7(b), where the differential element is now parallel with the X axis and

$$dA = x\,dy = \sqrt{4 - y}\,dy$$

Now we observe that the distance between the X axis and the differential element dA is y, so that

$$\bar{y} = \frac{1}{A}\int y\,dA = \frac{1}{A}\int y\sqrt{4 - y}\,dy$$

$$= \frac{1}{5.33}\int_0^4 y\sqrt{4 - y}\,dy$$

$$= \frac{1}{5.33}\left[\frac{-2(8 + 3y)(4 - y)^{3/2}}{15}\right]_0^4 = 1.60 \qquad \textit{Answer}$$

The important observation to be made here is that we must be careful when we apply Eq. (5.12) for a given differential element dA. Remember to use the centroidal location of dA for x and y in these equations.

Example 5.4

Find the centroid for the inverted circular cone shown in Fig. 5.8.

Since the Z axis is an axis of symmetry

$$\bar{x} = \bar{y} = 0 \qquad \textit{Answer}$$

To find $\bar{z}$, consider the element of volume as a circular disk of radius r and thickness dz. The element of volume is

$$dV = \pi r^2\, dz$$

From the geometry we can obtain r in terms of z. That is, by similar triangles,

$$\frac{r}{z} = \frac{2}{8} \qquad r = \frac{1}{4} z$$

Therefore

$$V = \int \pi r^2\, dz = \frac{\pi}{16} \int_0^8 z^2\, dz$$

$$= \frac{\pi}{16}\left(\frac{8^3}{3}\right) = 33.5 \text{ m}^3$$

Now

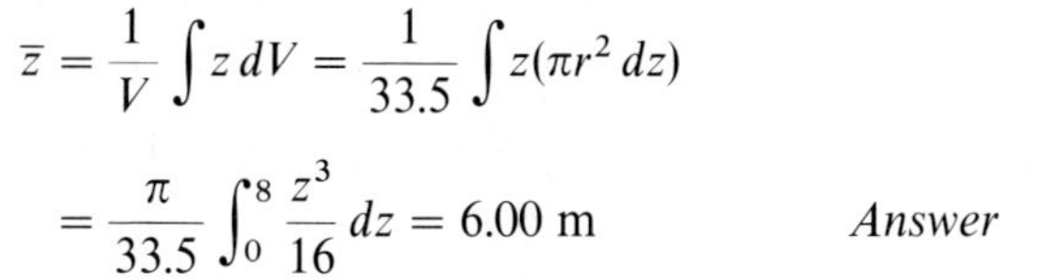

$$\bar{z} = \frac{1}{V} \int z\, dV = \frac{1}{33.5} \int z(\pi r^2\, dz)$$

$$= \frac{\pi}{33.5} \int_0^8 \frac{z^3}{16}\, dz = 6.00 \text{ m} \qquad \textit{Answer}$$

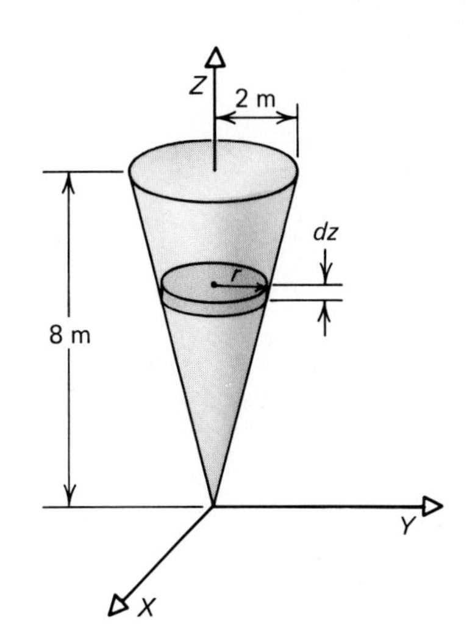

Figure 5.8

Note that this checks with the data in Table 5.3, where $\bar{z} = h/4 = 2$ m from the base of the cone.

PROBLEMS

5.1 Verify the centroid location of the quarter-circular arc shown in Table 5.1.

5.2 Verify the centroid location of the semicircular arc shown in Table 5.1.

5.3 Verify the centroid location of the arc of the circle shown in Table 5.1.

5.4 Find the centroid of the line in Fig. P5.4 from point A to point B.

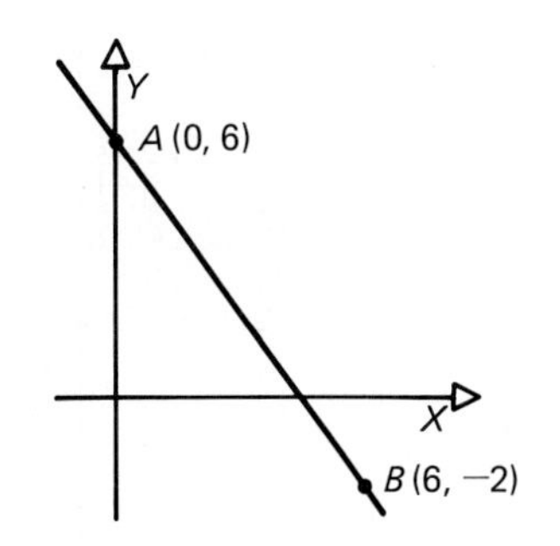

Figure P5.4

5.5 Obtain the centroid of the line $y = 2x + 1$ between $x = 0$ and $x = 2$ mm.

5.6 Obtain the centroid of the line $y = 3x + 2$ between $x = 0$ and $x = 10$ mm.

5.7 Find the centroid of the line $y = |x|$ between $x = -2$ in and $x = 2$ in.

5.8 Repeat Problem 5.7 between $x = -2$ mm and $x = 4$ mm.

5.9 Verify the centroid location of the area enclosed by a circular sector shown in Table 5.2.

5.10 Verify the centroid location of the area enclosed by a semicircle shown in Table 5.2.

5.11 Verify the centroid location of the area enclosed by the quarter-circular arc shown in Table 5.2.

5.12 Find the centroid for the area bounded by $x = 4$ mm, $y = x$, and $y = 0$.

5.13 Find the centroid for the area bounded by $y = 6$ in and $y = |x|$.

5.14 Find the centroid for the area bounded by $y = 0.5x$ and $y = x(x - 4)$.

5.15 Obtain the centroid for the area bounded by $y = 5 \sin x$, $x = 0$, and $x = \pi/2$.

5.16 Find the centroid of the area bounded by $y = 2x^2$, $y = 0$ and $x = 3$.

5.17 Find the centroid of the area bounded by $y = 4x^3$, $y = 0$, and $x = 4$.

5.18 Derive an expression for the centroid of the area bounded by $y = kx^n$, $y = 0$, $x = a$, and $n \geq 0$.

5.19 Verify the centroid location for the cone or pyramid shown in Table 5.3.

5.20 Verify the centroid location for the hemisphere shown in Table 5.3.

5.21 Obtain the centroid of a pyramid of height h meters and rectangular base a meters by b meters. Assume the origin of the X, Y, Z system located on the base directly under the apex of the pyramid.

5.22 Calculate the centroid of a pyramid of height 100 m and a rectangular base 20 m by 30 m.

5.23 Calculate the centroid of a hemisphere of radius 0.5 ft.

5.24 Calculate the centroid of a cone having a height of 0.5 ft and base radius of 0.2 ft.

5.3 Composite Bodies

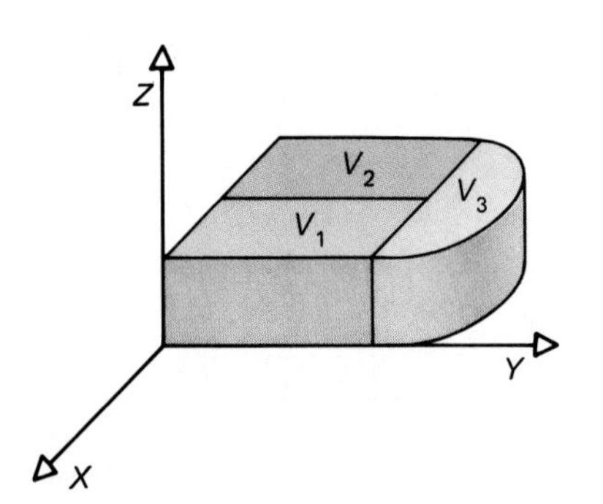

Figure 5.9

Often a complicated body can be represented as a combination of several common geometric shapes. When this is done, it is possible to obtain the centroid of thc total body by making use of the known centroids of the simpler shapes. Consider the body shown in Fig. 5.9 where the total volume is broken into three separate volumes V_1, V_2, and V_3. Furthermore, assume a different specific weight for each separate volume, namely γ_1, γ_2, and γ_3. If the total weight is concentrated at the centroid of the body, the magnitude of the moment about the Y axis is

$$M_y = \bar{x}(\gamma_1 V_1 + \gamma_2 V_2 + \gamma_3 V_3) \tag{5.13}$$

This moment must equal the sum of the moments produced by the individual volumes, namely

$$M_y = \bar{x}_1 \gamma_1 V_1 + \bar{x}_2 \gamma_2 V_2 + \bar{x}_3 \gamma_3 V_3 \tag{5.14}$$

where $\bar{x}_1$, $\bar{x}_2$, and $\bar{x}_3$ are the x centroids of V_1, V_2, and V_3, respectively. Equating Eqs. (5.13) and (5.14) and solving for the centroid of the entire body,

$$\bar{x} = \frac{\bar{x}_1\gamma_1 V_1 + \bar{x}_2\gamma_2 V_2 + \bar{x}_3\gamma_3 V_3}{\gamma_1 V_1 + \gamma_2 V_2 + \gamma_3 V_3}$$

$$= \frac{\sum \bar{x}_i \gamma_i V_i}{\sum \gamma_i V_i} \tag{5.15}$$

Similarly, we can develop the location of the center of gravity in the Y and Z directions to be

$$\bar{y} = \frac{\sum \bar{y}_i \gamma_i V_i}{\sum \gamma_i V_i} \tag{5.16}$$

$$\bar{z} = \frac{\sum \bar{z}_i \gamma_i V_i}{\sum \gamma_i V_i} \tag{5.17}$$

If the body is homogeneous then γ_i is constant, so that we have the following expressions for finding the centroid of a composite body:

$$\bar{x} = \frac{\sum \bar{x}_i V_i}{V} \qquad \bar{y} = \frac{\sum \bar{y}_i V_i}{V} \qquad \bar{z} = \frac{\sum \bar{z}_i V_i}{V} \tag{5.18}$$

where $V = \sum V_i$ is the total volume.

Equations having a form similar to Eqs. (5.18) can be obtained for areas. Consider a thin homogeneous plate lying in the XY plane that is divided into four areas, namely a semicircle, a rectangle, and two triangles as shown in Fig. 5.10. The magnitude of the moment about the Y axis caused by the entire plate is

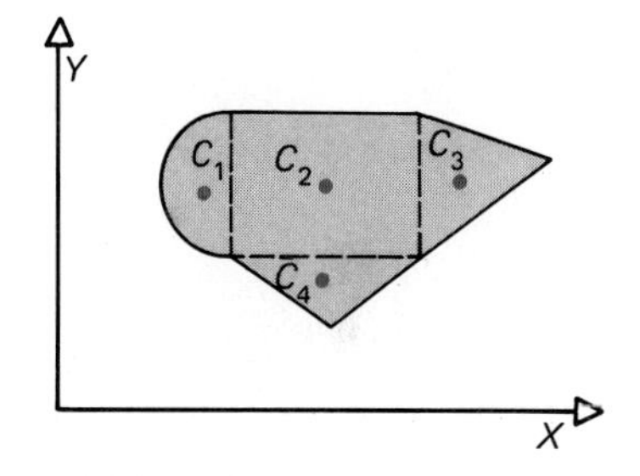

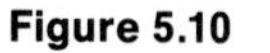

Figure 5.10

$$M_y = \gamma t A \bar{x} \tag{5.19}$$

where $$A = A_1 + A_2 + A_3 + A_4$$

and t is the plate thickness. This moment must equal the sum of the moments caused by the individual areas, namely

$$M_y = \bar{x}_1 A_1 \gamma t + \bar{x}_2 A_2 \gamma t + \bar{x}_3 A_3 \gamma t + \bar{x}_4 A_4 \gamma t \tag{5.20}$$

Equating the right side of Eqs. (5.19) and (5.20), we obtain

$$\bar{x} = \frac{\bar{x}_1 A_1 + \bar{x}_2 A_2 + \bar{x}_3 A_3 + \bar{x}_4 A_4}{A}$$

or $$\bar{x} = \frac{\sum \bar{x}_i A_i}{A} \tag{5.21a}$$

Similarly, we can show that

$$\bar{y} = \frac{\sum \bar{y}_i A_i}{A} \tag{5.21b}$$

In a similar manner, we can locate the centroid of a series of complicated lines by dividing them into a series of known line elements. It can be shown, following the above procedure for areas, that the centroid is given by

$$\bar{x} = \frac{\sum \bar{x}_i L_i}{L} \qquad \bar{y} = \frac{\sum \bar{y}_i L_i}{L} \tag{5.22}$$

where $L = \sum L_i$ = total length of the given line and $\bar{x}_i$ and $\bar{y}_i$ are the centroidal coordinates of each line element lying in the XY plane.

Example 5.5

Obtain the centroid of the line $y = |x|$ between $x = -2$ in and $x = 4$ in as shown in Fig. 5.11.

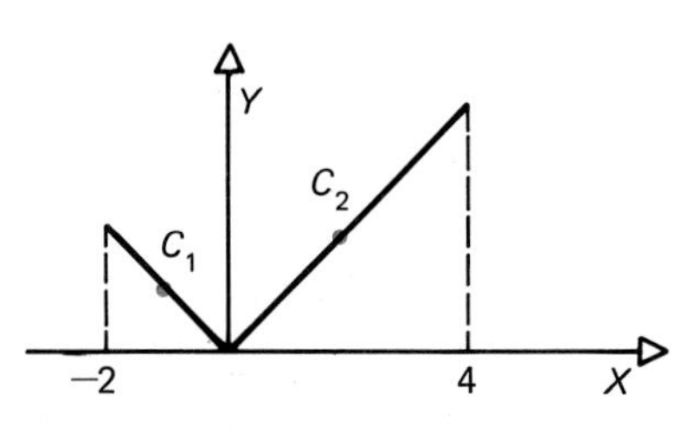

Figure 5.11

From Table 5.1 we note that the centroids C_1 and C_2 of the two lines are $(-1, 1)$ and $(2, 2)$, respectively. The composite centroid of the two lines can be obtained by carrying the calculations in tabular form.

i	L_i	$\bar{x}_i$	$\bar{y}_i$	$L_i\bar{x}_i$	$L_i\bar{y}_i$
1	$\sqrt{8}$	-1	1	$-\sqrt{8}$	$\sqrt{8}$
2	$2\sqrt{8}$	2	2	$4\sqrt{8}$	$4\sqrt{8}$

$$\sum L_i = 3\sqrt{8}$$

$$\sum L_i\bar{x}_i = -\sqrt{8} + 4\sqrt{8} = 3\sqrt{8},$$

$$\bar{x} = \frac{\sum L_i\bar{x}_i}{\sum L_i} = 1 \text{ in}$$ *Answer*

$$\sum L_i\bar{y}_i = \sqrt{8} + 4\sqrt{8} = 5\sqrt{8},$$

$$\bar{y} = \frac{\sum L_i\bar{y}_i}{\sum L_i} = 1.667 \text{ in}$$ *Answer*

Example 5.6

Locate the centroid for the area shown in Fig. 5.12.

The total area is divided into three composite areas as shown, one rectangle and two triangles. The total area is

$$\begin{aligned} A &= A_1 + A_2 + A_3 \\ &= 1(2) + \tfrac{1}{2}(0.5)(1) + \tfrac{1}{2}(1.5)(1) = 3 \text{ ft}^2 \end{aligned}$$

Using Eqs. (5.21), we find

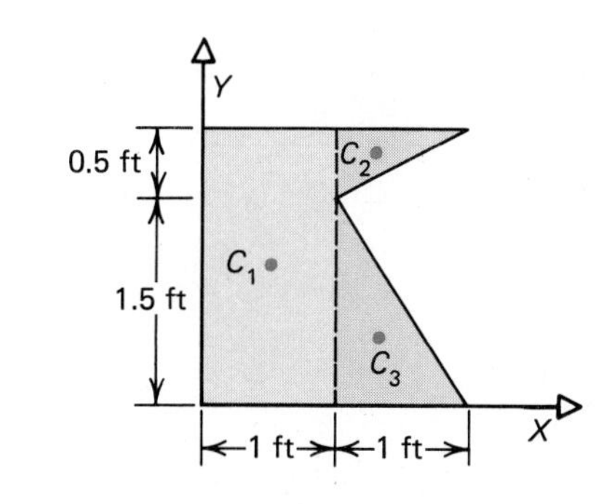

Figure 5.12

$$A\bar{x} = A_1\bar{x}_1 + A_2\bar{x}_2 + A_3\bar{x}_3$$
$$= 2(0.5) + 0.25(1 + 0.333)$$
$$+ 0.75(1 + 0.333) = 2.333$$

$$\bar{x} = 0.778 \text{ ft} \qquad \textit{Answer}$$

and

$$A\bar{y} = A_1\bar{y}_1 + A_2\bar{y}_2 + A_3\bar{y}_3$$
$$= 2(1) + 0.25[1.5 + \tfrac{2}{3}(0.5)]$$
$$+ 0.75[\tfrac{1}{3}(1.5)] = 2.833$$

$$\bar{y} = 0.944 \text{ ft} \qquad \textit{Answer}$$

The above calculations can be formalized in tabular form as follows:

i	A_i	$\bar{x}_i$	$\bar{y}_i$	$A_i\bar{x}_i$	$A_i\bar{y}_i$
1	2	0.5	1	1	2
2	0.25	1.333	1.833	0.333	0.458
3	0.75	1.333	0.5	1	0.375

$$\sum A_i = 3$$
$$\sum A_i\bar{x}_i = 1 + 0.333 + 1 = 2.333$$
$$\bar{x} = \frac{\sum A_i\bar{x}_i}{\sum A_i} = 0.778 \text{ ft}$$
$$\sum A_i\bar{y}_i = 2 + 0.458 + 0.375 = 2.833$$
$$\bar{y} = \frac{\sum A_i\bar{y}_i}{\sum A_i} = 0.944 \text{ ft}$$

Example 5.7

Obtain the centroid of the area shown in Fig. 5.13 neglecting the circular hole in the rectangle.

The total area is divided into the semicircular area A_1 and rectangular area A_2. The y centroid for the semicircular area is obtained from Table 5.2. The remaining information is tabulated below.

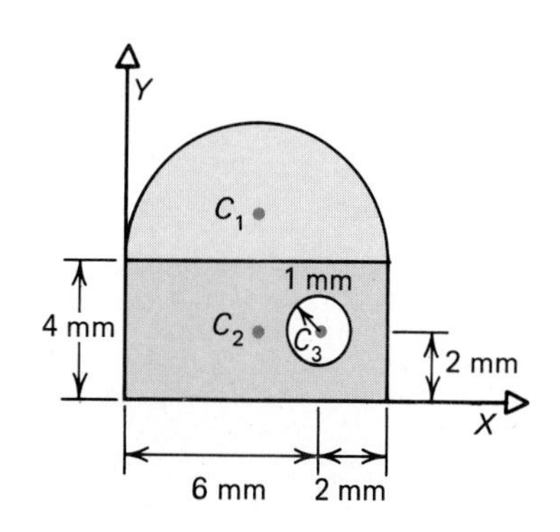

Figure 5.13

i	A_i	$\bar{x}_i$	$\bar{y}_i$	$A_i\bar{x}_i$	$A_i\bar{y}_i$
1	25.13	4	4 + 1.698	100.5	143.2
2	32	4	2	128	64

The total area is

$$A = \sum A_i = 57.13 \text{ mm}^2$$

The centroids become

$$\bar{x} = \frac{\sum A_i \bar{x}_i}{A} = \frac{228.5}{57.13} = 4.00 \text{ mm} \qquad \textit{Answer}$$

$$\bar{y} = \frac{\sum A_i \bar{y}_i}{A} = \frac{207.2}{57.13} = 3.63 \text{ mm} \qquad \textit{Answer}$$

Example 5.8

Locate the centroid for the area shown in Fig. 5.13 and compute the percentage error caused by neglecting the circular hole in Example 5.7.

In this example, we will find it convenient to divide the area into three parts: semicircular area A_1, a 4 mm by 8 mm rectangle A_2, and a negative area A_3 representing the circular hole of radius 1 mm. We include the hole in A_2 and subtract its contribution by introducing A_3. This procedure can save calculations in many practical problems.

i	A_i	$\bar{x}_i$	$\bar{y}_i$	$A_i\bar{x}_i$	$A_i\bar{y}_i$
1	25.13	4	5.698	100.5	143.2
2	32	4	2	128	64
3	−3.142	6	2	−18.85	−6.284

$$A = \sum A_i = 53.99 \text{ mm}^2$$

$$\sum A_i \bar{x}_i = 209.65 \qquad \bar{x} = \frac{\sum A_i \bar{x}_i}{A} = 3.88 \text{ mm} \qquad \textit{Answer}$$

$$\sum A_i \bar{y}_i = 200.92 \qquad \bar{y} = \frac{\sum A_i \bar{y}_i}{A} = 3.72 \text{ mm} \qquad \textit{Answer}$$

From Example 5.7 the centroid obtained by neglecting the circular hole is $\bar{x} = 4.0$ mm, $\bar{y} = 3.63$ mm, so the error is

$$\text{Error in } \bar{x} = \frac{3.88 - 4.00}{3.88} = -0.031 = -3.1\%$$

$$\text{Error in } \bar{y} = \frac{3.72 - 3.63}{3.72} = 0.024 = 2.4\%$$

Example 5.9

Locate the x coordinate of the centroid for the body shown in Fig. 5.14. Note that there are cylindrical holes of radius 0.05 in and 0.2 in deep.

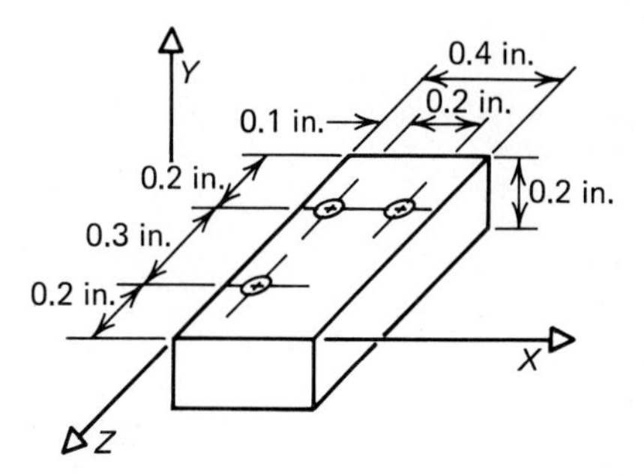

Figure 5.14

Let the body be represented by the 0.7 in by 0.2 in by 0.4 in block and three negative volumes, each being a circular cylinder of radius 0.05 in and height 0.2 in. Consequently, the tabulated data are as follows:

i	V_i	$\bar{x}_i$	$V_i\bar{x}_i$
1	0.056	0.2	0.0112
2	−0.00157	0.1	−0.000157
3	−0.00157	0.1	−0.000157
4	−0.00157	0.3	−0.000471

$$V = \sum V_i = 0.05129$$

$$\bar{x} = \frac{\sum V_i\bar{x}_i}{V} = \frac{0.01042}{0.05129} = 0.203 \text{ in} \qquad \textit{Answer}$$

PROBLEMS

5.25 Obtain the $\bar{y}$ and $\bar{z}$ centroid for the body in Example 5.9.

5.26 Obtain the centroid for the area shown in Fig. P5.26 where the total area is viewed as consisting of one rectangle and one triangle with a negative area.

5.27 Calculate the centroid of the shaded area shown in Fig. P5.27.

5.28 Find the centroid of the frustum of a cone shown in Fig. P5.28. Use the composite-body approach and let $h = 3$ in.

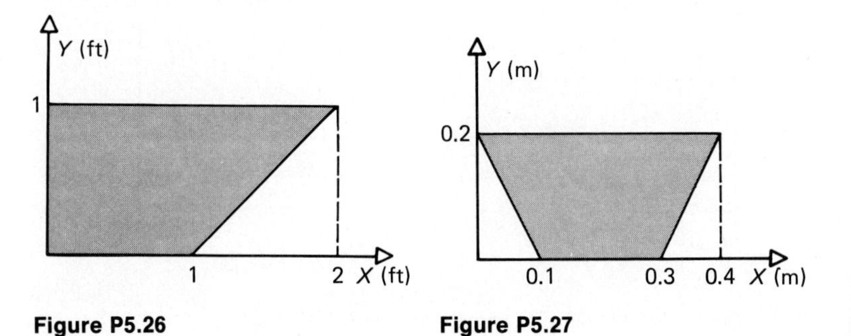

Figure P5.26

Figure P5.27

Figure P5.28

5.29 Find the centroid of the frustum of the cone in Problem 5.28 for $h = 6$ in.

5.30 Using the composite-body approach, find the centroid for the two lines shown in Fig. P5.30, i.e., the line $y = x$ and the quarter-circular arc.

5.31 Using the composite-body approach, find the centroid of the line segments forming the three-quarter circle shown in Fig. P5.31.

5.32 Using the composite-body approach, find the centroid for the four lines shown in Fig. P5.32.

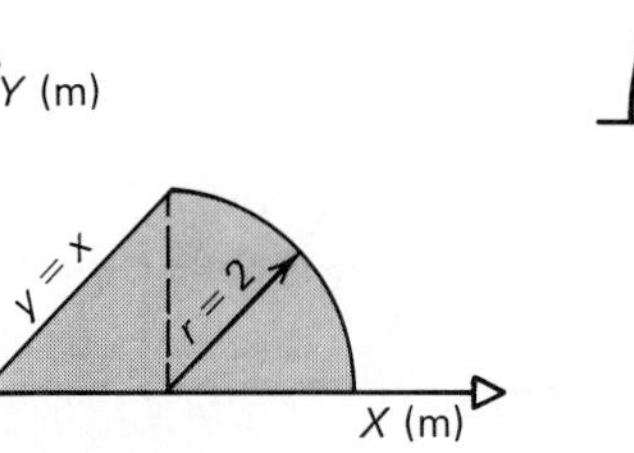

Figure P5.30

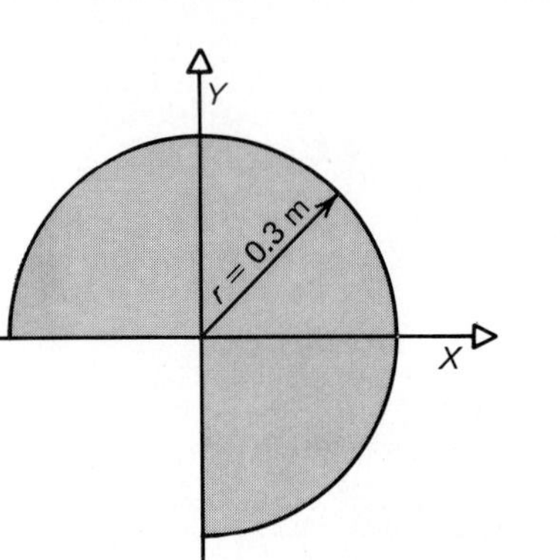

Figure P5.31

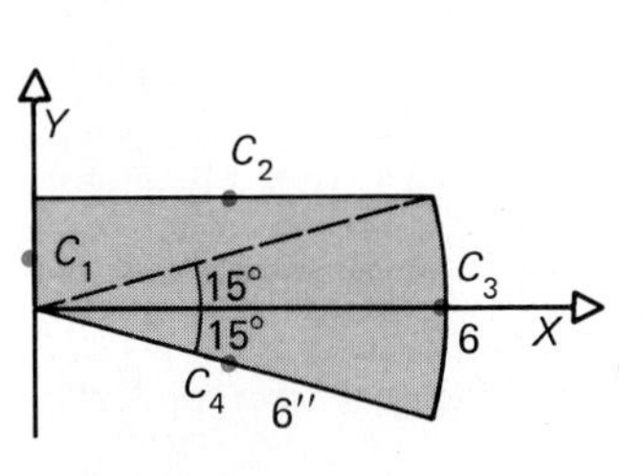

Figure P5.32

5.33 Using the composite-body approach, find the centroid for the two straight lines shown in Fig. P5.33.

5.34 Find the centroid of the enclosed area in Fig. P5.30.

5.35 Find the centroid of the enclosed area of Fig. P5.31.

5.36 Find the centroid of the enclosed area of Fig. P5.32.

5.37 Find the centroid of the enclosed area of Fig. P5.37.

5.38 Find the centroid of the enclosed area of Fig. P5.38.

5.39 Find the centroid of the enclosed area of Fig. P5.39.

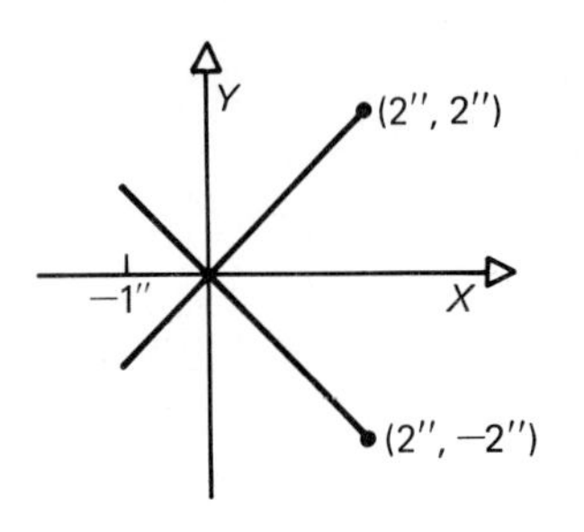

Figure P5.33

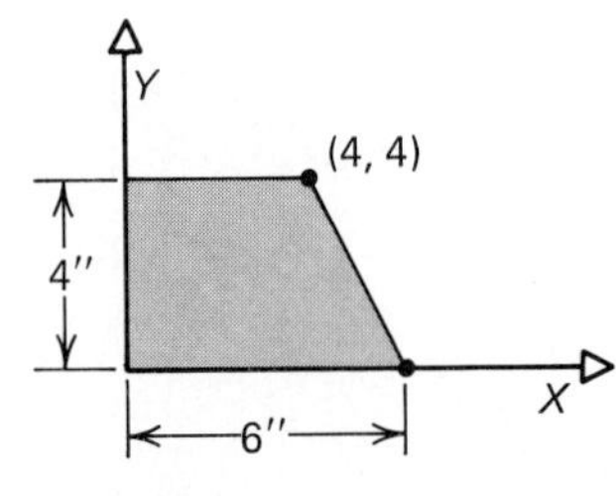

Figure P5.37

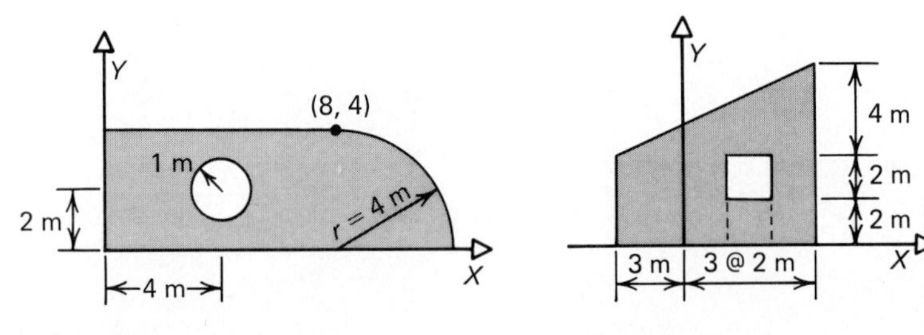

Figure P5.38

Figure P5.39

5.40 Find the centroid of the enclosed area of Fig. P5.40.

5.41 Find the centroid of the enclosed area of Fig. P5.41.

5.42 Locate the centroid for the area shown in Fig. P5.42 for $\theta = 15^\circ$.

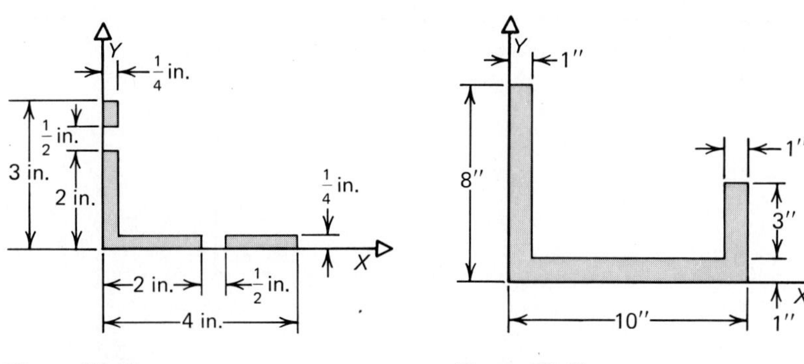

Figure P5.40

Figure P5.41

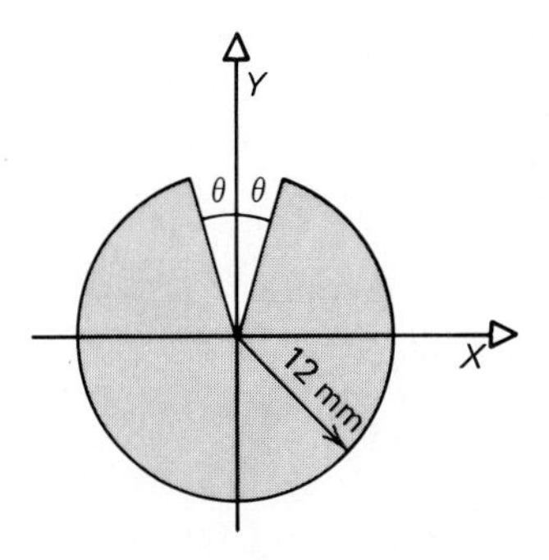

Figure P5.42

5.43 Find the centroid for the area in Problem 5.42 for $\theta = 45^\circ$.

5.44 Find the centroid for the quarter-circular strip shown in Fig. P5.44 for $r_1 = 7$ in and $r_2 = 10$ in.

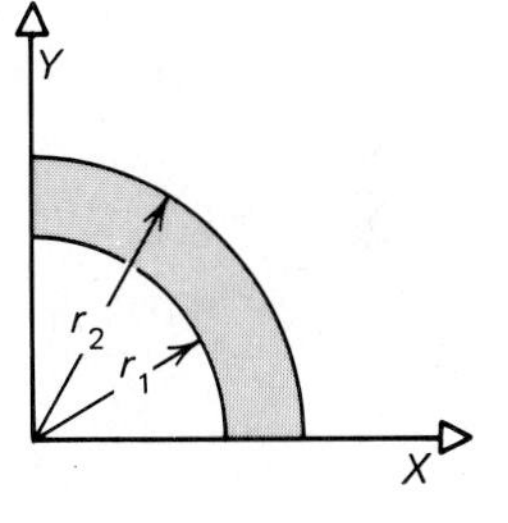

Figure P5.44

5.45 Find the centroid for the quarter-circular strip in Problem 5.44 for $r_2 = 2r_1 = 10$ in.

5.46 Find the centroid of the body composed of a cone and hemisphere as shown in Fig. P5.46. Let $h = 0.6$ m.

5.47 If $h = 0.4$ m, find the centroid of the body in Problem 5.46.

5.48 The hemispheric shell in Fig. P5.48 has an inner radius of $r_1 = 0.8$ ft and an outer radius of $r_2 = 1$ ft. Find the centroid of the body.

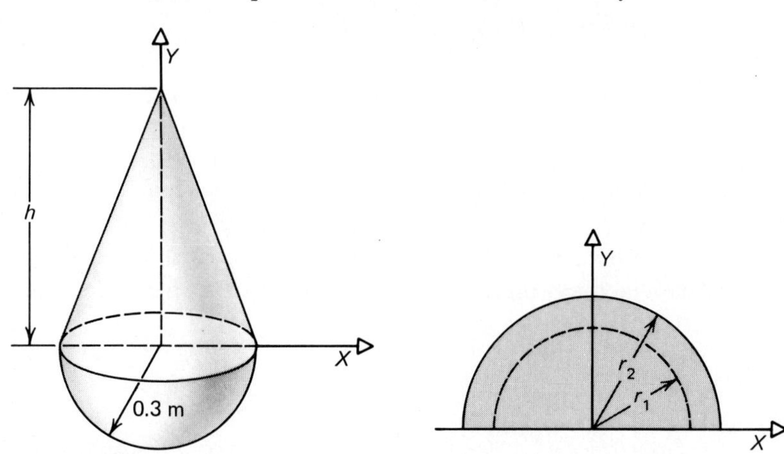

Figure P5.46

Figure P5.48

5.49 Find the centroid for the hemispheric shell in Problem 5.48 if $r_2 = 2r_1 = 1$ ft.

5.50 Find the centroid of the block containing two circular holes in Fig. P5.50. The diameter of each hole equals 10 mm.

5.51 Find the centroid of the two plates welded together in Fig. P5.51 for $w = 10$ mm.

5.52 Locate the centroid of the two plates welded together in Problem 5.51 for $w = 20$ mm.

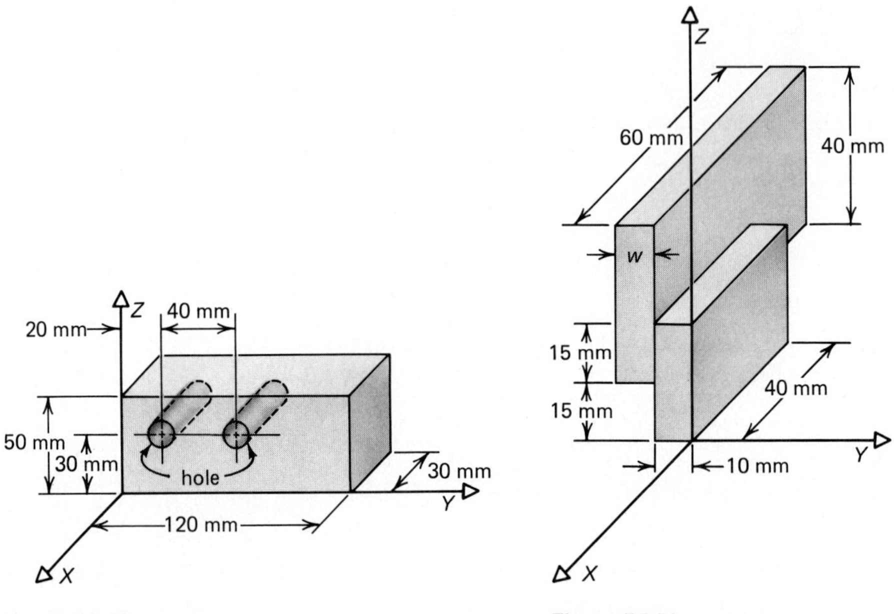

Figure P5.50

Figure P5.51

5.4 Theorems of Pappus and Guldinus

A Greek geometer named Pappus, and later a Swiss mathematician named Guldinus, formulated theorems that yield simple methods for obtaining surface areas and volumes generated by revolving a line or an area about a nonintersecting axis. By a nonintersecting axis we mean that the axis does not intersect any part of the line or area that is being revolved. As we shall see, these theorems take advantage of our knowledge of centroids for lines and areas.

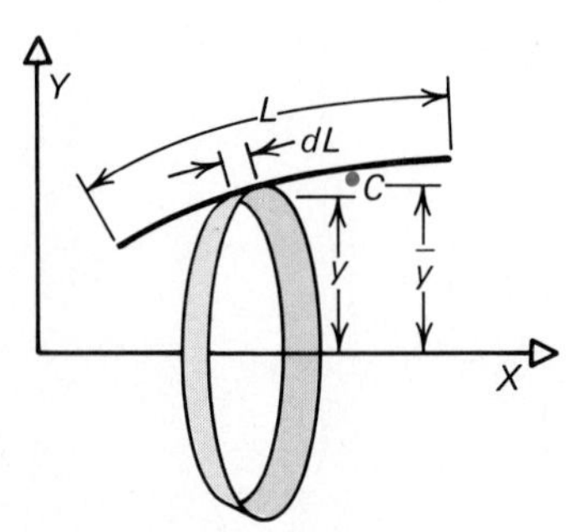

Figure 5.15

Theorem 1

The surface area obtained by revolving a curve of length L about a fixed, nonintersecting axis equals the curve length L times the distance traveled by the centroid of the generating line.

This theorem is proved in the following way. Consider the line of length dL in Fig. 5.15 that revolves about the nonintersecting X axis. The surface area dA of the ring, obtained by revolving dL, is

$$dA = 2\pi y\, dL$$

If the entire line of length L is revolved about the X axis, the total surface area is

$$A = 2\pi \int_0^L y\, dL \tag{5.23}$$

Since the centroid $\bar{y}$ of a line in the XY plane has been previously defined as

$$\bar{y} = \frac{1}{L} \int y\, dL$$

we can rewrite Eq. (5.23) as

$$A = 2\pi\bar{y}L \tag{5.24}$$

where $2\pi\bar{y}$ is the distance traveled by the centroid in generating the surface area. Consequently, the theorem is verified. Of course, if $\theta\bar{y}$ is the distance traveled by the centroid, then the generated surface area $A = \theta\bar{y}L$, where θ is in radians.

The consequences of this theorem may be used to compute the surface area for a variety of shapes. Examples are shown in Fig. 5.16. Here we see that the area of a cylinder, torus, sphere, disk, and cone may be generated by revolving the appropriate line about the X axis. The generated surface areas are found using the theorem.

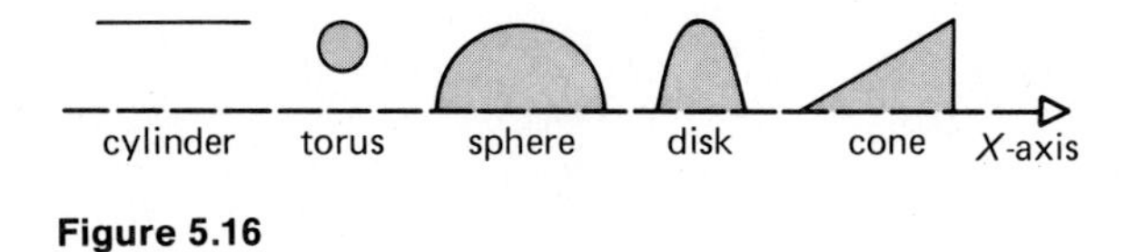

Figure 5.16

Theorem 2

The volume obtained by revolving a surface area A about a fixed nonintersecting axis equals the surface area A times the distance traveled by the centroid of the generating surface area.

Consider the area A in Fig. 5.17 that is to be revolved about the X axis. The element of volume obtained by revolving the area dA through 2π radians is

$$dV = 2\pi y\, dA$$

so that the total volume obtained by revolving A is

$$V = 2\pi \int y\, dA$$

But the centroid of the area has been previously defined as

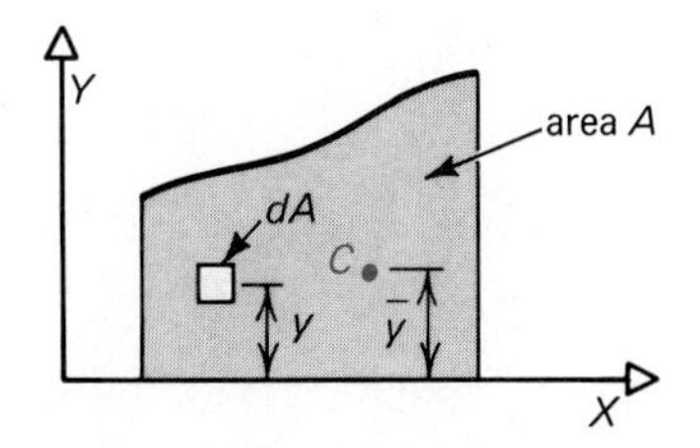

Figure 5.17

$$\bar{y} = \frac{1}{A}\int y\,dA$$

so that

$$V = 2\pi\bar{y}A \tag{5.25}$$

Since the term $2\pi\bar{y}$ is the distance traveled by the centroid relative to the fixed axis, the theorem is verified. Once again, if $\theta\bar{y}$ is the distance traveled by the centroid, then $V = \theta\bar{y}A$. Figure 5.18 shows how the volume of a cylinder, sphere, or cone can be generated.

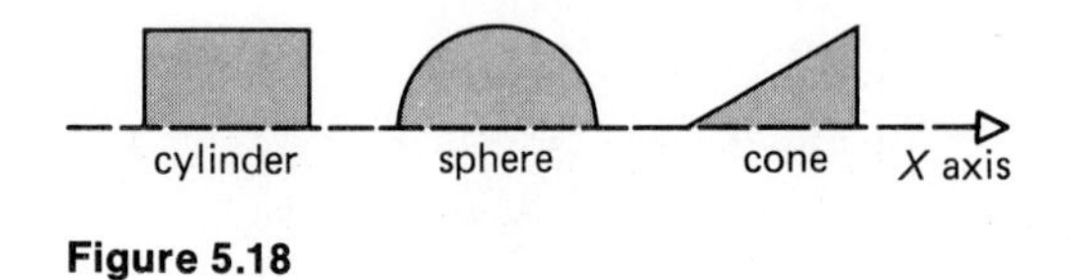

Figure 5.18

Example 5.10

Find the surface area obtained by completely revolving the triangle in Fig. 5.19 about (a) the X axis and (b) the Y axis.

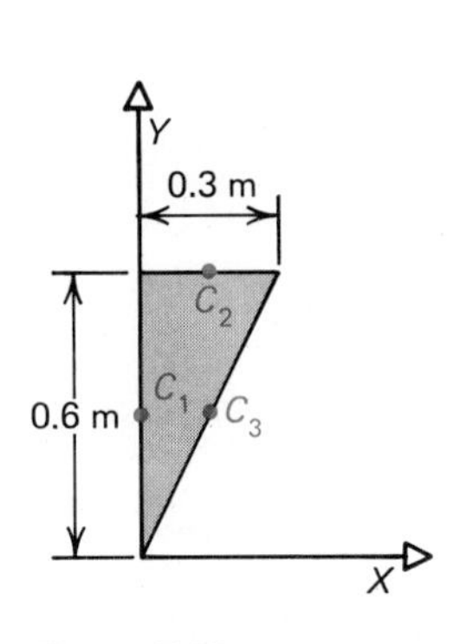

Figure 5.19

(a) We recall from the previous section that the centroid of a straight line is the midpoint. The location of the centroid of each line forming the triangle is labeled as $C_1(0, 0.3)$, $C_2(0.15, 0.6)$, and $C_3(0.15, 0.3)$. Now from the second expression in Eq. (5.22),

$$\begin{aligned}\bar{y}L &= \sum \bar{y}_i L_i \\ &= \bar{y}_1 L_1 + \bar{y}_2 L_2 + \bar{y}_3 L_3\end{aligned}$$

where

$$\bar{y}_1 = 0.3 \qquad \bar{y}_2 = 0.6 \qquad \bar{y}_3 = 0.3$$

$$L_1 = 0.6 \qquad L_2 = 0.3 \qquad L_3 = \sqrt{(0.3)^2 + (0.6)^2} = 0.671$$

$$\begin{aligned}\bar{y}L &= (0.3)(0.6) + (0.6)(0.3) + (0.3)(0.671) \\ &= 0.561\end{aligned}$$

The area is given by Eq. (5.24) as

$$\begin{aligned}A &= 2\pi\bar{y}L = 2\pi(0.561) \\ &= 3.52\ \text{m}^2\end{aligned}$$ *Answer*

(b) For the rotation about the Y axis, we have from the first expression in Eq. (5.22),

$$\begin{aligned}\bar{x}L &= \sum \bar{x}_i L_i \\ &= 0.15(0.3) + 0.15(0.671) = 0.146\end{aligned}$$

The area becomes

$$A = 2\pi\bar{x}L = 2\pi(0.146)$$
$$= 0.917\ \text{m}^2 \qquad \textit{Answer}$$

Example 5.11

Find the volume generated by rotating the triangle shown in Fig. 5.19 about (a) the X axis through π radians and (b) the Y axis through 2π radians.

(a) The area of the triangle is $A = \frac{1}{2}(0.3)(0.6) = 0.09\ \text{m}^2$
Equation (5.21b) reduces to

$$\bar{y}A = \sum \bar{y}_i A_i = \bar{y}_1 A_1$$

where $\bar{y}_1 = \frac{2}{3}(0.6) = 0.4$

Therefore, $\bar{y}A = (0.4)(0.09) = 0.036\ \text{m}^3$

The generated volume about the X axis by revolving the area through π radians is

$$V = \pi\bar{y}A = \pi(0.036) = 0.113\ \text{m}^3 \qquad \textit{Answer}$$

(b) To find the volume generated by revolving the triangle about the Y axis, apply Eq. (5.21a).

$$\bar{x}A = \sum \bar{x}_i A_i = \bar{x}_1 A_1$$

where $\bar{x}_1 = \frac{1}{3}(0.3) = 0.1$

Therefore, $\bar{x}A = (0.1)(0.09) = 0.009\ \text{m}^3$

The generated volume about the Y axis by revolving the area through 2π radians is

$$V = 2\pi\bar{x}A$$
$$= 2\pi(0.009) = 0.0565\ \text{m}^3 \qquad \textit{Answer}$$

Example 5.12

Find the volume generated by revolving through 2π radians the semicircular area shown in Fig. 5.20 (a) about the X axis, and (b) about the Y axis.

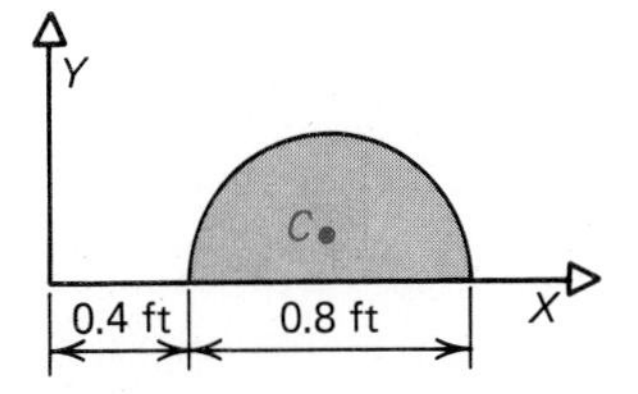

Figure 5.20

The area of the semicircle is

$$A = \frac{\pi r^2}{2} = \frac{\pi(0.4)^2}{2} = 0.2513\ \text{ft}^2$$

(a) The generated volume about the X axis is a sphere. Now

$$V = 2\pi\bar{y}A$$

where $$\bar{y} = \frac{4r}{3\pi} = \frac{4(0.4)}{3\pi} = 0.170 \text{ ft}$$

Therefore $$V = 2\pi(0.170)(0.2513)$$
$$= 0.268 \text{ ft}^3$$ *Answer*

It is interesting to note that we can derive the familiar expression of the volume of a sphere using the above information.

$$V = 2\pi\bar{y}A$$
$$= 2\pi\left(\frac{4r}{3\pi}\right)\frac{\pi r^2}{2}$$
$$= \tfrac{4}{3}\pi r^3$$

(b) By symmetry, the centroid is at $\bar{x} = 0.8$ ft so that the generated volume about the Y axis is

$$V = 2\pi(0.8)(0.2513)$$
$$= 1.263 \text{ ft}^3$$ *Answer*

Example 5.13

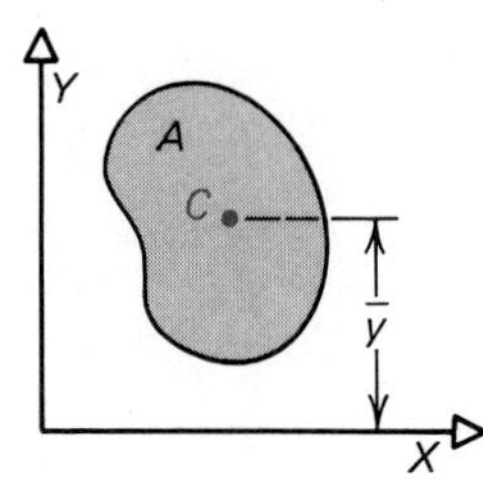

Figure 5.21

The area shown in Fig. 5.21 is measured to be 6.25 m^2. The area, revolved about the X axis, generates a body that is later completely submerged in a circular cylinder tank of water. The tank's radius measures 5 m and the water rises 1 m when the body is submerged. What is $\bar{y}$ in Fig. 5.21?

The volume of the body equals the volume of water displaced. The volume of the water displaced is

$$V = \pi r^2 \, \Delta h = \pi(25)(1) = 25\pi \text{ m}^3$$

By Theorem 2,

$$V = 2\pi\bar{y}A$$

Solving for $\bar{y}$ we find that

$$\bar{y} = \frac{V}{2\pi A} = \frac{25\pi}{2\pi(6.25)} = 2 \text{ m}$$ *Answer*

PROBLEMS

5.53 Find the surface area obtained by revolving the line segment in Fig. P5.53 about the X and Y axes for $\theta = 45°$.

5.54 Find the surface area obtained by revolving the line segment in Problem 5.53 about the X and Y axes for $\theta = 60°$.

5.55 Find the surface area if the three line segments that form the isosceles triangle in Fig. P5.55 is revolved about the X axis.

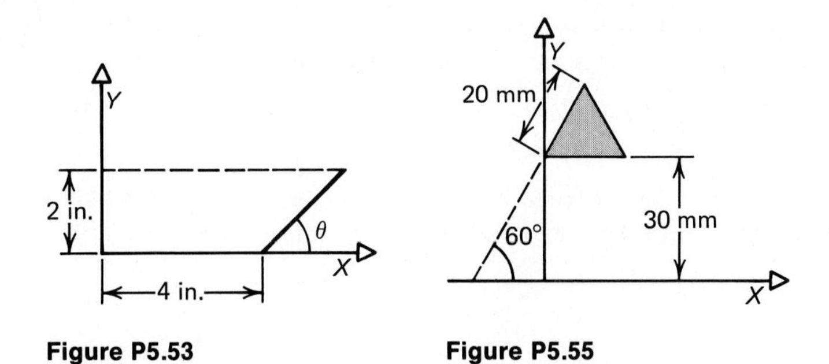

Figure P5.53

Figure P5.55

5.56 Find the surface area if the arc segment in Fig. P5.56 is revolved around the Y axis for $\theta = 10°$.

5.57 Find the surface area if the arc segment in Problem 5.56 is revolved around the Y axis through π radians and for $\theta = 45°$.

5.58 Find the surface area if the semicircular arc in Fig. P5.58 is revolved around, (a) the X axis, and (b) the Y axis for $r = 1$ ft.

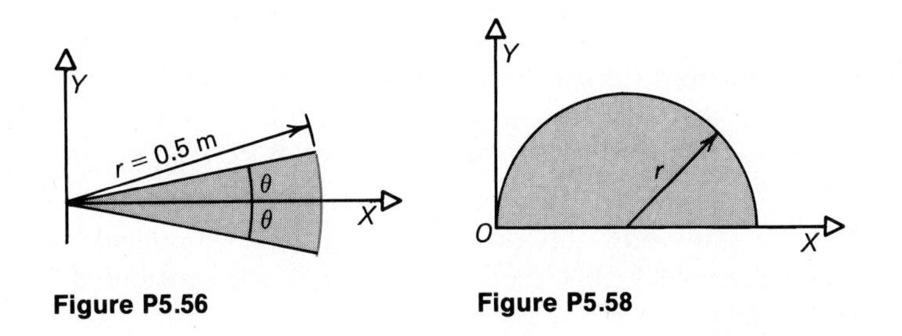

Figure P5.56

Figure P5.58

5.59 Find the surface area if the semicircular arc in Problem 5.58 is revolved through π radians about (a) the X axis and (b) the Y axis for $r = 0.5$ ft.

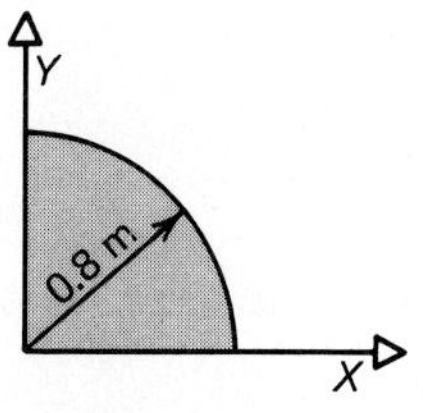

Figure P5.60

5.60 The circular arc in Fig. P5.60 is revolved around the X axis. Find the surface area of the generated hemisphere and check your result with the surface area of a sphere by doubling your answer.

5.61 Find the volume generated by rotating the area formed under the curve $y = 2x^2$ from $x = 0$ to $x = 4$ m about the Y axis.

5.62 Find the volume generated by revolving the enclosed area in Fig. P5.55 about the X axis.

5.63 Find the volume generated by revolving the area enclosed by the θ axis and the sine curve in Fig. P5.63 about the θ axis.

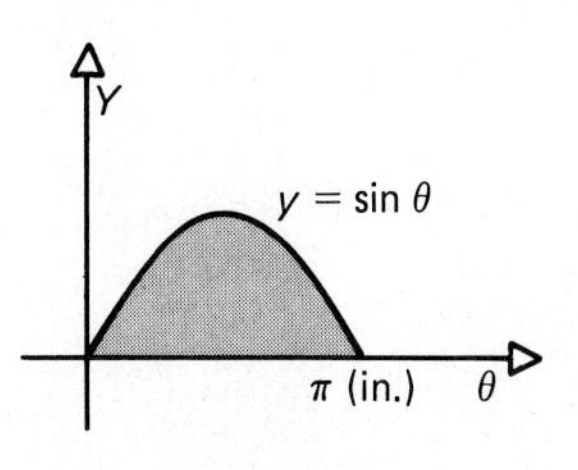

Figure P5.63

5.64 Find the volume generated by revolving the area enclosed by the X axis, Y axis, and the circular segment in Fig. P5.60 around the X axis.

5.65 The square shown in Fig. P5.65 is revolved around the Y axis. The volume generated is 30 ft^3. Find the dimension a.

5.66 Find the volume generated if the square in Problem 5.65 is revolved through π radians about the X axis for $a = 1.02$ ft.

5.67 Find the volume generated by revolving the shaded area of Fig. P5.67 about the Y axis.

5.68 The volume generated by revolving the area of Fig. P5.67 about an axis Y' parallel to the Y axis and through π radians is 261.8 m^3. Locate the Y' axis.

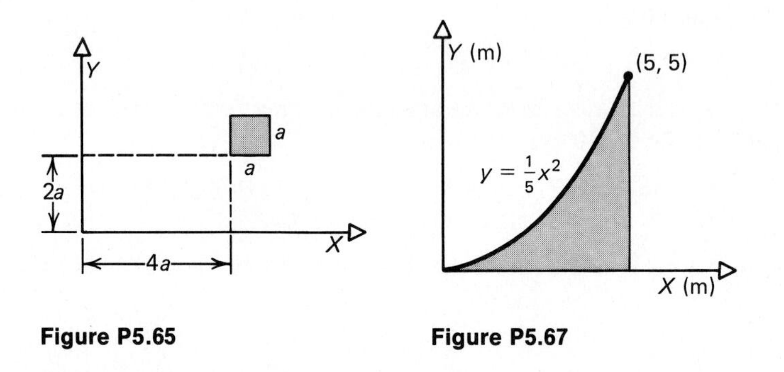

Figure P5.65

Figure P5.67

5.5 Distributed Loads

When a load is applied to a rigid body, it is often distributed over an area and is called a distributed or surface force. In this section we show how this distributed load can be replaced by a resultant force such that the resultant force and distributed load are equivalent.

Consider a beam in Fig. 5.22(a) subjected to the distributed load w whose magnitude varies with x and whose units are N/m or lb/ft. The magnitude of the differential force $d\mathbf{F}$ is given by

$$dF = w\,dx$$

The total magnitude of this resultant force is

$$F = \int_0^L w\,dx \tag{5.26}$$

that is, the magnitude of the resultant force is equal to the area under the w curve.

We now wish to place the resultant force $\mathbf{F}$ on the beam so that it produces the same moment about any point on the beam as the actual distributed load. To locate this point, let us take moments of the distributed load about point O. The magnitude of the moment due to the element of force $d\mathbf{F}$ is

$$dM_0 = x\,dF \tag{5.27}$$

where $dF = w\,dx$. The total magnitude of the moment is

$$M_0 = \int_0^L x\,dF = \int_0^L xw\,dx \tag{5.28}$$

Let the magnitude of the moment about point O be

$$M_0 = Fd \tag{5.29}$$

where d locates the position of the resultant force **F** as shown in Fig. 5.22(b). Equating the right sides of Eqs. (5.28) and (5.29) and using Eq. (5.26), we obtain

$$d = \frac{\int_0^L x\,dF}{F} = \frac{\int_0^L xw\,dx}{F} \tag{5.30}$$

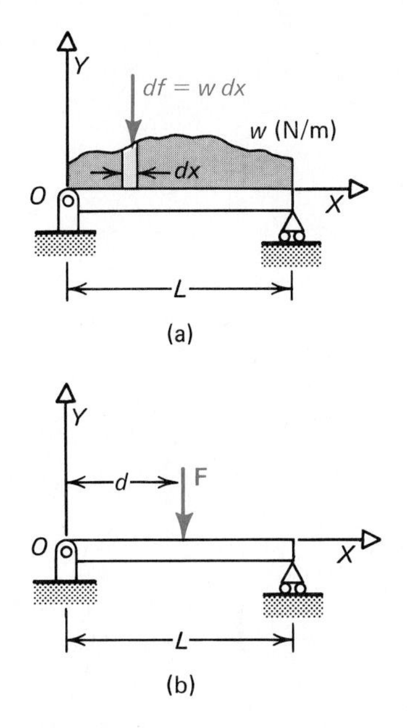

Figure 5.22

We note the similarity between Eq. (5.30) and Eq. (5.5a), where dF in Eq. (5.30) has replaced dw in Eq. (5.5a). Since the distributed load w is homogeneous, we conclude that the distance d in Eq. (5.30) is in fact the x centroid, that is $d = \bar{x}$. In this case we are only interested in the $\bar{x}$ coordinate which we call the **center of distributed force.** Note that this point generally does not coincide with the $\bar{x}$ for the beam, i.e., the centroid of the distributed force and the centroid of the beam are generally located at different points.

The location of the centroid of a distributed load allows us to replace the distributed load by a resultant force applied at the centroid. Consider, for example, the uniformly distributed load of w shown in Fig. 5.23(a). In this case w equals a constant across the beam. The magnitude of the resultant force of this load is obtained from Eq. (5.26) as

$$F = \int_0^L w\,dx = wL$$

The centroid of this force is obtained from Eq. (5.30) as

$$\bar{x} = \frac{\int_0^L wx\,dx}{F} = \frac{w\int_0^L x\,dx}{wL} = \frac{1}{L}\left[\frac{x^2}{2}\right]_0^L = \frac{L}{2}$$

The distributed load can therefore be replaced by a concentrated load located at $\bar{x} = L/2$ as shown in Fig. 5.23(a). If the load distribution is triangular as shown in Fig. 5.23(b), then

$$F = \int_0^L w\,dx = \int_0^L \frac{b}{L}x\,dx = \frac{bL}{2}$$

The centroid is located at

$$\bar{x} = \frac{\int_0^L wx\,dx}{F} = \frac{\int_0^L \frac{b}{L}x^2\,dx}{bL/2} = \frac{2L}{3}$$

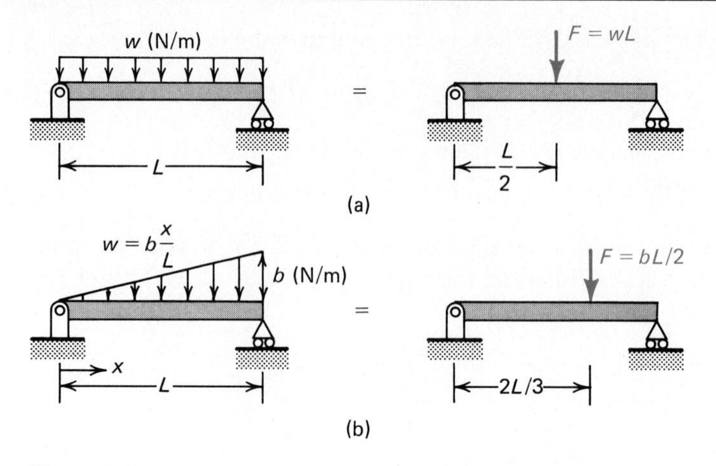

Figure 5.23

The equivalent resultant force of $bL/2$ applied at $\bar{x} = 2L/3$ is illustrated in Fig. 5.23(b).

Example 5.14

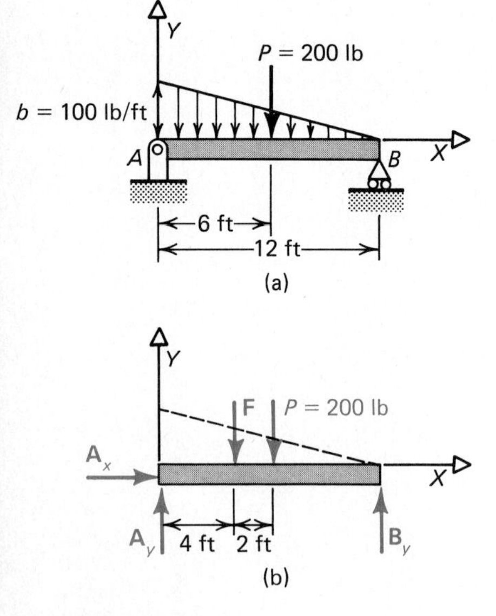

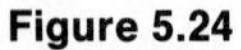
Figure 5.24

The beam shown in Fig. 5.24(a) has an applied load of 200 lb at its center and a triangular distributed load whose ordinate value at point A is 100 lb/ft. Find (a) the magnitudes of the reactions at the supports, and (b) the resultant force for the applied loads.

(a) Figure 5.24(b) shows the free-body diagram of the forces on the beam. Force **F** is the resultant force for the distributed load $w = 100(12 - x)/12$ (now shown by the dashed line) and is located at $\bar{x} = 12/3 = 4$ ft for the triangular load. Its magnitude is

$$F = \int_0^L w\,dx = \int_0^{12} 100\left(\frac{12 - x}{12}\right)dx$$

$$= \frac{1}{2}(100)(12) = 600 \text{ lb}$$

Note that for a triangular loading, F equals the area under the triangle whose base equals the length of loading and whose height equals the maximum magnitude of w. To find the magnitudes of the beam reactions we use the following equilibrium equations:

$$\sum M_A = 0 \qquad 12B_y - 4(600) - 6(200) = 0$$

$$B_y = 300 \text{ lb} \quad \textit{Answer}$$

$$\sum F_y = 0 \qquad A_y - 600 - 200 + 300 = 0$$

$$A_y = 500 \text{ lb} \qquad \textit{Answer}$$

and

$$\sum F_x = 0 \qquad A_x = 0 \qquad \textit{Answer}$$

(b) The magnitude of the resultant of the applied loads **F** and **P** is

$$R = 600 + 200 = 800$$

so that

$$\mathbf{R} = -800\mathbf{j} \text{ lb} \qquad \textit{Answer}$$

The location of the resultant force is

$$\bar{x} = \frac{600(4) + 200(6)}{800} = 4.5 \text{ ft} \qquad \textit{Answer}$$

Example 5.15

For the beam shown in Fig. 5.25(a) find (a) the magnitude of the reactions at the supports, and (b) the resultant force for the applied loads.

(a) Figure 5.25(b) shows the location of the two resultant forces, $\mathbf{F}_1$ and $\mathbf{F}_2$, where $\mathbf{F}_1$ is due to the triangular load on the left of $x = 3$ m and $\mathbf{F}_2$ is due to the triangular load to the right of $x = 3$ m. Their magnitudes are equal to the respective areas of the triangles.

$$F_1 = \tfrac{1}{2}(50)(3) = 75 \text{ N} \qquad F_2 = \tfrac{1}{2}(50)(12) = 300 \text{ N}$$

The magnitude of the beam reactions are computed as follows:

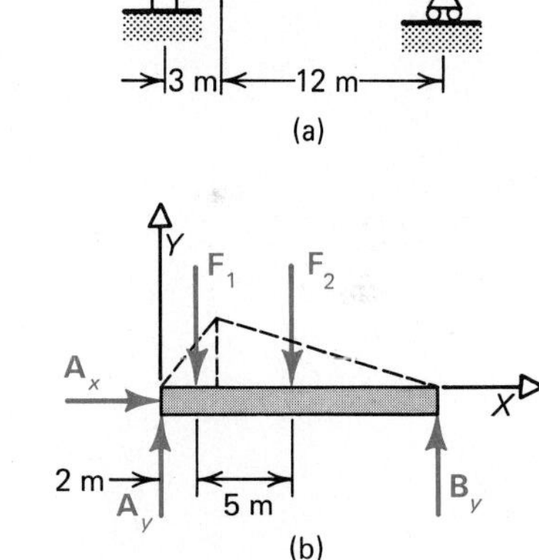

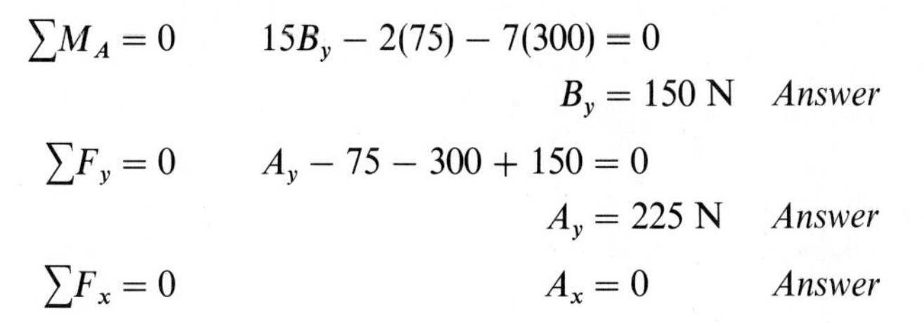

$$\sum M_A = 0 \qquad 15B_y - 2(75) - 7(300) = 0$$

$$B_y = 150 \text{ N} \qquad \textit{Answer}$$

$$\sum F_y = 0 \qquad A_y - 75 - 300 + 150 = 0$$

$$A_y = 225 \text{ N} \qquad \textit{Answer}$$

$$\sum F_x = 0 \qquad A_x = 0 \qquad \textit{Answer}$$

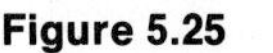

Figure 5.25

(b) The magnitude of the resultant of the applied loads $\mathbf{F}_1$ and $\mathbf{F}_2$ is

$$R = 75 + 300 = 375$$

so that

$$\mathbf{R} = -375\mathbf{j} \text{ N} \qquad \textit{Answer}$$

The location of the resultant force is

$$\bar{x} = \frac{75(2) + 300(7)}{375} = 6 \text{ m} \qquad \textit{Answer}$$

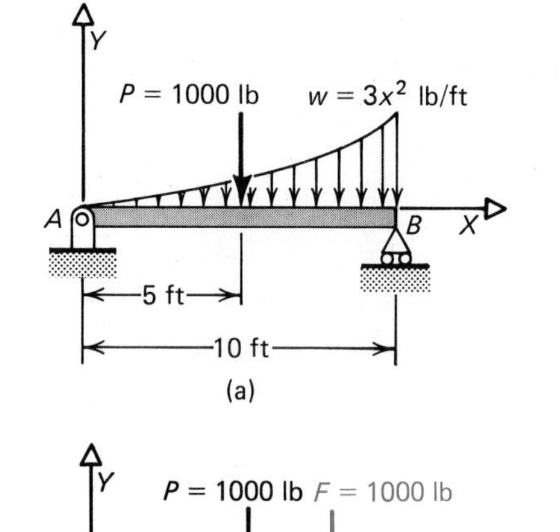

Figure 5.26

Example 5.16

A beam carries a concentrated load of 1000 lb as well as a distributed load as shown in Fig. 5.26(a). Find (a) the magnitude of the reactions at the supports and (b) the resultant force for the applied loads.

(a) The distributed load can be replaced by a concentrated load **F** as shown in Fig. 5.26(b) where

$$F = \int_0^L w\,dx = \int_0^{10} 3x^2\,dx$$

$$= [x^3]_0^{10} = 1000 \text{ lb}$$

The location of this force on the beam is

$$\bar{x} = \frac{1}{F}\int_0^L wx\,dx = \frac{1}{F}\int_0^{10} 3x^3\,dx$$

$$= \frac{1}{1000}\left[\frac{3x^4}{4}\right]_0^{10} = 7.5 \text{ ft}$$

The magnitudes of the reactions shown in Fig. 5.26(b) are obtained as follows:

$$\sum M_A = 0 \qquad -5(1000) - 7.5(1000) + 10B_y = 0$$

$$B_y = 1250 \text{ lb} \qquad \textit{Answer}$$

$$\sum F_y = 0 \qquad A_y + B_y - 1000 - 1000 = 0$$

$$A_y = 750 \text{ lb} \qquad \textit{Answer}$$

$$\sum F_x = 0 \qquad A_x = 0 \qquad \textit{Answer}$$

(b) The magnitude of the resultant of the applied loads is

$$R = 1000 + 1000 = 2000$$

so that

$$\mathbf{R} = -2000\mathbf{j} \text{ lb} \qquad \textit{Answer}$$

$$\bar{x} = \frac{(1000)5 + 1000(7.5)}{2000}$$

$$= 6.25 \text{ ft} \qquad \textit{Answer}$$

PROBLEMS

Find (a) the magnitude of the reactions at the supports and (b) the resultant force for the applied loads:

5.69 For the loaded beam in Fig. P5.69.

5.70 For the loaded beam in Fig. P5.70.

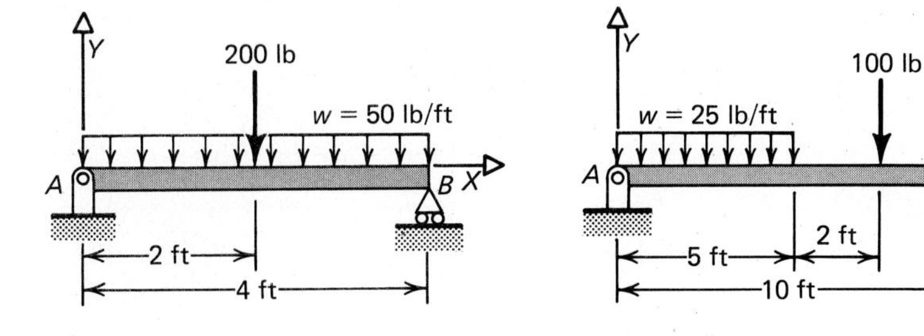

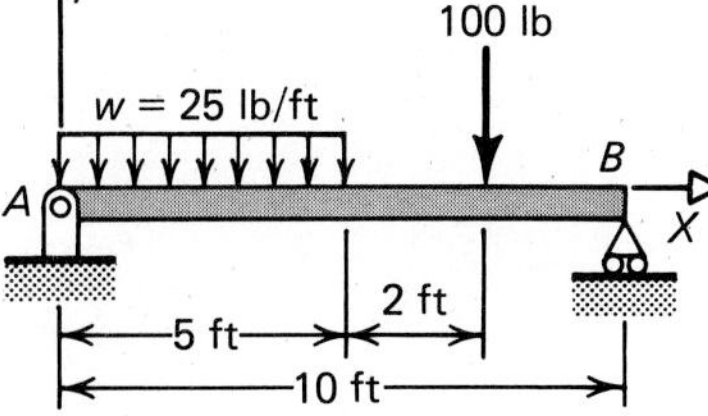

Figure P5.69

Figure P5.70

5.71 For the loaded beam in Fig. P5.71.

5.72 For the loaded beam in Fig. P5.72.

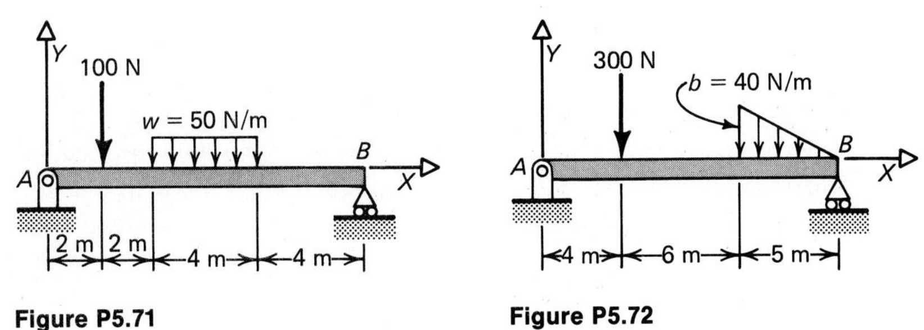

Figure P5.71

Figure P5.72

5.73 For the loaded beam in Fig. P5.73.

5.74 For the loaded beam in Fig. P5.74 for $L = 5$ ft.

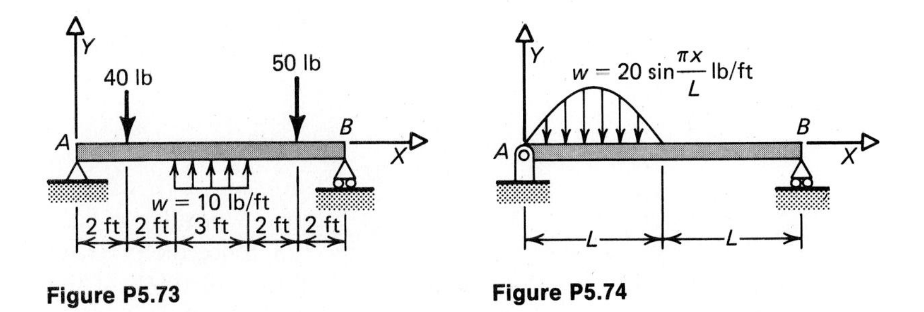

Figure P5.73

Figure P5.74

5.75 For the loaded beam in Problem 5.74 for $L = \pi$ ft.

5.76 For the loaded beam in Fig. P5.76.

5.77 For the loaded beam in Fig. P5.77 for $k = 6$, $n = 2$, and $L = 10$ m.

5.78 For the loaded beam in Problem 5.77 for $k = 8$, $n = 3$, and $L = 5$ m.

5.79 For the loaded beam in Fig. P5.79.

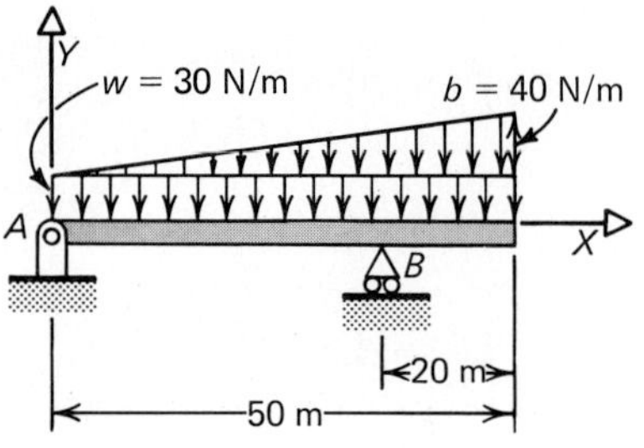

Figure P5.76

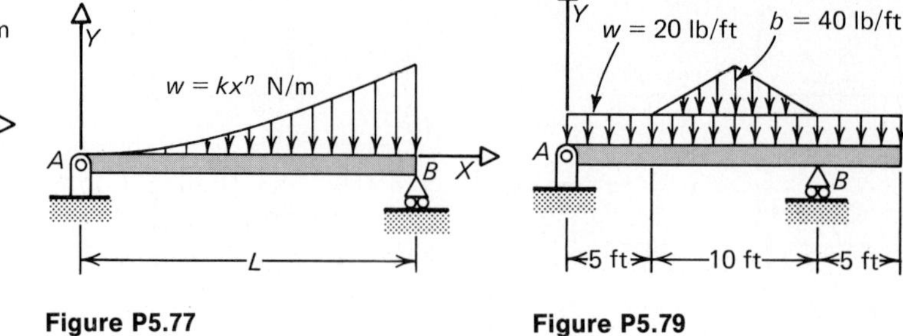

Figure P5.77

Figure P5.79

5.6 Fluid Pressure

When a body is submerged in a fluid, it is subjected to a pressure that is usually measured in N/m^2 or in lb/ft^2. Recall that 1 N/m^2 = 1 pascal where the pascal is abbreviated Pa. This pressure, called **hydrostatic pressure**, is due to the weight of the fluid above the point in question. The hydrostatic pressure varies linearly with the depth as shown in Fig. 5.27. Consequently, the absolute pressure p_a at a depth y is

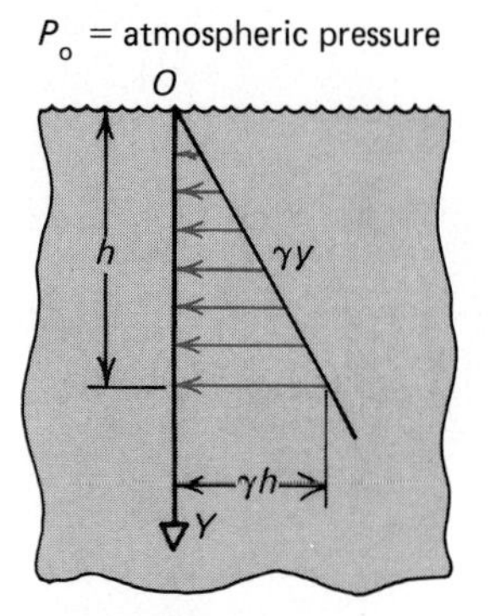

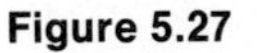

Figure 5.27

$$p_a = p_0 + \gamma y \tag{5.31}$$

where $\quad p_0$ = atmospheric pressure at the surface of the fluid

γ = specific weight of the fluid = ρg

where ρ is the density in kg/m^3 and $g = 9.81$ m/s^2. If the fluid is water $\rho = 1000$ kg/m^3. Usually we measure the gage pressure p, where

$$\begin{aligned} p &= p_a - p_0 \\ &= \gamma y = \rho g y \end{aligned} \tag{5.32}$$

This equation can be applied as follows. Consider a dam retaining a body of water as shown in Fig. 5.28. Since the pressure distribution varies in accord with Eq. (5.32), we can think of the hydrostatic pressure loading as a triangular distributed load acting on a vertical beam. Consequently, the resultant force associated with the fluid pressure is 4 m above ground level and its magnitude is

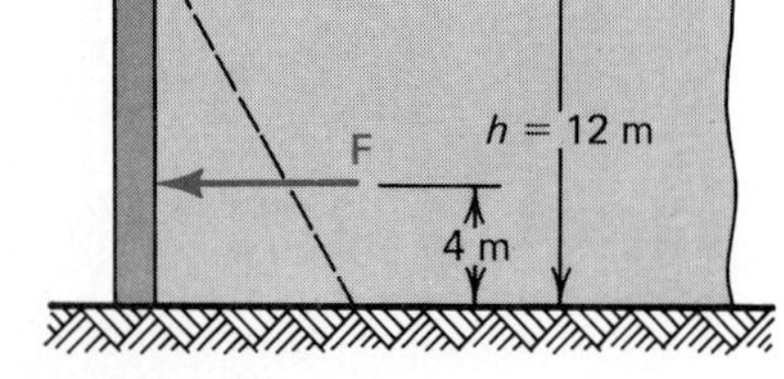

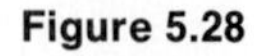

Figure 5.28

$$F = \tfrac{1}{2}(\rho g h)(h)(L) = \tfrac{1}{2}\rho g h^2 L = \tfrac{1}{2}\gamma h^2 L \tag{5.33}$$

where $\quad h$ = depth of the water

L = width of the wall

γ = specific weight of the fluid

Letting $L = 1$ m and noting that the density of water is 1000 kg/m^3, we have

$$F = \tfrac{1}{2}(1000)(9.81)(12)^2(1) = 706{,}320 \text{ N} = 706.3 \text{ kN per meter of width}$$

Should the wall be inclined as shown in Fig. 5.29, the most convenient approach to finding the resultant force due to the liquid is to partition the loads as shown in the figure. Here $\mathbf{F}_1$ is the weight of the water contained in the triangle ABC, while $\mathbf{F}_2$ is the force due to the hydrostatic pressure shown by the triangle ACD. Having these two forces, we can find their resultant by the techniques developed previously.

Since we include problems in the English gravitational system, we note that the specific weight of water is 62.4 lb/ft^3.

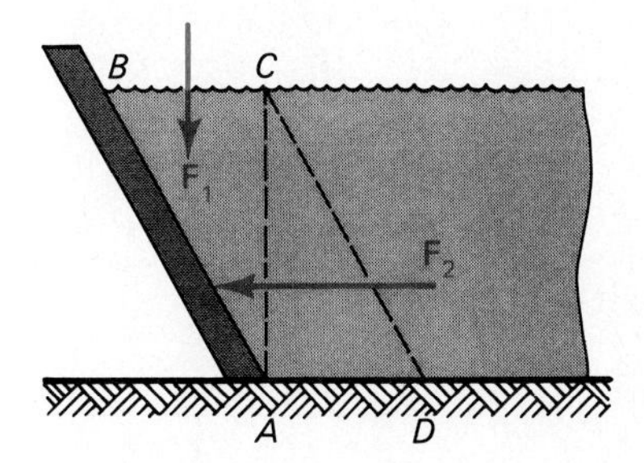

Figure 5.29

Example 5.17

Find the magnitude of the reactions at A and B in Fig. 5.30(a) assuming a smooth-pin support at A and no leakage of the restrained water. Consider only a 1-m section of the wall whose mass is 4000 kg.

Figure 5.30(b) shows the free-body diagram of the forces acting on the wall, where

$$F = \tfrac{1}{2}(1000)(9.81)(15)^2(1) = 1{,}104{,}000 \text{ N} = 1{,}104 \text{ kN}$$

The reactions are obtained as follows:

$$\sum M_A = 0 \qquad -B(18)\cos 30^\circ + 5F = 0$$

$$B = 354.1 \text{ kN} \qquad \textit{Answer}$$

$$\sum F_x = 0 \qquad A_x - F + B\cos 30^\circ = 0$$

$$A_x = 797.3 \text{ kN} \qquad \textit{Answer}$$

$$\sum F_y = 0 \qquad A_y - W + B\sin 30^\circ = 0$$

The weight of the wall is

$$W = 4000(9.81) = 39{,}240 \text{ N} = 39.24 \text{ kN}$$

$$A_y = 39.24 - 354.1 \sin 30^\circ$$

$$A_y = -137.8 \text{ kN} \qquad \textit{Answer}$$

The negative sign means that $\mathbf{A}_y$ acts downward.

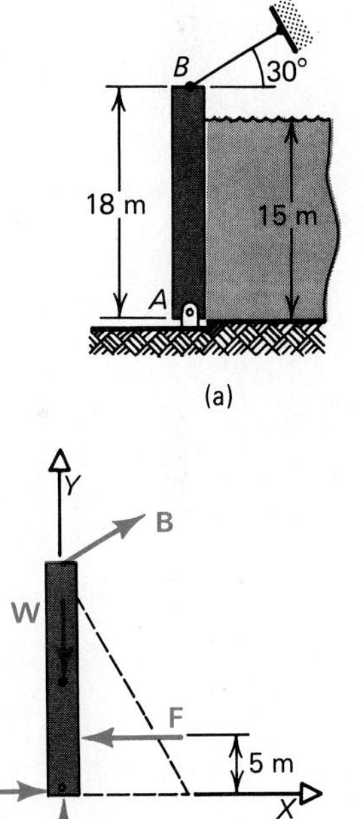

Figure 5.30

Example 5.18

Find the reactions of the concrete dam shown in Fig. 5.31(a) assuming that the specific weight of concrete is 150 lb/ft^3 and the restrained liquid is water. Treat only a 1-ft section of the dam and assume that

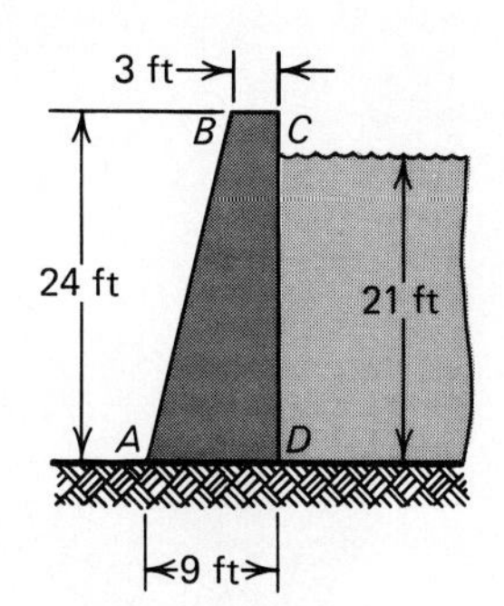

(a)

(b)

Figure 5.31

the dam behaves as a cantilever beam with reactions at point A. Let the specific weight of water equal 62.4 lb/ft^3.

Figure 5.31(b) shows a free-body diagram of the forces acting on the dam. The reactions at the support are labeled $\mathbf{A}_x$, $\mathbf{A}_y$ and the clamping moment $\mathbf{M}$. Force $\mathbf{F}$ is due to the fluid pressure acting on the dam. For convenience, the weight of the dam is divided into two values, $\mathbf{W}_1$ and $\mathbf{W}_2$. The magnitudes of the applied forces are found as follows:

$$\text{Volume } V_1 = \tfrac{1}{2}(6)(24)(1) = 72 \text{ ft}^3$$

$$\text{Volume } V_2 = (3)(24)(1) = 72 \text{ ft}^3$$

$$W_1 = \gamma V_1 = 150(72) = 10{,}800 \text{ lb}$$

$$W_2 = 10{,}800 \text{ lb}$$

$$F = \tfrac{1}{2}(62.4)(21)^2(1) = 13{,}760 \text{ lb}$$

The reactions are obtained as follows:

$$\sum F_x = 0 \qquad A_x - F = 0$$

$$A_x = 13{,}760 \text{ lb} \qquad \textit{Answer}$$

$$\sum F_y = 0 \qquad A_y - W_1 - W_2 = 0$$

$$A_y = 21{,}600 \text{ lb} \qquad \textit{Answer}$$

$$\sum M_A = 0 \qquad M - 4W_1 - 7.5W_2 + 7F = 0$$

$$M = 27{,}880 \text{ ft}\cdot\text{lb} \qquad \textit{Answer}$$

If we wish, we could replace A_y and M by an equivalent force of 21,600 lb placed at a distance

$$\bar{x} = \frac{M}{A_y} = \frac{27{,}880}{21{,}600} = 1.29 \text{ ft}$$

PROBLEMS

5.80 Draw the pressure distribution of the soil acting on the face AB of the retaining wall in Fig. P5.80. If the specific weight of the soil is 120 lb/ft^3, find the resultant force acting on a 1-ft section of the face AB of the retaining wall. Assume that the soil behaves like a liquid in its pressure effect on the retaining wall.

5.81 The retaining wall in Fig. P5.81 has water on one side, and soil on the other side with a density of 2000 kg/m^3. For a 1-m section of wall, draw the distributed forces on the wall AB, and find the resultant force acting on the wall if $h = 15$ m.

5.82 Find the resultant force acting on a 1-m section of the wall if the water level $h = 10$ m in Problem 5.81. Assume that the density of soil is 2000 kg/m^3.

5.83 The gate AB in Fig. P5.83 is 1 m wide. Find the magnitude of the force from the water acting on the gate.

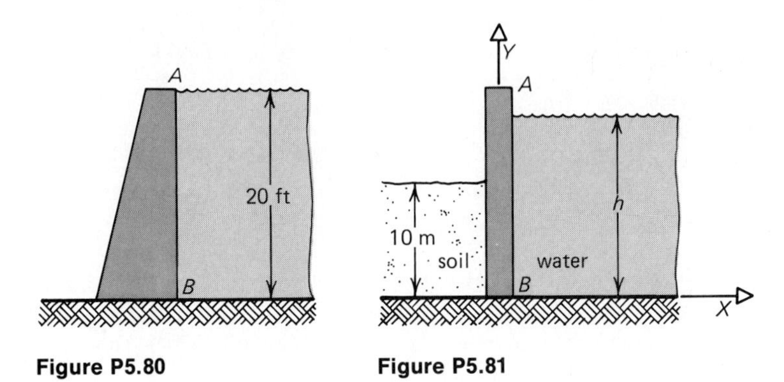

Figure P5.80 **Figure P5.81**

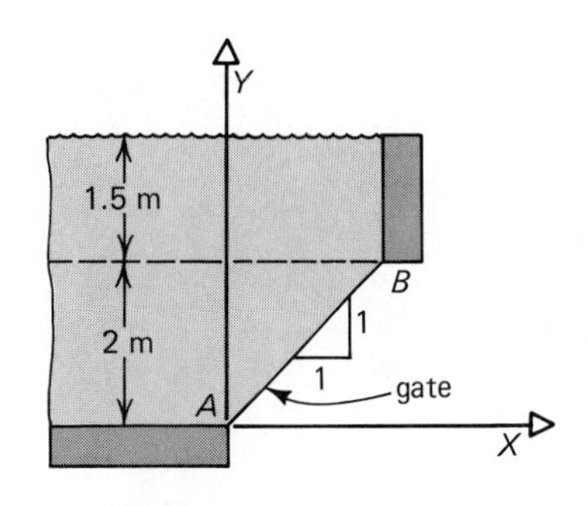

Figure P5.83

5.84 Find the tension in the cable used to support a 5-ft section of the gate shown in Fig. P5.84.

5.85 If the force necessary to break the cable shown in Fig. P5.85 is 500 lb, what is the allowable water level h? Assume that the gate is 1-ft wide.

5.86 Find the reactions of the concrete dam shown in Fig. P5.86 if the density for concrete is 2500 kg/m^3. Use a 1-m section of the dam and idealize the problem by assuming all the reactions occur at D. Also let $h = 4$ m.

Figure P5.84

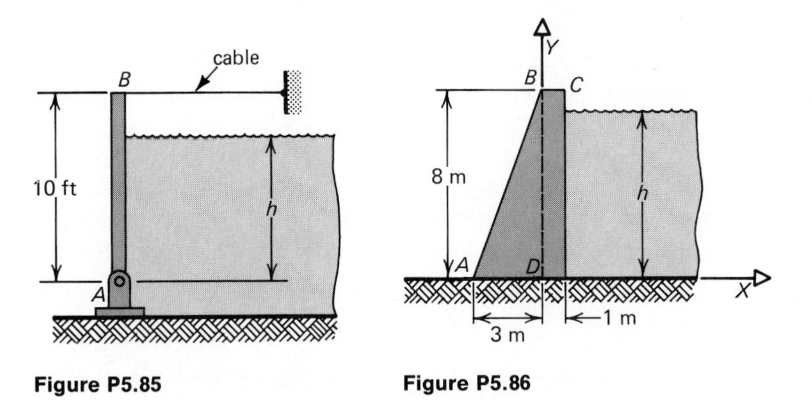

Figure P5.85 **Figure P5.86**

5.87 The earth dam in Fig. P5.87 has a density of 2000 kg/m^3. Due to seepage of water under the base AD, an uplifting pressure results which, for approximation purposes, varies linearly across the base starting with zero pressure at A to a maximum hydrostatic pressure at D. Find the reactions at A of the dam for a 1-m section of the dam.

5.88 Find the reactions of the concrete dam about the point A shown in Fig. P5.88. Use a 1-ft section of the dam and assume the specific weight of concrete to be 150 lb/ft^3. Let $h = 24$ ft.

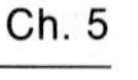

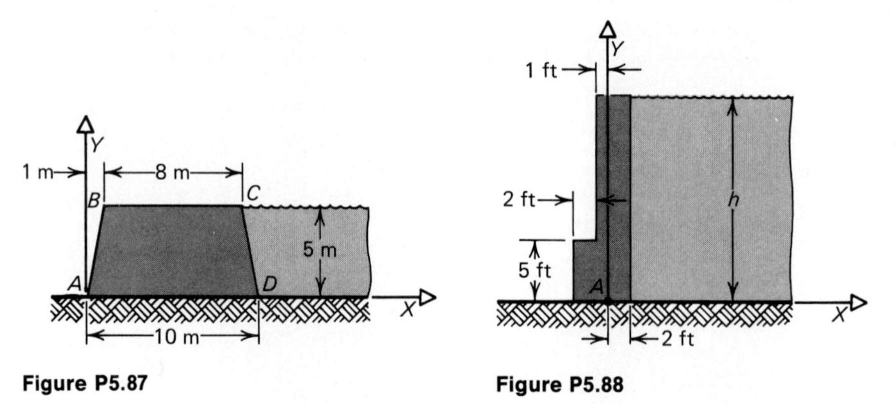

Figure P5.87 **Figure P5.88**

5.89 The dam shown in Fig. P5.88 is 24 ft high and is designed to have a maximum moment of 100 ft·kip clockwise at A. What is the height h of the water that provides this moment. Assume the specific weight of concrete as 150 lb/ft^3 and consider a 1-ft section of the dam.

5.7 Buoyancy

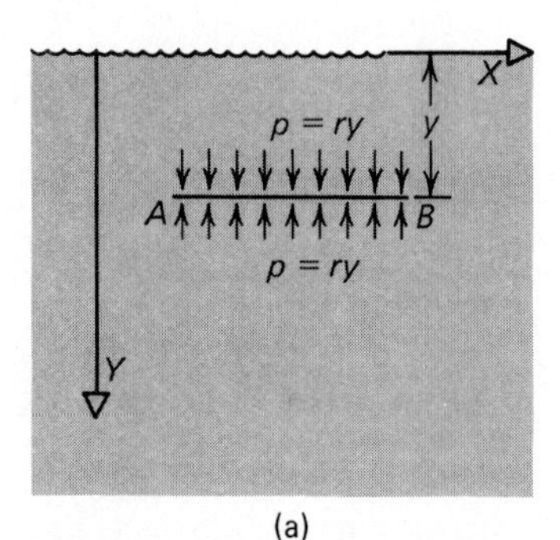

(a)

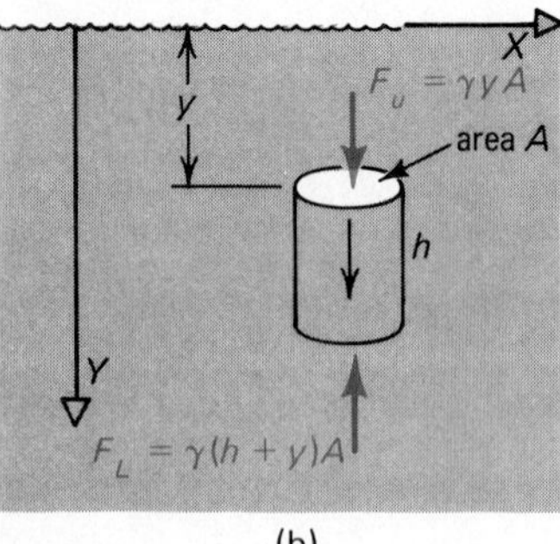

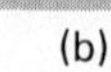

(b)

Figure 5.32

The pressure on a body under water increases as the body is submerged. If the body under consideration is a very thin sheet AB as shown in Fig. 5.32(a), then the pressure above and below the sheet is the same with the net pressure being zero. If, however, the body has a finite dimension such as the cylinder shown in Fig. 5.32(b), the net pressure and therefore the net force on the body is no longer zero. This can be easily understood if we observe that the force on the top surface is $\gamma y A$ while that at the lower surface is $\gamma(y + h)A$. The difference then gives rise to what is called the **buoyant force**.

A body submerged in a liquid undergoes a buoyant effect due to the increase of hydrostatic pressure with depth. For example, consider the submerged body shown in Fig. 5.33. For convenience, let the element of the vertical component of a force on a differential element of cross-sectional area dA at a depth y_1 be

$$d\mathbf{f}_1 = \gamma y_1 \, dA\mathbf{j}$$

At a depth y_2, let the vertical component of the force be

$$d\mathbf{f}_2 = -\gamma y_2 \, dA\mathbf{j}$$

The net differential force $d\mathbf{F}$ is

$$d\mathbf{F} = d\mathbf{f}_1 + d\mathbf{f}_2 = -\gamma(y_2 - y_1)\, dA\mathbf{j}$$

The term $(y_2 - y_1)\, dA$ may be viewed as a differential volume of a small cylinder dV. Now,

$$d\mathbf{F} = -\gamma \, dV\mathbf{j}$$

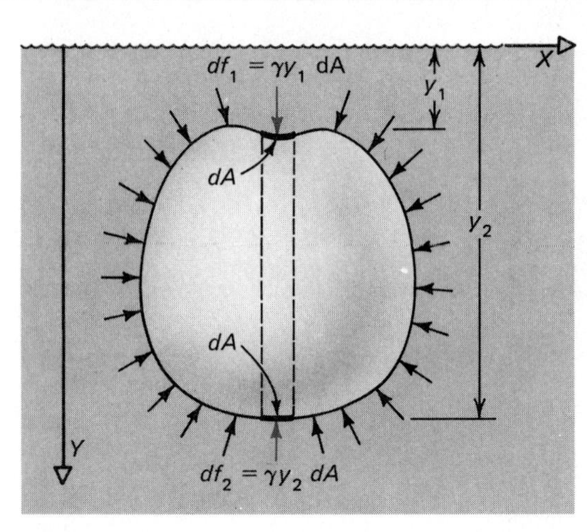

Figure 5.33

so that the total buoyant force is

$$\mathbf{F} = -\gamma \int dV \mathbf{j} = -\gamma V \mathbf{j} \tag{5.34}$$

Notice that the net resultant force on the submerged body in the X direction due to the hydrostatic pressure is zero. This is because each element of force in the X direction acting on the body due to hydrostatic pressure has an equal and opposite element of force, the summation of which is zero.

The resultant buoyant force given by Eq. (5.34) passes through the centroid of the body as in the case of the other examples of distributed forces.

Example 5.19

A sphere of radius 0.1 m is submerged in a water tank. What is the magnitude of the force required to keep the sphere submerged? Assume that the mass of the sphere is 0.621 kg and the density of water equals 1000 kg/m^3.

The sphere is shown in Fig. 5.34 with the resultant forces acting on it, where $\mathbf{F}_B$ is the buoyant force, $\mathbf{W}$ is the weight of the sphere, and $\mathbf{F}_A$ is the applied force to keep the sphere submerged. For equilibrium,

$$\sum F_y = F_A + W - F_B = 0$$

The magnitude of the buoyant force, equals $\gamma V = \rho g V$, where the volume of a sphere equals $4\pi r^3/3$. Therefore,

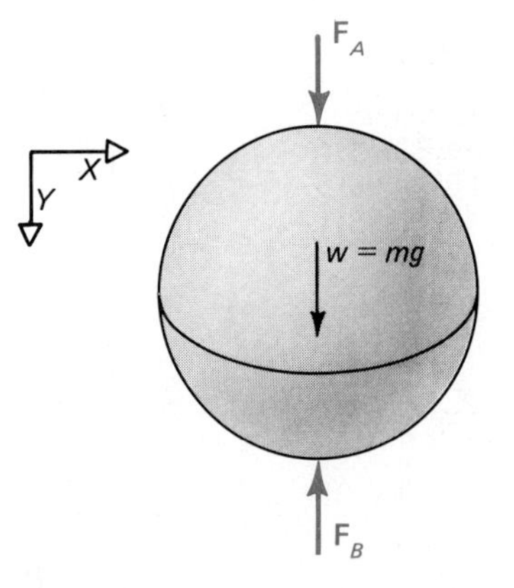

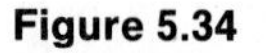
Figure 5.34

$$F_B = \rho g \tfrac{4}{3}\pi r^3$$
$$= (1000)(9.81)\tfrac{4}{3}\pi(0.1)^3 = 41.09 \text{ N}$$

The weight of the sphere is

$$W = mg$$
$$= (0.621)(9.81) = 6.09 \text{ N}$$

The magnitude of the applied force is

$$F_A = F_B - W = 35.0 \text{ N} \qquad \textit{Answer}$$

Example 5.20

Water is contained by the gate AB that is pinned at point A, and restrained by the wire at B and the counterweight D as shown in Fig. 5.35(a). The gate weighs 200 lb, and the counterweight is a sphere that weighs 300 lb. Neglecting the weight and volume of member AC, find the magnitude of the tensile force in the wire at B if the gate is 2 ft wide.

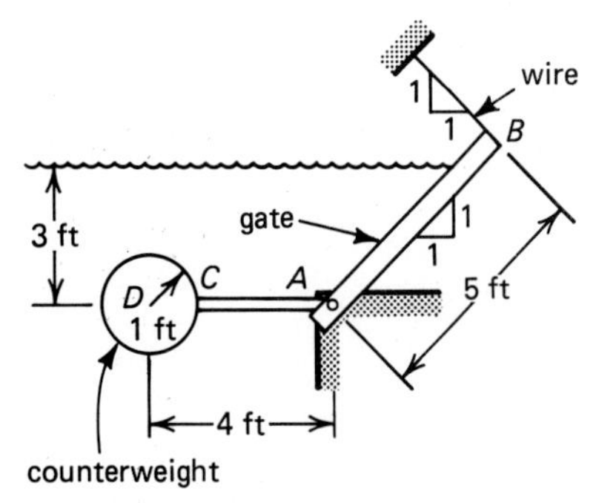

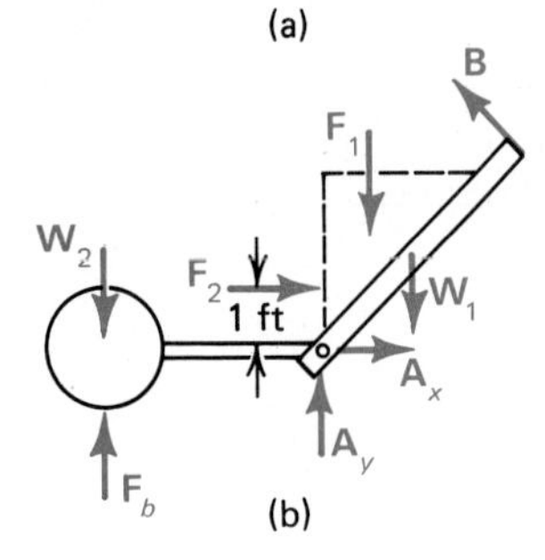

Figure 5.35

Figure 5.35(b) shows the free-body diagram of the forces. Force $\mathbf{F}_1$ represents the weight of the water above the gate, $\mathbf{F}_2$ is due to the hydrostatic pressure, and $\mathbf{F}_b$ is the buoyant force acting on the sphere. Since we are neglecting the volume of member AC, there is no buoyant force for this member. The magnitudes of these forces are:

$$F_1 = 62.4 \frac{(3)(3)(2)}{2} = 562 \text{ lb}$$
$$F_2 = \tfrac{1}{2}(62.4)(3)^2(2) = 562 \text{ lb}$$
$$F_b = 62.4(\tfrac{4}{3}\pi r^3) = 261 \text{ lb}$$
$$W_1 = 200 \text{ lb}$$
$$W_2 = 300 \text{ lb}$$

The magnitude of force **B** in the wire is found as follows:

$$\sum M_A = 0 \qquad 5B - W_1(2.5)\cos 45^\circ - F_1(1) + (W_2 - F_b)4 - F_2(1) = 0$$
$$B = 264 \text{ lb} \qquad \textit{Answer}$$

PROBLEMS

5.90 A freighter is designed such that when completely loaded it displaces 1000 m^3 of water. If the mass of the freighter is 500,000 kg, how much cargo can the freighter carry?

5.91 Two boys make a raft out of wood which has an average thickness of 50 mm and a density of 500 kg/m^3. If the boys have a combined mass of 125 kg, what is the minimum area of raft that will support the boys?

5.92 The float-valve arrangement in Fig. P5.92 consists of a plastic 3-in radius sphere that weighs 3 lb/ft^3. The rod AB weighs 2 lb and is uniform. The rod is pivoted at 0. Find the force transmitted to the spring at A. Let $L = 10$ in and assume that half the sphere is submerged.

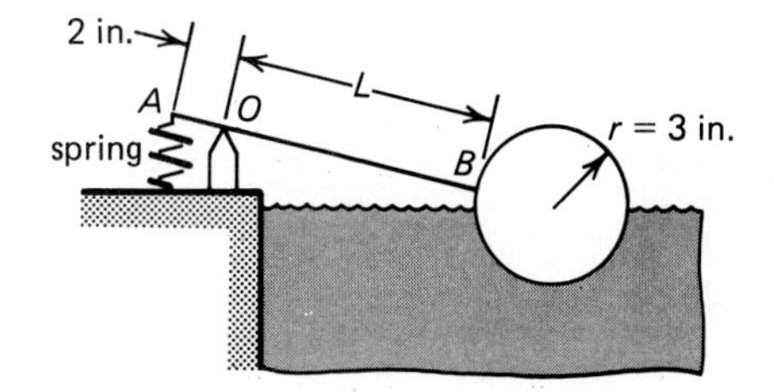

Figure P5.92

5.93 Calculate the force transmitted to the spring at A for the float-valve arrangement shown in Fig. P5.92. Assume that the sphere weighs 1 lb, rod AB weighs 2 lb, and $L = 15$ in. Assume that 60 percent of the sphere is submerged.

5.94 Find the force in the spring if the sphere is totally submerged in Problem 5.92.

5.95 A flat barge 10 m by 6 m is designed to support a load having a mass of 50,000 kg. The barge has a mass of 10,000 kg. If the water level must be at least 0.4 m from the top surface of the barge, what must be the thickness of the structure.

5.96 A package of scientific instruments having a mass of 75 kg is sent aloft by a spherical balloon containing helium. What is the minimum diameter of the balloon that will permit the package to be airborne? Assume the density of air as 1.28 kg/m^3 and that of helium as 0.176 kg/m^3.

5.97 The 8-ft-long block with a cross section 0.5 ft by 0.5 ft in Fig. P5.97 is secured to the floor of a tank by four strings, each having a breaking strength of 15 lb. How high must the water rise above the base of the block for the block to break away assuming $\gamma = 20$ lb/ft^3 for the block?

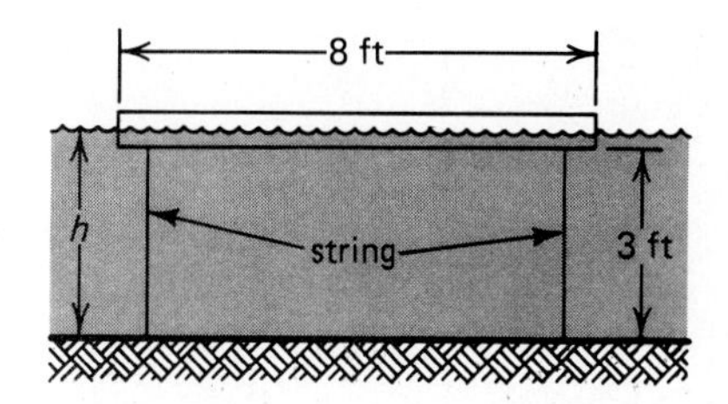

Figure P5.97

5.98 A cylindrical block measures 3 ft in diameter and is 0.5 ft thick. The circular rim of the cylinder is secured to the floor of a water tank by four strings, each of which will break if the tension exceeds 20 lb. If $\gamma = 15$ lb/ft^3 for the block, how high must the water rise above the base of the block for the block to break away?

5.8 Cables

Engineers use cables for many applications, such as suspension bridges, telephone lines, and electrical power lines. The successful use of cables requires a knowledge of what happens to a cable when a load is applied. This includes knowing the length of the cable, its tension, the reactions at its supports, its span, and its sag. For purposes of analysis, cables are assumed to have no resistance to bending, are inextensible, and carry tensile forces only.

Concentrated Load

By way of introduction to a study of cables, let us first treat the case of known concentrated loads $\mathbf{P}_1$ and $\mathbf{P}_2$ acting on a cable

as shown in Fig. 5.36(a). We will assume that the distances x_1, x_2, and L are known while y_1 and y_2 are unknowns. Figure 5.36(b) shows the free-body diagram of the cable, wherein there are four unknowns, A_x, A_y, B_x, and B_y. Note that we are presently neglecting the weight of the cable. Since we have only three equations of statics, the problem appears indeterminate. However, if we know the coordinates of some point on the cable, say point C in Fig. 5.36(b), the problem becomes solvable by the equation of statics.

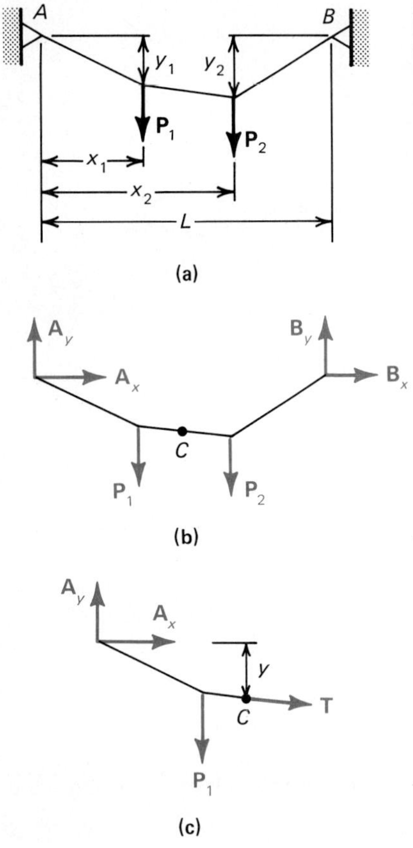

Consider the free-body diagram shown in Fig. 5.36(c). The figure shows the cable cut at the point C and a tension T acting on the cable. If we sum moments about point C, $\sum M_c = 0$, we obtain a relationship between the reaction components A_x and A_y. This equation, coupled with the equilibrium equations for Fig. 5.36(b), provide the necessary equations to obtain all the reactions. The force in each cable is found using the fact that the cable acts as a two-force member.

(a)

(b)

(c)

Figure 5.36

Example 5.21

A cable carries a concentrated load as shown in Fig. 5.37(a). Find the magnitude of the tension in section BC and the reactions at A and C.

From the free-body diagram of the entire cable shown in Fig. 5.37(b), we have

$$\sum M_A = 0 \qquad -10(100) + 35(C_y) = 0$$

$$C_y = 28.6 \text{ N}$$

$$\sum F_y = 0 \qquad A_y + C_y = 100$$

$$A_y = 71.4 \text{ N}$$

The free-body diagram of a part of the cable is shown in Fig. 5.37(c). The magnitude of the force in section BC is labeled T_{BC}. Now

$$\sum M_A = 0$$

$$2\left(\frac{25}{\sqrt{629}}\right)T_{BC} + 10\left(\frac{2}{\sqrt{629}}\right)T_{BC} - 10(100) = 0$$

$$T_{BC} = 358 \text{ N} \quad \textit{Answer}$$

$$\sum F_x = 0 \qquad A_x + \frac{25}{\sqrt{629}}T_{BC} = 0$$

$$A_x = -357 \text{ N} \quad \textit{Answer}$$

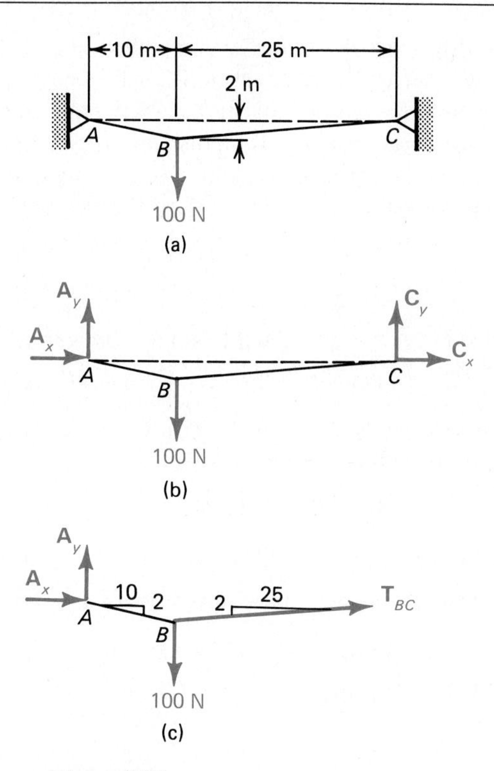

Figure 5.37

Now considering the entire cable shown in Fig. 5.37(b),

$$\sum F_x = 0 \qquad A_x + C_x = 0$$

$$C_x = 357 \text{ N} \qquad \textit{Answer}$$

The reactions are expressed as

$$\mathbf{A} = (-357\mathbf{i} + 71.4\mathbf{j}) \text{ N} \qquad \textit{Answer}$$

$$\mathbf{C} = (357\mathbf{i} + 28.6\mathbf{j}) \text{ N} \qquad \textit{Answer}$$

Distributed Loads

When the mass of the cable is important or the cable is carrying a load distributed along its length, it is necessary that we analyze a differential element of the cable and integrate over the length in order to determine the shape of the cable. Let us consider a cable carrying a uniformly distributed load w (N/m) across the span as shown in Fig. 5.38(a).

In order to determine the shape of the cable for this case, let

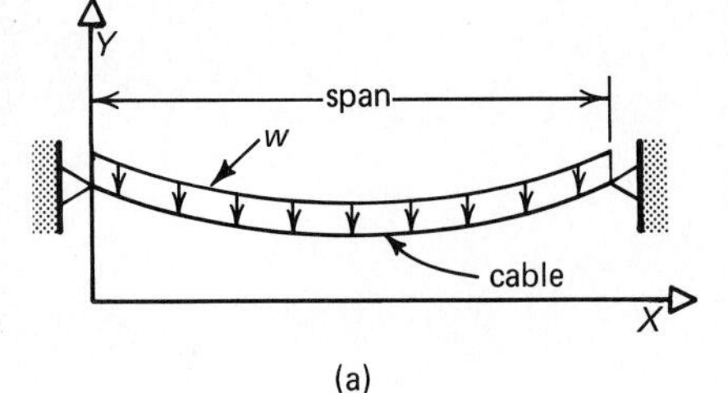

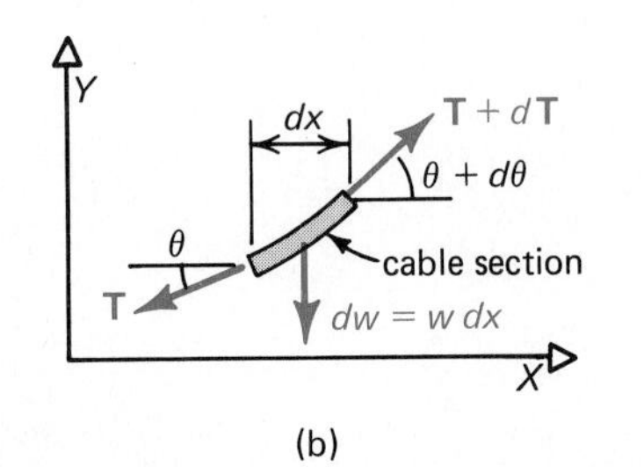

Figure 5.38

us consider the free-body diagram of a differential element of the cable in Fig. 5.38(b). Here we show the cable having an angle θ measured from the horizontal and $\theta + d\theta$ measured from the horizontal a distance dx away. The tension in the cable also changes from T to $T + dT$ over this increment. The vertical load equals $w\,dx$. We apply the equations of equilibrium to the element as follows.

$$\sum F_x = 0 \qquad (T+dT)\cos(\theta+d\theta) - T\cos\theta = 0$$

$$(T+dT)(\cos\theta\cos d\theta - \sin\theta\sin d\theta) - T\cos\theta = 0$$

$$\sum F_y = 0 \qquad (T+dT)\sin(\theta+d\theta) - T\sin\theta - w\,dx = 0$$

$$(T+dT)(\sin\theta\cos d\theta + \cos\theta\sin d\theta) - T\sin\theta - w\,dx = 0$$

For a small change in θ, $\sin d\theta \approx d\theta$ and $\cos d\theta \approx 1$, so that the above equations reduce to the following:

$$(T + dT)(\cos\theta - d\theta\sin\theta) - T\cos\theta = 0$$

$$(T + dT)(\sin\theta + d\theta\cos\theta) - T\sin\theta = w\,dx$$

If we neglect the product of differentials, these equations become

$$-T\,d\theta\sin\theta + dT\cos\theta = 0$$

$$T\,d\theta\cos\theta + dT\sin\theta = w\,dx$$

or

$$d(T\cos\theta) = 0 \tag{5.35}$$

$$d(T\sin\theta) = w\,dx \tag{5.36}$$

Since the derivative of a constant is zero, we can write the integral of Eq. (5.35) as

$$T\cos\theta = \text{constant} = T_0 \tag{5.37}$$

Now multiply both sides of Eq. (5.37) by $\sin\theta$ and rearrange to obtain

$$T\sin\theta = T_0\tan\theta \tag{5.38}$$

Substitute this into Eq. (5.36) to obtain

$$d(T_0\tan\theta) = w\,dx \tag{5.39}$$

But the slope $dy/dx = \tan\theta$, so that Eq. (5.39) becomes

$$d\left(T_0\frac{dy}{dx}\right) = w\,dx \qquad \text{or} \qquad \frac{d}{dx}\left(\frac{dy}{dx}\right) = \frac{w}{T_0}$$

therefore,

$$\frac{d^2y}{dx^2} = \frac{w}{T_0} \tag{5.40}$$

Equation (5.40) is a differential equation that governs the flexible cable loaded continuously by a uniformly distributed load w (N/m). The solution of the differential equation must also satisfy the boundary conditions at each end of the cable. In general, however, the equations governing the cable can become rather complex. In what follows we shall be content in analyzing two types of cables that have wide application and are reasonably simple to study in this first course.

Parabolic Cable

When the distributed load w is constant and the weight of the cable is neglected, the equilibrium shape of the cable is a parabola. Such a situation exists approximately in suspension bridges where w is the road load and the weight of the cable is small relative to the road load.

Consider a suspension cable, as shown in Fig. 5.39(a), that carries a uniform load w from the road. Now Eq. (5.40) is applicable with w and T_0 as constants. Let us integrate this equation twice as follows:

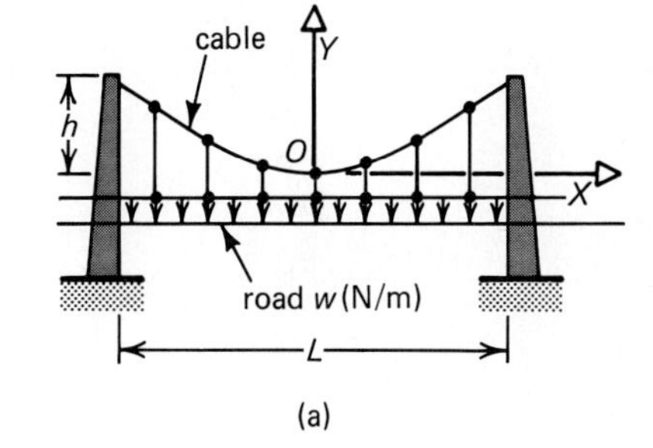

$$\frac{dy}{dx} = \frac{wx}{T_0} + C_1 \tag{5.41}$$

and

$$y = \frac{wx^2}{2T_0} + C_1 x + C_2 \tag{5.42}$$

where C_1 and C_2 are constants of integration. From Fig. 5.39(a) we observe that the boundary conditions are that

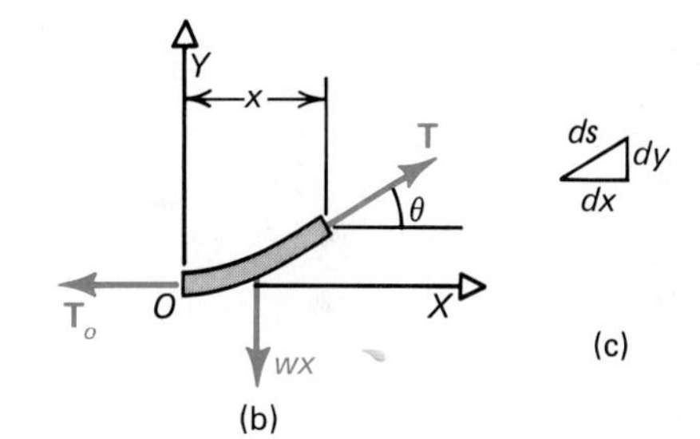

Figure 5.39

$$\text{at } x = 0 \begin{cases} \dfrac{dy}{dx} = 0 & (5.43) \\ y = 0 & (5.44) \end{cases}$$

Substituting Eq. (5.43) into Eq. (5.41) and Eq. (5.44) into Eq. (5.42), we obtain $C_1 = 0$ and $C_2 = 0$, respectively. Equation (5.42) reduces to

$$y = \frac{wx^2}{2T_0} \tag{5.45}$$

We can now evaluate T_0 since we notice that at $x = L/2$, $y = h$.

$$T_0 = \frac{wL^2}{8h}$$

so that

$$y = \frac{4h}{L^2} x^2 \tag{5.46}$$

which is the equation of a parabola.

The tension in the cable at any point can be found by considering the free-body diagram in Fig. 5.39(b). Note that T_0 acts in the horizontal direction at the origin where the slope of the cable is zero. The equilibrium equations are:

$$T\cos\theta = T_0 \qquad T\sin\theta = wx \tag{5.47}$$

Squaring both sides of each equation and adding, we obtain

$$T^2 = T_0^2 + w^2x^2 \qquad \text{or} \qquad T = \sqrt{T_0^2 + w^2x^2} \tag{5.48}$$

where

$$T_{max} = \sqrt{T_0^2 + \frac{w^2L^2}{4}}$$

Eliminating T from the two expressions in Eq. (5.47) gives

$$\tan\theta = \frac{wx}{T_0} \tag{5.49}$$

To determine the length of the cable, consider a differential length of cable ds in Fig. 5.39(c).

$$ds = \sqrt{(dx)^2 + (dy)^2} = dx\sqrt{1 + \left(\frac{dy}{dx}\right)^2} \tag{5.50}$$

Now the derivative of Eq. (5.45) is

$$\frac{dy}{dx} = \frac{wx}{T_0}$$

so that Eq. (5.50) can be written

$$ds = dx\sqrt{1 + \left(\frac{wx}{T_0}\right)^2} \tag{5.51}$$

The integral of this equation is

$$s = \int_0^{L/2} \sqrt{1 + \left(\frac{wx}{T_0}\right)^2}\, dx \tag{5.52}$$

which represents half the total length of the cable. This integration can be simplified if we use the binomial series expansion as follows:

$$\sqrt{1 + \left(\frac{wx}{T_0}\right)^2} = 1 + \frac{1}{2}\frac{w^2x^2}{T_0^2} - \frac{1}{8}\frac{w^4x^4}{T_0^4} + \cdots \tag{5.53}$$

Substituting this into Eq. (5.52) and integrating over one-half of the length, we obtain

$$\begin{aligned} s &= \int_0^{L/2} \left(1 + \frac{1}{2}\frac{w^2x^2}{T_0^2} - \frac{1}{8}\frac{w^4x^4}{T_0^4} + \cdots\right) dx \\ &= L\left(\frac{1}{2} + \frac{w^2L^2}{48T_0^2} - \frac{w^4L^4}{1280T_0^4} + \cdots\right) \end{aligned}$$

The total length of the cable is

$$S = 2s = L\left(1 + \frac{w^2L^2}{24T_0^2} - \frac{w^4L^4}{640T_0^4} + \cdots\right)$$

Since $w/T_0 = 8h/L^2$, we substitute this to obtain

$$S = L\left[1 + \frac{8}{3}\left(\frac{h}{L}\right)^2 - \frac{32}{5}\left(\frac{h}{L}\right)^4 + \cdots\right] \tag{5.54}$$

If $h/L < \frac{1}{4}$, this series converges. In most practical cases, h is much less than $L/4$ and a good approximation for the total length of the cable is

$$S \approx L\left[1 + \frac{8}{3}\left(\frac{h}{L}\right)^2\right]$$

Example 5.22

A cable is to span 200 ft across a canyon as shown in Fig. 5.40. Point B is known to be 30 ft higher than point A. If the force necessary to break the cable is 100,000 lb, find the maximum uniform load w (lb/ft) that the cable can carry.

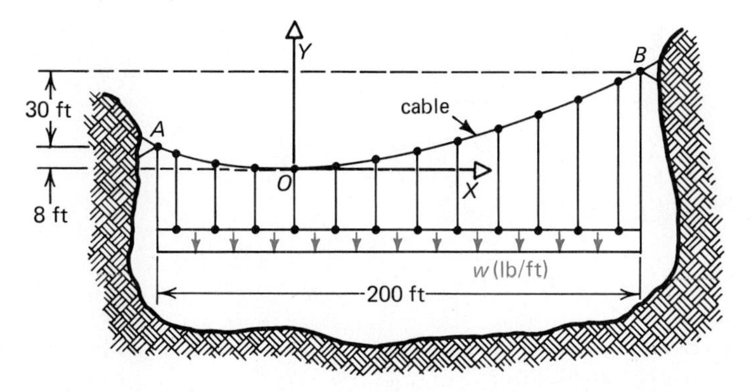

Figure 5.40

Since the distributed load is uniform, Eqs. (5.45) and (5.48) are applicable for this case. The first step in our solution is to locate the origin of the coordinate system at the lowest point along the cable from A to B. Label point A as $(x_A, 8)$ and point B as $(x_A + 200, 38)$. Using Eq. (5.45) for points A and B, we have

$$8 = \frac{wx_A^2}{2T_0}$$

and

$$38 = \frac{w(x_A + 200)^2}{2T_0}$$

Substituting the first equation into the second yields

$$38 = \frac{8}{x_A^2}(x_A + 200)^2$$

$$x_A = -63.0,\ 169.6$$

We choose $x_A = -63.0$ so that $x_B = 137.0$ ft. If we choose $x_A = 169.6$ then $x_B = 369.6$ ft which is an impossible solution for the given problem.

Now, at $x_B = 137$ ft, the maximum tension occurs in the cable. Rewriting Eq. (5.48) at this coordinate, we have

$$T_{\max} = \sqrt{T_0^2 + w^2 x_B^2}$$

Again from Eq. (5.45) we have

$$y_B = 38 = \frac{w x_B^2}{2T_0}$$

and substituting into the expression for $T_{\max}$ yields

$$T_{\max}^2 = w^2\left(\frac{x_B^2}{76}\right)^2 + w^2 x_B^2$$

$$= w^2 x_B^2\left[\left(\frac{x_B}{76}\right)^2 + 1\right]$$

Since $x_B = 137$ ft, this reduces to

$$T_{\max}^2 = 4.25 w^2 x_B^2$$

and, therefore,

$$w = \frac{T_{\max}}{2.06 x_B} = \frac{100{,}000}{2.06(137)} = 354 \text{ lb/ft} \qquad \textit{Answer}$$

Catenary Cable

A catenary cable is one that is loaded only by its own weight as shown in Fig. 5.41(a). By considering an element of length ds, we can derive expressions similar to Eqs. (5.35) and (5.36) except that ds now replaces dx, i.e.,

$$d(T\cos\theta) = 0$$

$$d(T\sin\theta) = w\,ds$$

so that the governing equation for the catenary cable is

$$\frac{d^2y}{dx^2} = \frac{w}{T_0}\frac{ds}{dx} \qquad (5.55)$$

Recall that

$$ds = dx\sqrt{1 + \left(\frac{dy}{dx}\right)^2}$$

so that Eq. (5.55) can be written

$$\frac{d^2y}{dx^2} = \frac{w}{T_0}\sqrt{1 + \left(\frac{dy}{dx}\right)^2}$$

To solve this differential equation, let $u = dy/dx$. Now $d^2y/dx^2 = du/dx$ so that

$$\frac{du}{dx} = \frac{w}{T_0}\sqrt{1 + u^2} \tag{5.56}$$

Since w is a constant, this equation can be integrated directly (see a table of integrals) to yield

$$\ln\left(u + \sqrt{1 + u^2}\right) = \frac{w}{T_0}x + C_1 \tag{5.57}$$

where C_1 is the constant of integration. The origin of the coordinate system is selected in Fig. 5.41(b) such that at $x = 0$,

$$\frac{dy}{dx} = u = 0$$

Using this boundary condition in Eq. (5.57), we find that $C_1 = 0$, and therefore

$$\ln\left(u + \sqrt{1 + u^2}\right) = \frac{wx}{T_0} \qquad \text{or} \qquad \sqrt{1 + u^2} = e^{wx/T_0} - u$$

Squaring both sides and simplifying, we obtain

$$u = \frac{e^{wx/T_0} - e^{-wx/T_0}}{2} = \sinh\left(\frac{wx}{T_0}\right)$$

Now,

$$u = \frac{dy}{dx} = \sinh\left(\frac{wx}{T_0}\right) \tag{5.58}$$

and we can integrate this equation to obtain

$$y = \frac{T_0}{w}\cosh\left(\frac{wx}{T_0}\right) + C_2 \tag{5.59}$$

The second boundary condition is that at $x = 0$, $y = 0$. Using this condition in Eq. (5.59), we have

$$C_2 = -\frac{T_0}{w}$$

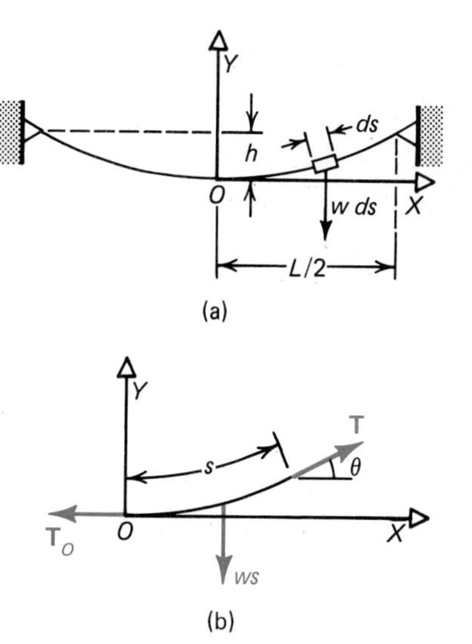

Figure 5.41

so that
$$y = \frac{T_0}{w}\left(\cosh\left(\frac{wx}{T_0}\right) - 1\right) \tag{5.60}$$

Equation (5.60) is the curve of the catenary cable. The sag h is given by the value of y at $x = L/2$, since the cable supports are at the same elevation as shown in Fig. 5.41.

$$y = h = \frac{T_0}{w}\left(\cosh\left(\frac{wL}{2T_0}\right) - 1\right)$$

To determine the length of the cable, we can consider a section of the catenary cable as shown in Fig. 5.41(b). For equilibrium,

$$T\cos\theta = T_0 \qquad \text{and} \qquad T\sin\theta = ws$$

or, by combining these equations,

$$\tan\theta = \frac{dy}{dx} = \frac{ws}{T_0} \tag{5.61}$$

Referring to Eq. (5.58), we can rewrite this equation as

$$\frac{ws}{T_0} = \sinh\left(\frac{wx}{T_0}\right)$$

and therefore

$$s = \frac{T_0}{w}\sinh\left(\frac{wx}{T_0}\right) \tag{5.62}$$

This provides the basic equation for the length of the catenary cable.

The tension T is obtained as before:

$$\begin{aligned} T &= \sqrt{T_0^2 + w^2s^2} = \sqrt{T_0^2 + T_0^2\sinh^2\left(\frac{wx}{T_0}\right)} \\ &= T_0\sqrt{1 + \sinh^2\left(\frac{wx}{T_0}\right)} \\ &= T_0\cosh\left(\frac{wx}{T_0}\right) \end{aligned} \tag{5.63}$$

As a final comment on the catenary cable, we observe that Eq. (5.60) is the basic expression for the curve. Notice that if we wish to find T_0, we need the appropriate x and y coordinate as well as the load w. With this information, however, we will require a trial-and-error process to find T_0. This is illustrated by the next example.

Example 5.23

A cable weighing 2 lb/ft is strung between two points at the same elevation across a horizontal span of 400 ft. If the sag is 100 ft, find the maximum tension in the cable.

Referring to Eq. (5.60), we have for $x = 200$, $y = 100$,

$$100 = \frac{T_0}{2}\left[\cosh\frac{2(200)}{T_0} - 1\right] \qquad \text{or} \qquad \cosh\left(\frac{400}{T_0}\right) - \frac{200}{T_0} = 1$$

A trial-and-error approach to finding T_0 is tabulated in Table 5.4. From the tabulated data, we find that $T_0 = 430$ lb. The maximum tension in the cable occurs when $x = 200$ ft. Using Eq. (5.63), we obtain

$$T_{max} = T_0 \cosh\frac{2(200)}{T_0}$$

$$= 430 \cosh (0.930)$$

$$= 630 \text{ lb} \qquad \textit{Answer}$$

TABLE 5.4

T_0	$\cosh\left(\frac{400}{T_0}\right)$	$\frac{200}{T_0}$	$\cosh\left(\frac{400}{T_0}\right) - \frac{200}{T_0}$
200	3.76	1	2.76
300	2.03	0.667	1.363
400	1.543	0.500	1.043
410	1.515	0.488	1.027
420	1.489	0.476	1.013
430	1.4648	0.465	0.9998 (close enough)

PROBLEMS

5.99 Find the reactions at A and D and the tension in the cable section AB in Fig. P5.99 for $L = 10$ m.

5.100 Find the reactions at A and D and the tension in the cable section AB in Fig. P5.99 for $L = 20$ m.

5.101 Find the tension in the cable section AB shown in Fig. P5.101.

5.102 Find the reactions at A and D for the cable shown in Fig. P5.102.

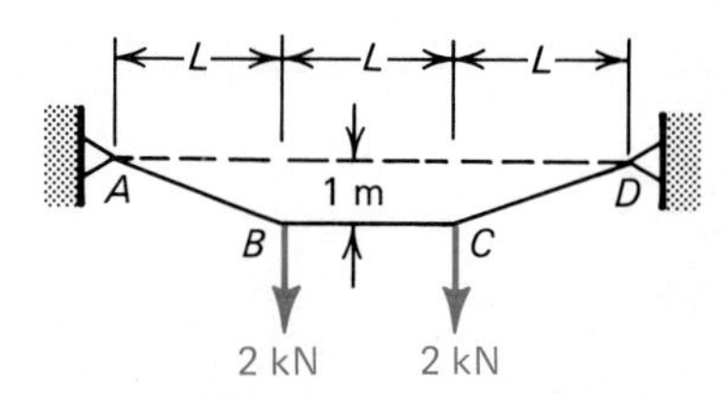

Figure P5.99

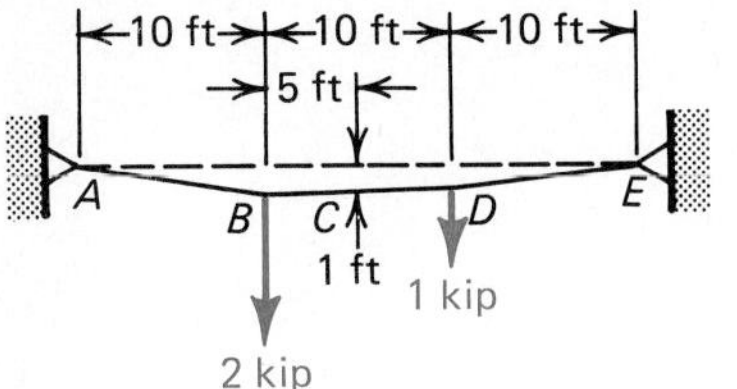
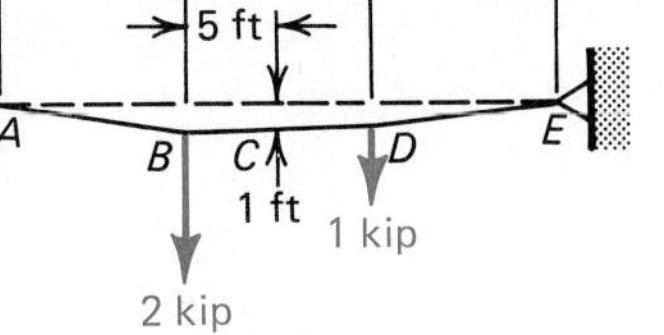

Figure P5.101

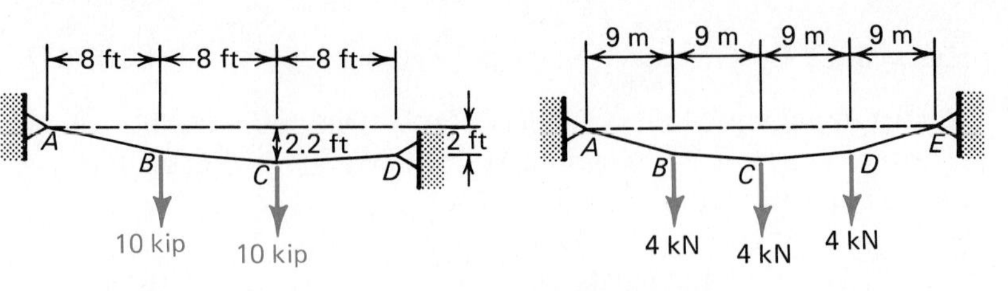

Figure P5.102

Figure P5.103

5.103 If the maximum deflection of the cable in Fig. P5.103 is $\frac{1}{2}$ m, find the reactions at *A* and *E*.

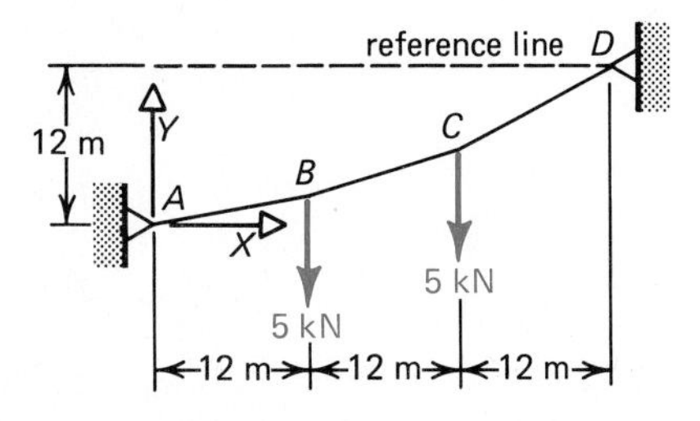

Figure P5.104

5.104 A force of 8 kN is known to be the vertical reaction at *D* for the cable shown in Fig. P5.104. Find the reaction at *A* and deflection at *C* from the reference line.

5.105 The cable in Fig. P5.105 carries a uniform load $w = 200$ lb/ft. Find the coordinates of the lowest point of the cable and the magnitude of the reactions at the supports *A* and *B*.

5.106 A cable whose maximum allowable force is 150 kip is to carry a uniform load $w = 100$ lb/ft across a span. If both the supports are at the same level, what is the maximum span if the sag must be limited to 10 ft?

5.107 A parabolic cable is designed to span 400 m and carry a uniform load of 100 N/m with the supports at the same level. If the maximum sag allowed is 20 m, determine the strength of the cable for this application.

5.108 The maximum allowable force of the cable in Fig. P5.108 is 120 kN. Find the maximum uniform load *w* that the cable can support.

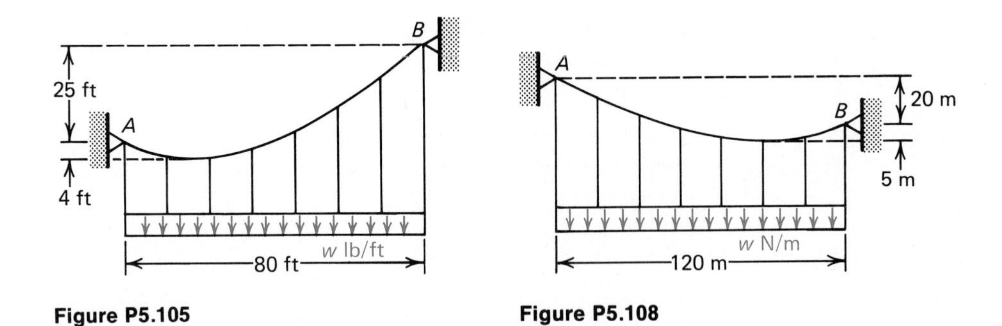

Figure P5.105

Figure P5.108

5.109 A cable weighing 50 lb/ft spans a river 200 ft wide. The cable supports are at the same elevation and the maximum allowable force of the cable is 10,000 lb. Find the sag of the cable.

5.110 An electric transmission line weighing 90 lb per 1000 ft spans 400 ft between towers at the same elevation. If the cable sags 15 ft, find the maximum tension in the cable.

5.9 Summary

When a force is applied over a specified area, it is called a distributed force. This fact, when included in our analysis, leads to the definition of the center of gravity and the centroid. We have shown how the centroids of lines, areas, and volumes may be obtained. Examples of distributed loads studied here included fluid pressure, buoyancy, and cables.

A. The important *ideas*:

1. The centroid is coincident with the center of gravity if the body is homogeneous.
2. The centroid of a composite body is found by treating the body as a combination of simpler geometric shapes.
3. The theorems of Pappus and Guldinus provide convenient formulas to find areas and volumes of bodies of revolution.
4. Distributed loads on beams, fluid pressure, and buoyancy are examples of distributed loads that can be replaced by a resultant force applied at the center of distributed force.
5. Complete analysis of cables subjected to distributed forces includes the solution of a differential equation with boundary conditions.

B. The important *equations*:

Centroids

Lines: $$\bar{x} = \frac{1}{L}\int x\,dL \qquad \bar{y} = \frac{1}{L}\int y\,dL \tag{5.9}$$

Areas: $$\bar{x} = \frac{1}{A}\int x\,dA \qquad \bar{y} = \frac{1}{A}\int y\,dA \tag{5.12}$$

Volumes: $$\bar{x} = \frac{1}{V}\int x\,dV \qquad \bar{y} = \frac{1}{V}\int y\,dV \tag{5.6}$$

$$\bar{z} = \frac{1}{V}\int z\,dV$$

Area of a surface of revolution:

$$A = 2\pi\bar{y}L \tag{5.24}$$

Volume of a body of revolution:

$$V = 2\pi\bar{y}A \tag{5.25}$$

Fluid pressure:

$$p = p_a - p_0 = \gamma y = \rho g y \tag{5.32}$$

Parabolic cable:

$$y = \frac{wx^2}{2T_0} \tag{5.45}$$

$$T = \sqrt{T_0^2 + w^2x^2} \tag{5.48}$$

Catenary cable:

$$y = \frac{T_0}{w}\left(\cosh\frac{wx}{T_0} - 1\right) \tag{5.60}$$

$$T = T_0 \cosh\frac{wx}{T_0} \tag{5.63}$$

6 Moments of Inertia

The area moment of inertia and the mass moment of inertia are two important properties that are used extensively in mechanics. For example, when we wish to evaluate the strength of structural beams and columns, we make use of the area moment of inertia, which, in effect, is a geometric property of the structural member. Likewise, we shall discover in dynamics that the rotational equations of motion for rigid bodies include the mass moment of inertia, a property that depends on the geometric shape and density of a body. In this chapter we shall derive general expressions for the area moment of inertia and the mass moment of inertia along with examples commonly found in engineering applications.

6.1 Area Moment of Inertia

Consider a flat surface $ABCD$ submerged in a fluid as shown in Fig. 6.1(a). We wish to evaluate the moment about the X axis due to the force on the area $ABCD$. The magnitude of the differential force $d\mathbf{F}$ due to the pressure p acting on the element of area dA is

$$dF = p\,dA$$

This force is shown in Fig. 6.1(b). The magnitude of the moment about the X axis due to $d\mathbf{F}$ is

$$dM_x = yp\,dA$$

The magnitude of the moment due to the pressure on the entire area $ABCD$ is expressed as

$$M_x = \int_{\text{area}} dM_x$$

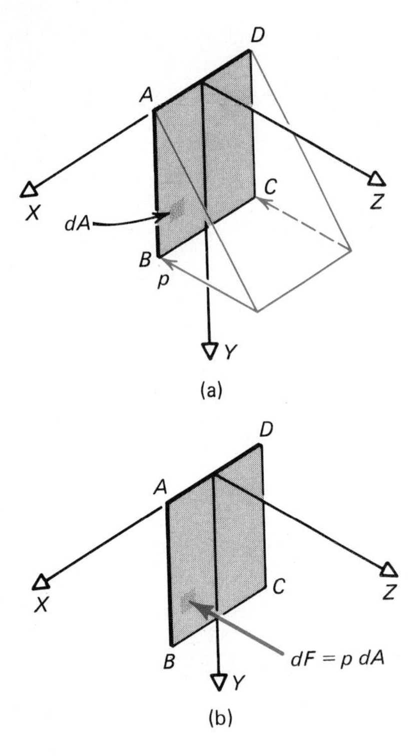

Figure 6.1

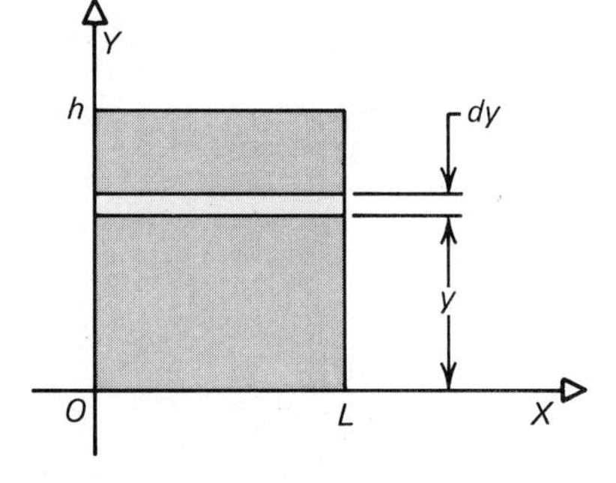

Figure 6.2

so that

$$M_x = \int_{\text{area}} yp\,dA$$

Since the pressure increases with the depth, we replace the pressure p with γy, so that

$$M_x = \int \gamma y^2\,dA \tag{6.1}$$

If γ is assumed to be constant, then

$$M_x = \gamma \int y^2\,dA$$

The integral $\int y^2\,dA$ is called a second moment of area about the X axis or the area moment of inertia. In this text, we shall call this integral the area moment of inertia. Since the area moment of inertia is about the X axis, the symbol I_x is used. If the axis were the Y axis, the area moment of inertia is written as I_y. The area moments of inertia, therefore, about the X and Y axes are defined as

$$I_x = \int y^2\,dA \qquad I_y = \int x^2\,dA \tag{6.2}$$

The units for the area moment of inertia are meter^4 or foot^4. As we shall see in what follows, the area moment of inertia depends on the shape of the area and the location of the axes chosen.

Rectangular Areas

If the area under consideration is rectangular, Eqs. (6.2) can be simplified. Consider the area shown in Fig. 6.2, where the differential area of width L and height dy is

$$dA = L\,dy$$

Substituting this into the first of Eqs. (6.2), we have

$$I_x = \int y^2\,dA = \int_0^h Ly^2\,dy$$

which, upon integration, yields

$$I_x = \frac{Lh^3}{3}$$

But we know that the area is given by $A = Lh$, so that

$$I_x = \frac{Ah^2}{3}$$

We can similarly show that the area moment of inertia about the Y axis is

$$I_y = \frac{hL^3}{3} = \frac{AL^2}{3}$$

Let us now replace the entire area of the rectangle in Fig. 6.2 by a fictitious thin strip of equal area A located at a distance k_x from the X axis as shown in Fig. 6.3. Since the fictitious area is a thin strip, the area moment of inertia about the X axis is

$$I_x = \int k_x^2 \, dA = k_x^2 A \tag{6.3}$$

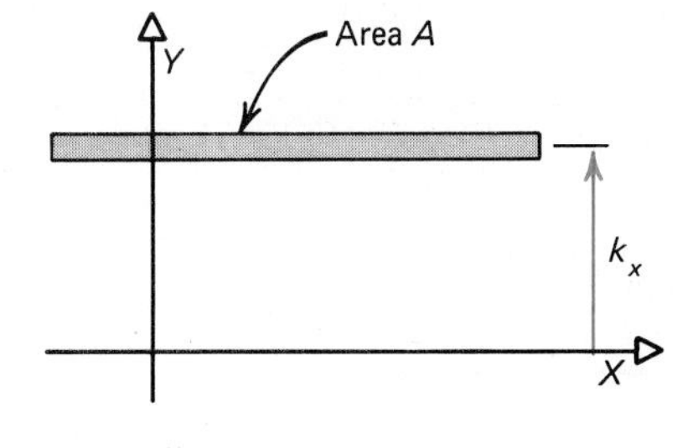

Figure 6.3

Here, k_x is called the *radius of gyration*. Rearranging Eq. (6.3),

$$k_x = \sqrt{\frac{I_x}{A}} \tag{6.4}$$

The expression for the radius of gyration is true for any surface. In the case of the rectangle in Fig. 6.2,

$$k_x = \sqrt{\frac{Ah^2/3}{A}} = \frac{h}{\sqrt{3}}$$

Likewise, the radius of gyration about the Y axis is

$$k_y = \sqrt{\frac{I_y}{A}}$$

In the case of the rectangle,

$$k_y = \sqrt{\frac{AL^2/3}{A}} = \frac{L}{\sqrt{3}}$$

The radius of gyration, which is measured in meters or feet, is a fictitious distance from the axis in question. If at this distance we envision the actual area to be represented by a thin strip of equal area, the area moment of inertia of this thin strip would be the same as the area moment of inertia of the actual area. This property is useful for computing the area moment of inertia for many irregular shapes commonly found in engineering and is listed in handbooks for various sizes so that the area moment of inertia can be quickly computed.

Area Polar Moment of Inertia

The area moment of inertia about the axis passing through the origin O and perpendicular to the XY plane in Fig. 6.4 is the area polar moment of inertia. It is designated as J and is defined mathematically as

$$J = \int r^2 \, dA \tag{6.5}$$

Since

$$r^2 = x^2 + y^2$$

we have

$$J = \int (x^2 + y^2) \, dA$$

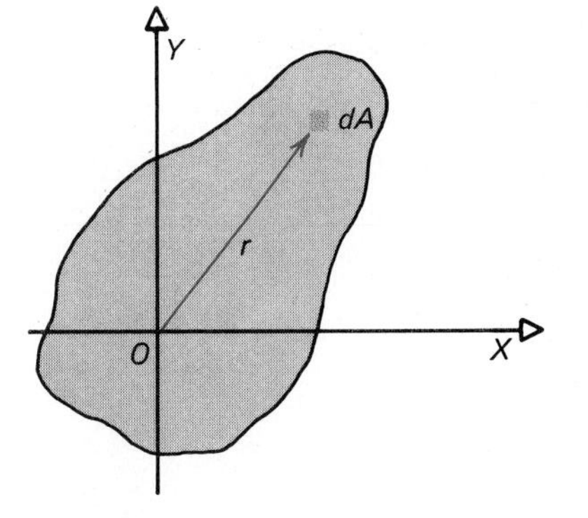

Figure 6.4

$$= \int x^2\,dA + \int y^2\,dA$$

$$= I_y + I_x \tag{6.6}$$

that is, the area polar moment of inertia is the sum of the two area moments of inertia I_x and I_y. The radius of gyration about the polar axis is expressed in the same form as Eq. (6.4), that is,

$$k_z = \sqrt{\frac{J}{A}} \tag{6.7}$$

in which k_z is the radius of gyration about the Z axis.

We can establish a relationship between k_z, k_x, and k_y by substituting

$$J = k_z^2 A \qquad I_x = k_x^2 A \qquad I_y = k_y^2 A$$

into Eq. (6.6) to obtain

$$k_z^2 A = k_x^2 A + k_y^2 A$$

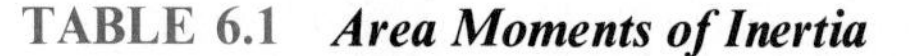

TABLE 6.1 *Area Moments of Inertia*

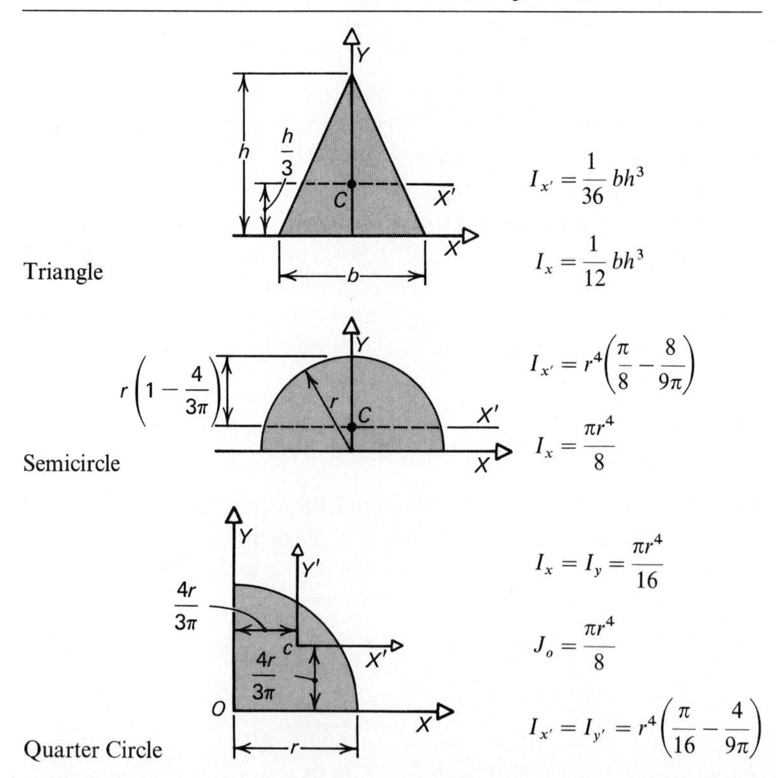

or
$$k_z = \sqrt{k_x^2 + k_y^2} \tag{6.8}$$

Table 6.1 lists the area moments of inertia as well as the area polar moments of inertia for some common geometric shapes. These geometric properties are important in the stress analysis of structures, which is studied in courses concerned with strength of materials, machine design, and theory of elasticity.

Example 6.1

Find the area moment of inertia about (a) the X axis and (b) the Y axis for the triangle shown in Fig. 6.5.

(a) The selection and arrangement of the element of area dA is very important in solving this problem. To find the area moment of inertia I_x, we choose the horizontal strip shown in Fig. 6.5(a). Note that each point on dA is located at a constant distance y

TABLE 6.1 ***Continued***

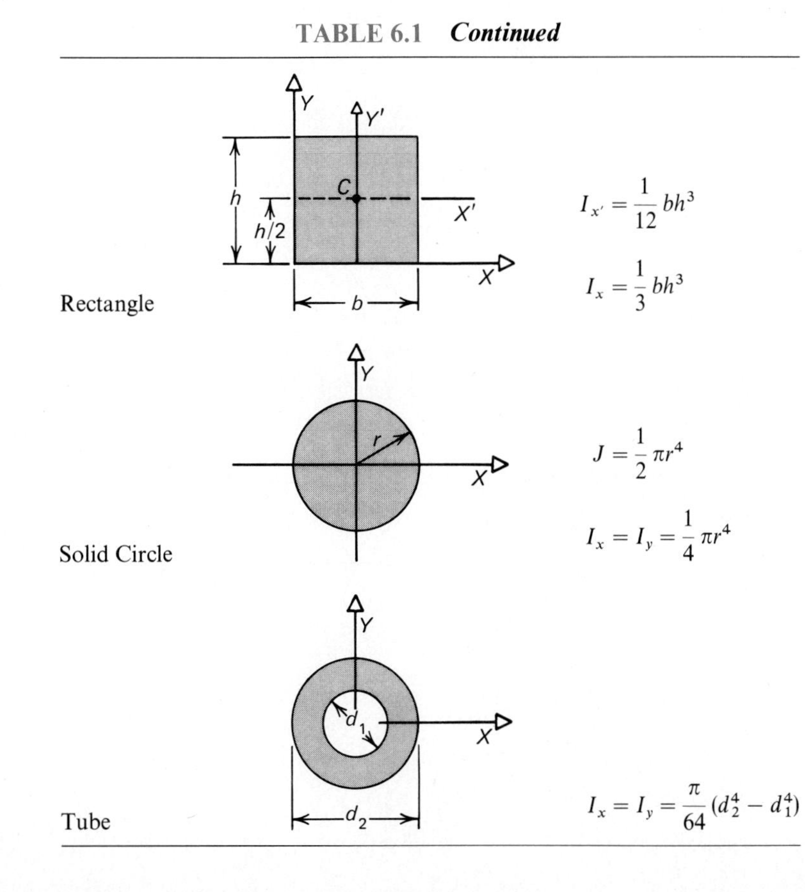

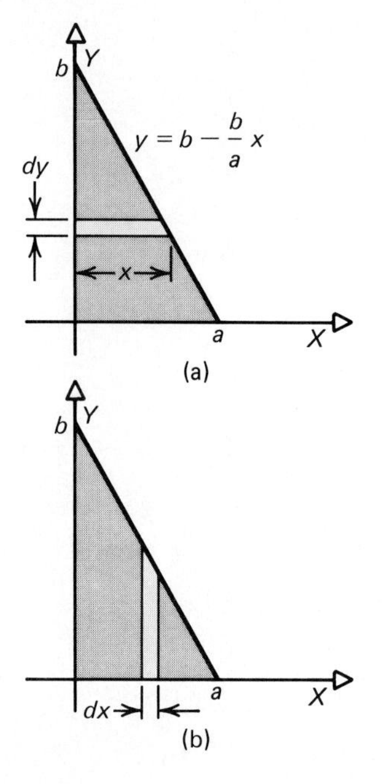

Figure 6.5

from the X axis. This arrangement of dA simplifies our derivation. Now

$$dA = x\,dy$$

Since
$$y = b - \frac{b}{a}x$$

then
$$x = \frac{a}{b}(b - y)$$

so that
$$dA = \frac{a}{b}(b - y)\,dy$$

The area moment of inertia about the X axis is

$$I_x = \int y^2\,dA$$

$$= \int_0^b \frac{a}{b}(b - y)y^2\,dy$$

$$= a\int_0^b y^2\,dy - \frac{a}{b}\int_0^b y^3\,dy$$

$$= a\left[\frac{y^3}{3}\right]_0^b - \frac{a}{b}\left[\frac{y^4}{4}\right]_0^b$$

$$= \frac{1}{3}ab^3 - \frac{1}{4}ab^3$$

$$= \frac{ab^3}{12}$$ *Answer*

(b) To find the area moment of inertia about the Y axis, the element of area is now drawn parallel to the Y axis as shown in Fig. 6.5(b). All points on dA are located at a constant distance x from the Y axis.

$$dA = y\,dx = \left(b - \frac{b}{a}x\right)dx$$

Now
$$I_y = \int x^2\,dA$$

$$= \int_0^a \left(b - \frac{b}{a}x\right)x^2\,dx$$

$$I_y = b\int_0^a x^2\,dx - \frac{b}{a}\int_0^a x^3\,dx$$

$$= b\left[\frac{x^3}{3}\right]_0^a - \frac{b}{a}\left[\frac{x^4}{4}\right]_0^a$$

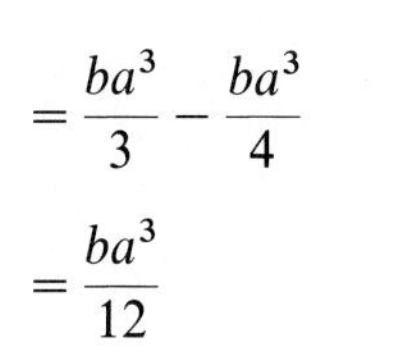

$$= \frac{ba^3}{3} - \frac{ba^3}{4}$$

$$= \frac{ba^3}{12} \qquad \textit{Answer}$$

Example 6.2

Calculate I_y for the shaded area of Fig. 6.6 for $a = 2$ m and $b = 1$ m.

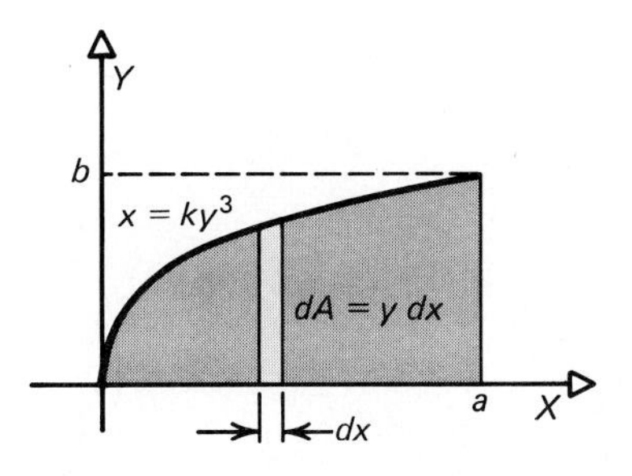

Figure 6.6

As in the previous example, we establish the differential area parallel to the Y axis since we wish to find I_y. Now $dA = y\,dx$, and since $x = ky^3$, we have

$$dx = 3ky^2\,dy$$

so that

$$dA = 3ky^3\,dy$$

The area moment of inertia about the Y axis is

$$I_y = \int x^2\,dA$$

$$= \int_0^b (k^2y^6)3ky^3\,dy$$

$$= \int_0^b 3k^3y^9\,dy$$

$$= \frac{3k^3b^{10}}{10}$$

Since $x = a$ when $y = b$, $k = a/b^3$, so that

$$I_y = \frac{3b^{10}}{10}\frac{a^3}{b^9} = \frac{3ba^3}{10}$$

Substituting $a = 2$ m and $b = 1$ m results in

$$I_y = 2.4 \text{ m}^4 \qquad \textit{Answer}$$

Example 6.3

Determine the area moment of inertia I_x and area polar moment of inertia J for the circular area shown in Fig. 6.7.

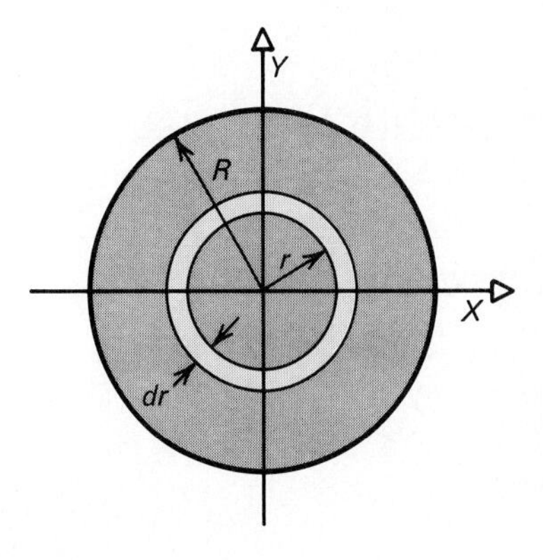

Figure 6.7

In this problem it is convenient to use polar coordinates. The differential area is the thin, shaded ring, so that

$$dA = 2\pi r\,dr$$

The polar moment of area becomes

$$J = \int r^2\,dA$$

$$= \int_0^R 2\pi r^3 \, dr$$

$$= \frac{\pi R^4}{2} \qquad \textit{Answer}$$

From symmetry, $I_x = I_y$, so that

$$J = I_x + I_y = 2I_x$$

and

$$I_x = \frac{J}{2}$$

$$= \frac{\pi R^4}{4} \qquad \textit{Answer}$$

Example 6.4

For the triangle shown in Fig. 6.8(a) determine (a) the area moments of inertia about the X and Y axes and the respective radii of gyration, and (b) the area polar moment of inertia about the origin O and the polar radius of gyration.

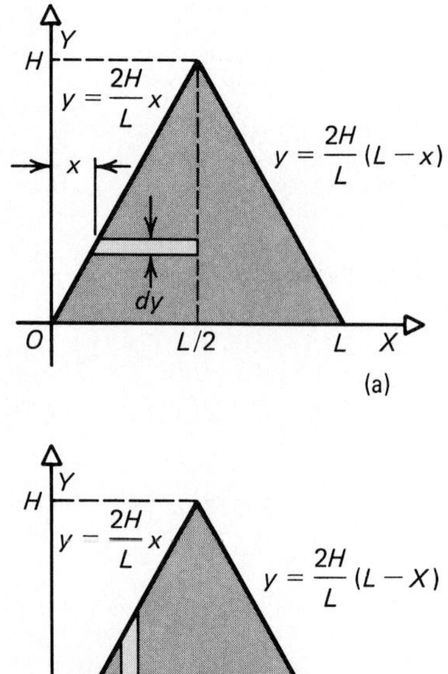

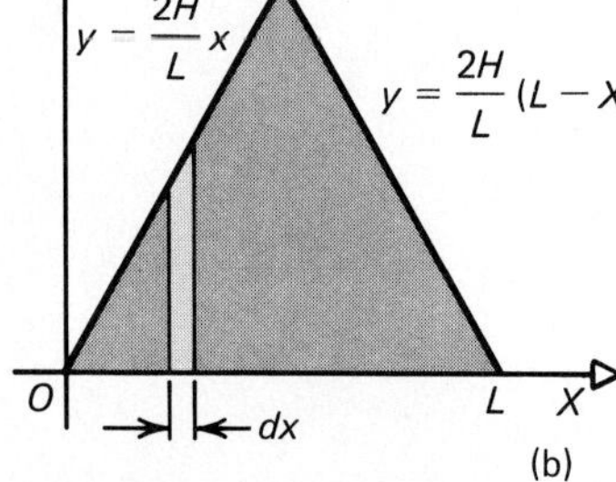

Figure 6.8

(a) To obtain the area moment of inertia about the X axis, it is convenient to treat half of the triangle by a differential element of area dA as shown in Fig. 6.8(a). Note that dA is arranged parallel to the X axis to find I_x. The differential area dA is

$$dA = \left(\frac{L}{2} - x\right) dy$$

Since $y = 2Hx/L$, we have $dy = 2H\,dx/L$. Now

$$I_x = \int y^2 \, dA$$

$$= 2\int_0^{L/2} \frac{4H^2}{L^2} x^2 \left(\frac{L}{2} - x\right) \frac{2H}{L} \, dx$$

Note that we multiply the integral by two in order to account for the entire area. The above integral becomes

$$I_x = \frac{16H^3}{L^3} \left[\frac{Lx^3}{6} - \frac{x^4}{4}\right]_0^{L/2}$$

$$= \frac{16H^3}{L^3} \frac{L^4}{192}$$

$$= \frac{H^3 L}{12} \qquad \textit{Answer}$$

The radius of gyration is

$$k_x = \sqrt{\frac{I_x}{A}}$$

Since the area of the triangle is $A = LH/2$

$$k_x = \sqrt{\frac{H^3L}{12}\frac{2}{LH}}$$

$$= \frac{H}{\sqrt{6}} \qquad \textit{Answer}$$

The area moment of inertia about the Y axis is

$$I_y = \int x^2\, dA \qquad dA = y\, dx$$

This element of area is now arranged parallel to the Y axis as shown in Fig. 6.8(b). Note that y is given by two equations depending upon the particular side of the triangle.

$$I_y = \int x^2\, dA = \int x^2 y\, dx$$

Now $\quad y = \frac{2H}{L}x \qquad$ for $\qquad 0 \le x \le L/2$

$$y = \frac{2H}{L}(L - x) \qquad \text{for} \qquad L/2 \le x \le L$$

Thus $\quad I_y = \int_0^{L/2} \frac{2H}{L}x^3\, dx + \int_{L/2}^{L} \frac{2H}{L}(L - x)x^2\, dx$

$$= \frac{HL^3}{32} + \frac{11}{96}HL^3$$

$$= \frac{7}{48}HL^3 \qquad \textit{Answer}$$

The radius of gyration is

$$k_y = \sqrt{\frac{7}{48}HL^3\frac{2}{HL}}$$

$$= L\sqrt{\frac{7}{24}} = 0.54L \qquad \textit{Answer}$$

(b) The area polar moment of inertia about the origin O is

$$J = I_x + I_y$$

$$= \frac{H^3L}{12} + \frac{7HL^3}{48}$$

$$J = \frac{HL}{12}\left(H^2 + \frac{7}{4}L^2\right) \qquad \textit{Answer}$$

and

$$k_z = \sqrt{\tfrac{1}{6}(H^2 + \tfrac{7}{4}L^2)} \qquad \textit{Answer}$$

PROBLEMS

6.1 Calculate the area moment of inertia about the X and Y axes for the rectangle shown in Fig. P6.1.

6.2 Calculate the area polar moment of inertia for the rectangle shown in Problem 6.1.

6.3 Calculate the radius of gyration, k_z, for the circular area shown in Fig. P6.3.

Obtain the area moment of inertia:

6.4 About the X axis for the shaded area of Fig. P6.4.

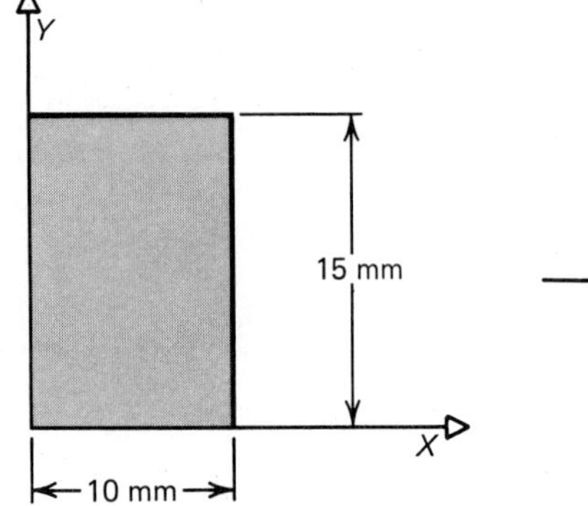

Figure P6.1

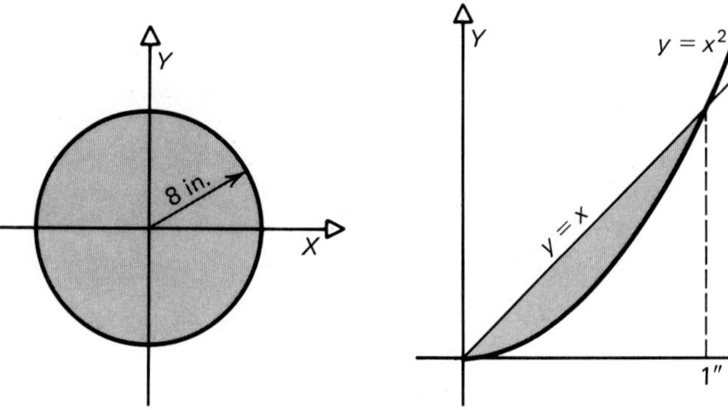

Figure P6.3 **Figure P6.4**

6.5 About the X axis for the shaded area of Fig. P6.5.

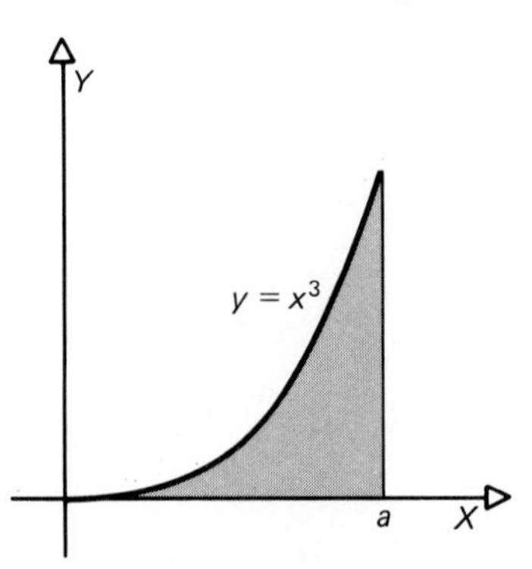

Figure P6.5

6.6 About the X axis for the shaded area of Fig. P6.6.

6.7 About the X axis for the shaded area of Fig. P6.7.

6.8 About the X axis for the shaded area of Fig. P6.8.

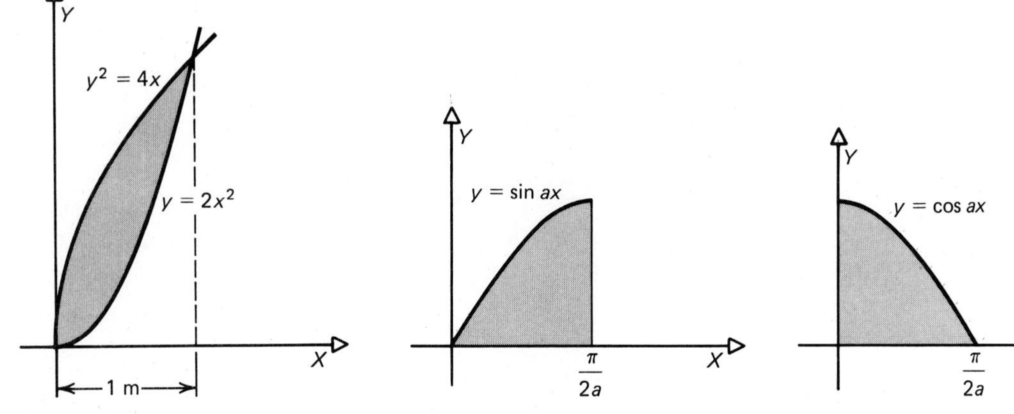

Figure P6.6 **Figure P6.7** **Figure P6.8**

6.9 About the X axis for the shaded area of Fig. P6.9.

6.10 About the Y axis for the shaded area in Problem 6.4.

6.11 About the Y axis for the shaded area in Problem 6.5.

6.12 About the Y axis for the shaded area in Problem 6.6.

6.13 About the Y axis for the shaded area in Problem 6.7.

6.14 About the Y axis for the shaded area in Problem 6.8.

6.15 About the Y axis for the shaded area in Problem 6.9.

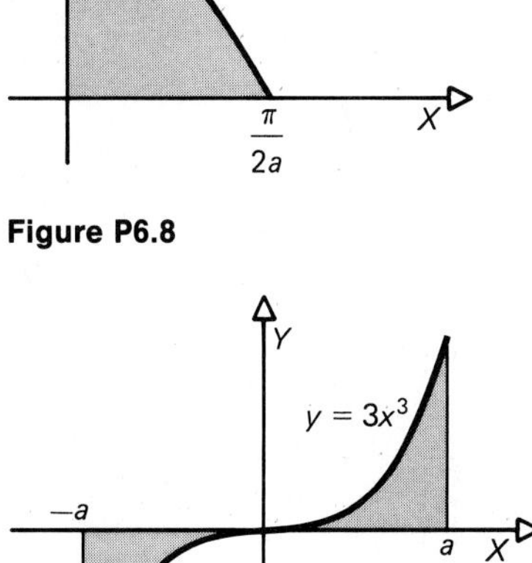

Figure P6.9

6.16 Find (a) the area polar moment of inertia for the quarter circle in Fig. P6.16. (b) What is the radius of gyration about the X axis?

6.17 Find (a) the area polar moment of inertia for the half circle in Fig. P6.17, and (b) the corresponding radius of gyration.

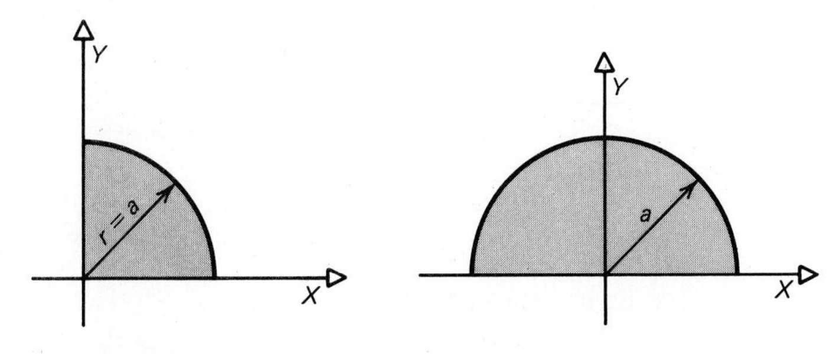

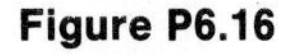
Figure P6.16 **Figure P6.17**

6.18 Find the area polar moment of inertia for the hollow tube shown in Fig. P6.18.

6.19 Find the area moment of inertia about the X axis for the ellipse shown in Fig. P6.19.

6.20 Find the polar radius of gyration for the hollow tube shown in Problem 6.18.

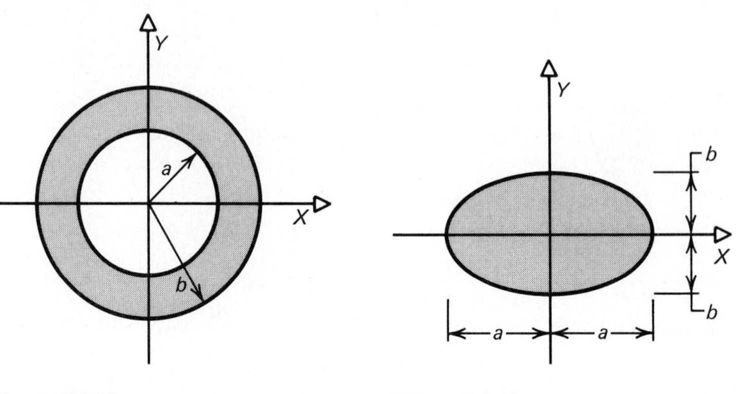

Figure P6.18 Figure P6.19

6.2 Parallel-Axis Theorem

We mentioned earlier that the value of the area moment of inertia depends upon the axes chosen. Here we shall derive a simple formula that enables us to obtain the area moment of inertia about any parallel axis once we obtain the area moment of inertia about the centroidal axis. Consider the area shown in Fig. 6.9 where the second moment of area about the Y axis is

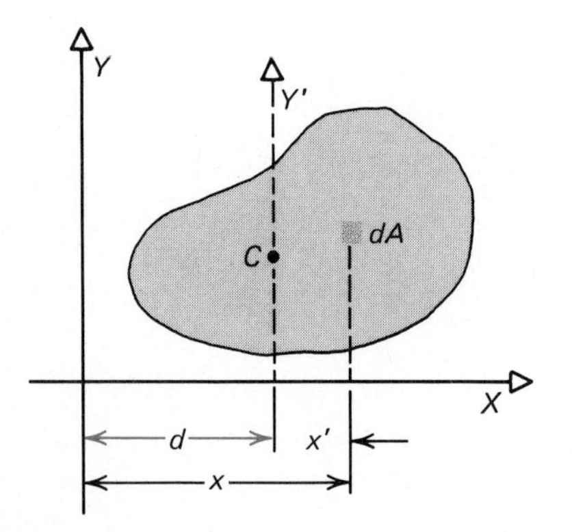

Figure 6.9

$$I_y = \int x^2\, dA$$

Let the Y' axis be drawn parallel to the Y axis at a distance d. We can therefore write $x = x' + d$, so that

$$I_y = \int (x' + d)^2\, dA = \int x'^2\, dA + d^2 \int dA + 2d \int x'\, dA$$

If we select the Y' axis to be a centroidal axis, i.e., it passes through the centroid of the area, then

$$\int x'\, dA = 0$$

and the previous expression reduces to

$$I_y = \int x'^2\, dA + d^2 \int dA$$

But the first integral is the area moment of inertia about the Y' axis and the second integral is simply the total area. Let $\bar{I}_{y'}$ be the area moment of inertia about the centroidal axis, that is,

$$\bar{I}_{y'} = \int x'^2 \, dA$$

then substituting, we have

$$I_y = \bar{I}_{y'} + Ad^2 \tag{6.9}$$

This is known as the parallel-axis theorem. It states that the area moment of inertia about an axis parallel to the centroidal axis is equal to the sum of the area moment of inertia about the centroidal axis and Ad^2, where d is the distance between the centroidal axis and the axis parallel to the centroidal axis.

If we introduce the radius of gyration into Eq. (6.9), we obtain

$$k_y^2 A = \bar{k}_{y'}^2 A + Ad^2 \qquad \text{then} \qquad k_y^2 = \bar{k}_{y'}^2 + d^2$$

We may, in a similar manner, show that for the area polar moment of inertia

$$J = \bar{J} + Ad^2 \qquad \text{and} \qquad k^2 = \bar{k}^2 + d^2$$

where d is the distance between the centroid and the origin of the parallel axis.

Composite Areas

The results of the parallel-axis theorem may be used to advantage for obtaining the area moment of inertia of composite areas. Consider the two areas A_1 and A_2 shown in Fig. 6.10. Let us assume the area moments of inertia of A_1 and A_2 about the centroidal axis parallel to the Y axis are $\bar{I}_{1y'}$ and $\bar{I}_{2y'}$, respectively. Then the area moments of inertia of areas A_1 and A_2 about the Y axis are

$$\bar{I}_{1y'} + A_1 d_1^2 \qquad \text{and} \qquad \bar{I}_{2y'} + A_2 d_2^2$$

so that the total area moment of inertia about the Y axis is

$$I_y = (\bar{I}_{1y'} + \bar{I}_{2y'}) + (A_1 d_1^2 + A_2 d_2^2) \tag{6.10}$$

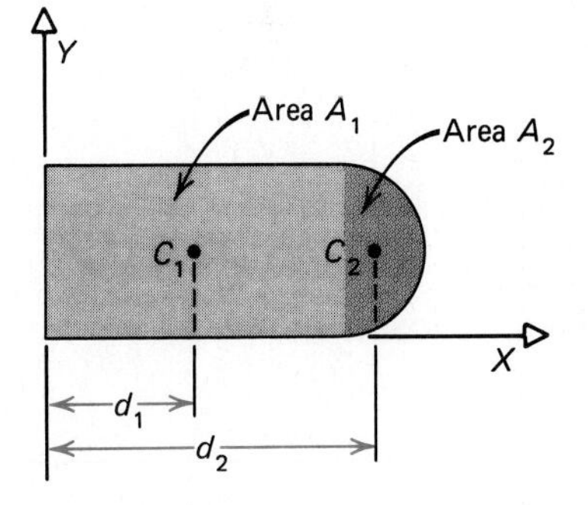

Figure 6.10

For the case of more than two areas, these results lead to the expression

$$I_y = \sum_i \bar{I}_{iy'} + \sum_i A_i d_{ix}^2 \tag{6.11a}$$

Likewise, the area moment of inertia about the X axis for a composite area is

$$I_x = \sum_i \bar{I}_{ix'} + \sum_i A_i d_{iy}^2 \tag{6.11b}$$

As we shall see in an example problem, the numerical computations for complicated geometric shapes are simplified by arranging the solution in tabular form.

The radius of gyration of a composite area cannot be obtained by adding the radii of gyration of individual areas. Rather, we must first compute the area moment of inertia by Eq. (6.11) and then substitute the result into Eq. (6.3). This is demonstrated in Example 6.7.

Example 6.5

Determine the area moment of inertia about the X' axis of the triangle shown in Fig. 6.11.

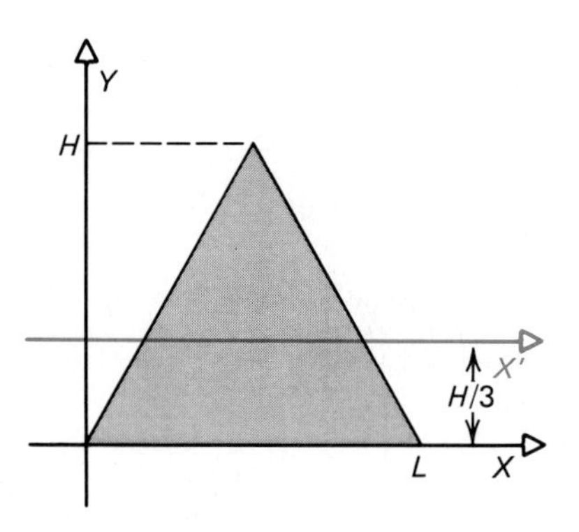

Figure 6.11

In this example, the X' axis passes through the centroid of the triangle so that we can make use of the parallel-axis theorem. From Example 6.4 we have calculated the area moment of inertia about the base of the triangle as

$$I_x = \frac{H^3L}{12}$$

The area of the triangle is

$$A = \frac{HL}{2}$$

Rearranging the parallel-axis theorem, we have the expression for the area moment of inertia about the centroidal axis.

$$\begin{aligned}\bar{I}_{x'} &= I_x - Ad^2 \\ &= \frac{H^3L}{12} - \frac{HL}{2}\frac{H^2}{9} \\ &= \frac{H^3L}{36}\end{aligned}$$

Answer

Example 6.6

For the semicircle in Table 6.1, establish that $I_x = \pi r^4/8$ using the given $\bar{I}_{x'}$.

By the parallel-axis theorem

$$I_x = \bar{I}_{x'} + Ad^2$$

in which

$$\bar{I}_{x'} = r^4\left(\frac{\pi}{8} - \frac{8}{9\pi}\right)$$

$$A = \frac{\pi r^2}{2}$$

$$d = \frac{4r}{3\pi}$$

Thus,

$$I_x = \frac{\pi r^4}{8} - \frac{8r^4}{9\pi} + \frac{\pi r^2}{2}\left(\frac{16r^2}{9\pi^2}\right)$$

$$= \frac{\pi r^4}{8} - \frac{8r^4}{9\pi} + \frac{8r^4}{9\pi}$$

$$= \frac{\pi r^4}{8}$$ *Answer*

Example 6.7

Calculate I_y and k_y for the shaded area in Fig. 6.12. The dimensions are given in inches.

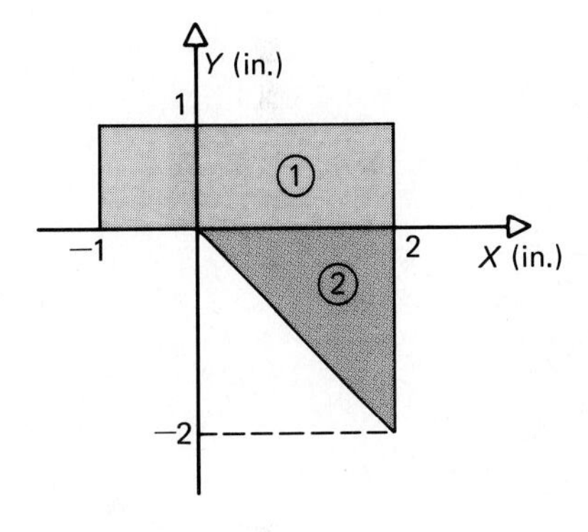

Figure 6.12

The solution for I_y is given by Eq. (6.11a) and is arranged in tabular form where area 1 refers to the rectangle while area 2 refers to the triangle.

$$\bar{I}_{1y'} = \frac{1}{12}(1)(3)^3 = 2.250$$

$$\bar{I}_{2y'} = \frac{1}{36}(2)(2)^3 = 0.444$$

The following table arranges the quantities for Eq. (6.11a).

i	A_i	d_{ix}	$\bar{I}_{iy'}$	$A_i d_{ix}^2$	$\bar{I}_{iy'} + A_i d_{ix}^2$
1	$(3)(1) = 3$	0.5	2.25	0.75	3.00
2	$\frac{1}{2}(2)(2) = 2$	1.33	0.44	3.56	4.00
					$I_y = 7.00$ in^4

For the radius of gyration,

$$k_y = \sqrt{\frac{I_y}{A}}$$

in which A equals the total area. For this case $A = 5.0$, so that

$$k_y = \sqrt{\frac{7.0}{5.0}}$$

$$= 1.18 \text{ in}$$ *Answer*

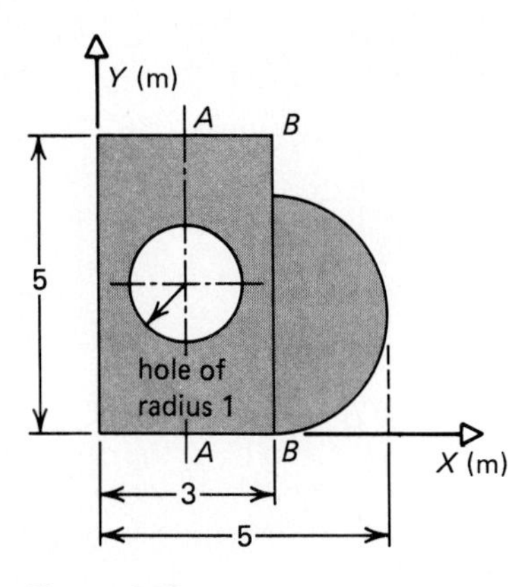

Figure 6.13

Example 6.8

What is the area moment of inertia of the composite area of Fig. 6.13 about the Y axis?

Since the rectangle contains a hole, we may use Eq. (6.11a) in which we have one area composed of a 5 by 3 rectangle. This means that we must subtract the contribution to I_y from the hole contained in the rectangle.

The area moment of inertia of the rectangle about the Y axis is

$$I_y\Big]_{\text{rectangle}} = \frac{bh^3}{3}$$

$$= \frac{(5)(3)^3}{3} = 45 \text{ m}^4$$

The area moment of inertia of the 1-m radius circle about the A axis is

$$\bar{I}_A = \frac{\pi(1)^4}{4} = \frac{\pi}{4} \text{ m}^4$$

Now we shift this to the Y axis, by the parallel-axis theorem.

$$I_y\Big]_{\text{circle}} = \frac{\pi}{4} + \pi(1)^2\left(\frac{3}{2}\right)^2$$

$$= 7.854 \text{ m}^4$$

The area moment of inertia of the semicircle about the centroidal axis is obtained by using the formula from Table 6.1.

$$\bar{I}\Big]_{\text{semicircle}} = (2)^4\left(\frac{\pi}{8} - \frac{8}{9\pi}\right)$$

$$= 2\pi - \frac{128}{9\pi}$$

$$= 1.756 \text{ m}^4$$

For shifting to the Y axis, we note that $A = 2\pi$ for the semicircle and the distance d between the Y axis and the semicircle centroid is

$$d = 3 + \frac{4r}{3\pi}$$

$$= 3 + \frac{4(2)}{3(\pi)}$$

$$= 3.849 \text{ m}$$

The area moment of inertia about the Y axis for the semicircle is

$$I_y]_{\text{semicircle}} = 1.756 + (3.849)^2(2\pi)$$
$$= 94.84 \text{ m}^4$$

The area moment of inertia of the composite area is obtained by the area moment of inertias for the rectangle and the semicircle and subtracting that due to the circle. Thus,

$$I_y = [I_y]_{\text{rectangle}} - [I_y]_{\text{circle}} + [I_y]_{\text{semicircle}}$$
$$= 45 - 7.854 + 94.84$$
$$= 132.0 \text{ m}^4 \qquad \textit{Answer}$$

PROBLEMS

6.21 Using the parallel-axis theorem, find the area moment of inertia about the X axis and the radius of gyration k_x for the rectangle shown in Fig. P6.21.

6.22 Using the parallel-axis theorem, find the area moment of inertia about the Y axis and the radius of gyration k_y for the rectangle shown in Problem 6.21.

6.23 Using the parallel-axis theorem, find the area moment of inertia about the X axis and the radius of gyration k_x for the triangle shown in Fig. P6.23.

6.24 Using the parallel-axis theorem and the data from Table 6.1, find the area moment of inertia about the u and v axes for the area shown in Fig. P6.24.

6.25 Using the parallel-axis theorem and the data from Table 6.1, find the area moment of inertia about the u and v axes for the area shown in Fig. P6.25 for $b = 1$ m and $h = 1.5$ m.

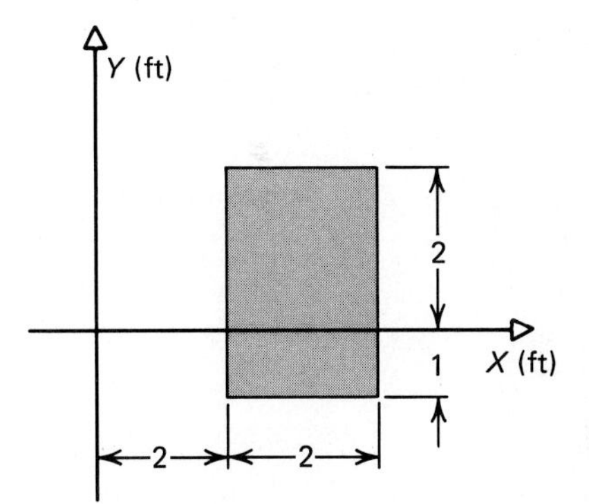

Figure P6.21

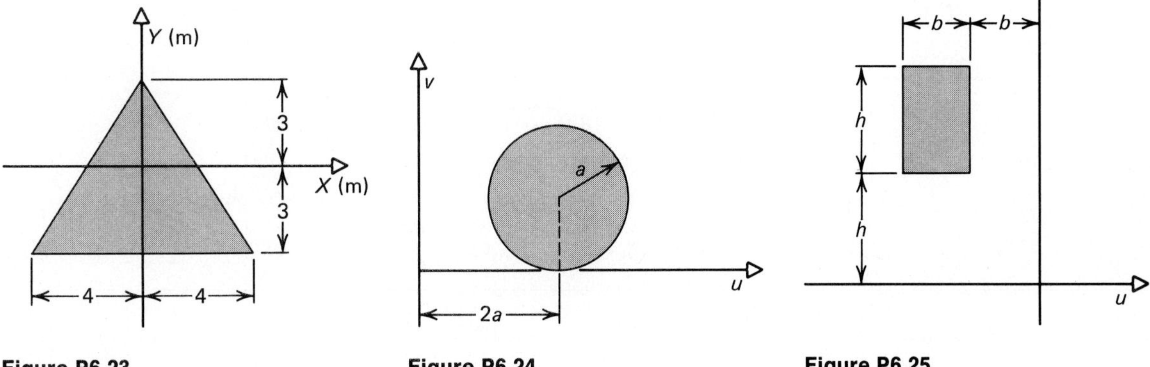

Figure P6.23 **Figure P6.24** **Figure P6.25**

6.26 Repeat Problem 6.25 for $b = 0.5$ m and $h = 0.8$ m.

6.27 Using the parallel-axis theorem and the data from Table 6.1, find the area moment of inertia about the u axis for the area shown in Fig. P6.27 for $b = 20$ mm and $h = 30$ mm.

6.28 Repeat Problem 6.27 for the v axis.

Using the composite area method, find the area moment of inertia about the X axis for the following:

6.29 The shaded area shown in Fig. P6.29.

6.30 The shaded area shown in Fig. P6.30.

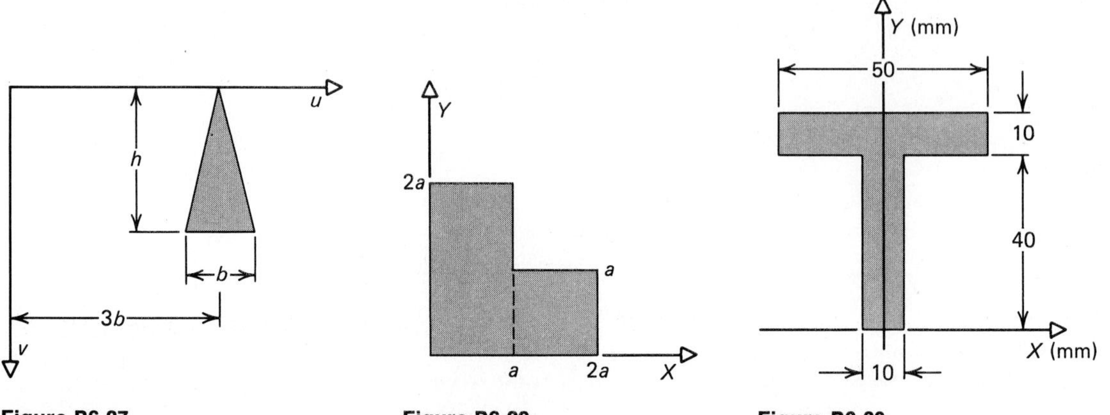

Figure P6.27 **Figure P6.29**

Figure P6.30

6.31 The shaded area shown in Fig. P6.31.

6.32 The shaded area shown in Fig. P6.32.

6.33 The shaded area shown in Fig. P6.33.

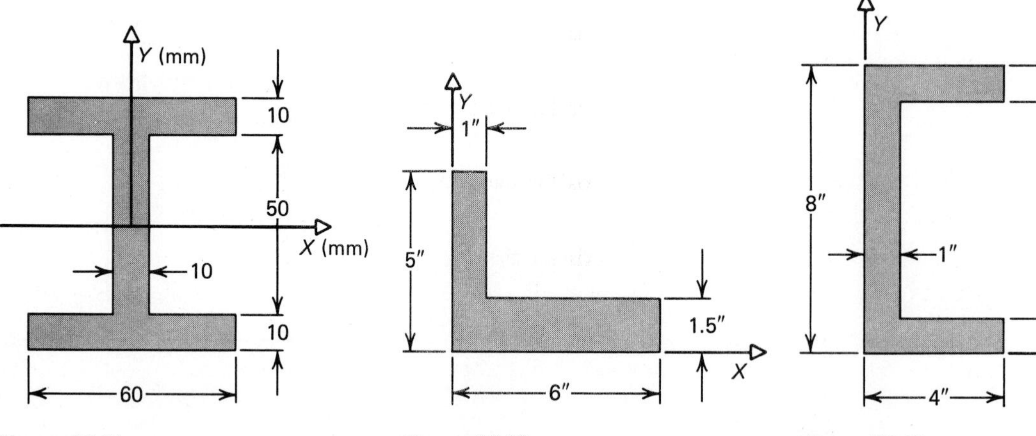

Figure P6.31

Figure P6.32 **Figure P6.33**

6.34 The shaded area shown in Fig. P6.34.

6.35 The shaded area shown in Fig. P6.35.

6.36 The shaded area shown in Fig. P6.36.

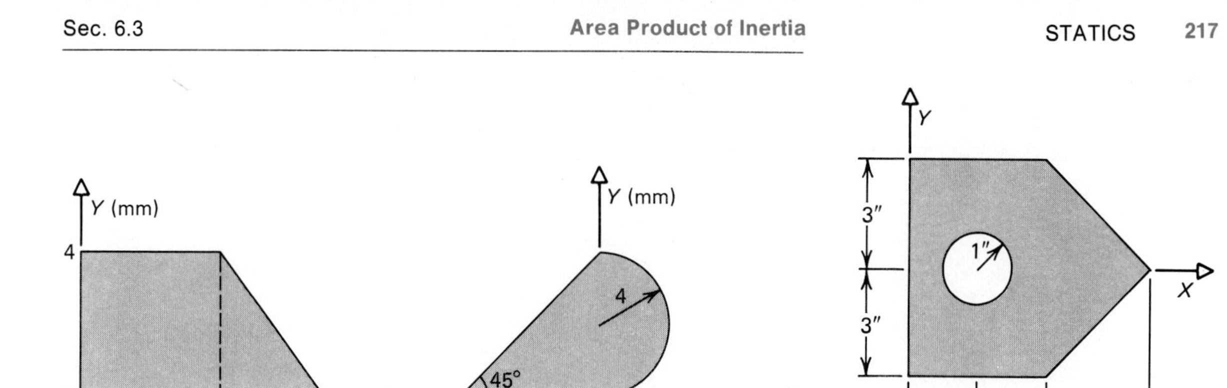

Figure P6.34

Figure P6.35

Figure P6.36

6.37 The shaded area shown in Fig. P6.37.

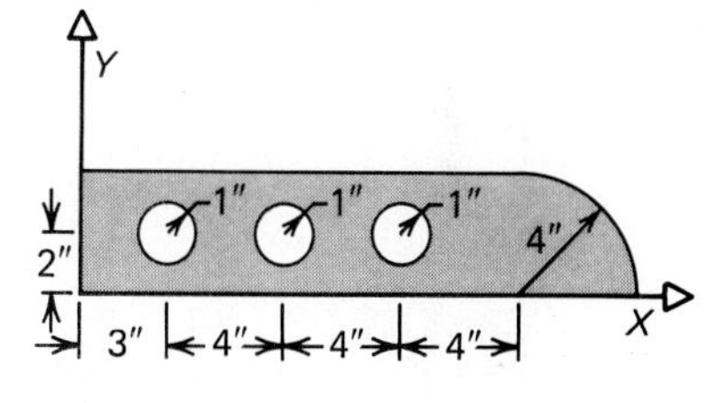

Figure P6.37

6.38 Obtain the area polar moment of inertia about the origin for the area in Problem 6.29.

6.39 Calculate the area polar moment of inertia about the origin for the area in Problem 6.30.

6.40 Calculate the area moment of inertia about the Y axis for the area in Problem 6.31.

6.41 Calculate the area moment of inertia about the Y axis for the area in Problem 6.32.

6.42 Find the area moment of inertia about the Y axis for the area in Problem 6.33.

6.43 Find the area moment of inertia about the Y axis for the area in Problem 6.34.

6.44 Find the area moment of inertia about the Y axis for the area in Problem 6.35.

6.45 Find the area moment of inertia about the Y axis for the area in Problem 6.36.

6.46 Find the area moment of inertia about the Y axis for the area in Problem 6.37.

6.47 Obtain the area polar moment of inertia about the center of the area shown in Fig. P6.47.

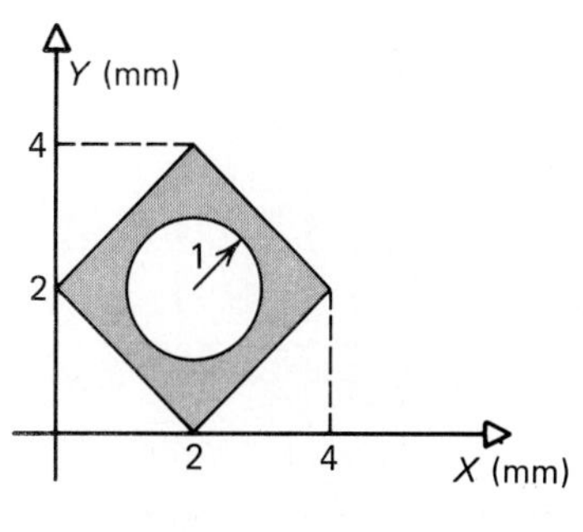

Figure P6.47

6.3 Area Product of Inertia

Let us return to the example in Section 6.1 where we computed the moment of the force about the X axis. Referring to Fig. 6.1, let us compute the magnitude of the moment about the Y axis where

$$M_y = \int_{\text{area}} xp\,dA$$

and if $p = \gamma y$, we have

$$M_y = \gamma \int_{\text{area}} xy\,dA$$

The term $\int xy\,dA$ is called the area product of inertia and is written as I_{xy}, i.e.,

$$I_{xy} = \int_{\text{area}} xy\,dA \tag{6.12}$$

Note that the units are normally meters^4 and feet^4 and that the sign for the results of integrating by Eq. (6.12) may be plus or minus.

By way of example, consider the rectangle shown in Fig. 6.14 First we locate the centroid of the differential area that we use in Eq. (6.12). Letting $dA = b\,dy$, and noting that the centroid of dA is at $(b/2, y)$ we substitute into Eq. (6.12) to obtain

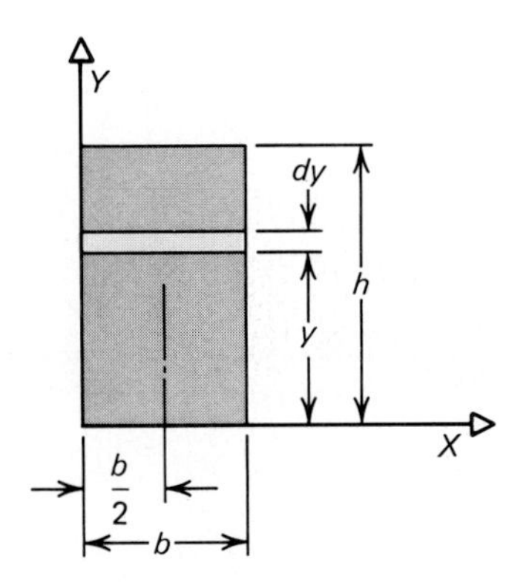

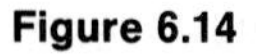
Figure 6.14

$$I_{xy} = \int_0^h \frac{b}{2} y(b\,dy) = \frac{b^2h^2}{4}$$

As a second example, let us evaluate the area product of inertia for the elliptical cross section shown in Fig. 6.15. We have placed the Y axis so that the ellipse is symmetrical about this axis. The area product of inertia becomes

$$I_{xy} = \int_{A1} xy\,dA + \int_{A2} (-x)y\,dA + \int_{A3} (-x)(-y)\,dA + \int_{A4} x(-y)\,dA$$

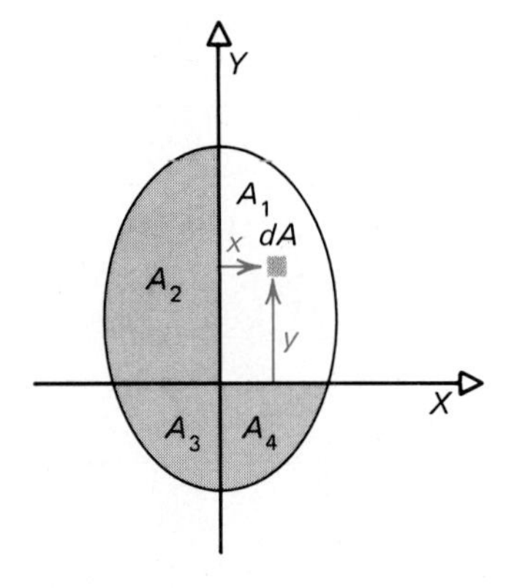

Figure 6.15

Since $A_1 = A_2$, $A_3 = A_4$, and these areas are symmetrical about the Y axis, the expression for I_{xy} goes to zero. The area product of inertia is zero if the X or the Y axis is a symmetrical axis of the body.

Let the area product of inertia about the centroidal axes X' and Y' be $\bar{I}_{x'y'}$ for the area in Fig. 6.16. The area product of inertia about the XY axes is

$$\begin{aligned} I_{xy} &= \int (x' + d_x)(y' + d_y)\,dA \\ &= \left(\int x'y'\,dA\right) + \left(d_x \int y'\,dA\right) + \left(d_y \int x'\,dA\right) + \left(d_x d_y \int dA\right) \end{aligned}$$

Since the X' and Y' axes are centroidal, the middle two terms go to zero and

$$I_{xy} = \bar{I}_{x'y'} + A d_x d_y \tag{6.13}$$

This is the parallel-axis theorem for the area product of inertia. This equation is very useful for calculating the area product of inertia of a symmetrically shaped body since d_x and d_y represent the location of the centroid of the body. Having these distances and the area, we immediately have I_{xy} since $\bar{I}_{x'y'} = 0$ for the symmetric body. For example, the rectangle is redrawn in Fig. 6.17 to include the centroidal axes $X'Y'$ parallel with the XY axes. Now $d_x = b/2$, $d_y = h/2$, $A = bh$, and $\bar{I}_{x'y'} = 0$. Referring to Eq. (6.13),

$$I_{xy} = \frac{b}{2}\left(\frac{h}{2}\right)(bh) = \frac{b^2h^2}{4}$$

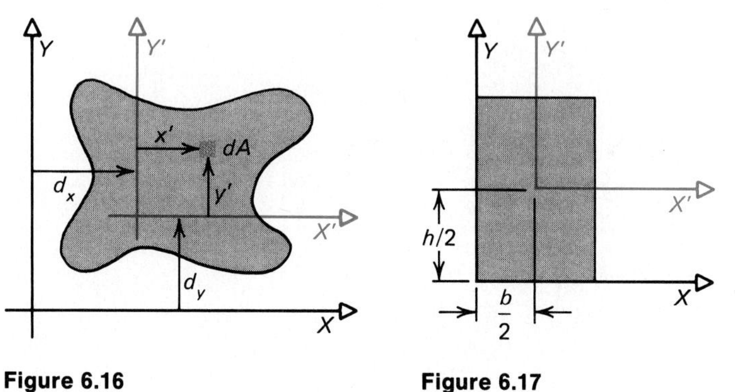

Figure 6.16

Figure 6.17

Example 6.9

Calculate the area product of inertia for the semicircle in Fig. 6.18 about the XY axes (a) by direct integration and (b) by use of Eq. (6.13).

(a) Figure 6.18 shows a differential element of area $dA = dr(r\,d\theta)$. Referring to Eq. (6.12),

$$I_{xy} = \int_{\text{area}} xy\,dA$$

Now $x = (a + r\cos\theta)$ and $y = r\sin\theta$, so that

$$\begin{aligned} I_{xy} &= \int_0^{\pi}\int_0^{a}(a + r\cos\theta)(r\sin\theta)(r\,dr\,d\theta) \\ &= \int_0^{\pi}\int_0^{a} ar^2\,dr(\sin\theta\,d\theta) \\ &\quad + \int_0^{\pi}\int_0^{a} r^3\,dr(\sin\theta\cos\theta\,d\theta) \\ &= \frac{a^4}{3}\int_0^{\pi}\sin\theta\,d\theta + \frac{a^4}{4}\int_0^{\pi}\sin\theta\cos\theta\,d\theta \\ &= \frac{a^4}{3}\Big[-\cos\theta\Big]_0^{\pi} + \frac{a^4}{4}\left[\frac{\sin^2\theta}{2}\right]_0^{\pi} \\ &= \frac{2a^4}{3} \end{aligned}$$

Answer

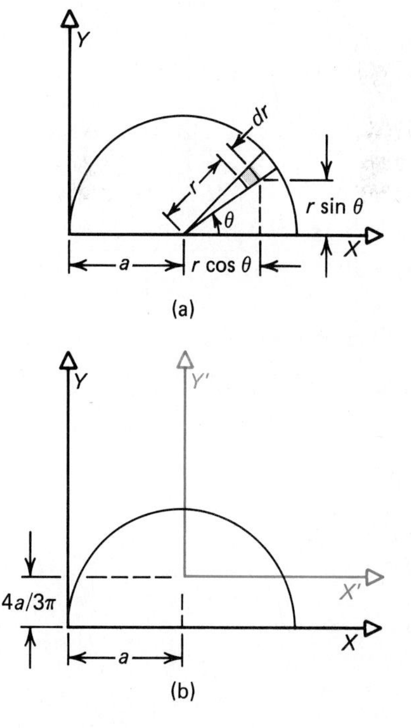

Figure 6.18

(b) The area product of inertia for the semicircle is obtained by Eq. (6.13) with the aid of Fig. 6.18(b).

$$I_{xy} = \bar{I}_{x'y'} + d_x d_y A$$

in which

$$\bar{I}_{x'y'} = 0$$

$$d_x = a$$

$$d_y = \frac{4a}{3\pi}$$

$$A = \frac{\pi a^2}{2}$$

Substituting into Eq. (6.13),

$$I_{xy} = \frac{4a^2}{3\pi}\left(\frac{\pi a^2}{2}\right) = \frac{2a^4}{3} \qquad \textit{Answer}$$

Example 6.10

For the triangle shown in Fig. 6.19, determine the area product of inertia about (a) the X and Y axes and (b) the X' and Y' axes.

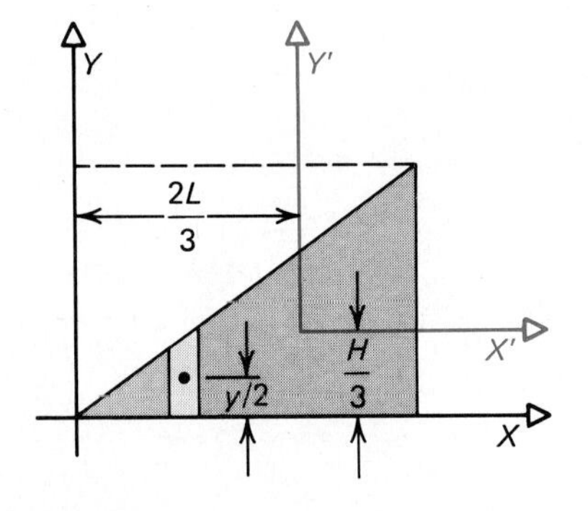

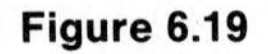
Figure 6.19

(a) Select a differential area as $dA = y\,dx$ so that the coordinate of the centroid of the element is $(x, y/2)$. When we substitute these coordinates for x and y into Eq. (6.12), we obtain

$$I_{xy} = \int \frac{xy^2}{2}\,dx$$

Substituting $y = (H/L)x$ and the limits for x gives us

$$I_{xy} = \int_0^L \frac{H^2}{2L^2}x^3\,dx$$

$$= \frac{H^2L^2}{8} \qquad \textit{Answer}$$

(b) The area product of inertia about the X' and Y' axes is

$$\bar{I}_{x'y'} = I_{xy} - d_x d_y A$$

$$= \frac{H^2L^2}{8} - \frac{2L}{3}\frac{H}{3}\frac{HL}{2}$$

$$= \frac{1}{72}H^2L^2 \qquad \textit{Answer}$$

PROBLEMS

6.48 Find the area product of inertia about the XY axes for the area shown in Fig. P6.48.

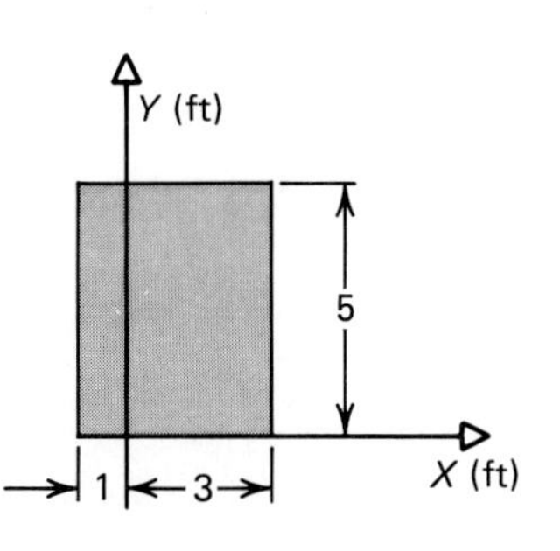

Figure P6.48

6.49 Find the area product of inertia about the XY axes for the area shown in Fig. P6.49.

6.50 Using the result of Problem 6.49, obtain the area product of inertia about the $X'Y'$ axes for the area shown in Fig. P6.49. Use the parallel-axis theorem.

6.51 Find the area product of inertia about the XY axes of the shaded area shown in Fig. P6.51.

6.52 Find the area product of inertia about the XY axes of the shaded area shown in Fig. P6.52.

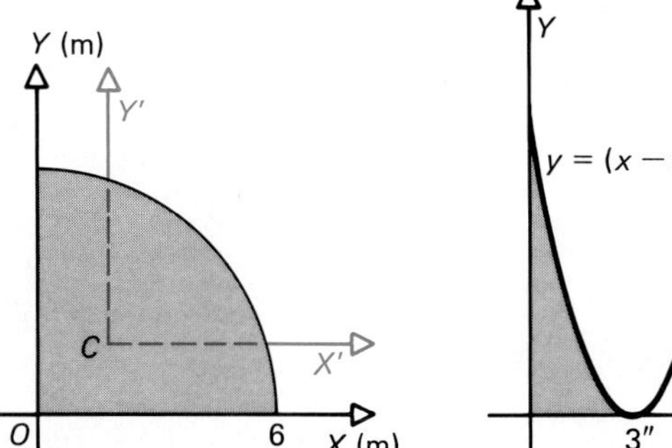

Figure P6.49

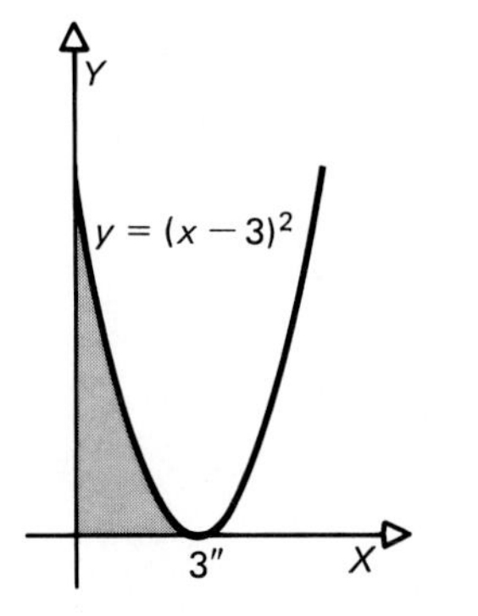

Figure P6.51

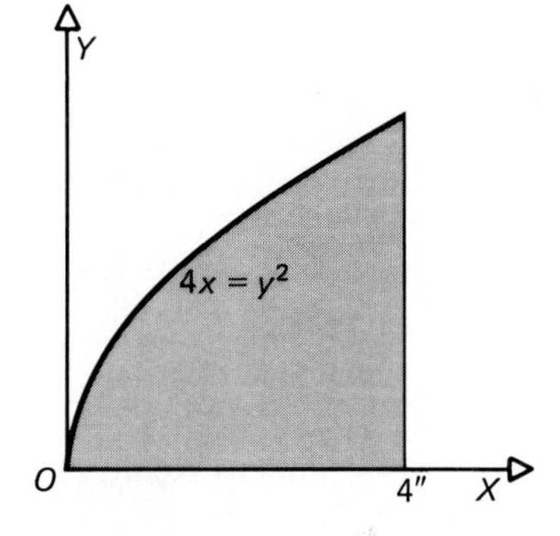

Figure P6.52

6.53 Obtain the area product of inertia about the XY axes for the shape shown in Fig. P6.53.

6.54 Obtain the area product of inertia about the XY axes for the shape shown in Fig. P6.54.

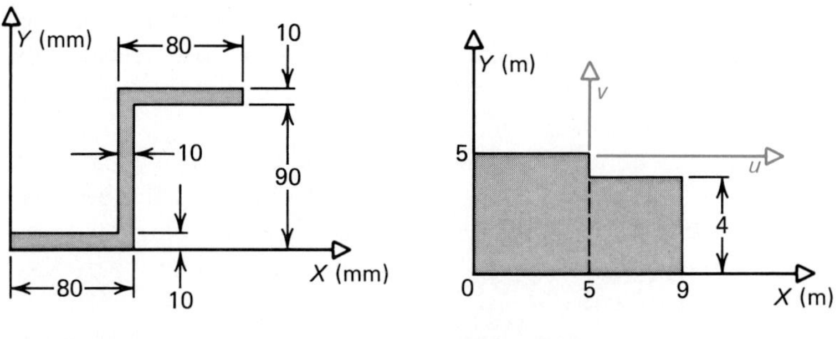

Figure P6.53

Figure P6.54

6.55 Obtain the area product of inertia about the uv axes for the shape shown in Fig. P6.54.

6.56 Obtain the area product of inertia about the XY axes for the shape shown in Fig. P6.56.

6.57 Obtain the area product of inertia about the uv axes for the shape shown in Fig. P6.56.

6.58 Obtain the area product of inertia about the XY axes for the shape shown in Fig. P6.58.

6.59 Obtain the area product of inertia about the uv axes for the shape shown in Fig. P6.58.

6.60 Obtain the area product of inertia about the XY axes for the shape shown in Fig. P6.60.

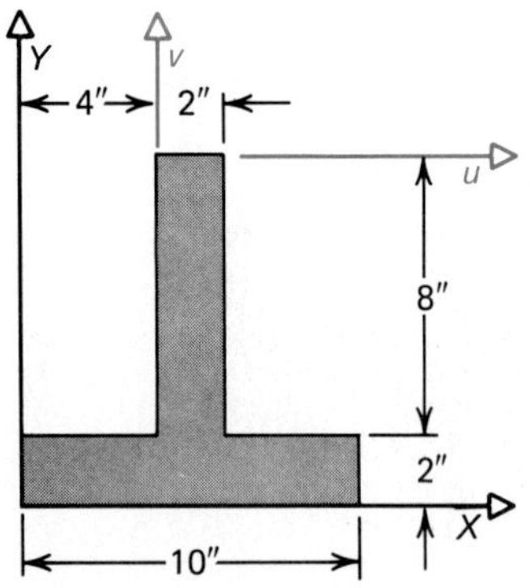

Figure P6.56

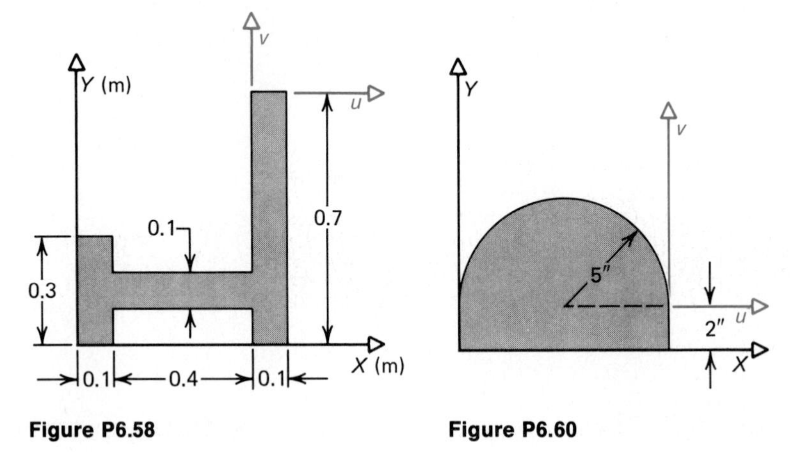

Figure P6.58

Figure P6.60

6.4 Principal Axes

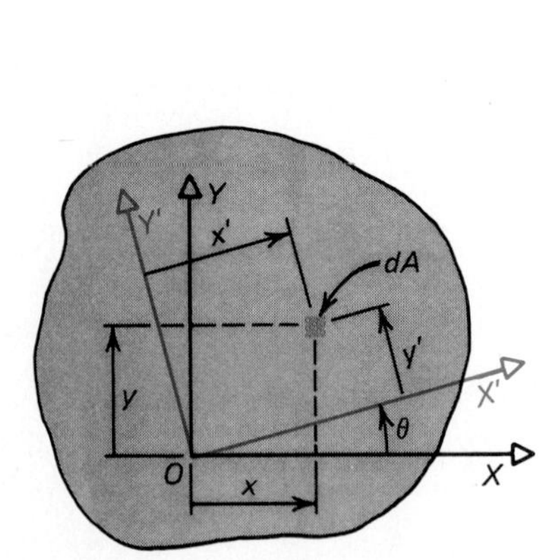

Figure 6.20

Consider the area in Fig. 6.20 where the origin of the XY axes is located at some arbitrary point O. We shall use the XY axes as our reference axes in what follows. Let us also consider a second set of axes $X'Y'$ that are inclined at an angle θ counterclockwise relative to the reference XY axes. We wish to determine the area moment of inertia about the X' axis and the Y' axis, respectively. We also wish to find the angles θ that give us the maximum and minimum area moments of inertia.

The axes about which the area moment of inertia is a maximum or a minimum are called the principal axes. To find the principal axes for the arbitrary area in Fig. 6.20, consider the following transformation equations between the two sets of coordinates that locate the element of area dA:

$$\begin{aligned} x' &= y \sin \theta + x \cos \theta \\ y' &= y \cos \theta - x \sin \theta \end{aligned} \tag{6.14}$$

Again, the problem is to find the rotation θ such that the area moment of inertia about the X' axis and Y' axis is an extremum. The area moments of inertia about the X' and Y' axis are defined as follows:

$$I_{x'} = \int y'^2 \, dA \qquad I_{y'} = \int x'^2 \, dA \qquad I_{x'y'} = \int x'y' \, dA$$

Substituting Eqs. (6.14), we obtain

$$I_{x'} = \int (y \cos\theta - x \sin\theta)^2 \, dA$$

$$I_{y'} = \int (y \sin\theta + x \cos\theta)^2 \, dA$$

$$I_{x'y'} = \int (y \sin\theta + x \cos\theta)(y \cos\theta - x \sin\theta) \, dA$$

Expanding these equations and using the following identities:

$$2 \sin^2\theta = 1 - \cos 2\theta$$

$$2 \cos^2\theta = 1 + \cos 2\theta$$

$$\sin 2\theta = 2 \sin\theta \cos\theta$$

$$I_x = \int y^2 \, dA$$

$$I_y = \int x^2 \, dA$$

$$I_{xy} = \int xy \, dA$$

we obtain

$$I_{x'} = \frac{I_x + I_y}{2} + \frac{I_x - I_y}{2} \cos 2\theta - I_{xy} \sin 2\theta \tag{6.15a}$$

$$I_{y'} = \frac{I_x + I_y}{2} - \frac{I_x - I_y}{2} \cos 2\theta + I_{xy} \sin 2\theta \tag{6.15b}$$

$$I_{x'y'} = \frac{I_x - I_y}{2} \sin 2\theta + I_{xy} \cos 2\theta \tag{6.15c}$$

Since we wish to determine the maximum or minimum of the area moment of inertia, we take the derivative of Eqs. (6.15a) or (6.15b) with respect to θ and set it equal to zero.

$$\frac{dI_{x'}}{d\theta} = 0 \qquad \text{or} \qquad \frac{dI_{y'}}{d\theta} = 0$$

Referring to Eq. (6.15a), we have

$$\frac{dI_{x'}}{d\theta} = -(I_x - I_y) \sin 2\theta - 2I_{xy} \cos 2\theta = 0$$

Rearranging terms, we obtain

$$\tan 2\theta = -\frac{2I_{xy}}{I_x - I_y} = q \tag{6.16}$$

Since $\tan 2\theta = \tan(2\theta - \pi)$, we have two solutions, one for the maximum and one for the minimum area moment of inertia.

$$2\theta_m = \tan^{-1} q \qquad \text{and} \qquad 2\theta_m - \pi = \tan^{-1} q$$

so that

$$\theta_m = \frac{1}{2}\tan^{-1} q \qquad \text{and} \qquad \theta_m = \frac{\pi}{2} + \frac{1}{2}\tan^{-1} q \tag{6.17}$$

where q is defined in Eq. (6.16). These two solutions for θ locate the principal axes relative to the X and Y axes. Note that positive θ is measured counterclockwise off the reference X axis as shown in Fig. 6.20.

If we substitute either value of θ from Eq. (6.17) into Eq. (6.15c), we find that $I_{x'y'} = 0$. Also, substituting into Eq. (6.15a) and (6.15b) we obtain the area moments of inertia about the principal axes that are the maximum and minimum values.

$$I_{max} = \frac{I_x + I_y}{2} + \sqrt{\left(\frac{I_x - I_y}{2}\right)^2 + I_{xy}^2} \tag{6.18a}$$

$$I_{min} = \frac{I_x + I_y}{2} - \sqrt{\left(\frac{I_x - I_y}{2}\right)^2 + I_{xy}^2} \tag{6.18b}$$

We observe from these results that if the original XY axes in Fig. 6.20 were symmetric axes, $I_{xy} = 0$ so that $I_{max} = I_x$ and $I_{min} = I_y$. Therefore, symmetric axes are principal axes, although the converse is not necessarily true. It should also be pointed out that the origin of the principal axes is often selected to be the centroid of the area. If this is done the axes are called centroidal axes. Although principal axes whose origin is at the centroid become centroidal axes, the converse is not true.

Example 6.11

For the triangle shown in Fig. 6.21, determine (a) the principal axes relative to the centroid c, and (b) the area moments of inertia about these principal axes.

(a) The origin of the reference XY axes is located at the centroid c of the triangle. From earlier results we can calculate the area moments of inertia and the area product of inertia about these centroidal axes as follows:

$$\bar{I}_x = \frac{1}{36}(6)(9)^3 = 121.5 \text{ m}^4$$

$$\bar{I}_y = \frac{1}{36}(9)(6)^3 = 54.0 \text{ m}^4$$

To find the area product of inertia, consider the element of area $dA = (y + 3)\,dx$ that is shown in Fig. 6.21(a). The centroid at the vertical strip has coordinates x and $(y - 3)/2$. Also, the relationship between y and x is

$$y = 3 - 1.5x$$

The area product of inertia relative to the centroidal axes is

$$\bar{I}_{xy} = \int x\left(\frac{y-3}{2}\right)(y + 3)\,dx$$

$$= \frac{1}{2}\int_{-2}^{4}(-9x^2 + 2.25x^3)\,dx$$

$$= -40.5 \text{ m}^4$$

Referring to Eq. (6.17),

$$\theta_m = \frac{1}{2}\tan^{-1} q = \frac{1}{2}\tan^{-1}\left[-\frac{2(-40.5)}{121.5 - 54}\right]$$

$$= 0.438 \text{ radians}$$

$$= 25.1^\circ$$ *Answer*

Figure 6.21(b) shows the principal axes $X'Y'$ rotated 25.1° counterclockwise relative to the XY axes.

(b) The area moments of inertia about the principal axes are given by Eq. (6.18).

$$I_{max} = \frac{121.5 + 54}{2} + \sqrt{\left(\frac{121.5 - 54}{2}\right)^2 + (-40.5)^2}$$

$$= 87.75 + 52.72 = 140.5 \text{ m}^4$$ *Answer*

$$I_{min} = 87.75 - 52.72 = 35.0 \text{ m}^4$$ *Answer*

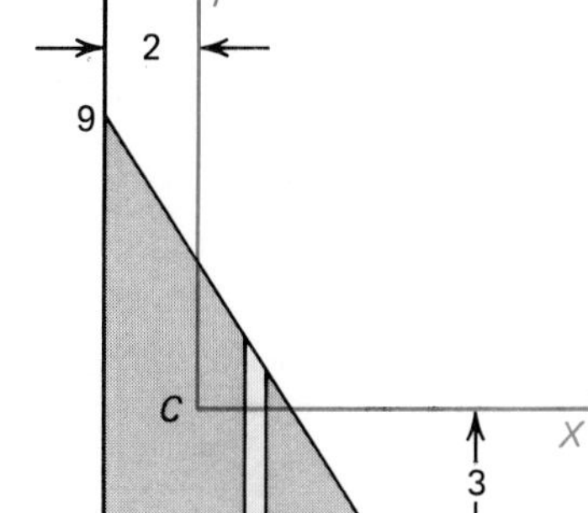

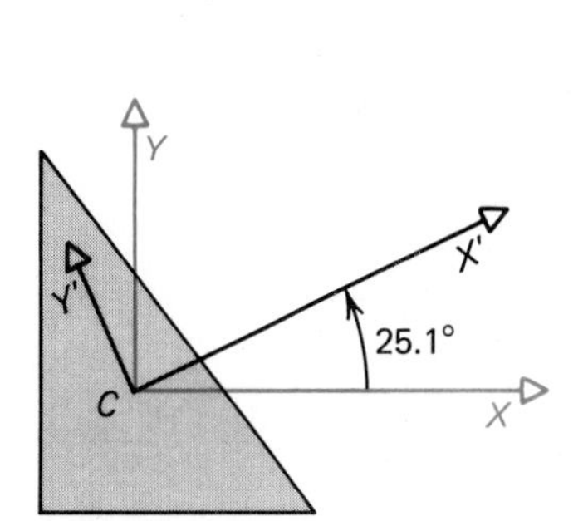

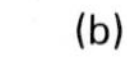

Figure 6.21

PROBLEMS

6.61 Verify the derivation of Eqs. (6.15a), (6.15b), and (6.15c).

6.62 Verify the derivation of Eqs. (6.18a) and (6.18b).

Find the principal axes relative to the centroidal XY axes and the area moments of inertia about these principal axes for the following:

6.63 The shaded area shown in Fig. P6.63.

6.64 The shaded area shown in Fig. P6.64.

6.65 The shaded area shown in Fig. P6.65.

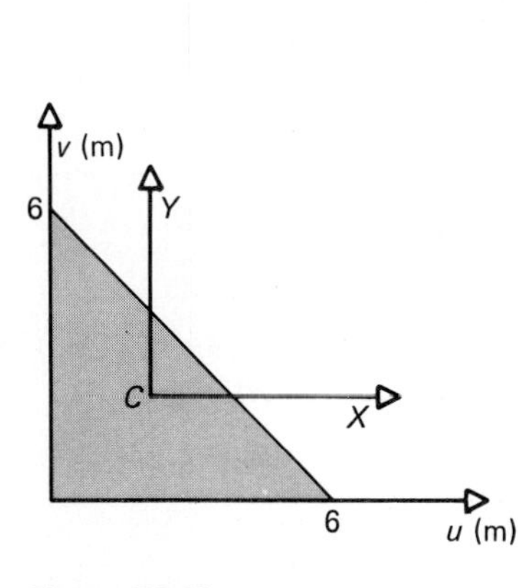

Figure P6.63

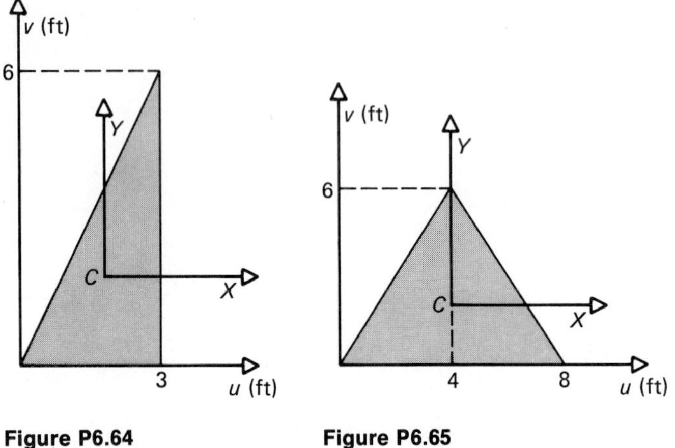

Figure P6.64

Figure P6.65

6.66 The shaded area shown in Fig. P6.66.

6.67 The shaded area shown in Fig. P6.67.

6.68 The shaded area shown in Fig. P6.68.

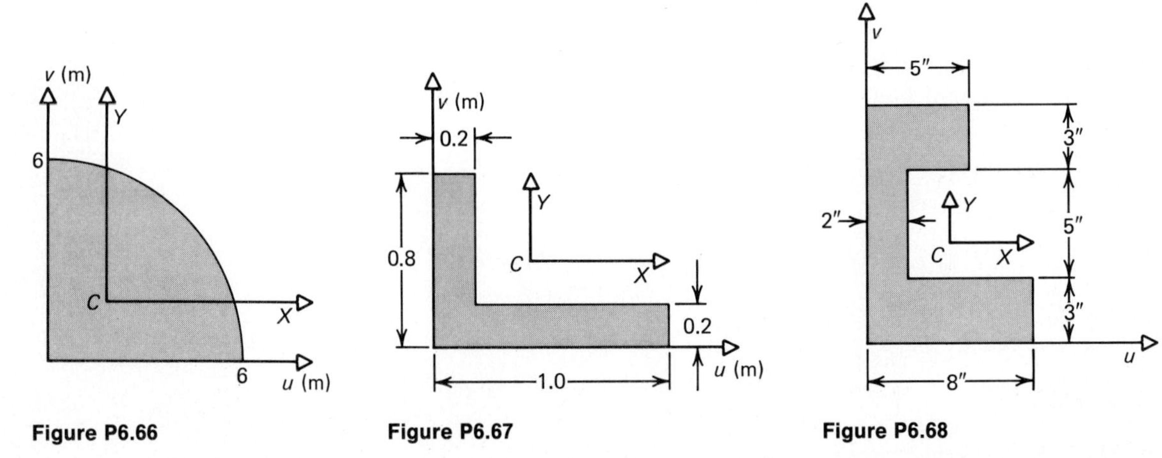

Figure P6.66

Figure P6.67

Figure P6.68

6.69 Find $\bar{I}_u$, $\bar{I}_v$, and $\bar{I}_{uv}$ for the structural beam shown in Fig. P6.69 whose properties are: $\bar{I}_x = 5454.2 \text{ in}^4$, $\bar{I}_y = 1986.0 \text{ in}^4$, and $\bar{I}_{xy} = 0$.

6.70 Find $\bar{I}_u$, $\bar{I}_v$, and $\bar{I}_{uv}$ for the American Standard Channel shown in Fig. P6.70 whose properties are: $\bar{I}_x = 128.1 \text{ in}^4$, $\bar{I}_y = 3.9 \text{ in}^4$, $\bar{I}_{xy} = -15 \text{ in}^4$.

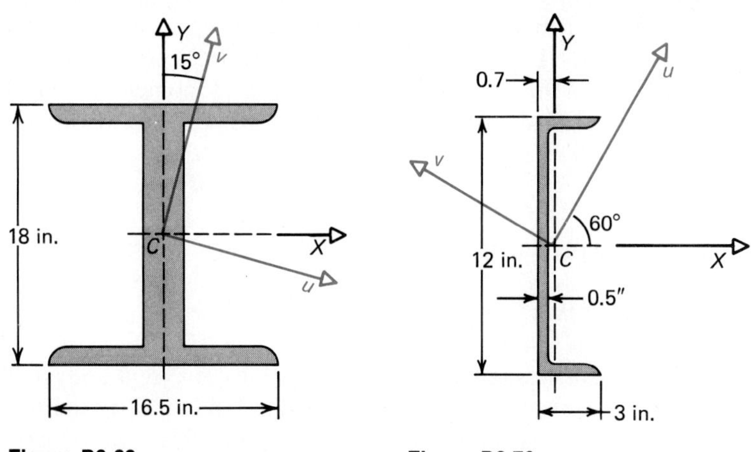

Figure P6.69 **Figure P6.70**

6.5 Mass Moment of Inertia

Like the area moment of inertia, the mass moment of inertia is a property that is important in our study of engineering mechanics. By way of introduction, consider the rigid body in Fig. 6.22(a) rotating at a constant rate of ω radians per second about the AA axis. The element of mass dm experiences a velocity of magnitude ωr acting in the tangential direction as shown in Fig. 6.22(b). The angular momentum of this element of mass about the AA axis is defined as the moment of the linear momentum about the AA axis. Since the linear momentum of the element equals $r\omega\, dm$, the magnitude of the angular momentum is $r^2\omega\, dm$. The magnitude of the angular momentum of the entire body is

$$\int r^2\omega\, dm = \omega \int r^2\, dm$$

since ω is constant. We define the mass moment of inertia, I, as

$$I = \int r^2\, dm \tag{6.19}$$

This equation is similar to the equation for area moment of inertia. Note, however, that the units for I are $\text{kg}\cdot\text{m}^2$, or $\text{lb}\cdot\text{ft}\cdot\text{sec}^2$ which is the same as $\text{slug}\cdot\text{ft}^2$.

While the mass moment of inertia plays an important role in studying the rotational motion of rigid bodies, here we shall focus on the computational procedures for calculating this property. As in the case of the area moment of inertia, there exists the definition of the radius of gyration, k, which is given by

$$k = \sqrt{\frac{I}{m}} \tag{6.20}$$

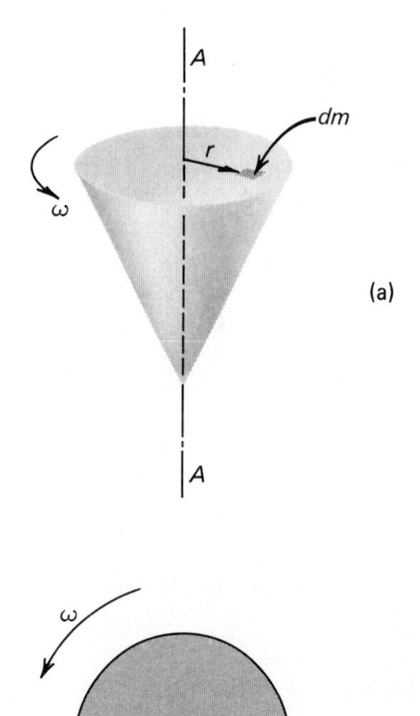

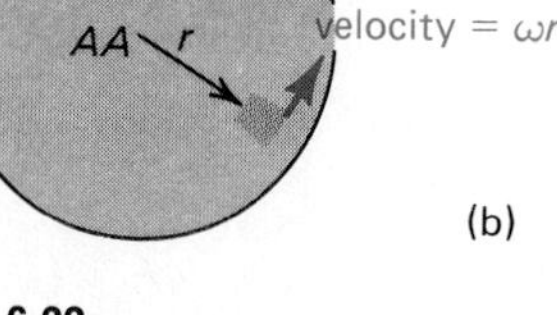

Figure 6.22

where I is determined from Eq. (6.19) and m represents the total mass of the body. Likewise, the parallel axis theorem applies for the mass moment of inertia property. If $\bar{I}$ is the mass moment of inertia about a centroidal axis of the body, then the mass moment of inertia about a parallel axis is

$$I = \bar{I} + md^2 \tag{6.21}$$

where d is the perpendicular distance between the two axes.

Mass Points

Sometimes we can neglect the dimensions of a rigid body and represent it simply as a mass concentrated at a point. For example, consider such a mass point shown in Fig. 6.23. For this case, the mass moment of inertia about the AA axis is simply

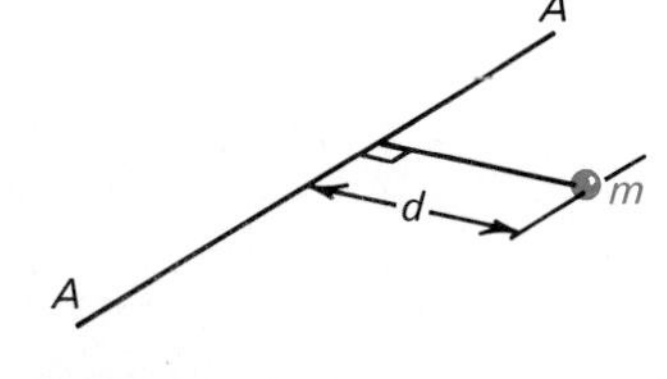

Figure 6.23

$$I = md^2 \tag{6.22}$$

If there are n mass points representing a body each at a perpendicular distance d_i from the AA axis, then

$$I = \sum_{i=1}^{n} m_i d_i^2 \tag{6.23}$$

Example 6.12 demonstrates the application of this equation to a system consisting of several mass points.

Thin Plates

The mass moments of inertia of thin plates are relatively simple to evaluate. Consider the plate shown in Fig. 6.24. We wish to obtain the mass moment of inertia about the X, Y, and Z axes, respectively. A differential mass of the thin plate is given by

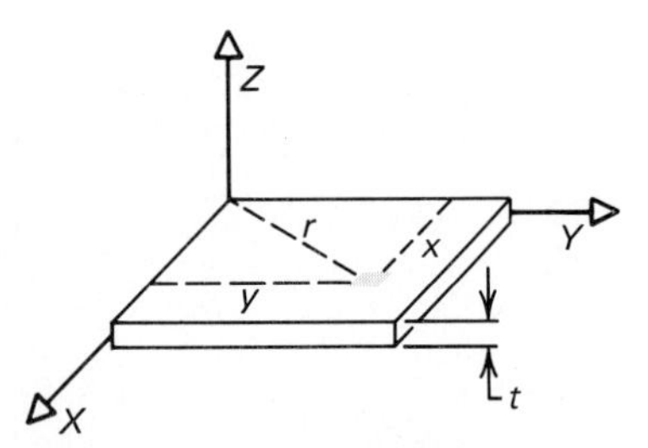

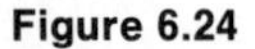
Figure 6.24

$$dm = \rho\, dV$$

where ρ is the density and is assumed to be constant. Since the plate thickness is t, the differential mass becomes

$$dm = \rho t\, dA$$

The mass moments of inertia about the X, Y, and Z axes are

$$I_x = \int y^2\, dm = \rho t \int y^2\, dA$$

$$I_y = \int x^2\, dm = \rho t \int x^2\, dA$$

$$I_z = \int (x^2 + y^2)\, dm = I_y + I_x$$

From the above equations we observe that for the case of thin plates, the expressions for the mass moment of inertia include the area moments of inertia. Thus, we may refer to Table 6.1 to compute the mass moment of inertia for thin plates of varying shapes. Simply multiply the area moment of inertia about the axis in question by the product of the density and plate thickness.

Likewise the mass products of inertia for the thin plates are

$$I_{xy} = I_{yx} = \int xy\,dm = \rho t \int xy\,dA$$

$$I_{xz} = I_{zx} = \int xz\,dm = 0$$

$$I_{yz} = I_{zy} = \int yz\,dm = 0$$

Again, we observe that the mass product of inertia of a thin plate is directly related to the area product of inertia. Since we are neglecting integration in the Z direction for a thin plate, two of the three sets of the product of inertia are zero.

Three-Dimensional Bodies

Consider the differential mass dm shown in Fig. 6.25. The mass moment of inertia about the X axis is

$$I_x = \int (y^2 + z^2)\,dm$$

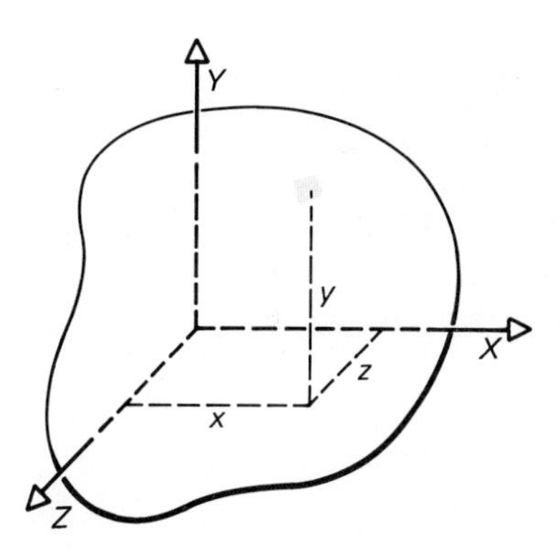

Figure 6.25

Letting ρ represent the density of the body, the differential mass becomes

$$dm = \rho\,dV$$

where dV is the volume of the differential mass dm. Assuming the density is constant throughout the body,

$$I_x = \rho \int_{\text{volume}} (y^2 + z^2)\,dV$$

Similarly,

$$I_y = \rho \int_{\text{volume}} (x^2 + z^2)\,dV$$

$$I_z = \rho \int_{\text{volume}} (x^2 + y^2)\,dV$$

The mass products of inertia are given by the following expressions:

$$I_{xy} = I_{yx} = \rho \int_{\text{volume}} xy\,dV$$

$$I_{xz} = I_{zx} = \rho \int_{\text{volume}} xz\,dV$$

$$I_{yz} = I_{zy} = \rho \int_{\text{volume}} yz\,dV$$

The mass moments of inertia for some common geometric bodies are given in Table 6.2. For composite bodies, the mass moments of inertia can be found by finding the mass moment of inertia of each member about its centroidal axis, and then using the parallel-axis theorem for each member. This approach is demonstrated below in Example 6.15.

TABLE 6.2 ***Mass Moments of Inertia of Common Geometric Bodies***

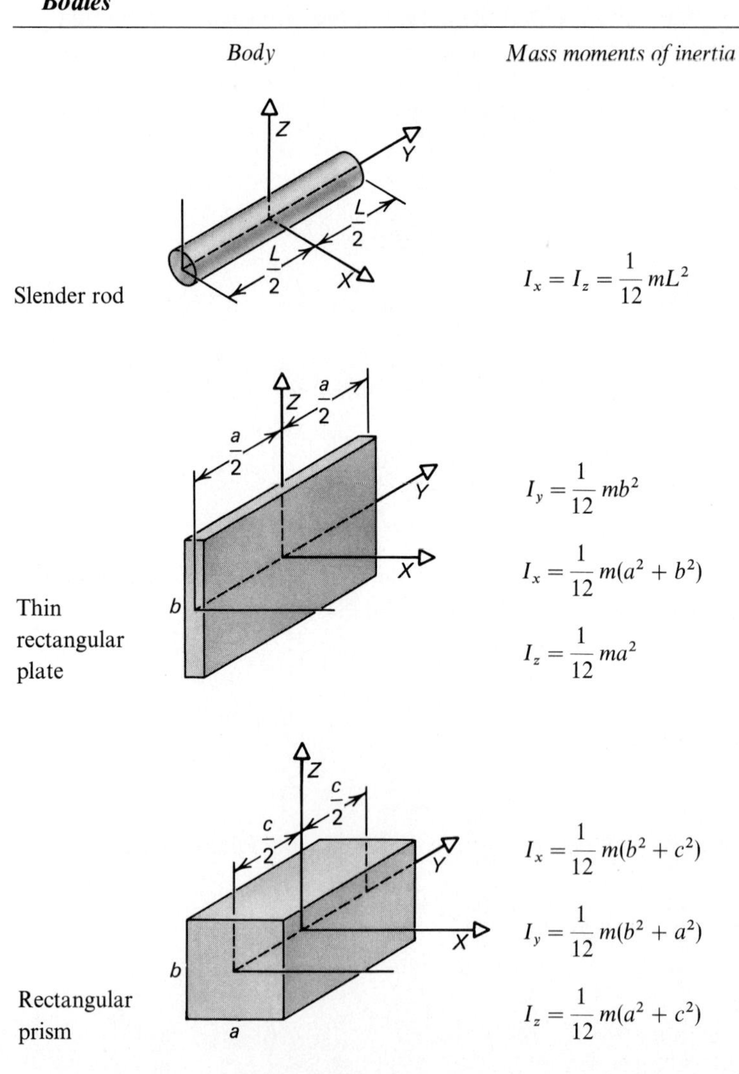

Body	Mass moments of inertia
Slender rod	$I_x = I_z = \frac{1}{12} mL^2$
Thin rectangular plate	$I_y = \frac{1}{12} mb^2$ $I_x = \frac{1}{12} m(a^2 + b^2)$ $I_z = \frac{1}{12} ma^2$
Rectangular prism	$I_x = \frac{1}{12} m(b^2 + c^2)$ $I_y = \frac{1}{12} m(b^2 + a^2)$ $I_z = \frac{1}{12} m(a^2 + c^2)$

Example 6.12

An unbalanced shaft contains two point masses as shown in Fig. 6.26. Determine the mass moment of inertia about the shaft axis due to the point masses only, if $w_1 = 25$ lb, $w_2 = 30$ lb, $d_1 = 2$ ft, and $d_2 = 1.5$ ft.

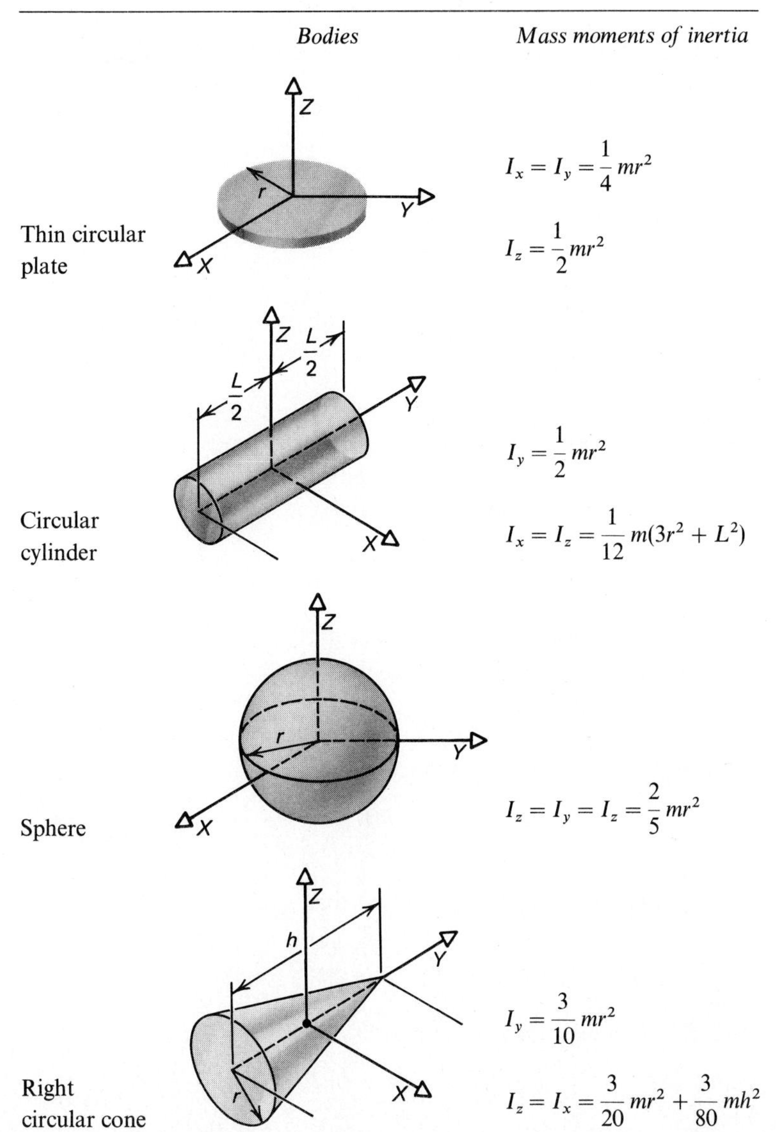

TABLE 6.2 ***Continued***

	Bodies	*Mass moments of inertia*
Thin circular plate		$I_x = I_y = \frac{1}{4} mr^2$ $I_z = \frac{1}{2} mr^2$
Circular cylinder		$I_y = \frac{1}{2} mr^2$ $I_x = I_z = \frac{1}{12} m(3r^2 + L^2)$
Sphere		$I_z = I_y = I_z = \frac{2}{5} mr^2$
Right circular cone		$I_y = \frac{3}{10} mr^2$ $I_z = I_x = \frac{3}{20} mr^2 + \frac{3}{80} mh^2$

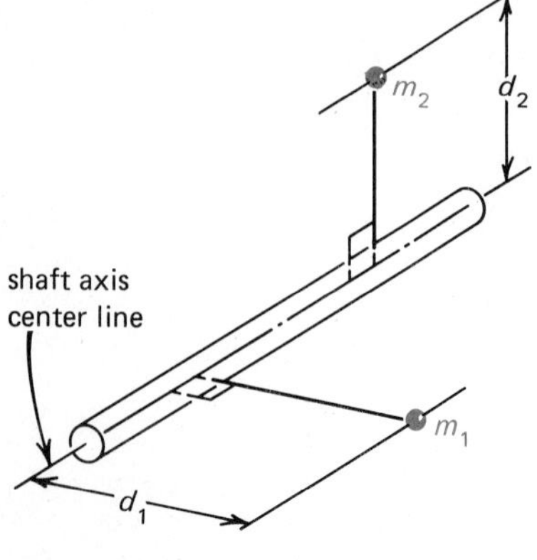

Figure 6.26

First compute the mass of each point mass using the relationship between mass and weight.

$$w = mg$$

where $g = 32.2$ ft/sec^2. Now

$$m_1 = \frac{w_1}{g} = \frac{25}{32.2} = 0.776 \text{ lb}\cdot\text{sec}^2/\text{ft}$$

$$m_2 = \frac{w_2}{g} = \frac{30}{32.2} = 0.932 \text{ lb}\cdot\text{sec}^2/\text{ft}$$

The mass moment of inertia due to the point masses is given by Eq. (6.23). For this case,

$$\begin{aligned} I &= m_1 d_1^2 + m_2 d_2^2 \\ &= 0.776(2)^2 + 0.932(1.5)^2 \\ &= 5.20 \text{ lb}\cdot\text{ft}\cdot\text{sec}^2 \end{aligned} \qquad \textit{Answer}$$

Example 6.13

Determine the mass moment of inertia about the X, Y, and Z axes, respectively, for the thin circular disk shown in Fig. 6.27.

Figure 6.27

From our earlier discussion on the mass moment of inertia of thin plates, we noted that this property was simply the product of the area moment of inertia multiplied by ρt. For the thin disk in Fig. 6.27, the mass moment of inertia is

$$I_x = \rho t \frac{\pi R^4}{4}$$

where $\pi R^4/4$ is obtained from Table 6.1 for a circle of radius R. Since the mass is given by

$$m = \rho V = \rho \pi R^2 t$$

the mass moment of inertia reduces to

$$I_x = \frac{mR^2}{4} \qquad \textit{Answer}$$

From symmetry we notice $I_y = I_x$, so that

$$I_y = I_x = \frac{mR^2}{4} \qquad \textit{Answer}$$

Similarly, the mass moment of inertia about the Z axis is

$$I_z = \rho t \frac{\pi R^4}{2} = \frac{mR^2}{2} \qquad \textit{Answer}$$

These expressions agree with the formulas in Table 6.2 for a thin circular plate.

Example 6.14

Determine the mass moment of inertia of the prismatic bar shown in Fig. 6.28 about (a) the X axis, and (b) the centroidal X' axis.

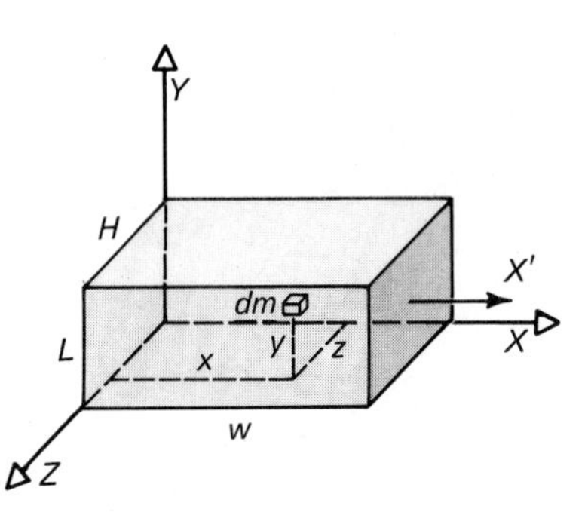

Figure 6.28

(a) The differential mass is,

$$dm = \rho\, dx\, dy\, dz$$

The mass moment of inertia about the X axis is

$$I_x = \int (y^2 + z^2)\, dm$$

Substituting for dm, we find that

$$I_x = \rho \int y^2\, dx\, dy\, dz + \rho \int z^2\, dx\, dy\, dz$$

Performing the integration, we obtain

$$I_x = \rho \frac{WHL^3}{3} + \rho \frac{H^3 LW}{3}$$

The mass of the prismatic bar is

$$m = LWH\rho$$

so that

$$I_x = \frac{m}{3}(L^2 + H^2) \qquad \textit{Answer}$$

(b) The mass moment of inertia about the centroidal axis is obtained using the parallel-axis theorem.

$$\bar{I}_{x'} = I_x - md^2$$

Now

$$d^2 = \left(\frac{L}{2}\right)^2 + \left(\frac{H}{2}\right)^2 = \frac{1}{4}(L^2 + H^2)$$

so that

$$\bar{I}_{x'} = \frac{m}{3}(L^2 + H^2) - \frac{m}{4}(L^2 + H^2)$$

$$= \frac{m}{12}(L^2 + H^2) \qquad \textit{Answer}$$

This result agrees with the data in Table 6.2 for a rectangular prism.

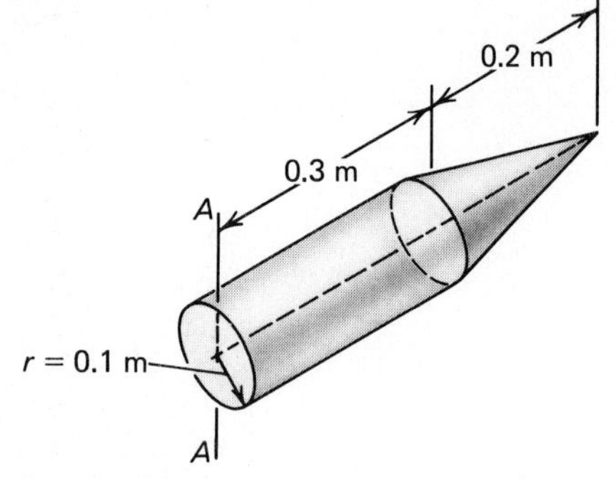

Figure 6.29

Example 6.15

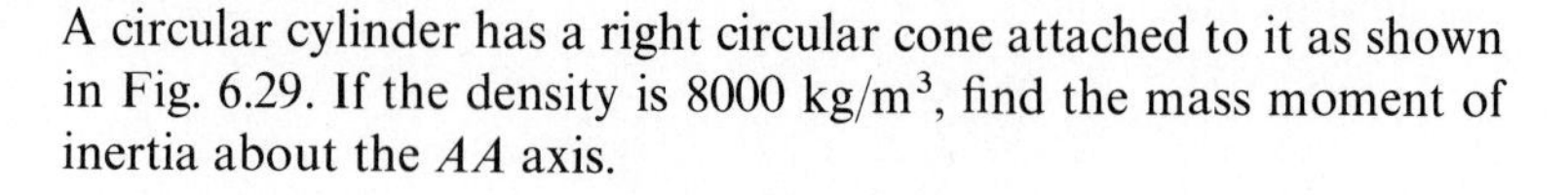

A circular cylinder has a right circular cone attached to it as shown in Fig. 6.29. If the density is 8000 kg/m³, find the mass moment of inertia about the AA axis.

This is an example of a composite body so that

$$I_{AA} = (I_{AA})_{\text{cylinder}} + (I_{AA})_{\text{cone}}$$

For the cylinder

$$\text{Volume:} \quad V = \pi(0.1)^2(0.3) = 9.42 \times 10^{-3}\ \text{m}^3$$

$$\text{Mass:} \quad m = \rho V = 8000(9.42 \times 10^{-3}) = 75.36\ \text{kg}$$

Referring to Table 6.2 for a circular cylinder and using the parallel-axis theorem,

$$(I_{AA})_{\text{cylinder}} = \frac{1}{12}m(3r^2 + L^2) + m\left(\frac{L}{2}\right)^2$$

$$= \frac{1}{12}m(3r^2 + 4L^2)$$

$$= \frac{75.36}{12}\left[3(0.1)^2 + 4(0.3)^2\right]$$

$$= 2.449\ \text{kg}\cdot\text{m}^2$$

For the circular cone,

$$\text{Volume:} \quad V = \frac{1}{3}\pi r^2 h = \frac{1}{3}\pi(0.1)^2(0.2) = 2.094 \times 10^{-3}\ \text{m}^3$$

$$\text{Mass:} \quad m = \rho V = 8000(2.094 \times 10^{-3}) = 16.75\ \text{kg}$$

Referring to Table 6.2 for a right circular cone we first obtain the mass moment of inertia about the centroidal axis of the circular cone.

$$I_x = \frac{3}{20}mr^2 + \frac{3}{80}mh^2$$

$$= \frac{3}{20}m\left(r^2 + \frac{h^2}{4}\right)$$

$$= \frac{3}{20}(16.75)\left[(0.1)^2 + \frac{(0.2)^2}{4}\right]$$

$$= 50.25 \times 10^{-3}\ \text{kg}\cdot\text{m}^2$$

Now use the parallel-axis theorem between the centroid of the circular cone and the AA axis.

$$(I_{AA})_{\text{cone}} = \bar{I}_x + md^2$$

$$= (50.25 \times 10^{-3}) + 16.75\left(0.3 + \frac{0.2}{4}\right)^2$$

$$= 2.102 \text{ kg}\cdot\text{m}^2$$

Finally, adding the mass moment of inertia for each body about the AA axis,

$$I_{AA} = 2.449 + 2.102 = 4.551 \text{ kg}\cdot\text{m}^2 \qquad \textit{Answer}$$

PROBLEMS

6.71 Verify the expressions for I_x and I_z for the slender rod shown in Table 6.2.

6.72 Verify the equations for I_x and I_z for the thin rectangular plate shown in Table 6.2.

6.73 Verify the equation for I_x for the sphere shown in Table 6.2.

6.74 Verify the equation for I_x for the circular cylinder shown in Table 6.2.

6.75 A 10-kg mass point is located as shown in Fig. P6.75. Obtain the mass moments of inertia about the X, Y, and Z axes.

6.76 Calculate the mass moment of inertia about the shaft axis AA due to the three mass points shown in Fig. P6.76. Assume $m_1 = m_2 = m_3 = 0.5$ kg and $d_1 = 30$ *mm*, $d_2 = 40$ *mm*, and $d_3 = 30$ *mm*.

6.77 The pendulum in Fig. P6.77 consists of a thin wire and a thin circular disk. If the weight of the disk is 5 lb and neglecting the weight of the wire, calculate the mass moment of inertia about the axis AA (a) due to the thin disk, and (b) treating the disk as a point mass.

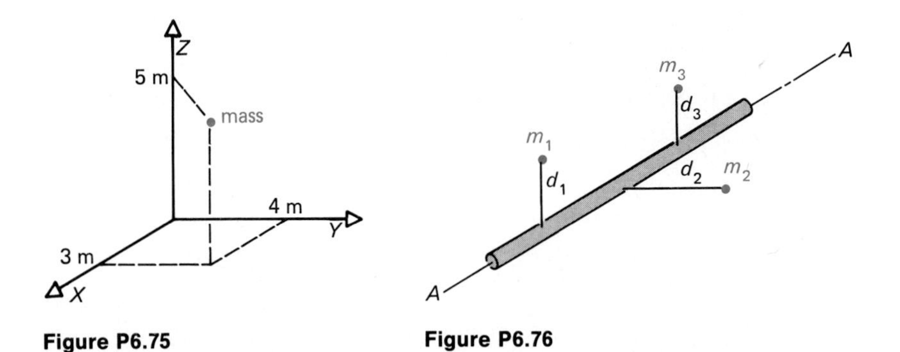

Figure P6.75 **Figure P6.76**

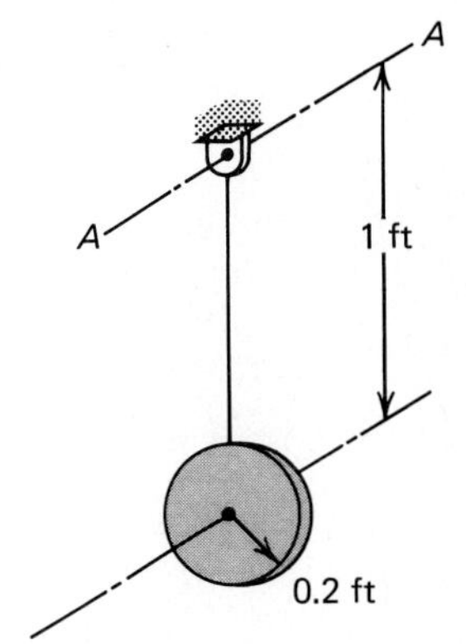

Figure P6.77

6.78 The specific weight of the thin steel plate in Fig. P6.78 is 490 lb/ft³. Calculate the mass moment of inertia about the X axis if the plate thickness is 1 in.

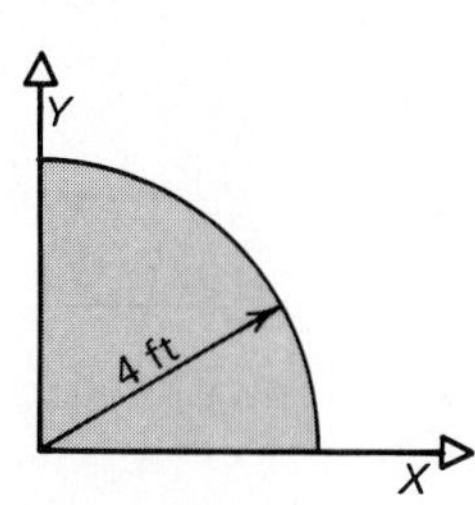

Figure P6.78

6.79 The thin triangular plate shown in Fig. P6.79 is made of glass whose density is 2.60 t/m^3. Assuming the plate thickness is 10 mm, calculate the mass moment of inertia about the X axis.

6.80 Calculate the mass moment of inertia about the Z axis for the prism shown in Fig. P6.80. Assume the prism is made of gold whose density is 19.3 t/m^3.

6.81 Calculate the mass moment of inertia about the X axis for the cylinder in Fig. P6.81 if the specific weight equals 100 lb/ft^3.

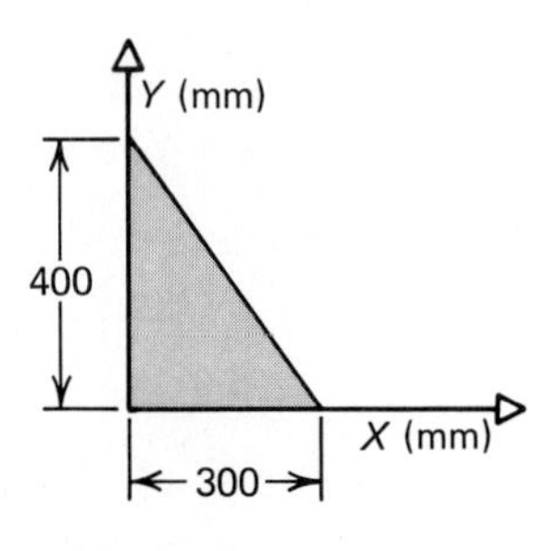

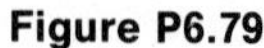
Figure P6.79

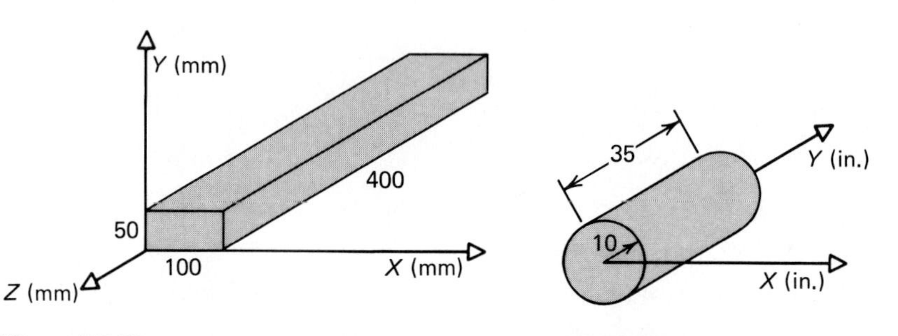

Figure P6.80

Figure P6.81

6.82 The brass pendulum shown in Fig. P6.82 consists of a thin rod 1 ft in length and 1 inch in diameter with a sphere of radius 3 in attached at the end. If the specific weight of brass is 534 lb/ft^3, find the mass moment of inertia about the axis through O.

6.83 A 10-kg mass point is located at (3, 4, 5). Obtain the mass moment of inertia of the mass about an axis coincident with the unit vector $\boldsymbol{\lambda} = 0.577\mathbf{i} + 0.577\mathbf{j} + 0.577\mathbf{k}$. Let the distance be in meters.

6.84 Find the mass moment of inertia for the spinning body about the AA axis as shown in Fig. P6.84. The body is made of aluminum which has a density of 2.25 t/m^3 and is assumed to consist of two similar thin circular disks and a slender rod. Let $d = 25$ mm be the diameter of the rod and $t = 20$ mm be the thickness of each disk.

6.85 A spinning satellite shown in Fig. P6.85 releases a pair of masses m

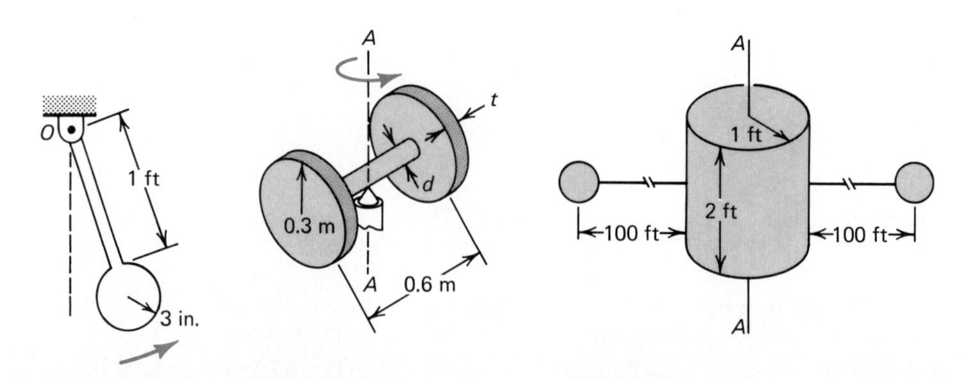

Figure P6.82

Figure P6.84

Figure P6.85

to reduce the amount of spin about the AA axis. The weight of the cylinder is 50 lb while each mass m equals 0.1 slugs. Neglecting the mass of the cord, find the mass moment of inertia of the body about the AA axis.

6.86 Calculate the mass moment of inertia about the longitudinal axis of a hollow cylinder with an inner radius of 0.3 m, outer radius of 1.0 m, and length of 2.0 m. Assume $\rho = 8.0 \text{ t/m}^3$.

6.87 Calculate the mass moment of inertia about an axis through the center of a hollow sphere with an inner radius of 75 mm and outer radius of 200 mm. Assume $\rho = 6.0 \text{ t/m}^3$.

6.6 Summary

A. The important *ideas*:

1. The area moment of inertia is used in the study of stress analysis of structures such as beams and columns.
2. The mass moment of inertia is applied in the study of rotational motion of rigid bodies.
3. The principal axes are the axes about which the area moment of inertia is either maximum or minimum.
4. The area products of inertia are zero about the symmetric axes of a body.
5. Symmetric axes are principal axes while principal axes are not necessarily symmetric axes.

B. The important *equations*:

$$I_x = \int y^2 \, dA \qquad I_y = \int x^2 \, dA \tag{6.2}$$

$$I_x = k_x^2 A \tag{6.3}$$

Area polar moment of inertia:

$$J = \int r^2 \, dA \tag{6.5}$$

$$J = I_x + I_y \tag{6.6}$$

$$k_z = \sqrt{k_x^2 + k_y^2} \tag{6.8}$$

Parallel-axis theorem:

$$I_y = \bar{I}_{y'} + Ad^2 \tag{6.9}$$

Area product of inertia:

$$I_{xy} = \int xy \, dA \tag{6.12}$$

Mass moment of inertia:

$$I = \int r^2 \, dm \tag{6.19}$$

7 Beam Analysis

Beams are perhaps the most important structural members used in engineering. It is therefore important that we understand fully the fundamentals of beam analysis. We analyzed beams for their support reactions to concentrated and distributed loads in Chapter 5. Here, we will extend our analysis to include the relationships between these applied loads and the consequential effects of shear and bending on the beam. This information is necessary in order to design beams of proper size for various applications. This design process, however, will be treated in the mechanics of materials course that generally follows a statics course.

7.1 Types of Beams

When the reactions of a beam can be found from the equations of statics, that beam is classified as statically determinate. If, however, there are more unknown reactions at the supports than there are equations of statics, a beam is classified as statically indeterminate. Complete analysis of this latter group must include the load-deformation properties of the beams along with the equilibrium equations of statics. In this text, we will limit ourselves to statically determinate beams. Table 7.1 shows some common beam supports and their classification.

Figure 7.1 shows selected beam loads with which we will be concerned in our study of beams; namely, the concentrated load, an applied couple, and the distributed load. Beam reactions due to distributed loads can be determined by placing the equivalent resultant load at the centroid of the distributed load and then applying the equations of equilibrium as discussed in Chapter 5.

TABLE 7.1 ***Types of Beams***

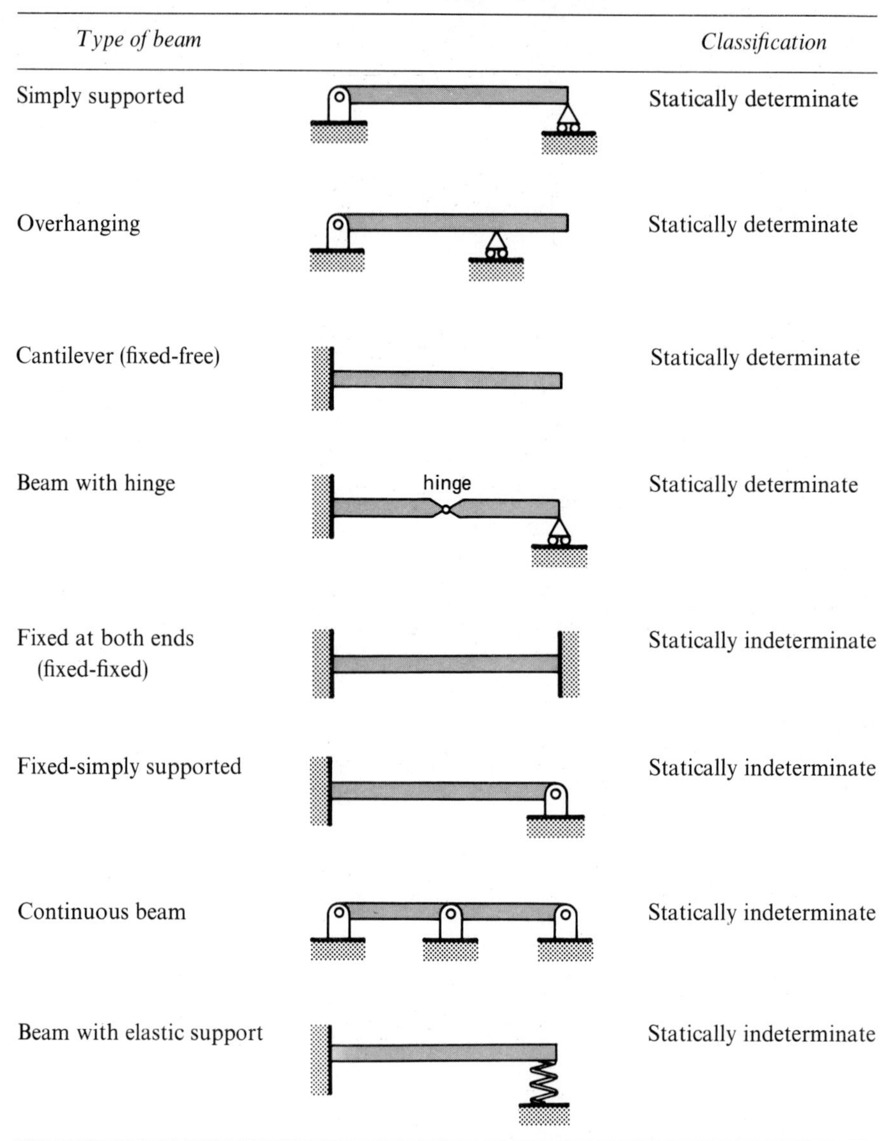

Type of beam	*Classification*
Simply supported	Statically determinate
Overhanging	Statically determinate
Cantilever (fixed-free)	Statically determinate
Beam with hinge	Statically determinate
Fixed at both ends (fixed-fixed)	Statically indeterminate
Fixed-simply supported	Statically indeterminate
Continuous beam	Statically indeterminate
Beam with elastic support	Statically indeterminate

When simple structures were analyzed earlier, a structural member was considered to be either in tension or compression. Such members we called two-force members. In the case of beams, the behavior of the internal forces is more complicated. For example, Fig. 7.2(a) shows a simply supported beam subject to a concentrated force **P**. Figure 7.2(b) shows two free-body diagrams if the beam is cut vertically at point C. The diagram to the left of the cut section

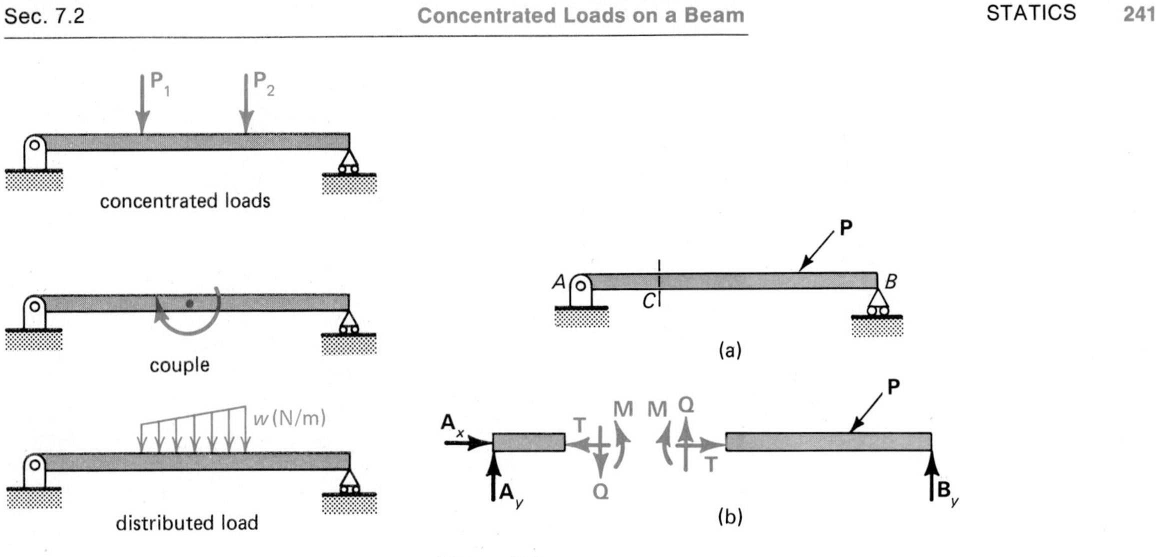

Figure 7.1

Figure 7.2

shows forces **Q** and **T** and a moment **M** which result internally when the beam supports the applied force. In the free-body diagram to the right of the cut section, we have the same forces **Q** and **T** and moment **M** applied in equal but opposite direction in accord with Newton's third law. The force **Q** is called the shear force, **T** the axial force, and **M** the bending moment. We shall consider the force **Q** and the moment **M** as being constant over the cross section of the beam. While the force **T** gives rise to stresses across the cross section of the beam, we are concerned primarily with vertically applied loads, whether concentrated or distributed. We shall therefore develop the relationships between vertical applied loads, bending moments, and shear forces.

7.2 Concentrated Loads on a Beam

Consider a beam subjected to concentrated loads $\mathbf{P}_1$ and $\mathbf{P}_2$ as shown in Fig. 7.3(a). The free-body diagram of the entire beam is shown in Fig. 7.3(b). Using this load diagram and the three equations of statics,

$$\sum F_x = 0 \qquad \sum F_y = 0 \qquad \sum M = 0$$

we can find the beam reactions $\mathbf{A}_x$, $\mathbf{A}_y$, and $\mathbf{B}_y$. Note that $\mathbf{A}_x = 0$ for this case.

To investigate the internal forces and moments, let us consider the free-body diagram in Fig. 7.3(c) for a section of the beam cut at point C. We note that the shear force, **Q**, is perpendicular to the beam axis. Since there is no load applied in the X direction, there

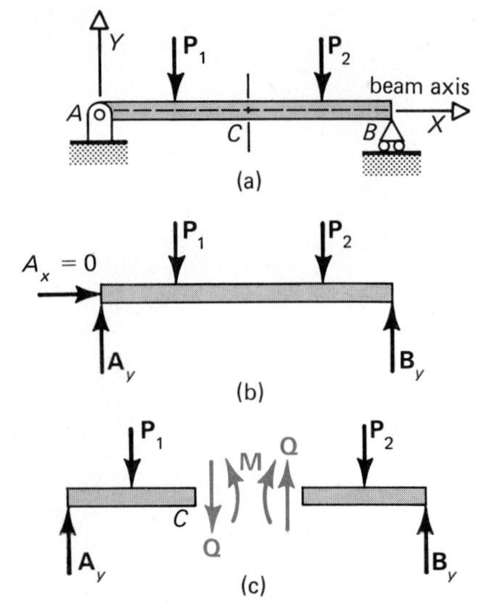

Figure 7.3

is no axial force acting on the beam. The bending moment **M** that is developed by the beam is also shown in the figure. Here **Q** tends to shear the beam while **M** has a tendency to bend the beam. In beam analysis, we are interested in obtaining the shear force **Q** and bending moment **M** throughout the length of the beam, and the relationships between the applied load, shear forces, and bending moments.

We can calculate the shear force **Q** and the bending moment **M** acting at the cross section under consideration by applying the equations of statics to either free-body diagram in Fig. 7.3(c). For example, considering the left section, $\sum F_y = 0$ will give us the shearing force **Q**. Having **Q**, we can sum moments about any point on the left section such that $\sum M = 0$. This allows us to find the bending moment **M**. Note that the magnitudes of **Q** and **M** depend upon the cross section selected. In other words, if we selected a section between point A and $\mathbf{P}_1$, we would calculate a new set of values for **Q** and **M** associated with the new cross section.

In the above example, we assumed the directions of the shear force **Q** and bending moment **M** when we drew Fig. 7.3(c). Let us adopt these assumed directions as the positive directions for the shear and the bending moment. This assumption identifies shear forces as positive when the external forces deform the beam as shown in Fig. 7.4(a), and negative when external forces deform the beam shown in Fig. 7.4(b). The bending moment is positive if it deforms the beam as indicated in Fig. 7.4(c) and negative as shown in Fig. 7.4(d). In problem solving, we assume positive shear and bending moment as shown in Fig. 7.3(c). We know these assumptions are correct when computation for the magnitudes at **Q** or **M** provides a positive value. If the magnitude of **Q** or **M** is computed

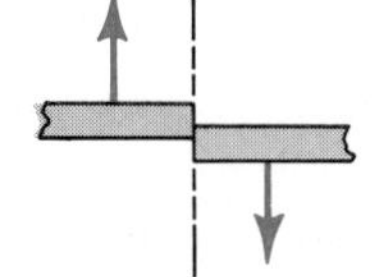
(a) Positive shear deformation

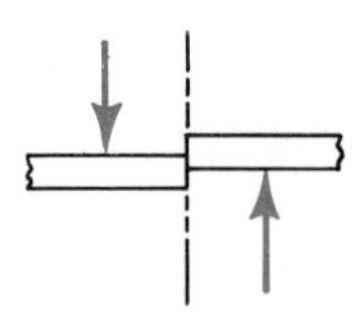

(b) Negative shear deformation

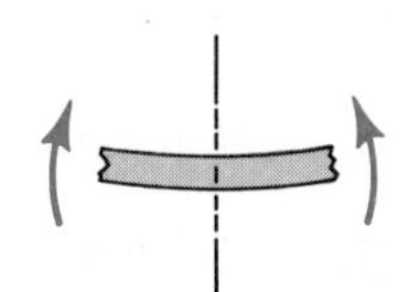
(c) Positive bending deformation

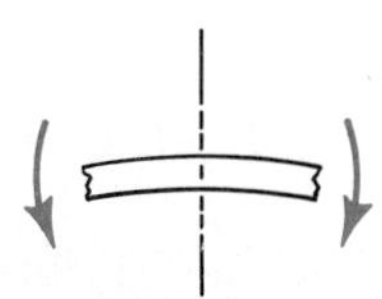
(d) Negative bending deformation

Figure 7.4

as a negative quantity, we know that the shear force, or bending moment, acts in the reverse direction.

In our example in Fig. 7.3 we noted that the shear force and bending moment are influenced by the section at which we cut the beam. In general, we wish to know how this force and moment vary throughout the length of the beam. It is, therefore, necessary that the shear and bending moment be computed everywhere and recorded in a convenient manner. To accomplish this, we draw a shear diagram and a bending-moment diagram. By way of example, consider the beam carrying the load **P** in Fig. 7.5(a). The free-body diagram for the entire beam is shown in Fig. 7.5(b). From the equations of statics, we obtain $A_x = 0$, $A_y = B_y = P/2$.

To establish the shear and bending moment in the beam, consider the free-body diagrams in Fig. 7.5(c) and (d). In Fig. 7.5(c), a section at a distance x from point A is taken to the left of the applied load P. From this diagram we can write the following: For $0 < x < L/2$,

$$\sum F_y = 0 \qquad \frac{P}{2} - Q = 0$$

$$Q = \frac{P}{2} \tag{7.1}$$

Now sum the moments about the cut section, letting positive moments act counterclockwise.

$$\sum M_x = 0 \qquad M - \frac{Px}{2} = 0$$

$$M = \frac{Px}{2} \tag{7.2}$$

Since Q and M are calculated as positive quantities, our assumed directions are correct.

Referring to the section to the right of the load **P** as shown in Fig. 7.5(d), we have the following: For $L/2 < x < L$,

$$\sum F_y = 0 \qquad \frac{P}{2} - P - Q = 0$$

$$Q = -\frac{P}{2} \tag{7.3}$$

Again sum moments about the cut section letting counterclockwise moments be positive.

$$\sum M_x = 0 \qquad M - \frac{Px}{2} + P\left(x - \frac{L}{2}\right) = 0$$

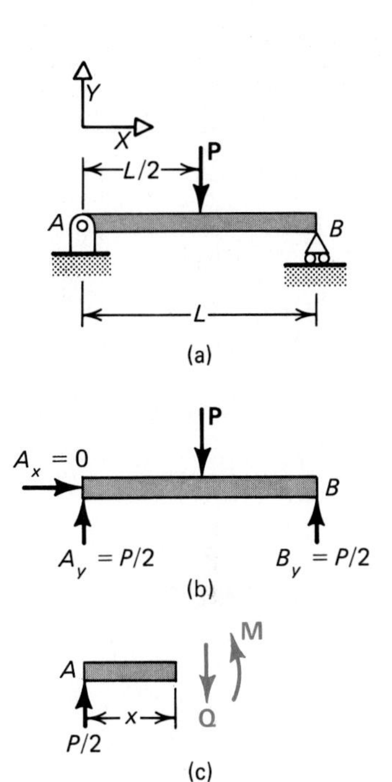

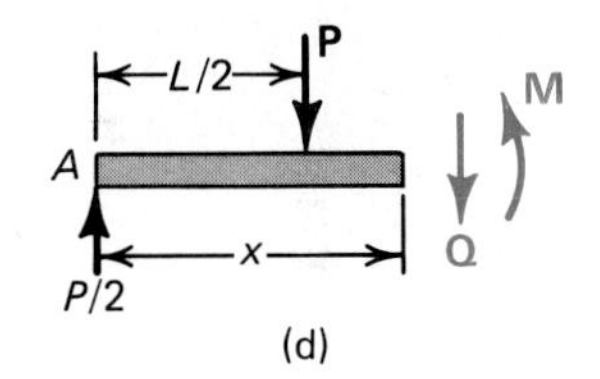

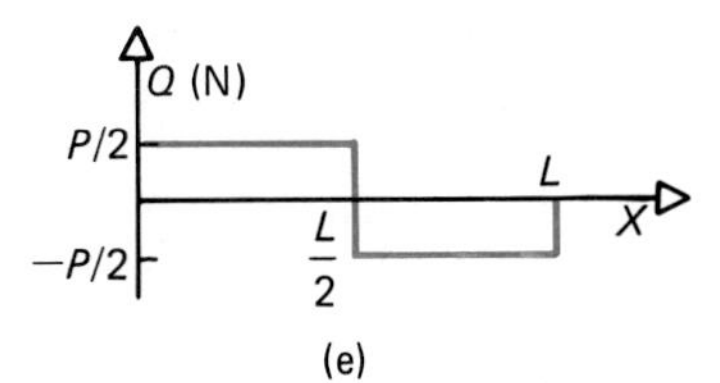

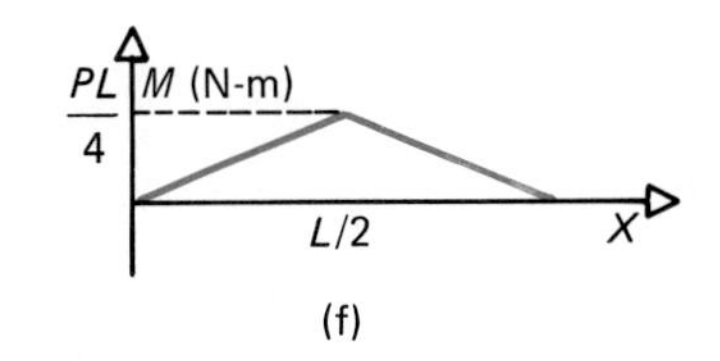

Figure 7.5

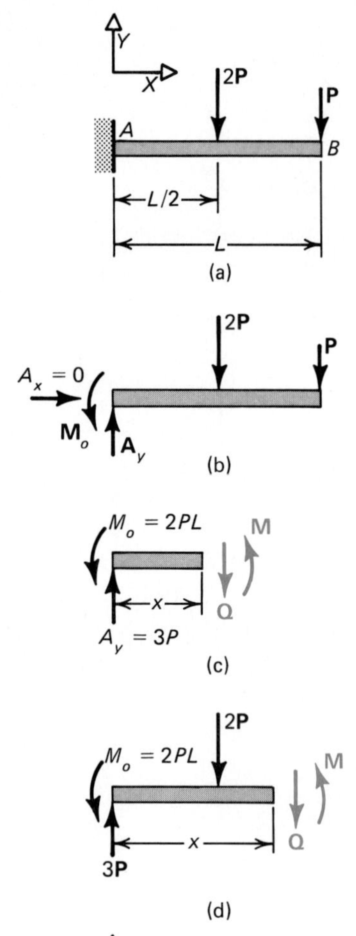

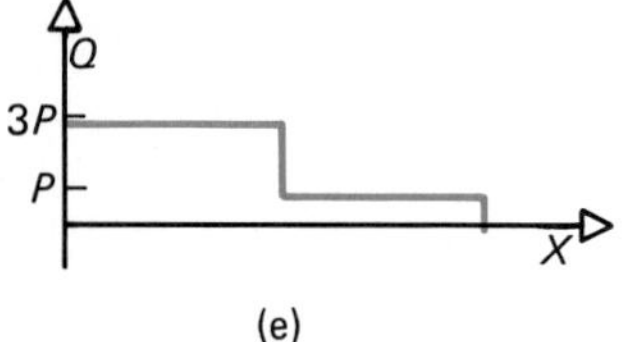

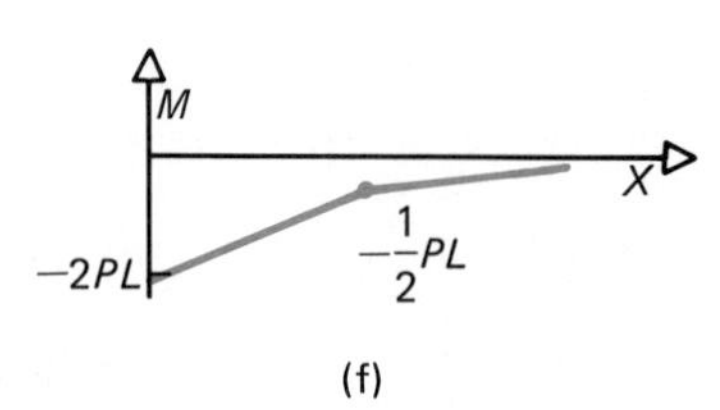

Figure 7.6

$$M = P\left(\frac{L}{2} - \frac{x}{2}\right) \tag{7.4}$$

In this section of the beam our assumed direction of Q was incorrect, while the direction on M is correct.

The results for the shear throughout the beam, given by Eqs. (7.1) and (7.3), are plotted as a shear diagram in Fig. 7.5(e). The shear force begins at A with a value $P/2$, until we reach the location of the applied load P, at which point the shear changes by an amount $(-P)$. The magnitude of this discontinuity is equal to the magnitude of the applied load at this section of the beam. Beyond $x = L/2$, the shear remains $(-P/2)$ throughout until we reach the reaction at B at which point the shear changes by an amount $(P/2)$. We note that the magnitude of the shear at both ends of the beam is equal to the reactions at these ends.

Equations (7.2) and (7.4) are plotted in Fig. 7.5(f), which is called the bending-moment diagram. Note that the boundary conditions for a simply supported beam are satisfied, that is, $M = 0$ at $x = 0$ and $x = L$. In this case, the maximum bending moment occurs at $x = L/2$ and has a magnitude equal to $PL/4$. We also note that in this example the bending moment has a maximum magnitude right under the point of applied load **P**.

Example 7.1

Find the equations for the shear and bending moment and draw the shear and bending-moment diagrams for the cantilever beam in Fig. 7.6(a).

The free-body diagram of the beam is shown in Fig. 7.6(b). Here we see a reaction $\mathbf{A}_y$ and a moment $\mathbf{M}_o$ due to the fixed-support condition. From $\sum F_x = 0$ we note that $A_x = 0$. Next we have

$$\sum F_y = 0 \qquad A_y = 3P$$

and

$$\sum M_A = 0 \qquad M_o - 2P\left(\frac{L}{2}\right) - P(L) = 0$$

$$M_o = 2PL$$

The free-body diagram to the left of the load $2P$ is shown in Fig. 7.6(c). For this case, where $0 < x < L/2$, we find

$$\sum F_y = 0 \qquad 3P - Q = 0$$

$$Q = 3P \qquad \textit{Answer}$$

and $\quad \sum M_x = 0 \qquad M + 2PL - 3Px = 0$

$$M = 3Px - 2PL \quad \textit{Answer}$$

The free-body diagram for a section cut between the applied loads is shown in Fig. 7.6(d). For this case, where $L/2 < x < L$, we find

$$\sum F_y = 0 \qquad 3P - 2P - Q = 0$$

$$Q = P \qquad \textit{Answer}$$

and summing the moments at the cut section x,

$$\sum M_x = 0 \qquad M + 2PL - 3Px + 2P\left(x - \frac{L}{2}\right) = 0$$

$$M = P(x - L) \quad \textit{Answer}$$

The Q equations are plotted to give Fig. 7.6(e) for the shear diagram. Again we note that wherever we have a concentrated load, including the reactions, we have a corresponding step change in the shear magnitude. In this case, it occurs at $x = 0$, $x = L/2$, and $x = L$. The bending-moment diagram in Fig. 7.6(f) is obtained from the moment equations. Again we note that the boundary conditions are met in the diagram, that is, $M_o = 2PL$ acting counterclockwise at $x = 0$, and $M = 0$ at $x = L$. The negative sign on M must be included in the bending-moment diagram to agree with the sign convention adopted in Fig. 7.4.

Example 7.2

Find the equations for the shear and bending moment and draw the shear and bending-moment diagrams for the overhanging beam in Fig. 7.7(a).

From the free-body diagram in Fig. 7.7(b) we have the following:

$$\sum M_A = 0 \qquad 10B_y - 10(20) = 0$$

$$B_y = 20 \text{ kip}$$

and $\quad \sum F_y = 0 \qquad -A_y + B_y - 10 = 0$

$$A_y = 10 \text{ kip}$$

For the free-body diagram of a cut section between A and B in Fig. 7.7(c), where $0 < x < 10$, we have the following:

$$\sum F_y = 0 \qquad -Q - 10 = 0$$

$$Q = -10 \text{ kip} \qquad \textit{Answer}$$

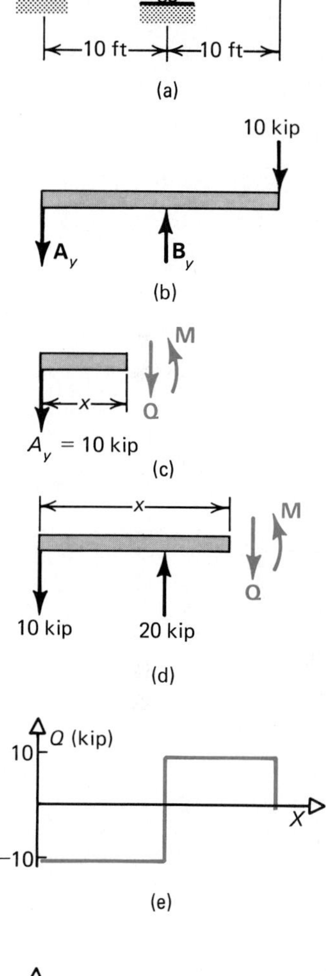

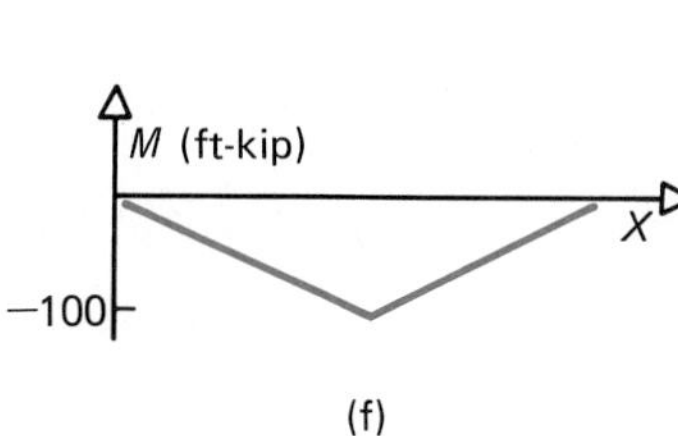

Figure 7.7

Summing moments about the cut section at x,

$$\sum M_x = 0 \qquad M + 10x = 0$$

$$M = -10x \text{ ft·kip} \qquad \textit{Answer}$$

From Fig. 7.7(d), where $10 < x < 20$, we have

$$\sum F_y = 0 \qquad -Q - 10 + 20 = 0$$

$$Q = 10 \text{ kip} \qquad \textit{Answer}$$

and summing moments about the cut section at x,

$$\sum M_x = 0 \qquad M + 10x - 20(x - 10) = 0$$

$$M = (10x - 200) \text{ ft·kip} \qquad \textit{Answer}$$

The Q equations are plotted in Fig. 7.7(e) for the shear diagram, while the M equations are plotted in Fig. 7.7(f) for the bending-moment diagram.

PROBLEMS

Derive the shear and bending-moment equations and draw the shear and bending-moment diagrams for the beams shown in:

7.1 Figure P7.1.

7.2 Figure P7.2.

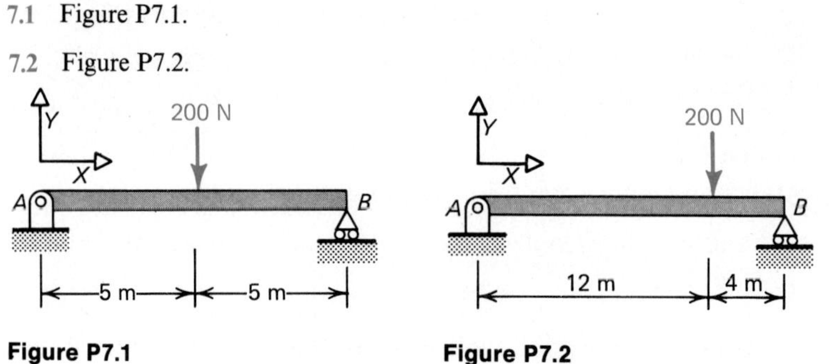

Figure P7.1 **Figure P7.2**

7.3 Figure P7.3 for $P = 100$ N, $L = 10$ m

7.4 Figure P7.3 for $P = 2$ kN, $L = 8$ m

7.5 Figure P7.5 for $2L_1 = L_2 = 4$ m

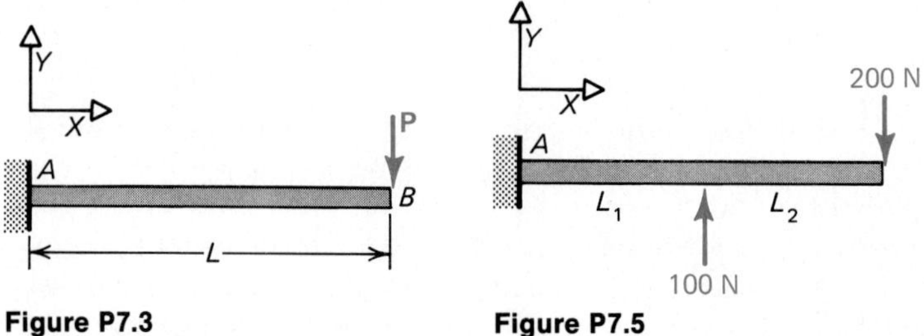

Figure P7.3 **Figure P7.5**

7.6 Figure P7.5 for $L_1 = 2$ m, $L_2 = 10$ m

7.7 Figure P7.7.

7.8 Figure P7.8.

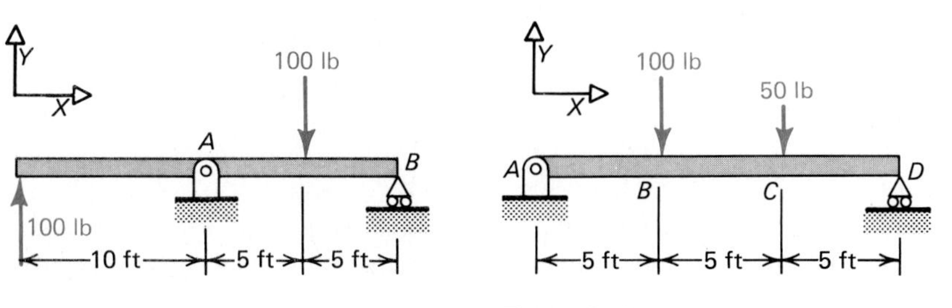

Figure P7.7

Figure P7.8

7.9 Figure P7.9.

7.10 Figure P7.10.

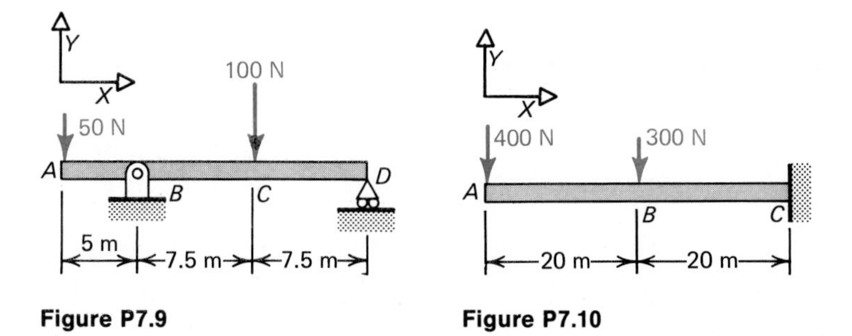

Figure P7.9

Figure P7.10

7.3 Applied Couple on a Beam

The second kind of special loading that is of interest is that due to an applied couple as discussed in Section 7.1. These couples may arise from applied forces that are equal and opposite, as shown in Fig. 7.8(a). Here the forces are applied to a rigid bracket spaced a distance d to produce a couple of magnitude Pd. Figure 7.8(b) is an equivalent loading diagram showing the couple applied at point C of the beam. Figure 7.9(a) shows another loading arrangement; its equivalent loading diagram is shown in Fig. 7.9(b). Note that the vertical load $\mathbf{P}$ is transferred to point C as well as the couple.

Once an equivalent loading diagram is drawn showing the applied couple and any other forces that may be present, the reactions can be obtained by considering the equilibrium of the entire beam. The shear and bending-moment diagrams are then obtained by analyzing different sections of the beam as discussed in Section

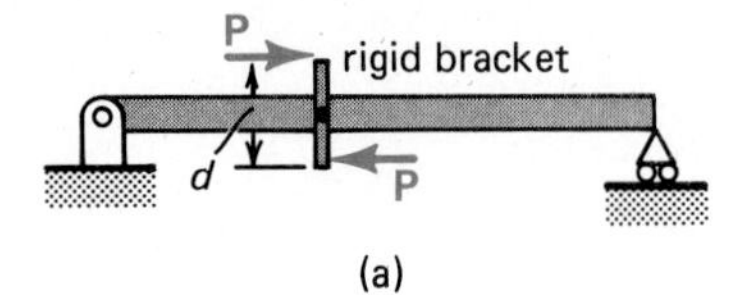

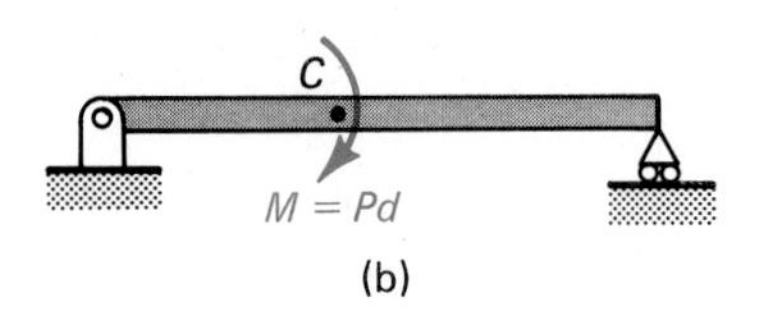

Figure 7.8

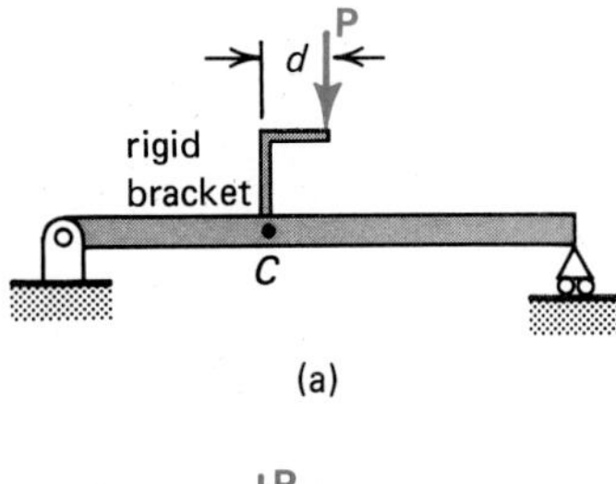

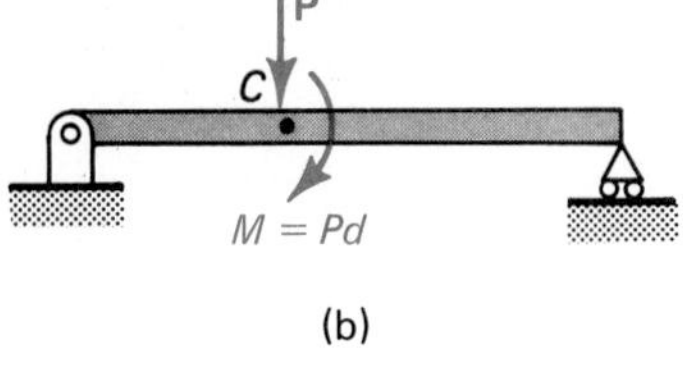

Figure 7.9

7.2. The following example problems show how to draw shear and moment diagrams for cases where applied couples are present.

Example 7.3

Find the equations for the shear and bending moment and draw the shear and bending-moment diagrams for the beam shown in Fig. 7.10(a). Let $P = 50$ kN.

The 50-kN load is replaced by an equivalent couple of magnitude 25 m·kN acting clockwise as shown in Fig. 7.10(b). The reactions can be obtained from the free-body diagram of the entire beam shown in Fig. 7.10(c).

$$\sum F_x = 0 \qquad A_x = 0$$

$$\sum F_y = 0 \qquad A_y = 50 \text{ kN}$$

$$\sum M_A = 0 \qquad M_o - 50(5) - 25 = 0$$

$$M_o = 275 \text{ m·kN}$$

Considering a section to the left of the vertical load P shown in Fig. 7.10(d) for $0 < x < 5$,

$$\sum F_y = 0 \qquad Q = A_y = 50 \text{ kN} \qquad \textit{Answer}$$

$$\sum M_x = 0 \qquad M_o + M - 50x = 0$$

$$M = (50x - 275) \text{ m·kN} \qquad \textit{Answer}$$

Now considering the beam section to the right of the vertical load P shown in Fig. 7.10(e) for $5 < x < 12$,

$$\sum F_y = 0 \qquad -Q + 50 - 50 = 0$$

$$Q = 0 \qquad \textit{Answer}$$

$$\sum M_x = 0 \qquad M + M_o - 50x + 50(x - 5) = 0$$

$$M = -25 \text{ m·kN} \qquad \textit{Answer}$$

The shear and bending-moment diagrams are shown in Fig. 7.10(f) and 7.10(g). Note the step change in the moment diagram at the end of the beam where the applied couple acts. This is equivalent to step changes in the shear diagram where concentrated loads act on the beam.

Example 7.4

Find the equations for the shear and bending moment and draw the shear and bending-moment diagrams for the beam in Fig. 7.11(a).

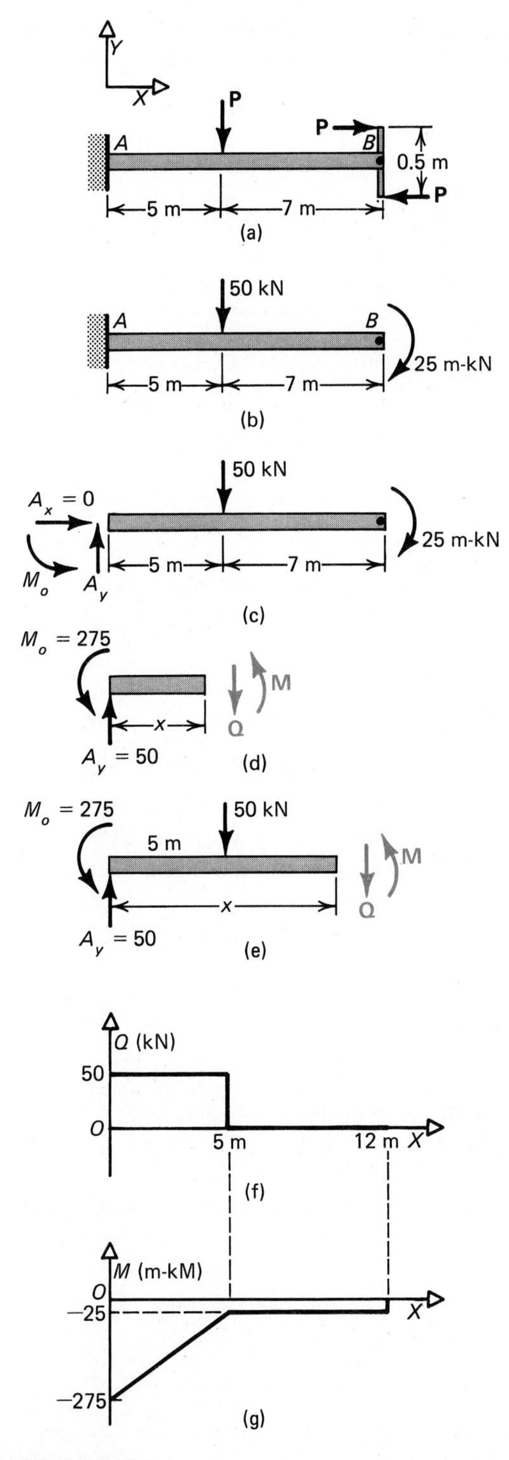
Y
X
P
P
B
A
0.5 m
P
5 m
7 m
(a)
50 kN
A
B
25 m-kN
5 m
7 m
(b)
50 kN
$A_x = 0$
25 m-kN
5 m
7 m
M_o
A_y
(c)
$M_o = 275$
M
x
Q
$A_y = 50$
(d)
$M_o = 275$
50 kN
5 m
M
x
Q
$A_y = 50$
(e)
Q (kN)
50
O
5 m
12 m X
(f)
M (m-kM)
O
−25
X
−275
(g)

Figure 7.10

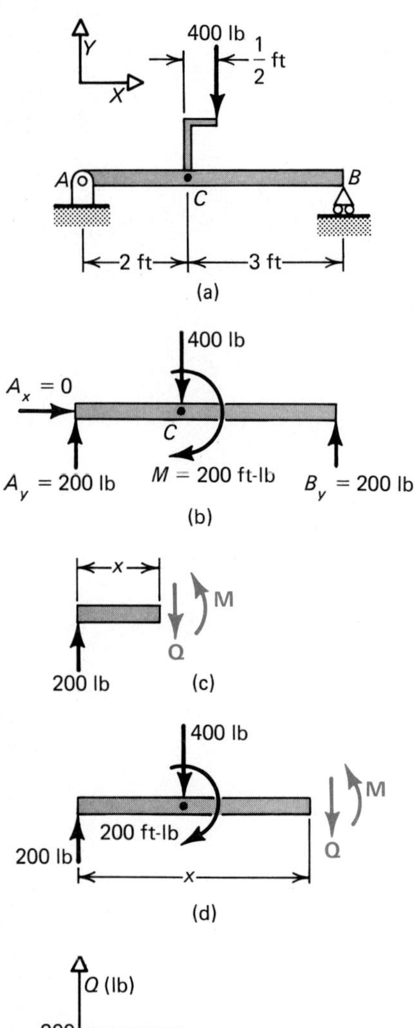

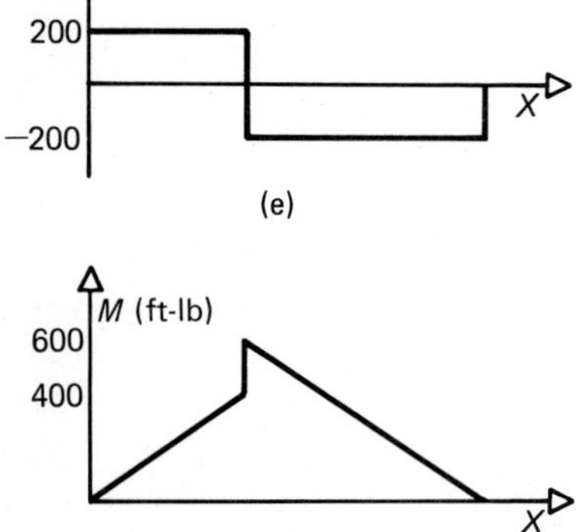

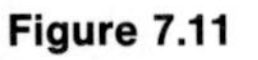
Figure 7.11

The beam reactions are obtained using the equivalent loads at point C for the free-body diagram in Fig. 7.11(b).

$$\sum M_A = 0 \qquad 5B_y - 400(2) - 200 = 0$$
$$B_y = 200 \text{ lb}$$

and

$$\sum F_y = 0 \qquad A_y + B_y - 400 = 0$$
$$A_y = 200 \text{ lb}$$

Referring to Fig. 7.11(c) and (d), we have the following. For $0 < x < 2$,

$$\sum F_y = 0 \qquad 200 - Q = 0$$
$$Q = 200 \text{ lb} \qquad \textit{Answer}$$

and

$$\sum M_x = 0 \qquad M - 200x = 0$$
$$M = 200x \text{ ft}\cdot\text{lb} \qquad \textit{Answer}$$

For $2 < x < 5$,

$$\sum F_y = 0 \qquad 200 - 400 - Q = 0$$
$$Q = -200 \text{ lb} \qquad \textit{Answer}$$

and

$$\sum M_x = 0 \qquad M - 200x + 400(x - 2) - 200 = 0$$
$$M = (1000 - 200x) \text{ ft}\cdot\text{lb} \qquad \textit{Answer}$$

The shear and bending-moment diagrams are shown in Fig. 7.11(e) and (f), respectively. Since we have an applied couple at point C, note the corresponding step change in the moment diagram. Again it is emphasized that this is similar to a step change in the shear diagram where we have a concentrated load.

PROBLEMS

Derive the shear and bending-moment equations and draw the shear and bending-moment diagrams for the beams shown in:

7.11 Figure P7.11 for $L_1 = 4$ m, $L_2 = 6$ m, $h = 2$ m and $P = 1000$ N.

7.12 Figure P7.11 for $L_1 = 5$ m, $L_2 = 5$ m, $h = 1$ m and $P = 1000$ N.

7.13 Figure P7.13.

7.14 Figure P7.14.

7.15 Figure P7.15.

7.16 Figure P7.16.

7.17 Figure P7.17.

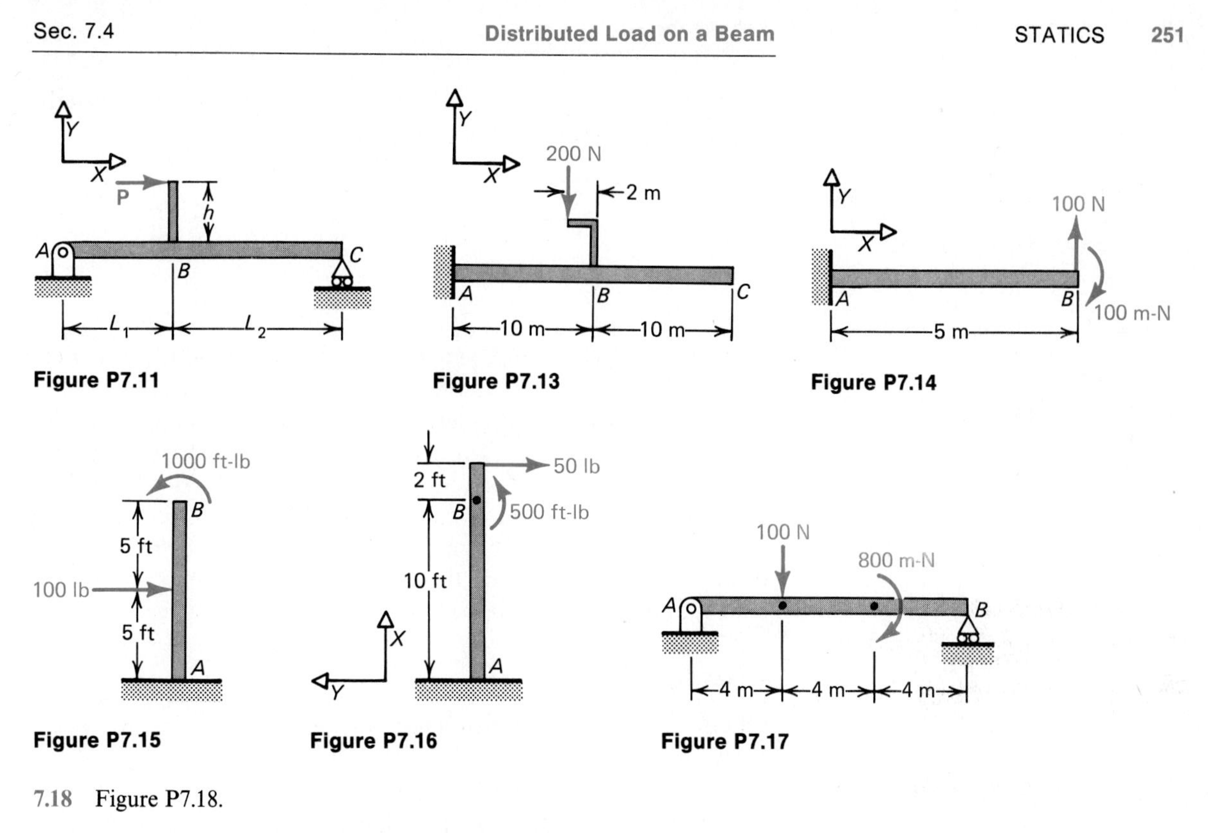

Figure P7.11

Figure P7.13

Figure P7.14

Figure P7.15

Figure P7.16

Figure P7.17

7.18 Figure P7.18.

7.19 Figure P7.19 for $M_A = 300$ m·N and $M_c = 200$ m·N.

7.20 Figure P7.19 for $M_A = 200$ m·N and $M_c = 100$ m·N.

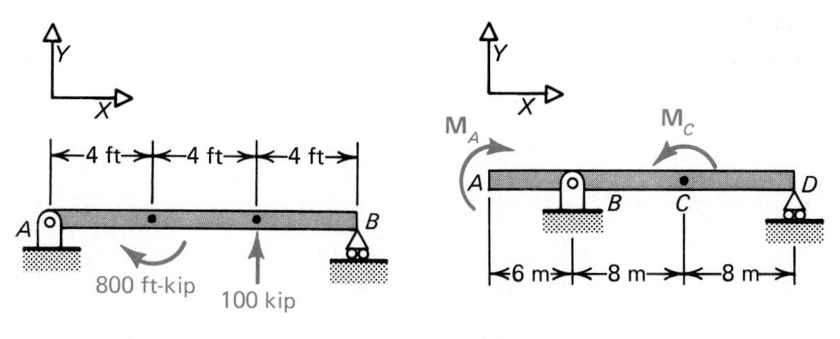

Figure P7.18

Figure P7.19

7.4 Distributed Load on a Beam

Before developing the shear and bending-moment diagram for the case of distributed loads, let us review the procedures for finding the beam reactions subject to distributed loads as discussed in Chapter 5. Figure 7.12(a) shows a beam supporting a distributed load. The free-body diagram of the beam is shown in Fig. 7.12(b),

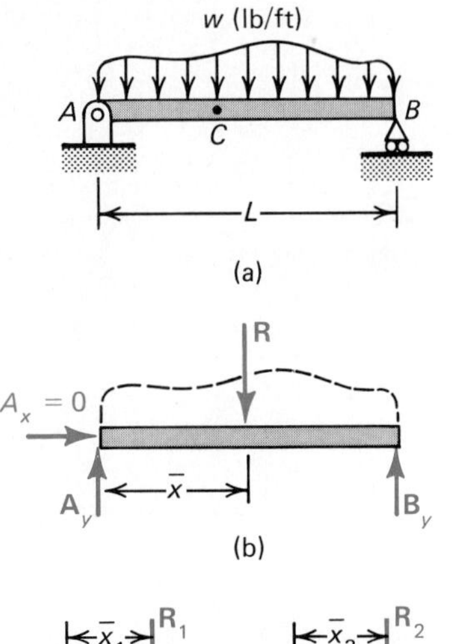

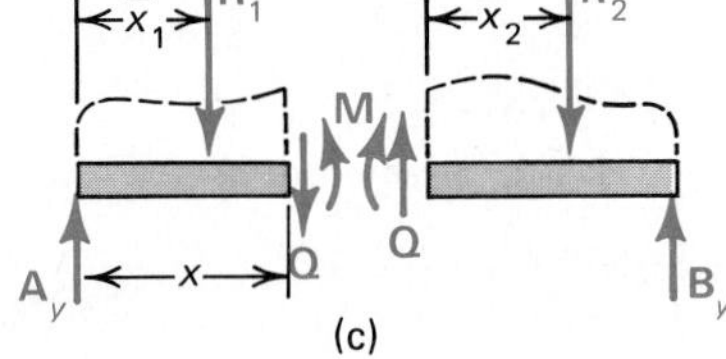

Figure 7.12

where we replace the distributed load w by the resultant load R at a distance $\bar{x}$ from point A such that

$$R = \int_0^L w\,dx \tag{7.5}$$

$$\bar{x} = \frac{\int_0^L wx\,dx}{\int_0^L w\,dx} \tag{7.6}$$

After the resultant load R and its location are obtained as shown in Fig. 7.12(b), we can find the reaction forces $\mathbf{A}_x$, $\mathbf{A}_y$, and $\mathbf{B}_y$. In this case Fig. 7.12(c) shows the free-body diagrams of the cut at point C. Now the distributed loads to either side of the section are replaced by R_1 and R_2 placed at the centroid of each portion of the distributed load. Note that $\bar{x}_2$ is measured to the right of the cut section as shown in Fig. 7.12(c). Having R_1, R_2, and their locations on the beam, we can now develop the expressions for the shear Q and the bending moment M at the section in question.

Example 7.5

Obtain the shear and bending-moment diagrams for the beam carrying the distributed load shown in Fig. 7.13(a).

We take the distributed load and replace it by its equivalent load R where

$$R = 50(5) = 250 \text{ N}$$

The load R is placed at the center of the uniformly distributed load as shown in Fig. 7.13(b). The reactions of the beam for the symmetrical loading are

$$A_y = B_y = 125 \text{ N}$$

We now consider the three sections shown in Fig. 6.13(c), (d), and (e). For $0 < x < 5$

$$\sum F_y = 0 \qquad Q = A_y = 125 \text{ N}$$

Summing moments about the cut section

$$\sum M_x = 0 \qquad M - 125x = 0$$

$$M = 125x \text{ m}\cdot\text{N}$$

For $5 < x < 10$ the distributed load acts over a portion of the beam as shown in Fig. 7.13(d). The equivalent load is R_1 which equals

$$R_1 = 50(x - 5)$$

and is located at the center of the uniformly distributed load such that

$$\bar{x}_1 = 5 + \frac{x-5}{2} = \frac{x+5}{2}$$

We now obtain the shear and bending moment.

$$\sum F_y = 0 \qquad A_y - Q - R_1 = 0$$

$$Q = 125 - 50(x-5)\ \text{N}$$

Summing moments about the cut section,

$$\sum M_x = 0 \qquad M + R_1(x - \bar{x}_1) - A_y x = 0$$

$$M = 125x - 50(x-5)\left(x - \frac{x+5}{2}\right)$$

$$= (-25x^2 + 375x - 625)\ \text{m}\cdot\text{N}$$

For the section of the beam shown in Fig. 7.13(e), where $10 < x < 15$,

$$\sum F_y = 0 \qquad A_y - 250 - Q = 0$$

$$Q = -125\ \text{N}$$

Summing moments about the cut section,

$$\sum M_x = 0 \qquad M + 250(x - 7.5) - 125x = 0$$

$$M = (-125x + 1875)\ \text{m}\cdot\text{N}$$

The shear and bending-moment diagrams are plotted in Fig. 7.13(f) and (g).

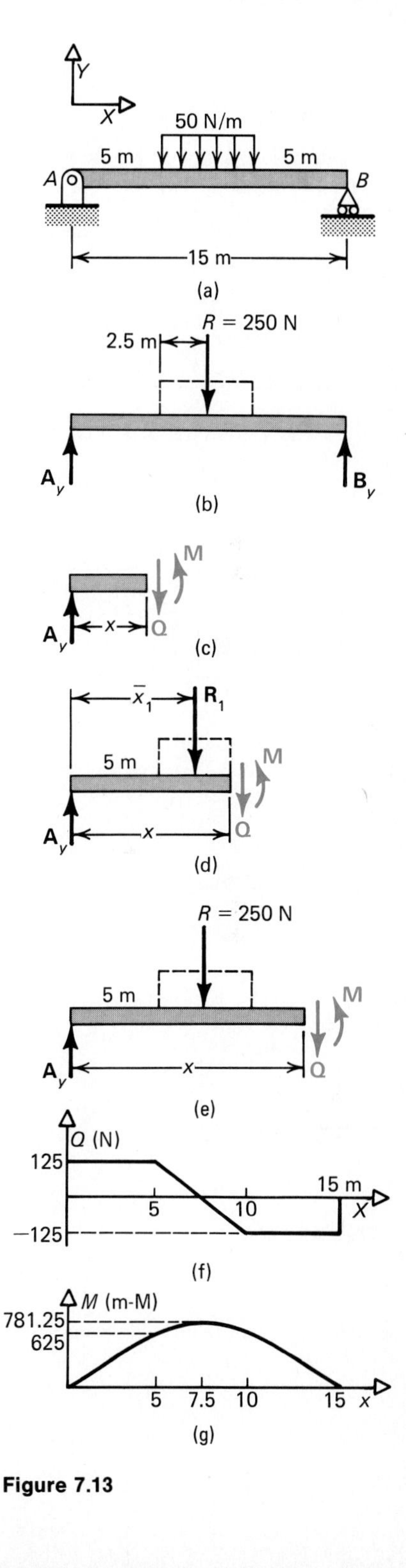

Figure 7.13

Example 7.6

Find the expressions for the shear and bending moment for the beam in Fig. 7.14(a) and draw the shear and bending-moment diagrams. Obtain the maximum value of the bending moment.

For the free-body diagram in Fig. 7.14(b),

$$R = wL$$

located at the beam center at $\bar{x} = L/2$. For the beam reactions,

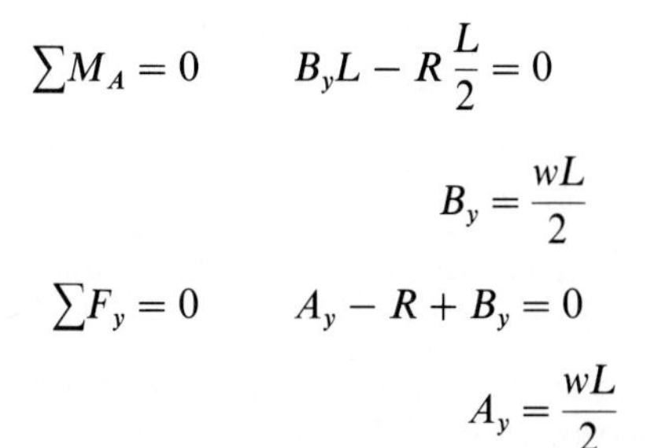

$$\sum M_A = 0 \qquad B_y L - R\frac{L}{2} = 0$$

$$B_y = \frac{wL}{2}$$

and $$\sum F_y = 0 \qquad A_y - R + B_y = 0$$

$$A_y = \frac{wL}{2}$$

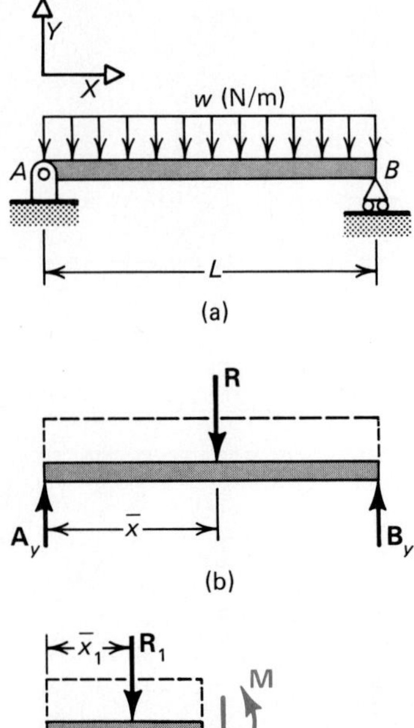

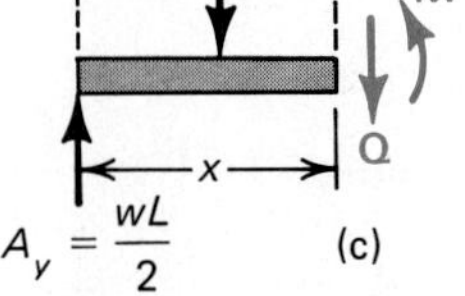

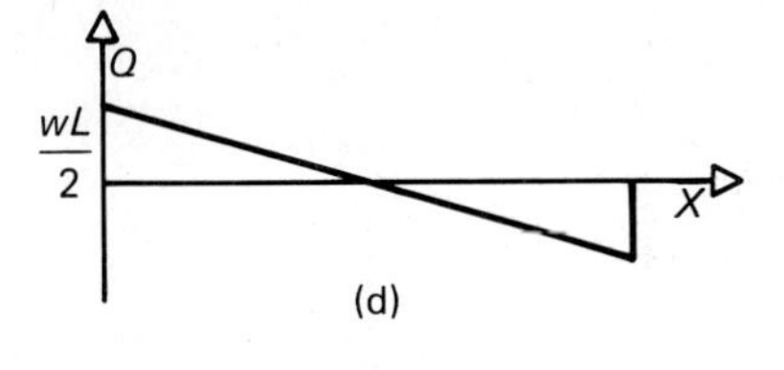

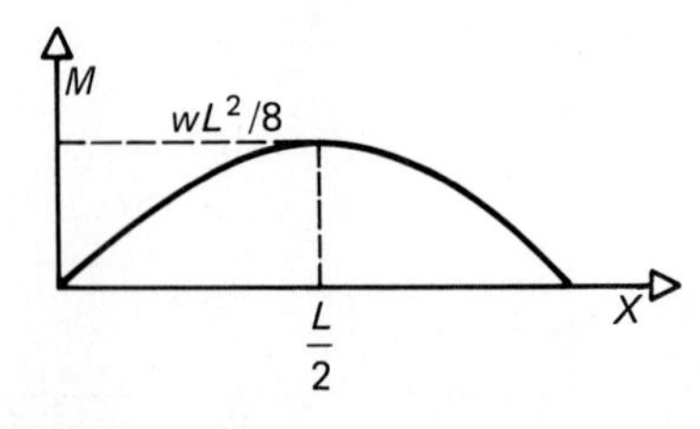

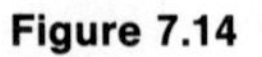

Figure 7.14

Fig. 7.14(c) shows the free-body diagram of a typical section of the beam. Now

$$R_1 = wx \qquad \text{and} \qquad \bar{x}_1 = \frac{x}{2}$$

For equilibrium, where $0 < x < L$, we find

$$\sum F_y = 0 \qquad \frac{wL}{2} - R_1 - Q = 0$$

$$Q = \frac{wL}{2} - wx \qquad \textit{Answer}$$

Summing moments about the cut section,

$$\sum M_x = 0 \qquad M - \frac{wL}{2}x + R_1\frac{x}{2} = 0$$

$$M = \frac{wLx}{2} - \frac{wx^2}{2} \qquad \textit{Answer}$$

Figures 7.14(d) and (e) show the shear and bending-moment diagrams. The maximum value of the bending moment can be obtained by differentiating the moment equation with respect to x and setting the equation equal to zero.

$$\frac{dM}{dx} = \frac{wL}{2} - wx = 0$$

This tells us where the maximum bending moment occurs. As we might guess,

$$x = \frac{L}{2}$$

Substituting this value of x into the bending-moment equation,

$$M_{\max} = \frac{wL}{2}\frac{L}{2} - \frac{w}{2}\left(\frac{L}{2}\right)^2 = \frac{wL^2}{8} \qquad \textit{Answer}$$

PROBLEMS

7.21 Obtain the shear and bending-moment equations and diagrams for the beam shown in Fig. P7.21.

7.22 Obtain the shear and bending-moment equations and diagrams for the beam shown in Fig. P7.22.

7.23 Obtain the shear and bending-moment equations for the beam with a distributed load shown in Fig. P7.23.

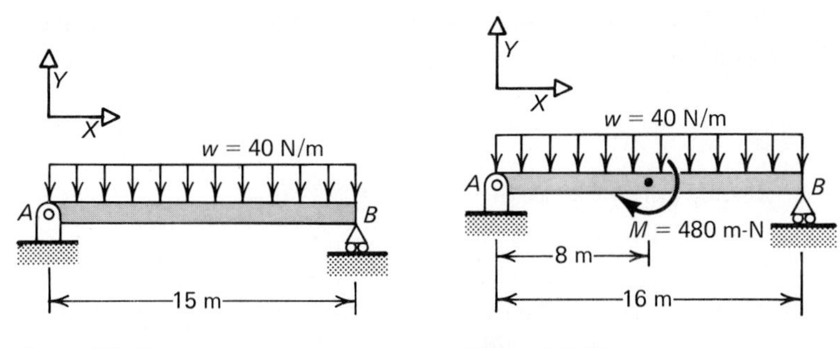

Figure P7.21

Figure P7.22

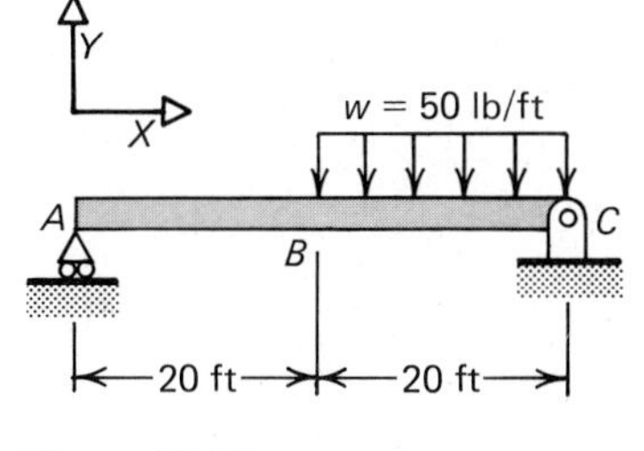

Figure P7.23

7.24 Obtain the shear and bending-moment equations for the beam with a distributed load shown in Fig. P7.24.

7.25 For the beam in Fig. P7.25, obtain the shear and bending-moment equations.

7.26 For the beam in Fig. P7.26, obtain the shear and bending-moment equations.

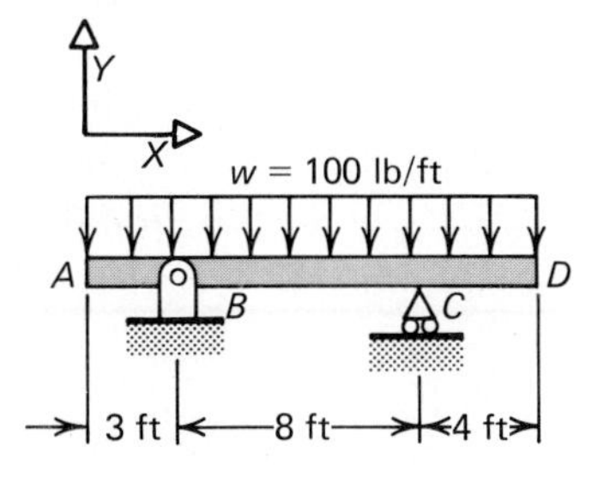

Figure P7.24

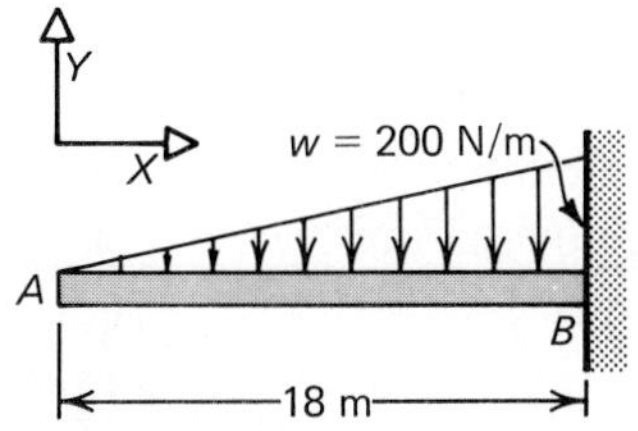

Figure P7.25

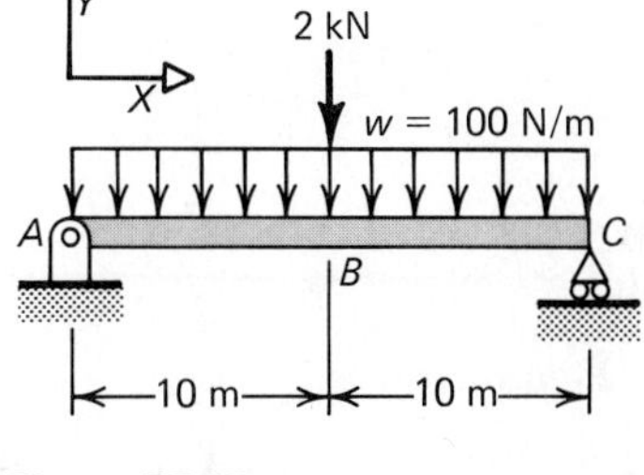

Figure P7.26

7.27 What are the shear and bending-moment equations for the beam shown in Fig. P7.27?

7.28 What are the shear and bending-moment equations for the beam shown in Fig. P7.28?

7.29 For beam in Fig. P7.29, obtain the shear and bending-moment equations for $P = 2000$ N.

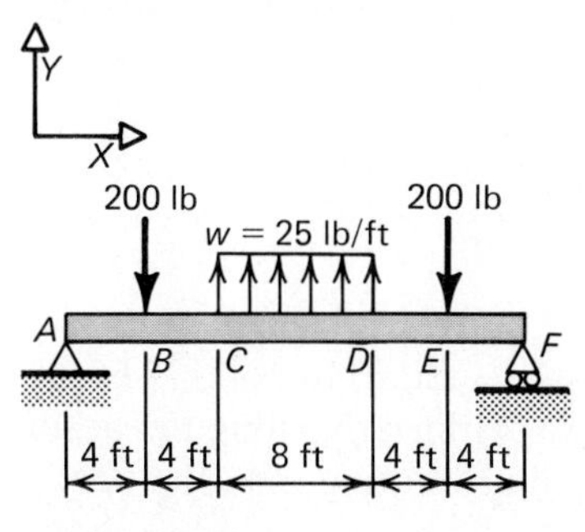

Figure P7.27

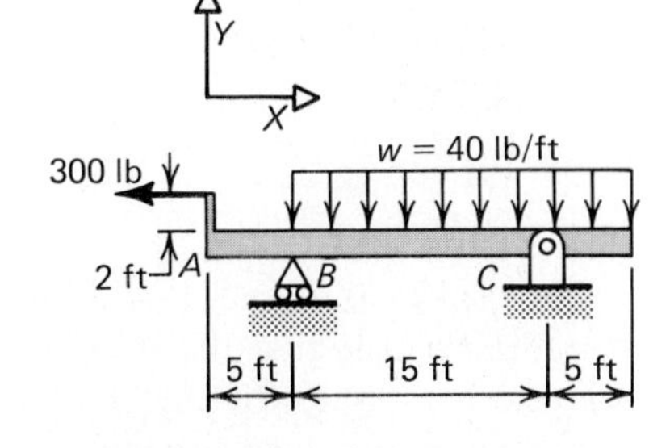

Figure P7.28

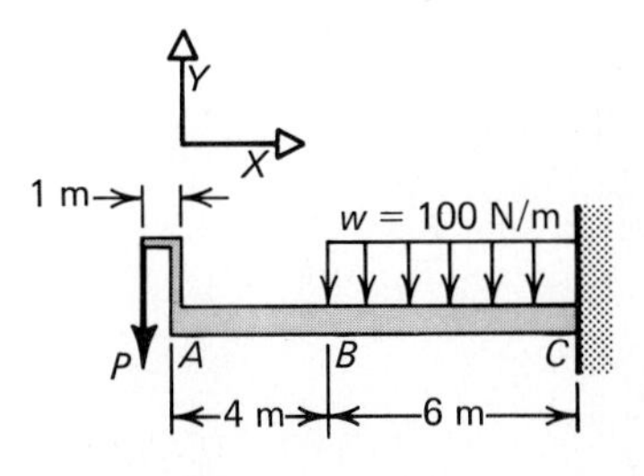

Figure P7.29

7.30 For the beam in Problem 7.29, obtain the shear and bending-moment equations for $P = 1000$ N.

7.5 Load, Shear, and Bending-Moment Relations

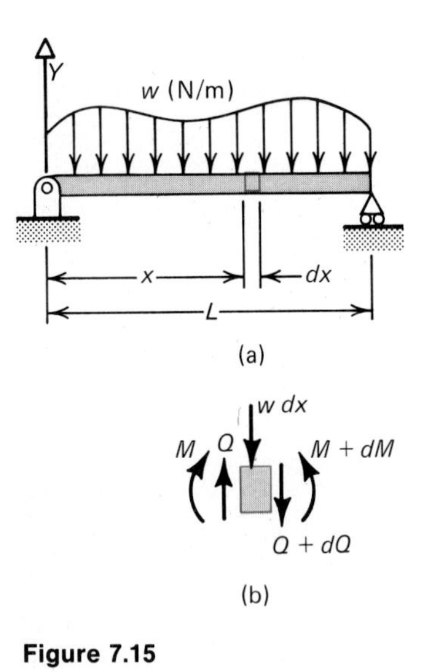

Figure 7.15

Consider the beam in Fig. 7.15(a) subject to a distributed load w (N/m). Let us isolate a differential length of beam dx and draw the free-body diagram as shown in Fig. 7.15(b). The magnitude of the shear Q and the bending moment M on the left side of the element increases by the amount $(Q + dQ)$ and $(M + dM)$, respectively, on the right side. Positive directions for the shear and bending moment are assumed. For equilibrium, we write

$$\sum F_y = 0 \qquad Q - (Q + dQ) - w\,dx = 0$$

$$dQ = -w\,dx$$

or

$$\frac{dQ}{dx} = -w \tag{7.7}$$

Integrating Eq. (7.7) between x_1 and x_2 where x_1 and x_2 are arbitrary distances from the left end of the beam, we obtain

$$\int_{Q_1}^{Q_2} dQ = -\int_{x_1}^{x_2} w\,dx$$

so that

$$Q_2 - Q_1 = -\int_{x_1}^{x_2} w\,dx \tag{7.8}$$

Equation (7.8) indicates that the difference of the shear between two arbitrary points of the beam equals the negative of the area under the load curve. Note that this expression holds only for distributed loads, and that the positive direction of w is downward. By way of example, consider the beam in Fig. 7.16. The shear between A and B is a constant 500 N since a free-body diagram of this part would show that the shear magnitude is equal to the reaction A_y. The change in shear between B and C equals the negative of the area under the load diagram. That is,

$$Q_2 = Q_1 - \int_{x_1}^{x_2} w\,dx$$

$$= Q_1 - w(x_2 - x_1)$$

where $x_1 = 10$ m and $x_2 = 20$ m as measured from point A of the beam. We note that the shear force varies linearly along the beam as the load is distributed uniformly. Substituting into the above expression, the shear becomes

$$Q_2 = 500 - 100(20 - 10) = 500 - 1000 = -500 \text{ N}$$

Between C and D the shear remains -500 N until point D is reached, which satisfies the boundary condition.

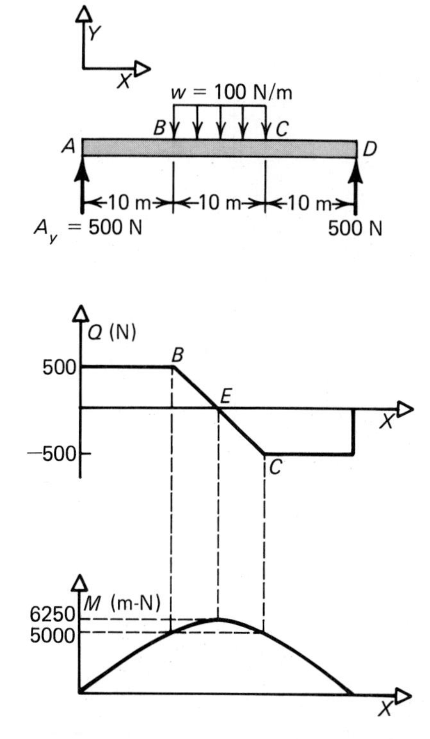

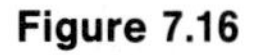
Figure 7.16

Returning to Fig. 7.15(b) for the beam element, we must also have equilibrium for the moments about any point on the element. Summing moments about the left edge,

$$\sum M_x = 0 \qquad (M + dM) - M - w\,dx\,\frac{dx}{2} - (Q + dQ)\,dx = 0$$

$$dM - \frac{w}{2}(dx)^2 - Q\,dx - dQ(dx) = 0$$

Since dM, dQ, and dx are differentials and, therefore, are small quantities, their products may be neglected so that the above expression reduces to

$$dM - Q\,dx = 0$$

or

$$\frac{dM}{dx} = Q \tag{7.9}$$

We integrate as before to obtain

$$\int_{M_1}^{M_2} dM = \int_{x_1}^{x_2} Q\,dx$$

so that

$$M_2 - M_1 = \int_{x_1}^{x_2} Q\,dx \tag{7.10}$$

This says that the difference between the bending moments at two points on the beam equals the area under the shear curve between these same two points. Returning to Fig. 7.16, the moment must be zero at A. Between A and B,

$$M_B = M_A + \int_0^{10} Q\,dx = 0 + 500(10) = 5000 \text{ m}\cdot\text{N}$$

Between B and E,

$$M_E = M_B + \int_{10}^{15} Q\,dx = 5000 + \tfrac{1}{2}(500)(5) = 6250 \text{ m}\cdot\text{N}$$

Between E and C,

$$M_C = M_E + \int_{15}^{20} Q\,dx = 6250 - \tfrac{1}{2}(500)(5) = 5000 \text{ m}\cdot\text{N}$$

Between C and D,

$$M_D = M_C + \int_{20}^{30} Q\,dx = 5000 - 500(10) = 0 \qquad \text{(Checks)}$$

Note that the maximum moment occurred at point E where the shear is zero. This agrees with Eq. (7.9) since the slope of the moment diagram is zero where $Q = 0$.

Consequently, when we have the shear diagram, we can locate the coordinates of the bending-moment diagram fairly rapidly. In the above example, we note that the shear is constant between points A and B and the corresponding curve between A and B is termed a zero-order curve. The integral of this curve is a straight line (first-order curve) as shown by the bending-moment curve between A and B. Similarly, the integral of the straight line between points B and C of the shear curve is represented by a parabola (second-order curve) on the bending-moment curve.

Example 7.7

Draw the shear and bending-moment diagrams for the beam in Fig. 7.17.

Beam reactions:

$$\sum M_A = 0 \qquad 20D_y - 2000(5) - 1000(15) = 0$$
$$D_y = 1250 \text{ lb}$$

and

$$\sum F_y = 0 \qquad A_y - 2000 - 1000 + D_y = 0$$
$$A_y = 1750 \text{ lb}$$

Shear diagram:

Having the reaction A_y, we begin construction of the shear diagram at point A. Between A and B, the area under the load diagram is 2000 lb, so that

$$Q_B = Q_A - \int_A^B w\,dx$$
$$= 1750 - 2000 = -250 \text{ lb}$$

Since the load diagram is zero-order, the shear diagram is first-order, and hence the straight line between A and B. Between B and C the shear is constant, and at C it experiences a step change in value of 1000 lb. Between C and D it remains constant at -1250 lb and checks at D with a reaction of $+1250$ lb.

Moment diagram:

We now construct the bending-moment diagram directly from the shear diagram. First we locate point E from the geometry. Using similar triangles, point E is 8.75 ft from the left end. At point A, $M = 0$, so that

$$M_E = \text{area under } Q \text{ curve between } A \text{ and } E$$
$$= \tfrac{1}{2}(1750)(8.75) = 7656 \text{ ft}\cdot\text{lb}$$

Between E and B,

$$M_B = M_E + \text{area under Q curve between } E \text{ and } B$$
$$= 7656 - \tfrac{1}{2}(250)(1.25) = 7500 \text{ ft}\cdot\text{lb}$$

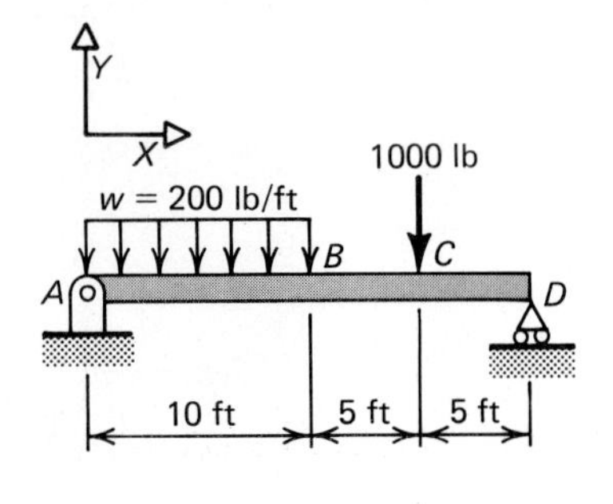

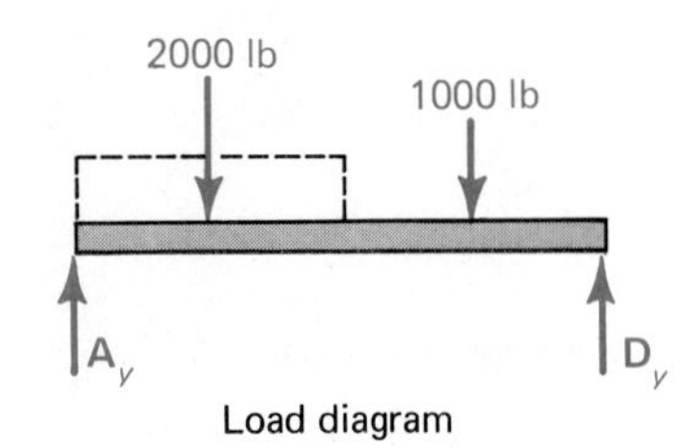

Load diagram

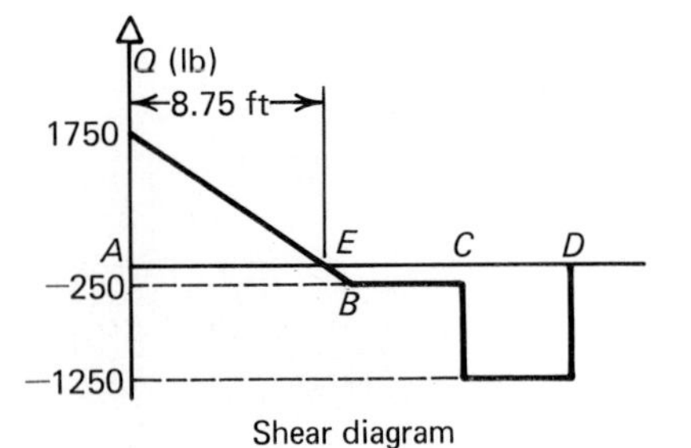

Shear diagram

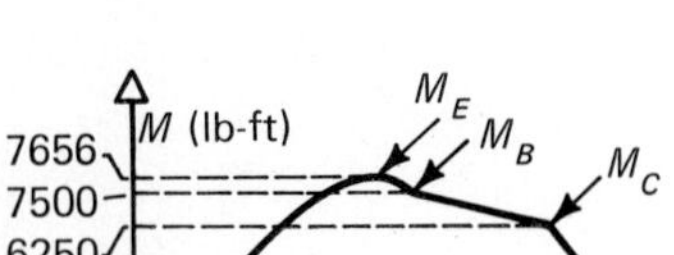
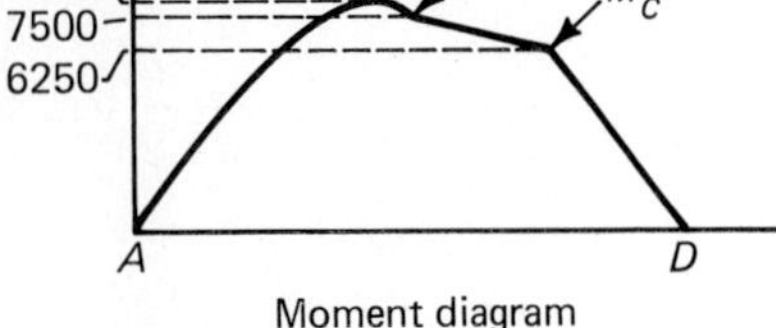

Moment diagram

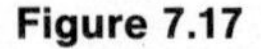
Figure 7.17

Between B and C,

$$M_C = M_B + \text{area under } Q \text{ curve between } B \text{ and } C$$
$$= 7500 - 250(5) = 6250 \text{ ft}\cdot\text{lb}$$

Between C and D,

$$M_D = M_C + \text{area under } Q \text{ curve between } C \text{ and } D$$
$$= 6250 - 1250(5) = 0 \qquad \text{(Checks)}$$

Again we observe that the maximum bending moment occurs at point E where $Q = 0$.

Example 7.8

Draw the shear and bending-moment diagrams for the beam in Fig. 7.18.

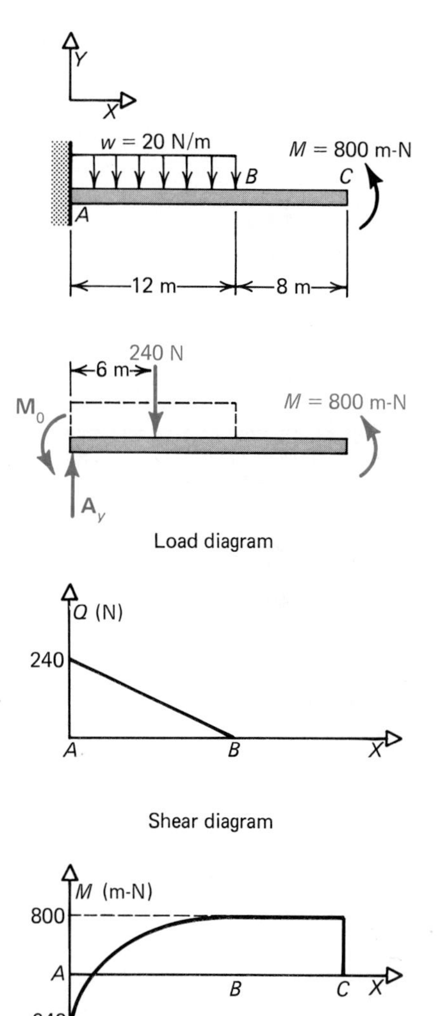

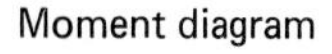

Figure 7.18

Beam reactions:

$$\sum M_A = 0 \qquad M_o - 240(6) + 800 = 0$$
$$M_o = 640 \text{ m}\cdot\text{N}$$

and

$$\sum F_y = 0 \qquad A_y - 240 = 0$$
$$A_y = 240 \text{ N}$$

Shear diagram:

Starting at point A with a value of 240 N, the shear changes linearly between A and B.

$$Q_B = Q_A - \int_A^B w\,dx$$
$$= 240 - 240 = 0$$

The shear remains zero for the remainder of the beam since there are no other forces present.

Moment diagram:

At A, $M_o = -640$ m · N, where the negative sign is used due to our sign convention for bending moments.

$$M_B = M_A + \text{area under the } Q \text{ curve between } A \text{ and } B$$
$$= -640 + \tfrac{1}{2}(240)(12) = 800 \text{ m}\cdot\text{N}$$

The moment between B and C remains 800 m·N and agrees with the couple at point C.

Example 7.9

Draw the shear and bending-moment diagrams for the beam in Fig. 7.19.

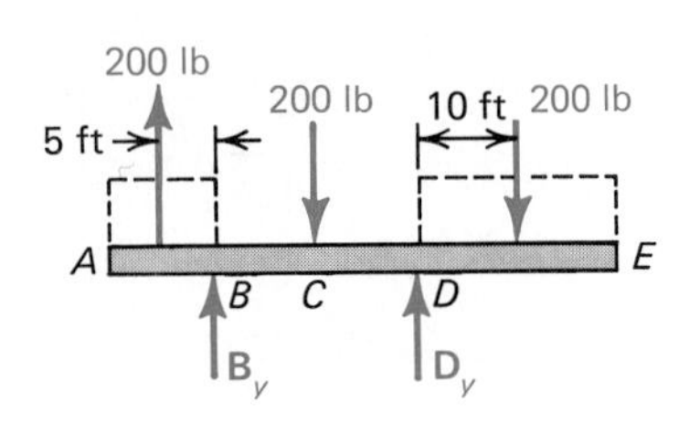

Load diagram

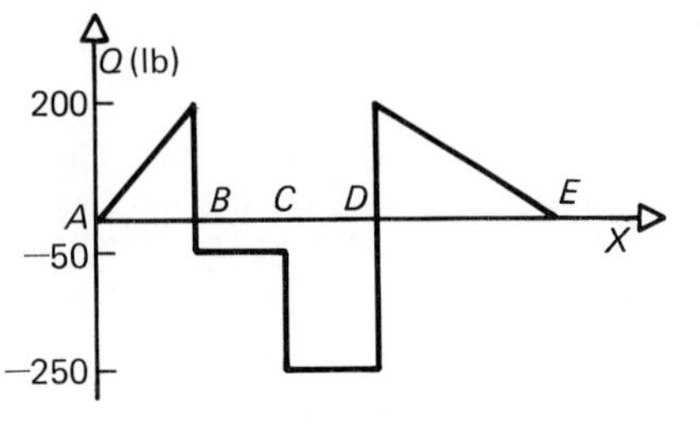

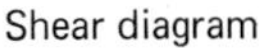
Shear diagram

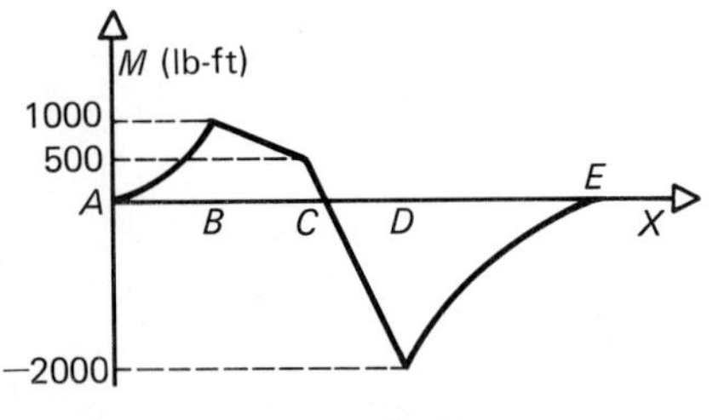

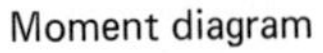
Moment diagram

Figure 7.19

Beam reactions:

$$\sum M_B = 0 \qquad 20D_y - 200(5) - 200(10) - 200(30) = 0$$

$$D_y = 450 \text{ lb}$$

and

$$\sum F_y = 0 \qquad 200 + B_y - 200 + 450 - 200 = 0$$

$$B_y = -250 \text{ lb}$$

Note that the reaction at B acts downward.

Shear diagram:

Starting at A the shear $Q = 0$ and increases linearly between A and B. At B,

$$Q_B = Q_A - \int_A^B w\,dx$$

Since w acts *upward* it is negative by the sign convention in Eq. (7.8). The shear just to the left of the reaction at B is

$$Q_B = 0 + 20(10) = 200 \text{ lb}$$

At B, there is a step change in shear due to the downward reaction at B. Between B and C, the shear is constant and undergoes a step change in shear at C and D. From D to E, the load is uniform so that

$$Q_E = Q_D - \int_D^E w\,dx$$

$$= 200 - 10(20) = 0 \qquad \text{(Checks)}$$

Moment diagram:

Between A and B,

$$M_B = M_A + \int_A^B Q\,dx$$

$$= 0 + \tfrac{1}{2}(200)(10) = 1000 \text{ ft}\cdot\text{lb}$$

Note the curvature of the second-order curve for the bending moment.

Between B and C,

$$M_C = M_B + \int_B^C Q\,dx$$

$$= 1000 + (-50)(10) = 500 \text{ ft}\cdot\text{lb}$$

Between C and D,

$$M_D = M_C + \int_C^D Q\,dx$$

$$= 500 + (-250)(10) = -2000 \text{ ft}\cdot\text{lb}$$

Between D and E,

$$M_E = M_D + \int_D^E Q\,dx$$

$$= -2000 + \tfrac{1}{2}(200)(20) = 0 \qquad \text{(Checks)}$$

Again, note the curvature for this portion of the bending-moment diagram.

Example 7.10

Draw the shear and bending-moment diagram for the beam in Fig. 7.20.

Beam reactions:

$$\sum M_A = 0 \qquad M_o - 300(8) - 100(20) = 0$$

$$M_o = 4400 \text{ m·N}$$

and

$$\sum F_y = 0 \qquad A_y - 300 - 100 = 0$$

$$A_y = 400 \text{ N}$$

Shear diagram:

Starting at A, $Q_A = 400$ N.

$$Q_B = Q_A - \int_A^B \frac{50}{12} x\,dx = 400 - \left[\frac{50}{24}x^2\right]_0^{12}$$

$$= 400 - 300 = 100 \text{ N}$$

The shape of the curve connecting Q_A and Q_B is second-order (parabola) since the loading curve between A and B is first-order (straight line). From B to C the shear remains constant at 100 N and checks with the load at C.

Moment diagram:

Between A and B,

$$M_B = M_A + \int_A^B Q\,dx$$

To find the area under the shear curve between A and B, divide the area into the area under the parabola DEF and the area $ABFE$ as shown in Fig. 7.20. We can either use the expression for Q and integrate to obtain an expression for M_B or simply calculate the areas from the diagram. The areas are

$$\text{Area under parabola } DEF = \tfrac{2}{3}(DE)(EF) = \tfrac{2}{3}(400 - 100)(12) = 2400$$

$$\text{Area under rectangle } ABFE = 100(12) = 1200$$

$$\text{Total area} = 2400 + 1200 = 3600 \text{ m·N}$$

$$M_B = -4400 + 3600 = -800 \text{ m·N}$$

If we had integrated we obtain

$$M_B = -4400 + \left[400x\right]_0^{12} - \left[\frac{50}{72}x^3\right]_0^{12} = -800 \text{ m·N}$$

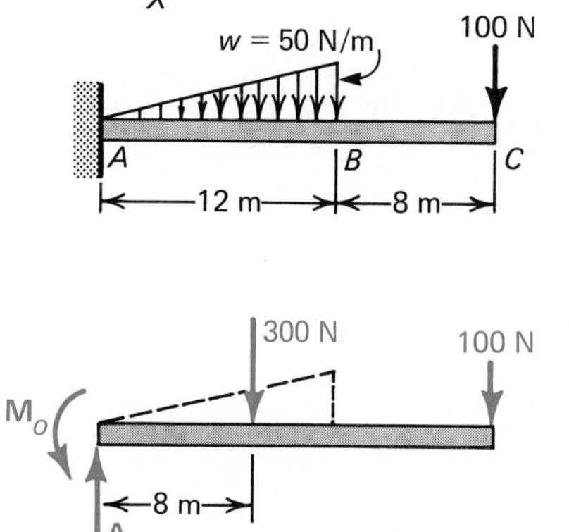

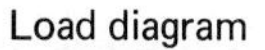

Load diagram

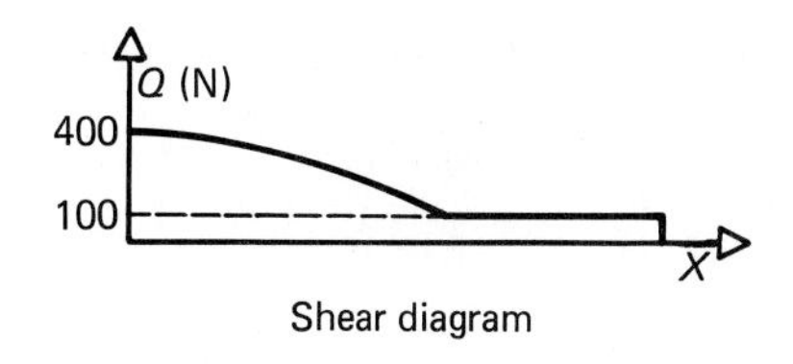

Shear diagram

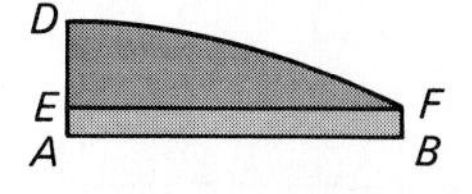

Area between A and B under shear diagram

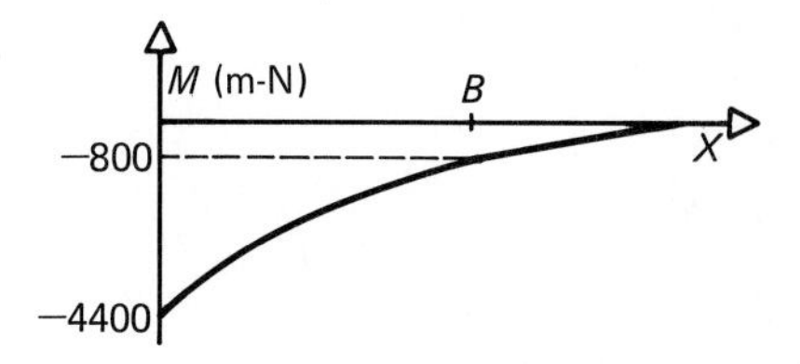

Moment diagram

Figure 7.20

Between B and C,

$$M_C = M_B + \int_B^C Q\,dx$$

$$= -800 + 100(8) = 0 \qquad \text{(Checks)}$$

Example 7.11

Draw the shear and bending-moment diagram for the beam whose load diagram is shown in Fig. 7.21.

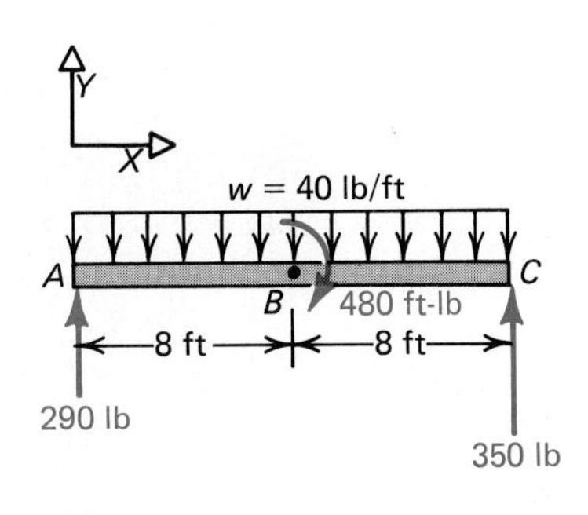

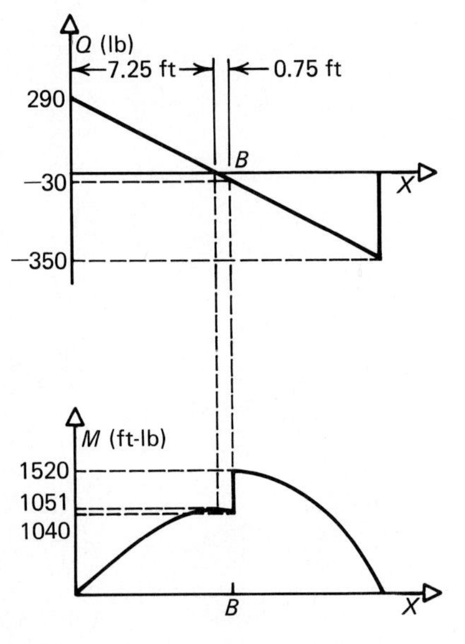

Figure 7.21

Shear diagram:

Starting at A, $Q_A = 290$ lb. Now,

$$Q_C = Q_A - \int_A^C w\,dx = Q_A - 40(16)$$

$$= 290 - 640 = -350 \text{ lb}$$

The shear force varies linearly between A and C and is zero at $x =$ 7.25 ft.

Moment diagram:

Between A and B,

$$M_B = M_A + \int_A^B Q\,dx$$

$$= 0 + \tfrac{1}{2}(290)(7.25) - \tfrac{1}{2}(30)(0.75)$$

$$= 1051 - 11$$

$$= 1040 \text{ ft}\cdot\text{lb}$$

We note that the moment has a maximum value of 1051 ft·lb at $x = 7.25$ ft after which the moment decreases since Q becomes negative. At B we add a couple of 480 ft·lb so that

$$M_C = M_B + \int_B^C Q\,dx$$

$$= (1040 + 480) + \int_B^C Q\,dx$$

$$= 1520 - \tfrac{1}{2}(30 + 350)8 = 0 \qquad \text{(Checks)}$$

The shape of the curve for the bending moment between B and C is second-order as shown in Fig. 7.21.

Example 7.12

Draw the shear and bending-moment diagram for the vertical beam whose load diagram is shown in Fig. 7.22.

The XY axes are oriented in Fig. 7.22 so that we are in effect looking at the beams studied earlier rotated through 90°.

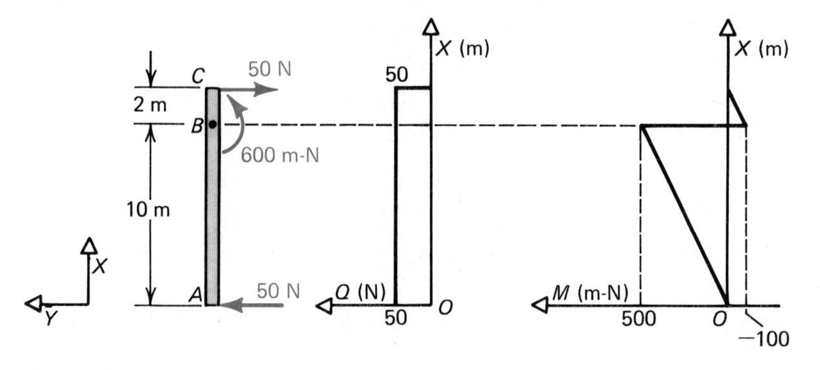

Figure 7.22

Shear diagram:

Starting at A, $Q_A = 50$ N and since there is no applied load between A and C, the shear force is constant over the entire beam.

Moment diagram:

Between A and B we have

$$M_B = M_A + \int_A^B Q\,dx$$

$$= 0 + 50(10) = 500 \text{ m}\cdot\text{N}$$

At B we subtract 600 m·N due to the applied couple, so that

$$M_C = (500 - 600) + \int_B^C Q\,dx$$

$$= -100 + 50(2) = 0 \qquad \text{(Checks)}$$

PROBLEMS

7.31 Draw the shear and bending-moment diagrams for the beam shown in Fig. P7.31.

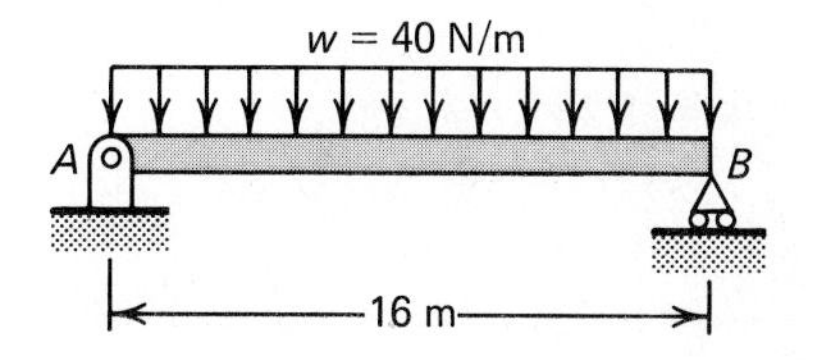

Figure P7.31

7.32 Obtain the shear and bending-moment diagrams for the beam shown in Fig. P7.22 on page 255.

7.33 Draw the shear and bending-moment diagrams for the beam in Problem 7.23 using the integral approach.

7.34 Draw the shear and bending-moment diagrams for the beam in Problem 7.24 using the integral approach.

7.35 Draw the shear and bending-moment diagrams for the beam in Problem 7.25 using the integral approach.

7.36 Draw the shear and bending-moment diagrams for the beam in Problem 7.26 using the integral approach.

7.37 Obtain the shear and bending-moment diagrams for the beam in Problem 7.27 using the integral approach.

7.38 Obtain the shear and bending-moment diagrams for the beam in Problem 7.28 using the integral approach.

7.39 Draw the shear and bending-moment diagrams for the beam in Problem 7.29 for $P = 2000$ N, using the integral approach.

7.40 Draw the shear and bending-moment diagrams for the beam in Problem 7.29 for $P = 1000$ N, using the integral approach.

7.41 Using the integral approach obtain the shear and bending-moment diagrams for the beam in Fig. P7.41.

7.42 Draw the shear and bending-moment diagrams for the beam in Fig. P7.42.

7.43 By using the integral approach, obtain the shear and bending-moment diagrams for the beam in Fig. P7.43.

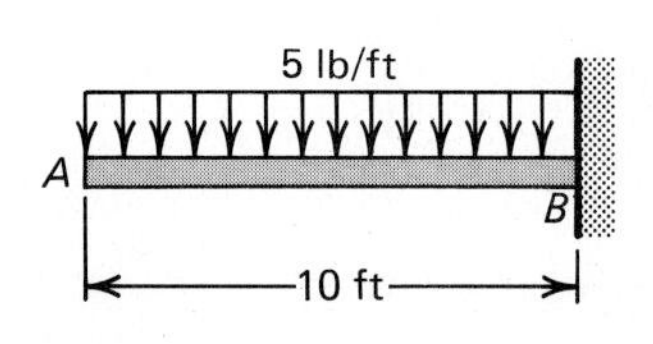

Figure P7.41

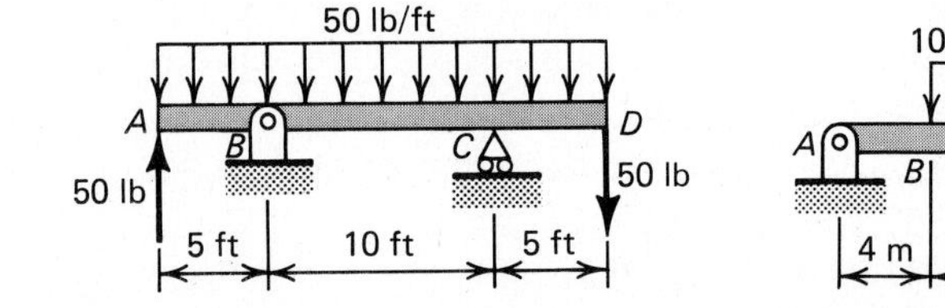

Figure P7.42

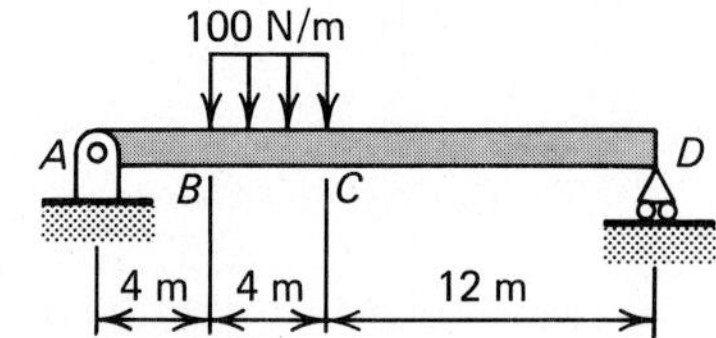

Figure P7.43

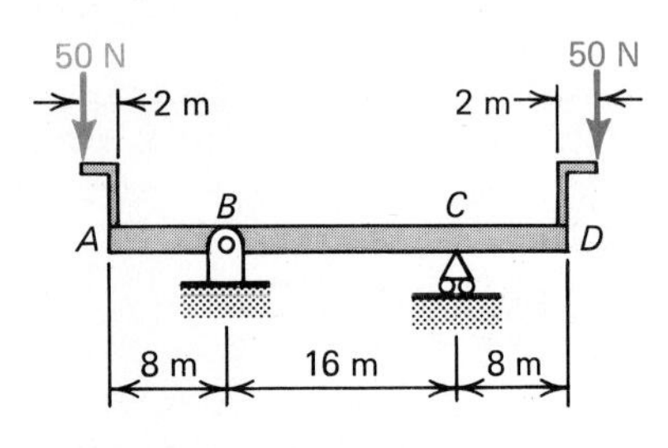

Figure P7.44

7.44 Sketch the shear and bending-moment diagrams for the beam in Fig. P7.44.

7.45 Find the magnitude and location of the maximum bending moment in the beam shown in Fig. P7.45. Sketch the shear and bending-moment diagrams for $w_1 = 2w_2 = 40$ lb/ft

7.46 If $w_1 = 4w_2 = 20$ lb/ft for the beam in Problem 7.45, then find the point where the shear force is zero.

7.47 Sketch the shear and bending-moment diagrams for the beam in Fig. P7.47.

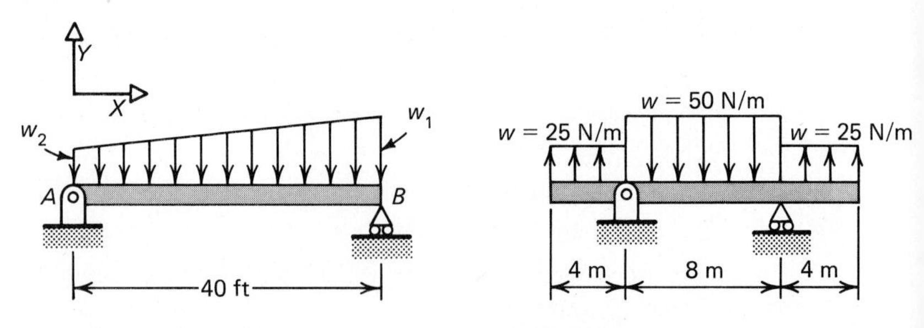

Figure P7.45

Figure P7.47

7.48 A mathematical model of a fin stabilizer for surface ships is shown in Fig. P7.48 subjected to two concentrated loads. Sketch the shear and bending-moment diagrams.

7.49 A trench 18-ft deep is supported on each side by sheet piling and braces as shown in Fig. P7.49. Let $\gamma = 120\ \text{lb/ft}^3$ for the soil and consider a 5-ft section between the braces. The braces apply 15-kip loads to the sheet piling. Sketch the shear diagram of the piling assuming it behaves as a cantilever beam. Neglect the mass of the piling.

7.50 Sketch the shear and bending-moment diagrams for the vertical beam shown in Fig. P7.50.

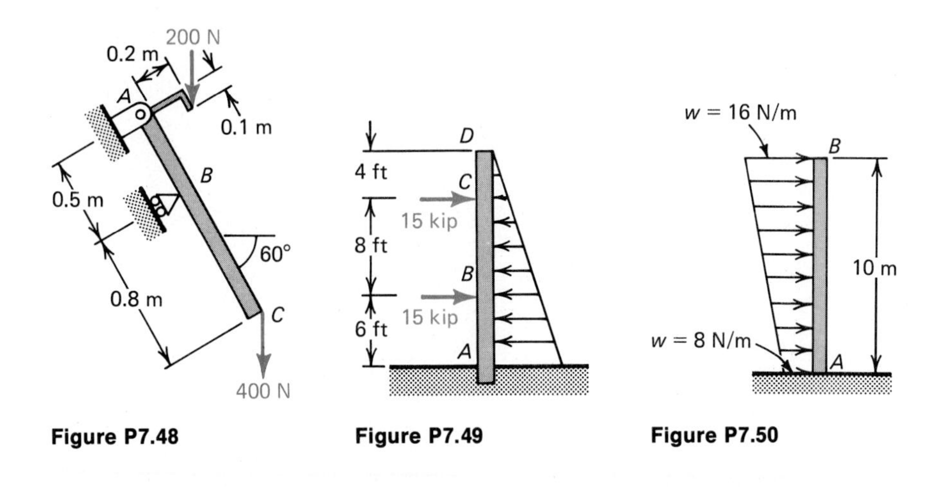

Figure P7.48 Figure P7.49 Figure P7.50

7.6 Summary

A. The important *ideas*:
1. A beam develops internal forces and moments represented by the shear force and bending moment, respectively.
2. The shear and bending-moment diagrams provide a graphic display of the variation of the shear forces and bending moments throughout the length of a beam.
3. The difference between the shear forces at two points of a beam equals the negative of the area under the load curve between these same two points.
4. The difference between the bending moments at two points on a beam equals the area under the shear curve between these same two points.
5. The slope of the shear diagram is zero where the distributed load w is zero.
6. The slope of the moment diagram is zero where the shear is zero.

B. The important *equations*:

Shear load:

$$\frac{dQ}{dx} = -w \tag{7.7}$$

$$Q_2 - Q_1 = -\int_{x_1}^{x_2} w\,dx \tag{7.8}$$

Shear and bending moment:

$$\frac{dM}{dx} = Q \tag{7.9}$$

$$M_2 - M_1 = \int_{x_1}^{x_2} Q\,dx \tag{7.10}$$

8 Friction

When a surface slides or rolls along another surface with which it is in contact, a force is developed that tends to oppose the motion of one body relative to the other. The force is tangential to the plane of the surfaces and is called friction. Fluid friction develops between layers of fluids, dry friction between rigid bodies in contact. If there is to be motion of one surface or body relative to another with which it is in contact, frictional force must be overcome by externally applied forces. Although in ideal systems friction is normally ignored, in real systems it must be accounted for and included in the analysis. While the occurrence of friction has many advantages, as in wedges, brakes, clutches, belt drives, and even walking, in some cases it is more detrimental than useful. In the reentry of space capsules, for example, friction is a serious problem; and the results of friction on gears, bearings, and the like are certainly undesirable.

Friction results in the dissipation of energy in the form of heat. The real nature of friction is very complex and not entirely understood. It is known to be a function of surface roughness and of microscopic phenomena at the interface. We shall be concerned here primarily with dry, or Coulomb, friction as it occurs between the surfaces of two rigid bodies.

8.1 Coulomb Friction

A simple example of Coulomb friction involves the sliding of a block along a rough surface. Assume a block to be at rest on the floor. Figure 8.1(a) shows the free-body diagram under this condition. Now assume a small force **P** applied as shown in Fig. 8.1(b). Although the block does not move, a force **f** appears that balances the applied force **P**. The force **f** is due to friction between the two

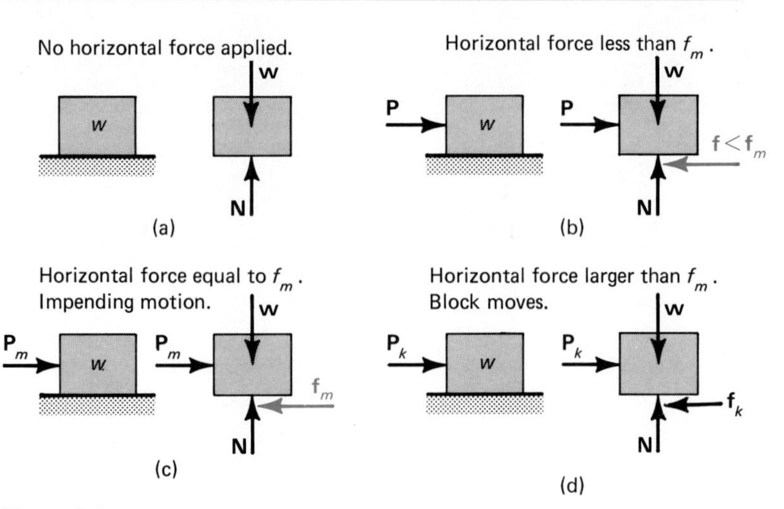

Figure 8.1

contacting surfaces. As **P** is increased in magnitude until the block just begins to move, **f** increases to its maximum value $\mathbf{f}_m$. This is the instant of **impending motion** and is represented by Fig. 8.1(c). Once the block does move as shown in Fig. 8.1(d), the friction force decreases to $\mathbf{f}_k$, i.e., the friction force during the motion of the block is less than that necessary to begin the motion. The maximum force $\mathbf{f}_m$ is known as the **static-friction force**, while $\mathbf{f}_k$ is the **kinetic-** or **sliding-friction force**. These forces are shown in Fig. 8.2. Here we observe that as the applied horizontal force **P** increases, the friction force increases linearly, until it reaches $\mathbf{f}_m$. With a further increase in **P**, motion begins and the friction force decreases suddenly to $\mathbf{f}_k$ and stays at this value as long as there is motion.

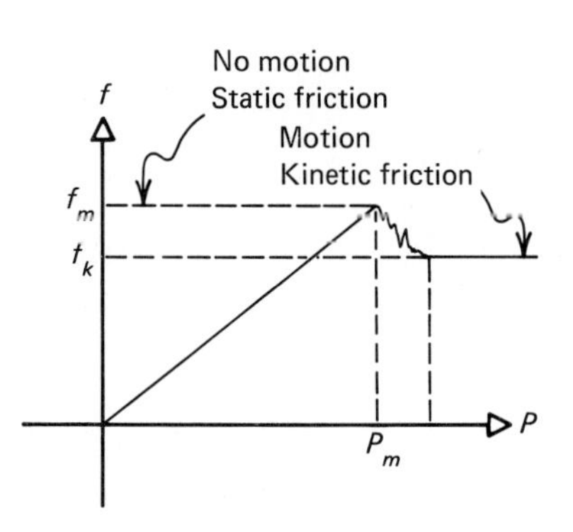

Figure 8.2

From experimentation, it has been found that the magnitude of the static-friction force is proportional to the magnitude of the normal force N,

$$f_m = \mu_0 N \tag{8.1}$$

where μ_0 is the **coefficient of static friction**. The magnitude of the kinetic-friction force is also proportional to the magnitude of the normal force. That is,

$$f_k = \mu N \tag{8.2}$$

where μ is the **coefficient of kinetic** or **sliding friction**. Notice that both of these forces are in a direction opposite to **P**, and therefore tend to oppose motion. Some commonly used values of μ_0 and μ appear in Table 8.1. It should be noted that the values given in Table 8.1 vary because of differing surface conditions. The wide range of values given in Table 8.1 is due mainly to the difficulty of measuring the coefficient accurately and the variability of surface finishes.

TABLE 8.1 ***Values of μ_0 and μ***

	μ_0	μ
Metal on stone	0.30–0.70	0.20–0.45
Metal on metal	0.15–0.60	0.10–0.40
Wood on wood	0.25–0.50	0.15–0.35
Wood on metal	0.20–0.60	0.15–0.40
Rubber on asphalt	0.70–0.90	0.50–0.60

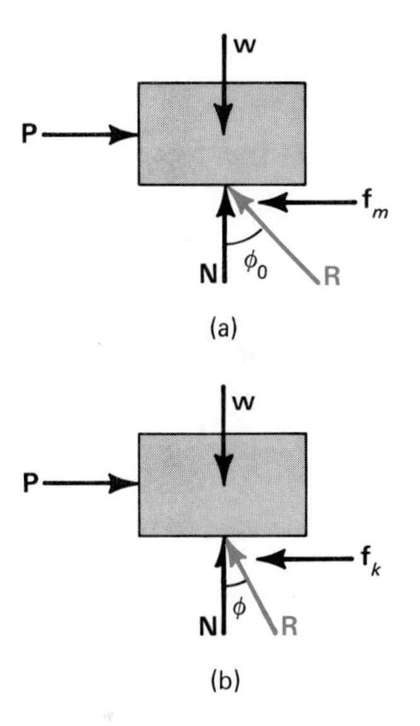

Figure 8.3

In the presence of friction forces, the magnitude of the resultant reaction force **R** is a combination of **N** and $\mathbf{f}_m$ as shown in Fig. 8.3(a). If the motion is impending, then their magnitudes are related by

$$R = \sqrt{N^2 + f_m^2} = N\sqrt{1 + \mu_0^2} \tag{8.3a}$$

$$\tan \phi_0 = \frac{f_m}{N} = \mu_0 \tag{8.3b}$$

The angle ϕ_0 is called the angle of static friction or angle of repose. As soon as motion occurs, the forces are as shown in Fig. 8.3(b), so that

$$R = \sqrt{N^2 + f_k^2} = N\sqrt{1 + \mu^2} \tag{8.4a}$$

$$\tan \phi = \frac{f_k}{N} = \mu \tag{8.4b}$$

where ϕ is the angle of kinetic friction. We note that $\phi \leq \phi_0$. The magnitude of the normal force **N** (not to be confused with N for newtons) can be computed by summing the forces in the vertical direction and in this case, $N = w$.

The externally applied force **P** need not be a horizontal force. The block in Fig. 8.4(a) is subjected to a force **P** that makes an angle θ with the horizontal. For the case of impending motion, a free-body diagram is drawn in Fig. 8.4(b). Summing the vertical forces gives us

$$N = w + P_y$$

and since $\quad f_m = \mu_0 N$

we have $\quad f_m = \mu_0(w + P_y)$

The magnitude of the reaction **R** is

$$R = \sqrt{N^2 + f_m^2} = (w + P_y)\sqrt{1 + \mu_0^2}$$

and the angle of repose is

$$\tan \phi_0 = \frac{f_m}{N} = \mu_0$$

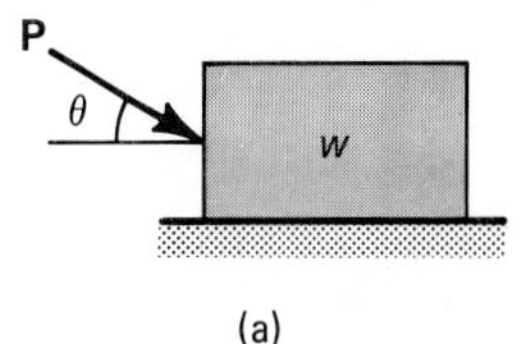

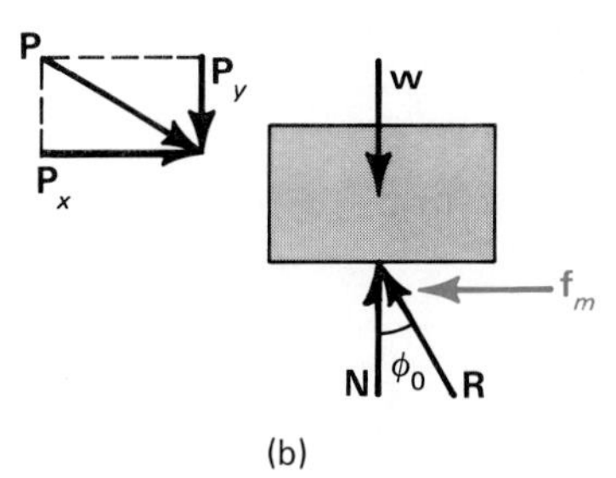

Figure 8.4

We note that although the magnitude of **R** changes for this case, its direction is the same as in the case where **P** is applied horizontally.

Static friction and its associated angle of repose may also be viewed with the aid of an inclined plane. Assume that a block is resting on a plane that is at an angle θ with the horizontal as shown in Fig. 8.5(a). Allow the plane to be tilted slowly upward. We would now like to establish the maximum angle θ before the block begins to slide. First, a free-body diagram is drawn as shown in Fig. 8.5(b). We then assume that the plane has been tilted so that there exists impending motion. Summing the forces in the Y and X directions, respectively, gives us

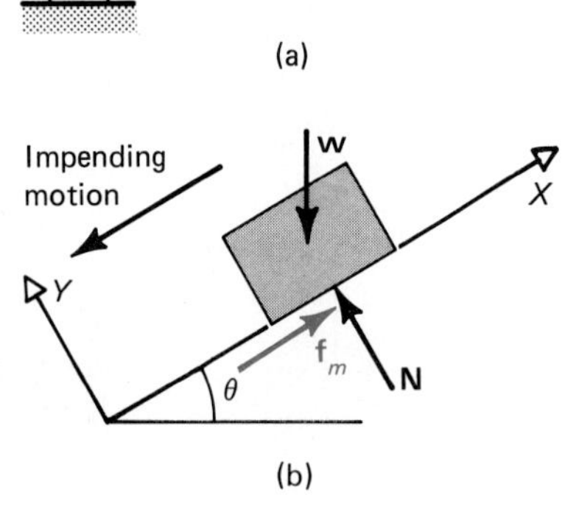

Figure 8.5

$$N = w \cos \theta$$

$$f_m = w \sin \theta$$

But we already know that

$$f_m = \mu_0 N$$
$$= \mu_0 w \cos \theta$$

so that $$f_m = w \sin \theta = \mu_0 w \cos \theta$$

which yields $$\tan \theta = \mu_0$$

We may conclude that by measuring θ experimentally, we obtain μ_0 between the block and the plane.

Problems involving dry friction can be conveniently divided into three categories as follows:

1. The condition of impending motion exists so that the friction force is static and $f_m = \mu_0 N$. The friction force must be included when drawing the free-body diagram.
2. The friction force **f** is less than the static-friction force $\mathbf{f}_m$. The friction force is assumed to exist in the free-body diagram and then computed from the conditions of equilibrium. It is usually of interest to know whether **f** is sufficient to maintain the body at rest.
3. Motion is not impending but has occurred, so that the friction force is kinetic and $f = f_k = \mu N$.

Example 8.1

The force **T** acting on the 100-lb block in Fig. 8.6(a) makes an angle of 45° with the horizontal. If the coefficient of static friction between the block and the surface is 0.6, determine the minimum magnitude of **T** to move the block.

The free-body diagram of the forces acting on the block is shown in Fig. 8.6(b). Since motion is impending, the static friction force $\mathbf{f}_m$ acts on the block. For static equilibrium,

$$\sum F_x = 0 \qquad T\cos 45^\circ - f_m = 0$$

$$T = 1.414 f_m = 1.414(\mu_0 N)$$

$$= 1.414(0.6)N = 0.848N \qquad \text{(a)}$$

Likewise,

$$\sum F_y = 0 \qquad N - w + T\sin 45^\circ = 0$$

$$N = 100 - 0.707T \qquad \text{(b)}$$

Substitute Eq. (b) into (a) and solve for T.

$$T = 0.848(100 - 0.707T)$$

$$T = 53.0 \text{ lb} \qquad \textit{Answer}$$

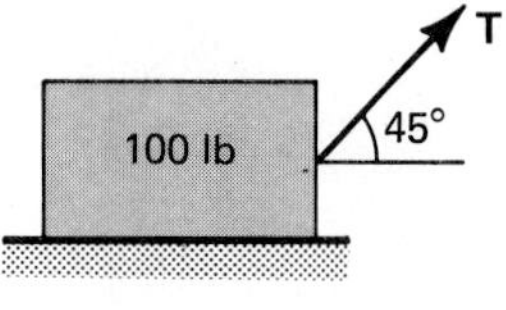

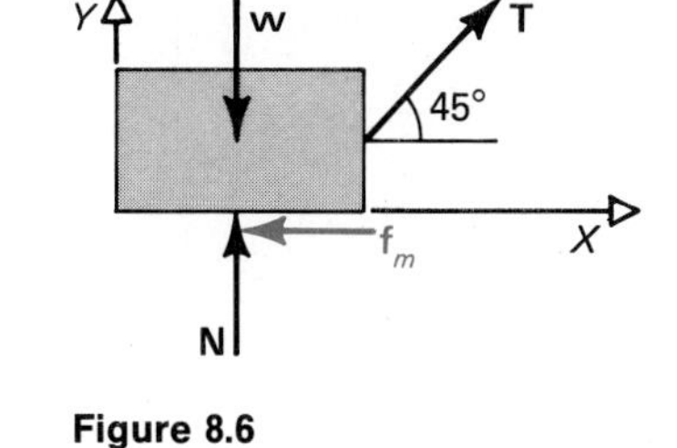

Figure 8.6

Example 8.2

The force $P = 20$ N acts on the 20-kg block initially at rest as shown in Fig. 8.7(a). If $\mu_0 = 0.8$ between the block and the inclined surface, does the block move down the plane?

The free-body diagram is shown in Fig. 8.7(b). We represent the friction force as $\mathbf{f}$ which we shall calculate and then compare with $\mathbf{f}_m$. If $f < f_m$, the block will not move.

$$\sum F_y = 0 \qquad N - 20(9.81)\cos 30^\circ = 0$$

$$N = 169.9 \text{ N}$$

$$\sum F_x = 0 \qquad -P + f - 20(9.81)\sin 30^\circ = 0$$

$$f = 98.1 + 20 = 118.1 \text{ N}$$

Now the maximum friction force that can be developed between the block and surface is

$$f_m = \mu_0 N = 0.8(169.9) = 135.9 \text{ N}$$

Since the actual friction force required for equilibrium is less than f_m, the block does not move.

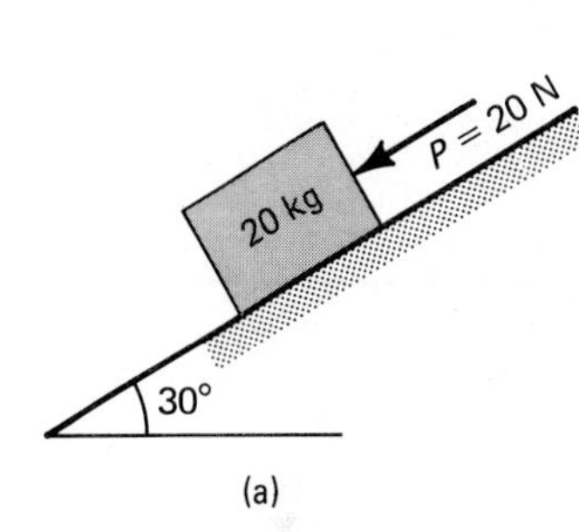

(a)

(b)

Figure 8.7

Example 8.3

The ladder of mass m rests against the wall as shown in Fig. 8.8(a). If $\mu_0 = 0.4$ between the ladder and both the floor and wall, find θ when motion is impending.

Figure 8.8(b) shows the free-body diagram of the forces acting on the ladder. Note that N_1 is the normal reaction between the

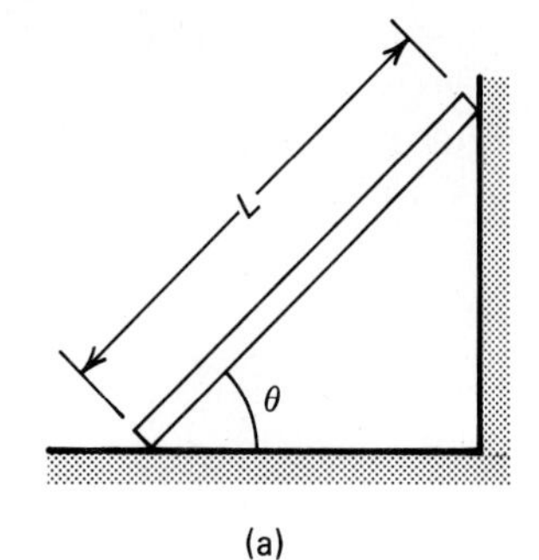

(a)

(b)

Figure 8.8

ladder and the floor while N_2 is the normal reaction between the ladder and the wall. Also, observe the direction of the friction forces that are opposing the impending motion of the ladder. For equilibrium,

$$\sum M_o = 0$$

$$\mu_0 N_2(L \cos \theta) + N_2(L \sin \theta) - mg \frac{L}{2} \cos \theta = 0$$

$$N_2(\mu_0 + \tan \theta) = 0.5mg \qquad \text{(a)}$$

Also $\quad \sum F_x = 0 \qquad \mu_0 N_1 - N_2 = 0$

$$N_2 = \mu_0 N_1 \qquad \text{(b)}$$

Finally, $\quad \sum F_y = 0 \qquad N_1 - mg + \mu_0 N_2 = 0 \qquad \text{(c)}$

Substitute Eq. (b) into (c) and solve for N_1.

$$N_1 - mg + \mu_0^2 N_1 = 0$$

$$N_1 = \frac{mg}{1 + \mu_0^2}$$

Substituting this result into Eq. (b),

$$N_2 = \frac{\mu_0 mg}{1 + \mu_0^2} \qquad \text{(d)}$$

Substitute Eq. (d) into (a).

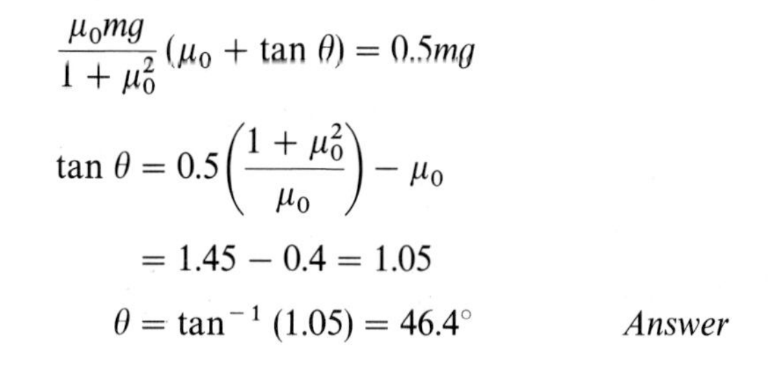

$$\frac{\mu_0 mg}{1 + \mu_0^2}(\mu_0 + \tan \theta) = 0.5mg$$

$$\tan \theta = 0.5\left(\frac{1 + \mu_0^2}{\mu_0}\right) - \mu_0$$

$$= 1.45 - 0.4 = 1.05$$

$$\theta = \tan^{-1}(1.05) = 46.4^\circ \qquad \textit{Answer}$$

PROBLEMS

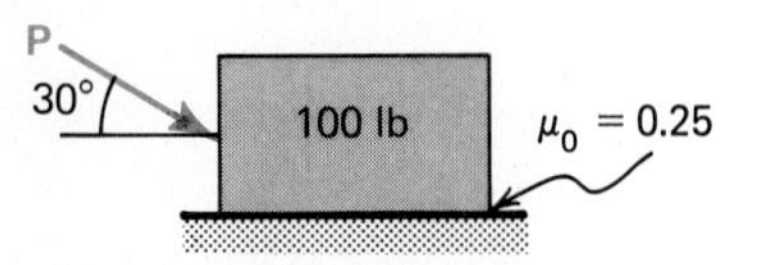

Figure P8.1

8.1 Determine P so that the block in Fig. P8.1 just begins to move.

8.2 Determine P necessary to begin moving the 100-lb block up the incline shown in Fig. P8.2.

8.3 Obtain the magnitude of the force **P** necessary to hold the 100-lb block on the incline shown in Problem 8.2.

8.4 The 100-kg block shown in Fig. P8.4 is to be pulled up the slope. Determine the angle θ and magnitude of the minimum force necessary to achieve this. The coefficient of static friction is 0.3.

8.5 For the block in Problem 8.4, determine the minimum force necessary to hold it in the position shown if θ is zero.

8.6 If $P = 50$ lb acts on the block in Problem 8.2, does the block move if $\mu_0 = 0.20$?

8.7 If $P = 200$ N and $\theta = 15^\circ$ acts on the block in Problem 8.4, does the block move if $\mu_0 = 0.30$?

8.8 If $\theta = 45^\circ$ for the ladder in Fig. 8.8(a), find μ_0 for the ladder to begin to move.

8.9 Find the angle θ for which the 25-kg block in Fig. P8.9 begins to move down the inclined plane if $P = 25$ N and $\mu_0 = 0.4$.

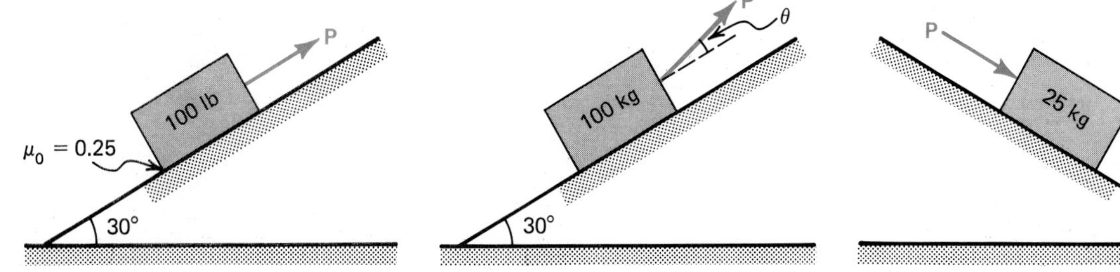

Figure P8.2 Figure P8.4 Figure P8.9

8.10 A 150-lb man has climbed 15 ft up a 50-lb ladder when the ladder begins to slide as shown in Fig. P8.10. What is the coefficient of friction between the ladder and the wall as well as the floor assuming they are equal? Assume the force acting on the ladder from the man's weight is applied at point A.

8.11 Determine the value of P in Fig. P8.11 necessary to (a) hold the block and (b) move the block up the incline. Assume $\mu_0 = 0.25$.

8.12 A 30-kg beam, shown in Fig. P8.12, is placed against a wall. If $\mu_0 = 0.2$ for all contacting surfaces and the magnitude of the horizontal restraining force is $P = 70$ N, determine the minimum value of θ when the beam just begins to slide.

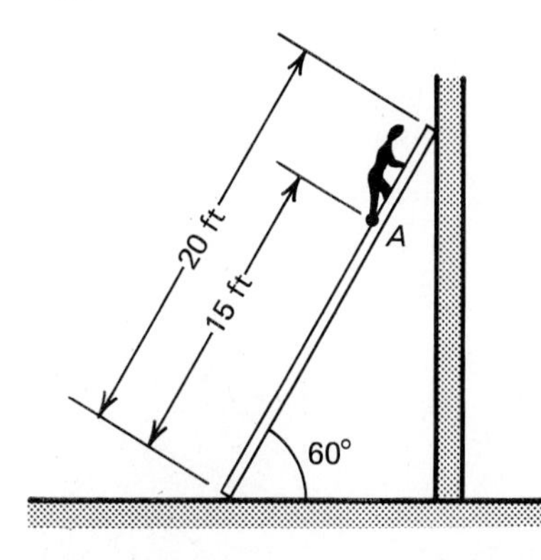

Figure P8.10

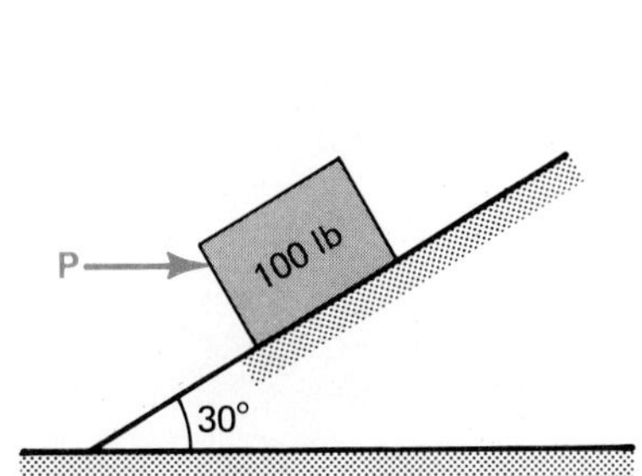

Figure P8.11

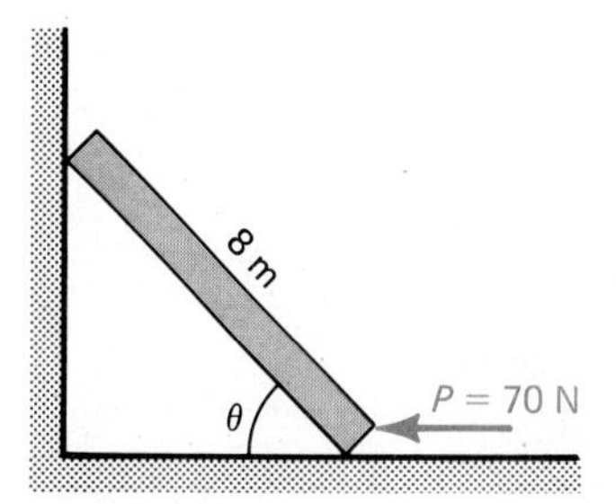

Figure P8.12

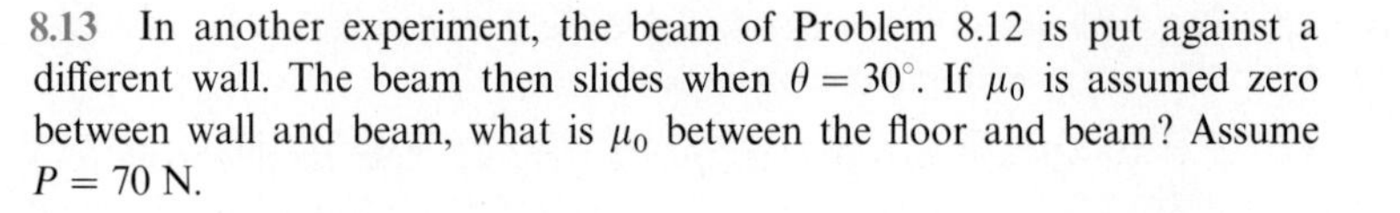

8.13 In another experiment, the beam of Problem 8.12 is put against a different wall. The beam then slides when $\theta = 30^\circ$. If μ_0 is assumed zero between wall and beam, what is μ_0 between the floor and beam? Assume $P = 70$ N.

8.14 A pair of tongs is used to hold a small steel cylinder as shown in Fig. P8.14. Determine the μ_0 that is necessary for the cylinder not to slip. Neglect the weight of the cylinder.

8.15 A 100-lb block is to be lifted by a pair of tongs as shown in Fig. P8.15. What must be the coefficient of static friction in order to prevent the block from slipping?

8.16 The homogeneous rectangular block of mass m shown in Fig. P8.16 is subjected to the force **P**. How large is the distance h so that the block slides without tipping? Assume $\mu_0 = 0.5$ and $a = 1$ m.

8.17 A 100-kg crate is subjected to the force **P** as shown in Fig. P8.17. Determine the direction and the magnitude of the force **P** in order to start the crate moving but without tipping about point A. Let $\mu_0 = 0.25$.

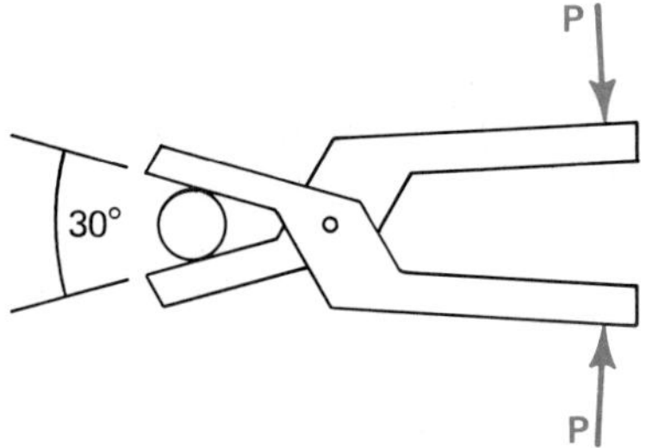

Figure P8.14

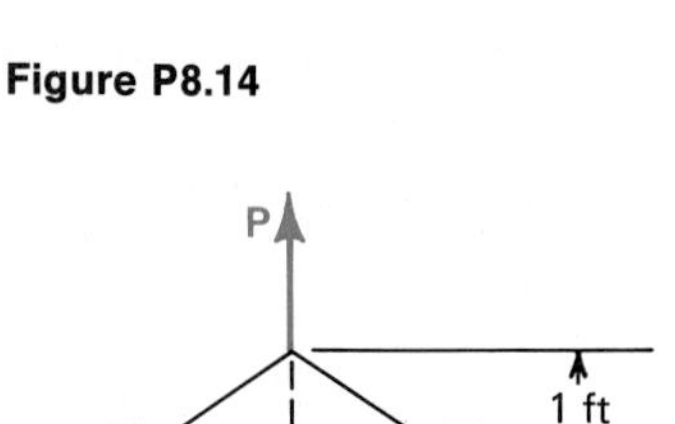

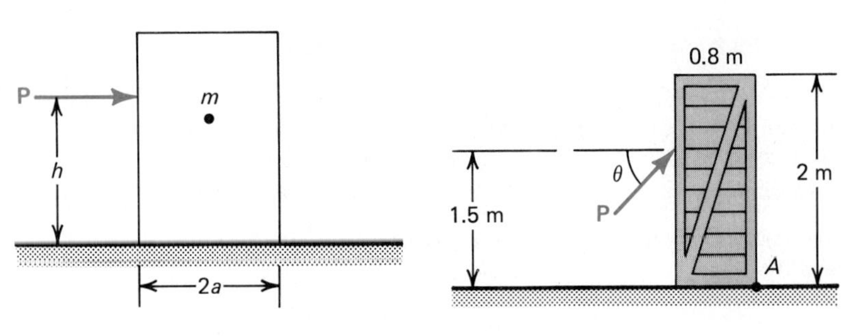

Figure P8.15

Figure P8.16

Figure P8.17

8.18 If the force **P** for Problem 8.17 is applied 1.8 m from the floor, determine values for θ and **P** that allow the crate to be pushed without tipping.

8.19 A 100-lb roller is supported as shown in Fig. P8.19. Determine the coefficient of static friction and the force **T** necessary to prevent it from moving.

8.20 The 50-kg wheel is supported by the force **P** as shown in Fig. P8.20. Determine μ_0 for the wheel to remain at rest.

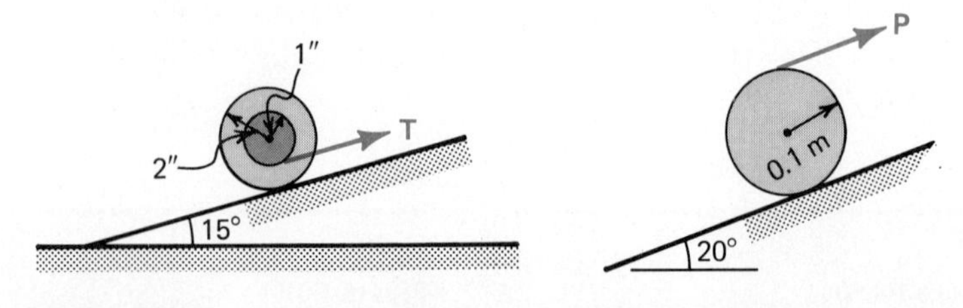

Figure P8.19

Figure P8.20

8.2 Wedges

A wedge is a simple device for transmitting large forces. For example, consider the wedge A in Fig. 8.9(a) that is used to lift block B. Knowing the static coefficients of friction between each surface, we wish to determine the magnitude of the force **P** as applied to the wedge to lift the block B. The free-body diagrams of the block B and the wedge are shown in Fig. 8.9(b) and (c), respectively. Note that each friction force acts in the direction that opposes the impending motion of the body to which it is applied. Also, the magnitude of each friction force f_A, f_B, and f_C equals the product of the coefficient of static friction and the appropriate normal reaction force.

The total number of unknowns are N_A, N_B, N_C, and P. These unknown forces are obtained from the conditions of equilibrium applied to the block and wedge separately, so that we obtain the following four independent equations.

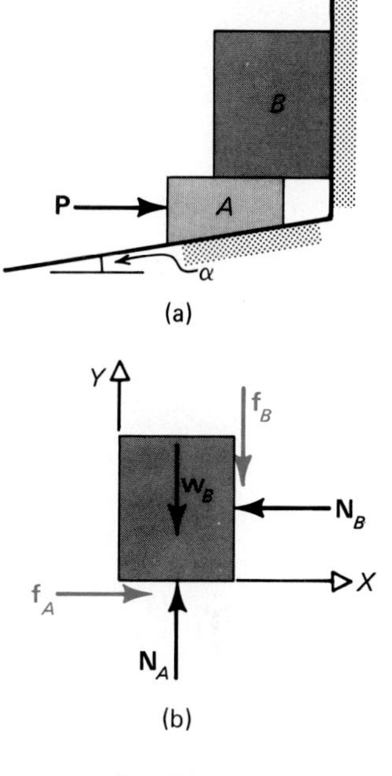

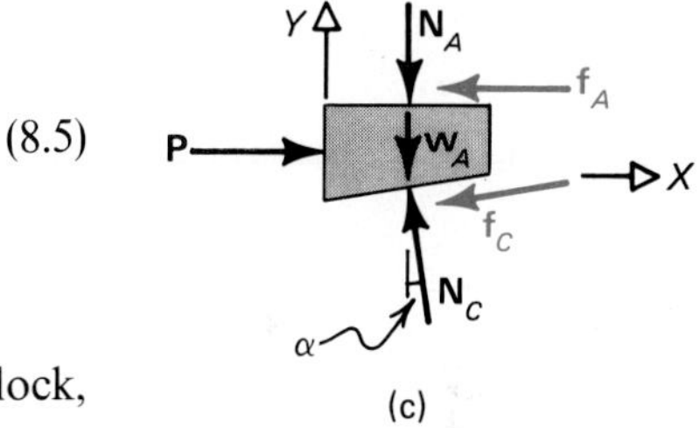

Figure 8.9

Block B:

$$\sum F_x = 0 \qquad f_A - N_B = 0$$

$$\sum F_y = 0 \qquad N_A - w_B - f_B = 0$$

Wedge A: (8.5)

$$\sum F_x = 0 \qquad P - f_A - f_C \cos\alpha - N_C \sin\alpha = 0$$

$$\sum F_y = 0 \qquad N_C \cos\alpha - N_A - f_C \sin\alpha - w_A = 0$$

Note that w_A and w_B are the weights of the wedge and the block, respectively.

Example 8.4

Determine the magnitude of the horizontal force **P** necessary to lift block B in Fig. 8.9(a), assuming $\mu_0 = 0.25$ for each contacting surface, $\alpha = 15°$, and the block B weighs 1962 lb. Neglect the weight of the wedge.

Since the weight of the wedge is neglected, $w_A = 0$ in Eqs. (8.5). Referring to these equations and noting that $f_A = 0.25\ N_A$ and $f_B = 0.25\ N_B$, we have the following:

Block B: $\sum F_x = 0.25N_A - N_B = 0$

$$\therefore N_B = 0.25N_A \qquad \text{(a)}$$

$$\sum F_y = N_A - 1962 - 0.25N_B = 0$$

$$N_A = 1962 + 0.25N_B \qquad \text{(b)}$$

Substituting Eq. (b) into (a),

$$N_B = 490.5 + 0.0625N_B$$

$$N_B = 523.2 \text{ lb}$$

Therefore, $$N_A = 2093 \text{ lb}$$

Wedge A:

$$\sum F_x = P - 0.25N_A - 0.25N_C \cos 15^\circ - N_C \sin 15^\circ = 0$$

$$P - 523.2 - 0.241N_C - 0.259N_C = 0$$

$$P = 523.2 + 0.500N_C \qquad \text{(c)}$$

$$\sum F_y = N_C \cos 15^\circ - 2093 - 0.25N_C \sin 15^\circ = 0$$

$$N_C = 2322 \text{ lb} \qquad \text{(d)}$$

Substituting Eq. (d) into (c),

$$P = 523.2 + 0.500(2322)$$

$$= 1684 \text{ lb}$$

Answer

Example 8.5

A 5-kg wedge is to be driven between two 50-kg plates as shown in Fig. 8.10(a). What is the magnitude of the force **P** necessary to move the plates? Assume that $\mu_0 = 0.5$ between all the surfaces.

First, a free-body diagram of the right plate is drawn as shown in Fig. 8.10(b). N_1 is the normal reaction between the plate and surface while N_2 is the normal reaction between the wedge and the plate. Summing the forces in the vertical and horizontal directions gives us

$$\sum F_y = N_1 - 50(9.81) - N_2 \sin 30^\circ - f_2 \cos 30^\circ = 0$$

$$\sum F_x = N_2 \cos 30^\circ - f_2 \sin 30^\circ - f_1 = 0$$

Since $f_1 = 0.5N_1$ and $f_2 = 0.5N_2$, the above equations become

$$N_1 - N_2(\sin 30^\circ + 0.5 \cos 30^\circ) = 490.5$$

$$0.5N_1 + N_2(0.5 \sin 30^\circ - \cos 30^\circ) = 0$$

Solving for N_1 and N_2, we obtain

$$N_1 = 2020 \text{ N} \qquad N_2 = 1640 \text{ N}$$

Now we can draw the free-body diagram of the wedge as shown in Fig. 8.10(c). Note that due to symmetry, we have the normal reaction N_2 acting on each side of the wedge surfaces.

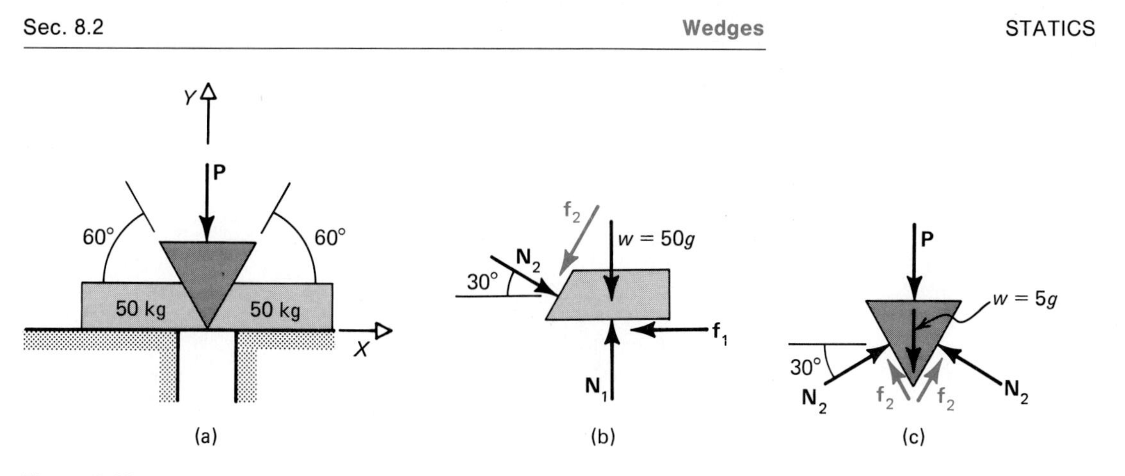

Figure 8.10

Summing the forces in the vertical direction gives us

$$\sum F_y = -P - 5(9.81) + 2N_2 \sin 30^\circ + 2\mu_0 N_2 \cos 30^\circ = 0$$

and substituting for N_2, we obtain

$$P = -49.1 + 2(1640) \sin 30^\circ + 2(0.5)(1640) \cos 30^\circ$$

$$P = 3011 \text{ N}$$ *Answer*

PROBLEMS

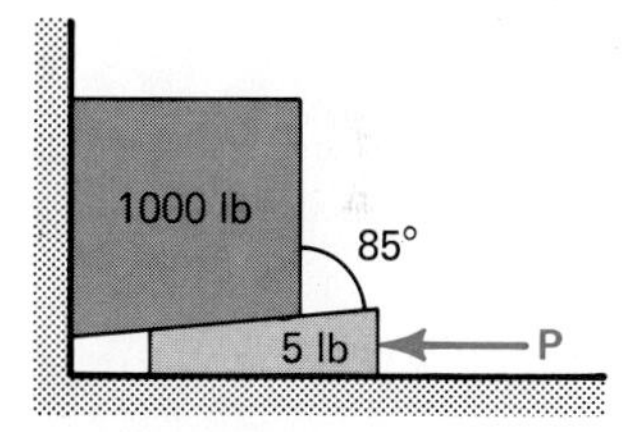

Figure P8.21

8.21 Determine the magnitude of the horizontal force **P** necessary to move the 1000-lb block shown in Fig. P8.21. The coefficient of static friction is 0.3 between all surfaces.

8.22 Repeat Problem 8.21 for $\mu_0 = 0.20$ between the block and the wedge, $\mu_0 = 0.25$ between the block and wall, and $\mu_0 = 0.30$ between the wedge and the floor.

8.23 A 10-kg wedge is used to lift the 5000-kg block shown in Fig. P8.23. The coefficient of static friction is 0.2 between all surfaces. Determine the value of P.

8.24 Repeat Problem 8.23 assuming $\mu_0 = 0$ between the wall and the 5000-kg block and 0.2 between all other surfaces.

8.25 A wedge is used to move the two 200-lb blocks shown in Fig. P8.25. Determine the value of P if $\mu_0 = 0.3$ between all surfaces shown.

8.26 Repeat Problem 8.25 assuming $\mu_0 = 0.4$ between the blocks and floor and $\mu_0 = 0$ between the wedge and the blocks.

8.27 A wedge is used to split wood as shown in Fig. P8.27. Determine the value of θ if it is desired that the wedge should not slip out once inserted. Assume $\mu_0 = 0.7$.

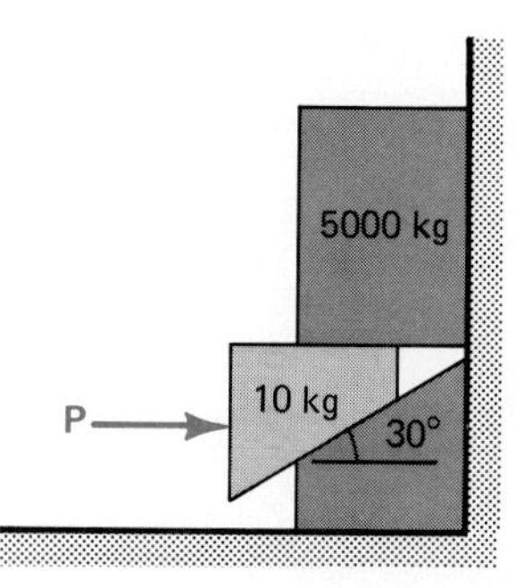

Figure P8.23

8.28 A 1000-kg block is moved up by a wedge as shown in Fig. P8.28. If $\mu_0 = 0.3$ for all surfaces, determine the necessary magnitude of the force **P**. Assume that the rollers are frictionless.

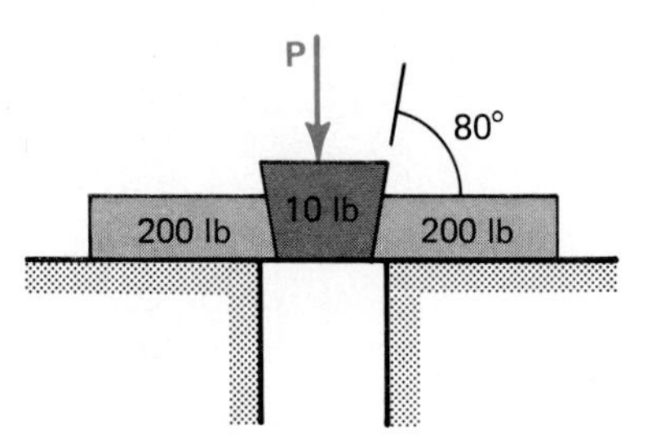

Figure P8.25

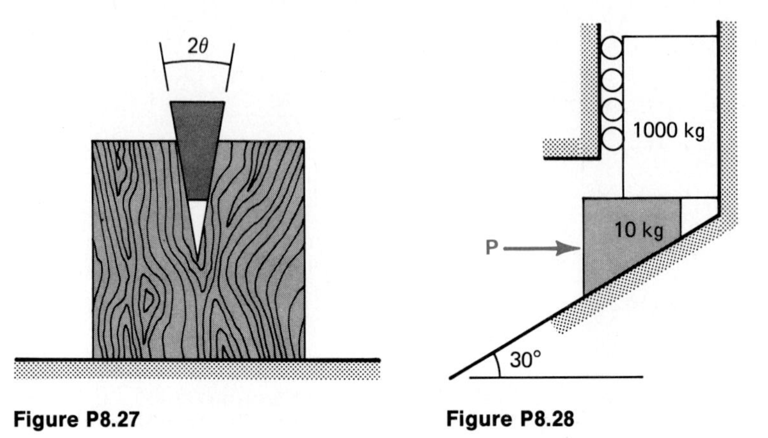

Figure P8.27

Figure P8.28

8.3 Screws

Machines such as presses, jackscrews, and screwdrivers utilize threaded screws to transmit loads. The screw can be analyzed by applying the method developed earlier for the case of a block sliding on an inclined plane. Consider the screw shown in Fig. 8.11(a), where an axial force **w**, that includes the weight of the screw, is applied. We isolate a small element of the screw and draw its free-body diagram as shown in Fig. 8.11(b). We note that this small element of the screw will slide on the nut strip that is stationary. The screw, when turned relative to the nut, will be subjected to a friction force **f**. Moving the screw up or down is analogous to raising or lowering the load w along an inclined plane as shown in Fig. 8.11(b) and (c). Consider the case of raising the load w for the condition of impending motion. Summing the forces in Fig. 8.11(b) in the vertical direction,

$$\sum F_y = -f_m \sin \alpha + N \cos \alpha - w = 0$$

Now $f_m = \mu_0 N$, so that the above equation becomes

$$N(\cos \alpha - \mu_0 \sin \alpha) - w = 0$$

and therefore

$$N = \frac{w}{\cos \alpha - \mu_0 \sin \alpha} \tag{8.6}$$

The net horizontal force is obtained by summing the forces in

Fig. 8.11(b) in the X direction, that is,

$$-f_m \cos \alpha - N \sin \alpha = -N(\mu_0 \cos \alpha + \sin \alpha)$$

and this force gives rise to a moment about the axis of the screw. Since the moment arm is r, which is the mean radius from the centerline of the screw, the magnitude of the moment involved in raising the load is

$$M_U = Nr(\mu_0 \cos \alpha + \sin \alpha)$$

Substituting for N from Eq. (8.6) we obtain

$$M_U = wr\left(\frac{\mu_0 \cos \alpha + \sin \alpha}{\cos \alpha - \mu_0 \sin \alpha}\right) \tag{8.7}$$

Since $\tan \phi_0 = \mu_0$, we have

$$M_U = wr\left(\frac{\tan \phi_0 \cos \alpha + \sin \alpha}{\cos \alpha - \tan \phi_0 \sin \alpha}\right)$$

$$= wr \tan (\phi_0 + \alpha) \tag{8.8}$$

Thus, we are able to calculate the required magnitude of the moment to lift the load w.

Figure 8.11(c) shows the free-body diagram of the screw element for impending downward motion. If no moment is applied to the screw, and $\phi_0 > \alpha$, the screw remains in place and is self-locking. To lower the screw, we sum the forces in the Y direction from Fig. 8.11(c),

$$\sum F_y = f_m \sin \alpha + N \cos \alpha - w = 0$$

Since $f_m = \mu_0 N$, we obtain

$$N = \frac{w}{\mu_0 \sin \alpha + \cos \alpha}$$

The net horizontal force becomes

$$(f_m \cos \alpha - N \sin \alpha) = N(\mu_0 \cos \alpha - \sin \alpha)$$

The magnitude of the moment to lower the load is

$$M_L = Nr(\mu_0 \cos \alpha - \sin \alpha)$$

Substituting for N and again noting that $\mu_0 = \tan \phi_0$,

$$M_L = rw\left(\frac{\tan \phi_0 \cos \alpha - \sin \alpha}{\tan \phi_0 \sin \alpha + \cos \alpha}\right)$$

which simplifies to

$$M_L = wr \tan (\phi_0 - \alpha) \tag{8.9}$$

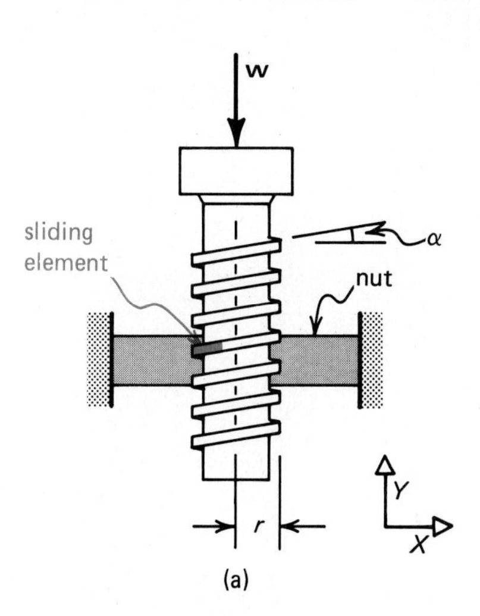

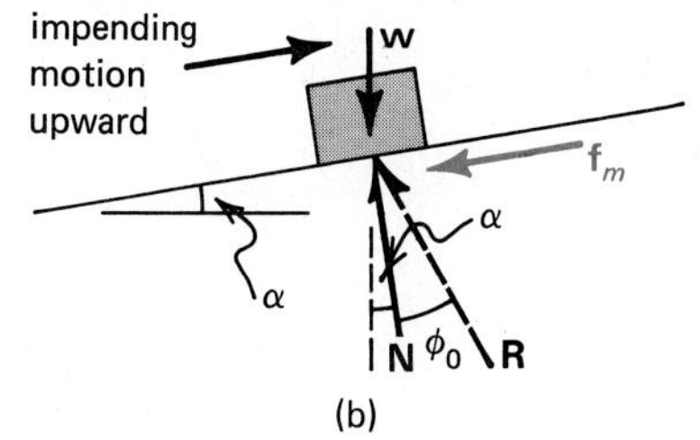

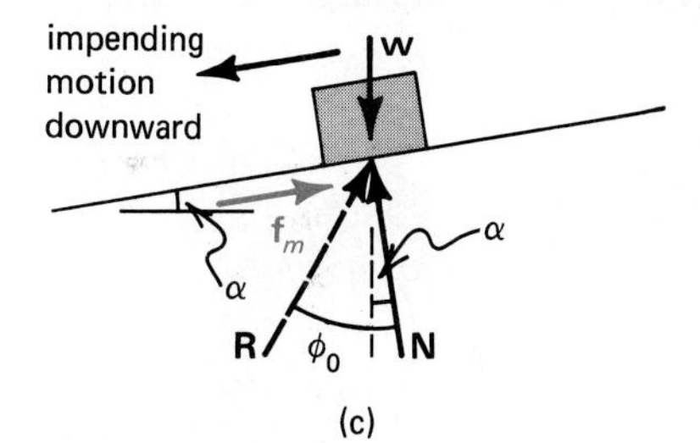

Figure 8.11

(a)

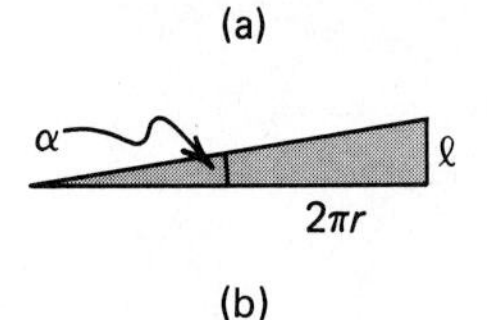

(b)

Figure 8.12

Example 8.6

The jackscrew having 5 threads per inch shown in Fig. 8.12(a) is used to lift a weight. The maximum magnitude of force **P** that produces the uplifting moment is 300 lb and the moment is directed along the positive Y axis. This force is applied at the edge of the jackscrew platform at a distance of 1 ft from the jackscrew centerline. If $\mu_0 = 0.3$, what maximum weight w can be lifted?

Since the screw moves through 5 turns and travels 1 in, it moves 0.2 in per turn. This is called the **pitch of the thread**, l. Now we can determine the angle α using Fig. 8.12(b),

$$\tan \alpha = \frac{l}{2\pi r} = \frac{0.2}{2\pi(0.5)} = 0.0637$$

$$\alpha = 3.64^\circ$$

Also,

$$\phi_0 = \tan^{-1} 0.3 = 16.70^\circ$$

Using Eq. (8.8),

$$300(12) = w(0.5) \tan(16.70^\circ + 3.64^\circ)$$

$$w = 19{,}422 \text{ lb}$$ *Answer*

Example 8.7

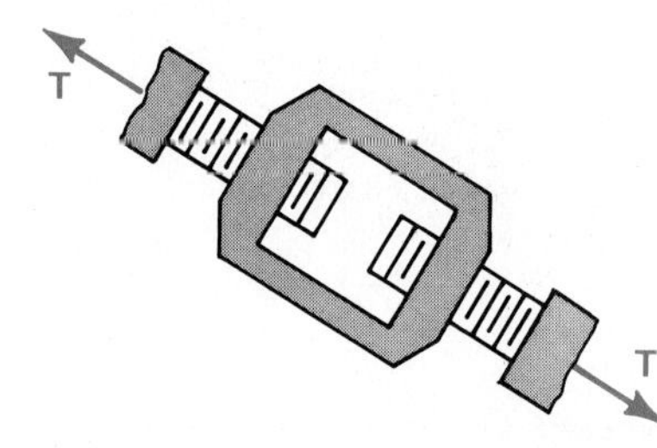

Figure 8.13

The turnbuckle shown in Fig. 8.13 supports a tensile force whose magnitude is $T = 300$ N. The screws have a pitch of 4.5 mm, mean radius of 20 mm, and $\mu_0 = 0.20$. Determine the torque that must be applied on the turnbuckle to loosen it.

Turning the turnbuckle to relieve the force T is analogous to lowering the weight in Fig. 8.11. Since there are two screws in this case, Eq. (8.9) is modified as follows:

$$M_L = 2wr \tan(\phi_0 - \alpha)$$

Now

$$\tan \alpha = \frac{l}{2\pi r} = \frac{4.5}{2\pi(20)}$$

$$\alpha = 2.05^\circ$$

$$\tan \phi_0 = \mu_0 = 0.20$$

$$\phi_0 = 11.31^\circ$$

Thus,

$$M_L = 2(300)(20) \tan(11.31^\circ - 2.05^\circ)$$

$$= 1956 \text{ mm} \cdot \text{N}$$

$$= 1.96 \text{ m} \cdot \text{N}$$ *Answer*

PROBLEMS

8.29 A jackscrew having 4 threads per inch is used to lift a weight. If a 10,000-lb weight must be lifted and $\mu_0 = 0.4$, what is the magnitude of the force **P** that must be applied to the end of the lever shown in Fig. P8.29.

8.30 If a 200 ft·lb moment is applied to a jackscrew of the type illustrated in Problem 8.29 in order to lift 25,000 lb, what is the coefficient of static friction?

8.31 One end of a car is to be lifted by the jackscrew shown in Fig. P8.31. The pitch is 4.5 mm, the mean radius is 20 mm, and $\mu_0 = 0.3$. If the weight to be lifted is 20 kN, what moment must be applied to the jackscrew?

8.32 The jackscrew in Problem 8.31 is slightly greased so that the coefficient of static friction is decreased by 20 percent. What is the moment required to lift the 20-kN weight?

8.33 The turnbuckle shown in Fig. 8.13 supports a tension of 500 lb. The screws have a pitch of 0.18 in, mean radius is 0.75 in, and $\mu_0 = 0.25$. Determine the moment that must be applied on the turnbuckle to loosen it.

8.34 For the turnbuckle of Problem 8.33, determine the moment that must be applied to loosen it if the tension is 400 lb and $\mu_0 = 0.3$.

8.35 The clamp shown in Fig. P8.35 holds two pieces of angle iron. Determine the forces exerted on the angles if a 50 m·N moment is applied to tighten the clamp. Assume that the pitch is 3 mm, mean radius is 10 mm, and $\mu_0 = 0.25$.

8.36 For the clamp of Problem 8.35, determine the torque required to loosen the clamp if the angles are pressed with a force of 1000 N.

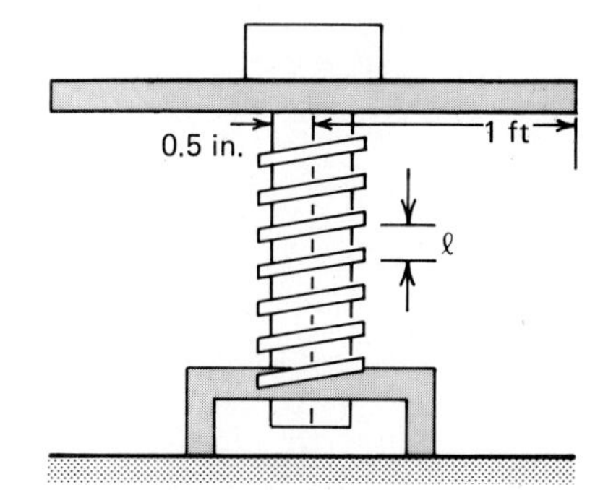

Figure P8.29

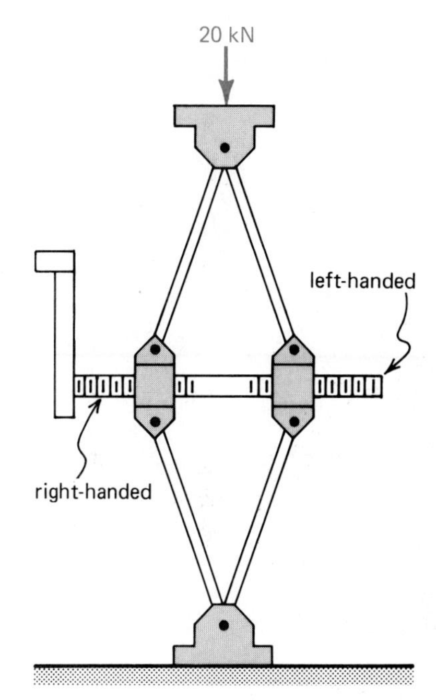

Figure P8.31

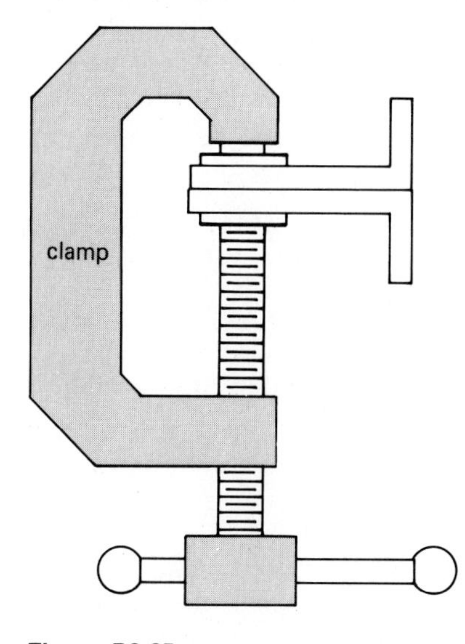

Figure P8.35

8.4 Belt Friction

Engineers make frequent use of ropes and pulleys in the design of driving mechanisms and brakes. Since these devices can slip over the surfaces with which they are in contact, friction forces must be considered. Such forces are called belt friction.

Consider a belt wrapped around a drum as shown in Fig. 8.14(a). The fixed drum is subjected to the two belt tensions T_1 and T_2 as well as a bearing reaction R and a moment of magnitude M that prevents rotation. We wish to establish the relationship between the two belt tensions at the instant of impending motion. Owing to the direction of **M**, we note that $T_2 > T_1$. A relationship between the tensions can be derived from the free-body diagram of the differential element of the belt shown in Fig. 8.14(b). Here, the tensile force T acts on one end of the belt element and $(T + dT)$ at the other end of the belt element. Also, the normal reaction force dN and friction force df act as shown in Fig. 8.14(b). Summing the forces in the X direction yields

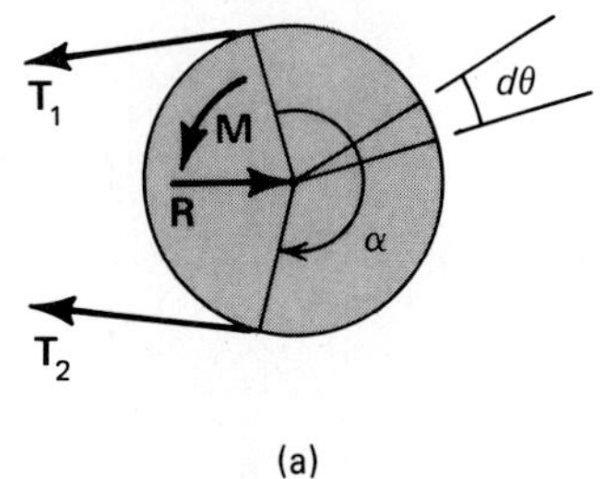

(a)

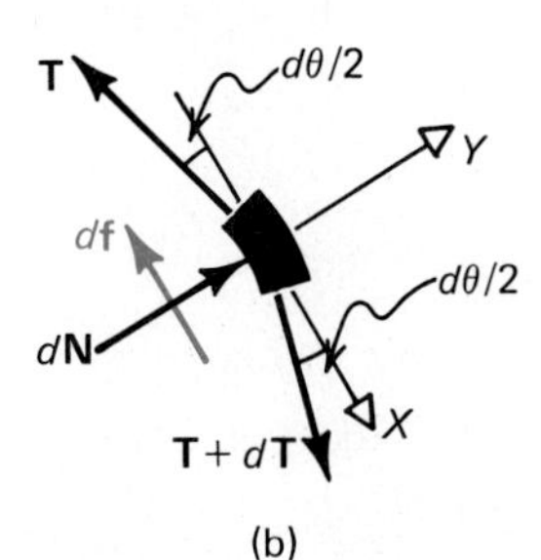

(b)

Figure 8.14

$$\sum F_x = -T\cos\frac{d\theta}{2} + (T + dT)\cos\frac{d\theta}{2} - df = 0$$

If $d\theta$ is small, $\cos(d\theta/2) \approx 1$, and the above equation yields

$$df = dT$$

Since

$$df = \mu_0\, dN$$

we have

$$dN = \frac{dT}{\mu_0} \tag{8.10}$$

Now we sum the forces in the Y direction, and noting that $\sin(d\theta/2) \approx d\theta/2$ for small $d\theta$, we obtain

$$\sum F_y = dN - T\sin\frac{d\theta}{2} - (T + dT)\sin\frac{d\theta}{2} = 0$$

$$dN = T\, d\theta \tag{8.11}$$

where the term $dT\, d\theta$ is negligible, and therefore set equal to zero. Equating the right sides of Eqs. (8.10) and (8.11) gives us

$$\frac{dT}{\mu_0} = T\, d\theta \qquad \text{or} \qquad \frac{dT}{T} = \mu_0\, d\theta$$

Integrating this expression over the length of belt in contact with the drum,

$$\int_{T_1}^{T_2} \frac{dT}{T} = \int_0^{\alpha} \mu_0\, d\theta$$

$$\log \frac{T_2}{T_1} = \mu_0 \alpha$$

or

$$\frac{T_2}{T_1} = e^{\mu_0 \alpha} \tag{8.12}$$

This gives the ratio of the two tensions when the rope or belt is about to slip. Note that α is expressed in radians and represents the contact angle. If the rope or belt is wrapped around n times, then $\alpha = 2\pi n$ radians. Another observation to be made is that the equation for tension is independent of the drum radius. For this reason the equation may be applied to noncircular cross sections as long as the correct angle is used. The analysis of vee belts is carried out in much the same way and is illustrated in Example 8.10.

Example 8.8

A rope is used to lift a 100-lb weight using the locked pulley shown in Fig. 8.15. If $\mu_0 = 0.2$, determine the magnitude of the force **P** necessary to begin lifting the weight.

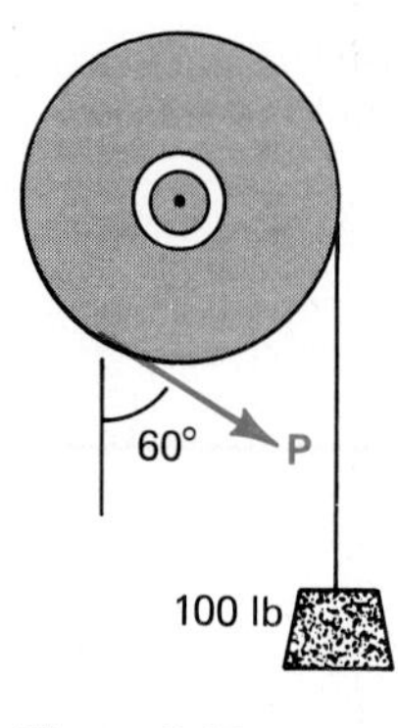

Figure 8.15

The tension P must be greater than the weight since frictional forces have to be overcome. The angle of contact is

$$\alpha = 240\left(\frac{\pi}{180}\right) = 4.19 \text{ radians}$$

Using Eq. (8.12) and letting $T_2 = P$ and $T_1 = 100$,

$$\frac{P}{100} = e^{(0.2)(4.19)} = 2.31$$

$$P = 231 \text{ lb} \qquad \textit{Answer}$$

Example 8.9

When $P = 80$ N, the brake drum in Fig. 8.16(a) is restrained from turning. What is the magnitude of this restraining moment if $\mu_0 = 0.3$?

The free-body diagrams of the brake and drum are shown in Fig. 8.16(b) and (c), respectively. Summing moments about point A on the brake we have

$$\sum M_A = T_2(0.5) - P(1.5) = 0$$

$$T_2 = \frac{80(1.5)}{0.5} = 240 \text{ N}$$

Now

$$T_2 = T_1 e^{\mu_0 \alpha} = T_1 e^{0.3\pi}$$

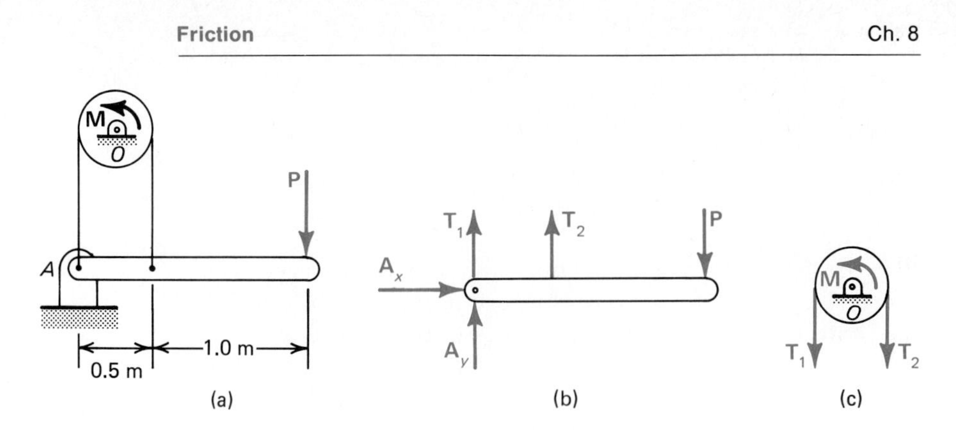

Figure 8.16

$$\therefore T_1 = 240e^{-0.3\pi} = 93.5 \text{ N}$$

Referring to Fig. 8.16(c),

$$\sum M_o = (T_2 - T_1)r - M = 0$$

$$M = (240 - 93.5)(0.25)$$

$$= 36.6 \text{ m} \cdot \text{N}$$ *Answer*

Example 8.10

If, in Example 8.8, a vee belt is used instead of a rope, as shown in Fig. 8.17(a), what force is necessary to lift the weight?

A free-body diagram of a differential element of the vee belt is shown in Fig. 8.17(b). Note that the friction force df acts on both sides of the belt where $df = \mu_0 \, dN$. Summing the forces in the X and Y directions, respectively, gives us

$$\sum F_x = -T \cos\frac{d\theta}{2} + (T + dT)\cos\frac{d\theta}{2} - 2\mu_0 \, dN = 0$$

$$\sum F_y = -T \sin\frac{d\theta}{2} - (T + dT)\sin\frac{d\theta}{2} + 2 \sin\beta \, dN = 0$$

Since $d\theta$ is small, we set $\cos(d\theta/2) \approx 1$, $\sin d\theta/2 \approx d\theta/2$, and we neglect terms like $dT\, d\theta$ to obtain,

$$dT = 2\mu_0 \, dN$$

$$T\, d\theta = 2\, dN \sin\beta$$

Solving these equations for dN and equating, we obtain

$$\frac{dT}{2\mu_0} = \frac{T\, d\theta}{2 \sin\beta} \quad \text{or} \quad \frac{dT}{T} = \left(\frac{\mu_0}{\sin\beta}\right) d\theta$$

We now define the *effective friction* as

$$\mu_{0v} = \frac{\mu_0}{\sin \beta}$$

so that

$$\frac{dT}{T} = \mu_{0v}\, d\theta$$

Integrating the differential equation,

$$\int_{T_1}^{T_2} \frac{dT}{T} = \int_0^{\alpha} \mu_{0v}\, d\theta$$

so that

$$\frac{T_2}{T_1} = e^{\mu_{0v}\alpha}$$

This result is similar to the flat-belt example except that we now use the effective friction μ_{0v}. For the values in Example 8.8 and assuming that $2\beta = 60^\circ$, we have

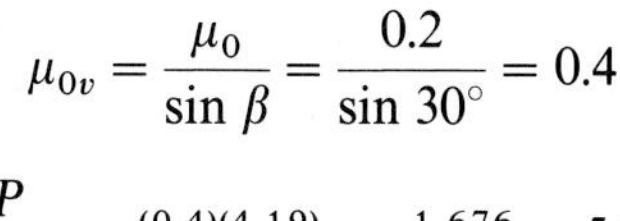

$$\mu_{0v} = \frac{\mu_0}{\sin \beta} = \frac{0.2}{\sin 30^\circ} = 0.4$$

and

$$\frac{P}{100} = e^{(0.4)(4.19)} = e^{1.676} = 5.34$$

$$P = 534 \text{ lb} \qquad \textit{Answer}$$

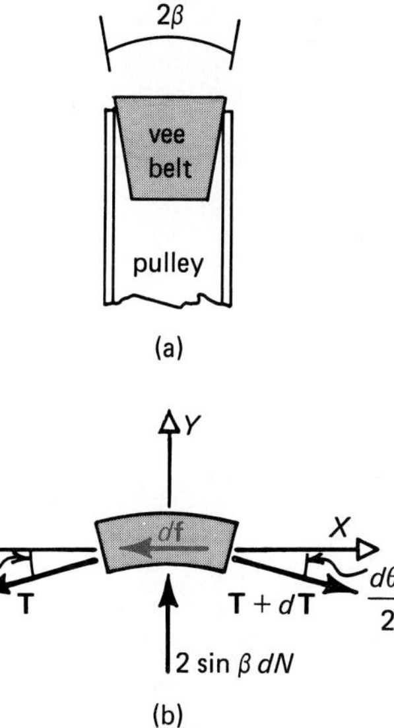

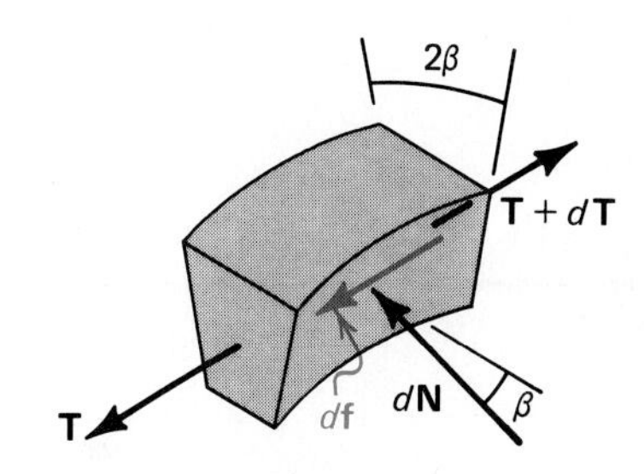

Figure 8.17

PROBLEMS

8.37 Determine the magnitude of the force **P** necessary to hold the 100-lb weight shown in Fig. P8.37. The coefficient of static friction is 0.4 between the belt and the locked pulley.

8.38 In Problem 8.37, determine the value of P necessary to raise the weight.

8.39 A flat belt is used to transmit the 50 m·N torque developed by the motor shown in Fig. P8.39. Determine the tensions in the belt. Assume that there is no slippage. The coefficient of static friction is 0.3.

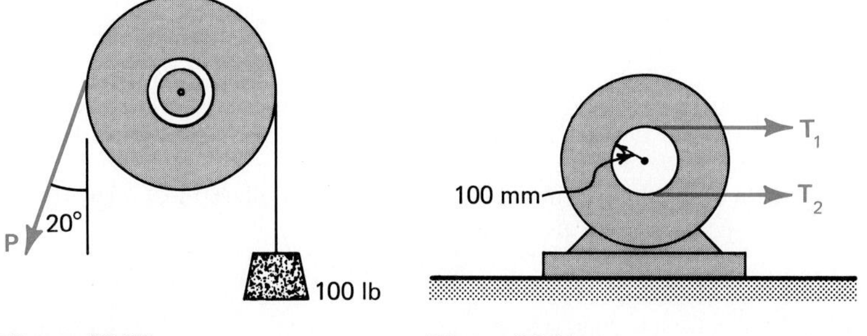

Figure P8.37 **Figure P8.39**

8.40 In Problem 8.39, if the maximum tension the belt can withstand is 1000 N, determine the largest torque that can be transmitted.

8.41 Determine the force P necessary (a) to hold and (b) to raise the 250-lb weight in Fig. P8.41. Each pulley is locked and μ_0 varies for each pulley as shown in the figure.

8.42 Determine the force P necessary to (a) hold and (b) raise the 100-lb weight shown in Fig. P8.42. Let the coefficient of static friction be 0.2 between the rope and each locked pulley.

8.43 The motor shown in Fig. P8.43 is used to transmit 100 m·N of torque. If $\mu_0 = 0.2$, determine the tensions in the belt. Assume that there is no belt slippage and the belt is flat. Let $\theta = 0°$.

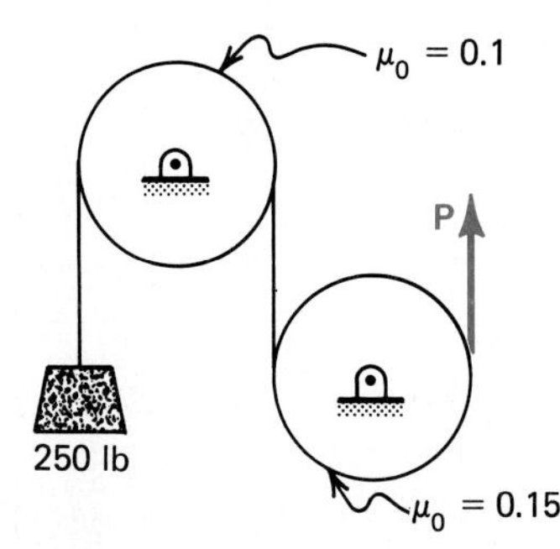

Figure P8.41

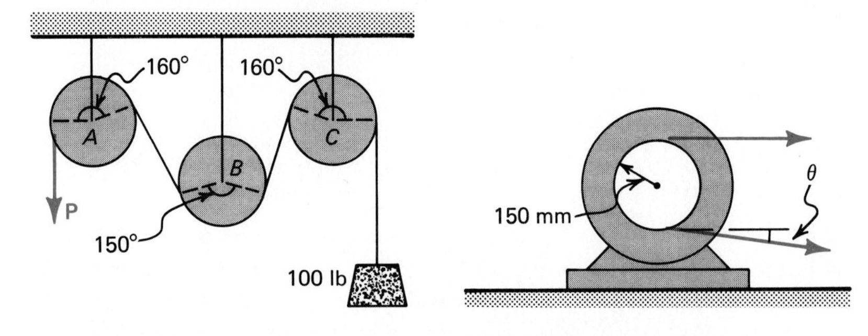

Figure P8.42

Figure P8.43

8.44 Determine the tensions in the belt for $\theta = 30°$ in Problem 8.43.

8.45 Find the braking moment that may be applied to the brake drum in Fig. P8.45 if $P = 40$ N, $a = 0.36$ m, and $\mu_0 = 0.35$.

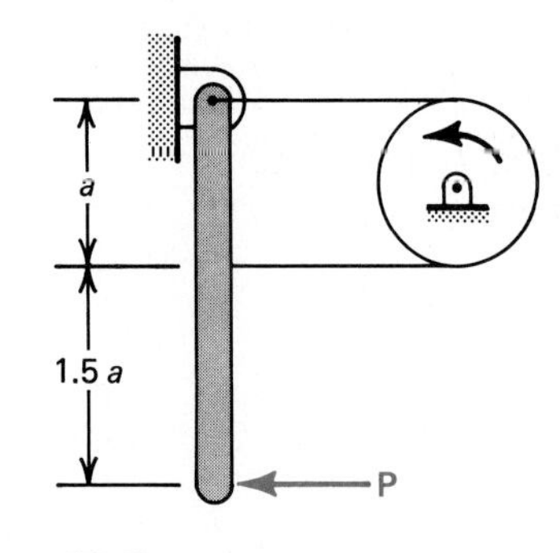

Figure P8.45

8.46 For the brake drum shown in Fig. P8.46, derive a relationship between the applied force **P** and the moment **M** due to the friction force applied to the drum. Let $\mu_0 = 0.5$ and neglect the weight of the beam.

8.47 For Problem 8.46, compute M if $P = 200$ lb, $r = 0.5$ ft, and $\mu_0 = 0.5$.

8.48 A 100-kg mass is suspended by a cord wrapped twice around the rod whose cross section is circular of radius 150 mm as shown in Fig. P8.48. If $\mu_0 = 0.25$, determine the force P necessary to (a) hold the weight and (b) raise the weight.

8.49 In Problem 8.48, if 2000 N is necessary to raise the weight, determine the coefficient of friction.

8.50 Obtain the value of P to (a) hold and (b) raise the 100-lb weight in Problem 8.37 if a vee belt of the type shown in Fig. P8.50 is used.

8.51 If the vee belt shown in Fig. P8.50 is used in place of the flat belt of Problem 8.46, determine the tensions in the belt.

8.52 Repeat Problem 8.39, using the vee belt shown in Fig. P8.50.

8.53 If the vee belt shown in Fig. P8.50 is used in place of the flat belt of Problem 8.43, determine the tension in the belt.

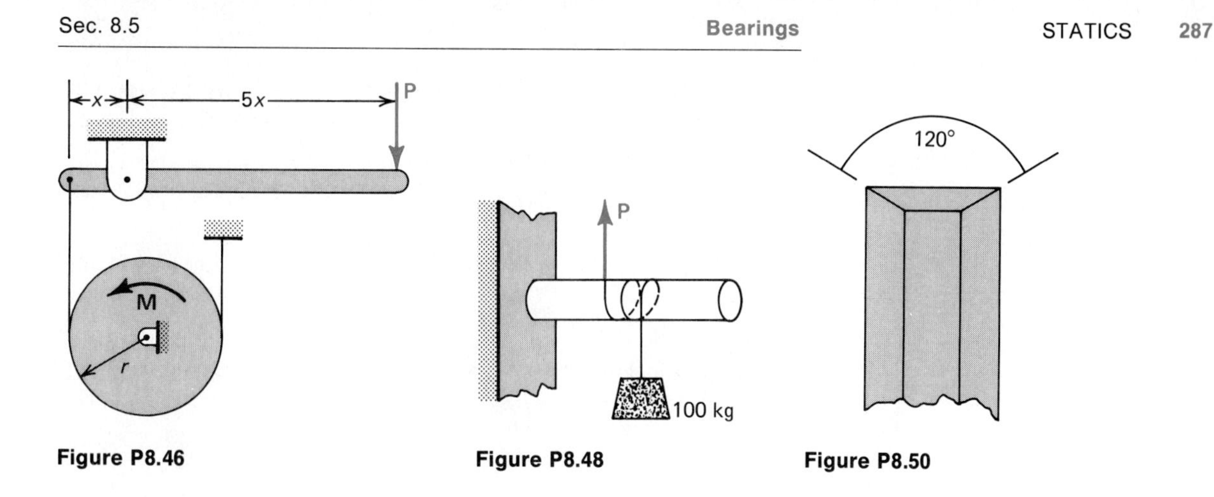

Figure P8.46

Figure P8.48

Figure P8.50

8.5 Bearings

Bearings are commonly used to support rotating machinery consisting of shafts and axles. Although a small amount of lubricant is often used, analyses using the methods developed for cases of dry friction provide satisfactory approximations. Bearings can be classified into journal bearings and thrust bearings. Journal bearings give lateral support to shafts; thrust bearings supply axial or thrust support.

Journal Bearings

Consider the journal bearing shown in Fig. 8.18(a), that supports a shaft that is rotating uniformly. This shaft requires a torque to first initiate and then maintain rotation. A radial load P causes the shaft to touch the bearing at a point C and thereby gives rise to reaction forces consisting of the normal force of magnitude N and the friction force $\mathbf{f}_k$. From the free-body diagram in Fig. 8.18(b), we have, for equilibrium,

$$\sum F_y = -P + N \cos\phi + f_k \sin\phi = 0$$

Substituting $f_k = \mu N$ and $\mu = \tan\phi$ we obtain

$$P = N\cos\phi + N\tan\phi\sin\phi$$

$$P = \frac{N}{\cos\phi}$$

Since R is the resultant of f_k and N,

$$R = \sqrt{\mu^2 N^2 + N^2} = N\sqrt{\mu^2 + 1} = \frac{N}{\cos\phi}$$

$$\therefore R = P$$

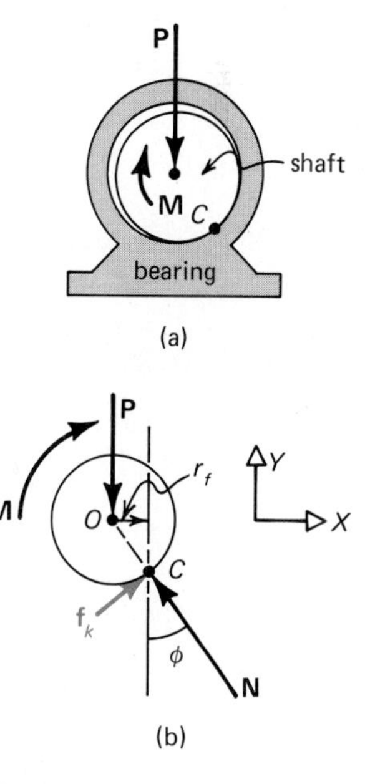

Figure 8.18

where the line of action of the force R is tangent to a circle with radius r_f. This circle is called the *friction circle*. Summing the moments about O gives us

$$\sum M_o = M_k - f_k r = 0$$

$$M_k = f_k r$$
$$= \mu N r$$

where μ is the coefficient of kinetic friction due to the existence of rotational motion and M_k is the magnitude of the applied moment. Substituting for N, we have

$$M_k = \mu Pr \cos \phi = Pr \sin \phi \tag{8.13}$$

This is the magnitude of the moment to maintain rotation. If ϕ is small because of small friction, this can be approximated to $M_k = \mu Pr$. To initiate rotation we use the coefficient of static friction, so that

$$M_o = \mu_0 Pr \cos \phi_0 = Pr \sin \phi_0 \tag{8.14}$$

Again, this can be approximated by $M_o = \mu_0 Pr$ if the friction is small.

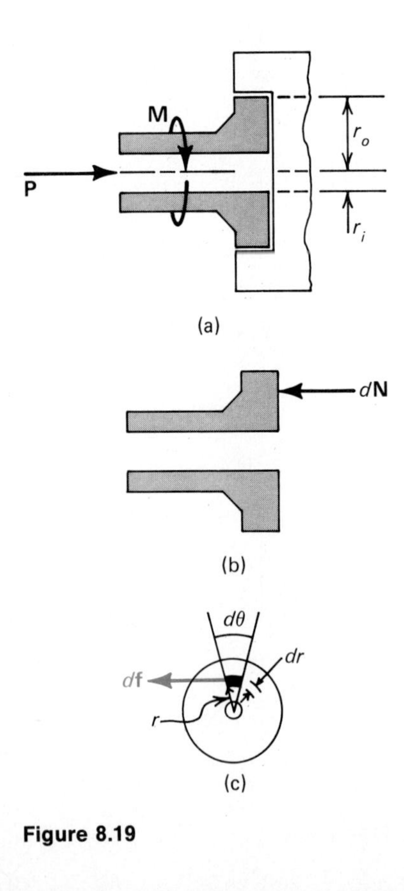

Figure 8.19

Thrust Bearings

Thrust bearings are used for providing axial support to shafts and axles. In general, there are two types of thrust bearings—end bearings and collar bearings. In the former, the shaft end rotates in a stationary collar so that friction is produced primarily between the shaft end and the collar. In collar bearings, friction occurs between two ring-shaped areas that are in contact. It is worth noting that friction developed in thrust bearings is also found in clutch plates and disk brakes.

By way of example of a thrust bearing, consider a hollow rotating shaft whose end is in contact with an end bearing as shown in Fig. 8.19(a). Let us assume that the hollow shaft is cylindrical with outer radius r_o and inner radius r_i. The thrust P keeps the surfaces in contact, while M_k is the torque necessary to sustain uniform rotation. The area in contact is

$$A = \pi(r_o^2 - r_i^2)$$

and, therefore, the pressure p between the contacting surfaces is

$$p = \frac{P}{A} = \frac{P}{\pi(r_o^2 - r_i^2)}$$

The normal force dN acting on the differential area dA in Fig. 8.19(b) is

$$dN = p\,dA$$

The frictional force df in Fig. 8.19(c) is

$$df = \mu\,dN = \mu p\,dA$$

and the moment produced by this frictional force is

$$dM_k = r\,df = r\mu p\,dA$$

We recall that the differential area $dA = r\,d\theta\,dr$, so that the total moment becomes

$$M_k = \int_0^{2\pi} \int_{r_i}^{r_o} \mu p r^2\,d\theta\,dr$$

Integrating this yields

$$M_k = \frac{2}{3}\mu P \frac{r_o^3 - r_i^3}{r_o^2 - r_i^2} \tag{8.15}$$

which is the moment necessary to sustain the rotation. If the disk is not hollow, then $r_i = 0$ and

$$M_k = \frac{2}{3}\mu P r_o \tag{8.16}$$

Example 8.11

The 2-m long shaft in Fig. 8.20 is supported between two journal bearings. A 1000-N load is transmitted to the 75-mm diameter shaft as shown. If the density of the shaft material is 4 t/m^3 and $\mu = 0.10$, find the torque necessary to maintain uniform motion of the shaft.

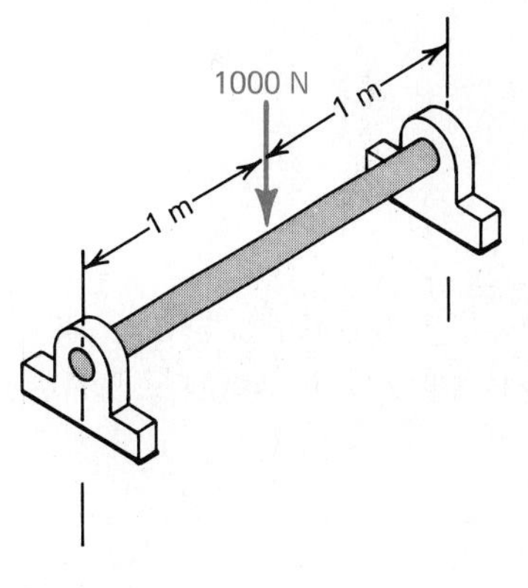

Figure 8.20

The load transmitted to each bearing is first calculated. The total load consists of the 1000-N load plus the shaft weight. Letting P represent the load transmitted to one bearing, the total load is

$$2P = 1000 + 4000\left[\frac{\pi(0.075)^2}{4}(2)\right](9.81)$$

$$= 1346.7$$

$$P = 673.4 \text{ N}$$

Since $\mu = 0.1 = \tan\phi$, we find $\phi = 5.7°$. Now referring to Eq. (8.13)

$$M_k = Pr\sin\phi = (673.4)\left(\frac{0.075}{2}\right)\sin 5.7°$$

$$M_k = 2.508 \text{ m}\cdot\text{N}$$

To account for the two journal bearings, we double this moment to obtain

$$M_k = 5.02 \text{ m}\cdot\text{N} \qquad \textit{Answer}$$

Example 8.12

Calculate the magnitude of the moment necessary to rotate the shaft in Fig. 8.19 if the axial thrust $P = 250$ N, $r_o = 75$ mm, $r_i = 25$ mm, and $\mu = 0.1$.

Referring to Eq. (8.15)

$$M_k = \frac{2}{3}\mu P \frac{r_o^3 - r_i^3}{r_o^2 - r_i^2}$$

$$= \frac{2}{3}(0.1)(250)\left[\frac{(0.075)^3 - (0.025)^3}{(0.075)^2 - (0.025)^2}\right]$$

$$= 1.354 \text{ m}\cdot\text{N} \qquad \textit{Answer}$$

PROBLEMS

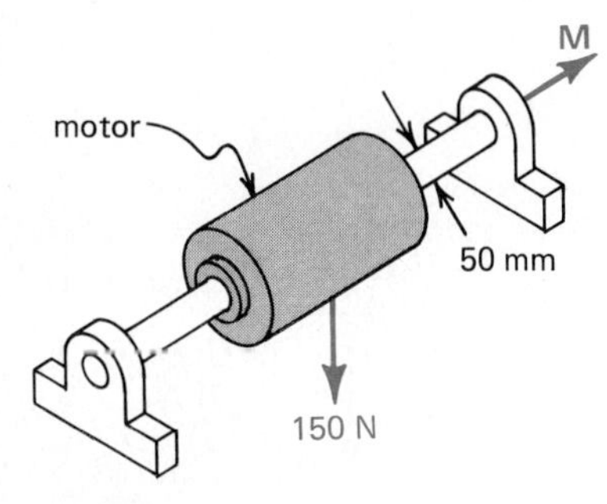

Figure P8.54

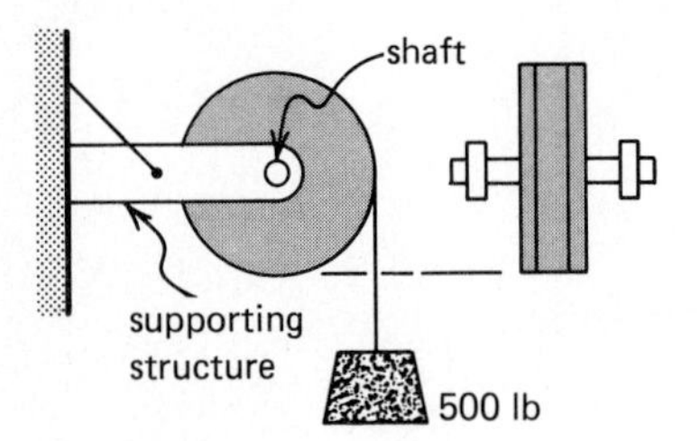

Figure P8.57

8.54 The shaft of a motor is fitted in a bearing as shown in Fig. P8.54. If the motor shaft exerts a moment of 1.5 m·N to start rotation, determine μ_0 between the shaft and bearing at each of the two supports. Assume that the motor and shaft weighs 150 N.

8.55 The shaft in Fig. 8.20 is replaced by a lighter material so that $\rho = 3.2$ t/m^3. Calculate the required torque to maintain uniform motion of the new shaft if $\mu = 0.12$.

8.56 Calculate the coefficient of friction between a shaft and journal bearing if the turning moment is 25 in·lb and $P = 100$ lb. Let the shaft radius be 2 in.

8.57 The drum of 1-ft radius shown in Fig. P8.57 is used to hoist a 500-lb block. The diameter of the shaft equals 3 in. The weight of the drum plus the shaft is 100 lb and $\mu = 0.3$. Determine the magnitude of the moment that must be applied to the drum shaft in order to raise the weight at uniform speed.

8.58 In Problem 8.57, if $\mu = 0.25$, what is the magnitude of the moment necessary to overcome friction at each of the two bearings?

8.59 Verify Eq. (8.15) by performing the double integration on the previous equation.

8.60 Derive an expression for the moment M_o necessary to begin turning a sander, shown in Fig. P8.60, under an axial load of P. Assume that the pressure p between the sander and floor is constant and the coefficient of friction is μ_0.

8.61 In Problem 8.60, obtain the moment necessary to begin turning the sander if $P = 300$ lb, $\mu_0 = 0.3$, $r = 3$ in.

8.62 Determine the inner diameter of a hollow shaft supported by a thrust bearing if the axial force $P = 250$ N, the torque $M_k = 3.5$ m·N, $\mu = 0.18$, and $r_0 = 100$ mm.

8.63 The thrust P is supported by the flat collar bearing shown in Fig. P8.63. The thrust is 400 lb and $\mu = 0.1$. Determine the moment necessary to maintain a constant speed of rotation.

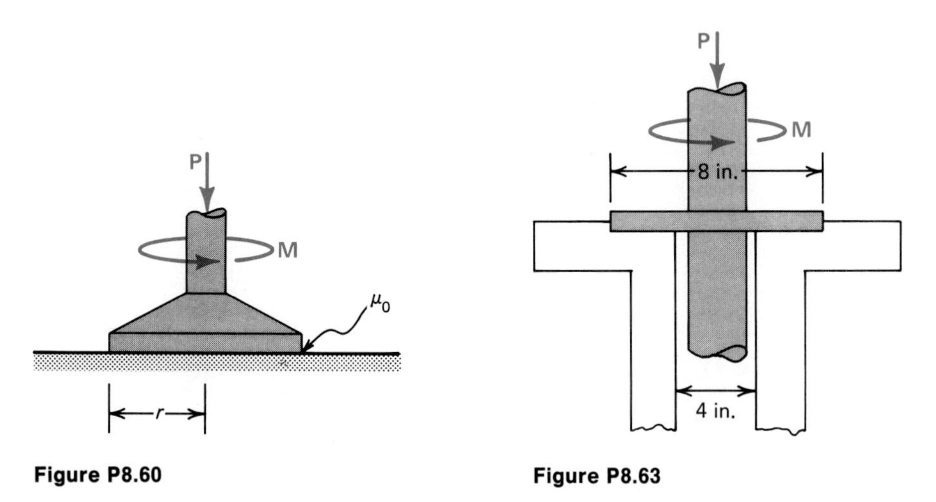

Figure P8.60 **Figure P8.63**

8.6 Summary

When one body moves relative to another with which it is in contact, a tangent frictional force tends to impede motion. Since this frictional force affects the conditions of equilibrium, we must include this force in writing the equations of equilibrium. The particular type of frictional force that concerned us in this chapter is called Coulomb friction. We examined the application of this friction to wedges, screws, and belts.

A. The important *ideas*:

1. Coulomb friction force is proportional to the normal force. The proportionality constant is the coefficient of friction.
2. The coefficient of static friction μ_0 is greater than the coefficient of kinetic friction μ.
3. For impending motion the friction force is equal to $\mu_0 N$.
4. The angle of repose is the maximum angle to which a plane can be tilted before a block on the plane will begin to slide.

B. The important *equations*:

$$f_m = \mu_0 N \tag{8.1}$$

$$R = N\sqrt{1 + \mu_0^2} \tag{8.3a}$$

$$\tan \phi_0 = \mu_0 \tag{8.3b}$$

$$M_U = wr \tan (\phi_0 + \alpha) \tag{8.8}$$

$$M_L = wr \tan (\phi_0 - \alpha) \tag{8.9}$$

$$\frac{T_2}{T_1} = e^{\mu_0 \alpha} \tag{8.12}$$

$$M_k = Pr \sin \phi \tag{8.13}$$

$$M_o = Pr \sin \phi_0 \tag{8.14}$$

$$M_k = \frac{2}{3} \mu P \frac{r_o^3 - r_i^3}{r_o^2 - r_i^2} \tag{8.15}$$

9 Virtual Work and Energy

To this point we have studied static equilibrium by means of the basic equations of statics, from which we have computed the forces and moments required to maintain the equilibrium of a body. The principle of virtual work is another method for studying static equilibrium. Its major advantage is found in treating complicated problems. As an introduction to this principle, we shall define what we mean by the concepts of work, virtual displacements, and virtual work and then use these concepts to derive the equations of static equilibrium.

Once the equilibrium of a body is established, it is often necessary to know whether the body is in a stable or unstable equilibrium position. We shall treat this special topic by considering the behavior of the potential energy of the body in the neighborhood of its static equilibrium positions.

The topics covered in this chapter are very useful when applied to problems in mechanics. However, it is worth noting that they are especially important since they form the foundation for the development of more advanced techniques found in dynamics.

9.1 The Concept of Work

Consider the block in Fig. 9.1 that moves through a displacement $\mathbf{r} = r\mathbf{i}$ due to a constant force $\mathbf{F} = F\mathbf{i}$. The **work** done by the force $\mathbf{F}$ on the block is

$$\text{Work} = W = \mathbf{F} \cdot \mathbf{r}$$
$$= F\mathbf{i} \cdot r\mathbf{i} = Fr \tag{9.1}$$

We see that the work is a scalar quantity and the usual units are N·m or lb·ft.

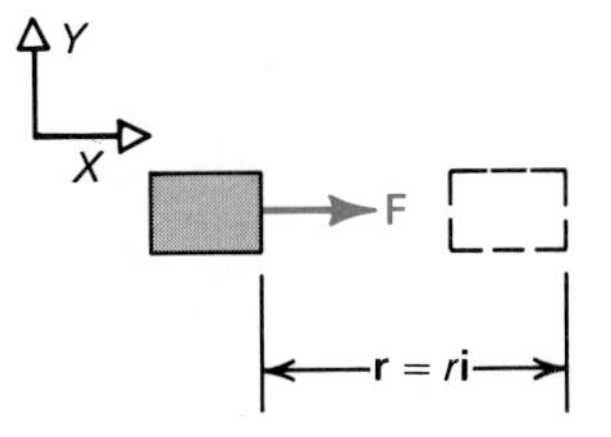

Figure 9.1

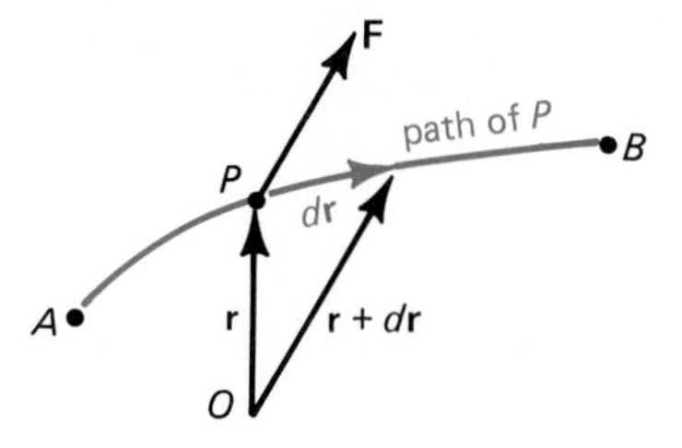

Figure 9.2

Let us now consider a particle P moving along some path under the action of a force $\mathbf{F}$ as shown in Fig. 9.2. The figure shows the particle P at some instant along its path and its differential displacement $d\mathbf{r}$. The work done by $\mathbf{F}$ in this differential displacement is

$$dW = \mathbf{F} \cdot d\mathbf{r}$$

This differential element of work is, therefore, the projection of $\mathbf{F}$ on the differential displacement $d\mathbf{r}$. Consequently, the total work done by the force $\mathbf{F}$ as P moves along the path from position A to position B is:

$$W = \int_A^B \mathbf{F} \cdot d\mathbf{r} \tag{9.2}$$

Equation (9.2) is an example of a line integral and is the general form of the expression for the work done by a force $\mathbf{F}$. We see that it reduces to Eq. (9.1) when the force is of constant magnitude and acts in the same direction as the displacement of the particle.

If there is more than one force acting on the particle, the total work is the summation of the work done by each force. If there are n forces acting on the particle,

$$\mathbf{F} = \mathbf{F}_1 + \mathbf{F}_2 + \cdots + \mathbf{F}_n$$

then the total work is given by

$$W = \int_A^B \mathbf{F} \cdot d\mathbf{r} = \int_A^B (\mathbf{F}_1 + \mathbf{F}_2 + \cdots + \mathbf{F}_n) \cdot d\mathbf{r} \tag{9.3}$$

The weight of a body, friction, and externally applied forces are examples of forces that do work when a body moves from one position to another. On the other hand, there are forces that, because of physical constraints, do no work. These forces are associated with what we call "workless constraints." Two examples of workless constraints are shown in Fig. 9.3. In Fig. 9.3(a), the normal force, which is the reaction force from the surface on the block, is always perpendicular to the path of the body, so that this force does no work. In the case of the rigid rod AB with the internal force $\mathbf{F}_{AB}$ acting at each joint, as in Fig. 9.3(b), it can be shown that since the rod does not deform, the net work done by these forces is zero even though the rod itself may move through a displacement.

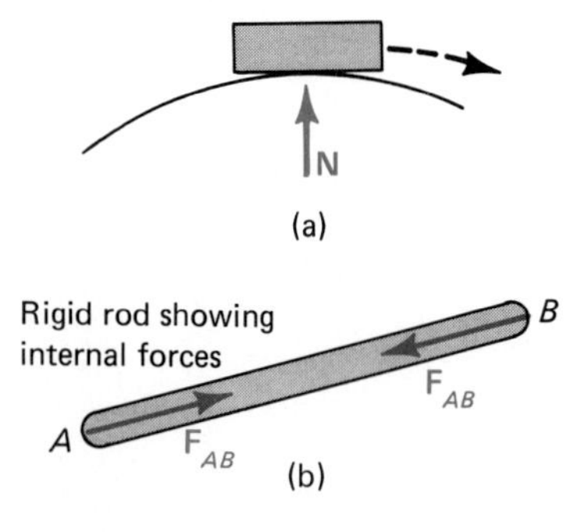

Figure 9.3

When a moment $\mathbf{M}$ is applied to a body that rotates through an infinitesimal angle $d\boldsymbol{\theta}$, then the differential work equals

$$dW = \mathbf{M} \cdot d\boldsymbol{\theta}$$

The total work done by the moment as the body rotates from position A to position B is

$$W = \int_A^B \mathbf{M} \cdot d\boldsymbol{\theta} \tag{9.4}$$

This equation is again an example of a line integral and is the general form of the expression for the work done by a moment **M**. Note that the units remain the same since the moment is in m·N or ft·lb while the angular displacement is in radians.

In addition to the work done by forces and moments, we also define the work done by couples. For example, consider the disk in Fig. 9.4 that rotates from position A to position B under the action of two equal and opposite forces of magnitude F. During this motion, each force travels along a circular arc through a distance $(a\theta)$. If we consider that these forces remain tangent to the arc, the work done by these forces is

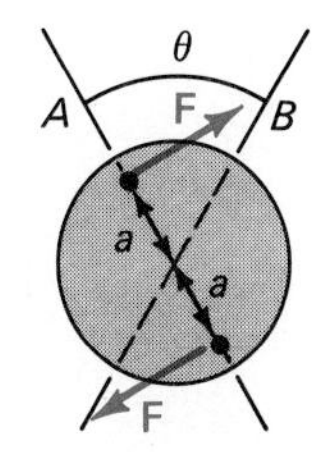

Figure 9.4

$$W = F(a\theta) + F(a\theta) = F(2a)\theta$$
$$= M\theta \tag{9.5}$$

where $M = F(2a)$ = magnitude of the moment of the couple formed by the forces. The work done by the couple is the product of the moment of the couple and the angle through which the body rotates.

As in the case of forces, the expression for the work done by a series of couples can be written as vectors in integral form, that is,

$$W = \int_A^B \mathbf{M} \cdot d\boldsymbol{\theta} \tag{9.6}$$

where **M** is the sum of the moments of all applied couples acting on the body as it changes from position A to position B. This expression is identical to Eq. (9.4).

Example 9.1

A block is sliding down an inclined plane as shown in Fig. 9.5(a). The coefficient of kinetic friction between the block and the plane is 0.20. Find the work done as the block slides a distance of 10 ft. The block weighs 50 lb.

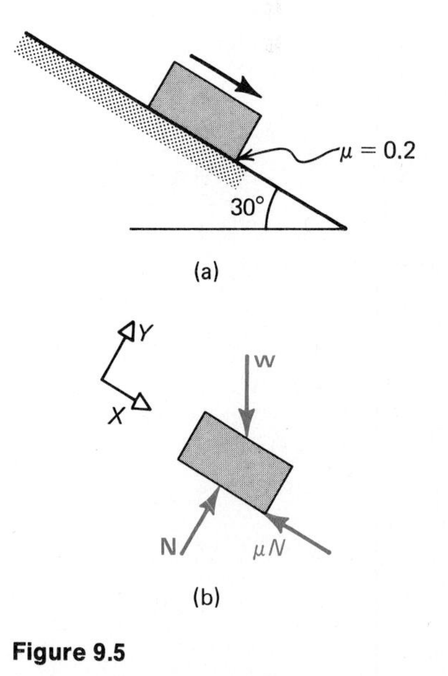

Figure 9.5

The free-body diagram of the block is shown in Fig. 9.5(b). The total force acting on the block is written as

$$\mathbf{F} = w(\sin 30°\mathbf{i} - \cos 30°\mathbf{j}) + N\mathbf{j} - \mu N\mathbf{i}$$

The differential displacement that the block may experience is

$$d\mathbf{r} = dx\mathbf{i}$$

Referring to Eq. (9.2)

$$W = \int_A^B \mathbf{F} \cdot d\mathbf{r}$$

$$W = \int_0^{10} [w(\sin 30°\mathbf{i} - \cos 30°\mathbf{j}) + N\mathbf{j} - \mu N\mathbf{i}] \cdot (dx\mathbf{i})$$

$$= \int_0^{10} (w \sin 30° - \mu N)\, dx$$

Note that the force $N\mathbf{j}$, which is a workless constraint, does not contribute to the work. Since the block does not leave the inclined plane,

$$\sum F_y = 0 \qquad N - w\cos 30^\circ = 0$$

so that

$$N = w\cos 30^\circ$$

Therefore

$$\begin{aligned} W &= \int_0^{10} (w\sin 30^\circ - \mu N)\,dx \\ &= \int_0^{10} (w\sin 30^\circ - \mu w\cos 30^\circ)\,dx \\ &= \int_0^{10} [50(0.5) - 0.2(50)(0.866)]\,dx \\ &= \int_0^{10} (16.34)\,dx = 163.4\ \text{lb}\cdot\text{ft} \end{aligned}$$

Answer

Example 9.2

A brake is applied to a spinning drum as shown in Fig. 9.6(a). The kinetic coefficient of friction is 0.2. If the drum stops spinning in 10 revolutions, what work is done on the drum?

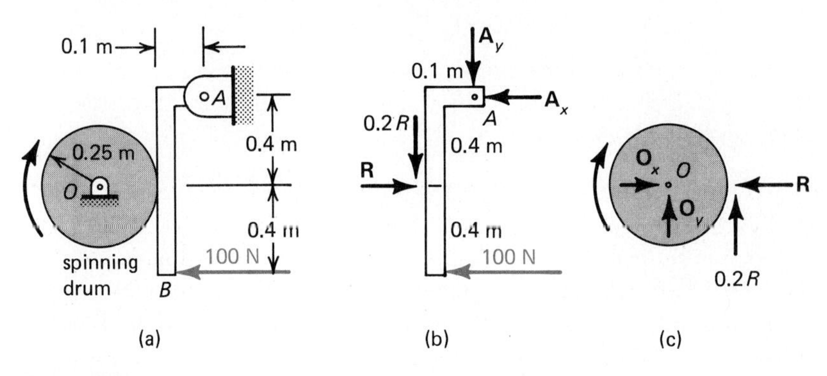

Figure 9.6

The free-body diagram of the brake is shown in Fig. 9.6(b) while the free-body diagram of the drum is shown in Fig. 9.6(c). Note the direction of the friction force $0.2R$ is opposing the direction of the rotating drum. This force acts in the opposite direction when placed on the free-body diagram of the brake. Referring to Fig. 9.6(b) we have the following equilibrium condition:

$$\sum M_A = 0 \qquad 0.1(0.2)R + 0.4R - 100(0.8) = 0$$

$$R = 190.5\ \text{N}$$

Now consider the spinning drum in Fig. 9.6(c) to calculate the moment acting on the drum.

$$M = (0.2R)0.25 = 0.2(190.5)(0.25) = 9.525$$

The work done is

$$W = -M\theta = -(9.525)(10)(2\pi)$$

$$W = -598.5 \text{ N}\cdot\text{m} \qquad \textit{Answer}$$

Note that the negative sign is included since the moment acts CCW while the motion of the drum is CW.

PROBLEMS

9.1 A block is moved on a trajectory defined by $\mathbf{r} = (5x\mathbf{i} + 6\mathbf{j})$ ft under a force $\mathbf{F} = (18\mathbf{i} - 10x\mathbf{j})$ lb. How much work is done on the block if x goes from 1 to 4 ft?

9.2 The block shown in Fig. P9.2 moves up the inclined plane under the action of a 30-lb force. If the kinetic coefficient of friction between the block and the plane is $\mu = 0.1$, find the work done on the block in moving it 5 ft.

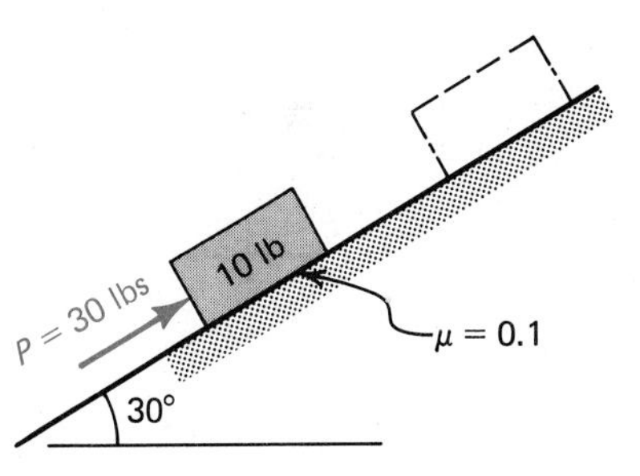

Figure P9.2

9.3 A particle of mass m kg is moved from A to B over a smooth curved surface as shown in Fig. P9.3 by the applied force $\mathbf{P} = C(\sin\theta\mathbf{i} + \cos\theta\mathbf{j})$ N where C is a constant. What is the work done on the particle for this change in the vertical position?

9.4 Repeat Problem 9.3 to include a coefficient of friction μ between the particle and surface.

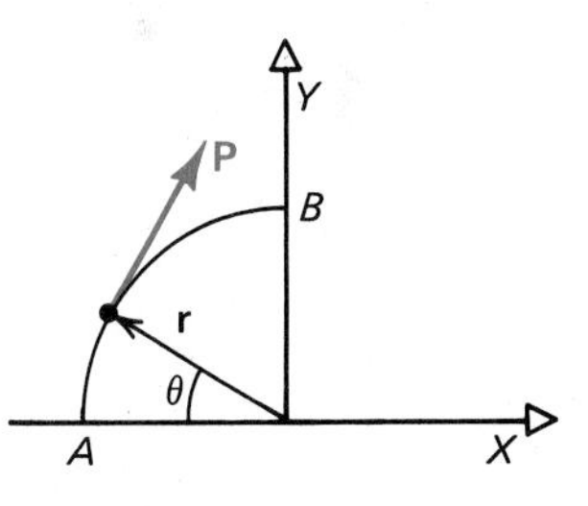

Figure P9.3

9.5 A particle travels along the curve $y = x^2$ from $x = 1$ ft to $x = 3$ ft. If $\mathbf{F} = (2\mathbf{i} - 4\mathbf{j})$ lb, find the work done on the particle.

9.6 Repeat Problem 9.5 for $\mathbf{F} = (10x\mathbf{i} + 4y\mathbf{j})$ lb.

9.7 A particle is to move from position A to position B over two possible paths, ACB or ADB in Fig. P9.7. If $\mathbf{F} = (20\mathbf{i} + 20\mathbf{j})$ N, find the work over each path.

9.8 If the particle in Problem 9.7 travels around the path $ACBDA$, find the total work done on the particle.

9.9 A body is subjected to a constant moment of 180 ft·lb for 50 revolutions. How much work was done on the body?

9.10 A moment of $6t\mathbf{k}$ ft·lb is applied to a wheel whose displacement is given by $\theta = 2t^2$ rad. If the moment is applied from $t = 1$ to $t = 4$ s, how much work is done on the wheel?

9.11 A satellite is accelerated over 100 revolutions by applying a constant moment of 50 m·N and then decelerated over 25 revolutions by applying an opposite moment of 150 m·N. What is (a) the net work and (b) total work done on the satellite?

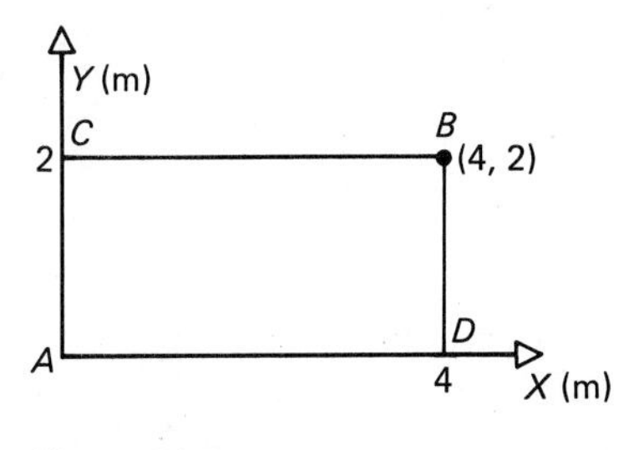

Figure P9.7

9.12 The cylinder in Fig. P9.12 turns through 90° under the action of the applied forces of magnitude $P = 6$ N. What work is done on the cylinder for this change in position if the forces remain perpendicular to the rods?

9.13 What is the total work done by the force **F** and the couple **M** shown in Fig. P9.13 as the body moves from position A to position B while it rotates 2 revolutions clockwise?

9.14 A brake AB is applied to the spinning drum shown in Fig. P9.14. If the kinetic coefficient of friction is 0.1 between the brake and the drum, what work is done on the drum for 20 revolutions?

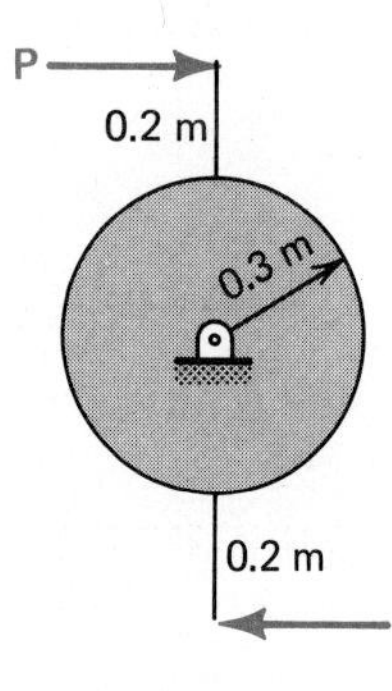

Figure P9.12

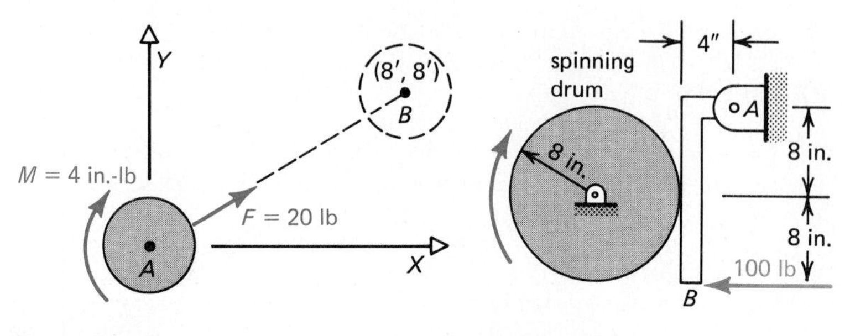

Figure P9.13

Figure P9.14

9.2 The Concept of Virtual Work

If a particle moves through a displacement $d\mathbf{r}$ under the action of a force **F** as described in the last section, the particle is not in a state of equilibrium. However, in studying the equilibrium of bodies, we sometimes find it useful to introduce fictitious displacements that are called **virtual displacements**. Virtual displacements are small displacements that occur without the passage of time and do not violate the constraints on the body. A virtual displacement is represented as $\delta\mathbf{r}$ rather than $d\mathbf{r}$. Consequently, the **virtual work** done by the net force **F** acting on a particle that experiences the virtual displacement $\delta\mathbf{r}$ is

$$\delta W = \mathbf{F} \cdot \delta\mathbf{r} \tag{9.7}$$

Likewise, the virtual work for a rigid body that may translate an amount $\delta\mathbf{r}$ and rotate an amount $\delta\boldsymbol{\theta}$ is

$$\delta W = \mathbf{F} \cdot \delta\mathbf{r} + \mathbf{M} \cdot \delta\boldsymbol{\theta} \tag{9.8}$$

The principle of virtual work states that if a body is in equilibrium, the virtual work done by all the applied forces and moments is zero.

Thus, for the case of a particle in equilibrium,

$$\delta W = \mathbf{F} \cdot \delta\mathbf{r} = 0 \tag{9.9}$$

and for a rigid body in equilibrium,

$$\delta W = \mathbf{F} \cdot \delta\mathbf{r} + \mathbf{M} \cdot \delta\boldsymbol{\theta} = 0 \tag{9.10}$$

We can summarize the principle of virtual work as follows:

1. A particle is in equilibrium if the virtual work of all applied forces is zero for all virtual displacements.
2. A rigid body is in equilibrium if the virtual work due to all forces and moments is zero for all virtual displacements.

We can demonstrate the principle of virtual work by finding the magnitude of the force $\mathbf{B}_y$ required to keep the rigid rod AB in Fig. 9.7(a) in a state of equilibrium. Since the bar is pinned at A, we allow it to rotate an amount $\delta\theta$ as shown in Fig. 9.7(b). Keep in mind that $\delta\theta$ is very small. The virtual work is therefore,

$$\delta W = -P_1(L_1\,\delta\theta) - P_2(L_2\,\delta\theta) + B_y(L\,\delta\theta)$$

For equilibrium, $\delta W = 0$ so that

$$(-P_1L_1 - P_2L_2 + B_yL)\,\delta\theta = 0$$

Since $\delta\theta$ is very small, but not zero,

$$-P_1L_1 - P_2L_2 + B_yL = 0$$

$$B_y = \frac{P_1L_1 + P_2L_2}{L}$$

It should be apparent that the above result could have been obtained using the equilibrium equation, $\sum M_A = 0$. However, as we shall discover in the next section, the principle of virtual work is a useful tool in studying the equilibrium of more complex systems.

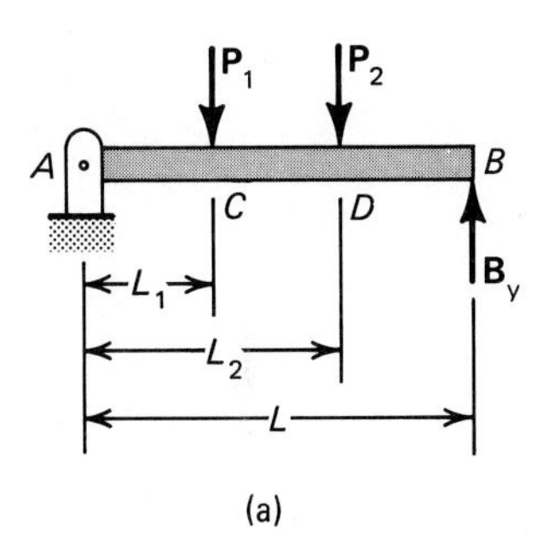

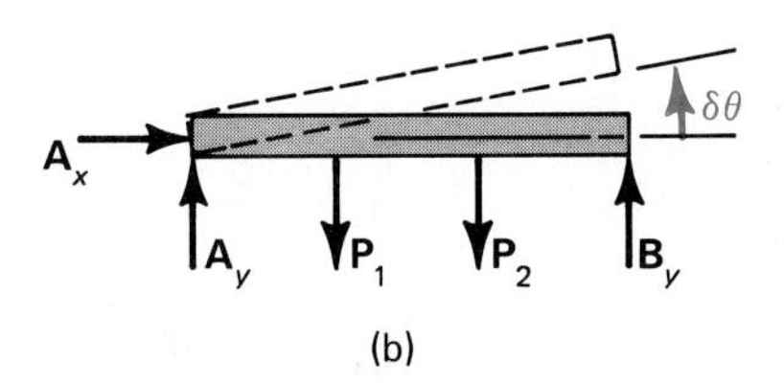

Figure 9.7

Example 9.3

Use the principle of virtual work to find the force **F** required to keep the rigid bar in Fig. 9.8(a) in static equilibrium if the magnitude of $M = 400$ ft·lb and (a) $P = 200$ lb, (b) $P = 100$ lb, and (c) $P = 50$ lb.

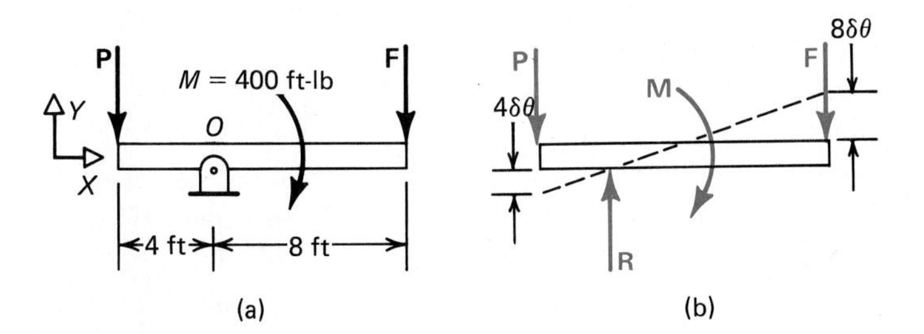

Figure 9.8

Allow the rod, which is pivoted at 0, to rotate by a small amount $\delta\theta$ as shown in Fig. 9.8(b). By the principle of virtual work,

$$\delta W = P(4\,\delta\theta) - M\,\delta\theta - F(8\,\delta\theta) = (4P - M - 8F)\,\delta\theta = 0$$

so that
$$F = \frac{4P - M}{8} = \frac{4P - 400}{8}$$

(a) If P = 200 lb	$F = 50$ lb	$\mathbf{F} = -50\mathbf{j}$ lb	*Answer*
(b) If $P = 100$ lb	$F = 0$		*Answer*
(c) If $P = 50$ lb	$F = -25$ lb	$\mathbf{F} = 25\mathbf{j}$ lb	*Answer*

Example 9.4

Use the principle of virtual work to find the magnitude of the force $\mathbf{B}_y$ to keep the frame subjected to the force **P** shown in Fig. 9.9(a) in static equilibrium.

Figure 9.9(b) shows the frame rotated through the virtual displacement $\delta\theta$. In this case, the position vector from the origin to the force **P** is

$$\mathbf{r}_c = r_c(\cos\theta\mathbf{i} + \sin\theta\mathbf{j})$$

Therefore
$$\delta\mathbf{r}_c = r_c(-\sin\theta\mathbf{i} + \cos\theta\mathbf{j})\,\delta\theta$$

For $\theta = 45^\circ$ and $r_c = 21.2$ m,

$$\delta\mathbf{r}_c = 15(-\mathbf{i} + \mathbf{j})\,\delta\theta$$

Note that we differentiate in the usual way except that we use the symbol δ instead of the familiar d since we are interested in virtual displacements.

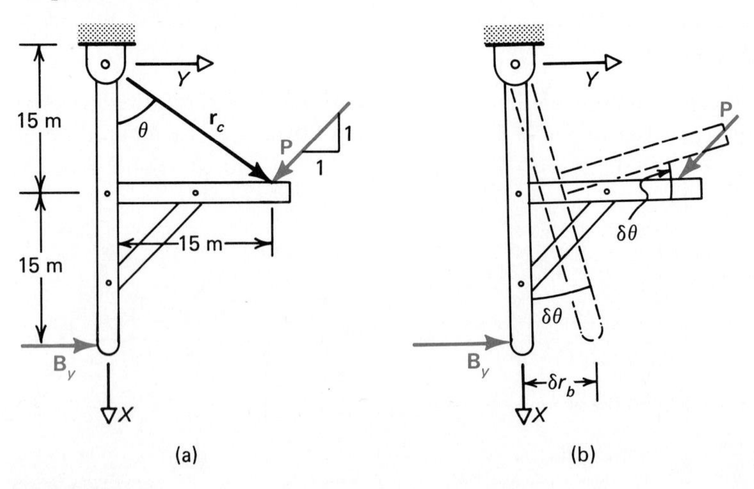

Figure 9.9

From Fig. 9.9(b) we see that $\delta \mathbf{r}_B = 30\,\delta\theta\mathbf{j}$. For equilibrium,

$$\delta W = (B_y\mathbf{j}) \cdot (30\,\delta\theta\mathbf{j}) + \mathbf{P} \cdot \delta\mathbf{r}_c = 0$$

$$30B_y\,\delta\theta + (0.707P\mathbf{i} - 0.707P\mathbf{j}) \cdot (-\mathbf{i} + \mathbf{j})15\,\delta\theta = 0$$

$$(30B_y - 21.2P)\,\delta\theta = 0$$

Since $\delta\theta$ does not equal zero,

$$B_y = 0.707P \qquad \textit{Answer}$$

PROBLEMS

9.15 Using the principle of virtual work, find an expression for P necessary to keep the block from sliding down an inclined plane shown in Fig. P9.15. Assume $\mu_0 = 0.2$ for the coefficient of friction, $\theta = 30^\circ$ and $m = 10$ kg.

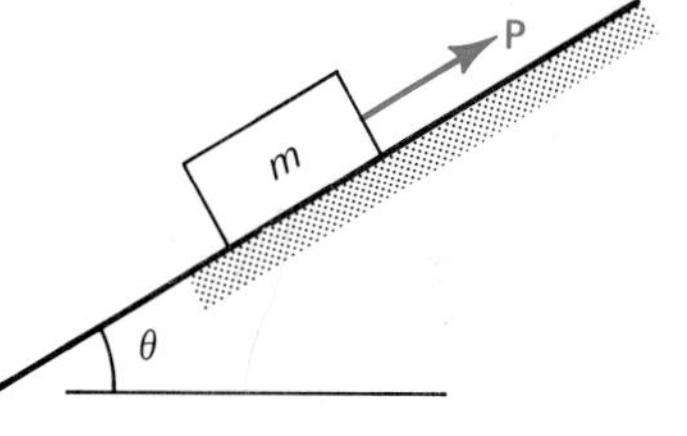

Figure P9.15

9.16 Using the principle of virtual work find the value of P for static librium of the beam shown in Fig. P9.16.

9.17 Find the virtual work done by P and the distributed load on the beam, and find P necessary for static equilibrium of the beam shown in Fig. P9.17.

9.18 Figure P9.18 shows a rod of length L and weight w at the position θ. What is the virtual work done on the rod by w for $\delta\theta$?

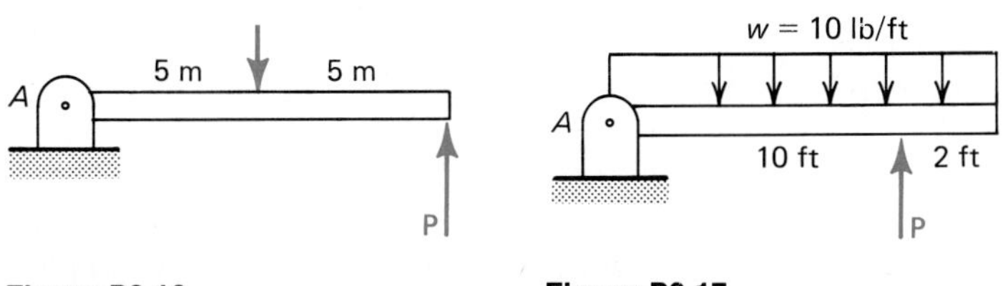

Figure P9.16

Figure P9.17

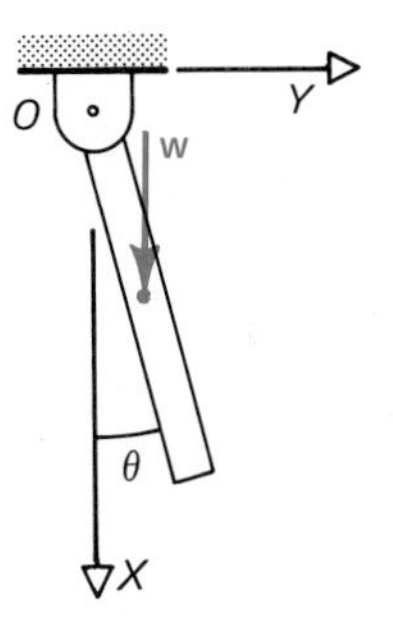

Figure P9.18

9.19 Repeat Problem 9.18 with an added resisting couple M acting clockwise on the rod.

9.20 Two rods are connected to form a double pendulum as shown in Fig. P9.20. For the forces shown, what is the virtual work (a) if θ_1 varies by $\delta\theta_1$ holding θ_2 fixed, and (b) if θ_2 varies by $\delta\theta_2$ holding θ_1 fixed?

9.21 What is the virtual work for Problem 9.20 if θ_1 and θ_2 are both allowed to vary simultaneously?

9.22 The rod of length L shown in Fig. P9.22 has a force $\mathbf{F}$ and couple $\mathbf{M}$ applied to it as shown. What is the virtual work due to $\mathbf{F}$ and $\mathbf{M}$?

9.23 Obtain the magnitude of the applied couple necessary for the static equilibrium of the beam shown in Fig. P9.23 by means of virtual work.

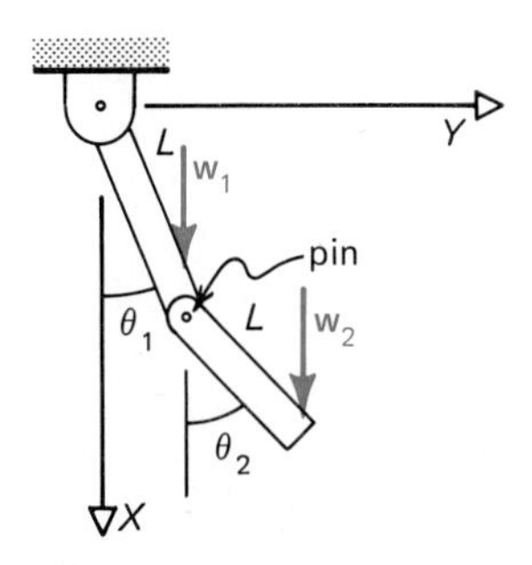

Figure P9.20

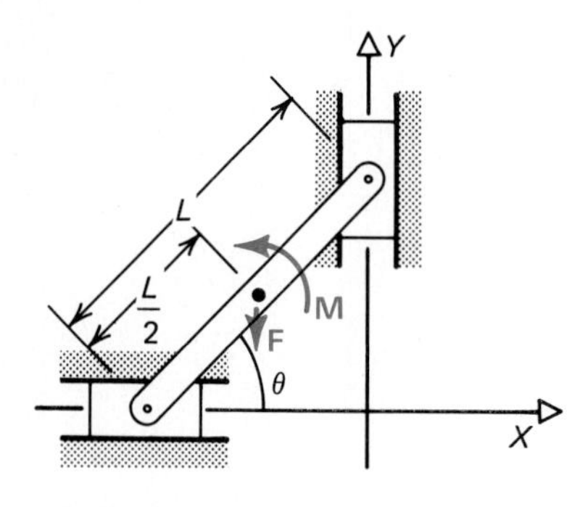

Figure P9.22

9.24 Use the principle of virtual work to find the magnitude of the unknown force **P** to maintain static equilibrium for the system shown in Fig. P9.24.

9.25 Obtain the magnitude of the force **P** to maintain static equilibrium for the system in Problem 9.24 for $a = 1$ ft, $b = 0.5$ ft, and $\theta = 45°$.

9.26 Use the principle of virtual work to find the magnitude of the moment of the unknown couple **M** to maintain static equilibrium for the system shown in Fig. P9.26.

9.27 Obtain the magnitude of the moment of the couple **M** necessary to maintain static equilibrium for the system in Problem 9.26 for $l = 1$ ft and $\theta = 30°$.

Figure P9.23

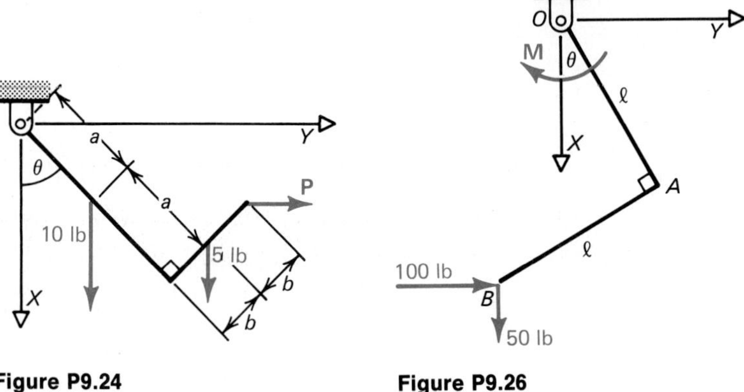

Figure P9.24

Figure P9.26

9.3 Connected Systems

In applying the principle of virtual work to bodies that are connected to each other, we shall treat those cases where we have ideal connectors, that is, where there is no friction at the joints. By way of example, consider the slider-crank mechanism in Fig. 9.10(a) that has a force **P** acting on the piston at C. Assuming that no friction is present, let us use the principle of virtual work to find the magnitude of the torque **M** required to maintain the system in equilibrium.

Our first step in dealing with connected systems is to establish the number of ways the system is free to move, that is, its number of degrees of freedom. From the figure we observe that there is only one degree of freedom that can be described by the coordinate x_c or the angle θ. It is important then to establish the relationship between these two coordinates since only one degree of freedom is present. Such a relationship is called an equation of constraint. In this example, we see from Fig. 9.10(a) that

$$x_c = r \sin \theta + l \cos \beta \tag{9.11}$$

where $$l \sin \beta = r \cos \theta$$

From Fig. 9.10(b) we observe that

$$\cos \beta = \frac{\sqrt{l^2 - r^2 \cos^2 \theta}}{l} \tag{9.12}$$

Substituting Eq. (9.12) into (9.11), we obtain the desired relationship between x_c and θ:

$$x_c = r \sin \theta + \sqrt{l^2 - r^2 \cos^2 \theta} \tag{9.13}$$

so that

$$\delta x_c = r \cos \theta \, \delta\theta + \frac{1}{2} [l^2 - r^2 \cos^2 \theta]^{-1/2}$$

$$\times [-2r^2 \cos \theta (-\sin \theta) \, \delta\theta]$$

$$= \left(r \cos \theta + \frac{r^2 \sin \theta \cos \theta}{\sqrt{l^2 - r^2 \cos^2 \theta}} \right) \delta\theta \tag{9.14}$$

To investigate the equilibrium requirement, we see in Fig. 9.10(c) that the mechanism is given a virtual displacement that we can view as a result of $\delta\theta$ or δx_c. The virtual work is

$$\delta W = M \, \delta\theta - P \, \delta x_c = 0$$

$$M \, \delta\theta = P \, \delta x_c \tag{9.15}$$

Using Eq. (9.14), we obtain

$$M = P \left(r \cos \theta + \frac{r^2 \sin \theta \cos \theta}{\sqrt{l^2 - r^2 \cos^2 \theta}} \right) \tag{9.16}$$

Note that when $\theta = 90^\circ$, $M = 0$ and the system is locked so that no moment is required for equilibrium. If $\theta = 0$, $M = Pr$ for equilibrium which should be obvious after sketching the free-body diagram of each member in this configuration.

The above example demonstrates the value of the principle of virtual work when we study involved mechanisms. Note that it is not necessary to draw a free-body diagram of each member of the system if each member is joined by frictionless pin connections. This is one of the advantages of the method.

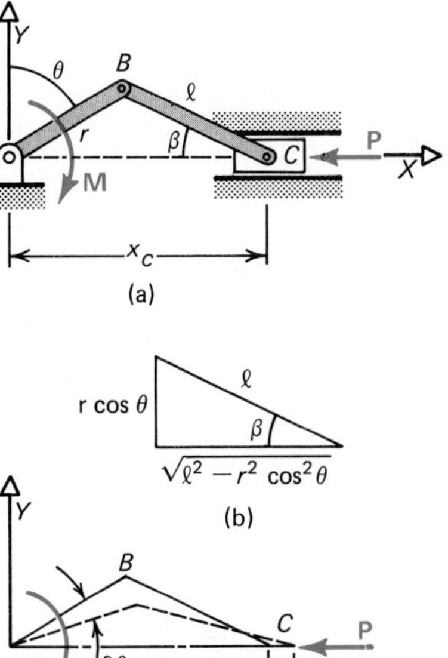

Figure 9.10

Example 9.5

For the two gear circles shown in Fig. 9.11 obtain the restraining moment M_A for static equilibrium if $M_B = 50$ m·N.

If gear A has a virtual displacement of $\delta\theta_A$ and gear B has a virtual displacement of $\delta\theta_B$, then the equation of constraint between

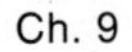

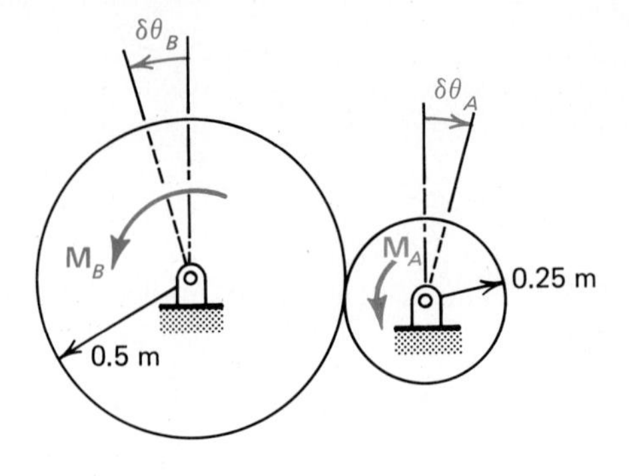

Figure 9.11

these coordinates is

$$r_A \delta\theta_A = r_B \delta\theta_B$$

where r_A and r_B are the radii of gear A and gear B, respectively.
The virtual work done on the system is

$$\delta W = -M_A \delta\theta_A + M_B \delta\theta_B = 0$$

$$\left(M_A - M_B \frac{r_A}{r_B}\right) \delta\theta_A = 0$$

$$M_A = \frac{r_A}{r_B} M_B = \frac{0.25}{0.6}(50)$$

$$= 20.8 \text{ m} \cdot \text{N} \qquad \textit{Answer}$$

Example 9.6

Find the value of P in terms of θ necessary to keep the mechanism in Fig. 9.12(a) in equilibrium.

Since the mechanism has one degree of freedom described by the angle θ, Fig. 9.12(b) shows the mechanism deformed due to the

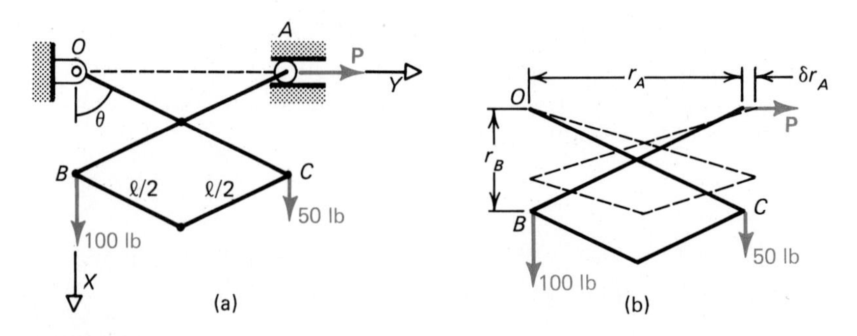

Figure 9.12

virtual displacement $\delta\theta$. The position vectors measured from the origin to each applied force are as follows:

$$\mathbf{r}_A = l \sin \theta \mathbf{j} \qquad \delta\mathbf{r}_A = l \cos \theta \, \delta\theta \mathbf{j}$$

$$\mathbf{r}_B = l \cos \theta \mathbf{i} \qquad \delta\mathbf{r}_B = -l \sin \theta \, \delta\theta \mathbf{i}$$

$$\mathbf{r}_C = l \cos \theta \mathbf{i} + l \sin \theta \mathbf{j} \qquad \delta\mathbf{r}_C = -l \sin \theta \, \delta\theta \mathbf{i} + l \cos \theta \, \delta\theta \mathbf{j}$$

The virtual work is

$$\delta W = P\mathbf{j} \cdot \delta\mathbf{r}_A + 100\mathbf{i} \cdot \delta\mathbf{r}_B + 50\mathbf{i} \cdot \delta\mathbf{r}_C = 0$$

$$P\mathbf{j} \cdot (l \cos \theta \, \delta\theta \mathbf{j}) + 100\mathbf{i} \cdot (-l \sin \theta \, \delta\theta \mathbf{i})$$
$$+ 50\mathbf{i} \cdot (-l \sin \theta \, \delta\theta \mathbf{i} + l \cos \theta \, \delta\theta \mathbf{j}) = 0$$

$$(Pl \cos \theta - 100l \sin \theta - 50l \sin \theta) \, \delta\theta = 0$$

Since $\delta\theta$ does not equal zero,

$$P = 150 \tan \theta \text{ lb} \qquad \textit{Answer}$$

PROBLEMS

9.28 Obtain a relationship between P and Q for equilibrium of the mechanism shown in Fig. P9.28.

9.29 What is the relationship between P and Q for equilibrium of the mechanism shown in Fig. P9.29?

9.30 What is the relationship between P and Q for the equilibrium of the mechanism shown in Fig. P9.30?

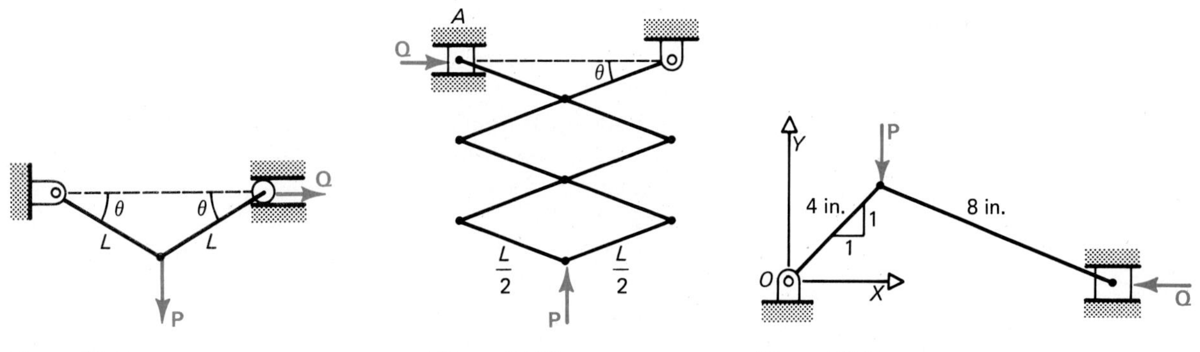

Figure P9.28 **Figure P9.29** **Figure P9.30**

9.31 For the mechanism in Problem 9.28, find P for static equilibrium if $\theta = 30°$ and $Q = 100$ lb.

9.32 Calculate the magnitude of P for the static equilibrium of the mechanism in Problem 9.29 if $\theta = 15°$ and $Q = 100$ lb.

9.33 Obtain the relationship between P and M for equilibrium of the system shown in Fig. P9.33.

9.34 Determine P for the equilibrium of the bar shown in Fig. P9.34.

9.35 Obtain the magnitude of the moment M for the static equilibrium of the system shown in Problem 9.33 if $P = 100$ N, $L = 0.5$ m, $r = 0.1$ m, and $\theta = 30°$.

9.36 Obtain the relationship between P and M for equilibrium of the system shown in Fig. P9.36.

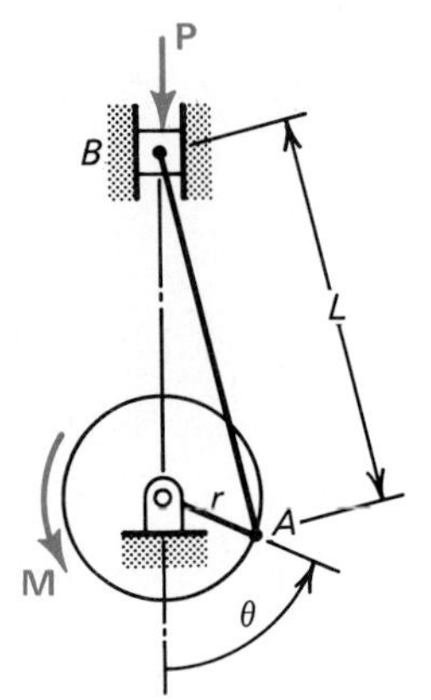

Figure P9.33

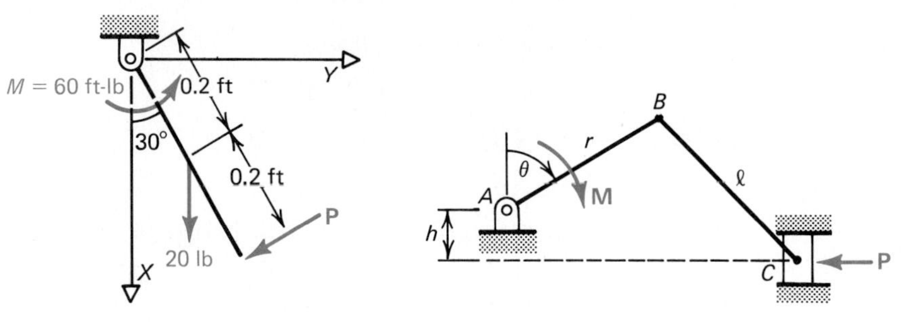

Figure P9.34

Figure P9.36

9.37 Find the magnitude of M for the static equilibrium of the system in Problem 9.36 if $r = 0.25$ m, $l = 0.6$ m, $h = 0.1$ m, $\theta = 30°$ and $P = 100$ N.

9.4 Potential Energy

On occasion, the elastic members with which bodies are connected to each other are idealized as springs. The spring modulus, k, is a measure of the resistance to deformation offered by a spring. The units of k are N/m or lb/ft for a linear spring such as that shown in Fig. 9.13. If a spring is compressed or stretched an amount x, where x is measured from the unstretched length of the spring, then the force that tends to restore the spring to its original shape is kx. For example, if $k = 1000$ N/m, then it takes 100 N to stretch the spring by 0.1 m. When a spring is deformed, the spring has energy stored in it. This energy is called **potential energy**. The amount of work required to deform a spring is equal to the change in the potential energy stored in the elastic member.

Consider, for example, a linear spring of unstretched length L, as shown in Fig. 9.13(a). Gradually apply a force $\mathbf{F}$ to the end of the spring so that it stretches by an amount x, as shown in Fig. 9.13(b). Figure 9.13(c) shows the free-body diagram of the spring; and, for static equilibrium,

$$\sum F_x = 0 \qquad F - kx = 0$$

Therefore, $$F = kx \tag{9.17}$$

The work done on the spring is

$$W = \int_A^B \mathbf{F} \cdot d\mathbf{r}$$

where from Eq. 9.17 and the free-body diagram shown in Fig. 9.13(c),

$$\mathbf{F} = kx\mathbf{i} \qquad d\mathbf{r} = dx\mathbf{i}$$

Therefore the work done in stretching the spring from x_A to x_B is

$$W = \int_{x_A}^{x_B} kx\mathbf{i} \cdot dx\mathbf{i}$$

$$= \frac{k(x_B)^2}{2} - \frac{k(x_A)^2}{2} \tag{9.18}$$

Note that the work is positive since $x_B > x_A$.

Next consider the spring connected to a block as shown in Fig. 9.13(d). We calculate the work done by the spring on the block as the block moves from x_A to x_B as shown in Fig. 9.13(d). From Eq. (9.2),

$$W = \int_A^B \mathbf{F} \cdot d\mathbf{r}$$

From the free-body diagram of the block shown in Fig. 9.13(e) and assuming a frictionless surface, we have,

$$\mathbf{F} = -kx\mathbf{i} \qquad d\mathbf{r} = dx\mathbf{i}$$

and therefore

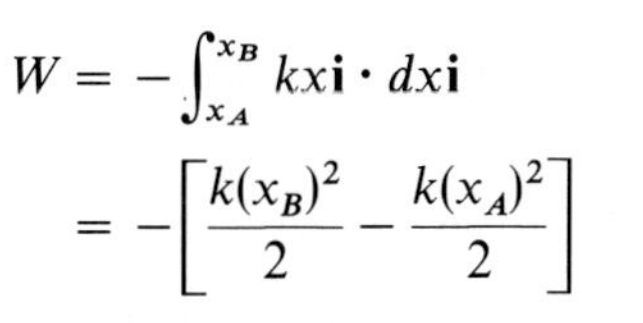

$$W = -\int_{x_A}^{x_B} kx\mathbf{i} \cdot dx\mathbf{i}$$

$$= -\left[\frac{k(x_B)^2}{2} - \frac{k(x_A)^2}{2}\right]$$

The term $kx^2/2$ is defined as the potential energy of a spring and is denoted by V. If we write

$$V_B = \tfrac{1}{2}kx_B^2 \qquad V_A = \tfrac{1}{2}kx_A^2$$

then the work done on the block by the spring is

$$W = -(V_B - V_A) \tag{9.19}$$

that is, the work done on the block by the spring is the negative of the potential energy change of the spring. The work done in stretching the spring is, however, positive and is equal to the potential energy stored in it. In general, we may say that the potential energy of a spring stretched or compressed by an amount x measured from its unstretched position is

$$V = \tfrac{1}{2}kx^2 \tag{9.20}$$

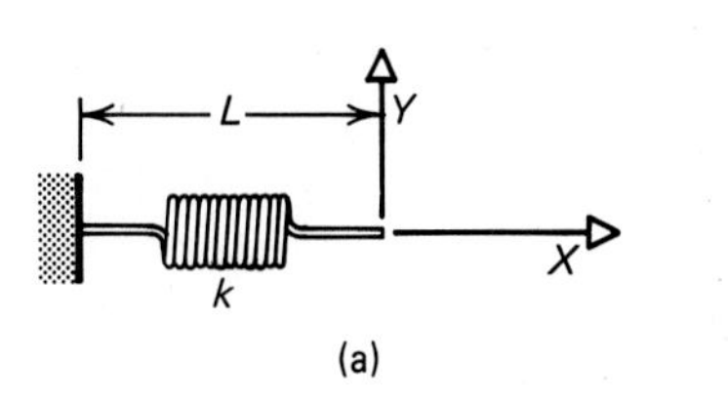

(a)

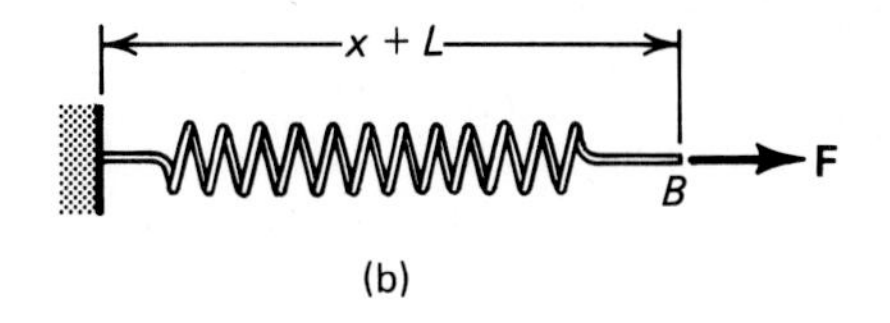

(b)

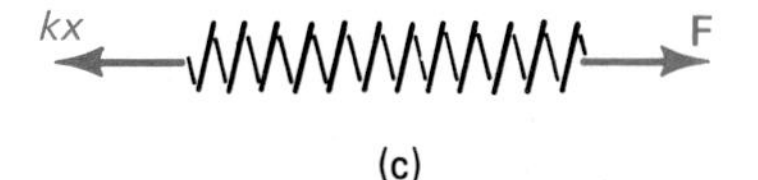

(c)

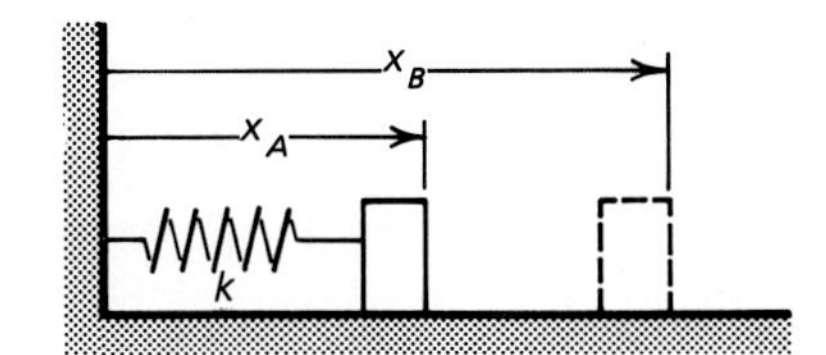

(d)

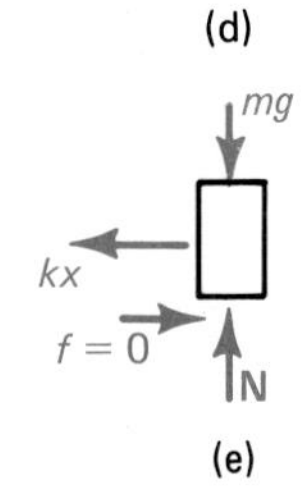

(e)

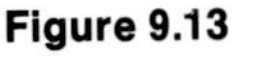

Figure 9.13

We can similarly write expressions for torsional springs that are used for resistance to angular deformation. The modulus k_t of a torsional spring has units equal to m·N/rad or ft·lb/rad. The work done in unwinding or winding a torsional spring is given by

$$W = \int_A^B M\,d\theta$$

where the moment M is given by

$$M = k_t\theta$$

so that

$$W = \int_A^B k_t\theta\,d\theta$$

The potential energy stored in the torsional spring is

$$V = \tfrac{1}{2}k_t\theta^2 \tag{9.21}$$

There is also a change in potential energy when a body changes its position relative to the surface of the earth. Consider, for example, a particle, shown in Fig. 9.14, having a mass of m. The work done on the particle by the weight is

Figure 9.14

$$W = \int_A^B \mathbf{F}\cdot d\mathbf{r}$$

where $\quad \mathbf{F} = -w\mathbf{j} \qquad d\mathbf{r} = dy\mathbf{j} \qquad w = mg$

so that,

$$W = -\int_{y_A}^{y_B} w\mathbf{j}\cdot dy\mathbf{j} = -(wy_B - wy_A) = -wh \tag{9.22}$$

Referring to Eq. (9.19), we see that

$$-(V_B - V_A) = -w(y_B - y_A) = -wh \tag{9.23}$$

By convention, we let $V_A = 0$, so that the potential energy of the weight at point B relative to point A is $V_B = wh$. Consequently, if a particle is a vertical distance h above a given reference point, its potential energy relative to the reference point is

$$V = mgh = wh \tag{9.24}$$

Note that if the body is a vertical distance below the reference point, $V = -wh$ relative to the reference point.

Example 9.7

What is the potential energy of a 2-kg particle in positions A, B, and C, respectively, relative to lines 1, 2, and 3 in Fig. 9.15?

$$w = mg = 2(9.81) = 19.62 \text{ N}$$

Line 1: at A, $\quad V = 0$

at B, $\quad V = 19.62(30) = 588.6$ N·m

at C, $\quad V = 19.62(10) = 196.2$ N·m

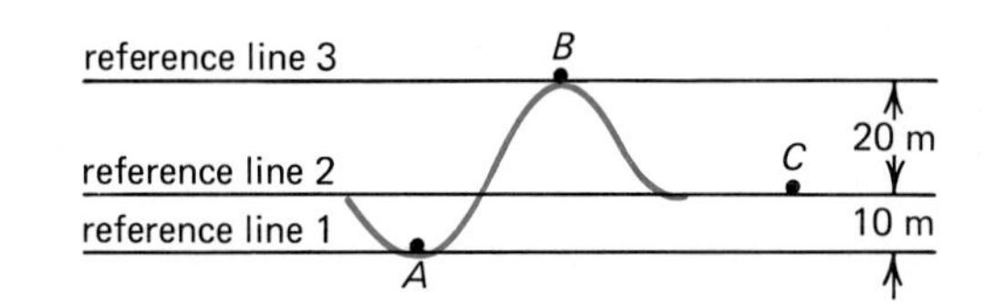

Figure 9.15

Line 2: at A, $V = 19.62(-10) = -196.2\ \text{N}\cdot\text{m}$

at B, $V = 19.62(20) = 392.4\ \text{N}\cdot\text{m}$

at C, $V = 0$

Line 3: at A, $V = 19.62(-30) = -588.6\ \text{N}\cdot\text{m}$

at B, $V = 0$

at C, $V = 19.62(-20) = -392.4\ \text{N}\cdot\text{m}$

Note that for each reference line, ΔV is the same between two points. For example, between points A and C,

$$\Delta V = V_C - V_A = 196.2\ \text{N}\cdot\text{m}$$

no matter which reference line is chosen. Since in mechanics we are always interested in *changes* in potential energy, we can choose the most convenient reference line.

Example 9.8

A 100-lb sphere strikes a spring as illustrated in Fig. 9.16. If the spring is compressed by 0.25 ft, how much potential energy is imparted to the spring by the sphere. After impact thc spring stretches to its original position sending the sphere up the frictionless chute. How far does the sphere rise? Assume $k = 7000$ lb/ft for the spring.

Figure 9.16

The potential energy stored in the spring when it is compressed 0.25 ft is

$$V = \tfrac{1}{2}kx^2 = \tfrac{1}{2}(7000)(0.25)^2$$

$$V = 218.8\ \text{lb}\cdot\text{ft} \qquad \textit{Answer}$$

Since there is no friction present, the potential energy is constant. This means that the compressed spring will impart its energy to the mass as it returns to its unstretched length. The particle travels up the chute such that

$$V = mgh = wh$$

$$h = \frac{218.8}{100}$$

$$= 2.19\ \text{ft} \qquad \textit{Answer}$$

Example 9.9

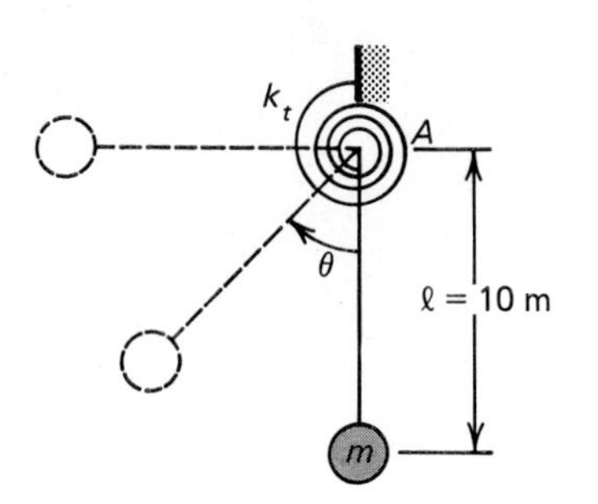

Figure 9.17

A simple pendulum has a 4-kg mass attached at its end and has a torsional spring with modulus $k_t = 1000$ m·N/rad as shown in Fig. 9.17. When $\theta = 0$, the torsional spring is undeformed. Neglecting the mass of the pendulum, find the potential energy of the system for (a) $\theta = 0$, (b) $\theta = 45°$, and (c) $\theta = 90°$.

Choose point A as the reference point for the potential energy so that the general expression for the potential energy of the mass is given by

$$V = -mgl \cos \theta$$

The weight is

$$w - mg = (4)(9.81) = 39.24 \text{ N}$$

The total potential energy consists of that due to the winding of the torsional spring and that due to the position of the mass.

$$V = \tfrac{1}{2}k_t\theta^2 - mgl \cos \theta = \tfrac{1}{2}k_t\theta^2 - 39.24l \cos \theta$$

Note that the reference position for calculating the potential energy of the mass is taken relative to point A. The potential energy becomes

(a) $\theta = 0$ $\quad V = -39.24(10) = -392.4$ N·m *Answer*

(b) $\theta = \dfrac{\pi}{4}$ $\quad V = \dfrac{1}{2}(1000)\left(\dfrac{\pi}{4}\right)^2 - 39.24(10 \cos 45°)$

$= 31.0$ N·m *Answer*

(c) $\theta = \dfrac{\pi}{2}$ $\quad V = \dfrac{1}{2}(1000)\left(\dfrac{\pi}{2}\right)^2 = 1234$ N·m *Answer*

PROBLEMS

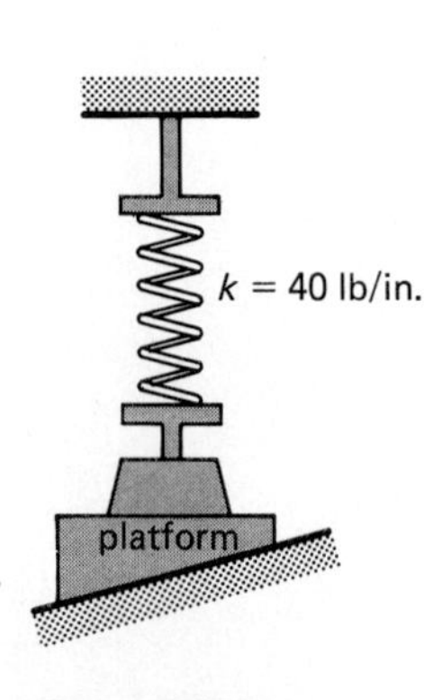

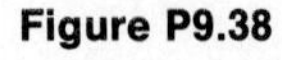
Figure P9.38

9.38 A linear spring, attached to a 20-lb weight as shown in Fig. P9.38, is unstretched as the weight rests on a platform. If the platform is slowly removed, find (a) the change in the total potential energy of the system and (b) the potential energy stored in the spring.

9.39 Find (a) the change in the total potential energy of the system in Problem 9.38 and (b) the potential energy stored in the spring if the platform is slowly removed. Let the weight be 50 lb.

9.40 Consider a linear spring, shown in Fig. P9.40, of modulus $k = 50{,}000$ N/m and unstretched length $L = 0.1$ m. What is the potential energy if $\bar{x}_A = -0.11\mathbf{i}$ m and $\bar{x}_B = 0.12\mathbf{i}$ m?

9.41 What is the potential energy stored in the spring shown in Fig. P9.41

when the rod rotates 10° counterclockwise if $k = 20$ lb/ft? The spring is in its unstretched position when the rod is vertical.

9.42 An inverted pendulum supports a 50-lb weight as shown in Fig. P9.42. The attached helical spring of modulus $k_t = 50$ ft·lb/rad is unstretched for the position shown. What is the change in the potential energy of the system if the pendulum rotates 20° clockwise?

9.43 A 5-kg rod is held in its position of static equilibrium by a torsional spring of modulus $k_t = 500$ m·N/rad as shown in Fig. P9.43. The torsional spring is undeformed when the rod is horizontal. If the rod rotates 30° clockwise, what is the change in the potential energy of the system?

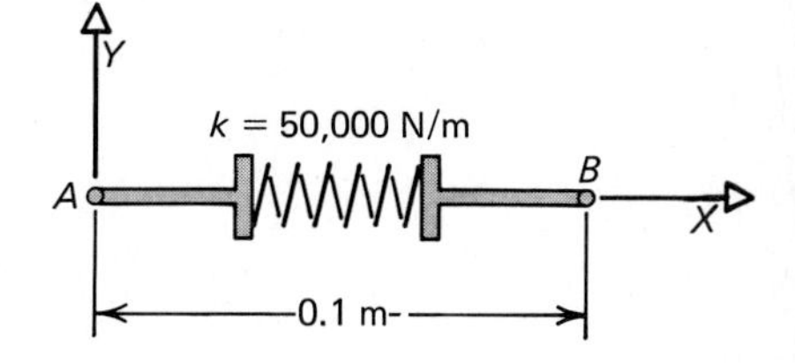

Figure P9.40

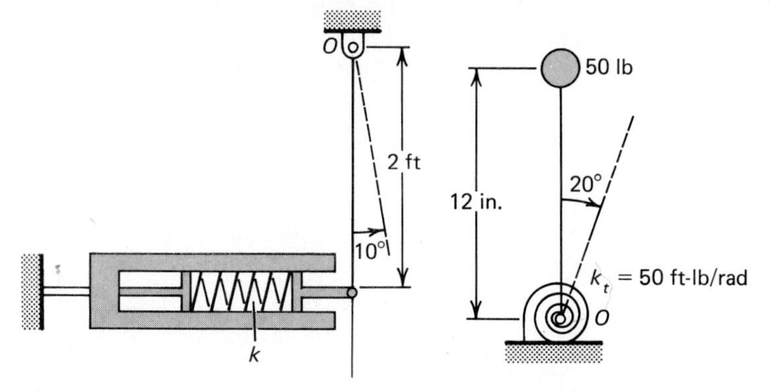

Figure P9.41

Figure P9.42

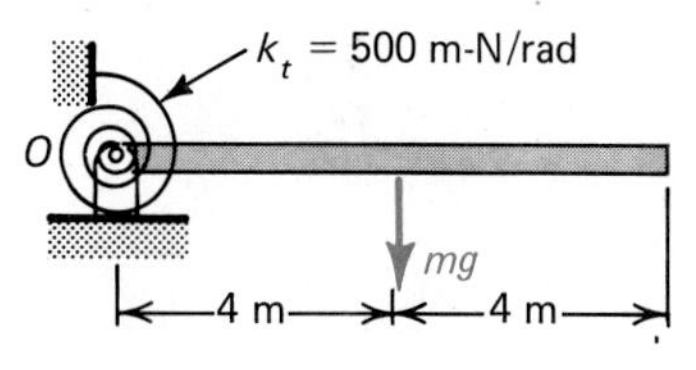

Figure P9.43

9.44 A beam, supported by springs, is shown in its static equilibrium position in Fig. P9.44. If point C deflects 0.2 m downward and the beam rotates 8° counterclockwise, what is the change in the potential energy? The mass of the beam equals 15 kg.

9.45 For Problem 9.44, find the angle through which the beam must rotate so that $\Delta V = 0$.

9.46 Repeat Problem 9.44 for point C moving 0.2 m upward and the beam rotating 8° clockwise.

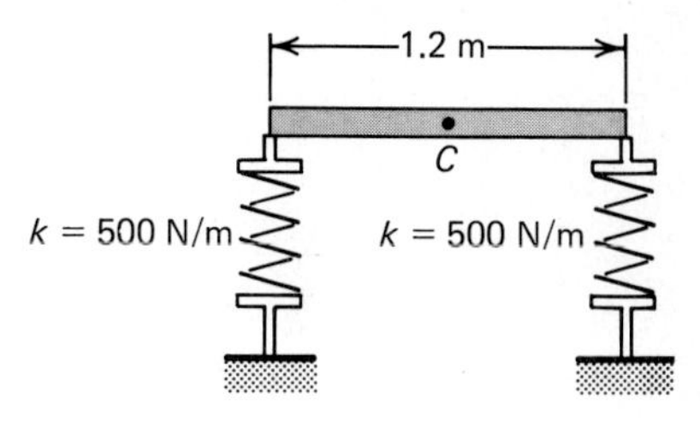

Figure P9.44

9.47 A particle rides along a frictionless rod and is restrained by a linear spring as shown in Fig. P9.47. The unstretched length of the spring is 0.2 ft. What is the change in the potential energy as the system moves from A to B if $r_A = 0.1$ ft, $\theta_A = 0°$, and $r_B = 0.3$ ft, $\theta_B = 45°$? The particle weighs 50 lb and $k = 100$ lb/ft for the spring.

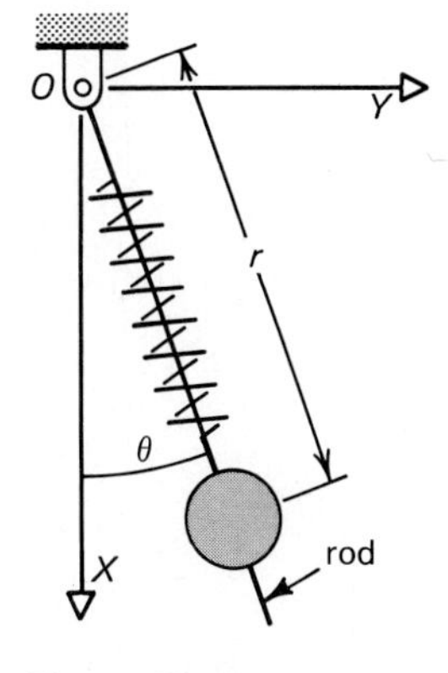

Figure P9.47

9.5 Stability of Equilibrium

Figure 9.18 shows a pendulum in static equilibrium in two different positions. Given a slight disturbance, the rod in Fig. 9.18(a) will oscillate back and forth about its position of static equilibrium, never moving far away. On the other hand, if we disturb, no matter how slightly, the inverted pendulum in Fig. 9.18(b), it will deviate an

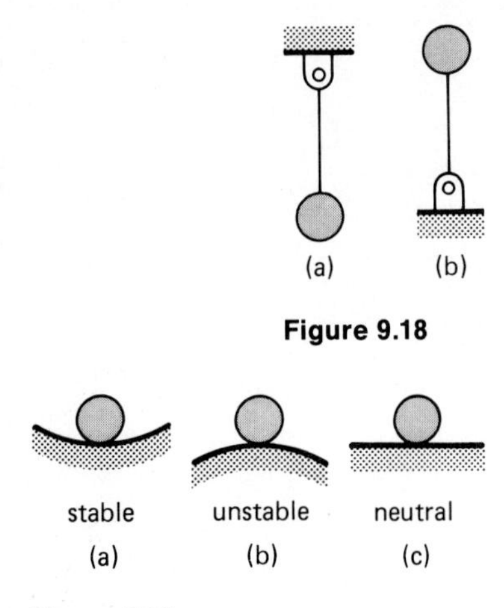

Figure 9.18

Figure 9.19

appreciable distance from its position of static equilibrium and will not return to its original position. We say that the pendulum in the first position is in a **stable** equilibrium position, while in the second position it is in an **unstable** equilibrium position.

The concept of stability is also demonstrated by the ball shown in Fig. 9.19. Again, a slight disturbance of the ball as shown in Fig. 9.19(a) would increase its potential energy and the ball would exhibit a tendency to return to its original position. Consequently, this is a position of stable equilibrium. On the other hand, the ball in Fig. 9.19(b) is in a position of unstable equilibrium because, given a slight disturbance, its potential energy would decrease as the ball leaves its equilibrium position. Figure 9.19(c) shows the ball in a position of neutral equilibrium. That is, a slight disturbance would neither increase nor decrease its potential energy. In Figures 9.19(a) and (b) we note that the potential energy is at an extremum in each case. That is, it is at a local minimum in Fig. 9.19(a) and at a local maximum in Fig. 9.19(b).

Suppose we assume that the potential energy is a function of one variable, say x.

$$V = V(x)$$

We wish to investigate the question of stability about the point of equilibrium, which we will assume is at $x = x_0$. Let us expand the potential energy function V about this equilibrium point in a Taylor series expansion,

$$V = V(x_0) + (x - x_0)\left[\frac{dV}{dx}\right]_{x=x_0} + \frac{(x - x_0)^2}{2!}\left[\frac{d^2V}{dx^2}\right]_{x=x_0} + \cdots$$

For equilibrium, we know the body is at an extremum position, so that

$$\left[\frac{dV}{dx}\right]_{x=x_0} = 0$$

The potential energy function becomes

$$V = V(x_0) + \frac{(x - x_0)^2}{2!}\left[\frac{d^2V}{dx^2}\right]_{x=x_0} + \cdots$$

The change ΔV in the potential energy of the body from the equilibrium position is

$$\Delta V = V(x) - V(x_0)$$

or

$$\Delta V = V - V(x_0) = \frac{(x - x_0)^2}{2!}\left[\frac{d^2V}{dx^2}\right]_{x=x_0} + \cdots \tag{9.25}$$

Generally, the higher-order terms in the series are small and so we will delete them.

Consequently, for the potential energy to increase as the system deviates about a stable equilibrium position, we must have

$$\Delta V > 0$$

which, by Eq. (9.25), requires that

$$\left[\frac{d^2V}{dx^2}\right]_{x=x_0} > 0$$

This is what happens when the pendulum in Fig. 9.18(a) swings. As it moves away from the equilibrium position, the potential energy increases with the minimum being at the equilibrium position. Likewise, for the potential energy to decrease as the system deviates about an unstable equilibrium position, we must have

$$\left[\frac{d^2V}{dx^2}\right]_{x=x_0} < 0$$

This occurs when the pendulum in Fig. 9.18(b) moves away from the equilibrium position, where the potential energy decreases with the maximum being at the equilibrium position.

We can summarize our findings for stability about the equilibrium point as follows:

$$\text{Stable equilibrium:} \quad \frac{dV}{dx} = 0 \qquad \frac{d^2V}{dx^2} > 0 \tag{9.26}$$

$$\text{Unstable equilibrium:} \quad \frac{dV}{dx} = 0 \qquad \frac{d^2V}{dx^2} < 0 \tag{9.27}$$

A word of caution: If the second derivative of V is also zero, then we must investigate higher-order derivatives. If all higher-order derivatives are zero, equilibrium will be neutral. To be stable, the higher-order derivatives must be of even order and positive. Otherwise, unstable equilibrium exists. The use of the potential energy of of system for the determination of static equilibrium and stability is illustrated in the following examples.

Example 9.10

Find the condition for equilibrium as well as the type of equilibrium for the system shown in Fig. 9.20. Assume that the system lies in a vertical plane and is pinned at the joint O.

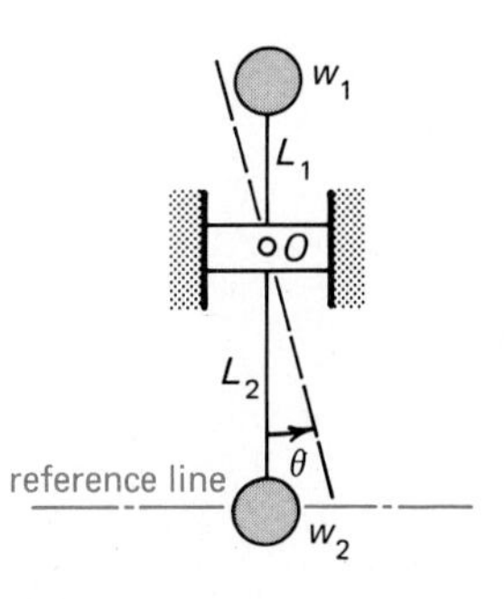

Figure 9.20

The potential energy relative to the reference line shown in Fig. 9.20 is given by

$$V = w_1(L_1 \cos\theta + L_2) + w_2(L_2 - L_2 \cos\theta)$$

For an extremum

$$\frac{dV}{d\theta} = -w_1 L_1 \sin\theta + w_2 L_2 \sin\theta = 0$$

so that the condition for equilibrium is

$$w_1 L_1 = w_2 L_2$$

For the type of equilibrium we obtain

$$\frac{d^2V}{d\theta^2} = -w_1 L_1 \cos\theta + w_2 L_2 \cos\theta$$
$$= 0$$

Since the second derivative equals zero at the equilibrium position, we must investigate higher-order derivatives. In fact all the higher-order derivatives are indeed zero so that we may conclude that the system is therefore in neutral equilibrium.

Example 9.11

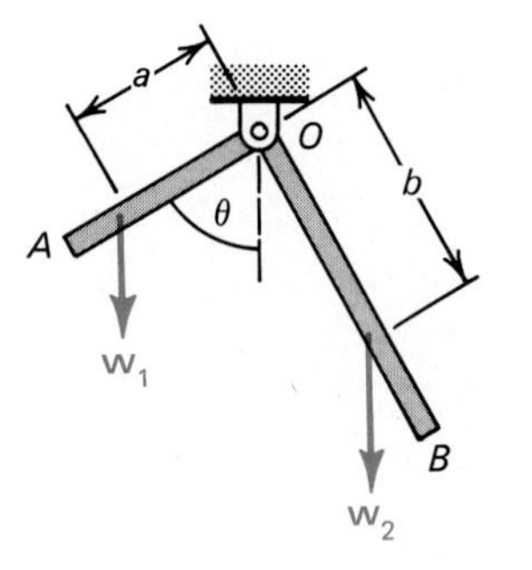

Figure 9.21

A right-angle member AOB is pivoted at O as shown in Fig. 9.21. Member OA has a mass of m_1 while member OB has a mass of m_2. Find the position of static equilibrium and determine if it is stable or unstable equilibrium.

The weights of OA and OB are given by

$$w_1 = m_1 g \qquad \text{and} \qquad w_2 = m_2 g$$

These forces are shown in Fig. 9.21 acting at their respective centers of gravity. The potential energy of the member relative to point O is

$$V = -w_1 a \cos\theta - w_2 b \cos(90^\circ - \theta)$$
$$= -w_1 a \cos\theta - w_2 b \sin\theta$$

Now
$$\frac{dV}{d\theta} = w_1 a \sin\theta - w_2 b \cos\theta = 0$$

$$\tan\theta = \frac{w_2 b}{w_1 a}$$

and therefore

$$\theta = \tan^{-1}\left(\frac{w_2 b}{w_1 a}\right) = \theta_0$$

is the position of static equilibrium.

To study the question of stability we establish $d^2V/d\theta^2$ first.

$$\frac{d^2V}{d\theta^2} = w_1 a \cos\theta + w_2 b \sin\theta$$

Note that at $\theta = \theta_0$,

$$\tan \theta_0 = \frac{w_2 b}{w_1 a}$$

so that

$$\sin \theta_0 = \frac{w_2 b}{\sqrt{(w_1 a)^2 + (w_2 b)^2}}$$

$$\cos \theta_0 = \frac{w_1 a}{\sqrt{(w_1 a)^2 + (w_2 b)^2}}$$

Now

$$\left[\frac{d^2V}{d\theta^2}\right]_{\theta=\theta_0} = \frac{(w_1 a)^2 + (w_2 b)^2}{\sqrt{(w_1 a)^2 + (w_2 b)^2}} = \sqrt{(w_1 a)^2 + (w_2 b)^2}$$

Since $(d^2V/d\theta^2) > 0$ for all θ, the position shown in Fig. 9.21 is a stable equilibrium position.

PROBLEMS

9.48 Show that the position of the inverted pendulum in Fig. P9.48 is that of unstable equilibrium.

9.49 Obtain the value of k for stable equilibrium of the pendulum in Fig. P9.49. The spring is unstretched for the position shown. Assume θ is small and the system lies in the vertical plane.

9.50 Obtain the maximum mass required for the system in Problem 9.49 to be stable. Assume small θ and let $b = 2a = 0.5$ m and $k = 1200$ N/m.

9.51 Find the condition as well as type of equilibrium for the system shown in Fig. P9.51. Assume point O is a pin connection and the system lies in the vertical plane.

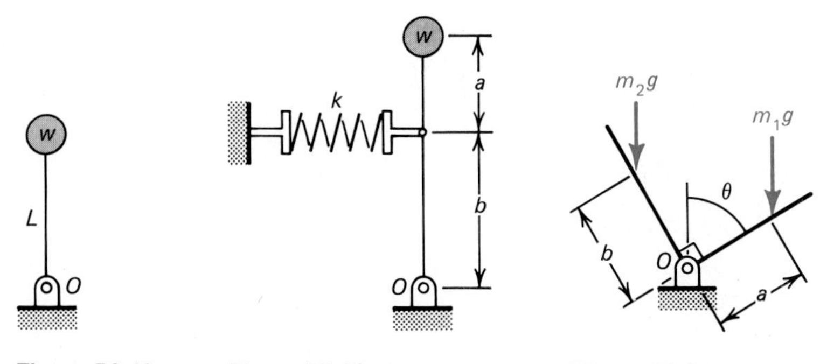

Figure P9.48 **Figure P9.49** **Figure P9.51**

9.52 The potential energy of a system is given by $V = x^3/3 + 5x^2 - 60x$. Obtain the conditions of static equilibrium and the type of stability.

9.53 Find (a) the position for equilibrium and (b) the type of stability for the rigid member in Fig. P9.53. Assume the member lies in the vertical plane.

9.54 Find the position of equilibrium for the system in Fig. P9.54 and the type of stability. w_2 is the weight of the rod OA. Assume small θ, no friction at the pulley B, and when $\theta = 0$, points A and B are at the same level.

9.55 A bar of weight w is restrained by a spring of modulus $k = 4w/L$ and unstretched length $s = L/2$ as shown in Fig. P9.55. What is the position of equilibrium?

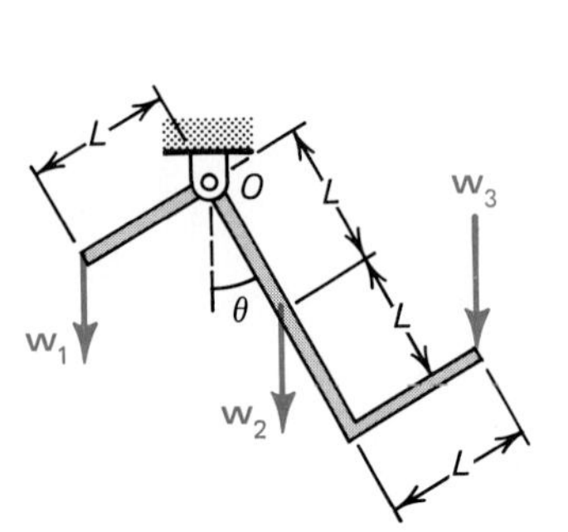

Figure P9.53

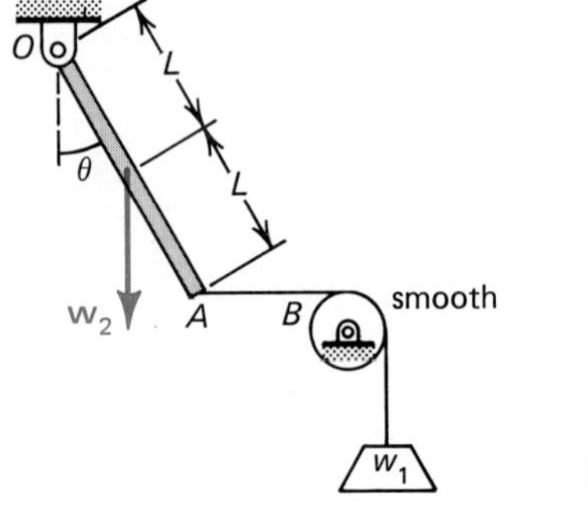

Figure P9.54

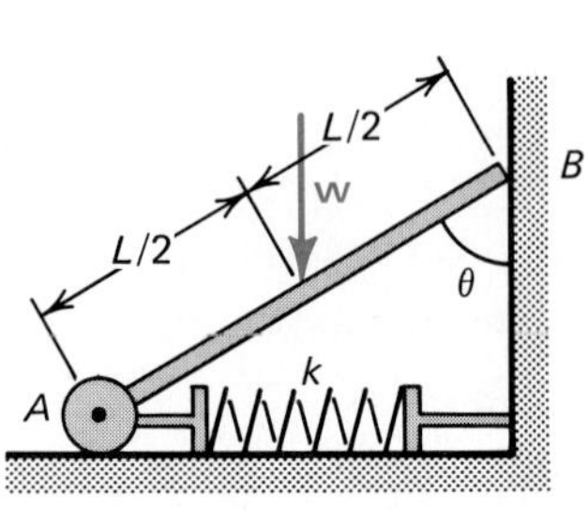

Figure P9.55

9.56 For the system in Problem 9.55 it is desired to have a position of equilibrium at $\theta = 60°$. If $L = 0.5$ m and $w = 100$ N, what spring modulus k is necessary to achieve this? Note that rod AB is connected to the roller by a smooth pin at A.

9.6 Summary

The principle of virtual work is another method for obtaining the equations of equilibrium of rigid bodies and structures. Once the equilibrium of the structure is established, the stability of equilibrium is determined by considering higher order derivatives of the potential energy function.

A. The important *ideas*:

1. Work is done on a body as it moves from one position to another under the action of forces and moments applied to it.
2. Virtual displacements are fictitious displacements that are very small in magnitude and do not violate the constraints of a system.
3. The principle of virtual work provides another technique for studying the equilibrium of single bodies as well as

connected bodies. A body is in equilibrium if the virtual work done by all the applied forces and moments is zero.

4. Potential energy is energy stored in a body.
5. Static equilibrium may be stable or unstable, depending upon the change in the potential energy of a system about the equilibrium position.

B. The important *equations*:

$$W = \int_A^B \mathbf{F} \cdot d\mathbf{r} \tag{9.2}$$

$$W = \int_A^B \mathbf{M} \cdot d\boldsymbol{\theta} \tag{9.4}$$

Principle of virtual work

Particle equilibrium: $\delta W = \mathbf{F} \cdot \delta\mathbf{r} = 0$ (9.9)

Rigid-body equilibrium: $\delta W = \mathbf{F} \cdot \delta\mathbf{r} + \mathbf{M} \cdot \delta\boldsymbol{\theta} = 0$ (9.10)

Potential energy

Spring: $V = \frac{1}{2}kx^2$ (9.20)

Weight: $V = wh$ (9.24)

Stable equilibrium: $\dfrac{dV}{dx} = 0 \qquad \dfrac{d^2V}{dx^2} > 0$ (9.26)

Unstable equilibrium: $\dfrac{dV}{dx} = 0 \qquad \dfrac{d^2V}{dx^2} < 0$ (9.27)

Answers to Even-Numbered Problems

Chapter 1

1.2 $\mathbf{i} + 14\mathbf{j}$, $9\mathbf{i} + 2\mathbf{j}$

1.4 $3\mathbf{i} - \mathbf{j}$, $-7\mathbf{i} + 7\mathbf{j}$

1.6 $2\mathbf{i} - 4\mathbf{j}$, $10\mathbf{i} - 8\mathbf{j}$

1.8 $8\mathbf{j}$

1.10 $\mathbf{i} - \mathbf{j}$

1.12 $-0.554\mathbf{i} - 0.831\mathbf{j}$

1.14 $-(V_r \cos\theta + V_\theta \sin\theta)\mathbf{i} - (V_r \sin\theta - V_\theta \cos\theta)\mathbf{j}$

1.16 $6.46\mathbf{i} + 5.46\mathbf{j}$

1.18 0.895, -0.448

1.20 0

1.22 0.548, $-0.182\mathbf{i} - 0.365\mathbf{j} + 0.912\mathbf{k}$

1.24 0.768, 0.384, -0.512

1.26 5.51

1.28 81.8°

1.32 0.136, -0.952, 0.272

1.36 (a) $-4\mathbf{k}$; (b) -33.6°

1.38 $-2\mathbf{i} - 4\mathbf{j} - 3\mathbf{k}$

1.40 $20\mathbf{i} - 18\mathbf{j} - 10\mathbf{k}$

1.42 $2\mathbf{i} - \mathbf{j} - \mathbf{k}$

1.46 $\pm(0.857\mathbf{i} + 0.429\mathbf{j} + 0.286\mathbf{k})$

Chapter 2

2.2 $R = 640$ N, $\theta = 261^\circ$

2.4 $\mathbf{R} = -3\mathbf{i}$ N

2.6 $R = 232$ lb, $\theta = 6^\circ$

2.8 $R = 201$ N, $\theta = 16.2^\circ$

2.10 $\mathbf{R} = -2\mathbf{j}$ N

2.12 $\mathbf{R} = (207\mathbf{i} + 6\mathbf{j})$ lb

2.14 $\alpha = 7^\circ$, $F_1 = 418$ lb

2.16 $\mathbf{T} = (-1424\mathbf{i} - 474\mathbf{j})$ N

2.18 $R = 69$ N, $\theta = 27.5^\circ$

2.20 (a) $\mathbf{F} = (-160\mathbf{i} + 192\mathbf{j})$ lb, (b) $\mathbf{F} = (160\mathbf{i} - 192\mathbf{j})$ lb

2.22 $F_x = 663$ lb, $F_y = 643$ lb, $F_z = 383$ lb

2.24 $\mathbf{F}_2 = (18.4\mathbf{i} + 14.6\mathbf{j} + 1.14\mathbf{k})$ N

2.26 $R = 1396$ lb; 0.286, 0.771, 0.569

2.28 $R = 455$ N; 0.374, 0.549, -0.747

2.30 $\mathbf{R} = (19.7\mathbf{i} + 1.40\mathbf{j} - 0.64\mathbf{k})$ lb

2.32 $T_1 = 63.5$ N, $T_2 = 16.0$ N

2.34 $T_{AC} = 14.3$ lb, $T_{BC} = 37.7$ lb

2.36 $T = 8.0$ N, $\alpha = 90^\circ$

2.38 $T_1 = 64.0$ lb, $T_2 = 16.8$ lb

2.40 $T_{AC} = 14.1$ lb, $T_{BC} = 38.1$ lb

2.42 $T = 8.05$ N, $\alpha = 90.3^\circ$

2.44 $T_{AB} = 110$ lb, $T_{BC} = 135$ lb

2.46 $F_1 = 117$ lb, $F_2 = 179$ lb

2.48 $T_{AB} = 213$ lb, $T_{BC} = 2110$ lb

2.50 $\mathbf{F}_3 = (-6.51\mathbf{i} - 11.1\mathbf{j} + 0.065\mathbf{k})$ N

2.52 $\mathbf{F}_3 = (-2.41\mathbf{i} + 15.2\mathbf{j} - 13.7\mathbf{k})$ lb

2.54 $T_{DA} = 393$ N, $T_{DB} = 690$ N, $T_{DC} = 0$

2.56 $T_{AB} = 2260$ N, $T_{AC} = 2190$ N, $T_{AD} = 612$ N

2.58 $\mathbf{M}_A = -29.2\mathbf{k}$ m·N, $\mathbf{M}_B = -21.7\mathbf{k}$ m·N

2.60 $\mathbf{M}_A = -152\mathbf{k}$ ft·lb, $\mathbf{M}_B = -52\mathbf{k}$ ft·lb

2.62 $\mathbf{M}_A = 17.3\mathbf{k}$ m·N, $\mathbf{M}_B = 0$

2.64 $\mathbf{M}_o = -140\mathbf{k}$ ft·lb

2.66 $\mathbf{M}_A = -74.8\mathbf{k}$ m·N

2.68 $\mathbf{M}_P = -177\mathbf{k}$ ft·lb

2.70 $\mathbf{M}_P = 21.2\mathbf{k}$ m·N

2.72 $\mathbf{M}_A = 49.2\mathbf{k}$ ft·lb

2.74 $M_A = 229$ ft·lb

2.76 $\mathbf{M}_o = (\mathbf{i} - \mathbf{j} - \mathbf{k})$ m·N

2.78 $\mathbf{M}_o = (24\mathbf{i} - 7\mathbf{j} - 10\mathbf{k})$ ft·lb

2.80 $\mathbf{M}_o = (-98.1\mathbf{i} - 294\mathbf{k})$ m·N

2.82 $36.0\mathbf{k}$ m·N

2.84 $49.2\mathbf{k}$ ft·lb

2.86 $(28.5\mathbf{i} + 9.48\mathbf{j})$ m·N

2.88 11.87 m·N; −0.590, 0.590, 0.590

2.90 $(-250\mathbf{i} + 200\mathbf{j} + 10\mathbf{k})$ mm·N

2.92 $\mathbf{M}_A = -29.2\mathbf{k}$ m·N, $\mathbf{R} = (86.6\mathbf{i} + 50\mathbf{j})$ N

2.94 $\mathbf{M}_C = -52.0\mathbf{k}$ ft·lb, $\mathbf{R} = 100\mathbf{j}$ lb

2.96 $\mathbf{M}_B = 0$, $\mathbf{R} = 173\mathbf{i}$ N

2.98 $\mathbf{M}_Q = -106\mathbf{k}$ ft·lb, $\mathbf{R} = 35.4(\mathbf{i} + \mathbf{j})$ lb

2.100 $\mathbf{M}_A = 162(\mathbf{i} - \mathbf{j})$ ft·lb,
$\mathbf{R} = (27.0\mathbf{i} + 27.0\mathbf{j} - 32.3\mathbf{k})$ lb

2.102 $\mathbf{M}_A = 14.0\mathbf{k}$ m·N, $\mathbf{R} = (10\mathbf{i} - 10\mathbf{j})$ N

2.104 $\mathbf{M}_A = 0$, $\mathbf{R} = (-32.6\mathbf{i} - 185\mathbf{j})$ lb

2.106 $\mathbf{M}_A = -216\mathbf{k}$ m·N, $\mathbf{R} = (50\mathbf{i} - 136.6\mathbf{j})$ N

2.108 $\mathbf{M}_B = 70.7(-2\mathbf{i} + 2\mathbf{j} + \mathbf{k})$ m·N,
$\mathbf{R} = 70.7(\mathbf{i} + \mathbf{j})$ N

2.110 $\mathbf{M}_B = (163\mathbf{i} - 81.6\mathbf{j})$ ft·lb, $\mathbf{R} = 40.8(\mathbf{i} + 2\mathbf{j} + \mathbf{k})$ lb

2.112 $\mathbf{M}_B = (-86.0\mathbf{i} - 29.4\mathbf{j} - 28.3\mathbf{k})$ m·N,
$\mathbf{R} = (-29.4\mathbf{i} + 36.0\mathbf{j} - 57.7\mathbf{k})$ N

2.114 $x = 1.33$ ft, $y = 1.33$ ft

2.116 $\mathbf{F}_1 = (-200\mathbf{i} + 150\mathbf{j})$ N, $\mathbf{F}_2 = 200\mathbf{i}$ N

Chapter 3

3.2 $\mathbf{A} = (106\mathbf{i} + 100\mathbf{j})$ N, $\mathbf{B} = 206\mathbf{j}$ N

3.4 $\mathbf{A} = 50\mathbf{j}$ N, $\mathbf{B} = 150\mathbf{j}$ N

3.6 $\mathbf{B} = -25\mathbf{j}$ lb, $\mathbf{M} = 100\mathbf{k}$ ft·lb

3.8 $\mathbf{A} = -18.8\mathbf{j}$ N, $\mathbf{B} = 68.8\mathbf{j}$ N

3.10 $T = 75$ lb

3.12 $T = 400$ lb

3.14 (a) 1730 lb, (b) $(300\mathbf{i} - 700\mathbf{j})$ lb

3.16 (a) 500 N (b) $(433\mathbf{i} + 150\mathbf{j})$ N

3.18 (a) $(294\mathbf{i} + 1490\mathbf{j})$ N, (b) 589 N

3.20 (a) 388 lb, (b) $(236\mathbf{i} - 94.0\mathbf{j})$ lb

3.22 $\mathbf{A} = (1750\mathbf{i} + 1000\mathbf{j})$ N, $T = 1750$ N

3.24 $x = 1.43$ ft

3.26 $\mathbf{A} = -50\mathbf{i}$ lb, $\mathbf{B} = -100\mathbf{i}$ lb

3.28 20 N

3.30 $\mathbf{A} = (68.4\mathbf{i} + 587.9\mathbf{j})$ N, $\mathbf{M} = 52.3\mathbf{k}$ m·N

3.32 $\mathbf{A} = (-1000\mathbf{i} - 2000\mathbf{j})$ lb, $\mathbf{B} = 4000\mathbf{j}$ lb

3.34 $T = \dfrac{w_1(a + b) + 2w_2 a}{2(a + b)\sin\theta}$,
$A = [T\cos\theta\mathbf{i} + (w_1 + w_2 - T\sin\theta)\mathbf{j}]$,
$T = 5.80$ kip, $\mathbf{A} = (5.02\mathbf{i} + 2.10\mathbf{j})$ kip

3.36 $T = 3913$ N, $\mathbf{A} = (3205\mathbf{i} + 2661\mathbf{j})$ N

3.38 $P = 86.6$ lb

3.40 $T = 21.7$ N, $\mathbf{A} = (-10.9\mathbf{i} + 31.2\mathbf{j})$ N

3.42 $\mathbf{A} = (-31.6\mathbf{i} - 4.0\mathbf{j})$ N, $\mathbf{M} = 2669\mathbf{k}$ mm·N

3.44 $\mathbf{A} = (100\mathbf{j} + 50\mathbf{k})$ lb, $\mathbf{M} = -600\mathbf{j}$ ft·lb

3.46 $\mathbf{A} = (-50\mathbf{j} + 100\mathbf{k})$ N, $\mathbf{M} = (-25\mathbf{i} - 90\mathbf{j} - 45\mathbf{k})$ m·N

3.48 $\mathbf{A} = (746\mathbf{i} + 400\mathbf{j} + 600\mathbf{k})$ lb,
$\mathbf{M} = (2517\mathbf{i} - 3000\mathbf{j})$ ft·lb

3.50 $\mathbf{C} = (86.7\mathbf{i} - 100\mathbf{j})$ N

3.52 $\mathbf{A} = 25\mathbf{k}$ lb, $\mathbf{M} = (175\mathbf{i} + 250\mathbf{j})$ in·lb

3.54 $\mathbf{A} = (263\mathbf{i} + 39\mathbf{j})$ N, $T = 397$ N

3.56 $\mathbf{A} = (-50\mathbf{i} - 50\mathbf{j})$ lb, $\mathbf{M} = (-100\mathbf{j} - 300\mathbf{k})$ ft·lb

3.58 $\mathbf{O} = (247\mathbf{j} - 34.7\mathbf{k})$ N,
$\mathbf{M} = (29.6\mathbf{i} + 13.9\mathbf{j} + 88.8\mathbf{k})$ m·N

Chapter 4

4.2 $F_{AB} = 539.4$ N C, $F_{AC} = 202.3$ N T,
$F_{BC} = 539.4$ N C

4.4 $F_{AB} = 1250$ lb C, $F_{AH} = 224$ lb T, $F_{BC} = 1250$ lb C,
$F_{BH} = 0$

4.6 $F_{BF} = 1222$ N C, $F_{BC} = 1351$ N C

4.8 $F_{DE} = 282.8$ lb C, $F_{DC} = 200.0$ lb T,
$F_{CE} = 141.4$ lb C, $F_{BE} = 141.4$ lb C

4.10 $F_{AB} = 158.1$ N C, $F_{AE} = 150.0$ N C,
$F_{BE} = 158.1$ N T, $F_{BC} = 100.0$ N T

4.12 $F_{AF} = 0$, $F_{AE} = 707.1$ lb T, $F_{AB} = 1000$ lb T,
$F_{BC} = 1414$ lb T, $F_{BE} = 1000$ lb C

4.14 $F_{AG} = 901.4$ N T, $F_{AB} = 500$ N C,
$F_{GB} = 901.4$ N C, $F_{GF} = 1000$ N T

4.16 $F_{BC} = 19.5$ kip C

4.18 $F_{AD} = 0.787$ kN T, $F_{AE} = 6.70$ kN C, $F_{BC} = 3.146$ kN C, $F_{BD} = 3.196$ kN T

4.20 $F_{AF} = F_{DE} = 2404$ lb T, $F_{BC} = F_{CD} = 1700$ lb C, $F_{CF} = 0$, $F_{EF} = 1700$ lb T, $F_{BF} = F_{CE} = 1000$ lb C

4.22 $F_{DE} = F_{BD} = 390$ lb C, $F_{CE} = F_{AC} = 310$ lb T, $F_{CD} = F_{AD} = 0$, $F_{AB} = 150$ lb T

4.24 $F_{AF} = 2.000$ kN T, $F_{FG} = 2.828$ kN T, $F_{BF} = 0$

4.36 $F_{CF} = 0.292$ kip C

4.38 $F_{EG} = 960$ N T

4.40 $F_{BC} = 8.75$ kN C, $F_{KL} = 12.23$ kN T, $F_{CL} = 4.16$ kN C

4.42 $F_{DC} = F_{DE} = 12.0$ kN C

4.44 $F_{GI} = 5000$ lb T, $F_{GH} = 3000$ lb C

4.46 $F_{IL} = 5.25$ kip T, $F_{HJ} = 5.25$ kip C

4.48 $F_{FD} = 1.204$ kN C, $F_{ED} = 4.179$ kN C, $F_{CE} = 3.291$ kN T

4.50 $F_{GE} = 2.828$ kN T, $F_{BC} = 4.00$ kN C

4.52 $F_{BD} = 3.233$ kip C

4.54 $F_{HG} = 3.010$ kip T

4.56 $F_{EL} = 3.391$ kip T

4.58 $F_{BC} = 2.828$ N T, $F_{CD} = 1.580$ N C, $F_{AC} = 0.500$ N C

4.60 $F_{AB} = 0$, $F_{AC} = 141$ lb C, $F_{AD} = 0$

4.62 $F_{BC} = 6.403$ kN T

4.64 $\mathbf{A} = (2.667\mathbf{i} + 0.846\mathbf{k})$ kip, $\mathbf{B} = (-4\mathbf{j} + 5.615\mathbf{k})$ kip, $\mathbf{D} = (-2.667\mathbf{i} - 3.461\mathbf{k})$ kip

4.66 $F_{AE} = 1.436$ kN T

4.68 $F_{DE} = 0.833$ kip T, $F_{BF} = 0.833$ kip C

4.70 $\mathbf{A} = (-354\mathbf{i} - 41.4\mathbf{j})$ N, $\mathbf{D} = 395\mathbf{j}$ N, $\mathbf{C} = -619\mathbf{j}$ N, $T = 442$ N

4.72 160 N

4.74 $T = 312.5$ N, $\mathbf{C} = 620.5\mathbf{j}$ N

4.76 $T = 80$ kip, $\mathbf{A} = (-80\mathbf{i} + 30\mathbf{j})$ kip

4.78 $\mathbf{D} = (-3.954\mathbf{i} - 17.47\mathbf{j})$ kN, A = 12**j** kN

4.80 $\mathbf{A} = 70.7(-\mathbf{i} - \mathbf{j})$ lb, $\mathbf{B} = 0$

4.82 $\mathbf{B} = (275\mathbf{i} + 550\mathbf{j})$ N, $\mathbf{G} = -275\mathbf{j}$ N, $\mathbf{E} = 275(-\mathbf{i} - \mathbf{j})$ N

4.84 1800 N

4.86 $\mathbf{A} = (-99.0\mathbf{i} + 250\mathbf{j})$ lb, $\mathbf{B} = (28.3\mathbf{i} + 321\mathbf{j})$ lb

4.88 960 N

4.90 $P = 25$ N

4.92 $M = 10{,}500$ in·lb

4.94 $M_R = 40$ m·N

Chapter 5

5.4 $\bar{x} = 3$, $\bar{y} = 2$

5.6 $\bar{x} = 5$ mm, $\bar{y} = 17$ mm

5.8 $\bar{x} = 1$ mm, $\bar{y} = 1.667$ mm

5.12 $\bar{x} = 2.667$ mm, $\bar{y} = 1.333$ mm

5.14 $\bar{x} = 2.25$, $\bar{y} = 2.03$

5.16 $\bar{x} = 2.25$, $\bar{y} = 5.40$

5.18 $\bar{x} = a(n+1)/(n+2)$, $\bar{y} = a^n k(n+1)/[2(2n+1)]$

5.22 $\bar{x} = \bar{y} = 0$, $\bar{z} = 25$ m

5.24 $\bar{x} = \bar{y} = 0$, $\bar{z} = 0.125$ ft

5.26 $\bar{x} = 0.778$ ft, $\bar{y} = 0.555$ ft

5.28 $\bar{x} = \bar{y} = 0$, $\bar{z} = 1.18$ in

5.30 $\bar{x} = 2.196$ m, $\bar{y} = 1.144$ m

5.32 $\bar{x} = 3.203$ in, $\bar{y} = 0.337$ in

5.34 $\bar{x} = 2.26$ m, $\bar{y} = 0.778$ m

5.36 $\bar{x} = 3.30$ in, $\bar{y} = 0.335$ in

5.38 $\bar{x} = 5.728$ m, $\bar{y} = 1.908$ m

5.40 $\bar{x} = 1.245$ in, $\bar{y} = 0.658$ in

5.42 $\bar{x} = 0$, $\bar{y} = -0.719$ mm

5.44 $\bar{x} = \bar{y} = 5.467$ in

5.46 $\bar{x} = \bar{z} = 0$, $\bar{y} = 0.01875$ m

5.48 $\bar{x} = \bar{z} = 0$, $\bar{y} = 0.454$ ft

5.50 $\bar{x} = -15.0$ mm, $\bar{y} = 60.5$ mm, $\bar{z} = 24.9$ mm

5.52 $\bar{x} = -28.0$ mm, $\bar{y} = -17.0$ mm, $\bar{z} = 31.0$ mm

5.54 14.51 in^2, 66.40 in^2

5.56 0.546 m^2

5.58 (a) 12.57 ft^2, (b) 19.74 ft^2

5.60 4.022 m^2

5.62 38,930 mm^3

5.64 1.072 m^3

5.66 8.335 ft^3

5.68 $x = -6.25$ m

5.70 (a) $\mathbf{A} = 123.8\mathbf{j}$ lb, $\mathbf{B} = 101.3\mathbf{j}$ lb, (b) $\mathbf{R} = -225\mathbf{j}$ lb, $\bar{x} = 4.5$ ft

5.72 (a) $\mathbf{A} = 242.2\mathbf{j}$ N, $\mathbf{B} = 157.8\mathbf{j}$ N, (b) $\mathbf{R} = -400\mathbf{j}$ N, $\bar{x} = 5.92$ m

5.74 (a) $\mathbf{A} = 47.75\mathbf{j}$ lb, $\mathbf{B} = 15.92\mathbf{j}$ lb, (b) $\mathbf{R} = -63.66\mathbf{j}$ lb, $\bar{x} = 2.50$ ft

5.76 (a) $\mathbf{A} = 139\mathbf{j}$ N, $\mathbf{B} = 2361\mathbf{j}$ N, (b) $\mathbf{R} = -2500\mathbf{j}$ N, $\bar{x} = 28.3$ m

5.78 (a) $\mathbf{A} = 250\mathbf{j}$ N, $\mathbf{B} = 1000\mathbf{j}$ N,
(b) $\mathbf{R} = -1250\mathbf{j}$ N, $\bar{x} = 4.0$ m

5.80 $\mathbf{R} = -24{,}000\mathbf{i}$ lb

5.82 $\mathbf{R} = 490.5\mathbf{i}$ kN, $\bar{y} = 3.33$ m

5.84 $T = 1404$ lb

5.86 $\mathbf{D} = (78.5\mathbf{i} + 490.5\mathbf{j})$ kN, $\mathbf{M} = -300.9\mathbf{k}$ m·kN

5.88 $\mathbf{A} = (17{,}970\mathbf{i} + 12{,}300\mathbf{j})$ lb, $\mathbf{M} = -141{,}400\mathbf{k}$ ft·lb

5.90 500,000 kg

5.92 8.00 lb

5.94 21.3 lb

5.96 5.062 m

5.98 0.301 ft

5.100 $\mathbf{A} = (-40\mathbf{i} + 2\mathbf{j})$ kN, $\mathbf{D} = (40\mathbf{i} + 2\mathbf{j})$ kN,
$T_{AB} = 40.1$ kN

5.102 $\mathbf{A} = (-92.3\mathbf{i} + 17.7\mathbf{j})$ kip, $\mathbf{D} = (92.3\mathbf{i} + 2.3\mathbf{j})$ kip

5.104 $\mathbf{A} = (-9.0\mathbf{i} + 2.0\mathbf{j})$ kN, $y_c = 10.7$ m

5.106 345 m

5.108 747 N/m

5.110 121.4 lb

Chapter 6

6.2 $J = 16{,}250$ mm^4

6.4 $I_x = \frac{1}{28}$ in^4

6.6 $I_x = 0.685$ m^4

6.8 $I_x = \dfrac{2}{9a}$

6.10 $I_y = \frac{1}{20}$ in^4

6.12 $I_y = 0.171$ m^4

6.14 $I_y = 0.467a^{-3}$

6.16 (a) $J = \pi a^4/8$, (b) $k_x = a/2$

6.18 $J = (b^4 - a^4)\pi/2$

6.20 $k_z = [0.5(a^2 + b^2)]^{1/2}$

6.22 $I_y = 56$ft^4, $k_y = 3.06$ ft

6.24 $I_u = 3.927a^4$, $I_v = 13.35a^4$

6.26 $I_u = 0.597$ m^4, $I_v = 0.233$ m^4

6.28 $I_v = 1.085 \times 10^6$ mm^4

6.30 $I_x = 1.230 \times 10^6$ mm^4

6.32 $I_x = 47.30$ in^4

6.34 $I_x = 101.3$ mm^4

6.36 $I_x = 84.7$ in^4

6.38 $J = 6.0a^4$

6.40 $I_y = 3.640 \times 10^5$ mm^4

6.42 $I_y = 44.67$ in^4

6.44 $I_y = 441.8$ mm^4

6.46 $I_y = 7453$ in^4

6.48 $I_{xy} = 50$ ft^4

6.50 $\bar{I}_{x'y'} = -21.25$ m^4

6.52 $I_{xy} = 42.7$ in^4

6.54 $I_{xy} = 380.3$ m^4

6.56 $I_{xy} = 580$ in^4

6.58 $I_{xy} = 0.0155$ m^4

6.60 $I_{xy} = 909.4$ in^4

6.64 $\theta_m = -16.8°$, $\bar{I}_{max} = 19.36$ ft^4, $\bar{I}_{min} = 3.14$ ft^4

6.66 $\theta_m = 45°$, $\bar{I}_{max} = 92.37$ m^4, $\bar{I}_{min} = 49.87$ m^4

6.68 $\theta_m = 15.8°$, $\bar{I}_{max} = 684.1$ in^4, $\bar{I}_{min} = 193.7$ in^4

6.70 $\bar{I}_v = 47.94$ in^4, $\bar{I}_u = 84.06$ in^4, $\bar{I}_{uv} = 61.28$ in^4

6.76 $I = 1700$ kg·mm^2

6.78 $I_x = 63.74$ lb·ft·sec^2

6.80 $I_z = 0.1608$ kg·m^2

6.82 $I_o = 1.753$ lb·ft·sec^2

6.84 $I_{AA} = 2.73$ kg·m^2

6.86 $I = 24{,}930$ kg·m^2

Chapter 7

7.2 $0 < x < 12$: $Q = 50$ N,
$M = 50x$ m · N
$12 < x < 16$: $Q = -150$ N,
$M = (2400 - 150x)$ m·N

7.4 $0 < x < 8$: $Q = 2$ kN, $M = (2x - 16)$ m·kN

7.6 $0 < x < 2$: $Q = 100$ N, $M = (100x - 2200)$ m·N
$2 < x < 12$: $Q = 200$ N, $M = (200x - 2400)$ m·N

7.8 $0 < x < 5$: $Q = 83.3$ lb,
$M = 83.3x$ ft·lb
$5 < x < 10$: $Q = -16.7$ lb,
$M = (-16.7x + 500)$ ft·lb
$10 < x < 15$: $Q = -66.7$ lb,
$M = (-66.7x + 1000)$ ft·lb

7.10 $0 < x < 20$: $Q = -400$ N,
$M = -400x$ m·N
$20 < x < 40$: $Q = -700$ N,
$M = (-700x + 6000)$ m·N

7.12 $0 < x < 5$: $Q = -100$ N,
$M = -100x$ m·N
$5 < x < 10$: $Q = -100$ N,
$M = (1000 - 100x)$ m·N

7.14 $0 < x < 5$: $Q = -100$ N, $M = (400 - 100x)$ m·N

7.16 $0 < x < 10$: $Q = 50$ lb, $M = (50x - 100)$ ft·lb
$10 < x < 12$: $Q = 50$ lb, $M = (50x - 600)$ ft·lb

7.18 $0 < x < 4$: $Q = -100$ kip, $M = -100x$ ft·kip
$4 < x < 8$: $Q = -100$ kip, $M = (800 - 100x)$ ft·kip
$8 < x < 12$: $Q = 0$, $M = 0$

7.20 $0 < x < 6$: $Q = 0$, $M = 200$ m·N
$6 < x < 14$: $Q = -6.25$ N, $M = (237.5 - 6.25x)$ m·N
$14 < x < 22$: $Q = -6.25$ N, $M = (137.5 - 6.25x)$ m·N

7.22 $0 < x < 8$: $Q = (290 - 40x)$ N, $M = (290x - 20x^2)$ m·N
$8 < x < 16$: $Q = (290 - 40x)$ N, $M = (-20x^2 + 290x + 480)$ m·N

7.24 $0 < x < 3$: $Q = -100x$ lb, $M = -50x^2$ ft·lb
$3 < x < 11$: $Q = (656 - 100x)$ lb, $M = (-50x^2 + 656x - 1968)$ ft·lb
$11 < x < 15$: $Q = (1500 - 100x)$ lb, $M = (-50x^2 + 1500x - 11250)$ ft·lb

7.26 $0 < x < 10$: $Q = (2 - 0.1x)$ kN, $M = (-0.05x^2 + 2x)$ m·kN
$10 < x < 20$: $Q = -0.1x$ kN, $M = (-0.05x^2 + 20)$ m·kN

7.28 $0 < x < 5$: $Q = 0$, $M = -600$ ft·lb
$5 < x < 20$: $Q = (-40x + 506.7)$ lb, $M = (-20x^2 + 506.7x - 2634)$ ft·lb
$20 < x < 25$: $Q = (-40x + 1000)$ lb, $M = (-20x^2 + 1000x - 12{,}500)$ ft·lb

7.30 $0 < x < 4$: $Q = -1000$ N, $M = (-1000x - 1000)$ m·N
$4 < x < 10$: $Q = (-100x - 600)$ N, $M = (-50x^2 - 600x - 1800)$ m·N

7.46 $x = 21.9$ ft

Chapter 8

8.2 $P = 71.7$ lb
8.4 $\theta = 16.7°$, $P = 714$ N
8.6 no
8.8 $\mu_0 = 0.414$
8.10 $\mu_0 = 0.372$
8.12 $\theta = 44.6°$
8.14 $\mu_0 = 0.268$
8.16 $h = 2$ m
8.18 $\theta = -26.6°$, $P = 313$ N
8.20 $\mu_0 = 0.182$
8.22 $P = 641$ lb
8.24 $P = 53.0$ kN
8.26 $P = 20.4$ lb
8.28 $P = 14.8$ kN
8.30 $\mu_0 = 0.111$
8.32 $M_U = 223$ m·N
8.34 $M_L = 12.9$ ft·lb
8.36 $M_L = 2.00$ m·N
8.38 $P = 306$ lb
8.40 $M = 61.0$ m·N
8.42 (a) $P = 19.4$ lb, (b) $P = 516$ lb
8.44 971 N, 1639 N
8.46 $M = 3.96Pr$
8.48 (a) 42.4 N, (b) 22,700 N
8.50 $P = 27.5$ lb
8.52 254 N, 754 N
8.54 $\mu_0 = 0.8$
8.56 $\mu_0 = 0.125$
8.58 $M = 9.38$ ft·lb
8.60 $M = \frac{2}{3}\mu Pr$
8.62 $r_i = 50$ mm

Chapter 9

9.2 $W = 120.7$ lb·ft
9.4 $W = r\left(\dfrac{c\pi}{2} - mg - \mu mg\right)$
9.6 $W = 200$ lb·ft
9.8 $W = 0$
9.10 $W = 504$ N·m
9.12 $W = 9.42$ N·m
9.14 $W = -1596$ lb·ft
9.16 $P = 50$ N
9.18 $\delta W = -\dfrac{wL}{2}\sin\theta\,\delta\theta$
9.20 (a) $\delta W = -(w_1 + w_2)L\sin\theta_1\,\delta\theta_1$
(b) $\delta W = -w_2 L\sin\theta_2\,\delta\theta_2$

9.22 $\delta W = \left(-\frac{FL}{2}\cos\theta + M\right)\delta\theta$

9.24 $P = (10a \sin\theta + 2.5b \cos\theta)/(a \cos\theta - b \sin\theta)$

9.26 $M = 50l(3 \cos\theta + \sin\theta)$

9.28 $P = 2Q \tan\theta$

9.30 $P = 1.378Q$

9.32 $P = 10.7$ lb

9.34 $P = 145$ lb

9.36 $M = Pr\left[\cos\theta + \frac{\sin\theta(r\cos\theta + h)}{\sqrt{L^2 - (r\cos\theta + h)^2}}\right]$

9.38 (a) $\Delta V = 0$, (b) $V = 1.67$ lb·ft

9.40 $V = 423$ N·m

9.42 $\Delta V = 0.0312$ lb·ft

9.44 $\Delta V = -5.95$ N·m

9.46 $\Delta V = 52.9$ N·m

9.50 $m = 40.8$ kg

9.52 $x = -14.22$, unstable equilibrium;
$x = 4.22$, stable equilibrium

9.54 $\theta = \tan^{-1}\left(\frac{2w_1}{w_2}\right)$, stable equilibrium

9.56 $k = 473$ N/m

Basic Concepts 1

When we talk about the motion of physical objects, whether they are the planets in our solar system, the roll-pitch-yaw motions of a spacecraft, the acceleration of an automobile, the motion of a charged particle in an electric field, the curving of a baseball, or the swinging of a pendulum, we are talking about the general area of mechanics, and more specifically, about dynamics. Some of the problems from dynamics are of direct importance to engineering students who study them in great detail in upper-level courses. For all students, however, the subject of dynamics provides challenging problems that require appreciable analytical reasoning.

Early studies in dynamics were carried on by Galileo (1564–1642), who applied the concept of virtual work and formulated the parallelogram law for motion. Around the same time, the renowned scientist Kepler (1571–1630) stated his famous laws governing planetary motion. This remarkable achievement was made by astronomical observations alone, without the help of calculus.

The classical mechanics that we know today began when Newton (1642–1727) introduced the law of gravitation and explained the concepts of force, momentum, and acceleration. Although much of Newton's work was geometric, it was restated using calculus by Euler (1707–1783), who also extended the original ideas of Newton to include rigid-body motions. In honor of his work, the angles that describe the rotational motion of a rigid body are called Euler angles.

The concept of equilibrium was first postulated by D'Alembert (1717–1783), who introduced what is called the inertia force. This concept, coupled with the results of Newton's work, yields an equation stating that the sum of the forces on a body in motion is zero if the inertia force is included. This seemingly simple fact has very important implications in applied mechanics.

Up to this time, work in mechanics had its foundations in astronomical data and geometrical concepts, but involved very little calculus. This work was formalized by Lagrange (1736–1813), who analytically derived the generalized equations of motion by using energy concepts. His work was one of the most notable milestones in the development of classical mechanics, since it was derived independently of newtonian mechanics. The method of deriving equations of motion from energy considerations is now known as the lagrangian approach, and the equations themselves are called Lagrange's equations.

The work done by Newton and Lagrange was rederived by another towering scientist, Hamilton (1805–1865), who based his work on the principle of least action. He introduced the concept of generalized momentum, which is useful in more advanced concepts of dynamics as well as in the study of the stability of motion.

Many other scientists were engaged in much research; the period from the late eighteenth century to the late nineteenth century saw a rapid advance in mechanics. The torque-free motion of a rigid body was explained via geometrical representation by Poinsot (1777–1859). Coriolis (1792–1843) showed how the introduction of additional terms validated Newton's laws when the reference frame is moving. Foucault (1819–1868) developed the equations for the pendulum and the gyroscope. These are only a few examples of this era's rapid advancement. The great progress made up to this time was reported by Routh (1831–1907) in his *Treatise on Rigid Dynamics* (1860) and *Dynamics of the Particle* (1898).

Einstein's (1878–1955) introduction of the theory of relativity in the twentieth century marked another major jump in the development of mechanics. Although Einstein's work has had a profound effect on mechanics, newtonian mechanics persists for all engineering applications where the magnitude of the velocities under consideration are less than the speed of light.

The first course in dynamics will be restricted to the consideration of particles and rigid bodies. Specifically, we will deal with the kinematics and then the kinetics of particles and rigid bodies. In this chapter we begin our study by first reviewing some basic concepts important in formulating problems in dynamics. These include a review of vectors, coordinate systems, some fundamental laws, and finally a discussion of dimensions and units.

1.1 Vectors

Historically, much of the early work in dynamics was done via astronomical data or geometrical concepts. Later, many of the laws and equations of motion were derived using calculus. The field of dynamics has grown immensely, however, and it is now necessary that we employ a notation that is concise, simple, and complete. Such a notation is to be found in vectors and vector mathematics. When either a physical law or mathematical manipulation is expressed in vector notation, it is independent of the coordinate axes. While greater detail was devoted to the subject of vectors in *Statics*, we shall present a brief review of the highlights on this topic.

A **vector** is a variable that has two properties, magnitude and direction. These properties are independent of any particular coordinate system. If a variable has magnitude but no direction, then it is called a scalar. Velocity and force are two examples of vectors. Both of these quantities are characterized by a magnitude and a direction. On the other hand, the speed of an object, as well as its volume, are two examples of scalar quantities. A vector shall be denoted by $\mathbf{V}$, $\mathbf{A}$, etc., and shall be said to have a magnitude of V, A, etc. Some sample vectors are shown in Fig. 1.1.

A vector may also be represented by means of a **unit vector** whose magnitude is unity and whose direction lies along the direction of the represented vector.

$$\mathbf{A} = A\mathbf{e}_A \tag{1.1}$$

is a vector of magnitude A in the direction of unit vector $\mathbf{e}_A$. The symbol $\mathbf{e}$ together with an appropriate subscript will normally be used for unit vectors. The vectors in Fig. 1.1 are represented using their unit vectors as shown in Fig. 1.2. Unit vectors associated with cartesian coordinates, X, Y, and Z are indicated in Fig. 1.3 by the conventional unit vectors $\mathbf{i}$, $\mathbf{j}$, and $\mathbf{k}$.

Two vectors are defined to be equal if they have the same magnitude *and* the same direction. The sum of two vectors can be obtained by the parallelogram rule or the triangle rule of addition of vectors, as shown in Fig. 1.4. The difference of two vectors can be obtained by using the triangle rule as shown in Fig. 1.5. In adding or subtracting vectors the following rules should be remembered.

(a) Vector addition is commutative since the result is independent of the order of vector addition, i.e., $\mathbf{A} + \mathbf{B} = \mathbf{B} + \mathbf{A}$.

(b) Vector addition and subtraction are associative since vectors may be added or subtracted in any order, i.e., $\mathbf{A} + (\mathbf{B} + \mathbf{C}) = (\mathbf{A} + \mathbf{B}) + \mathbf{C}$.

(c) If k is a scalar quantity, then

$$k(\mathbf{A} + \mathbf{B}) = k\mathbf{A} + k\mathbf{B}$$

and therefore multiplication of a vector by a scalar is distributive.

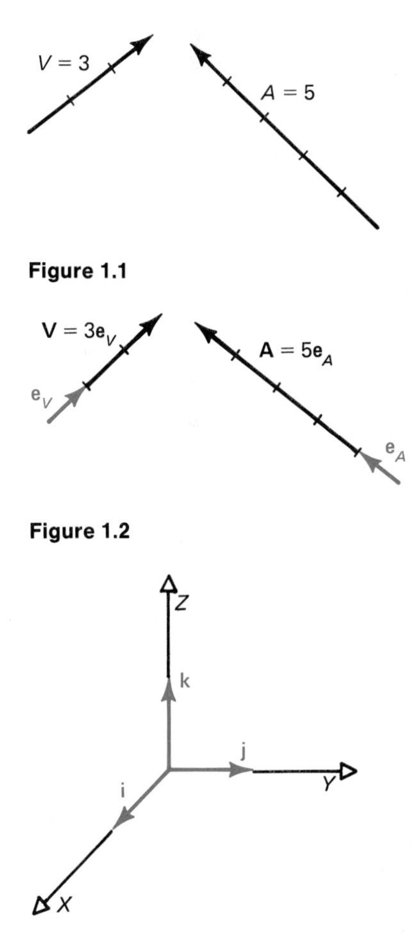

Figure 1.1

Figure 1.2

Figure 1.3

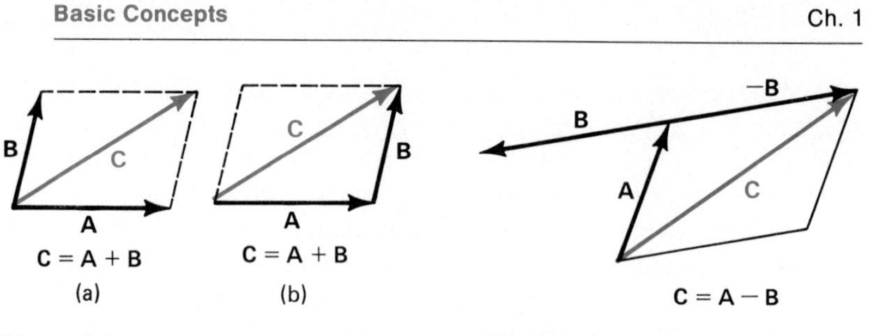

Figure 1.4

Figure 1.5

A vector can be broken into several components according to convenience. Consider the vector **V** shown in Fig. 1.6(a). The components of **V** are also shown in this figure, which illustrates a cartesian coordinate system whose origin coincides with the tail of the vector. Conversely, the two component vectors $\mathbf{V}_x$ and $\mathbf{V}_y$ can be added to yield **V** as shown in Fig. 1.6(b).

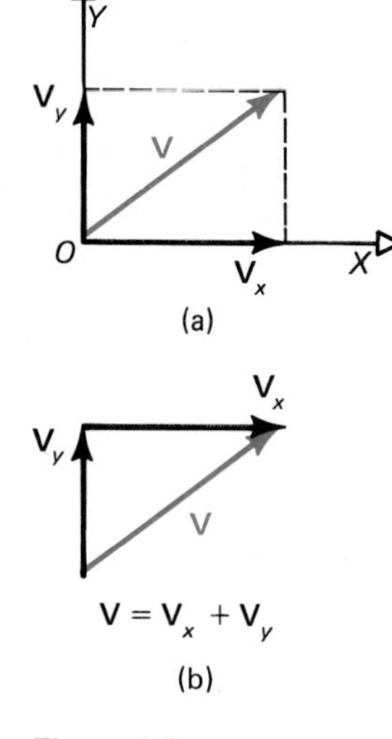

Figure 1.6

To add or subtract vectors, it is convenient to use the unit vector representation with a suitable coordinate system. For example, if

$$\mathbf{A} = A_x\mathbf{i} + A_y\mathbf{j} \qquad \text{and} \qquad \mathbf{B} = B_x\mathbf{i} + B_y\mathbf{j}$$

then

$$\mathbf{A} + \mathbf{B} = (A_x + B_x)\mathbf{i} + (A_y + B_y)\mathbf{j}$$

while

$$\mathbf{A} - \mathbf{B} = (A_x - B_x)\mathbf{i} + (A_y - B_y)\mathbf{j}$$

There are two types of multiplication involving vector quantities: a scalar or dot product and a vector or cross product. The scalar product of **A** and **B** is written as $\mathbf{A} \cdot \mathbf{B}$, and is defined as

$$\mathbf{A} \cdot \mathbf{B} \equiv AB \cos \theta \tag{1.2}$$

where θ is the angle between the vectors **A** and **B**. If

$$\mathbf{A} \cdot \mathbf{B} = 0$$

and $A \neq 0$, $B \neq 0$, then **A** and **B** are orthogonal, i.e., the vectors are perpendicular to each other. If

$$\mathbf{A} \cdot \mathbf{B} = AB$$

then **A** and **B** are collinear and lie in the same direction. If $\mathbf{A} \cdot \mathbf{B} = -AB$, then **A** and **B** are collinear but lie in opposite directions, i.e., $\theta = 180°$.

The scalar product can be used to obtain the projection of a vector in a given direction. In Fig. 1.7 the projection of the vector **A** on the X axis can be obtained by

$$\mathbf{A} \cdot \mathbf{i} = A \cos \theta$$

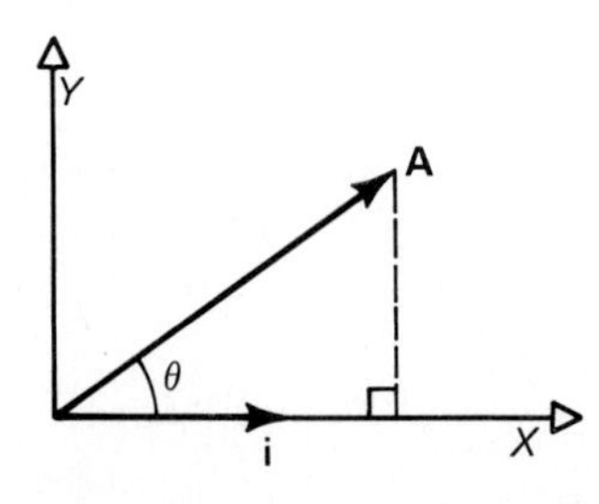

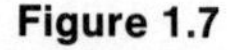
Figure 1.7

Similarly, in Fig. 1.8, the projection of vector **A** along the same direction as vector **B** can be obtained as,

$$\mathbf{A} \cdot \mathbf{e}_B = A \cos \beta$$

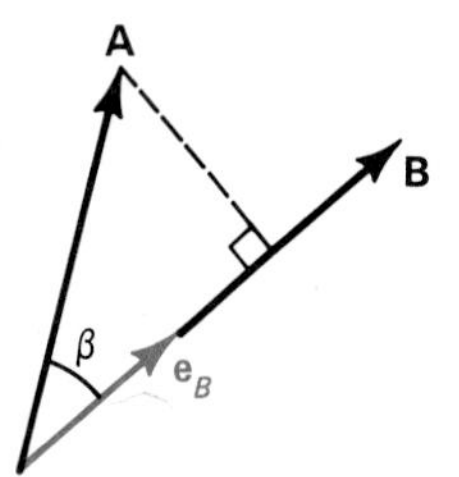

Figure 1.8

Using the definition of a unit vector from Eq. (1.1), we can write $\mathbf{e}_B$ as

$$\mathbf{e}_B = \frac{\mathbf{B}}{B}$$

so that the projection of **A** on the line of action of **B** becomes $(\mathbf{A} \cdot \mathbf{B})/B$.

The scalar product is also used to define the direction cosines of a vector. Consider vector **A** in the cartesian system shown in Fig. 1.9. Here,

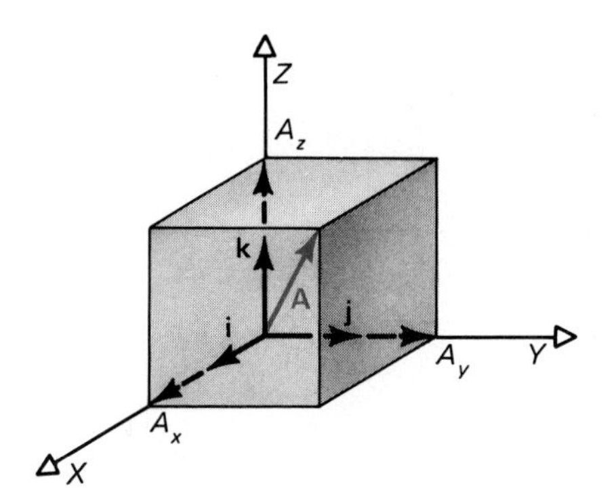

Figure 1.9

$$\mathbf{A} = A_x\mathbf{i} + A_y\mathbf{j} + A_z\mathbf{k} \tag{1.3}$$

where **i**, **j**, **k** are unit vectors along the X, Y, Z axes, and A_x, A_y, A_z are scalar components of **A** along the X, Y, Z axes. Let us form the scalar product of both sides with **i**,

$$\mathbf{A} \cdot \mathbf{i} = A_x(\mathbf{i} \cdot \mathbf{i}) + A_y(\mathbf{j} \cdot \mathbf{i}) + A_z(\mathbf{k} \cdot \mathbf{i}) = A_x$$

since $\mathbf{i} \cdot \mathbf{i}$ is unity and $\mathbf{j} \cdot \mathbf{i} = \mathbf{k} \cdot \mathbf{i} = 0$. Therefore, A_x is the scalar product of **A** and **i**. And, from the definition of a scalar product,

$$\mathbf{A} \cdot \mathbf{i} = A_x = A \cos \theta_x \tag{1.4a}$$

where θ_x is the angle between **A** and **i**. Similarly, we obtain

$$\mathbf{A} \cdot \mathbf{j} = A_y = A \cos \theta_y \tag{1.4b}$$

$$\mathbf{A} \cdot \mathbf{k} = A_z = A \cos \theta_z \tag{1.4c}$$

Substituting into Eq. (1.3), we have

$$\mathbf{A} = A \cos \theta_x\mathbf{i} + A \cos \theta_y\mathbf{j} + A \cos \theta_z\mathbf{k} \tag{1.5}$$

However, we have established that

$$\mathbf{A} = A\mathbf{e}_A \tag{1.1}$$

Substituting Eq. (1.1) into the left side of Eq. (1.5) and canceling the A, we obtain

$$\mathbf{e}_A = \cos \theta_x\mathbf{i} + \cos \theta_y\mathbf{j} + \cos \theta_z\mathbf{k} \tag{1.6}$$

The three cosines are called the **direction cosines** and represent the cosines of the angles between **A** and the coordinate axes. If we take the scalar product of both sides with $\mathbf{e}_A$, we then have

$$\cos^2 \theta_x + \cos^2 \theta_y + \cos^2 \theta_z = 1$$

The direction cosines are

$$\cos \theta_x = \frac{A_x}{A} \qquad \cos \theta_y = \frac{A_y}{A} \qquad \cos \theta_z = \frac{A_z}{A}$$

Here A is the magnitude of **A**, which is obtained by forming the dot product of **A** with itself.

$$\mathbf{A} \cdot \mathbf{A} = A^2 = A_x^2 + A_y^2 + A_z^2$$

and the magnitude A becomes

$$A = \sqrt{A_x^2 + A_y^2 + A_z^2} \tag{1.7}$$

The scalar product of two vectors **A** and **B** was written above as

$$\mathbf{A} \cdot \mathbf{B} = AB \cos \theta$$

But, if the vectors are written in cartesian coordinates as

$$\mathbf{A} = A_x\mathbf{i} + A_y\mathbf{j} + A_z\mathbf{k} \qquad \text{and} \qquad \mathbf{B} = B_x\mathbf{i} + B_y\mathbf{j} + B_z\mathbf{k}$$

then the scalar product becomes

$$\begin{aligned} \mathbf{A} \cdot \mathbf{B} &= (A_x\mathbf{i} + A_y\mathbf{j} + A_z\mathbf{k}) \cdot (B_x\mathbf{i} + B_y\mathbf{j} + B_z\mathbf{k}) \\ &= A_xB_x + A_yB_y + A_zB_z \end{aligned}$$

The angle between the vectors **A** and **B** can be obtained by substituting into Eq. (1.2) and solving for $\cos \theta$.

$$\cos \theta = \frac{A_xB_x + A_yB_y + A_zB_z}{AB}$$

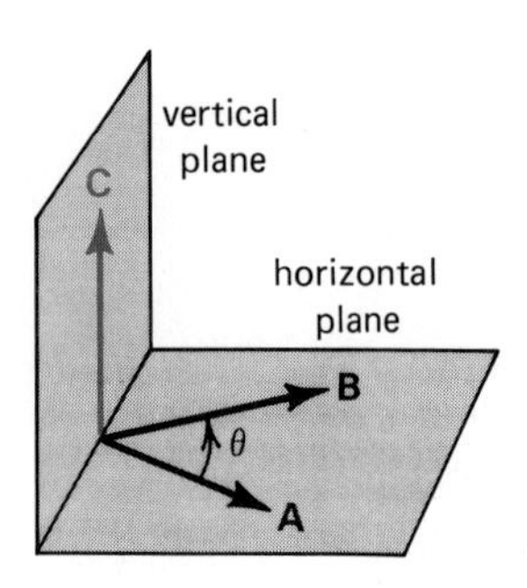

Figure 1.10

where A is the magnitude of **A** and B is the magnitude of **B**.

Let us now consider the vector or cross product of two vectors, written $\mathbf{A} \times \mathbf{B}$, that is defined as

$$\mathbf{C} = \mathbf{A} \times \mathbf{B} = AB \sin \theta \ \mathbf{e}_c \tag{1.8}$$

where $0 \leq \theta \leq 180°$ and $\mathbf{e}_c$ is a unit vector perpendicular to the plane containing the vectors **A** and **B** as shown in Fig. 1.10. The direction of **C** is obtained by the right-hand rule, i.e., if the vector **A** is rotated by the smallest angle that brings it into coincidence with **B**, then the direction of **C** is that of an advancing right-handed screw that turns in the same direction as the vector **A**, as shown in Fig. 1.11. From the definition of a cross product, if

$$\mathbf{A} \times \mathbf{B} = 0$$

and $A \neq 0$, $B \neq 0$, then **A** and **B** are collinear. If **A** and **B** are orthogonal, then

$$\mathbf{A} \times \mathbf{B} = AB\mathbf{e}_c$$

It can be shown that the cross product is not commutative since

$$\mathbf{A} \times \mathbf{B} \neq \mathbf{B} \times \mathbf{A}$$

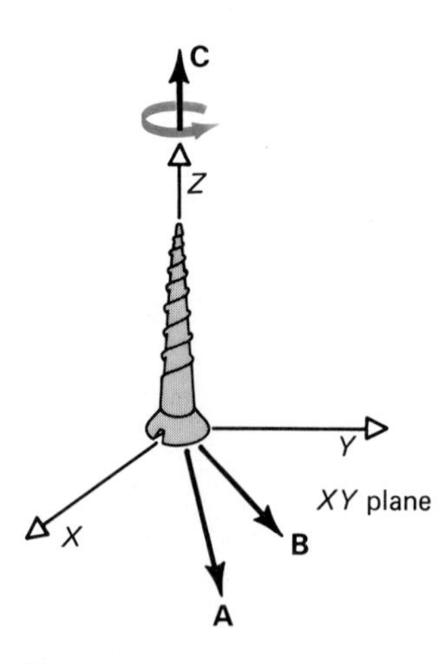

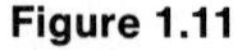
Figure 1.11

Let us consider **A** and **B** in cartesian coordinates and form the cross product,

$$\mathbf{A} \times \mathbf{B} = (A_x\mathbf{i} + A_y\mathbf{j} + A_z\mathbf{k}) \times (B_x\mathbf{i} + B_y\mathbf{j} + B_z\mathbf{k}) \qquad (1.9)$$

It can be shown that the cross product is distributive so that the above expression can be computed term by term. Recalling that **i**, **j**, **k** are unit vectors perpendicular to each other, we have the following relationships for their cross products:

$$\mathbf{i} \times \mathbf{j} = \mathbf{k} = -\mathbf{j} \times \mathbf{i}$$

$$\mathbf{j} \times \mathbf{k} = \mathbf{i} = -\mathbf{k} \times \mathbf{j}$$

$$\mathbf{k} \times \mathbf{i} = \mathbf{j} = -\mathbf{i} \times \mathbf{k}$$

$$\mathbf{i} \times \mathbf{i} = \mathbf{j} \times \mathbf{j} = \mathbf{k} \times \mathbf{k} = 0$$

Using these relationships, Eq. (1.9) can be reduced to

$$\mathbf{A} \times \mathbf{B} = (A_yB_z - A_zB_y)\mathbf{i} + (A_zB_x - A_xB_z)\mathbf{j} + (A_xB_y - A_yB_x)\mathbf{k}$$

This result can be compactly written as a **determinant**:

$$\mathbf{A} \times \mathbf{B} = \begin{vmatrix} \mathbf{i} & \mathbf{j} & \mathbf{k} \\ A_x & A_y & A_z \\ B_x & B_y & B_z \end{vmatrix} \qquad (1.10)$$

The triple vector product, **A** × (**B** × **C**), is frequently used in dynamics. To evaluate this product, we first calculate the terms inside the parentheses. The vector (**B** × **C**) is then premultiplied by the vector **A**. These steps are demonstrated in Example 1.4.

Example 1.1

A vector **A** is given by

$$\mathbf{A} = 5\mathbf{i} + \mathbf{j} + 2\mathbf{k}$$

What is (a) the magnitude of **A**, (b) each direction cosine of **A**, and (c) the unit vector $\mathbf{e}_A$ in terms of the unit vectors **i**, **j**, and **k**?

(a) Using Eq. (1.7), we obtain

$$A = \sqrt{5^2 + 1 + 2^2} = 5.48 \qquad \textit{Answer}$$

(b) Using Eqs. (1.4), we obtain

$$\cos\theta_x = \frac{\mathbf{A}\cdot\mathbf{i}}{A} = \frac{5}{5.48} = 0.912 \qquad \textit{Answer}$$

$$\cos\theta_y = \frac{\mathbf{A}\cdot\mathbf{j}}{A} = \frac{1}{5.48} = 0.182 \qquad \textit{Answer}$$

$$\cos\theta_z = \frac{\mathbf{A}\cdot\mathbf{k}}{A} = \frac{2}{5.48} = 0.365 \qquad \textit{Answer}$$

(c) If we substitute the above into Eq. (1.6) we find that

$$\mathbf{e}_A = 0.912\mathbf{i} + 0.182\mathbf{j} + 0.365\mathbf{k} \qquad \textit{Answer}$$

The unit vector $\mathbf{e}_A$ is coincident with $\mathbf{A}$ and can be obtained by dividing $\mathbf{A}$ by its magnitude. Therefore, as an alternate approach,

$$\mathbf{e}_A = \frac{\mathbf{A}}{A} = \frac{1}{5.48}(5\mathbf{i} + \mathbf{j} + 2\mathbf{k})$$

$$\mathbf{e}_A = 0.912\mathbf{i} + 0.182\mathbf{j} + 0.365\mathbf{k}$$

Example 1.2

Obtain the projection of **A** on **B** if

$$\mathbf{A} = 2\mathbf{i} + 3\mathbf{j} \qquad \text{and} \qquad \mathbf{B} = 5\mathbf{i} + 2\mathbf{j} + \mathbf{k}$$

The projection of **A** on **B** is $A \cos \theta$. The dot product of **A** and **B** is

$$\mathbf{A} \cdot \mathbf{B} = AB \cos \theta$$

Therefore,

$$A \cos \theta = \frac{\mathbf{A} \cdot \mathbf{B}}{B} = \frac{(2\mathbf{i} + 3\mathbf{j}) \cdot (5\mathbf{i} + 2\mathbf{j} + \mathbf{k})}{\sqrt{5^2 + 2^2 + 1}}$$

$$= \frac{16}{\sqrt{30}} = 2.92 \qquad \textit{Answer}$$

Example 1.3

Obtain $\mathbf{A} \times \mathbf{B}$ for the above problem.

$$\mathbf{A} \times \mathbf{B} = \begin{vmatrix} \mathbf{i} & \mathbf{j} & \mathbf{k} \\ 2 & 3 & 0 \\ 5 & 2 & 1 \end{vmatrix} = 3\mathbf{i} - 2\mathbf{j} - 11\mathbf{k} \qquad \textit{Answer}$$

Example 1.4

Find $\mathbf{w} \times (\mathbf{w} \times \mathbf{r})$ if $\mathbf{w} = 2\mathbf{i} - 3\mathbf{j} + 4\mathbf{k}$ and $\mathbf{r} = 2\mathbf{i} + \mathbf{j}$.

First find the vector $(\mathbf{w} \times \mathbf{r})$ by forming the determinant.

$$\mathbf{w} \times \mathbf{r} = \begin{vmatrix} \mathbf{i} & \mathbf{j} & \mathbf{k} \\ 2 & -3 & 4 \\ 2 & 1 & 0 \end{vmatrix} = -4\mathbf{i} + 8\mathbf{j} + 8\mathbf{k}$$

Now cross **w** into $(\mathbf{w} \times \mathbf{r})$ to obtain

$$\mathbf{w} \times (\mathbf{w} \times \mathbf{r}) = \begin{vmatrix} \mathbf{i} & \mathbf{j} & \mathbf{k} \\ 2 & -3 & 4 \\ -4 & 8 & 8 \end{vmatrix} = -56\mathbf{i} - 32\mathbf{j} + 4\mathbf{k} \qquad \textit{Answer}$$

PROBLEMS

1.1 If $\mathbf{A} = 5\mathbf{i} + 8\mathbf{j}$ and $\mathbf{B} = 6\mathbf{i} - 8\mathbf{j} + 2\mathbf{k}$, find
(a) $\mathbf{A} + \mathbf{B}$ (c) $\mathbf{A} + 2\mathbf{B}$
(b) $\mathbf{A} - \mathbf{B}$ (d) $3\mathbf{A} - \mathbf{B}$

1.2 If $\mathbf{V}_1 = 8\mathbf{j} + 10\mathbf{k}$ and $\mathbf{V}_2 = -6\mathbf{j} + 15\mathbf{k}$, find
(a) $\mathbf{V}_1 + \mathbf{V}_2$ (c) $5\mathbf{V}_1 + 2\mathbf{V}_2$
(b) $\mathbf{V}_2 - \mathbf{V}_1$ (d) $3\mathbf{V}_1 - 4\mathbf{V}_2$

1.3 Show that $\mathbf{A} \cdot \mathbf{B} = \mathbf{B} \cdot \mathbf{A}$, if $\mathbf{A} = A_x\mathbf{i} + A_y\mathbf{j}$ and $\mathbf{B} = B_x\mathbf{i} + B_y\mathbf{j}$.

1.4 If $\mathbf{C} = 10\mathbf{i} + 4\mathbf{j}$ and $\mathbf{D} = 4\mathbf{i} - 10\mathbf{j}$, find $\mathbf{C} \cdot \mathbf{D}$.

1.5 If $\mathbf{V}_1 = 2\mathbf{i} + 3\mathbf{k}$ and $\mathbf{V}_2 = -4\mathbf{k}$, find $\mathbf{V}_1 \cdot \mathbf{V}_2$.

1.6 If $\mathbf{A} = 5\mathbf{i} + 8\mathbf{j}$ and $\mathbf{B} = 6\mathbf{i} - 8\mathbf{j} + 2\mathbf{k}$, find $(\mathbf{A} + \mathbf{B}) \cdot (3\mathbf{A} - \mathbf{B})$.

1.7 If $\mathbf{C} = -2\mathbf{i} + 4\mathbf{j}$ and $\mathbf{D} = 3\mathbf{i} - \mathbf{j}$, find $(\mathbf{C} - \mathbf{D}) \cdot (\mathbf{C} + 2\mathbf{D})$.

1.8 Obtain a unit vector that gives the direction of $\mathbf{B} - \mathbf{A}$ if

$$\mathbf{A} = 3\mathbf{i} + \mathbf{j} \qquad \mathbf{B} = \mathbf{i} + 2\mathbf{j} + 3\mathbf{k}$$

where $\mathbf{i}$, $\mathbf{j}$, $\mathbf{k}$ are unit vectors in the cartesian system.

1.9 If $\mathbf{C} = \mathbf{A} - \mathbf{B}$, then using the dot product show that $C^2 = A^2 + B^2 - 2AB\cos\theta$.

1.10 If $\mathbf{A} = 5\mathbf{i} + 5\mathbf{j} + \mathbf{k}$ and $\mathbf{B} = 2\mathbf{i} + 4\mathbf{k}$, find $\mathbf{A} \times \mathbf{B}$.

1.11 If $\mathbf{V}_1 = 2\mathbf{i} + \mathbf{j} + \mathbf{k}$ and $\mathbf{V}_2 = -\mathbf{j} + 2\mathbf{k}$, find $\mathbf{V}_1 \times \mathbf{V}_2$.

1.12 If $\mathbf{A} = 5\mathbf{i} + 8\mathbf{j}$ and $\mathbf{B} = 6\mathbf{i} - 8\mathbf{j} + 2\mathbf{k}$, find $(\mathbf{A} + \mathbf{B}) \times (3\mathbf{A} - \mathbf{B})$.

1.13 If $\mathbf{C} = -2\mathbf{i} + 4\mathbf{j}$ and $\mathbf{D} = 3\mathbf{i} - \mathbf{j}$, find $(\mathbf{C} - \mathbf{D}) \times (\mathbf{C} + 2\mathbf{D})$.

1.14 Using $\mathbf{A} = A_1\mathbf{i} + A_2\mathbf{j}$ and $\mathbf{B} = B_1\mathbf{i} + B_2\mathbf{j}$, show that $\mathbf{A} \times \mathbf{B} = -\mathbf{B} \times \mathbf{A}$.

1.15 Using the definition of a cross product, show that

$$\frac{\sin\theta_1}{B} = \frac{\sin\theta_2}{C}$$

where A, B, and C are the magnitudes of the vectors $\mathbf{A}$, $\mathbf{B}$, and $\mathbf{C}$ that form a triangle, and θ_1 is the angle between $\mathbf{A}$ and $\mathbf{C}$ while θ_2 is the angle between $\mathbf{A}$ and $\mathbf{B}$.

1.16 For Problem 1.10 obtain the magnitude of $\mathbf{A} \times \mathbf{B}$ by computing $AB \sin\theta$ and using the scalar product for evaluating $\sin\theta$.

1.17 If $\mathbf{A} = 3\mathbf{i}$ and $\mathbf{B} = \mathbf{i} - \mathbf{j}$, find $\mathbf{B} \times (\mathbf{A} \times \mathbf{B})$.

1.18 If $\mathbf{w} = \mathbf{j} - \mathbf{k}$ and $\mathbf{r} = \mathbf{i} + 2\mathbf{j}$, find $\mathbf{w} \times (\mathbf{w} \times \mathbf{r})$.

1.19 If $\mathbf{w} = 6\mathbf{k}$ and $\mathbf{r} = \mathbf{i} + 2\mathbf{j} + 3\mathbf{k}$, find $\mathbf{w} \times (\mathbf{w} \times \mathbf{r})$ and $(\mathbf{w} \times \mathbf{r}) \times \mathbf{w}$.

1.2 Coordinate Systems

While the properties of a vector are independent of any particular coordinate system, there are special coordinate systems that we frequently use to express vectors. The choice of a particular coordinate system is made based upon the physical details of a given problem. We now introduce four commonly used coordinate systems that we shall need for future use.

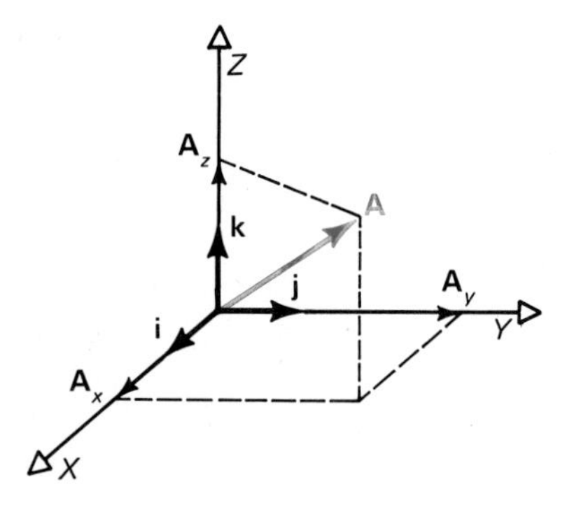

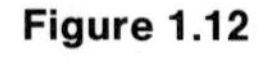
Figure 1.12

1. *Cartesian Coordinate System.* Figure 1.12 shows the vector **A** in three dimensions. Using the **i**, **j**, **k** unit vectors,

$$\mathbf{A} = A_x\mathbf{i} + A_y\mathbf{j} + A_z\mathbf{k} \tag{1.11}$$

Note that the unit vectors form a right-hand system, i.e., $\mathbf{i} \times \mathbf{j} = \mathbf{k}$.

2. *Polar Coordinate System.* This system uses the unit vectors $\mathbf{e}_r$ and $\mathbf{e}_\theta$ in Fig. 1.13 to represent the vector in two dimensions. Now

$$\mathbf{A} = A\mathbf{e}_r \tag{1.12}$$

where the unit vectors are related as follows:

$$\mathbf{e}_r = \cos\theta\mathbf{i} + \sin\theta\mathbf{j} \tag{1.13a}$$

$$\mathbf{e}_\theta = -\sin\theta\mathbf{i} + \cos\theta\mathbf{j} \tag{1.13b}$$

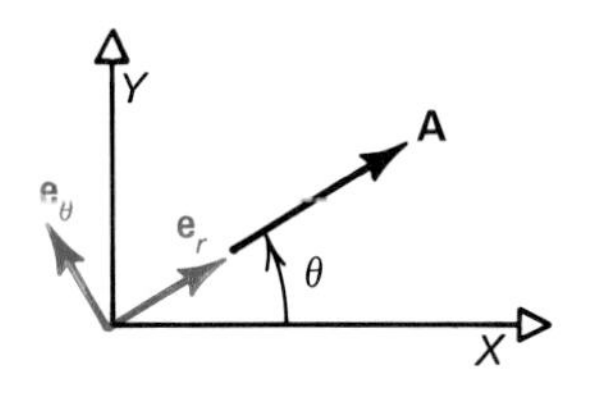

Figure 1.13

While $\mathbf{e}_\theta$ does not appear in Eq. (1.12), we shall see later that it does appear in the derivative of **A**.

3. *Cylindrical Coordinate System.* The cylindrical coordinate system represents a vector by an angle and two lengths, as shown in Fig. 1.14. The three unit vectors $\mathbf{e}_r$, $\mathbf{e}_\theta$, and **k** associated with cylindrical coordinates are mutually perpendicular, such that $\mathbf{e}_r \times \mathbf{e}_\theta = \mathbf{k}$. The vector **A** is given by

$$\mathbf{A} = A_r\mathbf{e}_r + A_z\mathbf{k} \tag{1.14}$$

where A_r and A_z are the components of **A** on the r and Z axes, respectively.

It can be seen from Fig. 1.14 that the cylindrical coordinate unit vectors are related to the cartesian system by the following relationships:

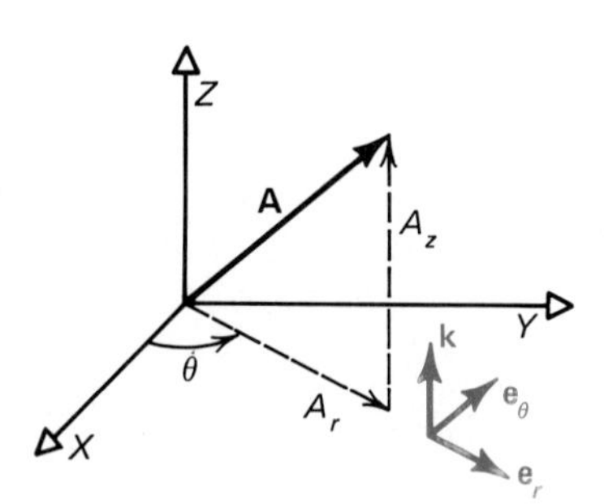

Figure 1.14

$$\mathbf{e}_r = \cos\theta\mathbf{i} + \sin\theta\mathbf{j} \tag{1.15a}$$

$$\mathbf{e}_\theta = -\sin\theta\mathbf{i} + \cos\theta\mathbf{j} \tag{1.15b}$$

$$\mathbf{k} = \mathbf{k} \tag{1.15c}$$

4. *Spherical Coordinate System.* This system represents a vector by two angles and one length, as shown in Fig. 1.15. The three unit vectors $\mathbf{e}_r$, $\mathbf{e}_\phi$, and $\mathbf{e}_\theta$ associated with spherical coordinates are mutually perpendicular, such that $\mathbf{e}_r \times \mathbf{e}_\phi = \mathbf{e}_\theta$. The vector **A** is given by

$$\mathbf{A} = A\mathbf{e}_r \tag{1.16}$$

where the relationships between the unit vectors in the spherical coordinate system and the cartesian coordinate system are as follows:

$$\mathbf{e}_r = \sin\phi(\cos\theta\mathbf{i} + \sin\theta\mathbf{j}) + \cos\phi\mathbf{k} \tag{1.17a}$$

$$\mathbf{e}_\phi = \cos\phi(\cos\theta\mathbf{i} + \sin\theta\mathbf{j}) - \sin\phi\mathbf{k} \tag{1.17b}$$

$$\mathbf{e}_\theta = -\sin\theta\mathbf{i} + \cos\theta\mathbf{j} \tag{1.17c}$$

Again, while $\mathbf{e}_\phi$ and $\mathbf{e}_\theta$ do not appear in Eq. (1.16), they are used to represent the derivatives of **A**.

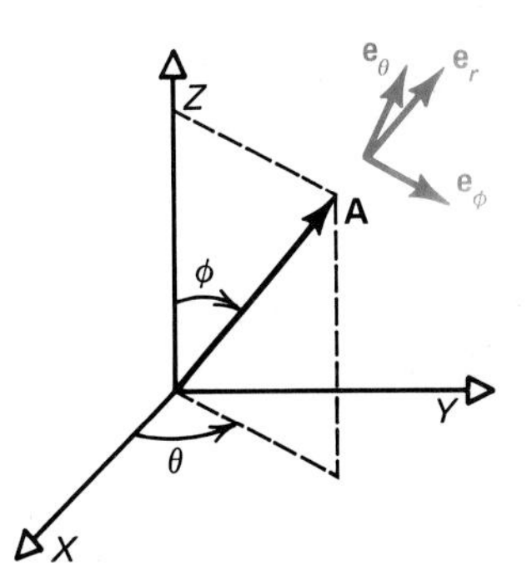

Figure 1.15

Example 1.5

The following vectors are expressed in cylindrical coordinates:

$$\mathbf{A} = 3\mathbf{e}_r + 2\mathbf{e}_\theta + \mathbf{k} \qquad \mathbf{B} = -4\mathbf{e}_r + 3\mathbf{k}$$

Find (a) $\mathbf{A} \cdot \mathbf{B}$ and (b) $\mathbf{A} \times \mathbf{B}$.

(a) Since the unit vectors $\mathbf{e}_r$, $\mathbf{e}_\theta$, and **k** form an orthogonal set of unit vectors,

$$\begin{aligned}\mathbf{A} \cdot \mathbf{B} &= (3\mathbf{e}_r + 2\mathbf{e}_\theta + \mathbf{k}) \cdot (-4\mathbf{e}_r + 3\mathbf{k}) \\ &= -12 + 3 = -9 \qquad \textit{Answer}\end{aligned}$$

(b) In finding the cross product, notice that the order of the unit vectors in the first row of the determinant follows the right-hand system, so that

$$\begin{aligned}\mathbf{A} \times \mathbf{B} &= \begin{vmatrix} \mathbf{e}_r & \mathbf{e}_\theta & \mathbf{k} \\ 3 & 2 & 1 \\ -4 & 0 & 3 \end{vmatrix} \\ &= 6\mathbf{e}_r - 13\mathbf{e}_\theta + 8\mathbf{k} \qquad \textit{Answer}\end{aligned}$$

Example 1.6

For the case of the unit vectors in a spherical coordinate system, use Eqs. (1.17) to show that $\mathbf{e}_r \times \mathbf{e}_\phi = \mathbf{e}_\theta$.

Referring to Eqs. (1.17a) and (1.17b),

$$
\begin{aligned}
\mathbf{e}_r \times \mathbf{e}_\phi &= \begin{vmatrix} \mathbf{i} & \mathbf{j} & \mathbf{k} \\ \sin\phi\cos\theta & \sin\phi\sin\theta & \cos\phi \\ \cos\phi\cos\theta & \cos\phi\sin\theta & -\sin\phi \end{vmatrix} \\
&= \mathbf{i}(-\sin^2\phi\sin\theta - \cos^2\phi\sin\theta) \\
&\quad - \mathbf{j}(-\sin^2\phi\cos\theta - \cos^2\phi\cos\theta) \\
&\quad + \mathbf{k}(\sin\phi\cos\phi\sin\theta\cos\theta - \sin\phi\cos\phi\sin\theta\cos\theta) \\
&= -\sin\theta\mathbf{i} + \cos\theta\mathbf{j} \\
&= \mathbf{e}_\theta
\end{aligned}
$$

PROBLEMS

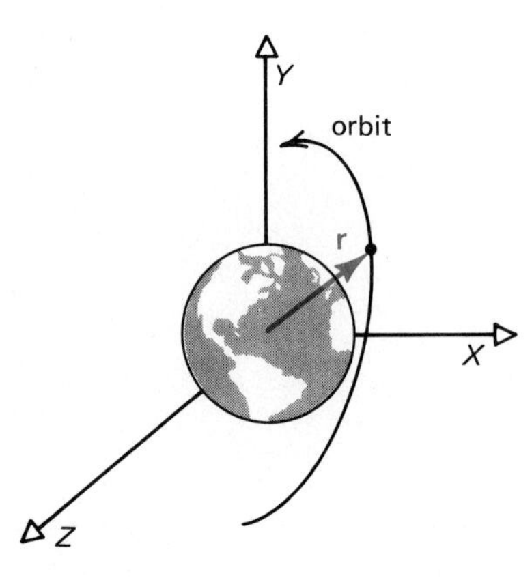

Figure P1.20

1.20 The position of a satellite in Fig. P1.20 is given by

$$\mathbf{r} = (5\mathbf{i} + 3\mathbf{j} + 4\mathbf{k})R$$

where $R = 6.38 \times 10^6$ m is the earth's radius. What is the magnitude of $\mathbf{r}$? Obtain a unit vector coincident with $\mathbf{r}$.

1.21 Find $\mathbf{A} \cdot \mathbf{B}$ for the vectors expressed in polar coordinates, $\mathbf{A} = 10\mathbf{e}_r - 5\mathbf{e}_\theta$, $\mathbf{B} = -15\mathbf{e}_\theta$.

1.22 Find $\mathbf{C} \times \mathbf{D}$ for the vectors expressed in polar coordinates, $\mathbf{C} = -4\mathbf{e}_r + \mathbf{e}_\theta$, $\mathbf{D} = 5\mathbf{e}_r - 3\mathbf{e}_\theta$.

1.23 Using the transformation equations for cylindrical coordinates given by Eqs. (1.15), find $\mathbf{e}_r \cdot \mathbf{e}_\theta$, $\mathbf{e}_r \cdot \mathbf{k}$, and $\mathbf{e}_\theta \cdot \mathbf{k}$.

1.24 A missile is to be fired with velocity $\mathbf{V}_1$ so that it is headed in the direction shown in Fig. P1.24. However, it misfires so that its velocity is $\mathbf{V}_2$. Obtain the magnitude of the velocity difference $\mathbf{V}_2 - \mathbf{V}_1$.

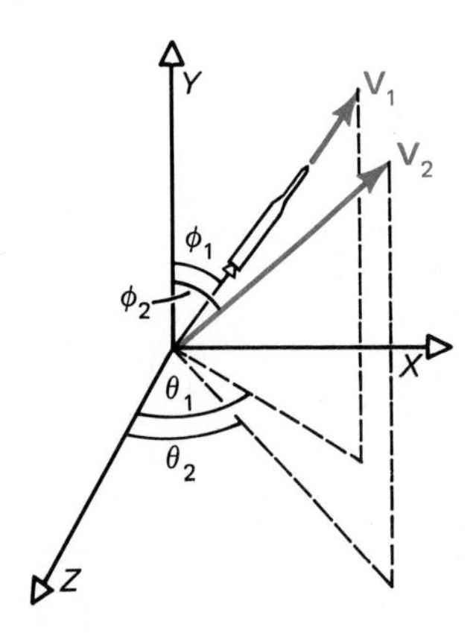

Figure P1.24

1.25 Using the transformation equations for cylindrical coordinates given by Eq. (1.15), find $\mathbf{e}_r \times \mathbf{e}_\theta$, $\mathbf{e}_\theta \times \mathbf{k}$, and $\mathbf{k} \times \mathbf{e}_r$.

1.26 Using the transformation equations for spherical coordinates given by Eq. (1.17), find $\mathbf{e}_\phi \times \mathbf{e}_\theta$.

1.27 An airplane is flying at a speed of V horizontally as shown in Fig. P1.27. Express the velocity $\mathbf{V}$ in terms of polar coordinates $\mathbf{e}_r$ and $\mathbf{e}_\theta$ at the instant shown.

1.28 The positions of two cars in radio contact in Fig. P1.28 are measured from a fixed point O and are given by

$$\mathbf{r}_B = 10(\cos\pi t\mathbf{i} + \sin\pi t\mathbf{j})$$

$$\mathbf{r}_A = 10\cos 3\pi t\mathbf{i} + 10(2 + \sin 3\pi t)\mathbf{j}$$

What is the location of car A relative to B at $t = 1$ s?

1.29 Two vectors from the origin to points A and B, respectively, are given in cylindrical coordinates as

$$\mathbf{A} = 2\mathbf{e}_{r1} + 7\mathbf{k} \qquad \mathbf{B} = \mathbf{e}_{r2} + 6\mathbf{k}$$

where $\mathbf{e}_{r1} = \cos\theta_1\mathbf{i} + \sin\theta_1\mathbf{j} \qquad \mathbf{e}_{r2} = \cos\theta_2\mathbf{i} + \sin\theta_2\mathbf{j}$

Obtain $\mathbf{A} \cdot \mathbf{B}$ and $\mathbf{A} \times \mathbf{B}$. What is the magnitude of the distance between A and B?

1.30 Verify the relation between cylindrical and cartesian unit vectors given by Eqs. (1.15).

1.31 Verify the relations between the spherical and cartesian unit vectors given by Eq. (1.17).

1.32 The velocity of an aircraft is seen to consist of two components, $\mathbf{V}_r$ and $\mathbf{V}_\theta$ as shown in Fig. P1.32. Obtain the components of the resultant velocity in cartesian coordinates.

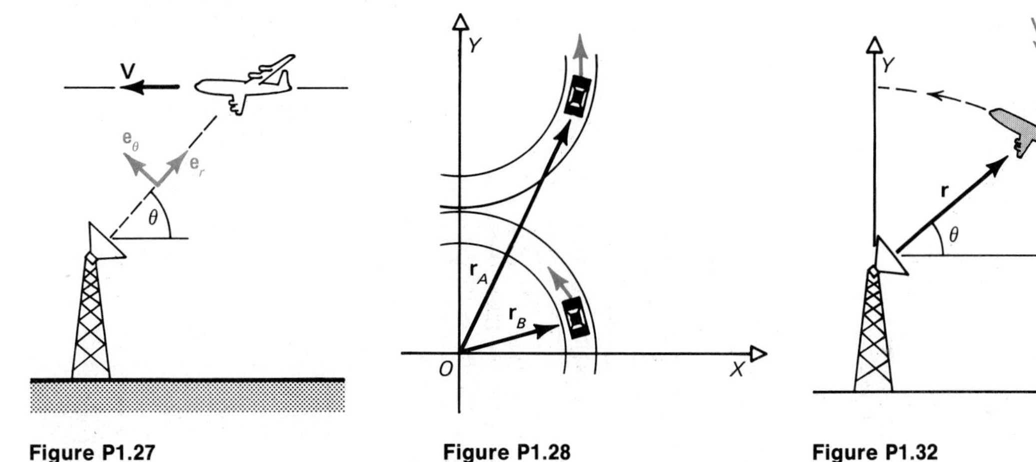

Figure P1.27 Figure P1.28 Figure P1.32

1.3 Derivative of a Vector

The derivative of a vector plays a very important role when we derive the equations of velocity and acceleration in the next chapter, and therefore must be clearly understood.

Consider the vector $\mathbf{A}$ that varies with time t. Referring to Eq. (1.1),

$$\mathbf{A} = A\mathbf{e}_A$$

where we assume that both the magnitude A and the unit vector $\mathbf{e}_A$ vary with time t. The derivative of $\mathbf{A}$ with respect to time t is

$$\frac{d\mathbf{A}}{dt} = \frac{dA}{dt}\mathbf{e}_A + A\frac{d\mathbf{e}_A}{dt} \tag{1.18}$$

This shows that the vector **A** undergoes a change in magnitude represented by dA/dt and a change in direction represented by $d\mathbf{e}_A/dt$. We shall see in the next chapter that $d\mathbf{e}_A/dt$ is determined by the rate of rotation of the vector **A**.

If the vector **A** is written in cartesian coordinate form,

$$\mathbf{A} = A_x\mathbf{i} + A_y\mathbf{j} + A_z\mathbf{k}$$

then the derivative becomes

$$\frac{d\mathbf{A}}{dt} = \left(\frac{dA_x}{dt}\mathbf{i} + \frac{dA_y}{dt}\mathbf{j} + \frac{dA_z}{dt}\mathbf{k}\right) + \left(A_x\frac{d\mathbf{i}}{dt} + A_y\frac{d\mathbf{j}}{dt} + A_z\frac{d\mathbf{k}}{dt}\right)$$

Note that the last three terms in this equation go to zero if the unit vectors **i**, **j**, and **k** are fixed, i.e., their directions do not vary with time. When the unit vectors are fixed, the coordinate system itself is fixed. If the unit vectors are functions of time, the particular functional relationship must be known in order to obtain the derivatives. When the unit vectors are moving, the particular coordinate system is called a moving coordinate system.

Example 1.7

Given the vector $\mathbf{B} = t^2\mathbf{e}_B$ where $\mathbf{e}_B = \sin 3t\mathbf{i} - \cos 3t\mathbf{j}$, find the derivative of **B** with respect to time. Assume the unit vectors **i** and **j** are fixed.

$$\frac{d\mathbf{B}}{dt} = 2t\mathbf{e}_B + t^2\frac{d\mathbf{e}_B}{dt}$$

where

$$\frac{d\mathbf{e}_B}{dt} = 3\cos 3t\mathbf{i} + 3\sin 3t\mathbf{j}$$

so that

$$\frac{d\mathbf{B}}{dt} = 2t(\sin 3t\mathbf{i} - \cos 3t\mathbf{j}) + 3t^2(\cos 3t\mathbf{i} + \sin 3t\mathbf{j})$$

$$= (2t\sin 3t + 3t^2\cos 3t)\mathbf{i} + (-2t\cos 3t + 3t^2\sin 3t)\mathbf{j}$$

Answer

Example 1.8

The position of a point A describes a spiral such that

$$\mathbf{A} = R(t)\mathbf{e}_R(t)$$

where $R(t) = 5t^2$ and $\mathbf{e}_R(t) = \cos 2t\mathbf{i} + \sin 2t\mathbf{j}$.

Assume that **i** and **j** are fixed. (a) What is the rate of change of **A** with respect to time? (b) Find **A** and its derivative at $t = 0$, $\pi/4$, $\pi/2$.

$$\mathbf{A} = 5t^2(\cos 2t\mathbf{i} + \sin 2t\mathbf{j})$$

$$\frac{d\mathbf{A}}{dt} = 10t(\cos 2t\mathbf{i} + \sin 2t\mathbf{j}) + 10t^2(-\sin 2t\mathbf{i} + \cos 2t\mathbf{j}) \qquad \textit{Answer}$$

At $t = 0$, $\quad \mathbf{A} = 0 \qquad \dfrac{d\mathbf{A}}{dt} = 0$ *Answer*

At $t = \dfrac{\pi}{4}$, $\quad \mathbf{A} = \dfrac{5\pi^2}{16}\mathbf{j} = 3.08\mathbf{j} \qquad \dfrac{d\mathbf{A}}{dt} = \dfrac{5\pi}{2}\mathbf{j} - \dfrac{5\pi^2}{8}\mathbf{i} = -6.17\mathbf{i} + 7.85\mathbf{j}$ *Answer*

At $t = \dfrac{\pi}{2}$, $\quad \mathbf{A} = -\dfrac{5\pi^2}{4}\mathbf{i} = -12.34\mathbf{i} \qquad \dfrac{d\mathbf{A}}{dt} = -5\pi\mathbf{i} - \dfrac{5\pi^2}{2}\mathbf{j} = -15.71\mathbf{i} - 24.67\mathbf{j}$ *Answer*

PROBLEMS

1.33 If $\mathbf{A} = (3 \sin 2t)(\mathbf{i} + \mathbf{j})$, find $d\mathbf{A}/dt$ using $d\mathbf{i}/dt = -5\mathbf{j}$ and $d\mathbf{j}/dt = -5\mathbf{i}$.

1.34 If $\mathbf{A} = 2t^2(\mathbf{i} - \mathbf{j})$ and $\mathbf{B} = e^t(\mathbf{i} + \mathbf{j})$, find $d(\mathbf{A} + \mathbf{B})/dt$ using $d\mathbf{i}/dt = 2\mathbf{j}$ and $d\mathbf{j}/dt = -3\mathbf{i}$.

1.35 Using the vectors in Problem 1.34, find $d(\mathbf{A} \times \mathbf{B})/dt$ assuming $d\mathbf{k}/dt = 0$.

1.36 A particle is moving in such a way that its velocity is given by

$$\mathbf{v} = e^t(3t\mathbf{i} + \mathbf{j} + t^2\mathbf{k})$$

What is $d\mathbf{v}/dt$ if **i**, **j**, and **k** are fixed? What happens if the coordinate system is moving?

1.37 If $\mathbf{w} = (2\mathbf{j} + \mathbf{k})$, find $\mathbf{w} \times \mathbf{i}$.

1.38 In polar coordinates, $\mathbf{e}_r = (\cos\theta\mathbf{i} + \sin\theta\mathbf{j})$. Find $d\mathbf{e}_r/dt$ assuming **i** and **j** are fixed.

1.39 In spherical coordinates, $\mathbf{e}_r = \sin\phi(\cos\theta\mathbf{i} + \sin\theta\mathbf{j}) + \cos\phi\mathbf{k}$. Find $d\mathbf{e}_r/dt$ assuming **i**, **j**, and **k** are fixed.

1.40 In the cylindrical coordinate system, it can be shown that $d\mathbf{e}_\theta/dt = \mathbf{w} \times \mathbf{e}_\theta$ where $\mathbf{w} = \dot{\theta}\mathbf{k}$. Verify that this is true, assuming the **i**, **j**, and **k** unit vectors are fixed.

1.4 Fundamental Concepts

The concepts fundamental to a study of dynamics are those of space, time, inertia, and force. In addition we must also define a particle and a body. We define these terms in a simple manner leaving a more sophisticated presentation for advanced texts.

Space. A geometrical region in which physical events occur. It can have one or more dimensions although we shall be concerned with at most three dimensions. Position in space is established relative to some reference system. The basic reference system necessary for newtonian mechanics is one that is nonaccelerating.

Inertia. The ability of a body to resist a change in its state of motion.

Mass. A quantitative measure of the inertia of a body.

Particle. A body is said to be a particle if the dimensions of the body can be treated as negligible. Thus, the mass of a particle is assumed to be concentrated at a point. A particle is sometimes called a mass point.

Rigid Body. A body characterized by the condition that any two points of the body remain at a fixed distance relative to each other for all time.

Deformable Body. A body that can be deformed so that any two points of the body can move relative to each other as a function of time.

Force. The action of one body on another. This action may exist because of contact between the bodies, which is called the push-pull effect, or this action may exist with the bodies apart, which is called the field-of-force effect.

Time. Essentially a measure of the orderly succession of events occurring in space. Time is considered as an absolute quantity. The standard for time is established by the frequency of oscillation of a cesium atom.

Having considered certain basic definitions, we shall now introduce some fundamental laws as simply as possible, reserving the complexities for later.

Newton's Laws

In 1687, Sir Isaac Newton published the following laws:

I. *Every particle remains at rest or continues to move in a straight line with a uniform motion if there is no unbalanced force acting upon it.*

II. *The time rate of change of the linear momentum of a particle is proportional to the unbalanced force acting upon it and occurs in the direction in which the force acts.*

III. *To every action there is an equal and opposite reaction. The mutual forces of two bodies acting upon each other are equal in magnitude, opposite in direction, and colinear.*

The third law is useful for drawing free-body diagrams that show the forces acting at common connecting points. These laws have been verified experimentally over the years. It is important to realize that the first two laws hold for measurements made in a reference frame that is essentially fixed.

The second law, written as an equation, forms the basis for most of the analysis in dynamics. Since the **linear momentum** of a particle is defined as the product of the particle mass and its velocity, it follows from Newton's second law that

$$\mathbf{F} = k\frac{d}{dt}(m\mathbf{v}) = km\frac{d\mathbf{v}}{dt} = km\mathbf{a} \tag{1.19}$$

in which k is the proportionality constant and mass m is assumed constant. By setting $k = 1$, the units of force, mass and acceleration are no longer independent. If $k = 1$, Eq. (1.19) simplifies to

$$\mathbf{F} = m\mathbf{a} \tag{1.20}$$

in which $\mathbf{F}$ is the sum of the external forces acting on the particle of mass m and $\mathbf{a}$ is the absolute acceleration of the particle. In the SI system the units of force are newtons (N), mass are kilograms (kg), and acceleration are meters per second per second (m/s^2). In the English gravitational system the units for mass are slugs, force are pounds (lb), and acceleration are feet per second per second (ft/sec^2).

Newton also formulated the **law of gravitation** that governs the mutual attraction between two isolated bodies. This law is expressed mathematically as

$$F = G\frac{m_1 m_2}{r^2} \tag{1.21}$$

where F = magnitude of the mutual force of attraction between the two bodies

$$G = \text{universal constant}$$
$$m_1, m_2 = \text{masses of the bodies}$$
$$r = \text{distance between the centers of mass of the bodies}$$

The mutual forces between the two bodies obey Newton's third law so that they are equal in magnitude and opposite in direction along the line joining the centers of mass of the two bodies.

D'Alembert's Approach

Equation (1.20) can be rearranged to read

$$\mathbf{F} - m\mathbf{a} = 0 \tag{1.22}$$

Recall that $\mathbf{F}$ represents the sum of the external forces acting on the particle, m is its mass, and $\mathbf{a}$ is the particle acceleration measured relative to a fixed frame of reference. The term $-m\mathbf{a}$ is called the **inertia force** or reversed force. Equation (1.22) states that the sum of external forces plus the inertia force equals zero. This statement is similar to the summation of forces used in statics, so that the methods of statics can be applied to problems in dynamics. The laws of static equilibrium apply to a system if the inertia force is added to the externally applied forces. This approach is attributed to D'Alembert who derived it by using the principle of virtual work. D'Alembert's approach has far-reaching effects in that the virtual-work principle is used in developing advanced methods such as Hamilton's principle and Lagrange's equations of motion. We shall discuss some of these ideas later, although more rigorous analysis is relegated to advanced texts.

This is an appropriate time to reiterate the importance of what we have learned in statics, viz., the free-body diagram. In applying Newton's second law we must draw a free-body diagram showing all the forces acting on the body or particle. The sum of the forces in any particular direction will be equal to the mass multiplied by the acceleration in that same direction. For example, if a particle of mass m is moving in the x and y direction, then

$$\sum F_x = ma_x \qquad \sum F_y = ma_y$$

where $\sum F_x$ and $\sum F_y$ must be obtained from the free-body diagram.

1.5 Dimensions and Units

When we measure the length of a string, we may choose to record its length in centimeters, inches, or perhaps feet. The concept of length is called a **dimension**; the terms centimeters, inches, and feet are called **units**. The concept of time is also a dimension, for which the units may be seconds, minutes, hours, etc.

In the field of newtonian mechanics, three dimensions are necessary for all units. It has been agreed that length and time are fundamental to all systems and that mass or force may be selected as convenient. In other words, since we set $k = 1$ in Eq. (1.19), if the dimension of mass is selected as the third fundamental dimension, then the dimension of force is found from Newton's second law. A system in which mass, length, and time are fundamental dimensions is called an **absolute system**. If the dimension of force is used, we have a **gravitational system** composed of force, length, and time.

In this text we shall use both the International System (SI) of units as well as the English gravitational system. In the SI system of units the mass is in kilograms (kg), the length is in meters (m) and the time is in seconds (s). To obtain the units of force in the SI system, consider Newton's second law of motion for a particle,

$$\text{Force} = \text{Mass} \times \text{Acceleration}$$

In the SI system we define the newton (N) as a unit of force, in which 1 N causes a mass of 1 kg to accelerate 1 m/s^2.

In the gravitational system the fundamental dimensions are force, length, and time. The unit of force is the pound (lb), the length is the foot (ft), and time is the second (sec). To obtain the unit of mass we use Newton's second law of motion:

$$\text{Force} = \text{Mass} \times \text{Acceleration}$$

If the force is in lb and acceleration in ft/sec^2, then the derived units of mass from Newton's second law become $lb \cdot sec^2/ft$. This unit of mass is called a slug. The units and dimensions of quantities generally useful in the study of dynamics are given in Table 1.1 for the English gravitational system and in Table 1.2 for the SI system. Table 1.3 gives some useful conversion factors in converting from one system to the other.

To be valid, a given equation must be dimensionally homogeneous. That is, the fundamental dimensions must be equal on both sides of the equation. As an example, suppose we are working in the SI system. In this case the fundamental dimensions are mass (M), length (L), and time (T). From Newton's second law, the dimension for force in the SI system is

$$F = MLT^{-2}$$

We can check the dimensional validity of the equation

$$Ft = mv$$

where F is a force, t is time, m is mass, and v is velocity, by substituting

TABLE 1.1 ***English Gravitational Units and Dimensions***

Dimension	*Quantity*	*English units*	*Abbreviations*
L	Length	foot	ft
T	Time	second	sec
LT^{-1}	Velocity	foot/second	ft/sec
LT^{-2}	Acceleration	foot/second^2	ft/sec^2
FT^2L^{-1}	Mass	pound-second^2/foot = slug	lb·sec^2/ft
F	Force	pound	lb
F	Weight	pound	lb
FT^2L^{-4}	Density	pound-second^2/foot^4 = slug/ft^3	(lb·sec^2)/ft^4
F	Ton	2000 lb	ton
FL	Work	pound-foot	lb·ft
FL	Energy	pound-foot	lb·ft
FLT^{-1}	Power	pound-foot/second	lb·ft/sec
FL^{-2}	Pressure	pound/foot^2	lb/ft^2
FL	Moment, torque	foot-pound	ft·lb
L^4	Area moment of inertia	foot^4	ft^4
FLT^2	Mass moment of inertia	pound-foot-second^2 = slug-foot^2	lb·ft·sec^2
—	Plane angle	radian	rad
FT	Linear momentum	pound-second = slug-foot/second	lb·sec
FT	Linear impulse	pound-second	lb·sec
T^{-1}	Angular velocity	radians/second	rad/sec
T^{-2}	Angular acceleration	radians/second^2	rad/sec^2
FTL	Angular momentum	pound-foot-second	lb·ft·sec
FTL	Angular impulse	pound-foot-second	lb·ft·sec

the dimensional symbols of the above dimension for force into the equation. Thus

$$MLT^{-2}T = MLT^{-1}$$

From this we see that the equation $Ft = mv$ is dimensionally correct although this is not sufficient to assure its correctness as an equation.

Next we consider the gravitational force of attraction exerted on a body by the earth. This gravitational force is known as the **weight** (w) of the body and its unit is the newton in the SI system and the pound in the English gravitational system. On the surface of the earth let $a = g_e$ in Eq. (1.20); in Eq. (1.21) let $m_1 = m =$ the mass of the body, $m_2 = m_e =$ the mass of the earth, and $r = R =$ the

TABLE 1.2 ***SI Units and Dimensions***

Dimensions	*Quantity*	*SI units*	*Abbreviations*
L	Length	meter	m
T	Time	second	s
LT^{-1}	Velocity	meter/second	m/s
LT^{-2}	Acceleration	meter/second^2	m/s^2
M	Mass	kilogram	kg
MLT^{-2}	Force	newton	N
MLT^{-2}	Weight	newton	N
ML^{-3}	Density	kilogram/meter^3	kg/m^3
M	Metric ton	1000 kilograms	t
ML^2T^{-2}	Work	newton-meter = joule	N·m = J
ML^2T^{-2}	Energy	newton-meter = joule	N·m = J
ML^2T^{-3}	Power	newton-meter/second = watt	N·m/s = W
$ML^{-1}T^{-2}$	Pressure	newton/meter^2 = pascal	N/m^2 = Pa
ML^2T^{-2}	Moment, torque	meter-newton	m·N
L^4	Area moment of inertia	meter^4	m^4
ML^2	Mass moment of inertia	kilogram-meter^2	kg·m^2
—	Plane angle	radian	rad
MLT^{-1}	Linear momentum	newton-second	N·s
MLT^{-1}	Linear impulse	newton-second = kg·m/s	N·s
T^{-1}	Angular velocity	radian/second	rad/s
T^{-2}	Angular acceleration	radian/second^2	rad/s^2
ML^2T^{-1}	Angular momentum	kilogram-meter^2/second	kg·m^2/s
ML^2T^{-1}	Angular impulse	newton-meter-second = kg·m^2/s	N·m·s

TABLE 1.3 ***Conversion from the English Gravitational Units to the International Standard of Units***

To convert	*to*	*Multiply by*
foot	meter	0.3048
pound-foot	joule	1.356
mile	meter	1,609
pound	newton	4.448
slug	kilogram	14.59
kip (1000 lb)	newton	4,448

radius of the earth. Now, with this notation, equate the right sides of Eqs. (1.20) and (1.21).

$$mg_e = \frac{Gmm_e}{R^2}$$

$$g_e = G\frac{m_e}{R^2} = 9.81 \text{ m/s}^2 \text{ or } 32.2 \text{ ft/sec}^2 \tag{1.23}$$

Actually, this value for the absolute acceleration due to gravity varies slightly due to the geometry of the earth. It is, however, adequate for our purposes, and we shall use 9.81 m/s^2 or 32.2 ft/sec^2 for the acceleration of gravity on the earth's surface throughout this text.

In general then, the weight w on the earth's surface is related to the mass m by the following:

$$w = mg_e \tag{1.24}$$

Therefore, a mass of 100 kg in a gravitational field having $g_e = 9.81$ m/s^2 weighs 981 N. On the other hand, a mass of 10 slugs at $g_e =$ 32.2 ft/sec^2 weighs 322 lb.

To show how the acceleration of gravity varies with the height h above the earth's surface, consider Fig. 1.16. The magnitude of the force of attraction acting on the particle is mg, where g is now the acceleration of gravity at the height h above the earth's surface. Again, combining the right sides of Newton's second law and the law of gravitation,

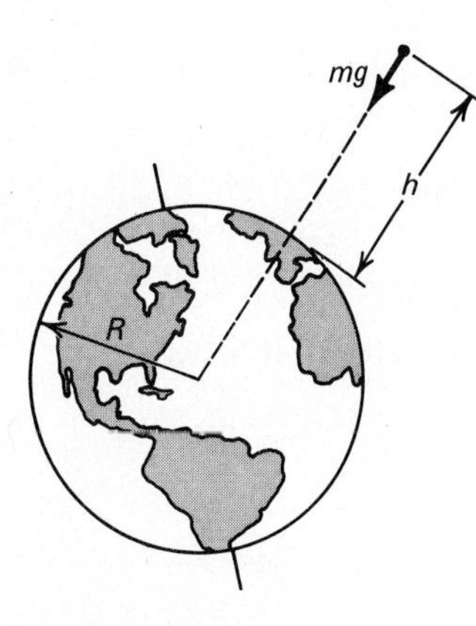

Figure 1.16

$$mg = \frac{Gm_e m}{(R+h)^2} \tag{1.25}$$

But, from Eq. (1.23) $Gm_e = g_e R^2$. Consequently,

$$g = g_e \frac{R^2}{(R+h)^2} \tag{1.26}$$

If we measure gravity at one earth radius above the surface of the earth, we see that its value is one-fourth the measured value on the earth's surface.

In the following chapters we shall use the symbol g for g_e when we are dealing with bodies close to the earth. Instead of using Eq. (1.24), we shall say that the weight on the earth's surface is mg, where $g = 9.81$ m/s^2 or 32.2 ft/sec^2.

Example 1.9

A block of 100 kg is subjected to a force of 1000 N as shown in Fig. 1.17(a). If the coefficient of kinetic friction is 0.2, what is the acceleration of the block?

All the forces on the block are shown in Fig. 1.17(b). Now

$$\sum F_x = ma_x \qquad \text{and} \qquad \sum F_y = ma_y$$

Since there is no motion in the y direction, $a_y = 0$ and

$$\sum F_y = 0$$

Referring to the free-body diagram in Fig. 1.17(b),

$$\mathbf{N} - mg = 0$$

$$\mathbf{N} = 100(9.81) = 981 \text{ N}$$

For the x direction, again refer to Fig. 1.17(b).

$$\sum F_x = ma_x$$

$$1000 - \mu N = ma_x$$

$$1000 - 0.2(981) = 100a_x$$

$$a_x = 8.04$$

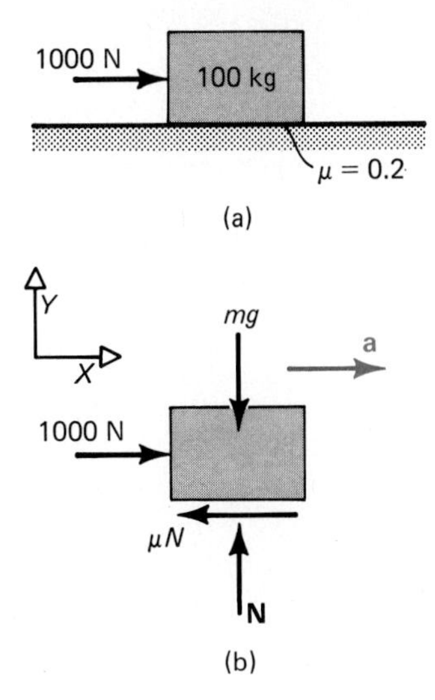

Figure 1.17

The acceleration of the block is

$$\mathbf{a} = 8.04\mathbf{i} \text{ m/s}^2 \qquad \textit{Answer}$$

Example 1.10

Determine if the following equations are dimensionally correct.
(a) Torque = Mass moment of inertia × Angular acceleration
(b) Work = Mass × Velocity

The dimensions from Table 1.2 are used to verify the above equations.

(a) Torque = Force × Distance = $MLT^{-2} \times L = ML^2T^{-2}$

Mass moment of inertia × Angular acceleration $= ML^2 \times T^{-2}$
$= ML^2T^{-2}$

The equation is dimensionally correct.

(b) Work $= ML^2T^{-2}$

Mass × Velocity $= M \times LT^{-1}$
$= MLT^{-1}$

The equation is dimensionally incorrect.

Example 1.11

What is the weight of a 100-kg mass at an altitude of 2×10^6 m above the surface of the earth.

The value of g is given by

$$g = \frac{g_e R^2}{(R + h)^2}$$

The radius of the earth is $R = 6378$ km, so that

$$g = \frac{(9.81)(6378)^2}{(6378 + 2000)^2} = 5.685$$

The weight is

$$w = mg = 100(5.685)$$
$$= 568.5 \text{ N} \qquad \textit{Answer}$$

PROBLEMS

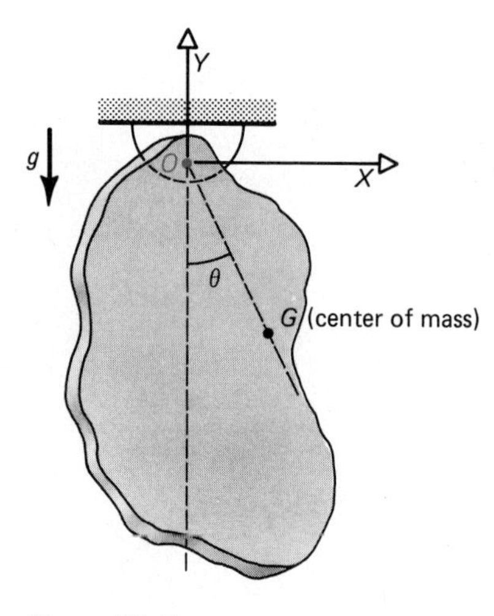

Figure P1.41

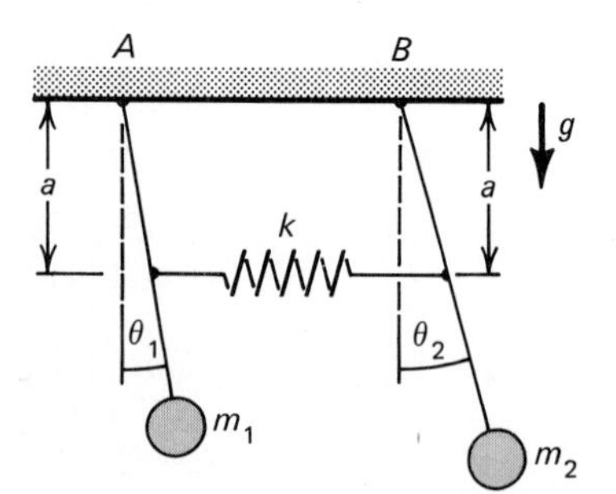

Figure P1.42

1.41 The compound pendulum of mass m pinned at O shown in Fig. P1.41 oscillates in the XY plane. Its motion about the static equilibrium position is described by the angle θ. Draw the free-body diagram of the forces acting on the pendulum.

1.42 Draw the free-body diagram for each moving pendulum and connecting spring shown in Fig. P1.42.

1.43 Draw the free-body diagram of the forces acting on mass m_1 and mass m_2 shown in Fig. P1.43. Mass m_1 is displaced a distance x from the unstretched length of the spring. The mass m_2 is attached to an assumed massless rigid rod that is rotated through the angular displacement θ. Assume μ is the kinetic coefficient of friction between m_1 and the horizontal surface.

1.44 A 100-kg block is accelerated along a horizontal surface. If the kinetic coefficient of friction between the block and surface is 0.4 and a 2500-N force is applied, determine the acceleration of the block.

1.45 Determine the acceleration of the block of Problem 1.44 if the surface is inclined upwards by 30° and the block is pulled up the slope by the 2500-N force.

1.46 A mass m is subjected to a force $\mathbf{F}$ such that $\mathbf{F} = F_1\mathbf{i} + F_2\mathbf{j}$, where F_1 and F_2 are constants. What is the resultant velocity of this mass?

1.47 In an experiment, an investigator claims that the readings from a force gauge and an accelerometer are given, respectively, as

$$\mathbf{F} = 10\mathbf{i} + 20\mathbf{j} + 40\mathbf{k} \text{ lb} \qquad \text{and} \qquad \mathbf{a} = 2\mathbf{i} + 4\mathbf{j} + 7\mathbf{k} \text{ ft/sec}^2$$

Are these readings consistent with Newton's second law?

1.48 Upon further investigation of the claims made by the investigator in Problem 1.47, it was found that the accelerometer readings are accurate and the component of the force in the $\mathbf{k}$ direction is in error. Taking this into account, what must the resultant force be?

1.49 The position vector of a particle of mass m is given by $\mathbf{r} = 6t^2\mathbf{i} + \sin \pi t\mathbf{j}$, where $\mathbf{i}$ and $\mathbf{j}$ are fixed. What is the force on the particle, expressed in the $\mathbf{i}$ and $\mathbf{j}$ components?

1.50 Express the answer to Problem 1.49 in polar coordinates.

1.51 Determine whether the following equations are dimensionally correct:
(a) $\frac{1}{2}mv^2$ = Force × Distance
(b) Torque = Time rate of change of angular momentum
(c) $\mathscr{I}$ = Linear impulse = mv^2

1.52 Calculate the acceleration of gravity of a body at heights of 637, 6378, and 12,756 km above the earth's surface.

1.53 What is the earth's force of attraction on the body when it is 12,756 km above the earth's surface if the mass of the body is 200 kg?

1.54 A body has a weight of 1500 lb at an altitude of one earth radius. What is its weight and mass on the planet Mars assuming that $g_{\text{Mars}} = \frac{1}{2}g_e$?

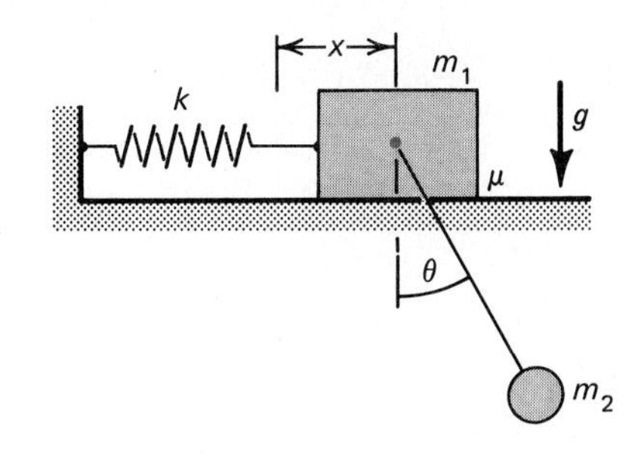

Figure P1.43

1.6 Summary

A. The important *ideas*:

1. The use of vector notation in dynamics allows us considerable flexibility since the equations in dynamics can then be developed independently of any specific coordinate system.
2. There are four commonly used coordinate systems in dynamics: cartesian, polar, cylindrical, and spherical.
3. Our study of dynamics is based on newtonian mechanics that requires nonaccelerating reference frames.
4. Reference frames can be fixed or moving and must be specified.
5. The derivative of a vector with respect to time includes two terms: the rate of change of the magnitude and the rate of change of the direction.

B. The important *equations*:

Vector:	$\mathbf{A} = A\mathbf{e}_A$	(1.1)
Unit vector:	$\mathbf{e}_A = \cos\theta_x\mathbf{i} + \cos\theta_y\mathbf{j} + \cos\theta_z\mathbf{k}$	(1.6)
Scalar product:	$\mathbf{A}\cdot\mathbf{B} = AB\cos\theta$	(1.2)
Vector product:	$\mathbf{A}\times\mathbf{B} = AB\sin\theta\,\mathbf{e}_c$	(1.8)
Cartesian components:	$\mathbf{A} = A_x\mathbf{i} + A_y\mathbf{j} + A_z\mathbf{k}$	(1.11)
Polar components:	$\mathbf{A} = A\mathbf{e}_r$	(1.12)

Cylindrical components: $\mathbf{A} = A_r\mathbf{e}_r + A_z\mathbf{k}$ (1.14)

Spherical components: $\mathbf{A} = A\mathbf{e}_r$ (1.16)

Derivative of a vector: $\dfrac{d\mathbf{A}}{dt} = \dfrac{dA}{dt}\mathbf{e}_A + A\dfrac{d\mathbf{e}_A}{dt}$ (1.18)

Newton's law: $\mathbf{F} = m\mathbf{a}$ (1.20)

Law of gravitation: $F = G\dfrac{m_1m_2}{r^2}$ (1.21)

Weight: $w = mg_e$ (1.24)

Acceleration of gravity: $g = g_e\dfrac{R^2}{(R + h)^2}$ (1.26)

2 Kinematics of a Particle

Before examining the relations between forces and the motion that they produce, it is necessary to understand motion in the abstract sense and be able to describe it mathematically. Kinematics is the branch of mechanics that deals with the displacement of a particle over time without reference to the forces that cause or change the motion. Kinematics is concerned with the position, velocity, and acceleration of moving bodies as functions of time.

The motion of a particle moving in a straight line is called **rectilinear motion**; when a particle moves along a curved path, its motion is known as **curvilinear motion**. When the curved path lies on a plane, the motion is referred to as **plane** curvilinear motion.

We begin our study of kinematics by relating the concepts of displacement, velocity, and acceleration to linear motion and angular motion. The specific cases of rectilinear motion of a particle and the angular motion of a line in a plane are examined. We shall develop the general expression for the velocity and acceleration of a particle in terms of a moving coordinate system. Finally, we introduce normal-tangential coordinates and polar coordinates as special coordinate systems used to describe plane curvilinear motion.

2.1 Displacement, Velocity, and Acceleration

Linear Motion

Suppose a particle P moves along the space curve in Fig. 2.1 from position A to position B during some time interval. The position vector $\mathbf{r}$ locates the particle P relative to the reference frame

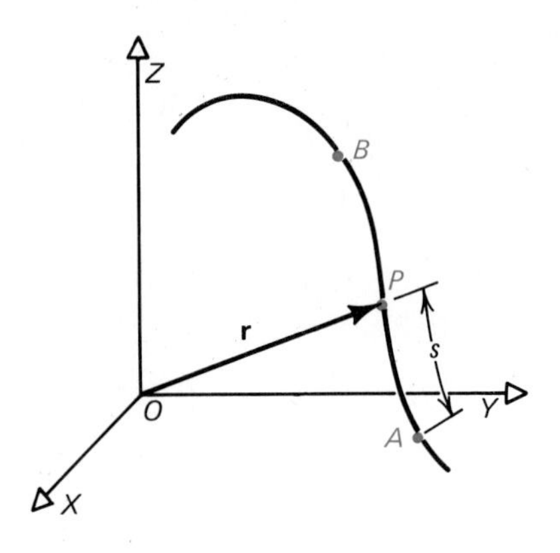

Figure 2.1

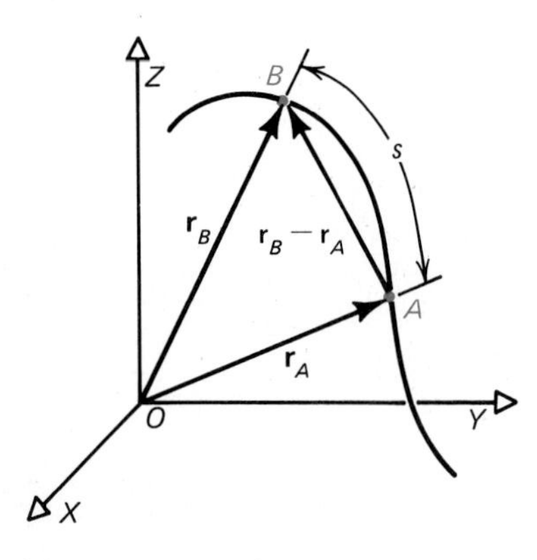

Figure 2.2

XYZ, while the distance traversed by the particle is measured along the curve and is designated by the symbol s. The change in position of the particle during the time interval is its **linear displacement**, which is represented by the vector $(\mathbf{r}_B - \mathbf{r}_A)$ as shown in Fig. 2.2. If the particle moves from A to B and then returns to A during some time interval, the distance traveled is $2s$ while its displacement is zero over the time interval.

Imagine that the particle is at position A_1 at time t_1 and at position B_1 and time $t_1 + \Delta t$, as shown in Fig. 2.3. The displacement vector from A_1 to B_1 is represented by $(\mathbf{r}_{B1} - \mathbf{r}_{A1}) = \Delta\mathbf{r}$. As the time interval Δt grows smaller, the magnitude and direction of $\Delta\mathbf{r}$ changes. We are usually interested in letting $\Delta t \to 0$, which leads to the definition of the **instantaneous linear velocity**. Mathematically, this is expressed as

$$\mathbf{v} = \lim_{\Delta t \to 0} \frac{\Delta\mathbf{r}}{\Delta t} = \frac{d\mathbf{r}}{dt} = \dot{\mathbf{r}} \tag{2.1}$$

The magnitude of this vector is called the **speed**. The units of velocity are meters per second (m/s) or feet per second (ft/sec). As $\Delta t \to 0$, the direction of the $\Delta\mathbf{r}$ vector becomes tangent to the space curve at position A_1. Thus, the direction of the velocity $\mathbf{v}$ is tangential to the space curve at position A_1.

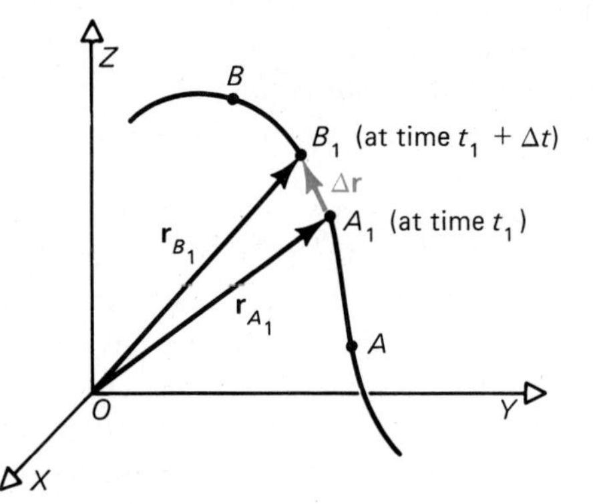

Figure 2.3

The derivative with respect to time of the velocity in Eq. (2.1) is the **instantaneous linear acceleration** of the particle.

$$\mathbf{a} = \lim_{\Delta t \to 0} \frac{\Delta\mathbf{v}}{\Delta t} = \frac{d\mathbf{v}}{dt} = \dot{\mathbf{v}} = \ddot{\mathbf{r}} \tag{2.2}$$

The units of acceleration are meters per second per second (m/s^2) or feet per second per second (ft/sec^2). The acceleration vector, also, is composed of terms that reflect both change in magnitude and change in direction.

Angular Motion

The concept of angular motion is necessary to understand the dynamic behavior of rigid bodies and particles. The motion of a rigid body may involve a change in orientation as well as a change in location. The following describes a special case of this motion in which the motion of a body remains in a plane. During a very short time interval Δt, the motion of a rigid body can be viewed as occurring in two stages—a small translation plus a small rotation—the order of occurrence being immaterial. Suppose points A and B are points on a thin circular plate that moves only on the XY plane as shown in Fig. 2.4(a). If translation of this body occurs through the linear displacement $\mathbf{\Delta r}$ as shown in Fig. 2.4(b), point A moves to A' and point B moves to B'.

To conceptualize the small rotation that occurs, imagine an axis perpendicular to the XY plane passing through point A'. Envision the body, with A' fixed, rotating through a small angle $\Delta\theta$ counterclockwise (CCW) about the axis as shown in Fig. 2.4(c), where, for clarity, only point B' is shown moving to B''. Since the angular rotation $\Delta\theta$ occurs during Δt, which again approaches zero, the angular rotation becomes an infinitesimal vector $\boldsymbol{\Delta\theta}$ which is shown in Fig. 2.4(d) as a vector along the axis of rotation in the XZ plane. It is important to note that we can treat infinitesimal rotations as vectors, but we cannot so treat finite rotations. (The commutative law of vector addition is violated for vector representation of finite rotations.) The direction of the vector $\boldsymbol{\Delta\theta}$ is found using the right-hand rule, i.e., the four fingers of the right hand will follow the direction of rotation while the thumb points in the direction of the rotation vector.

Consequently, the infinitesimal angular displacement $\boldsymbol{\Delta\theta}$ is a vector whose magnitude is equal to the angle of rotation and whose direction lies along the axis of rotation, i.e., the axis determined by those points not displaced by the infinitesimal rotation. The rate of change of the angular rotation is defined as the angular velocity, $\boldsymbol{\omega}$.

$$\boldsymbol{\omega} = \lim_{\Delta t \to 0} \frac{\boldsymbol{\Delta\theta}}{\Delta t} = \frac{d\boldsymbol{\theta}}{dt} \tag{2.3}$$

Note that the direction of this vector also follows the right-hand rule. The rate of change of the angular velocity is the **angular acceleration**,

$$\boldsymbol{\alpha} = \lim_{\Delta t \to 0} \frac{\Delta\boldsymbol{\omega}}{\Delta t} = \frac{d\boldsymbol{\omega}}{dt} \tag{2.4}$$

The units for angular velocity are radians per second (rad/s) and for angular acceleration are radians per second per second (rad/s^2).

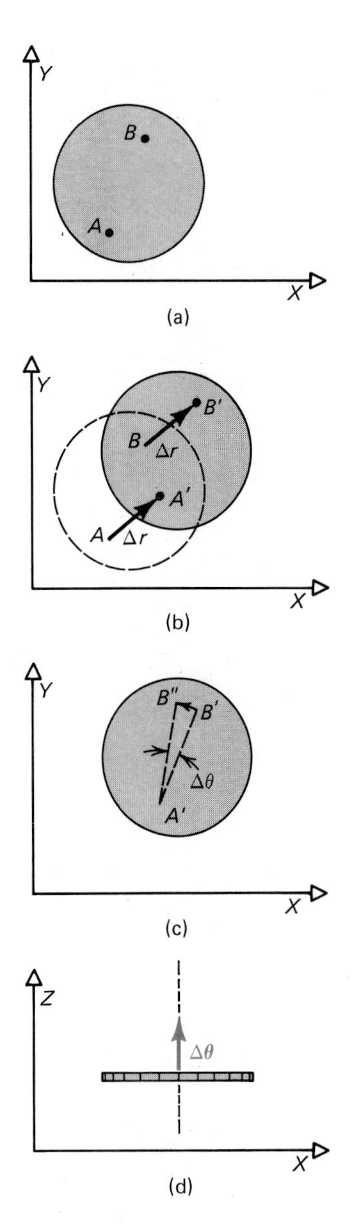

Figure 2.4

Linear and Angular Velocity

Finally, let us develop a relationship between linear velocity and angular velocity. As a model, consider a circular cone rotating about the axis AA as shown in Fig. 2.5(a). Let $\boldsymbol{\omega}$ be the angular velocity of the cone, $\mathbf{r}$ the position vector from the fixed point O to some point P on the cone, and θ the angle between the AA axis and $\mathbf{r}$. In Fig. 2.5(b) we are looking down on the point P that is moving in a circular path of radius $r \sin \theta$. Here, the magnitude of the linear velocity of P is $v = \omega r \sin \theta$ and its direction is tangent to the circular path of travel. The velocity of P is shown as $\mathbf{v}$ in Fig. 2.5(b). From this information and the definition given by Eq. (1.8), we see that $\mathbf{v}$ can be formed by taking the vector product between $\boldsymbol{\omega}$ and $\mathbf{r}$, that is,

$$\mathbf{v} = \boldsymbol{\omega} \times \mathbf{r} \tag{2.5}$$

Although Eq. (2.5) was derived for the special case of a rotating circular cone, it is independent of the geometric shape of the rotating body, and can be applied to any rigid body rotating with an angular velocity $\boldsymbol{\omega}$ where $\mathbf{r}$ is the position vector from a fixed point O to some point P on the rigid body.

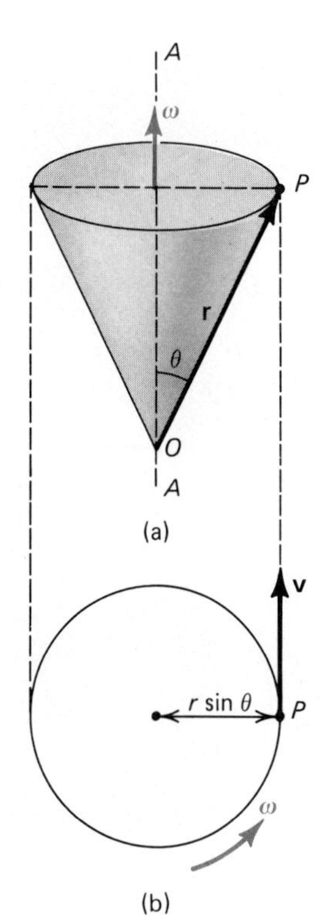

(b)

Figure 2.5

Example 2.1

The position of a particle, moving in the XY plane, is given by

$$\mathbf{r} = 5t^2\mathbf{i} + 2t^2\mathbf{j}$$

where r is in meters and t in seconds. Find (a) the velocity and acceleration; (b) obtain the magnitude of $\mathbf{v}$ and $\mathbf{a}$ at $t = 1$ s.

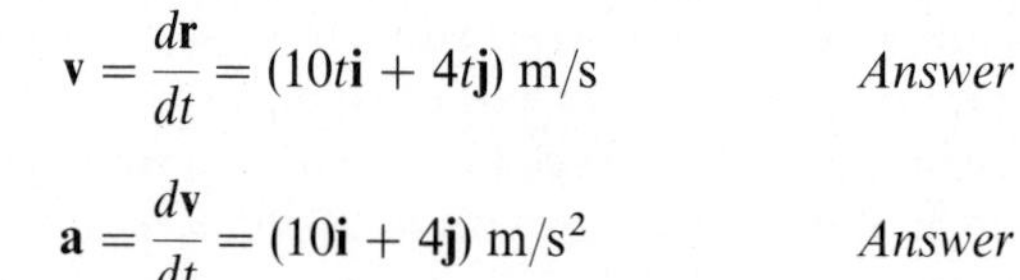

(a) $$\mathbf{v} = \frac{d\mathbf{r}}{dt} = (10t\mathbf{i} + 4t\mathbf{j}) \text{ m/s} \qquad \textit{Answer}$$

$$\mathbf{a} = \frac{d\mathbf{v}}{dt} = (10\mathbf{i} + 4\mathbf{j}) \text{ m/s}^2 \qquad \textit{Answer}$$

(b) At $t = 1$ s,

$$\mathbf{v} = (10\mathbf{i} + 4\mathbf{j}) \qquad \mathbf{a} = (10\mathbf{i} + 4\mathbf{j})$$

$$v = 10.77 \text{ m/s} \qquad a = 10.77 \text{ m/s}^2 \qquad \textit{Answer}$$

Example 2.2

Obtain the linear velocity $\mathbf{v}$ of a point P as it rotates about a point O with angular velocity $\boldsymbol{\omega} = 10t\mathbf{k}$. The position of P relative to O is given by $\mathbf{r} = 3\mathbf{i} + 4\mathbf{j}$. The units of $\mathbf{r}$ are in feet and t is in seconds. Obtain the magnitude of the velocity at $t = 8$ sec.

The velocity of P is given by Eq. (2.5).

$$\mathbf{v} = \boldsymbol{\omega} \times \mathbf{r}$$

Substituting for $\boldsymbol{\omega}$ and $\mathbf{r}$,

$$\begin{aligned}\mathbf{v} &= (10t\mathbf{k}) \times (3\mathbf{i} + 4\mathbf{j}) \\ &= (30t\mathbf{j} - 40t\mathbf{i}) \text{ ft/sec} \qquad \textit{Answer}\end{aligned}$$

At $t = 8$ sec,

$$\mathbf{v} = 240\mathbf{j} - 320\mathbf{i}$$

The magnitude of $\mathbf{v}$ is

$$v = 400 \text{ ft/sec} \qquad \textit{Answer}$$

PROBLEMS

2.1 The position of a particle as it moves in the XY plane is given by

$$\mathbf{r} = 3t^2\mathbf{i} - 4t\mathbf{j}$$

where the magnitude of $\mathbf{r}$ is in meters and t is in seconds. (a) Find the velocity $\mathbf{v}$ and the acceleration $\mathbf{a}$. (b) What is the magnitude of $\mathbf{v}$ and $\mathbf{a}$ when $t = \frac{1}{2}$ s?

2.2 What is the magnitude of the velocity at $t = \pi/8$ s of a particle described by the position vector

$$\mathbf{r} = t(\sin 2t\mathbf{i} + \cos 2t\mathbf{j}) \text{ m}$$

2.3 A particle travels in a straight line such that

$$\mathbf{r} = (A + 5 \sin 2\pi t)\mathbf{k}$$

where the magnitude of $\mathbf{r}$ is in feet, $2\pi t$ is in radians, and t is in seconds. Find the velocity and acceleration when $t = 0, \frac{1}{4}$, and $\frac{1}{2}$ sec.

2.4 The angular position of a particle is given by

$$\theta = 5t^3$$

where θ is in radians and t in seconds. What is the magnitude of the angular velocity and acceleration at $t = 2$ s.

2.5 The angular acceleration in rad/s^2 is given as

$$\boldsymbol{\alpha} = 3t^2\mathbf{i} + 2t\mathbf{j}$$

where t is measured in seconds. If at $t = 0$, $\boldsymbol{\omega} = 2\mathbf{i} - \mathbf{j}$, find $\boldsymbol{\omega}$ for all values of t.

2.6 The angular velocity in rad/s is

$$\boldsymbol{\omega} = e^{-t}t^2 \sin t\mathbf{k}$$

Find the angular acceleration $\boldsymbol{\alpha}$, assuming $\mathbf{k}$ is constant.

2.7 A wheel spins in the XY plane about an axis through O as shown in Fig. P2.7. The magnitude of the angular velocity is 1 rad/s and the wheel rotates in the clockwise (CW) direction. Find the velocity $\mathbf{v}$ of the points on the wheel labeled A, B, and C.

2.8 A unit vector $\mathbf{e}$ of length 1 m is given by

$$\mathbf{e} = \sin\theta \sin\phi\mathbf{i} + \sin\theta\cos\phi\mathbf{j} + \cos\theta\mathbf{k}$$

If θ is fixed and $\dot{\phi}$ is given by ω rad/s, derive an expression for the velocity $\mathbf{v}$ at the tip of $\mathbf{e}$. What is the magnitude of the velocity if $\theta = 30^\circ$ and $\omega = 4$ rad/s?

2.9 A wheel used as an energy storage device shown in Fig. P2.9 is spun up so that the angular velocity is

$$\boldsymbol{\omega} = (C_1 t + C_2 t^2)\mathbf{k}$$

where ω is in rad/s. It is known that $\omega = 5$ at $t = 1$ s and $\omega = 30$ at $t = 3$ s. What is (a) the angular velocity at $t = 4$ s, (b) the angular acceleration at $t = 4$ s, (c) the linear velocity at the tip of the wheel at point A at $t = 4$ s?

2.10 A car is traveling at 60 mi/hr in the direction shown in Fig. P2.10. What is the angular velocity of the radius vector of 100 ft about the point O at the instant shown? Assume $\mathbf{v}$ is perpendicular to the radius vector.

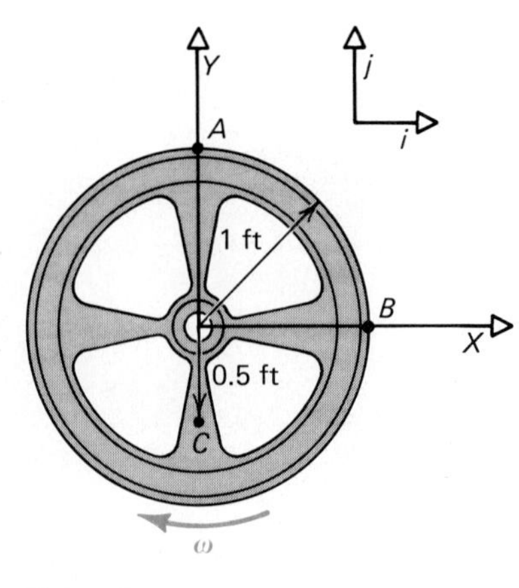

Figure P2.7

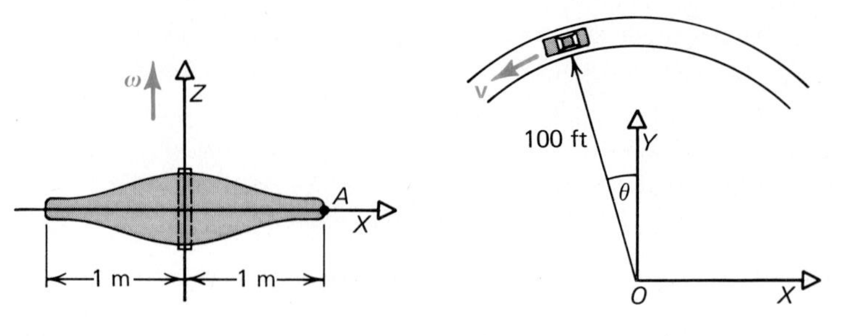

Figure P2.9

Figure P2.10

2.11 If the magnitude of the angular velocity of the car relative to O, in Fig. P2.10, is 0.8 rad/s, then what is the component of the velocity in the X and Y directions if $\theta = 30^\circ$.

2.2 Rectilinear Motion

A particle traveling along a straight line experiences rectilinear motion. In Fig. 2.6, the X axis lies along the line of motion, and the particle's position vector is simply

$$\mathbf{r} = s\mathbf{i}$$

where s is the distance measured along the X axis. Examples of rectilinear motion are shown in Fig. 2.7, where, in each case, the X axis is arbitrarily taken along the line of motion.

Figure 2.8 shows a particle in rectilinear motion at some time t and a short time later, at $t + \Delta t$, as it travels along a straight line. The ratio of $\Delta\mathbf{r}/\Delta t$ is called the average velocity; i.e.,

$$\mathbf{v}_{av} = \frac{\Delta\mathbf{r}}{\Delta t} = \frac{\Delta s}{\Delta t}\mathbf{i}$$

If we let $\Delta t \to 0$, we obtain the instantaneous linear velocity $\mathbf{v}$ as

$$\mathbf{v} = \lim_{\Delta t \to 0} \frac{\Delta\mathbf{r}}{\Delta t} = \frac{d\mathbf{r}}{dt}$$

or

$$\mathbf{v} = \lim_{\Delta t \to 0} \frac{\Delta s}{\Delta t}\mathbf{i} = \frac{ds}{dt}\mathbf{i} \tag{2.6}$$

If the velocity of the particle in Fig. 2.8 changes by an amount $\Delta\mathbf{v}$ during the time interval Δt, its average acceleration is defined as

$$\mathbf{a}_{av} = \frac{\Delta\mathbf{v}}{\Delta t}$$

Taking the limit as $\Delta t \to 0$, we obtain the instantaneous linear acceleration as:

$$\mathbf{a} = \lim_{\Delta t \to 0} \frac{\Delta\mathbf{v}}{\Delta t} = \frac{d\mathbf{v}}{dt} \tag{2.7}$$

These equations for rectilinear motion may be simplified. First, by letting v be the speed, i.e., the magnitude of the velocity, Eq. (2.6) can be written

$$\mathbf{v} = v\mathbf{i} = \frac{ds}{dt}\mathbf{i}$$

or

$$v = \frac{ds}{dt} \tag{2.8}$$

Second, letting a be the magnitude of the acceleration, Eq. (2.7) can be written as

$$\mathbf{a} = a\mathbf{i} = \frac{dv}{dt}\mathbf{i} = \frac{d^2s}{dt^2}\mathbf{i}$$

or

$$a = \frac{dv}{dt} = \frac{d^2s}{dt^2} \tag{2.9}$$

A third equation, that is sometimes useful, is obtained by eliminating the differential dt between Eqs. (2.8) and (2.9). This leads to

$$v\,dv = a\,ds \tag{2.10}$$

Equations (2.8), (2.9), and (2.10) are the scalar differential equations relevant to the displacement, velocity, and acceleration

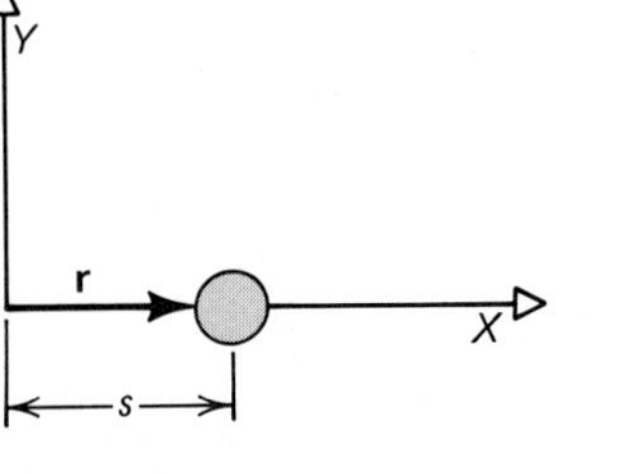

Figure 2.6

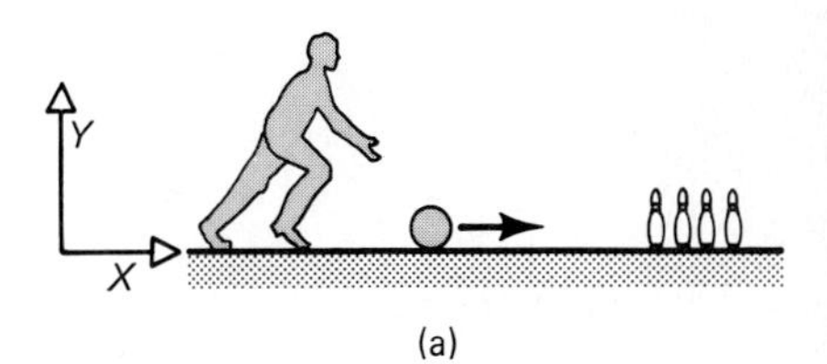

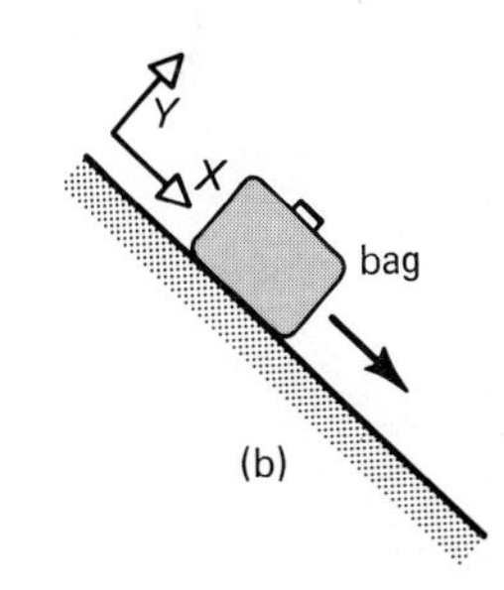

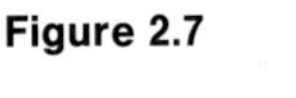
Figure 2.7

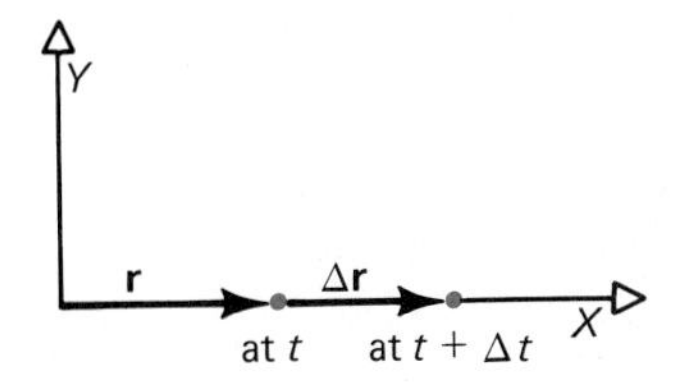

Figure 2.8

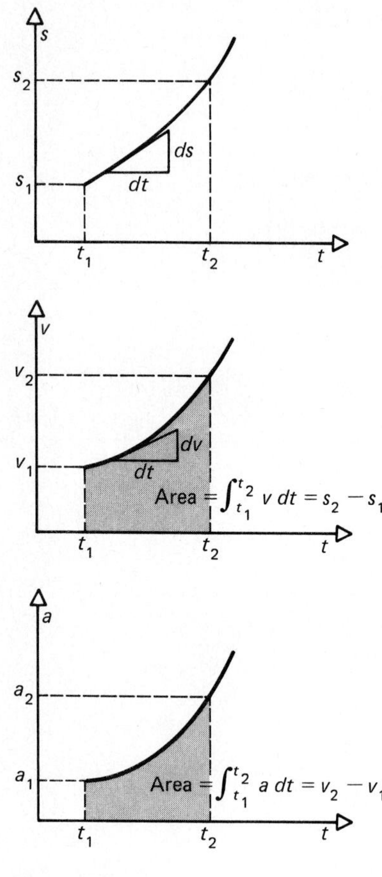

Figure 2.9

of a particle in rectilinear motion. Let us rewrite these three differential equations in integral form. We can rearrange Eqs. (2.8) and (2.9) in the following manner:

$$ds = v\,dt \qquad dv = a\,dt \tag{2.11}$$

If at some time t_1, $s = s_1$, $v = v_1$, and at time t_2, $s = s_2$, $v = v_2$, then the integrals of Eqs. (2.11) are, respectively,

$$\int_{s_1}^{s_2} ds = s_2 - s_1 = \int_{t_1}^{t_2} v\,dt \tag{2.12}$$

$$\int_{v_1}^{v_2} dv = v_2 - v_1 = \int_{t_1}^{t_2} a\,dt \tag{2.13}$$

Similarly, the integral form of Eq. (2.10) is

$$\int_{v_1}^{v_2} v\,dv = \int_{s_1}^{s_2} a\,ds \tag{2.14}$$

If the acceleration is a known function of time, Eq. (2.13) should be used; Eq. (2.14) is applicable if the acceleration is a known function of the displacement s.

Further insight into the meaning of these equations for rectilinear motion can be obtained by considering curves for some general rectilinear motion as shown in Fig. 2.9. First, we note that the slope of the st plot is the speed v, while the slope of the vt plot is the magnitude of the acceleration a. A second observation is that the area under the at curve equals the corresponding change in the speed, while the area under the vt curve equals the change in displacement. Sometimes, the solution to a problem is found more readily by sketching the given information in this graphical form, rather than by solving the differential equations directly.

Example 2.3

A car, moving in a straight line, starts from rest and accelerates uniformly at 20 m/s^2 for the first 2 s, travels at uniform speed for the next 2 s, and decelerates uniformly at 10 m/s^2 for the next 3 s. Find how far the car has traveled after 7 s.

The manner in which the given information is plotted, as shown in Fig. 2.10, is as follows:

(a) On the at plane we plot a constant value of 20 m/s^2 from 0 to 2 s, zero (for uniform speed) from 2 to 4 s, and -10 m/s^2 from 4 to 7 s.

(b) On the vt plane we start at zero velocity (car starts from rest) and draw a straight line to 40 m/s at $t = 2$ s. This is obtained by calculating the area under the at curve from zero to 2 s. Next draw a constant value of 40 m/s for the uniform velocity from 2 to 4 s; and a negative slope for the deceleration from 4 to 7 s. The value of v at $t = 7$ s is found by equating the area under the at curve between 4 and 7 s with the *change* in the speed, i.e.,

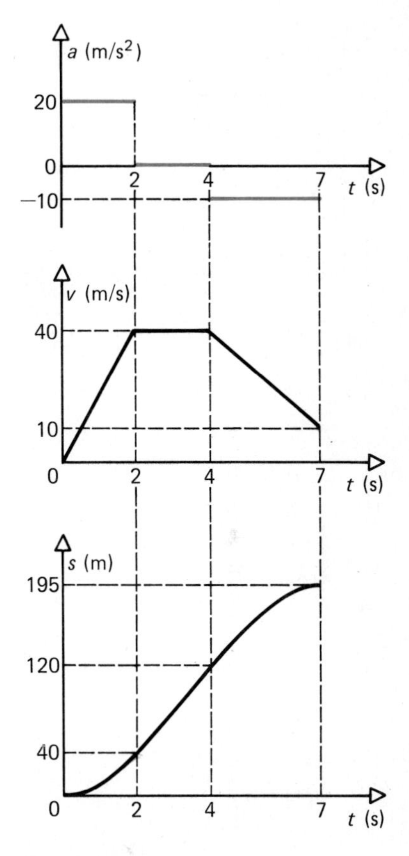

Figure 2.10

$$v_2 - v_1 = \int_{t_1}^{t_2} a\,dt$$

$$= \text{area under the } at \text{ curve between } t_1 \text{ and } t_2$$

$$v_2 - 40 = 3(-10)$$

$$v_2 = 10 \text{ m/s}$$

Note that the notation from Eq. (2.13) is used where in this case $t_1 = 4$ s, $t_2 = 7$ s, and $v_1 = 40$ m/s.

(c) Having the vt curve, we may sketch the st curve with the use of Eq. (2.12). For the first 2 s, $t_1 = 0$, $t_2 = 2$ s, $s_1 = 0$, so that

$$s_2 - s_1 = \int_{t_1}^{t_2} v\,dt$$

$$= \text{area under the } vt \text{ curve between } t_1 \text{ and } t_2$$

$$s_2 - 0 = \tfrac{1}{2}(40)(2)$$

$$s_2 = 40 \text{ m}$$

Between s_1 and s_2 we sketch a second-order curve starting with a zero slope at the origin since $v = 0$ at $t = 0$.

For the next 2 s, let $t_1 = 2$ s, $t_2 = 4$ s, and $s_1 = 40$ m in applying Eq. (2.12) over this interval. Now

$$s_2 - 40 = 40(2)$$

$$s_2 = 120 \text{ m}$$

A straight line is sketched over this interval since its slope must be constant for the constant velocity of 40 m/s.

For the last increment, let $t_1 = 4$ s, $t_2 = 7$ s, $s_1 = 120$ m, so that

$$s_2 - 120 = \tfrac{1}{2}(40 + 10)(3)$$

$$s_2 = 195 \text{ m}$$

Again, a second-order curve is drawn over the interval. Note that it begins with a positive slope at $t = 4$ s, and continuously decreases since the velocity is decreasing over the interval from 4 s to 7 s.

Example 2.4

A projectile travels rectilinearly through three different media, each 1 ft thick. The resulting relationship between the acceleration and

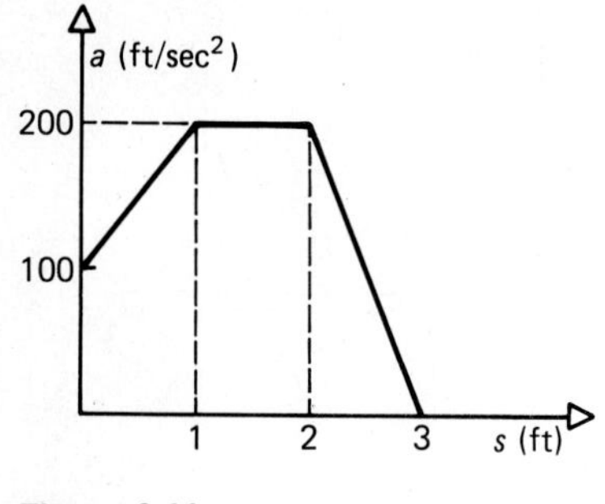

Figure 2.11

the distance traveled is shown in Fig. 2.11. If the initial velocity of the projectile is 10 ft/sec, find the velocity after the body has traveled through all three media.

By observing Fig. 2.11, the following equations relate the acceleration and displacement:

Between 0 and 1 ft $\quad a = 100 + 100s$

Between 1 and 2 ft $\quad a = 200$

Between 2 and 3 ft $\quad a = 600 - 200s$

Now in general,

$$\int_{v_1}^{v_2} v\,dv = \int_{s_1}^{s_2} a\,ds$$

Between 0 and 1 ft, $v_1 = 10$, $s_1 = 0$, $s_2 = 1$

$$\int_{10}^{v_2} v\,dv = \int_0^1 (100 + 100s)\,ds$$

$$\frac{v_2^2 - 100}{2} = [100s + 50s^2]_0^1 = 150$$

$$v_2^2 = 400$$

$$v_2 = 20 \text{ ft/sec}$$

Between 1 and 2 ft, the initial velocity equals 20 ft/sec, so that $v_1 = 20$, $s_1 = 1$, and $s_2 = 2$.

$$\int_{20}^{v_2} v\,dv = \int_1^2 200\,ds$$

$$\frac{v_2^2 - 400}{2} = 200(1) = 200$$

$$v_2^2 = 800$$

$$v_2 = 28.3 \text{ ft/sec}$$

Between 2 and 3 ft, the initial velocity now equals 28.3 ft/sec, so that $v_1 = 28.3$, $s_1 = 2$, and $s_2 = 3$.

$$\int_{28.3}^{v_2} v\,dv = \int_2^3 (600 - 200s)\,ds$$

$$\frac{v_2^2 - 800}{2} = [600s - 100s^2]_2^3$$

$$v_2^2 = 1000$$

$$v_2 = 31.6 \text{ ft/sec} \qquad \textit{Answer}$$

Example 2.5

Given the acceleration-time history of a particle in Fig. 2.12(a), sketch the velocity-time and displacement-time curves. Assume $v = 0$ and $s = 0$ at $t = 0$.

We first observe that over the first 10 s the vt curve is second-order and the st curve is third-order. The equation for vt over this increment is

$$\int_0^v dv = \int_0^t a\,dt = \int_0^t \frac{20t}{10}\,dt$$

$$v = t^2 \qquad \text{for } 0 \le t \le 10$$

We use this equation to find the general equation of the displacement over the first 10 s.

$$\int_0^s ds = \int_0^t v\,dt = \int_0^t t^2\,dt$$

$$s = \frac{t^3}{3} \qquad \text{for } 0 \le t \le 10$$

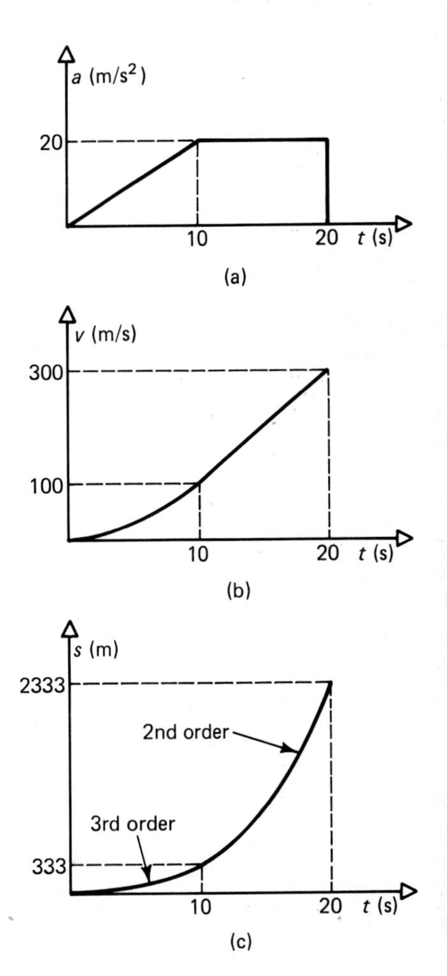

Figure 2.12

We can now sketch the vt and st curves up to $t = 10$ s as shown in Figs. 2.12(b) and (c). Note that $s = 333$ m at $t = 10$ s.

Having the velocity and displacement at $t = 10$ s, we complete the problem by either directly applying Eqs. (2.12) and (2.13) or using the curves in Fig. 2.12. Letting $t_1 = 10$, $t_2 = 20$, and $v_1 = 100$, we calculate the velocity v_2 at t_2 directly from the curve as follows:

$$v_2 - v_1 = \text{area under the } at \text{ curve between } t_1 \text{ and } t_2$$

$$v_2 - 100 = 20(10)$$

$$v_2 = 300 \text{ m/s}$$

Draw a straight line between $t = 10$ and $t = 20$ s as shown in Fig. 2.12(b).

Finally, let $t_1 = 10$, $t_2 = 20$, and $s_1 = 333$, so that the displacement s_2 at t_2 is

$$s_2 - s_1 = \text{area under the } vt \text{ curve between } t_1 \text{ and } t_2$$

$$s_2 - 333 = \tfrac{1}{2}(100 + 300)(10)$$

$$s_2 = 2333 \text{ m}$$

PROBLEMS

2.12 The displacement of a particle is given by $x = t(t^2 + 1)$ ft, where t is in seconds. Plot the velocity and acceleration as a function of time.

2.13 The displacement of a particle is given by $x = (t - t^2)$ m, where t is in seconds. Plot the velocity and acceleration as a function of time.

2.14 A vehicle, traveling in a straight line, starts from rest and accelerates uniformly at 10 m/s^2 for the first 3 s, and decelerates uniformly at 5 m/s^2 for the next 4 s. Plot the *at*, *vt*, and *st* curves and label the ordinate values at $t = 3$ s and $t = 7$ s.

2.15 A particle travels in a straight line starting with an initial velocity of 10 m/s. It accelerates uniformly at 5 m/s^2 for the first 4 s, and decelerates uniformly at 5 m/s^2 for the next 4 s. Plot the *at*, *vt*, and *st* curves and label the ordinate values at $t = 4$ s and $t = 8$ s.

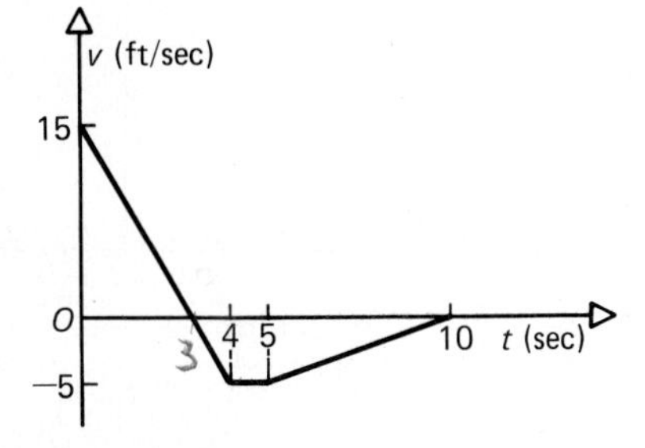

Figure P2.16

2.16 For the velocity-time history in Fig. P2.16 of a particle moving in a straight line, plot the *at* and *st* curves and label all significant ordinates.

2.17 A projectile travels through a 2-ft medium where its acceleration is described by $a = 10e^{-s}$, where a is in ft/sec^2 and s is in ft. Throughout a second medium, the acceleration is assumed to be constant at -5 ft/sec^2. Assuming that the velocity is 6 ft/sec when it enters the first medium, how far does the projectile travel until it comes to rest?

2.18 A car accelerates uniformly from rest to a speed of 60 km/h in 300 m. It travels at this speed for 1000 m and then decelerates uniformly to rest in 150 m. How much time has elapsed?

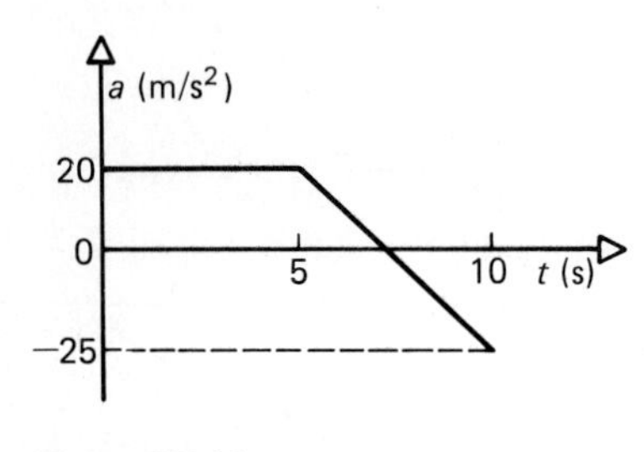

Figure P2.19

2.19 For the acceleration-time history in Fig. P2.19 of a particle moving in a straight line, plot the *vt* and *st* curves and label all significant ordinates.

2.20 The acceleration of a point P is given by $a = 8 \sin \pi t$ ft/sec^2. It is known that when $t = 0$, $x = 6$ ft and when $t = 1$ sec, $x = 20$ ft. (a) What is the velocity of P? (b) What is the displacement when $t = 5$ sec?

2.21 The acceleration of a particle P is related to its velocity by $a = -Kv^2$. Express the velocity as a function of time assuming v_0 is the initial velocity. (This kind of equation occurs when a particle moves in a resisting medium, e.g., a satellite moving in the upper atmosphere.)

2.22 When a train, traveling at 80 km/h, is one kilometer away from the next station, the train engineer applies the brakes so that the train decelerates uniformly in coming to rest at the station. What is the magnitude of the deceleration in m/s^2 and how long does it take for the train to come to rest?

2.23 A rocket blasts off vertically from the earth's surface. The thrust from the engines provides a constant acceleration of $10g$, where $g = 32.2$ ft/sec^2, that lasts for 20 sec. How high is the rocket above the earth's surface when the engines cut off?

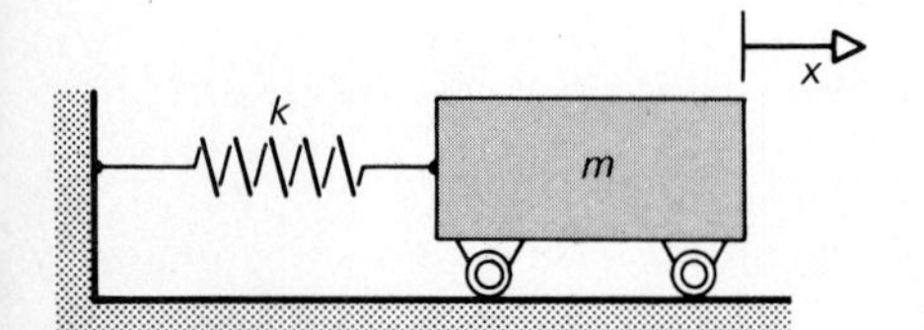

Figure P2.24

2.24 The mass m in Fig. P2.24 oscillates harmonically about the unstretched length of the spring. It can be shown from Newton's second law that the acceleration of the mass is $\ddot{x} = -kx/m$. If the mass is initially displaced an amount x_0 and then released, find the velocity after the mass travels $x_0/2$ and when the spring is unstretched.

2.25 A ball is dropped from rest vertically downward from a tower as shown in Fig. P2.25. It travels 20 ft when it enters a resisting medium that causes the acceleration to change linearly with time from $g = 32.2$ ft/sec^2, when it reached the medium, to $0.9g$ when it leaves the medium 10 sec later. What is the velocity of the ball when it reaches the ground?

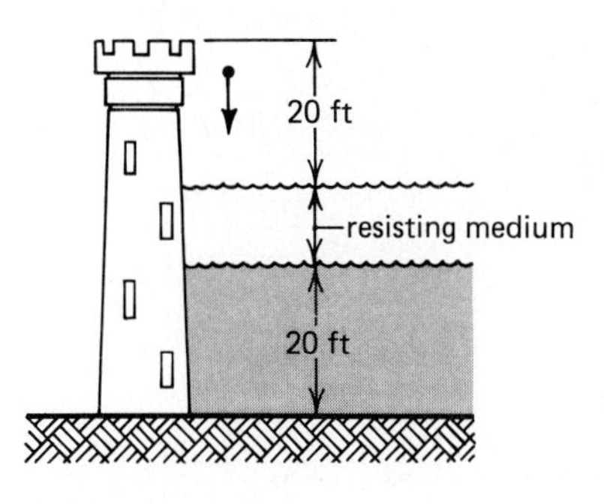

Figure P2.25

2.26 The touchdown speed of an airplane is 165 km/h and it must decelerate to a stop on a runway whose length is 1 km. Compute the deceleration assuming that it is uniform.

2.27 It is proposed to connect two cities 450 km apart by a new air cushion vehicle (ACV). This vehicle accelerates and decelerates uniformly at 20 m/s^2 and its top speed is 275 km/h. How much time does it take to make the trip?

2.28 If the ACV discussed in Problem 2.27 is to make one stop midway for 3 min, then what is the total time for the complete trip?

2.29 An airplane is flying at a constant velocity of 400 mi/hr when it suddenly comes across an air pocket that causes a deceleration as shown in Fig. P2.29. The plane recovers its cruising velocity in 1 min. How much delay does this cause in the plane's landing time?

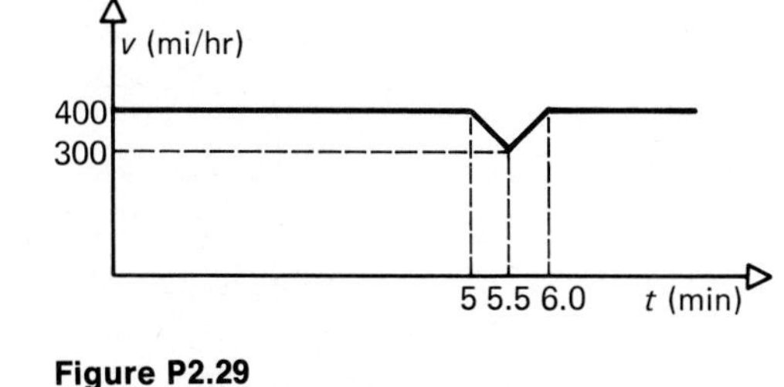

Figure P2.29

2.3 Angular Motion of a Line

Suppose a circular wheel is pinned at O and is free to rotate in the XY plane as shown in Fig. 2.13(a). Consider any line AB on the wheel that rotates through a small rotation $\Delta\theta$ in the counterclockwise direction in a time interval Δt as seen in Fig. 2.13(b). If we let $\Delta t \to 0$, the magnitude of the instantaneous angular velocity is obtained as

$$\omega = \lim_{\Delta t \to 0} \frac{\Delta\theta}{\Delta t} = \frac{d\theta}{dt} \tag{2.15}$$

The angular velocity also has direction in accord with the right-hand rule. Since the wheel rotates CCW, our thumb points in the Z direction so that

$$\boldsymbol{\omega} = \frac{d\theta}{dt}\mathbf{k} \tag{2.16}$$

Figure 2.13(c) is a view of the wheel in the XZ plane showing the $\boldsymbol{\omega}$

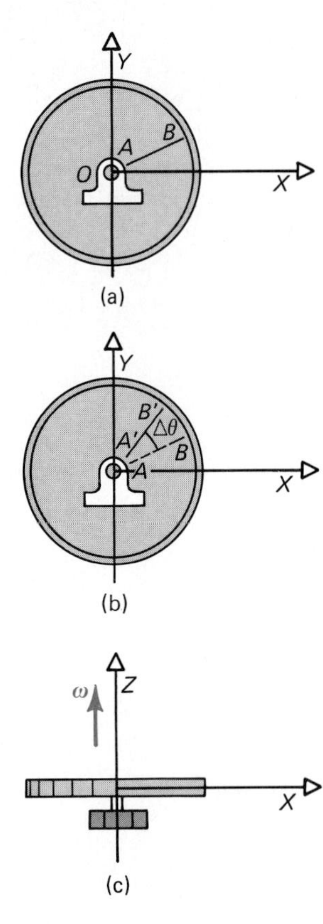

Figure 2.13

vector. As we shall discover below, $\boldsymbol{\omega}$ is a free vector so that it need not pass through any particular point on the plate.

If we consider the ratio of $\Delta\omega/\Delta t$ and let $\Delta t \to 0$, we obtain the magnitude of the instantaneous angular acceleration as

$$\alpha = \lim_{\Delta t \to 0} \frac{\Delta\omega}{\Delta t} = \frac{d\omega}{dt} \tag{2.17}$$

If we eliminate dt simultaneously from Eqs. (2.15) and (2.17), we obtain the following expression:

$$\omega\, d\omega = \alpha\, d\theta \tag{2.18}$$

By comparing Eqs. (2.15), (2.17), and (2.18) with Eqs. (2.8), (2.9), and (2.10), respectively, we see that they are identical in form, whereby θ, the angular displacement, replaces s, the linear displacement; the angular velocity ω replaces the linear velocity v; and the angular acceleration α replaces the linear acceleration a. The equations for the angular motion of a line in a plane can be integrated and represented graphically the same way as we did for the rectilinear motion of a particle. Keep in mind that θ is measured in radians, ω in rad/s, and α in rad/s^2.

A special comment about the angular velocity is useful at this point. Suppose the wheel in Fig. 2.14 is free to roll along a plane as shown in Fig. 2.14(a). Now the body no longer experiences pure rotation; rather it translates and rotates as it moves from one location to another. Suppose we follow the motion of lines AB and CD during a time interval Δt. We observe in Fig. 2.14(b) that each line rotates through the *same* angle $\Delta\theta$. Consequently, when we let $\Delta t \to 0$, the angular velocity of line AB is the same as that of line CD. In other words, the angular velocity $\boldsymbol{\omega}$ of any line painted on the rolling body is the same. This means that we no longer have to be concerned about the $\boldsymbol{\omega}$ vector passing through some fixed point. We say that $\boldsymbol{\omega}$ is a *free* vector, in that it has magnitude and direction but need not pass through a particular point. We shall find this idea useful when we study the motion of rigid bodies.

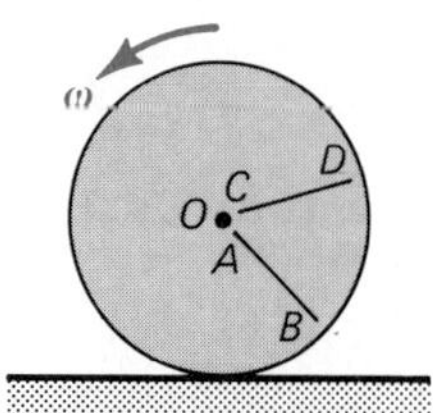

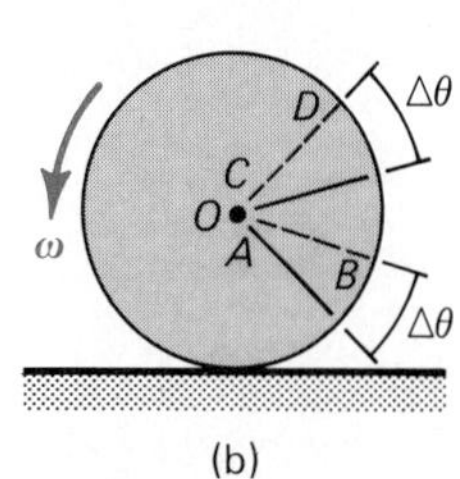

Figure 2.14

Example 2.6

A wheel initially at rest is subjected to the angular acceleration shown in Fig. 2.15(a). Sketch the ωt and θt curves.

Figures 2.15(b) and (c) show the ωt and θt curves. For the first 10 s, the angular velocity varies linearly and the angular displacement follows a second-order curve. For this interval ω varies with t as follows:

$$\omega = 2t$$

and $$\theta = \int_0^t 2t\,dt = t^2$$

At $t = 10$ s, $\omega = 20$ rad/s and $\theta = 100$ rad.

During the next 10 s,

$$d\omega = \alpha\,dt$$

For convenience introduce τ as being measured from $t = 10$ s as shown in Fig. 2.15(a). Since

$$\alpha = 2 - 0.2\tau$$

we may obtain an expression for the angular velocity in terms of τ.

$$\int_{20}^{\omega} d\omega = \int_0^{\tau} (2 - 0.2\tau)\,d\tau$$

$$\omega = 20 + 2\tau - 0.1\tau^2 \qquad \text{(a)}$$

At $\tau = 10$, $\omega = 30$ rad/s.

Also, we now obtain an expression for the angular displacement in terms of τ.

$$d\theta = \omega\,dt = \omega\,d\tau$$

Substituting Eq. (a) and forming the integrals, we have

$$\int_{100}^{\theta} d\theta = \int_0^{\tau} (20 + 2\tau - 0.1\tau^2)\,d\tau$$

$$\theta = 100 + 20\tau + \tau^2 - \frac{0.1\tau^3}{3} \qquad \text{(b)}$$

At $\tau = 10$, $\theta = 366.7$ rad

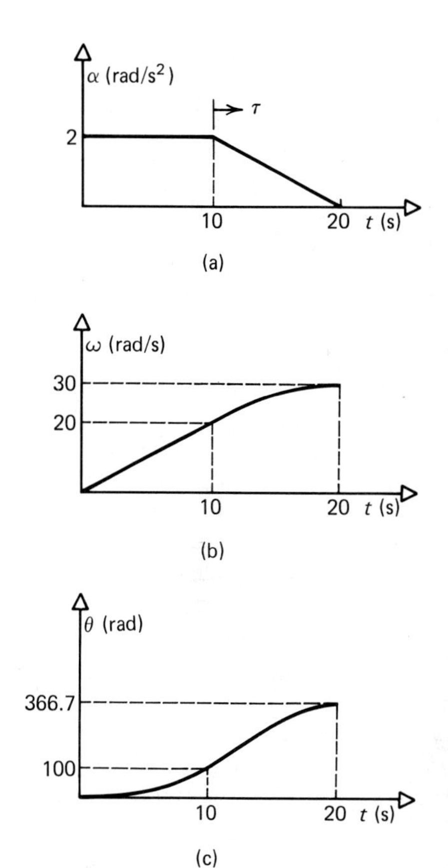

Figure 2.15

The last 10 s sketched in Figs. 2.15(b) and (c) show a second-order curve for ω and a third-order curve for θ in accord with Eqs. (a) and (b). Note that we could transform equations (a) and (b) from τ to t by using the relationship $\tau = t - 10$ and reworking the problem. This adds to the algebra but should yield the same answer and is suggested as an exercise.

Example 2.7

The angular velocity of a spinning wheel for the first 4 s is

$$\omega = (2t^2 - 4t - 6) \text{ rad/s}$$

where a positive value of ω refers to the counterclockwise direction and t is measured in seconds. After 4 s, the angular velocity is uniform at 10 rad/s CCW. Find the angular displacement at $t = 6$ s, and the total angle of rotation in 6 s.

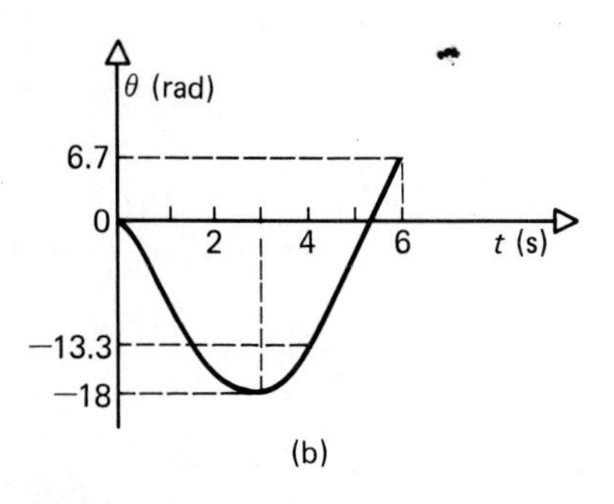

Figure 2.16

At $t = 0$, let $\theta = 0$. A plot of the given data on the ωt plane is shown in Fig. 2.16(a). Since the total angle turned during the first 6 s is required, let us first find the maximum rotation in the clockwise direction (plotted negatively). From the ωt plot we see this occurs at $t = 3$ s. Thus,

$$\int_0^{\theta_2} d\theta = \int_0^3 \omega\, dt$$

$$\theta_2 = \int_0^3 (2t^2 - 4t - 6)\, dt$$

$$= \left(\frac{2t^3}{3} - 2t^2 - 6t\right)_0^3$$

$$= -18 \text{ rad}$$

Figure 2.16(b) shows the θt curve. Note that the slope is zero at $t = 3$ s. Now integrate over the next second. Let $t_1 = 3$, $t_2 = 4$, and $\theta_1 = -18$ for this increment and note that θ_2 occurs at $t_2 = 4$ s.

$$\int_{-18}^{\theta_2} d\theta = \int_3^4 (2t^2 - 4t - 6)\, dt$$

$$\theta_2 + 18 = [(42.7 - 32 - 24) - (18 - 18 - 18)]$$

$$\theta_2 = -13.3 \text{ rad}$$

This completes the data for plotting the θt curve over the first 4 s. For the last 2 s, let $t_1 = 4$, $t_2 = 6$, and $\theta_1 = -13.3$. Note that $\omega = 10$ rad/s and θ_2 occurs at $t = 6$ s.

$$\int_{-13.3}^{\theta_2} d\theta = \int_4^6 10\, dt$$

$$\theta_2 + 13.3 = 20$$

$$\theta_2 = 6.7 \text{ rad} \qquad \textit{Answer}$$

The total angle turned during the 6 s period is

$$|\theta| = 18 + 18 + 6.7 = 42.7 \text{ rad} \qquad \textit{Answer}$$

PROBLEMS

2.30 The angular displacement of a wheel is given by $\theta = (t^3 - 2t^2 + 4)$ rad. Plot the angular velocity and angular acceleration up to $t = 2$ s.

2.31 The arm of a piston-crankshaft mechanism has an angular velocity of $\omega = 3t^2$ rad/s for the first 10 s, after which time the angular velocity remains constant at 300 rad/s. How many revolutions has the arm experienced when $t = 20$ s?

2.32 A particle undergoes angular motion such that the angular acceleration is given by $\alpha = C\omega$, where C is a constant. If $\omega = \omega_0$ at $t = 0$, obtain an expression for the angular velocity as a function of time.

2.33 Obtain an expression for the angular displacement for Problem 2.32 if $\theta = \theta_0$ at $t = 0$.

2.34 A drum is rotating at a constant angular velocity of 1200 rev/min when a brake is applied to bring it to rest. If the braking action is to produce a constant deceleration over a period of 2 min, what must be the magnitude of the deceleration and how many revolutions does the drum turn during this period?

2.35 The time history for the angular acceleration of a spinning wheel is in Fig. P2.35. Plot the ωt and θt curves and label all significant ordinate values.

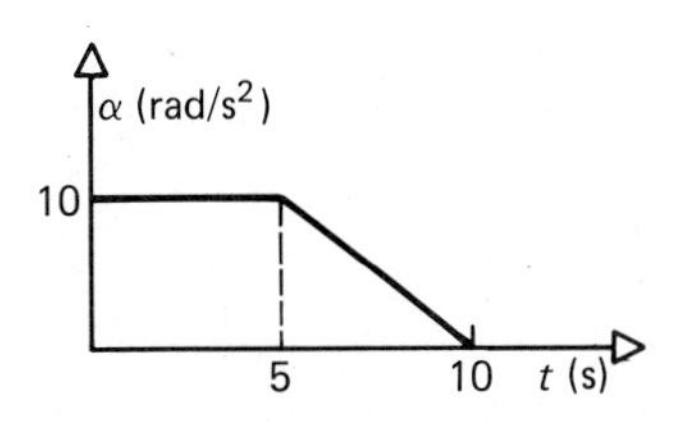

Figure P2.35

2.36 In an experiment, a motor shaft is rotating at a constant rate of 3600 rev/min when the power goes off. The manufacturer claims that in the initial deceleration mode the angular velocity is given by $\omega = 3600e^{-\sigma t}$ rev/min, where t is in min. What must σ be for ω to reduce to 2 rev/min in 0.9 min?

2.37 In Problem 2.36, how many revolutions will occur in the first 0.2 min after the power goes off?

2.38 The rotation of a wheel is such that its angular velocity varies as shown in Fig. P2.38. What is the angular acceleration when $\omega = 150$ rad/s?

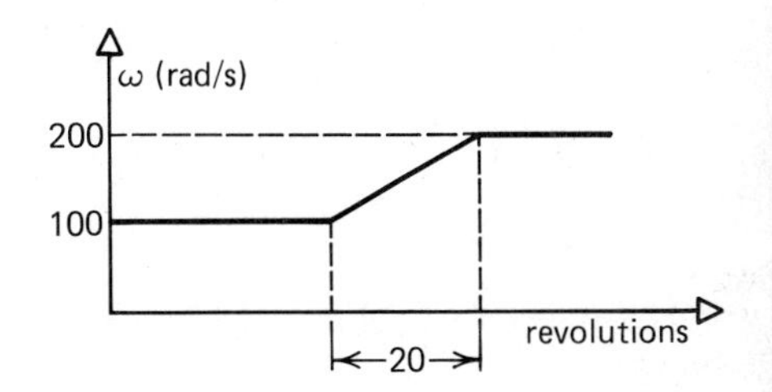

Figure P2.38

2.39 The angular acceleration of a spacecraft is approximated by $\alpha = A\theta$, where A is a constant. Obtain an expression for the angular velocity and displacement of the vehicle where $\omega = 0$ and $\theta = \theta_0$ at $t = 0$. Assume $\theta > \theta_0$.

2.40 A point on a rim of radius r undergoes the angular motion shown in Fig. P2.40. Plot its (a) angular velocity and (b) angular acceleration as a function of time.

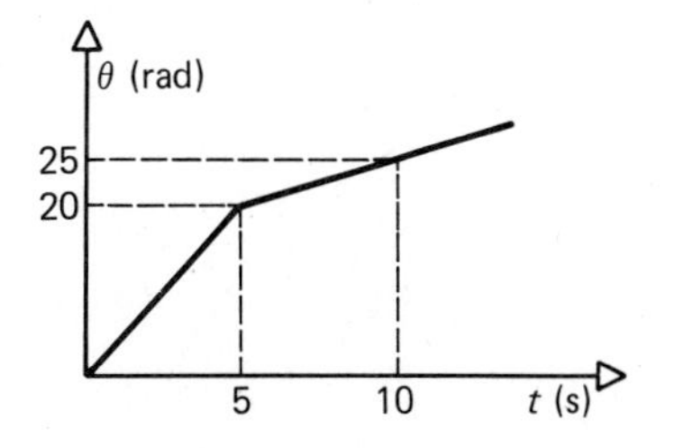

Figure P2.40

2.41 The angular velocity of a point P on a rim is observed using a stroboscope and is recorded as a function of time as shown in Fig. P2.41. Obtain a plot of the angular acceleration as a function of time.

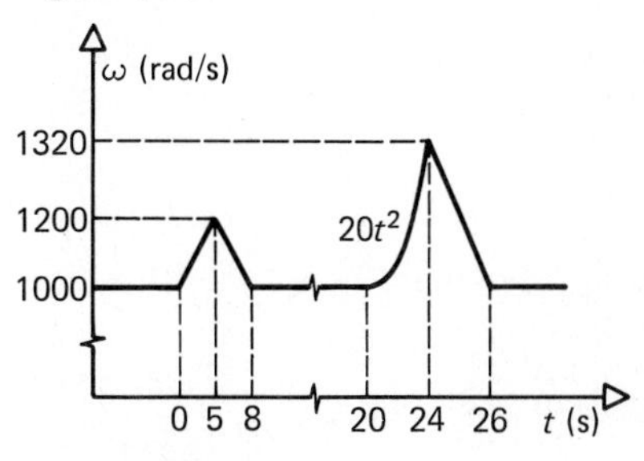

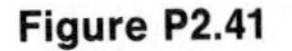
Figure P2.41

2.42 The angular velocity of a rotating flywheel changes uniformly from 10 rad/s CW to 20 rad/s CCW in 12 s. During this time interval, what is (a) the angular displacement and (b) the total angle change of the flywheel?

2.43 The angular acceleration of a rotating line is given by

$$\alpha = -\lambda^2 \beta \cos \lambda t \text{ rad/s}^2$$

When $t = 0$, α is clockwise, the angular velocity of the line is A rad/s CCW, and the angular displacement is C rad CW. Find the general expression for θ.

2.44 The rotor of a motor undergoes a uniform change in angular velocity from 1200 rev/min CW to 3600 rev/min CCW during a 4-min period. What is (a) the angular displacement and (b) the total angle change of the rotor?

2.45 The blades of the turbine start turning from rest under an angular acceleration of 8 rad/s^2 CCW that lasts 5 s, undergoes uniform angular velocity for the next 5 s, and decelerates uniformly to rest after an additional 10 s. Draw the θt, ωt, and αt curves.

2.46 The angular velocity of a wheel for the first 3 s is

$$\omega = (t^2 - 2t - 1.25) \text{ rad/s}$$

where t is measured in seconds. After 3 s the angular velocity suddenly shifts to a uniform rate of 4 rad/s. Find the angular displacement at $t = 5$ s, and the total angle of rotation in 5 s.

2.47 Repeat Problem 2.46 if $\omega = (t^2 - 2t)$ rad/s. Note that $\omega = 4$ rad/s from $t = 3$ s onward.

2.4 Absolute and Relative Motion

In studying dynamics it is important to distinguish between absolute and relative motion. Whether any given motion is considered to be absolute or relative depends on the choice of the reference frame. For example, when the speed of the astronauts was as high as 17,000 mi/hr relative to the earth, it was zero relative to their capsule and a few miles per hour relative to the last stage of the rocket. Clearly, then, it is necessary to know the reference, i.e., is the reference frame moving or is it assumed fixed? It is even more important to know whether the reference frame is accelerating, since nonaccelerating reference frames are used in newtonian mechanics.

The coordinate systems presented in Chapter 1 may be used as reference frames, and they may be fixed or they may be moving. When we speak of a reference frame, we are really speaking about the dependency of a coordinate system on time: a reference frame attached to a ball that has just been pitched is an accelerating reference frame with respect to the earth, whereas a reference frame fixed in the earth may be assumed as not accelerating. (Actually, a preponderance of the calculations in the field of dynamics ignores the motion of the earth and treats such a reference frame as not moving.)

A fixed reference frame is often called an **inertial frame**. It is also referred to as a newtonian or galilean frame of reference. A true inertial frame is very difficult to obtain. Indeed, if our entire solar system moves about the sun, a true nonaccelerating system should be fixed in the sun. Such a system is called a **heliocentric** system. A reference frame fixed in the earth is called a **geocentric** system, and suffices as an inertial system for most engineering applications.

When the motion of a particle or rigid body is referred to a fixed frame of reference, the motion is called **absolute motion**. If the reference frame is not fixed, the motion is called **relative motion**. Throughout the text, the frame of reference and the kind of motion will be specified.

2.5 Derivatives of Moving Vectors

Rotating Unit Vector

Consider the unit vector **e** rotating in the XY plane with an angular velocity $\boldsymbol{\omega}$ as shown in Fig. 2.17. Express this vector in its XY components as

$$\mathbf{e} = \cos\theta\mathbf{i} + \sin\theta\mathbf{j} \tag{2.19}$$

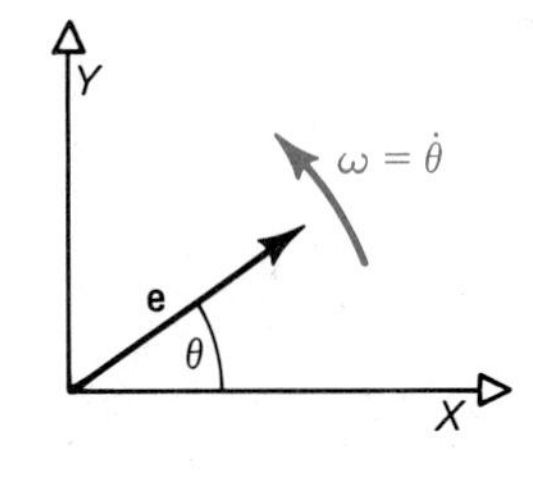

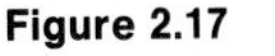
Figure 2.17

where the unit vectors **i** and **j** are fixed, so that

$$\frac{d\mathbf{i}}{dt} = \frac{d\mathbf{j}}{dt} = 0 \tag{2.20}$$

We now take the derivative of Eq. (2.19) using the results of Eq. (2.20) and obtain

$$\frac{d\mathbf{e}}{dt} = \dot{\theta}(-\sin\theta\mathbf{i} + \cos\theta\mathbf{j}) \tag{2.21}$$

where $\dot{\theta} = d\theta/dt$. The magnitude of this vector is

$$\left|\frac{d\mathbf{e}}{dt}\right| = \dot{\theta}\sqrt{\sin^2\theta + \cos^2\theta} = \dot{\theta}$$

which is the rate of rotation of **e** in the XY plane. The vector $d\mathbf{e}/dt$ is shown in Fig. 2.18 and is seen to be perpendicular to **e**. In the last section we called the magnitude of the angular velocity $\boldsymbol{\omega}$ the rate of rotation of a line in a plane; recalling Eq. (2.16) we have

$$\boldsymbol{\omega} = \dot{\theta}\mathbf{k} \tag{2.16}$$

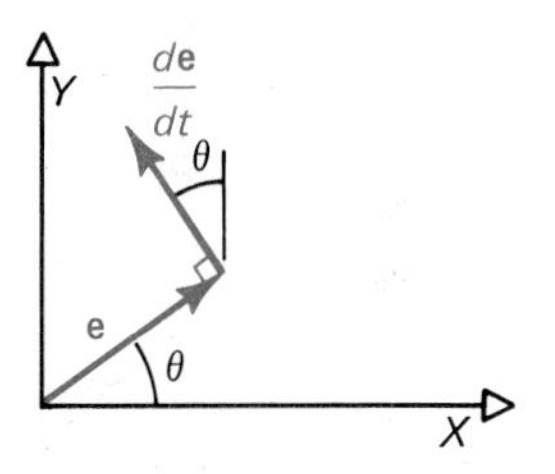

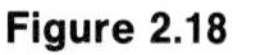
Figure 2.18

This vector is perpendicular to the XY plane. Therefore, the vectors representing **e**, $d\mathbf{e}/dt$, and $\boldsymbol{\omega}$ are all perpendicular to each other, i.e., they form an orthogonal set as shown in Fig. 2.19. Since the magnitudes of $d\mathbf{e}/dt$ and $\boldsymbol{\omega}$ are equal and the magnitude of **e** is unity, we may express the orthogonality shown in Fig. 2.19 by using the vector product, that is,

$$\frac{d\mathbf{e}}{dt} = \boldsymbol{\omega} \times \mathbf{e} \tag{2.22}$$

As an exercise, you may want to substitute Eqs. (2.19) and (2.16) into Eq. (2.22) to obtain Eq. (2.21).

We conclude that the derivative of a unit vector is equal to the rate of rotation of the unit vector crossed into the unit vector itself. The derivative of the unit vector represents the velocity of the tip of the unit vector. Note that Eq. (2.22) could be derived directly by applying Eq. (2.5).

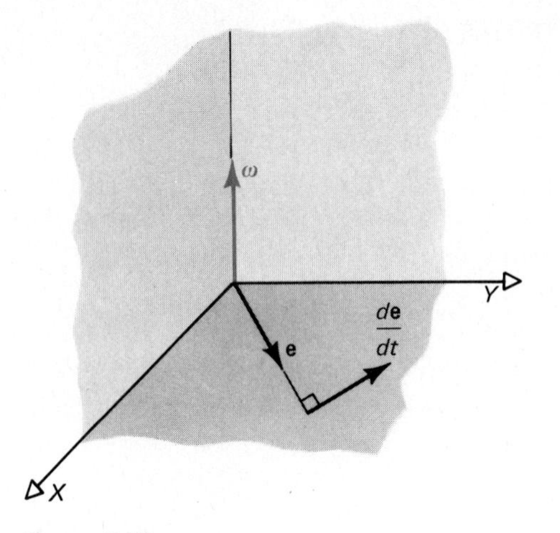

Figure 2.19

Vector in a Moving Reference Frame

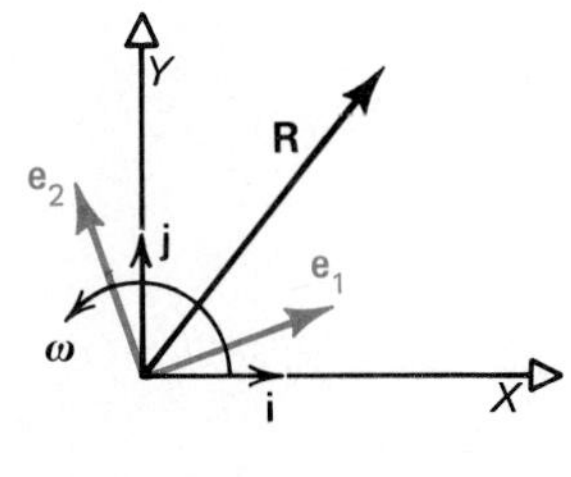

Figure 2.20

Consider the vector $\mathbf{R}$ in the XY plane in Fig. 2.20. Assume that the position of $\mathbf{R}$ varies with time. Let the $\mathbf{i}$, $\mathbf{j}$ unit vectors remain fixed while $\mathbf{e}_1$, $\mathbf{e}_2$ rotate at an absolute rate $\boldsymbol{\omega}$ rad/s. Suppose we express $\mathbf{R}$ in terms of the $\mathbf{i}$, $\mathbf{j}$ unit vectors:

$$\mathbf{R} = R_x\mathbf{i} + R_y\mathbf{j} \tag{2.23}$$

where R_x and R_y are the components of $\mathbf{R}$ on the X and Y axes, respectively. Now the derivative of this expression with respect to time is

$$\dot{\mathbf{R}} = \dot{R}_x\mathbf{i} + \dot{R}_y\mathbf{j} \tag{2.24}$$

where $\dot{R}_x = dR_x/dt$ and $\dot{R}_y = dR_y/dt$. We could just as well express $\mathbf{R}$ in terms of the $\mathbf{e}_1$, $\mathbf{e}_2$ unit vectors:

$$\mathbf{R} = R_1\mathbf{e}_1 + R_2\mathbf{e}_2 \tag{2.25}$$

where R_1 and R_2 are the components of $\mathbf{R}$ on the $\mathbf{e}_1$ and $\mathbf{e}_2$ axes, respectively. The derivative of Eq. (2.25) with respect to time is

$$\dot{\mathbf{R}} = \dot{R}_1\mathbf{e}_1 + R_1\dot{\mathbf{e}}_1 + \dot{R}_2\mathbf{e}_2 + R_2\dot{\mathbf{e}}_2 \tag{2.26}$$

where $\dot{\mathbf{e}}_1$ and $\dot{\mathbf{e}}_2$ are included since they are rotating at an angular velocity $\boldsymbol{\omega}$. Using Eq. (2.22),

$$R_1\dot{\mathbf{e}}_1 = R_1\boldsymbol{\omega} \times \mathbf{e}_1 = \boldsymbol{\omega} \times R_1\mathbf{e}_1$$

$$R_2\dot{\mathbf{e}}_2 = R_2\boldsymbol{\omega} \times \mathbf{e}_2 = \boldsymbol{\omega} \times R_2\mathbf{e}_2$$

Rewriting Eq. (2.26),

$$\dot{\mathbf{R}} = \dot{R}_1\mathbf{e}_1 + \dot{R}_2\mathbf{e}_2 + \boldsymbol{\omega} \times R_1\mathbf{e}_1 + \boldsymbol{\omega} \times R_2\mathbf{e}_2$$

$$= \dot{\mathbf{R}}_r + \omega \times (R_1\mathbf{e}_1 + R_2\mathbf{e}_2)$$
$$= \dot{\mathbf{R}}_r + \omega \times \mathbf{R} \tag{2.27}$$

where $\dot{\mathbf{R}}_r = \dot{R}_1\mathbf{e}_1 + \dot{R}_2\mathbf{e}_2$
= Rate of change of **R** as measured by an observer attached to the $\mathbf{e}_1$, $\mathbf{e}_2$ rotating reference frame

We conclude that while we can represent a vector in terms of different components such as Eqs. (2.23) and (2.25), the form of the derivative of the vector will depend on whether the chosen unit vectors are rotating or not. As we shall see later, there are advantages to representing a vector in terms of a moving coordinate system in certain applications.

2.6 Velocity of a Particle

Consider the particle P that moves in the XY plane, as shown in Fig. 2.21, and has a position vector of

$$\mathbf{S} = \mathbf{R} + \mathbf{r} \tag{2.28}$$

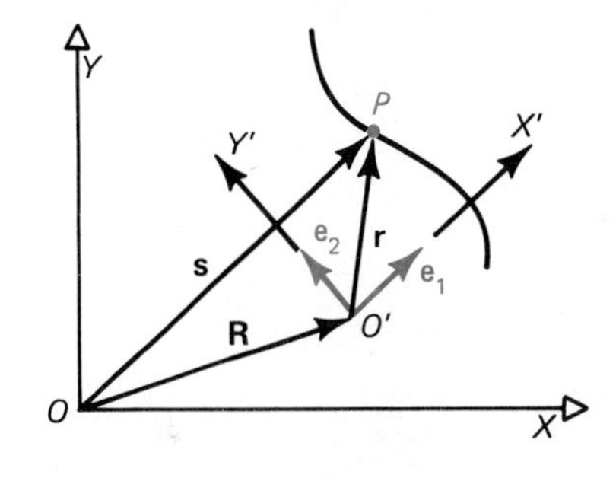

Figure 2.21

where **R** locates O' relative to O in the fixed XY system and **r** locates the particle P in the moving $X'Y'$ system. This means that the unit vectors $\mathbf{e}_1$ and $\mathbf{e}_2$ that are attached to the $X'Y'$ system also rotate at ω rad/s. Although not shown in Fig. 2.21, the unit vector $\mathbf{e}_3$ is present at O' such that $\mathbf{e}_3 = \mathbf{e}_1 \times \mathbf{e}_2$. For convenience we shall refer to the moving system in terms of the unit vectors $\mathbf{e}_1$ and $\mathbf{e}_2$ rather than the X' and Y' axes.

The derivative of Eq. (2.28) is the absolute velocity of the particle P.

$$\mathbf{v} = \dot{\mathbf{S}} = \dot{\mathbf{R}} + \dot{\mathbf{r}} \tag{2.29}$$

where $\dot{\mathbf{R}}$ is the velocity of O' relative to O. Since **r** is expressed in terms of a rotating reference frame, we may apply Eq. (2.27) to obtain

$$\dot{\mathbf{r}} = \dot{\mathbf{r}}_r + \omega \times \mathbf{r} \tag{2.30}$$

Substituting Eq. (2.30) into (2.29)

$$\mathbf{v} = \dot{\mathbf{R}} + \dot{\mathbf{r}}_r + \omega \times \mathbf{r} \tag{2.31}$$

where $\dot{\mathbf{R}}$ = Absolute velocity of the moving origin O'

ω = Absolute angular velocity of the moving reference frame $\mathbf{e}_1$, $\mathbf{e}_2$

r = The position vector of the point P measured in the $\mathbf{e}_1$, $\mathbf{e}_2$ system

$\dot{\mathbf{r}}_r$ = The velocity of the particle P as measured by an observer attached to the moving reference frame $\mathbf{e}_1, \mathbf{e}_2$

Equation (2.31) is fundamental to our treatment of the motion of a particle. While we have developed this equation for the motion of a particle in a plane, we shall see later that it also applies to the motion of a particle in three dimensions.

Equation (2.31) is simplified if the moving reference frame translates only, so that $\omega = 0$. For this special case,

$$\mathbf{v} = \dot{\mathbf{R}} + \dot{\mathbf{r}}_r \tag{2.32}$$

For example, suppose a person walks inside a train at a rate of 2 m/s relative to the train. Let the train travel at 500 m/s. Place the origin O' of the moving reference frame on the train so that $\dot{\mathbf{R}} = 500\mathbf{e}_1$. If the person moves in the same direction as the train, $\dot{\mathbf{r}}_r = 2\mathbf{e}_1$. Applying Eq. (2.32), the absolute velocity of the person is

$$\mathbf{v} = 500\mathbf{e}_1 + 2\mathbf{e}_1 = 502\mathbf{e}_1 \text{ m/s}$$

If the person moves in the opposite direction of the train's motion, $\dot{\mathbf{r}}_r = -2\mathbf{e}_1$, so that $\mathbf{v} = 498\mathbf{e}_1$ m/s. The term given by $\dot{\mathbf{r}}_r$ can be thought of as relative velocity.

Example 2.8

A table turns at a rate of 2 rad/sec CCW as shown in Fig. 2.22. As the table turns, a marble moves outward along a groove at a rate of 3 in/sec relative to the table. What is the absolute velocity of the marble at the instant shown?

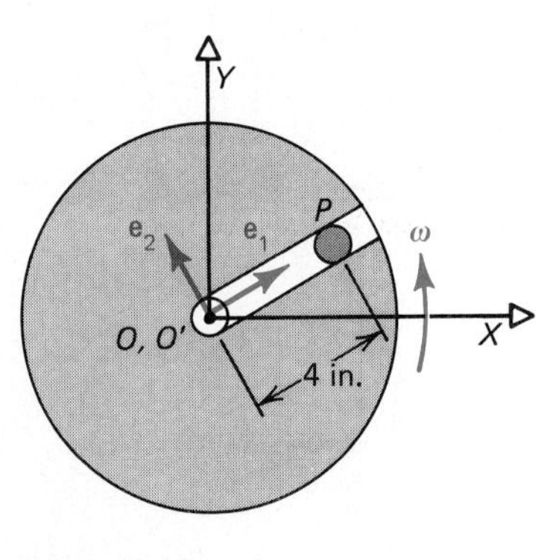

Figure 2.22

Attach the unit vectors $\mathbf{e}_1, \mathbf{e}_2, \mathbf{e}_3$ to the table as shown in the figure. Note that the origin O' of the rotating reference frame coincides with the fixed reference point O. Now from Eq. (2.31)

$$\mathbf{v} = \dot{\mathbf{R}} + \dot{\mathbf{r}}_r + \boldsymbol{\omega} \times \mathbf{r}$$

where $\quad \mathbf{R} = \dot{\mathbf{R}} = 0 \qquad \boldsymbol{\omega} = 2\mathbf{e}_3$

$\qquad \dot{\mathbf{r}}_r = 3\mathbf{e}_1 \qquad \mathbf{r} = 4\mathbf{e}_1$

Now $\quad \boldsymbol{\omega} \times \mathbf{r} = 2\mathbf{e}_3 \times 4\mathbf{e}_1 = 8\mathbf{e}_2$

The absolute velocity of the marble at the instant shown in Fig. 2.22 is

$$\mathbf{v} = (3\mathbf{e}_1 + 8\mathbf{e}_2) \text{ in/sec} \qquad \textit{Answer}$$

Alternate Approach

Suppose we place O' at P and again attach the moving unit vector system $\mathbf{e}_1, \mathbf{e}_2, \mathbf{e}_3$ to the table. For this case

$$\mathbf{R} = 4\mathbf{e}_1 \qquad \omega = 2\mathbf{e}_3$$

$$\dot{\mathbf{r}}_r = 3\mathbf{e}_1 \qquad \mathbf{r} = 0$$

Since O' is a point on the table that is rotating, its absolute velocity is

$$\dot{\mathbf{R}} = \omega \times \mathbf{R} = 2\mathbf{e}_3 \times 4\mathbf{e}_1 = 8\mathbf{e}_2$$

The absolute velocity of the marble remains

$$\mathbf{v} = (3\mathbf{e}_1 + 8\mathbf{e}_2)\text{ in/sec} \qquad \textit{Answer}$$

From the above we observe that we normally have a choice in placing the origin O' and, as we shall see later, how we attach the moving reference frame to the body.

Example 2.9

Assume that the rotating table in Example 2.8 is attached by a bearing connection to a rotating arm as shown in Fig. 2.23. The arm rotates at an absolute angular velocity of 3 rad/sec CW, and the table, which is free to rotate relative to the arm, still rotates at an absolute angular velocity of 2 rad/sec CCW. If the marble moves outward along the groove at a rate of 3 in/sec relative to the table, what is the absolute velocity of the marble at the instant shown?

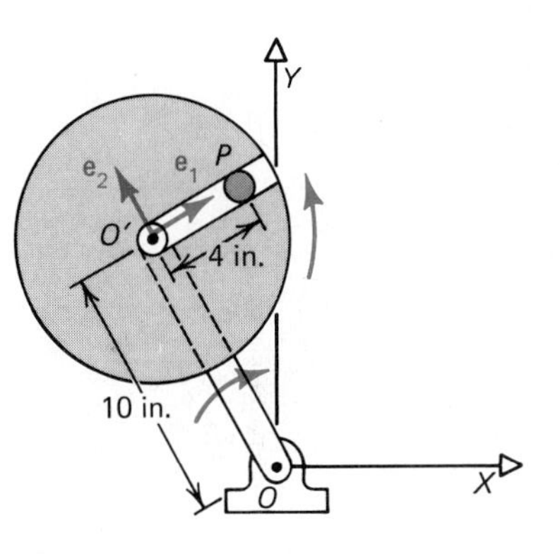

Figure 2.23

The fixed origin O and the moving origin O' are shown in Fig. 2.23. Attach $\mathbf{e}_1, \mathbf{e}_2, \mathbf{e}_3$ to the table so that $\omega = 2\mathbf{e}_3$ for the rotating unit vectors. Since O' is a point on the rotating arm, its absolute velocity is

$$\dot{\mathbf{R}} = \omega_a \times \mathbf{R}$$

where ω_a = the absolute angular velocity of the rotation arm = $-3\mathbf{e}_3$, and $\mathbf{R} = 10\mathbf{e}_2$ so that

$$\dot{\mathbf{R}} = -3\mathbf{e}_3 \times 10\mathbf{e}_2 = 30\mathbf{e}_1$$

The remaining terms are now identified in order to apply Eq. (2.31):

$$\omega = 2\mathbf{e}_3 \qquad \mathbf{r} = 4\mathbf{e}_1 \qquad \dot{\mathbf{r}}_r = 3\mathbf{e}_1 \qquad \omega \times \mathbf{r} = 2\mathbf{e}_3 \times 4\mathbf{e}_1 = 8\mathbf{e}_2$$

The absolute velocity of the marble at the instant shown in Fig. 2.23 is

$$\begin{aligned}\mathbf{v} &= \dot{\mathbf{R}} + \dot{\mathbf{r}}_r + \omega \times \mathbf{r} \\ &= 30\mathbf{e}_1 + 3\mathbf{e}_1 + 8\mathbf{e}_2 \\ &= (33\mathbf{e}_1 + 8\mathbf{e}_2)\text{ in/sec} \qquad \textit{Answer}\end{aligned}$$

Example 2.10

A particle travels around a circular groove in a plate as shown in Fig. 2.24. At the instant shown, the velocity of the particle P measured relative to the plate is 5 m/s in the CW direction. At the same instant, the plate which is rigidly attached to the 2-m arm is rotating in the XY plane at an absolute angular velocity of 3 rad/s CCW. Find the absolute velocity of the particle.

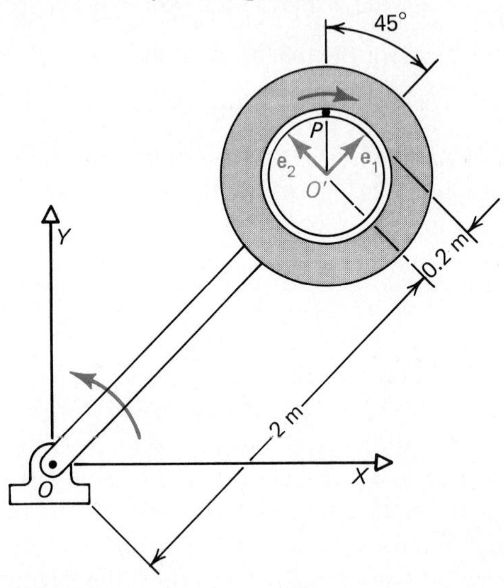

Figure 2.24

Figure 2.24 shows O and O'. We attach $\mathbf{e}_1, \mathbf{e}_2, \mathbf{e}_3$ to the assembly which is rotating at 3 rad/s CCW. The velocity of the particle is measured by an observer attached to the moving reference frame. It will appear as a vector tangent to the circular path with a magnitude equal to 5 m/s. The vectors for Eq. (2.31) are as follows:

$$\mathbf{R} = 2\mathbf{e}_1$$

$$\dot{\mathbf{R}} = 3\mathbf{e}_3 \times 2\mathbf{e}_1 = 6\mathbf{e}_2$$

$$\boldsymbol{\omega} = 3\mathbf{e}_3$$

$$\mathbf{r} = 0.2(\cos 45^\circ \mathbf{e}_1 + \sin 45^\circ \mathbf{e}_2) = 0.141(\mathbf{e}_1 + \mathbf{e}_2)$$

$$\dot{\mathbf{r}}_r = 5(\cos 45^\circ \mathbf{e}_1 - \sin 45^\circ \mathbf{e}_2) = 3.536(\mathbf{e}_1 - \mathbf{e}_2)$$

$$\boldsymbol{\omega} \times \mathbf{r} = 3\mathbf{e}_3 \times 0.141(\mathbf{e}_1 + \mathbf{e}_2) = 0.423(\mathbf{e}_2 - \mathbf{e}_1)$$

Substituting into Eq. (2.31),

$$\begin{aligned} \mathbf{v} &= \dot{\mathbf{R}} + \dot{\mathbf{r}}_r + \boldsymbol{\omega} \times \mathbf{r} \\ &= 6\mathbf{e}_2 + 3.536(\mathbf{e}_1 - \mathbf{e}_2) + 0.423(\mathbf{e}_2 - \mathbf{e}_1) \\ &= (3.113\mathbf{e}_1 + 2.887\mathbf{e}_2) \text{ m/s} \end{aligned}$$

Answer

Example 2.11

Cars A and B have velocities of 10 and 15 m/s, respectively, in the directions shown in Fig. 2.25. What is the velocity of car A relative to car B? What is the velocity of car B relative to car A?

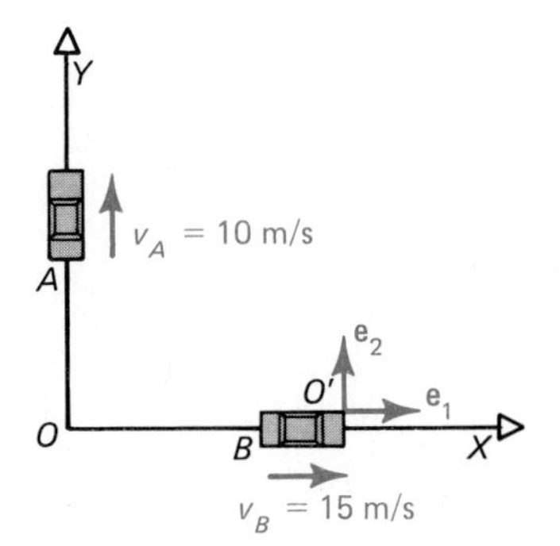

Figure 2.25

This is an example in which Eq. (2.32) may be applied. To find the velocity of car A relative to car B, place O' on car B as shown in Fig. 2.25. Note that the moving unit vectors do not rotate. Now the absolute velocity of the moving reference frame is

$$\dot{\mathbf{R}} = 15\mathbf{e}_1$$

The absolute velocity of the point in question, namely car A, is

$$\mathbf{v} = 10\mathbf{e}_2$$

so that the velocity of car A relative to car B is

$$\dot{\mathbf{r}}_r = \mathbf{v} - \dot{\mathbf{R}} = (10\mathbf{e}_2 - 15\mathbf{e}_1) \text{ m/s} \qquad \textit{Answer}$$

To find the velocity of car B relative to car A, place O' on car A and let the unit vectors $\mathbf{e}_1$ and $\mathbf{e}_2$ have the same orientation as above. Now

$$\dot{\mathbf{R}} = 10\mathbf{e}_2 \qquad \mathbf{v} = 15\mathbf{e}_1$$

so that the velocity of car B relative to car A is

$$\dot{\mathbf{r}}_r = (15\mathbf{e}_1 - 10\mathbf{e}_2) \text{ m/s} \qquad \textit{Answer}$$

We observe that the relative velocity of car A relative to car B is the negative of the relative velocity of car B relative to car A.

PROBLEMS

2.48 The unit vector $\mathbf{e} = (\cos\theta\mathbf{i} + \sin\theta\mathbf{j})$, whose magnitude equals one meter, rotates at 4 rad/s CCW. When $\theta = 135°$, what is the velocity of the tip of the vector?

2.49 The velocity of the tip of a unit vector $\mathbf{e}_2$ equals $-10\mathbf{e}_1$ m/s. What is the angular velocity of the vector?

2.50 A cylinder of diameter 80 mm rotates at a constant rate of 3 rad/s thereby allowing a rope wrapped around it to unwind. What is the magnitude of the velocity of the unwinding rope at the point where it leaves the cylinder? Neglect the changes of cylinder diameter during unwinding.

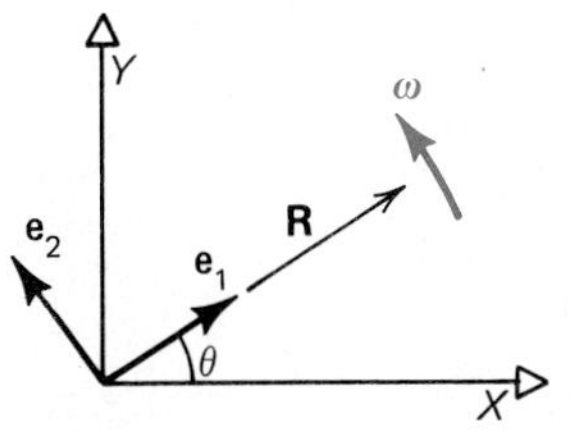

Figure P2.51

2.51 The vector $\mathbf{R} = t^2\mathbf{e}_1$ ft rotates in the XY plane at an angular velocity of $\omega\mathbf{e}_3$ rad/sec as shown in Fig. P2.51. What is the velocity of the tip of the vector $\mathbf{R}$?

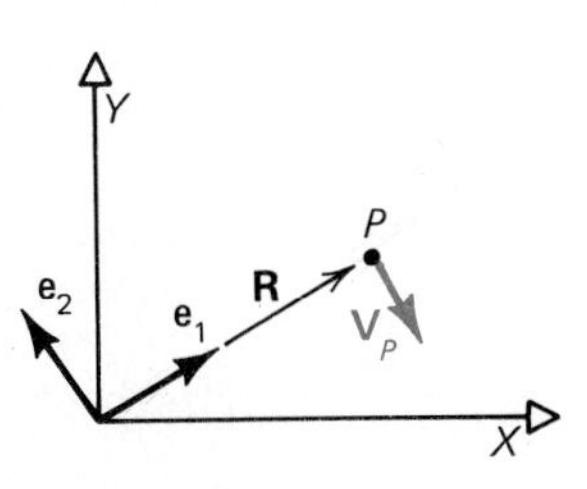

Figure P2.52

2.52 The position vector of a particle P moving in the XY plane is given by $\mathbf{R} = 4t\mathbf{e}_1$ ft where t is in seconds. When $t = \frac{1}{2}$ sec, the velocity of P is

$4\mathbf{e}_1 - 6\mathbf{e}_2$ ft/sec as shown in Fig. P2.52. What is the angular velocity of the unit vector $\mathbf{e}_1$?

2.53 A flat pan with a 1-m radius slides across the ice in Fig. P2.53 at a uniform speed of 4 m/s and spins such that its angular velocity is 3 rad/s CW. What is the velocity of point A on the pan?

2.54 A table turns at a rate of 3 rad/s CW as shown in Fig. P2.54. As the table turns, a particle moves inward along a groove at a uniform rate of 5 m/s relative to an observer attached to the table. What is the absolute velocity of the particle at the instant shown using the rotating unit vectors $\mathbf{e}_1$, $\mathbf{e}_2$?

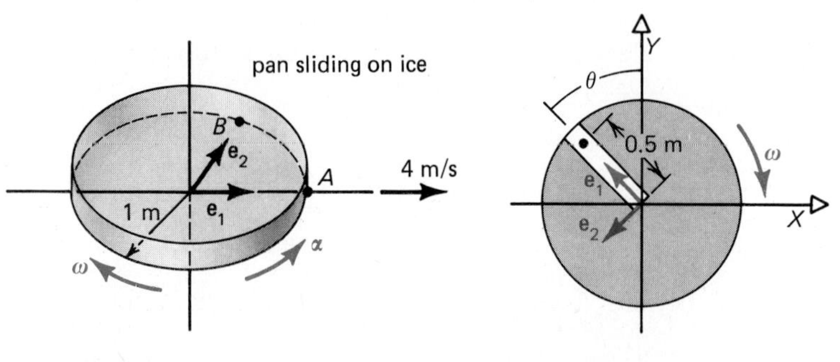

Figure P2.53

Figure P2.54

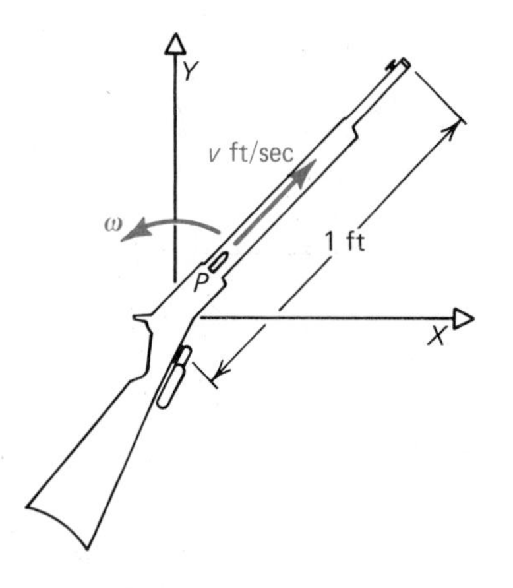

Figure P2.56

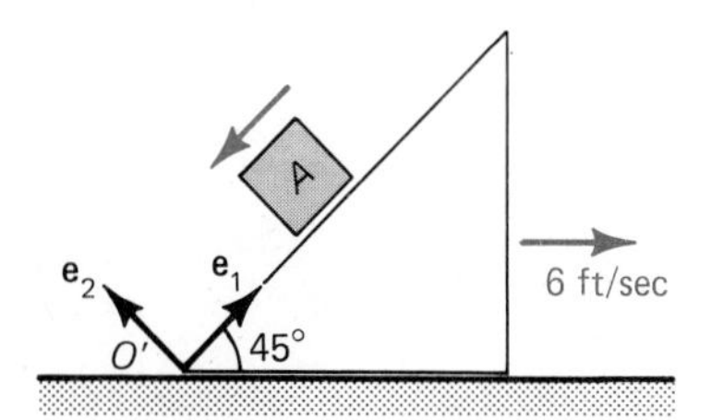

Figure P2.57

2.55 Repeat Problem 2.54 for the case where the origin of the XY axes also translates at 3 m/s in the positive X direction and $\theta = 45^\circ$ at the instant shown.

2.56 A projectile is fired from a gun that is rotating at ω rad/s CCW as shown in Fig. P2.56. The speed of the projectile relative to the muzzle is a constant v ft/sec. What is the magnitude of the velocity of the projectile at the instant it leaves the muzzle?

2.57 The wedge in Fig. P2.57 slides along a horizontal plane at 6 ft/sec while block A slides downward at a rate of 2 ft/sec relative to the wedge. Using the unit vectors $\mathbf{e}_1$, $\mathbf{e}_2$ that are attached to the wedge as shown, find the absolute velocity of the block.

2.58 The ring P slides along the bar in Fig. P2.58 at 7 m/s downward relative to the bar. Simultaneously, the bar lies in the XY plane and is rotating at 5 rad/s CCW. At the instant shown, find the absolute velocity of the ring.

2.59 A particle P travels around a circular groove in a plate as shown in Fig. P2.59. At the instant shown, the velocity measured relative to the plate is 20 m/s in the CW direction. At the same time, the plate is rigidly attached to the arm that is rotating at a uniform angular rate of 2 rad/s CW. Using the unit vectors $\mathbf{e}_1$, $\mathbf{e}_2$ attached to the plate, find the magnitude of the absolute velocity of the particle.

2.60 A hamster exercises in its cage as shown in Fig. P2.60. If the cage turns at a rate of π rad/sec, what is the velocity of the hamster relative to the cage? What is its absolute velocity?

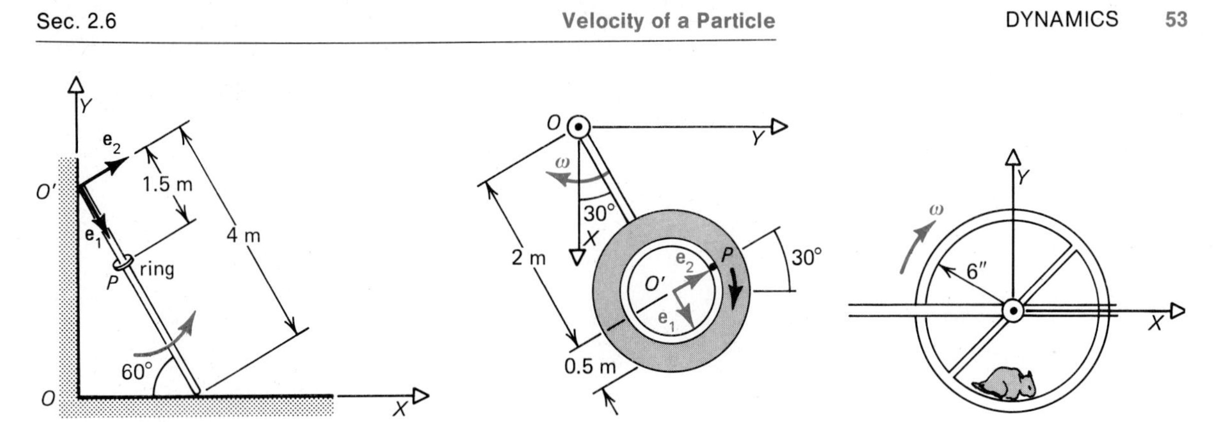

Figure P2.58

Figure P2.59

Figure P2.60

2.61 A waiter walks to a table with a velocity equal to 3 ft/sec relative to the floor of the restaurant as shown in Fig. P2.61. The restaurant turns through 5 revolutions each hour. What is the absolute velocity of the waiter at the instant shown?

2.62 An ice skater does a figure 8 on a movable stage as shown in Fig. P2.62. At the instant shown, the skater is traveling across the ice at a uniform rate of 5 m/s, and the stage is revolving at 1 rev/min CW. Find the magnitude of the skater's absolute velocity.

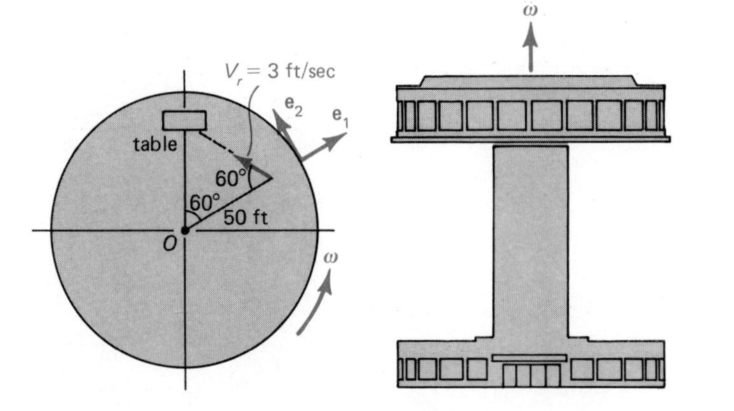

Figure P2.61

Figure P2.62

2.63 A cable of negligible diameter is employed to lower a weight as shown in Fig. P2.63. The cable wraps around the frustum of a cone. What is the magnitude of the velocity of the weight when the cone rotates at a constant ω rad/s?

2.64 A cam rotates with constant angular velocity $\dot{\theta}$ as shown in Fig. P2.64. The distance r is defined by $r = 2 - \cos\theta$. Since the point P is restrained to move in the vertical direction, what is the absolute velocity of P?

2.65 Particle P travels along a groove of a plate as shown in Fig. P2.65. At the instant shown, a velocity of 2 in/sec to the right is measured by an

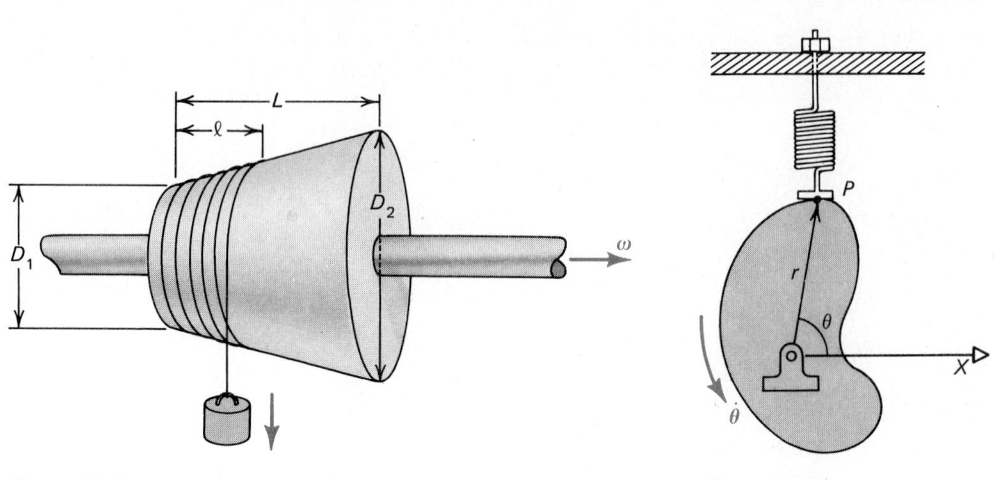

Figure P2.63

Figure P2.64

observer attached to the plate. If the angular velocity of the plate is 1 rad/sec CCW, find the absolute velocity of P.

2.66 Repeat Problem 2.65 for the case in which the origin of the XY axes is also traveling 4 in/sec in the positive Y direction.

2.67 A child, sitting in a wagon moving 1 m/s to the left, rolls marbles on the wagon in a straight line as shown in Fig. P2.67. Marble A rolls 1 m/s to the right, marble B moves 2 m/s to the right, and marble C moves 1 m/s to the left after hitting the back of the wagon (each motion relative to the child). What is the absolute velocity of each marble?

2.68 Obtain the velocity of car A relative to car B at the instant shown Fig. P2.68.

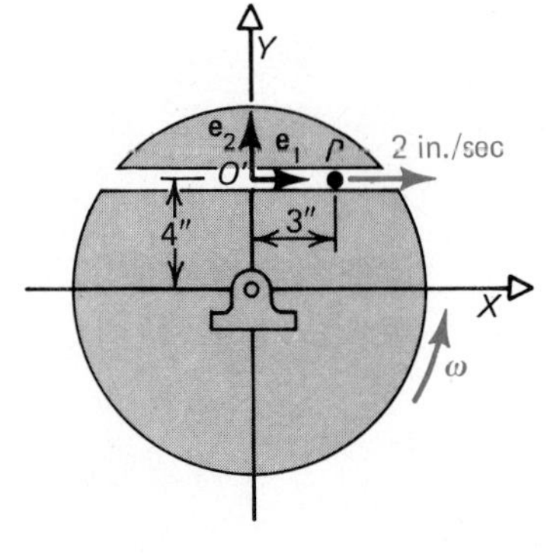

Figure P2.65

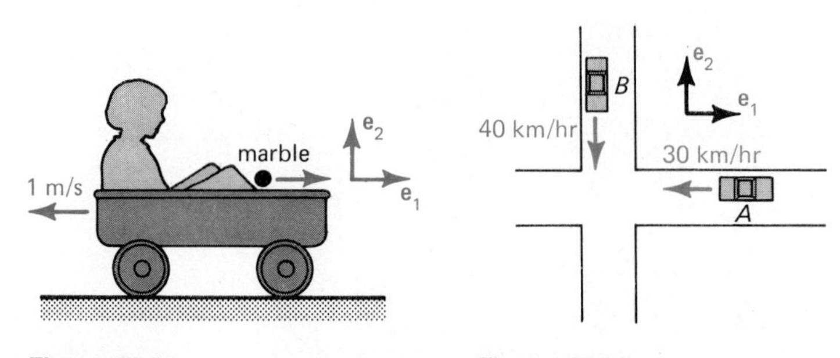

Figure P2.67

Figure P2.68

2.69 Boats A and B have velocities of 20 and 25 mi/hr, respectively, in the directions shown in Fig. P2.69. What is the velocity of boat B relative to boat A? What is the velocity of boat A relative to boat B?

2.70 Ship A in Fig. P2.70 travels in a northerly direction at a velocity of 10 mi/hr, while ship B travels northwesterly at 8 mi/hr. What is the velocity of ship B as measured by a sailor on ship A?

2.71 Obtain the absolute velocity of block 3 in Fig. P2.71 at the instant shown. Assume that the pulleys are free to rotate and that there is no slippage between the ropes and the pulleys.

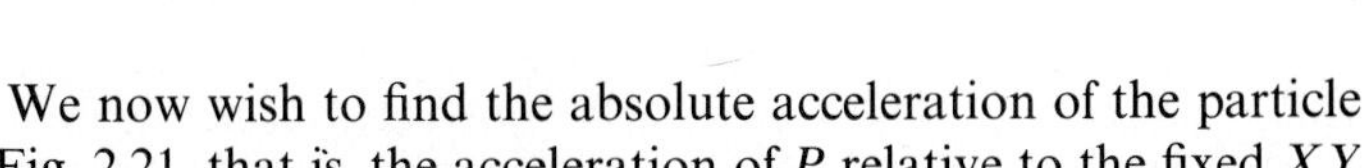

Figure P2.69

Figure P2.70

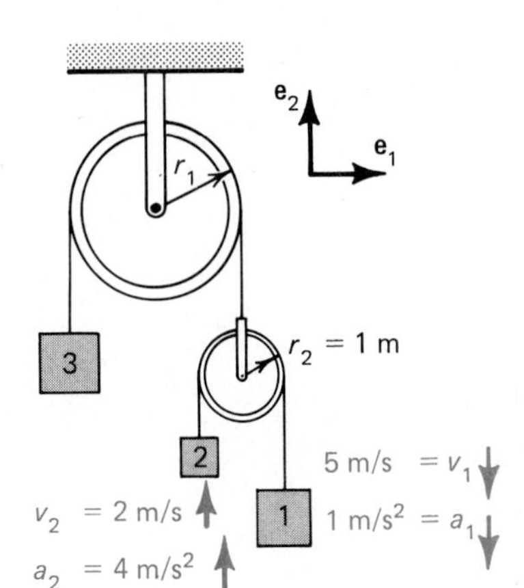

Figure P2.71

2.7 Acceleration of a Particle

We now wish to find the absolute acceleration of the particle P in Fig. 2.21, that is, the acceleration of P relative to the fixed XY coordinate system. Differentiating Eq. (2.28) twice with respect to time, we have

$$\mathbf{a} = \ddot{\mathbf{S}} = \ddot{\mathbf{R}} + \ddot{\mathbf{r}} \tag{2.33}$$

where $\ddot{\mathbf{R}}$ is the absolute acceleration of the moving origin O' relative to the fixed origin O. Since $\mathbf{r}$ is measured in a moving reference frame, its derivatives follow the rule given by Eq. (2.27), that is,

$$\dot{\mathbf{r}} = \dot{\mathbf{r}}_r + \boldsymbol{\omega} \times \mathbf{r}$$

$$\ddot{\mathbf{r}} = \frac{d}{dt}(\dot{\mathbf{r}}_r + \boldsymbol{\omega} \times \mathbf{r}) = \frac{d\dot{\mathbf{r}}_r}{dt} + \frac{d}{dt}(\boldsymbol{\omega} \times \mathbf{r})$$

Now

$$\frac{d\dot{\mathbf{r}}_r}{dt} = \ddot{\mathbf{r}}_r + \boldsymbol{\omega} \times \dot{\mathbf{r}}_r$$

$$\frac{d}{dt}(\boldsymbol{\omega} \times \mathbf{r}) = \dot{\boldsymbol{\omega}} \times \mathbf{r} + \boldsymbol{\omega} \times (\dot{\mathbf{r}}_r + \boldsymbol{\omega} \times \mathbf{r})$$

$$= \dot{\boldsymbol{\omega}} \times \mathbf{r} + \boldsymbol{\omega} \times \dot{\mathbf{r}}_r + \boldsymbol{\omega} \times (\boldsymbol{\omega} \times \mathbf{r})$$

Therefore,

$$\ddot{\mathbf{r}} = \ddot{\mathbf{r}}_r + \dot{\boldsymbol{\omega}} \times \mathbf{r} + \boldsymbol{\omega} \times (\boldsymbol{\omega} \times \mathbf{r}) + 2\boldsymbol{\omega} \times \dot{\mathbf{r}}_r \tag{2.34}$$

Letting $\boldsymbol{\alpha} = \dot{\boldsymbol{\omega}}$ represent the absolute angular acceleration of the moving reference frame, substitute Eq. (2.34) into (2.33) to obtain the final expression for the absolute acceleration of the particle P.

$$\mathbf{a} = \ddot{\mathbf{R}} + \ddot{\mathbf{r}}_r + \boldsymbol{\alpha} \times \mathbf{r} + \boldsymbol{\omega} \times (\boldsymbol{\omega} \times \mathbf{r}) + 2\boldsymbol{\omega} \times \dot{\mathbf{r}}_r \tag{2.35}$$

where

$\ddot{\mathbf{R}}$ = The absolute acceleration of the moving origin O'

$\ddot{\mathbf{r}}_r$ = The acceleration of P as measured by an observer attached to the moving reference frame $\mathbf{e}_1$, $\mathbf{e}_2$

$\boldsymbol{\alpha}$ = The absolute angular acceleration of the moving reference frame

$\boldsymbol{\omega}$ = The absolute angular velocity of the moving reference frame

$\mathbf{r}$ = The position vector of the particle P measured in the moving reference frame

$\boldsymbol{\alpha} \times \mathbf{r}$ = The **tangential** acceleration due to the angular acceleration of the moving reference frame. If the magnitude of $\boldsymbol{\omega}$ is constant, this term goes to zero.

$\boldsymbol{\omega} \times (\boldsymbol{\omega} \times \mathbf{r})$ = The acceleration due to the angular velocity of the moving axes. This is called the **centripetal** acceleration or **normal** component of acceleration

$2\boldsymbol{\omega} \times \dot{\mathbf{r}}_r$ = The acceleration due to the interaction of the angular velocity and the relative velocity term. This term is called the **Coriolis** acceleration

If the moving reference frame translates only, then $\boldsymbol{\omega} = \boldsymbol{\alpha} = 0$ and Eq. (2.35) simplifies to

$$\mathbf{a} = \ddot{\mathbf{R}} + \ddot{\mathbf{r}}_r \tag{2.36}$$

This equation says that to obtain the absolute acceleration of the particle P, we need the absolute acceleration of the moving origin O' and the relative acceleration or the acceleration of P as measured by an observer attached to the nonrotating moving reference frame.

We call the reader's attention, without further elaboration here, to the fact that the acceleration of a point in space has the same mathematical form as Eq. (2.35), which was derived for motion in a plane. The $\mathbf{r}$, $\boldsymbol{\omega}$, and $\boldsymbol{\alpha}$ vectors for motion in space can have components about all three axes instead of the restrictions for planar motion.

Having used vectors to develop general expressions for the velocity and acceleration of a particle, it is well to emphasize that these expressions are valid for all coordinate systems, in as much as a vectorial expression can be derived without regard to any specific coordinate system.

Example 2.12

Obtain the magnitude of the acceleration of car B relative to car A at the instant shown in Fig. 2.26

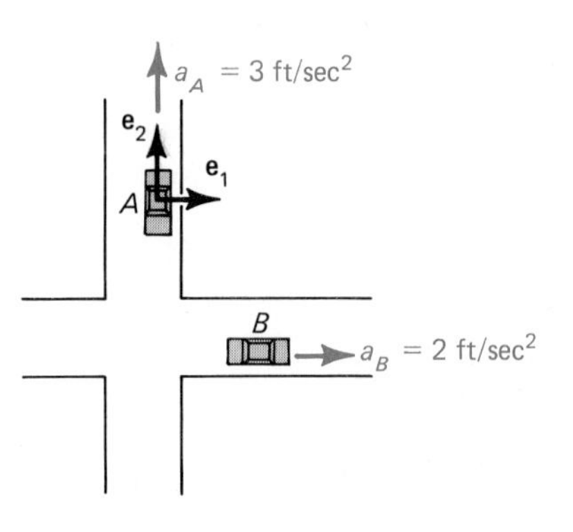

Figure 2.26

To find the acceleration of car B relative to car A place the moving reference frame on car A as shown in Fig. 2.26. Now the acceleration of the moving reference frame is

$$\ddot{\mathbf{R}} = 3\mathbf{e}_2$$

while the acceleration of the particle (represented by car A) is

$$\mathbf{a} = 2\mathbf{e}_1$$

so that the acceleration of car B as measured by an observer sitting in car A is

$$\ddot{\mathbf{r}}_r = \mathbf{a} - \ddot{\mathbf{R}} = 2\mathbf{e}_1 - 3\mathbf{e}_2$$

Its magnitude is

$$|\ddot{\mathbf{r}}_r| = \sqrt{(2)^2 + (-3)^2} = 3.61 \text{ ft/sec}^2 \qquad \textit{Answer}$$

Example 2.13

The table in Fig. 2.27 rotates at an angular velocity of 4 rad/s CCW and angular acceleration of 6 rad/s² CW. At the instant shown, the particle P moves along the groove at a uniform rate of 8 m/s outward relative to the table. Find the absolute acceleration of the particle.

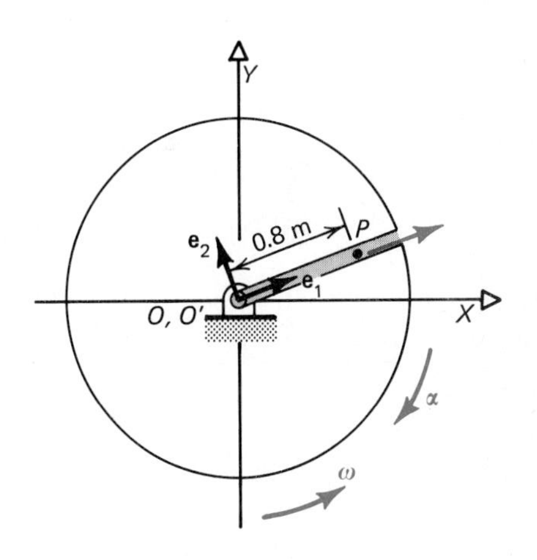

Figure 2.27

For this example, it is convenient to place O' at O and attach the moving reference frame $\mathbf{e}_1$, $\mathbf{e}_2$, $\mathbf{e}_3$ to the table as shown in the figure. Now

$$\boldsymbol{\omega} = 4\mathbf{e}_3 \qquad \boldsymbol{\alpha} = -6\mathbf{e}_3$$

$$\mathbf{r} = 0.8\mathbf{e}_1 \qquad \dot{\mathbf{r}}_r = 8\mathbf{e}_1$$

$\ddot{\mathbf{r}}_r = 0$ since the particle moves at a uniform rate relative to the table

$\ddot{\mathbf{R}} = 0$ since O' is coincident with O

We now develop the terms for Eq. (2.35).

$$\boldsymbol{\alpha} \times \mathbf{r} = -6\mathbf{e}_3 \times 0.8\mathbf{e}_1 = -4.8\mathbf{e}_2$$

To obtain $\boldsymbol{\omega} \times (\boldsymbol{\omega} \times \mathbf{r})$ first we form

$$\boldsymbol{\omega} \times \mathbf{r} = 4\mathbf{e}_3 \times 0.8\mathbf{e}_1 = 3.2\mathbf{e}_2$$

Now $\quad \boldsymbol{\omega} \times (\boldsymbol{\omega} \times \mathbf{r}) = 4\mathbf{e}_3 \times 3.2\mathbf{e}_2 = -12.8\mathbf{e}_1$

For the Coriolis acceleration,

$$2\boldsymbol{\omega} \times \dot{\mathbf{r}}_r = 8\mathbf{e}_3 \times 8\mathbf{e}_1 = 64\mathbf{e}_2$$

Recalling Eq. (2.35),

$$\begin{aligned} \mathbf{a} &= \ddot{\mathbf{R}} + \ddot{\mathbf{r}}_r + \boldsymbol{\alpha} \times \mathbf{r} + \boldsymbol{\omega} \times (\boldsymbol{\omega} \times \mathbf{r}) + 2\boldsymbol{\omega} \times \dot{\mathbf{r}}_r \\ &= -4.8\mathbf{e}_2 - 12.8\mathbf{e}_1 + 64\mathbf{e}_2 \\ &= (-12.8\mathbf{e}_1 + 59.2\mathbf{e}_2)\ \text{m/s}^2 \end{aligned}$$ *Answer*

Example 2.14

Suppose that a freight car travels along a flat, straight track at a speed of 30 ft/sec and is accelerated at 10 ft/sec² as shown in Fig. 2.28(a). A rod is pinned to the car's ceiling and at the instant shown rotates at an angular velocity of 2 rad/sec CCW and an angular acceleration of 10 rad/sec² CW. At the same time a weight P slides outward along the rod at a uniform speed of 4 ft/sec relative to the rod. Find the absolute velocity and acceleration of P at the instant shown.

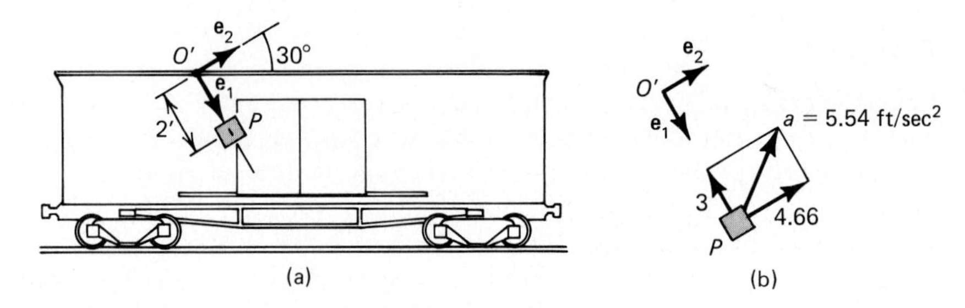

Figure 2.28

Place the moving reference frame $\mathbf{e}_1$, $\mathbf{e}_2$, $\mathbf{e}_3$ at O' and attach it to the rotating rod. Since O' is a point on the car and the car is moving,

$$\dot{\mathbf{R}} = 30(\sin 30^\circ \mathbf{e}_1 + \cos 30^\circ \mathbf{e}_2)$$

$$\ddot{\mathbf{R}} = 10(\sin 30^\circ \mathbf{e}_1 + \cos 30^\circ \mathbf{e}_2)$$

Notice that these vectors are in terms of the unit vectors of the moving reference frame. This is done for convenience so that all vectors will be in terms of the same unit vectors. From the statement of the problem, we can write the following:

$$\boldsymbol{\omega} = 2\mathbf{e}_3 \qquad \boldsymbol{\alpha} = -10\mathbf{e}_3 \qquad \mathbf{r} = 2\mathbf{e}_1 \qquad \dot{\mathbf{r}}_r = 4\mathbf{e}_1 \qquad \ddot{\mathbf{r}}_r = 0$$

The absolute velocity is found by using Eq. (2.31):

$$\begin{aligned}\mathbf{v} &= \dot{\mathbf{R}} + \dot{\mathbf{r}}_r + \boldsymbol{\omega} \times \mathbf{r}\\ &= 30(0.5\mathbf{e}_1 + 0.866\mathbf{e}_2) + 4\mathbf{e}_1 + (2\mathbf{e}_3 \times 2\mathbf{e}_1)\\ &= 15\mathbf{e}_1 + 26\mathbf{e}_2 + 4\mathbf{e}_1 + 4\mathbf{e}_2\\ &= (19\mathbf{e}_1 + 30\mathbf{e}_2)\text{ ft/sec} \qquad \textit{Answer}\end{aligned}$$

The acceleration is found with Eq. (2.35):

$$\mathbf{a} = \ddot{\mathbf{R}} + \ddot{\mathbf{r}}_r + \boldsymbol{\alpha} \times \mathbf{r} + \boldsymbol{\omega} \times (\boldsymbol{\omega} \times \mathbf{r}) + 2\boldsymbol{\omega} \times \dot{\mathbf{r}}_r$$

Since

$$\boldsymbol{\omega} \times \mathbf{r} = 4\mathbf{e}_2$$

$$\boldsymbol{\omega} \times (\boldsymbol{\omega} \times \mathbf{r}) = 2\mathbf{e}_3 \times 4\mathbf{e}_2 = -8\mathbf{e}_1$$

and

$$\boldsymbol{\alpha} \times \mathbf{r} = -10\mathbf{e}_3 \times 2\mathbf{e}_1 = -20\mathbf{e}_2$$

and

$$2\boldsymbol{\omega} \times \dot{\mathbf{r}}_r = 4\mathbf{e}_3 \times 4\mathbf{e}_1 = 16\mathbf{e}_2$$

we find that

$$\begin{aligned}\mathbf{a} &= 10(0.5\mathbf{e}_1 + 0.866\mathbf{e}_2) - 20\mathbf{e}_2 - 8\mathbf{e}_1 + 16\mathbf{e}_2\\ &= (-3\mathbf{e}_1 + 4.66\mathbf{e}_2)\text{ ft/sec}^2 \qquad \textit{Answer}\end{aligned}$$

The acceleration of P is plotted in Fig. 2.28(b) which shows the magnitude of the acceleration, 5.54 ft/sec^2, and its direction.

Example 2.15

The rod that is pin-connected to the edge of a revolving table in Fig. 2.29, is rotating at an absolute angular velocity of 2 rad/s CCW and an absolute angular acceleration of 8 rad/s CCW. A sliding ring P is moving inward at a uniform velocity of 5 m/s relative to the rod. Find the absolute velocity and acceleration of P at the instant shown, assuming the velocity of O' is uniform at 10 m/s downward and its acceleration is 10 m/s^2 to the left.

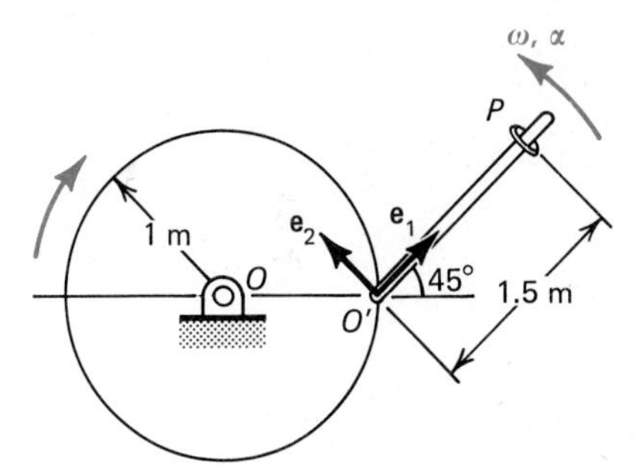

Figure 2.29

Place the origin O' at the point of connection between the rod and the table. Attach $\mathbf{e}_1$, $\mathbf{e}_2$, $\mathbf{e}_3$ to the rod. Now

$$\boldsymbol{\omega} = 2\mathbf{e}_3$$

$$\boldsymbol{\alpha} = 8\mathbf{e}_3$$

$$\mathbf{r} = 1.5\mathbf{e}_1$$

$$\dot{\mathbf{r}}_r = -5\mathbf{e}_1$$

$$\ddot{\mathbf{r}}_r = 0$$

$$\dot{\mathbf{R}} = 10(-\cos 45^\circ\mathbf{e}_1 - \sin 45^\circ\mathbf{e}_2) = -7.07\mathbf{e}_1 - 7.07\mathbf{e}_2$$

$$\ddot{\mathbf{R}} = 10(-\sin 45^\circ\mathbf{e}_1 + \cos 45^\circ\mathbf{e}_2) = -7.07\mathbf{e}_1 + 7.07\mathbf{e}_2$$

$$\boldsymbol{\omega} \times \mathbf{r} = 2\mathbf{e}_3 \times 1.5\mathbf{e}_1 = 3\mathbf{e}_2$$

The absolute velocity of P is

$$\begin{aligned} \mathbf{v} &= \dot{\mathbf{R}} + \dot{\mathbf{r}}_r + \boldsymbol{\omega} \times \mathbf{r} \\ &= -7.07\mathbf{e}_1 - 7.07\mathbf{e}_2 - 5\mathbf{e}_1 + 3\mathbf{e}_2 \\ &= (-12.07\mathbf{e}_1 - 4.07\mathbf{e}_2)\ \text{m/s} \end{aligned}$$ *Answer*

To obtain the absolute acceleration, we calculate

$$\boldsymbol{\alpha} \times \mathbf{r} = 8\mathbf{e}_3 \times 1.5\mathbf{e}_1 = 12\mathbf{e}_2$$

$$\boldsymbol{\omega} \times (\boldsymbol{\omega} \times \mathbf{r}) = 2\mathbf{e}_3 \times 3\mathbf{e}_2 = -6\mathbf{e}_1$$

$$2\boldsymbol{\omega} \times \dot{\mathbf{r}}_r = 4\mathbf{e}_3 \times (-5\mathbf{e}_1) = -20\mathbf{e}_2$$

Now

$$\begin{aligned} \mathbf{a} &= \ddot{\mathbf{R}} + \ddot{\mathbf{r}}_r + \boldsymbol{\alpha} \times \mathbf{r} + \boldsymbol{\omega} \times (\boldsymbol{\omega} \times \mathbf{r}) + 2\boldsymbol{\omega} \times \dot{\mathbf{r}}_r \\ &= -7.07\mathbf{e}_1 + 7.07\mathbf{e}_2 + 12\mathbf{e}_2 - 6\mathbf{e}_1 - 20\mathbf{e}_2 \\ &= (-13.07\mathbf{e}_1 - 0.93\mathbf{e}_2)\ \text{m/s}^2 \end{aligned}$$ *Answer*

PROBLEMS

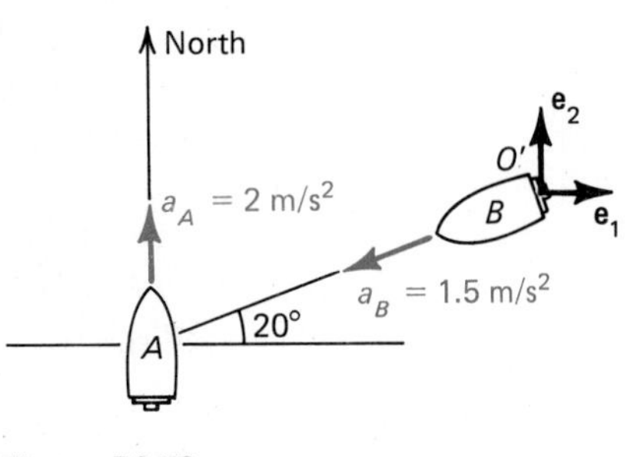

Figure P2.72

2.72 Boats A and B have accelerations of 2 m/s^2 and 1.5 m/s^2, respectively, in the directions shown in Fig. P2.72. What is the acceleration of A relative to B?

2.73 Suppose the wedge in Problem 2.57 slides along with the horizontal plane at an acceleration of 5 ft/sec^2 to the right, while the block slides downward at an acceleration of 3 ft/sec^2 relative to the wedge. Find the absolute acceleration of the block.

2.74 Cars A and B travel along a common straight path. If the acceleration of car A is $\mathbf{a}_A = 7\mathbf{e}_1$ m/s^2 and for car B, $\mathbf{a}_B = 8\mathbf{e}_1$ m/s^2, what is the acceleration of car B measured relative to an observer sitting in car A?

2.75 If the pan in Problem 2.53 has an angular acceleration of 2 rad/sec^2 CCW, then what is the acceleration of points A and B?

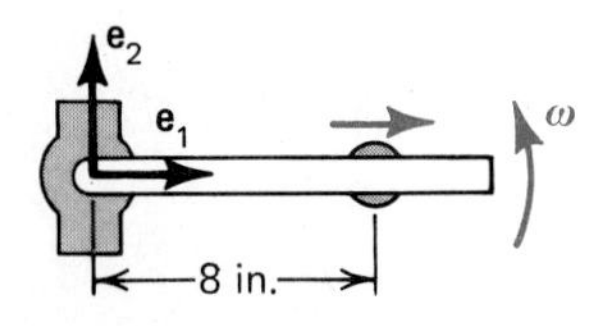

Figure P2.76

2.76 The arm of a mechanism in Fig. P2.76 rotates in the plane of the page at a uniform rate of 10 rad/sec CCW while a particle slides uniformly outward at 5 in/sec relative to the arm. What is the acceleration of the particle when it is 8 in from the pinned end of the rod? Sketch the Coriolis acceleration vector.

2.77 Find the absolute acceleration of the particle in Problem 2.54 if, in addition to the data given, the table rotates at an angular acceleration of 6 rad/s^2 CCW.

2.78 A particle of water, represented by point P in Fig. P2.78, is moving inside a water sprinkler at a constant velocity of v relative to the sprinkler. The sprinkler rotates at constant ω rad/s. What is the acceleration of P at the instant shown?

2.79 Find the absolute acceleration of the ring in Problem 2.58 if, in addition to the data given, the bar rotates at an angular acceleration of 5 rad/s^2 CCW and the ring accelerates along the bar at 4 m/s^2 downward relative to the bar.

2.80 Find the absolute acceleration of the particle in Problem 2.59 assuming the particle's velocity relative to the plate is a constant 20 m/s.

2.81 If the particle P in Problem 2.65 has an acceleration of 3 in/sec^2 to the left relative to the plate, while the angular acceleration of the plate is 2 rad/sec^2 CW, then what is the absolute acceleration of P?

2.82 A satellite at an altitude of $0.15R$, where R is the earth's radius, has constant angular velocity of 10^{-3} rad/s as it goes around the earth. What is the magnitude of the acceleration of this satellite?

2.83 At a given instant, the satellite of Problem 2.82 is given an angular acceleration of 10^{-6} rad/s^2. What is the magnitude of the acceleration of the satellite?

2.84 Obtain the acceleration of the particle P of Problem 2.54 at the instant shown. Assume the table rotates at a uniform rate of 3 rad/s CW.

2.85 Assume that the table of Problem 2.54 begins to decelerate at a rate of 2 rad/s^2. What is the acceleration of P at that instant?

2.86 Obtain the acceleration of the hamster relative to the cage in Problem 2.60 if the cage begins to accelerate at 0.1 rad/sec^2.

2.87 What is the magnitude of the acceleration of the ice skater of Problem 2.62 if the stage begins to decelerate at a rate of 0.05 rad/s^2?

2.88 Derive an expression for the absolute acceleration of point P in the cam in Problem 2.64.

2.89 The ring P slides down the bar in Fig. P2.89 at a speed of 4 ft/sec and an acceleration of 1 ft/sec^2, both measured relative to the bar. At the same instant, the angle between the bar and the wall is 30°. The bar is rotating in the XY plane at an angular velocity of 2 rad/sec CCW and an angular acceleration of 3 rad/sec^2 CCW. What is the magnitude of the acceleration of the ring at the instant shown?

2.90 Compute the acceleration of block 3 in Problem 2.71.

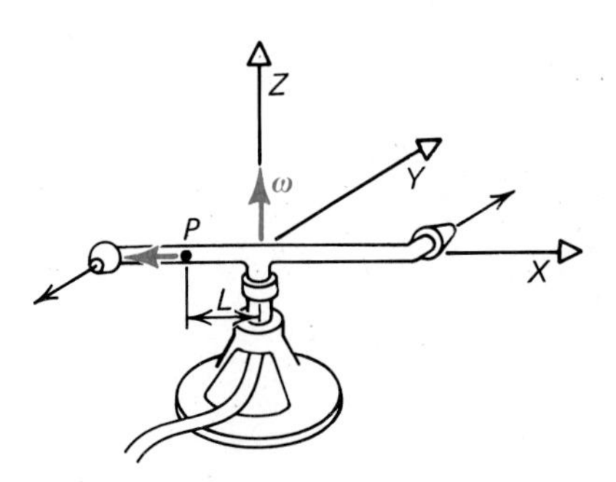

Figure P2.78

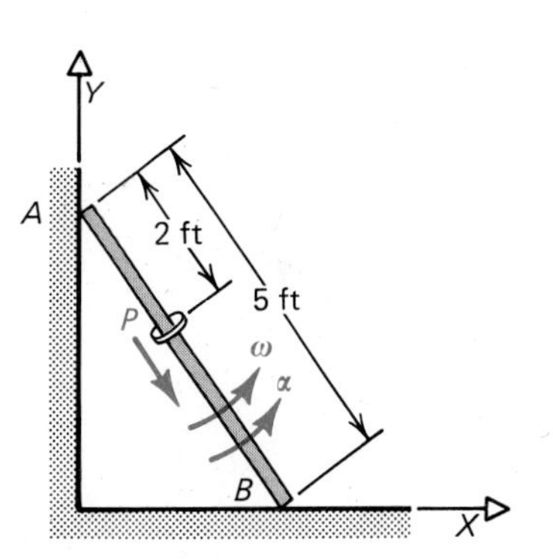

Figure P2.89

2.8 Motion of a Particle—Polar Coordinates

When the motion of a particle is expressed in a coordinate system whose coordinate directions are fixed, the motion is said to be expressed in a fixed coordinate system. If, however, the components are expressed along coordinates that move, then the motion is said to be represented in a moving coordinate system. This is precisely the approach taken with the results of Eqs. (2.31) and (2.35). The expressions for velocity and acceleration were expressed in terms of a general moving coordinate system represented by the unit vector system $\mathbf{e}_1$, $\mathbf{e}_2$, and $\mathbf{e}_3$. Now we shall consider the polar coordinate system as a special moving coordinate system to describe the motion of a particle in a plane.

Consider the point P moving in the XY plane in Fig. 2.30. Let $\mathbf{e}_r$, $\mathbf{e}_\theta$, and $\mathbf{k}$ be the unit vectors of the polar coordinate system. The position vector to point P is given by

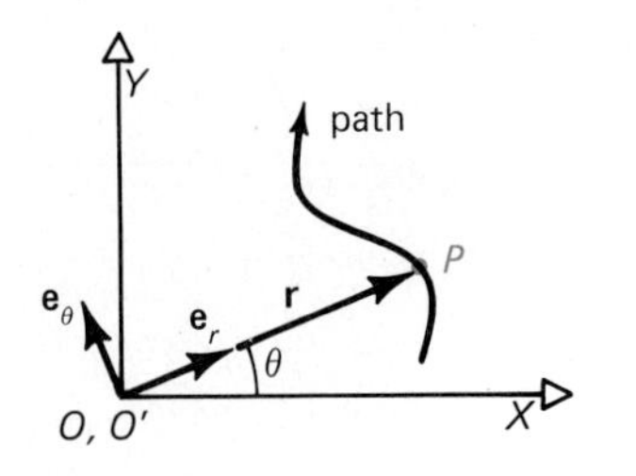

Figure 2.30

$$\mathbf{r} = r\mathbf{e}_r$$

If we assume that $\mathbf{r}$ rotates about O in the XY plane at the rate ω, then

$$\omega = \dot{\theta}\mathbf{k}$$

where positive θ is measured CCW from the X axis.

The velocity of P can be obtained by substituting into Eq. (2.31), so that

$$\mathbf{v} = \dot{\mathbf{R}} + \dot{\mathbf{r}}_r + \omega \times \mathbf{r} \qquad (2.31)$$

where $\qquad \dot{\mathbf{R}} = 0 \qquad$ and $\qquad \dot{\mathbf{r}}_r = \dot{r}\mathbf{e}_r$

Substituting into Eq. (2.31) we obtain

$$\mathbf{v} = \dot{r}\mathbf{e}_r + (\dot{\theta}\mathbf{k} \times r\mathbf{e}_r)$$

$$\mathbf{v} = \dot{r}\mathbf{e}_r + r\dot{\theta}\mathbf{e}_\theta \qquad (2.37)$$

Therefore, the velocity of P in polar coordinates consists of one component in the $\mathbf{e}_r$ direction of magnitude $\dot{r}$ and another component in the $\mathbf{e}_\theta$ direction of magnitude $r\dot{\theta}$.

To express the acceleration of the particle in polar coordinates, recall Eq. (2.35).

$$\mathbf{a} = \ddot{\mathbf{R}} + \ddot{\mathbf{r}}_r + \alpha \times \mathbf{r} + \omega \times (\omega \times \mathbf{r}) + 2\omega \times \dot{\mathbf{r}}_r \qquad (2.35)$$

Now $\qquad \ddot{\mathbf{R}} = 0 \qquad \ddot{\mathbf{r}}_r = \ddot{r}\mathbf{e}_r \qquad \alpha = \ddot{\theta}\mathbf{k}$

Upon substituting into Eq. (2.35) and arranging terms, we obtain the following:

$$\mathbf{a} = (\ddot{r} - r\dot{\theta}^2)\mathbf{e}_r + (2\dot{r}\dot{\theta} + r\ddot{\theta})\mathbf{e}_\theta \qquad (2.38)$$

As an example, consider the motion of a particle along the spiral path shown in Fig. 2.31. Here, $r = At$. Assuming $\dot{\theta} = \Omega =$ constant, we can express the velocity and acceleration of the particle in polar coordinates by substituting directly into Eqs. (2.37) and (2.38).

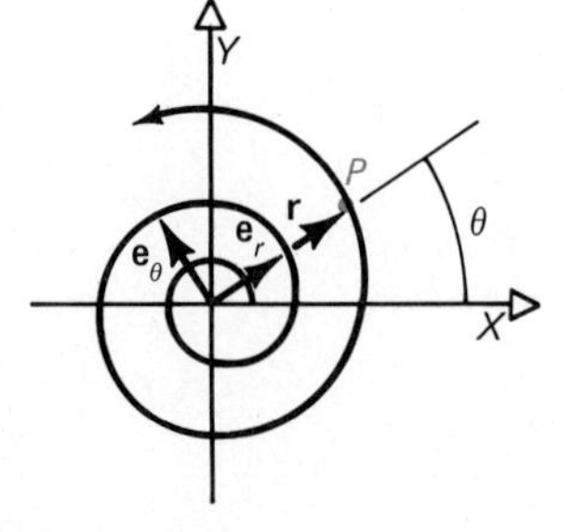

Figure 2.31

$$r = At \qquad \dot{\theta} = \Omega$$

$$\dot{r} = A \qquad \ddot{\theta} = 0$$

$$\ddot{r} = 0$$

Therefore,

$$\mathbf{v} = A\mathbf{e}_r + A\Omega t\mathbf{e}_\theta$$

$$\mathbf{a} = -At\Omega^2\mathbf{e}_r + 2A\Omega\mathbf{e}_\theta$$

These last two equations express the absolute velocity and acceleration of the particle P in this special moving coordinate system.

Example 2.16

The particle P in Fig. 2.32 is constrained by a spring inside the uniformly rotating tube. The particle's position relative to the tube is $r = (A + B \sin \beta t)$ where A, B, and β are constants. Express the velocity of the particle P in polar coordinates and find the time when the particle's speed first reaches a maximum. Let $A = 4$ in, $B = 2$ in, $\beta = 20$ rad/sec, and $\dot{\theta} = 10$ rad/sec.

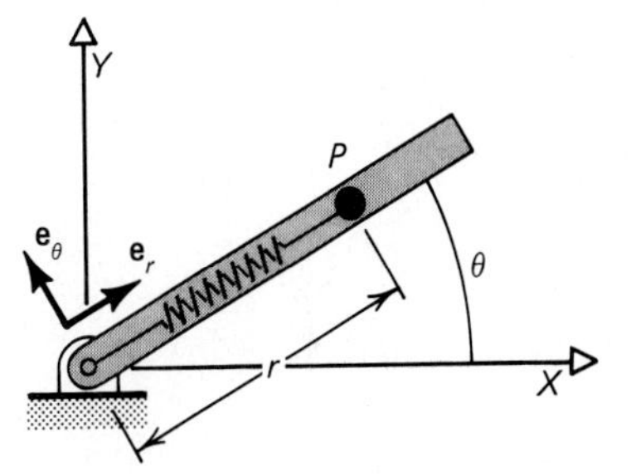

Figure 2.32

From the statement of the problem we have

$$r = A + B \sin \beta t$$

$$\dot{r} = \beta B \cos \beta t$$

$$\ddot{r} = -\beta^2 B \sin \beta t$$

Substituting into Eq. (2.37),

$$\begin{aligned}\mathbf{v} &= \beta B \cos \beta t\mathbf{e}_r + (A + B \sin \beta t)\dot{\theta}\mathbf{e}_\theta \\ &= 40 \cos 20t\mathbf{e}_r + (40 + 20 \sin 20t)\mathbf{e}_\theta \text{ in/sec} \quad \textit{Answer}\end{aligned}$$

To find the time when the speed is a maximum, consider the general Eq. (2.37):

$$\mathbf{v} = \dot{r}\mathbf{e}_r + r\dot{\theta}\mathbf{e}_\theta$$

The speed is

$$|\mathbf{v}| = v = \sqrt{\dot{r}^2 + (r\dot{\theta})^2}$$

Differentiate with respect to time and set the derivative of the speed equal to zero for obtaining the value of t when the speed is a maximum.

$$\frac{dv}{dt} = 0 = \frac{2\dot{r}\ddot{r} + 2(r\dot{\theta})(r\ddot{\theta} + \dot{r}\dot{\theta})}{2\sqrt{\dot{r}^2 + (r\dot{\theta})^2}}$$

or

$$\dot{r}\ddot{r} + r\dot{\theta}(r\ddot{\theta} + \dot{r}\dot{\theta}) = 0$$

Since the tube rotates uniformly, $\ddot{\theta} = 0$. Now

$$\dot{r}(\ddot{r} + r\dot{\theta}^2) = 0$$

Since $\dot{r}$ does not equal zero at all times,

$$\ddot{r} + r\dot{\theta}^2 = 0$$

or $$-\beta^2 B \sin \beta t + (A + B \sin \beta t)\dot{\theta}^2 = 0$$

$$(\beta^2 - \dot{\theta}^2)B \sin \beta t = A\dot{\theta}^2$$

$$(400 - 100)(2) \sin \beta t = 4(100)$$

$$\beta t = \sin^{-1}\left(\tfrac{2}{3}\right) = 0.7297 \text{ rad}$$

$$t = \frac{0.7297}{20} = 0.0365 \text{ sec} \qquad \textit{Answer}$$

Example 2.17

The moving gear A having a radius of r_2 rotates about the fixed gear B of radius r_1 as shown in Fig. 2.33. The angular velocity of the connecting rod of length L is ω_1 CCW. What is the velocity of the point P on gear A expressed in the $\mathbf{e}_r$, $\mathbf{e}_\theta$ coordinate system? Neglect the gear teeth.

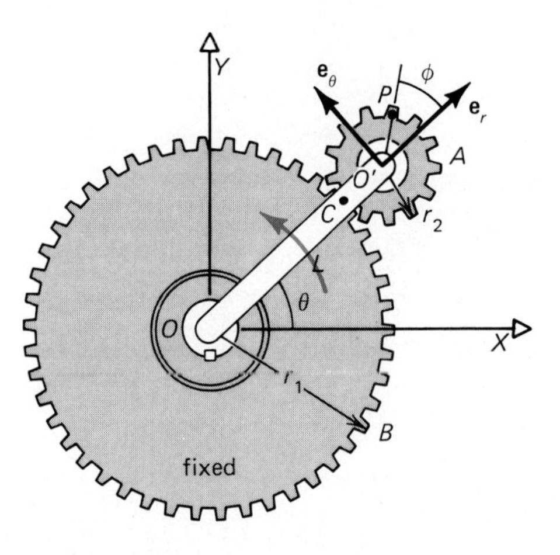

Figure 2.33

We shall use Eq. (2.31) to find the velocity of P. First place the moving origin O' as shown in Fig. 2.33 at the end of the connecting rod and attach the moving coordinate system $\mathbf{e}_r$, $\mathbf{e}_\theta$ to the rod. The velocity $\dot{\mathbf{R}}$ of the moving origin O' is established immediately by Eq. (2.37). That is,

$$\dot{\mathbf{R}} = \dot{r}\mathbf{e}_r + r\,\dot{\theta}\mathbf{e}_\theta$$

Now $\dot{r} = 0$ since the rod is rigid. Therefore, since $r = r_1 + r_2 = L$,

$$\dot{\mathbf{R}} = L\dot{\theta}\mathbf{e}_\theta$$

Recall Eq. (2.31), to find the velocity of P.

$$\mathbf{v} = \dot{\mathbf{R}} + \dot{\mathbf{r}}_r + \boldsymbol{\omega} \times \mathbf{r}$$

where

$$\mathbf{r} = r_2(\cos\phi\,\mathbf{e}_r + \sin\phi\,\mathbf{e}_\theta)$$

$$\boldsymbol{\omega} = \dot{\theta}\mathbf{k}$$

$$\dot{\mathbf{r}}_r = r_2\dot{\phi}(-\sin\phi\,\mathbf{e}_r + \cos\phi\,\mathbf{e}_\theta)$$

$$\boldsymbol{\omega} \times \mathbf{r} = r_2\dot{\theta}\cos\phi\,\mathbf{e}_\theta - r_2\dot{\theta}\sin\phi\,\mathbf{e}_r$$

Therefore,

$$\begin{aligned}\mathbf{v} &= L\dot{\theta}\mathbf{e}_\theta - r_2\dot{\phi}\sin\phi\,\mathbf{e}_r + r_2\dot{\phi}\cos\phi\,\mathbf{e}_\theta \\ &\quad + r_2\dot{\theta}\cos\phi\,\mathbf{e}_\theta - r_2\dot{\theta}\sin\phi\,\mathbf{e}_r \\ &= -r_2\sin\phi(\dot{\phi} + \dot{\theta})\mathbf{e}_r + (L\dot{\theta} + r_2\dot{\phi}\cos\phi \\ &\quad + r_2\dot{\theta}\cos\phi)\mathbf{e}_\theta\end{aligned}$$

We may simplify this expression by observing the geometrical restraint that $r_1\theta = r_2\phi$. Therefore, differentiating and rearranging,

$$\dot{\phi} = \frac{r_1}{r_2}\dot{\theta}$$

Now,

$$\begin{aligned}\mathbf{v} &= -(r_1 + r_2)\dot{\theta}\sin\phi\,\mathbf{e}_r + [L + (r_1 + r_2)\cos\phi]\dot{\theta}\mathbf{e}_\theta \\ &= -L\dot{\theta}\sin\phi\,\mathbf{e}_r + L\dot{\theta}(1 + \cos\phi)\mathbf{e}_\theta \qquad \textit{Answer}\end{aligned}$$

If the moving reference frame is fixed on the small wheel then $\omega = \dot{\phi} + \dot{\theta}$ and $\dot{r}_r = 0$. Substitution into the equation for **v** will of course yield the same result.

PROBLEMS

2.91 The particle in Fig. 2.30 travels in the XY plane such that $r = e^t$ and $\theta = t^2$ rad where r is in meters and t is in seconds. Find the magnitude of its velocity and acceleration when $t = 1$ s.

2.92 A particle moves on a circle of radius 10 m. The distance measured along the circle from a fixed point is given by $s = t^3$, where s is in meters and t is in seconds. Obtain the acceleration at $t = 1$ s in polar coordinates.

2.93 The position vector of a point is given by

$$\mathbf{r} = 5e^{2t}(\cos\theta\mathbf{i} + \sin\theta\mathbf{j})\text{ ft}$$

where $\theta = 2t$ rad, t is in seconds, and **i**, **j** are fixed unit vectors along the XY axes. Express the velocity and acceleration in polar coordinate form.

2.94 Rework Problem 2.93 to express the velocity in terms of the fixed coordinate system.

2.95 The rocket in Fig. P2.95 is fired so that its trajectory is defined in a given plane by $\mathbf{r} = r_1(t)\mathbf{i} + r_2(t)\mathbf{j}$, where **i**, **j** are unit vectors along a fixed coordinate system. Express the position and velocity of the rocket, as tracked by radar, in polar form.

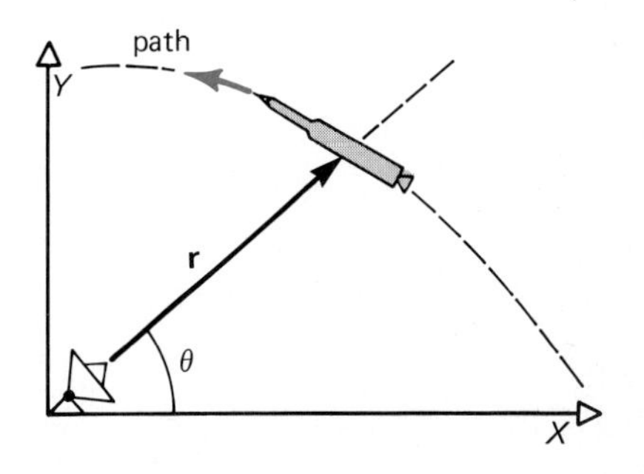

Figure P2.95

2.96 The position vector of a point P moving in the XY plane is given by $\mathbf{r} = (2t^3\mathbf{i} + 5t\mathbf{j})$ where r is in meters and t is in seconds. Express the velocity in polar coordinates when $t = 2$ s.

2.97 The radius vector of a point P is given by $6t^2\mathbf{e}_r$ and the angle θ measured CCW with the X axis is $6t$ rad. What is the velocity and acceleration of P expressed in the moving polar coordinates?

2.98 A fluid particle moves along an impeller as shown in Fig. P2.98. Let the path of the particle be given by $r = 0.75e^{2\theta}$ in. The pump is rotating at 2000 rev/min. What is the velocity and acceleration of P when $\theta = \pi/4$ radians?

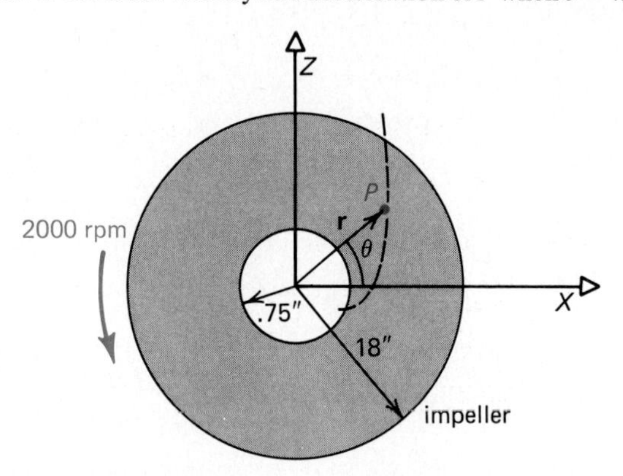

Figure P2.98

2.99 Express the velocity of point P of Problem 2.64 in polar coordinates.

2.100 A carnival ride is designed as shown in Fig. P2.100. A person represented by the particle P moves around the small circle of radius r at a uniform rate ω_2 rad/s, while the large wheel of radius R is spinning at a uniform rate ω_1 rad/s. What is the velocity of P expressed in the moving coordinate system $\mathbf{e}_r$, $\mathbf{e}_\theta$?

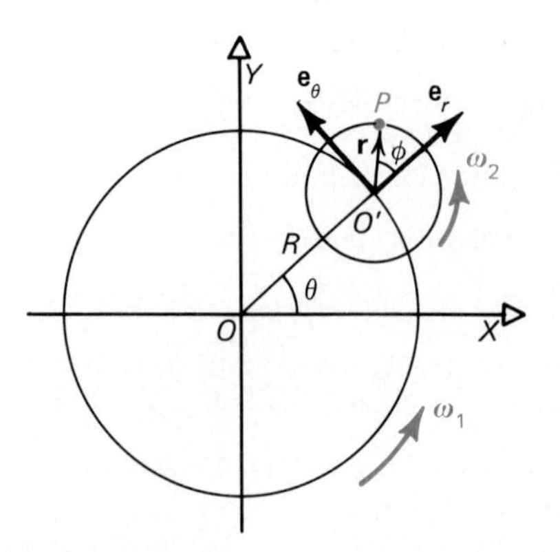

Figure P2.100

2.101 Repeat Problem 2.100 to find the acceleration of P assuming constant angular velocity rates.

2.102 The bead P in Fig. P2.102 travels around the circular wire of radius $a = 0.3$ m at a uniform rate of 10 m/s relative to the wire. The wire is rigidly attached to the rod that is rotating uniformly at $\dot{\theta} = 8$ rad/s CCW. Express the velocity of the bead in polar coordinates at the instant shown.

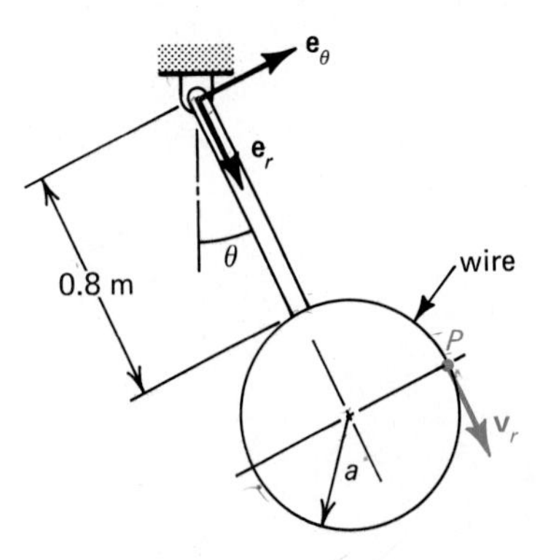

Figure P2.102

2.103 Repeat Problem 2.102 to find the acceleration of P assuming constant angular velocity rates.

2.9 Motion of a Particle—Normal-Tangential Coordinates

The normal and tangential components are frequently used to describe curvilinear motion in a moving coordinate system. These components are very useful if the path of the particle is already known. Let us first, however, define a set of moving unit vectors that can be used to describe the components of velocity and acceleration. Consider the particle P moving along a path in the XY plane as shown in Fig. 2.34(a). Let's measure the distance traveled by the particle along the curved path. Consider the particle moving a very short distance, say to point P_1. The change of $\mathbf{r}$ with respect to s is

$$\frac{d\mathbf{r}}{ds} = \lim_{\Delta s \to 0} \frac{\Delta \mathbf{r}}{\Delta s}$$

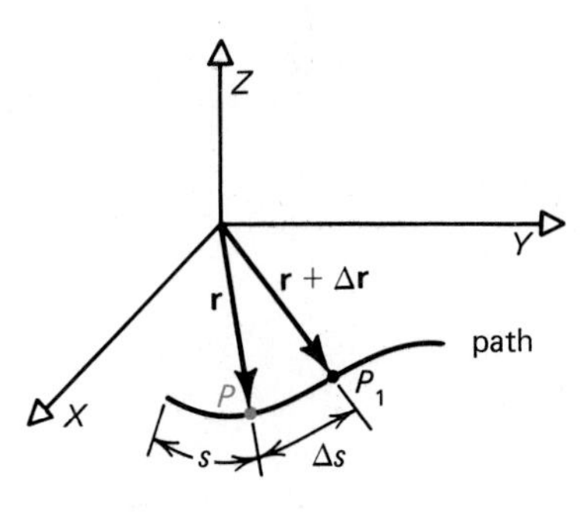

Now as Δs becomes small, P_1 approaches P, the ratio $|\Delta\mathbf{r}/\Delta s|$ approaches unity, and the direction of $\Delta\mathbf{r}$ approaches that of the tangent at P. In the limit, therefore, $d\mathbf{r}/ds$ has a magnitude of unity and a direction tangent to the curve at P. Let

$$\frac{d\mathbf{r}}{ds} = \mathbf{e}_\tau \tag{2.39}$$

The variation of $\mathbf{e}_\tau$ with s can be obtained by noting that

$$\mathbf{e}_\tau \cdot \mathbf{e}_\tau = 1 \qquad \text{and} \qquad \frac{d}{ds}(\mathbf{e}_\tau \cdot \mathbf{e}_\tau) = 2\mathbf{e}_\tau \cdot \frac{d\mathbf{e}_\tau}{ds} = 0$$

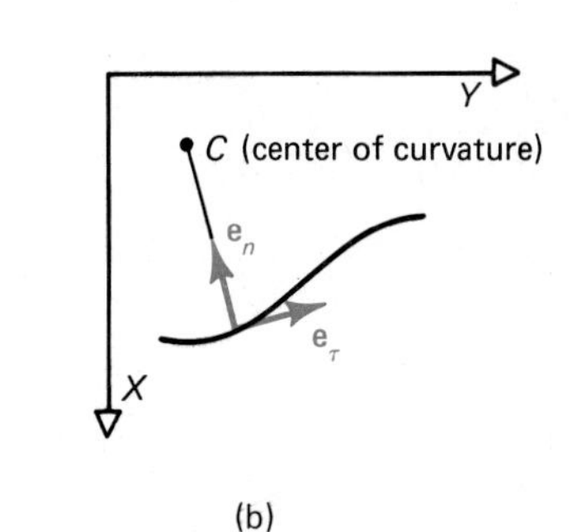

Figure 2.34

which implies that $\mathbf{e}_\tau$ and $d\mathbf{e}_\tau/ds$ are perpendicular. (We reject the trivial case where $d\mathbf{e}_\tau/ds = 0$.) Now, since there can be an infinite number of vectors perpendicular to $\mathbf{e}_\tau$ at P, we select a unit vector $\mathbf{e}_n$ lying in the XY plane that is perpendicular to $\mathbf{e}_\tau$ and directed toward the center of curvature as shown in Fig. 2.34(b). This is called the normal unit vector.

Figures 2.35(a) and (b) will help in determining the magnitude of $d\mathbf{e}_\tau/ds$. Here we show $\mathbf{e}_\tau$ and $(\mathbf{e}_\tau + \Delta\mathbf{e}_\tau)$ at the points P and P_1 in Fig. 2.35(a). Figure 2.35(b) shows the vector $(\mathbf{e}_\tau + \Delta\mathbf{e}_\tau)$ formed by adding vectorially $\Delta\mathbf{e}_\tau$ to $\mathbf{e}_\tau$. Now

(a)

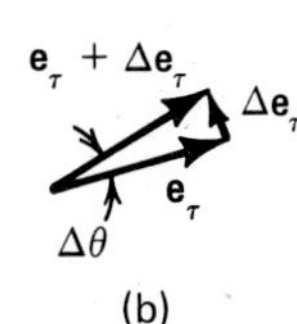

(b)

Figure 2.35

$$\left|\frac{d\mathbf{e}_\tau}{ds}\right| = \lim_{\Delta s \to 0} \left|\frac{\Delta \mathbf{e}_\tau}{\Delta\theta}\right| \frac{\Delta\theta}{\Delta s} \tag{2.40}$$

We note from Fig. 2.35(b) that, since the magnitude of $\mathbf{e}_\tau$ is unity,

$$|\Delta \mathbf{e}_\tau| = \Delta\theta$$

Also, from Fig. 2.35(a),

$$\Delta s = \rho\, \Delta\theta$$

where ρ is the radius of curvature. Substituting into Eq. (2.40), we obtain

$$\left|\frac{d\mathbf{e}_\tau}{ds}\right| = \frac{1}{\rho}$$

We can now define the normal unit vector at P as

$$\mathbf{e}_n = \rho \frac{d\mathbf{e}_\tau}{ds} \tag{2.41}$$

In general, the motion of a particle requires three directions; in the case of normal-tangential coordinates, the third direction is the binormal unit vector $\mathbf{e}_b$ defined such that

$$\mathbf{e}_b = \mathbf{e}_\tau \times \mathbf{e}_n$$

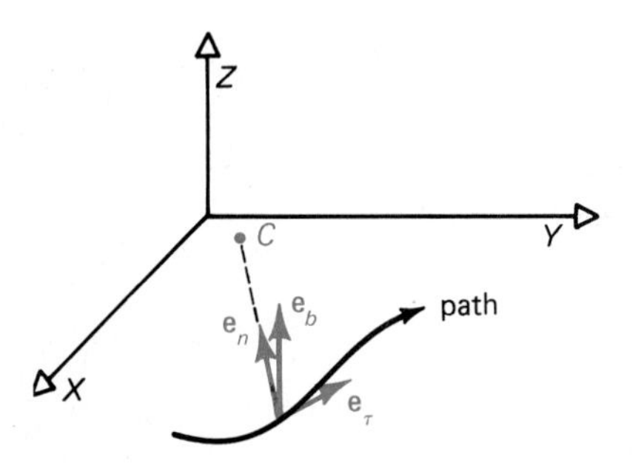

Figure 2.36

The set of vectors $\mathbf{e}_b$, $\mathbf{e}_n$, $\mathbf{e}_\tau$ is often referred to as a triad vector set and is shown in Fig. 2.36.

Having defined the unit vectors that move with the particle P, we can now resolve the velocity and acceleration along these directions. For the velocity,

$$\mathbf{v} = \frac{d\mathbf{r}}{dt} = \frac{d\mathbf{r}}{ds}\frac{ds}{dt}$$

where ds/dt is simply the speed v of the particle P. Since $d\mathbf{r}/ds$ is defined as a unit vector that is tangent to the path, we obtain the equation for the velocity of particle P as

$$\mathbf{v} = v\mathbf{e}_\tau \tag{2.42}$$

The equation for the acceleration of the particle P is

$$\mathbf{a} = \frac{d\mathbf{v}}{dt} = \dot{v}\mathbf{e}_\tau + v\frac{d\mathbf{e}_\tau}{dt} = \dot{v}\mathbf{e}_\tau + v\frac{d\mathbf{e}_\tau}{ds}\frac{ds}{dt}$$

When we substitute from Eq. (2.41), this becomes

$$\mathbf{a} = \dot{v}\mathbf{e}_\tau + \frac{v^2}{\rho}\mathbf{e}_n \tag{2.43}$$

The acceleration of a particle, therefore, has two components, one of which is tangent to the path while the other is normal to the path and directed toward the center of curvature. Figure 2.37(a) shows the velocity vector while Fig. 2.37(b) shows the acceleration components in terms of their normal and tangential components.

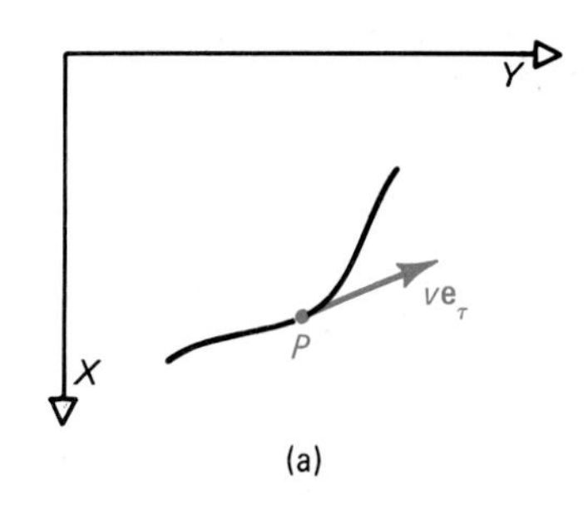

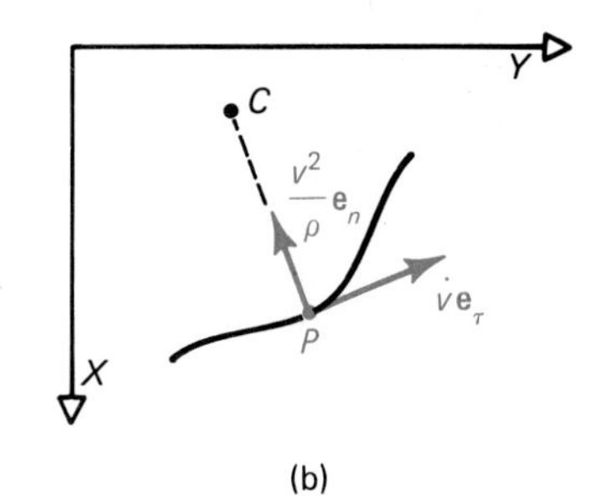

Figure 2.37

Example 2.18

Express the velocity and acceleration of the particle P traveling in a circular path in Fig. 2.38(a) in normal-tangential components.

Since the particle travels along the circular path, the radius of curvature equals the radius of the circle and the center of curvature remains at the origin. Now

$$\mathbf{v} = v\mathbf{e}_\tau$$

where
$$v = b\dot{\theta}$$

Therefore
$$\mathbf{v} = b\dot{\theta}\mathbf{e}_\tau \qquad \textit{Answer}$$

Also, the acceleration is given by Eq. (2.43).

$$\mathbf{a} = \dot{v}\mathbf{e}_\tau + \frac{v^2}{\rho}\mathbf{e}_n$$

where
$$\dot{v} = b\ddot{\theta}$$
$$\rho = b$$

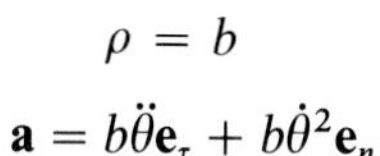

so that
$$\mathbf{a} = b\ddot{\theta}\mathbf{e}_\tau + b\dot{\theta}^2\mathbf{e}_n \qquad \textit{Answer}$$

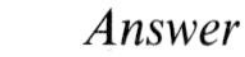

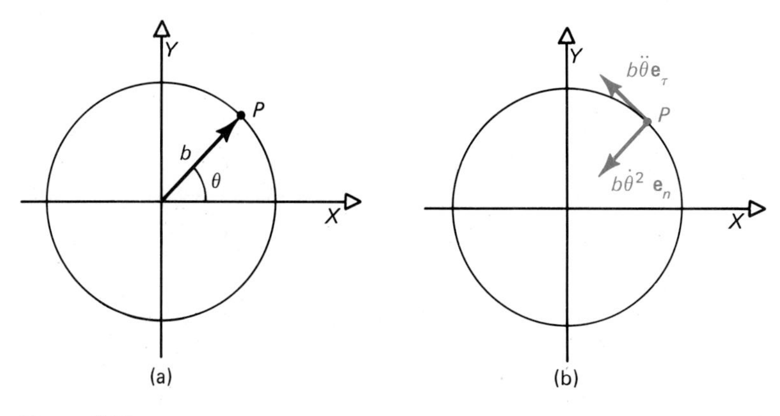

Figure 2.38

Figure 2.38(b) shows the acceleration components. Note that the normal component of acceleration is always directed toward the center of curvature and exists even if the particle travels along the circular path at a constant speed.

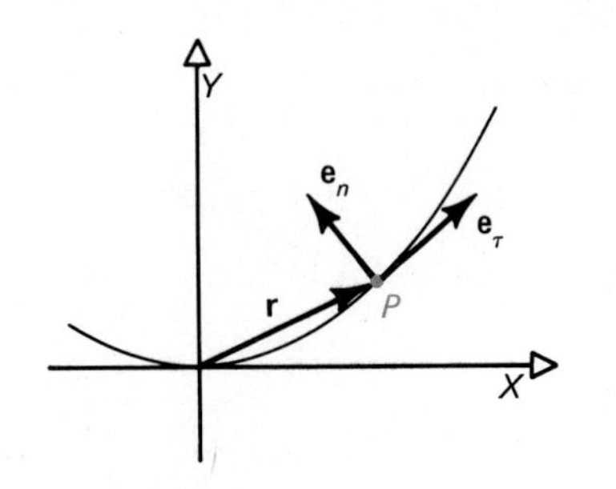

Figure 2.39

Example 2.19

Obtain the velocity and acceleration of particle P as shown in Fig. 2.39 in the fixed XY coordinate system as well as in the moving $\mathbf{e}_n$, $\mathbf{e}_\tau$ system. Assume that the motion of P is defined such that

$$x = At \qquad y = ct^2$$

For the fixed system

$$\mathbf{r} = At\mathbf{i} + ct^2\mathbf{j}$$

$$\mathbf{v} = A\mathbf{i} + 2ct\mathbf{j} \qquad \textit{Answer}$$

$$\mathbf{a} = 2c\mathbf{j} \qquad \textit{Answer}$$

To express the velocity and acceleration in tangential and normal components, we must obtain the radius of curvature. From calculus, the magnitude of the radius of curvature is

$$\rho = \frac{[1 + (dy/dx)^2]^{3/2}}{d^2y/dx^2}$$

and since

$$\frac{dy}{dx} = \frac{dy}{dt}\frac{dt}{dx} = \frac{\dot{y}}{\dot{x}}$$

$$\frac{d^2y}{dx^2} = \frac{d}{dt}\left(\frac{dy}{dx}\right)\frac{dt}{dx} = \frac{1}{\dot{x}}\frac{d}{dt}\left(\frac{\dot{y}}{\dot{x}}\right)$$

$$= \frac{1}{\dot{x}}\frac{\dot{x}\ddot{y} - \dot{y}\ddot{x}}{\dot{x}^2}$$

the radius of curvature can be expressed as

$$\rho = \frac{(\dot{x}^2 + \dot{y}^2)^{3/2}}{|\dot{x}\ddot{y} - \dot{y}\ddot{x}|}$$

For the problem under consideration, the derivatives are

$$\dot{x} = A \qquad \ddot{x} = 0 \qquad \dot{y} = 2ct \qquad \ddot{y} = 2c$$

Therefore

$$\rho = \frac{(A^2 + 4c^2t^2)^{3/2}}{2Ac}$$

The speed v at any time is found by

$$v = \sqrt{\dot{x}^2 + \dot{y}^2} = \sqrt{A^2 + 4c^2t^2}$$

The tangential acceleration is determined by

$$\dot{v} = \frac{dv}{dt} = \frac{4c^2t}{\sqrt{A^2 + 4c^2t^2}}$$

Substituting into Eq. (2.42) and Eq. (2.43),

$$\mathbf{v} = v\mathbf{e}_\tau = \sqrt{A^2 + 4c^2t^2}\,\mathbf{e}_\tau \qquad \textit{Answer}$$

$$\mathbf{a} = \dot{v}\mathbf{e}_\tau + \frac{v^2}{\rho}\mathbf{e}_n$$

$$= \frac{2c}{\sqrt{A^2 + 4c^2t^2}}(2ct\mathbf{e}_\tau + A\mathbf{e}_n) \qquad \textit{Answer}$$

It is interesting that at $t = 0$

$$\mathbf{v}(0) = A\mathbf{e}_\tau \qquad \text{and} \qquad \mathbf{a}(0) = 2c\mathbf{e}_n$$

but as t gets large, both the velocity and acceleration have the same direction, i.e., $\mathbf{e}_\tau$. The magnitude of the velocity continues to increase, the tangential component of acceleration goes to $2c$, and the normal component of acceleration goes to zero.

PROBLEMS

2.104 The motion of a particle is given by

$$\mathbf{r} = 5\cos t\mathbf{i} + \sin t\mathbf{j}$$

What is the velocity and acceleration of the particle expressed in tangential and normal components?

2.105 The position of a particle P is described by $x = \sin 2t$ and $y = e^t$. Express the velocity and acceleration in tangential and normal components.

2.106 The position vector for a particle P moving in the XY plane is given by $\mathbf{r} = 2t^3\mathbf{i} + 5t\mathbf{j}$. Express the velocity and acceleration in tangential and normal components.

2.107 A particle moves along a curve $y = x^2$ and $x = 2t$ where the distance is expressed in meters. What is the tangential and normal velocity and acceleration of the particle at $t = 1$ s?

2.108 A particle moves along a parabola $y = x^2$. Its distance measured along the path from the origin is given by $s = 4t^2$ and is measured in feet. Obtain the tangential and normal components of the velocity and acceleration.

2.109 A particle traveling along a curved path in a plane experiences an acceleration $\mathbf{a} = 10\mathbf{i} + 15\mathbf{j}$ m/s² as shown in Fig. P2.109. If the particle speed is 10 m/s, what is the radius of curvature at the instant shown?

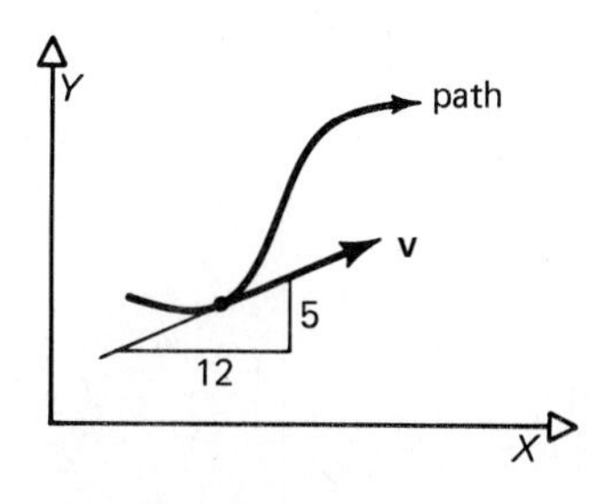

Figure P2.109

2.110 A train is going around a circular path described by a 2000-ft radius. The train moves at 25 mi/hr when the speed is seen to increase at the rate of 2 ft/sec² for a period of 10 sec. What are the normal and tangential components of acceleration after 10 sec have elapsed?

2.111 A particle is moving at a constant speed of 10 m/s along a parabola

$x = y^2$ where x is in meters. Assuming that the particle is moving up and to the right, compute the normal and tangential acceleration at the point $y = 3$.

2.112 A particle moves on a circular path of radius 10 ft in such a way that $v = 2t^2 + 4$ where v is in ft/sec and t is in seconds. What are the normal and tangential components of acceleration at $t = 4$ s? How much distance has been traversed?

2.113 A particle moves along a circle of radius 3 m. The distance measured along the curve from a fixed point is given by $s = t^3$, where s is in meters. Obtain the tangential and normal components of acceleration when $t = 2$ s.

2.10 Three-Dimensional Motion

In deriving the general equations for velocity and acceleration it was assumed that the motion under consideration was limited to the XY plane. This led to the restriction that the absolute velocities and accelerations had components in the XY plane and the angular velocities and angular accelerations had components only along the Z axis. The purpose of these assumptions was to simplify the mathematics and to develop a better grasp of the physical meaning of the various components of velocity and acceleration. In many real problems, however, the particle or body under consideration can have components of linear and angular velocity and acceleration about all three axes. This section, therefore, considers general three-dimensional motion.

If the angular velocity $\boldsymbol{\omega}$ and angular acceleration $\boldsymbol{\alpha}$ have components in all three directions and the position vector is represented in three-dimensional space, the general expressions derived earlier, namely Eq. (2.31) for the velocity and Eq. (2.35) for the acceleration, can be applied without modification. The difficulty arises not in the governing equations but in the increased complexity required in understanding the physical motion that occurs with three-dimensional motion.

Example 2.20

A disk shown in Fig. 2.40 is rotating about the Y axis at Ω rad/s while the arm to which it is pinned is rotating about the Z axis at ω_0 rad/s. Obtain the velocity of point A on the disk at the instant shown.

Attach O' at the center of the disk and let $\mathbf{e}_1$, $\mathbf{e}_2$, $\mathbf{e}_3$ be the moving unit vectors. Then the absolute angular velocity of the moving unit vectors is

$$\boldsymbol{\omega} = \Omega \mathbf{e}_2 + \omega_0 \mathbf{e}_3$$

Also, using the notation developed for Eq. (2.31)

$$\mathbf{R} = L\mathbf{e}_2 \qquad \mathbf{r} = r\mathbf{e}_1$$

$$\dot{\mathbf{R}} = \omega_0 \mathbf{e}_3 \times L\mathbf{e}_2 = -\omega_0 L\mathbf{e}_1$$

$$\dot{\mathbf{r}}_r = 0$$

$$\boldsymbol{\omega} \times \mathbf{r} = (\Omega\mathbf{e}_2 + \omega_0\mathbf{e}_3) \times (r\mathbf{e}_1) = -\Omega r\mathbf{e}_3 + \omega_0 r\mathbf{e}_2$$

Substituting into Eq. (2.31),

$$\mathbf{v} = -\omega_0 L\mathbf{e}_1 + \omega_0 r\mathbf{e}_2 - \Omega r\mathbf{e}_3 \qquad \textit{Answer}$$

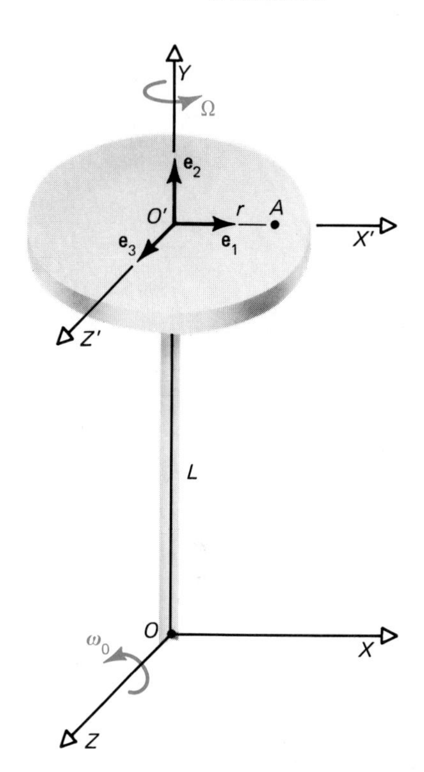

Figure 2.40

Example 2.21

Obtain the acceleration of point A at $t = 1$ s in Example 2.20 if $\omega_0 = 5t$ rad/s and Ω is constant. The position of point A at $t = 1$ s is shown in Fig. 2.40.

Since $\qquad \boldsymbol{\omega} = \Omega\mathbf{e}_2 + 5t\mathbf{e}_3$

the absolute angular acceleration of the moving reference frame is

$$\dot{\boldsymbol{\omega}} = \boldsymbol{\alpha} = 5\mathbf{e}_3$$

Since the moving reference frame is attached to the plate as in Example 2.20, the acceleration of O' is obtained by noting that it moves along a circular path of radius L. Expressing this acceleration in terms of its normal and tangential components,

$$\ddot{\mathbf{R}} = -5L\mathbf{e}_1 - 25L\mathbf{e}_2$$

Using the notation in Eq. (2.35),

$$\ddot{\mathbf{r}}_r = 0 \qquad \dot{\mathbf{r}}_r = 0$$

$$\boldsymbol{\alpha} \times \mathbf{r} = (5\mathbf{e}_3 \times r\mathbf{e}_1) = 5r\mathbf{e}_2$$

$$\boldsymbol{\omega} \times (\boldsymbol{\omega} \times \mathbf{r}) = (\Omega\mathbf{e}_2 + 5\mathbf{e}_3) \times [(\Omega\mathbf{e}_2 + 5\mathbf{e}_3) \times r\mathbf{e}_1]$$

$$= -(\Omega^2 + 25)r\mathbf{e}_1$$

Substituting into Eq. (2.35),

$$\mathbf{a} = -5L\mathbf{e}_1 - 25L\mathbf{e}_2 + 5r\mathbf{e}_2 - (\Omega^2 + 25)r\mathbf{e}_1$$

$$= -(\Omega^2 r + 25r + 5L)\mathbf{e}_1 - (25L - 5r)\mathbf{e}_2 \qquad \textit{Answer}$$

Example 2.22

A particle P moves along the inside surface of the tube shown in Fig. 2.41. The particle slides along the tube such that $r = t^2$, where r is in feet and t is in seconds. The tube rotates such that its angular velocity is $\boldsymbol{\omega} = 10t(\mathbf{e}_2 + \mathbf{e}_3)$ rad/sec where the unit vector system $\mathbf{e}_1$, $\mathbf{e}_2$, and $\mathbf{e}_3$ is attached to the tube as shown in the figure. Find the

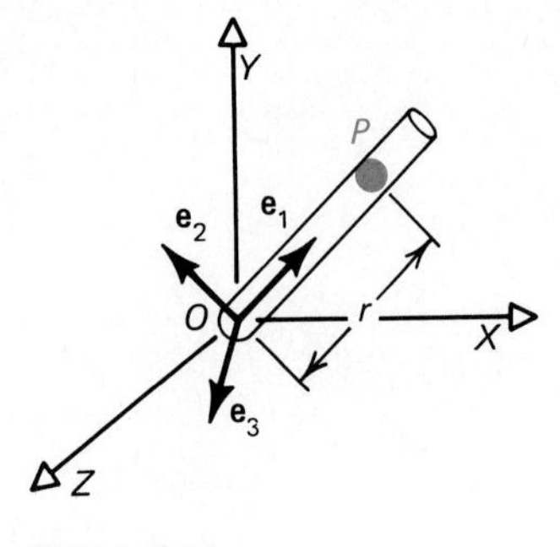

Figure 2.41

absolute velocity and acceleration of the particle when $t = 1$ s. Express the results in terms of $\mathbf{e}_1$, $\mathbf{e}_2$, and $\mathbf{e}_3$.

Since the moving unit vector system is attached to the tube, its angular velocity is also $\boldsymbol{\omega} = 10t(\mathbf{e}_2 + \mathbf{e}_3)$. Now

$$\mathbf{v} = \dot{\mathbf{R}} + \dot{\mathbf{r}}_r + \boldsymbol{\omega} \times \mathbf{r}$$

Since the origin O' is coincident with O,

$$\dot{\mathbf{R}} = 0$$

Also,
$$\mathbf{r} = t^2\mathbf{e}_1$$
$$\dot{\mathbf{r}}_r = 2t\mathbf{e}_1$$
$$\boldsymbol{\omega} \times \mathbf{r} = 10t(\mathbf{e}_2 + \mathbf{e}_3) \times (t^2\mathbf{e}_1) = -10t^3\mathbf{e}_3 + 10t^3\mathbf{e}_2$$

Therefore
$$\mathbf{v} = 2t\mathbf{e}_1 + 10t^3\mathbf{e}_2 - 10t^3\mathbf{e}_3$$

At $t = 1$ sec,

$$\mathbf{v} = (2\mathbf{e}_1 + 10\mathbf{e}_2 - 10\mathbf{e}_3) \text{ ft/sec} \qquad \textit{Answer}$$

To find the acceleration,

$$\mathbf{a} = \ddot{\mathbf{R}} + \ddot{\mathbf{r}}_r + \boldsymbol{\alpha} \times \mathbf{r} + \boldsymbol{\omega} \times (\boldsymbol{\omega} \times \mathbf{r}) + 2\boldsymbol{\omega} \times \dot{\mathbf{r}}_r$$

where
$$\ddot{\mathbf{R}} = 0$$
$$\ddot{\mathbf{r}}_r = 2\mathbf{e}_1$$
$$\boldsymbol{\alpha} = 10(\mathbf{e}_2 + \mathbf{e}_3)$$
$$\boldsymbol{\alpha} \times \mathbf{r} = 10(\mathbf{e}_2 + \mathbf{e}_3) \times t^2\mathbf{e}_1 = -10t^2\mathbf{e}_3 + 10t^2\mathbf{e}_2$$
$$\boldsymbol{\omega} \times (\boldsymbol{\omega} \times \mathbf{r}) = 10t(\mathbf{e}_2 + \mathbf{e}_3) \times (10t^3)(\mathbf{e}_2 - \mathbf{e}_3)$$
$$= 100t^4(-\mathbf{e}_1 - \mathbf{e}_1) = -200t^4\mathbf{e}_1$$
$$2\boldsymbol{\omega} \times \dot{\mathbf{r}}_r = 20t(\mathbf{e}_2 + \mathbf{e}_3) \times 2t\mathbf{e}_1 = -40t^2\mathbf{e}_3 + 40t^2\mathbf{e}_2$$

Therefore

$$\mathbf{a} = 2\mathbf{e}_1 - 10t^2\mathbf{e}_3 + 10t^2\mathbf{e}_2 - 200t^4\mathbf{e}_1 - 40t^2\mathbf{e}_3 + 40t^2\mathbf{e}_2$$

At $t = 1$ sec,

$$\mathbf{a} = (-198\mathbf{e}_1 + 50\mathbf{e}_2 - 50\mathbf{e}_3) \text{ ft/sec}^2 \qquad \textit{Answer}$$

PROBLEMS

2.114 Obtain the velocity and acceleration of a point A at $t = 2$ s in Example 2.20 if $2r = L = 1$ m, $\omega_0 = 10t$ rad/s and $\Omega = 8$ rad/s.

2.115 Rework Example 2.22 for $\mathbf{r} = t^2(2\mathbf{e}_1 + 3\mathbf{e}_3)$ and $\boldsymbol{\omega} = 5t(\mathbf{e}_2 + \mathbf{e}_3)$ at $t = 2$ s.

2.116 The arm of the mechanism in Problem 2.76 rotates at a uniform rate of 10 rad/sec about the $\mathbf{e}_3$ axis and also at a uniform rate of 5 rad/sec about the $\mathbf{e}_2$ axis so that the absolute angular velocity is $5\mathbf{e}_2 + 10\mathbf{e}_3$ rad/sec. If a particle slides uniformly outward at 5 in/sec relative to the arm, what is the velocity and acceleration of the particle at the instant shown?

2.117 The absolute angular velocity of the large wheel in Problem 2.100 is $\omega_0\mathbf{j} + \omega_1\mathbf{k}$ and the angular velocity of the small wheel relative to the large wheel is $\omega_2\mathbf{k}$. What is the velocity of point P expressed in the moving coordinate system $\mathbf{e}_r$, $\mathbf{e}_\theta$?

2.118 A body in a circular orbit around the earth has an absolute angular velocity of $\omega_1\mathbf{e}_1 + \omega_2\mathbf{e}_2 + \omega_3\mathbf{e}_3$. The position of the body at a given instant is given by $R\mathbf{e}_1$ and the absolute angular acceleration by $\alpha_1\mathbf{e}_1 + \alpha_2\mathbf{e}_2 + \alpha_3\mathbf{e}_3$. What is the absolute acceleration of the body at this instant?

2.11 Summary

A. The important *ideas*:

1. Absolute velocity and acceleration can be represented by fixed or moving reference frames.
2. The particular coordinate system selected is a matter of convenience.
3. The angular velocity vector $\boldsymbol{\omega}$ is a free vector, i.e., it has magnitude and direction but need not pass through a particular point.
4. The rate of change of the vector $\mathbf{R}$ measured in a moving reference frame includes the term $\dot{\mathbf{R}}_r$ which is the rate of change as measured by an observer attached to the moving reference frame.

B. The important *equations*:

Rectilinear motion of a particle:

$$ds = v\,dt \tag{2.8}$$

$$dv = a\,dt \tag{2.9}$$

$$v\,dv = a\,ds \tag{2.10}$$

Angular motion of a line:

$$d\theta = \omega\,dt \tag{2.15}$$

$$d\omega = \alpha\,dt \tag{2.17}$$

$$\omega\,d\omega = \alpha\,d\theta \tag{2.18}$$

Derivative of a unit vector **e** rotating at a rate of $\boldsymbol{\omega}$:

$$\frac{d\mathbf{e}}{dt} = \boldsymbol{\omega} \times \mathbf{e} \tag{2.22}$$

General expression for the absolute velocity:

$$\mathbf{v} = \dot{\mathbf{R}} + \dot{\mathbf{r}}_r + \boldsymbol{\omega} \times \mathbf{r} \tag{2.31}$$

General expression for the absolute acceleration:

$$\mathbf{a} = \ddot{\mathbf{R}} + \ddot{\mathbf{r}}_r + \boldsymbol{\alpha} \times \mathbf{r} + \boldsymbol{\omega} \times (\boldsymbol{\omega} \times \mathbf{r}) + 2\boldsymbol{\omega} \times \dot{\mathbf{r}}_r \tag{2.35}$$

Velocity and acceleration representation using moving polar coordinates:

$$\mathbf{v} = \dot{r}\mathbf{e}_r + r\dot{\theta}\mathbf{e}_\theta \tag{2.37}$$

$$\mathbf{a} = (\ddot{r} - r\dot{\theta}^2)\mathbf{e}_r + (2\dot{r}\dot{\theta} + r\ddot{\theta})\mathbf{e}_\theta \tag{2.38}$$

Velocity and acceleration representation using moving normal and tangential components:

$$\mathbf{v} = v\mathbf{e}_\tau \tag{2.42}$$

$$\mathbf{a} = \dot{v}\mathbf{e}_\tau + \frac{v^2}{\rho}\mathbf{e}_n \tag{2.43}$$

Kinetics of a Particle 3

Having studied the kinematics of a particle in motion, we are now ready to consider the behavior of a particle subject to the forces that produce the motion. This area of mechanics is called **kinetics**. To study the kinetics of a particle we make use of Newton's second law. This law is useful in calculating the acceleration of a particle for a given set of forces, or, if the acceleration is known, in calculating the resultant force acting on the particle. In all cases, the relationships hold *for instantaneous values only.*

By integrating Newton's second law spatially, we shall develop the work-energy relationship. This integral form of Newton's law is particularly useful in deriving the change in velocity as well as finding displacements that occur as a body changes position. The work-energy relationship is modified to include the change in potential energy, which accounts for the work done by the conservative forces.

A time integration of Newton's second law will give us the impulse-momentum relationships for a particle. These equations are useful for finding velocity changes when one or more of the forces acting on the particle are explicit functions of time. We shall see in Chapter 7 that these equations are essential to the study of the behavior of impacting bodies.

The equations developed in this chapter are applicable to a single particle, where we define a particle as a body with mass but no dimensions. Later, these equations will be extended to the case of a rigid body.

3.1 Equations of Motion of a Particle

The governing equation for the motion of a particle subjected to external forces was introduced in Chapter 1. This equation is a consequence of Newton's second law and is written

$$\mathbf{F} = m\mathbf{a} \tag{3.1}$$

where $\mathbf{F}$ is the resultant external force acting on the particle of mass m, and $\mathbf{a}$ is the absolute acceleration of the particle. This equation relates the instantaneous values of the variables $\mathbf{F}$ and $\mathbf{a}$. Since acceleration is the time rate of change of the instantaneous velocity, i.e., $\mathbf{a} = d\mathbf{v}/dt$, Eq. (3.1) can also be written in differential equation form:

$$\mathbf{F} = m\frac{d\mathbf{v}}{dt} \tag{3.2}$$

We shall find that forces can be of different types, that is, some forces are constant, others vary with position, time, and velocity. In working with Eqs. (3.1) or (3.2), it is sometimes desirable to know what force produced a given acceleration. On the other hand, it may be that the forces acting on the body are known and we wish to find the resulting instantaneous acceleration, or an integral of the acceleration.

The motion of a particle that is subjected to various types of forces is considered next. The equations used are in scalar form, so that forces and motions along each direction are considered separately.

CASE 1: $F = C = \text{constant}$

$$ma = F = C$$

$$m\frac{dv}{dt} = C$$

$$dv = \frac{C}{m}\,dt \tag{3.3}$$

Since C and m are constants, Eq. (3.3) can be integrated if the initial velocity of the particle is known. With v as a function of t, $ds = v\,dt$ can be integrated to find s, provided the initial displacement of the particle is known.

CASE 2: $F = F(t) = \text{Force as a function of time } t$

$$m\frac{dv}{dt} = F(t)$$

$$dv = \frac{1}{m}F(t)\,dt \tag{3.4}$$

When $F(t)$ is known, this differential equation can be integrated to find v; and if desired, the displacement s can then be found, again using $ds = v\,dt$. It is understood that both the initial displacement and the initial velocity must be known to obtain a complete solution.

CASE 3: $F = F(v) =$ Force as a function of velocity v

$$m\frac{dv}{dt} = F(v)$$

$$dt = m\frac{dv}{F(v)} \tag{3.5}$$

The ease with which this equation is integrated obviously depends on the complexity of $F(v)$.

CASE 4: $F = F(s) =$ Force as a function of displacement s

$$ma = F(s)$$

But

$$a\,ds = v\,dv$$

So that

$$v\,dv = \frac{1}{m}F(s)\,ds \tag{3.6}$$

Upon integrating this equation, we obtain $v = v(s)$. Since $ds = v\,dt$, this equation is written in the form

$$dt = \frac{ds}{v(s)}$$

for the next integration.

An interesting example of the case where force is a function of distance is the linear spring. When a spring is stretched or compressed by a distance x measured from the undeformed position, the restoring force is

$$F = kx$$

where k is the spring modulus having units of newtons per meter (N/m) or pounds per foot (lb/ft). The larger the value of k the stiffer the spring. We shall discuss this in greater detail in the next section.

Two special points should be emphasized before we consider example problems. First, before writing any equations for a problem, a free-body diagram of the forces acting on the particle should be drawn. Secondly, an appropriate reference frame must be included in every problem in order to assure that the absolute acceleration of the particle is used in Eq. (3.1).

Example 3.1

A particle, fired from a gun at an inclination of γ shown in Fig. 3.1(a), leaves the gun barrel with a velocity $\mathbf{V}_0$. Neglecting air resistance, determine the trajectory path of the particle.

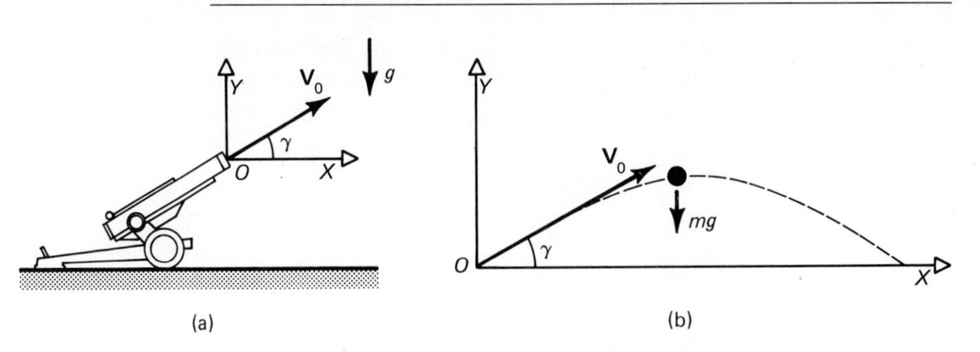

Figure 3.1

Figure 3.1(a) shows the inertial reference with its origin O at the exit of the gun barrel. The free-body diagram in Fig. 3.1(b) shows that the gravitational force is the only force acting on the particle. Therefore, for the force acting on the particle,

$$\mathbf{F} = -mg\mathbf{j}$$

Using the fixed cartesian coordinates to represent acceleration, Eq. (3.1) is written

$$-mg\mathbf{j} = m(\ddot{x}\mathbf{i} + \ddot{y}\mathbf{j})$$

Since the trajectory curve is desired, it is convenient to work with the scalar equations before integrating. The following equations are examples of Case 1, discussed above:

$$\ddot{y} = -g \qquad \ddot{x} = 0$$

$$\dot{y} = -gt + C_1 \qquad \dot{x} = C_3$$

$$y = \frac{-gt^2}{2} + C_1t + C_2 \qquad x = C_3t + C_4$$

The constants of integration are found from the following initial conditions:

$$y(0) = 0 \qquad x(0) = 0$$

$$\dot{y}(0) = V_0 \sin\gamma \qquad \dot{x}(0) = V_0 \cos\gamma$$

If the initial conditions are substituted into the above expressions for x, $\dot{x}$, y, and $\dot{y}$, four equations in the four unknown constants C_1, C_2, C_3, and C_4 are obtained. They can be solved to give

$$C_1 = V_0 \sin\gamma \qquad C_3 = V_0 \cos\gamma$$

$$C_2 = 0 \qquad C_4 = 0$$

In the present example, this can be done almost by inspection. Therefore, the position of the particle is given by

$$y = -\frac{gt^2}{2} + V_0 t \sin \gamma$$

$$x = V_0 t \cos \gamma$$

The equation for the trajectory is found by eliminating t from the above parametric equations. Doing this, we obtain

$$y = -\frac{gx^2}{2V_0^2 \cos^2 \gamma} + x \tan \gamma \qquad \textit{Answer}$$

so that the path of the particle is a parabola.

Example 3.2

Set up the differential equations of motion for the mass m shown in Fig. 3.2(a). The mass is part of a simple pendulum attached to a box which is accelerating. For the instant under consideration, the box acceleration is **a**, the pendulum is θ rad measured counterclockwise off the vertical line, and $\boldsymbol{\omega}$ and $\boldsymbol{\alpha}$ are the angular velocity and angular acceleration of the pendulum, respectively, both positive in the CCW direction.

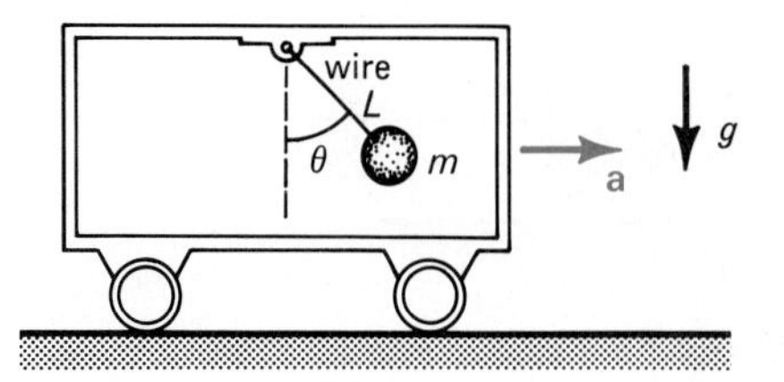

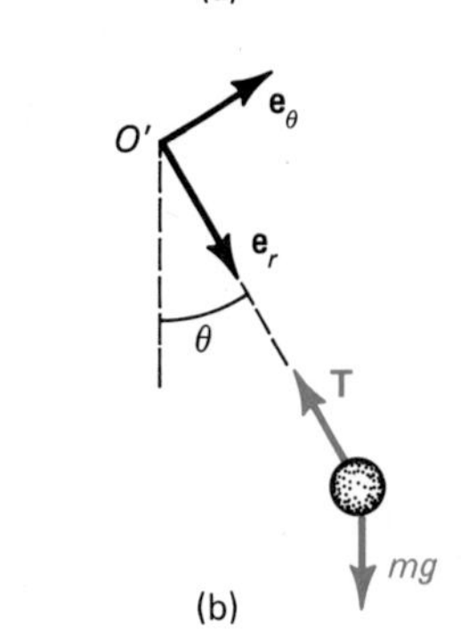

Figure 3.2

Figure 3.2(b) shows the origin O' of the moving reference frame and the free-body diagram of the mass. Assume that the wire transmits only the axial force **T**. The moving reference frame is attached to the pendulum, so that the absolute acceleration of the mass is

$$\mathbf{a} = \ddot{\mathbf{R}} + \boldsymbol{\alpha} \times \mathbf{r} + \boldsymbol{\omega} \times (\boldsymbol{\omega} \times \mathbf{r}) + \ddot{\mathbf{r}}_r + 2\boldsymbol{\omega} \times \dot{\mathbf{r}}_r$$

where
$$\ddot{\mathbf{R}} = a(\sin \theta \mathbf{e}_r + \cos \theta \mathbf{e}_\theta)$$
$$\boldsymbol{\alpha} = \alpha \mathbf{k} = \ddot{\theta} \mathbf{k}$$
$$\boldsymbol{\omega} = \omega \mathbf{k} = \dot{\theta} \mathbf{k}$$
$$\mathbf{r} = L\mathbf{e}_r$$
$$\dot{\mathbf{r}}_r = \ddot{\mathbf{r}}_r = 0$$

Therefore $\quad \mathbf{a} = (a \sin \theta - L\dot{\theta}^2)\mathbf{e}_r + (a \cos \theta + L\ddot{\theta})\mathbf{e}_\theta$

Applying Newton's law, we obtain

$$m[(a \sin \theta - L\dot{\theta}^2)\mathbf{e}_r + (a \cos \theta + L\ddot{\theta})\mathbf{e}_\theta]$$
$$= -T\mathbf{e}_r + mg(\cos \theta \mathbf{e}_r - \sin \theta \mathbf{e}_\theta)$$

Note that all vectors are represented in terms of the chosen reference frame $(\mathbf{e}_r, \mathbf{e}_\theta, \mathbf{k})$ at the instant under consideration. Writing the vector equation as two scalar equations gives the following differential equations:

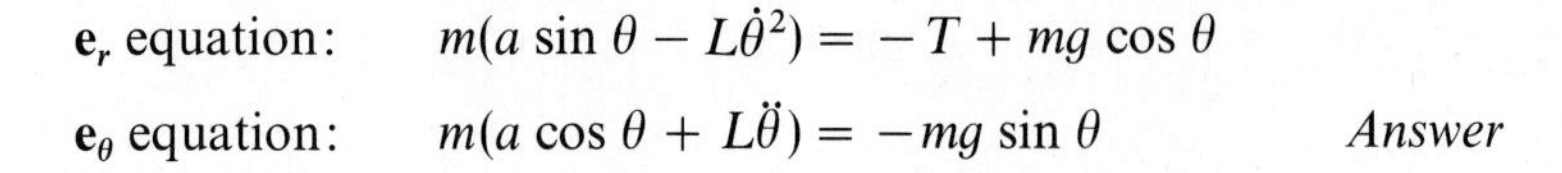

$\mathbf{e}_r$ equation: $m(a \sin\theta - L\dot{\theta}^2) = -T + mg\cos\theta$

$\mathbf{e}_\theta$ equation: $m(a\cos\theta + L\ddot{\theta}) = -mg\sin\theta$ *Answer*

Example 3.3

A mass m is attached to a collapsible telescope as shown in Fig. 3.3(a). At the instant under consideration, the telescope makes a positive angle θ from the X axis, its clockwise angular velocity $\boldsymbol{\omega} = -\dot{\theta}\mathbf{k}$ where $\dot{\theta}$ is constant, and the opening telescope is moving axially such that $L = (4t + 2\cos\lambda t)$, L is in meters and t is in seconds. Find the axial and shear forces acting on the mass if $m = 1$ kg, $\theta = 60°$, $\dot{\theta} = 5$ rad/s, $\lambda = 2$ rad/s and $t = 3$ s. Neglect the mass of the telescope.

(a)

(b)

Figure 3.3

A rotating reference frame represented by $\mathbf{e}_r$, $\mathbf{e}_\theta$, $\mathbf{k}$ is attached to the telescope as shown in Fig. 3.3(b), along with the free-body diagram of the forces acting on the mass. It is assumed that an equal and opposite shear force F_θ and axial force F_r are transmitted to the telescope. The acceleration of the mass is

$$\mathbf{a} = \ddot{\mathbf{R}} + \boldsymbol{\alpha} \times \mathbf{r} + \boldsymbol{\omega} \times (\boldsymbol{\omega} \times \mathbf{r}) + \ddot{\mathbf{r}}_r + 2\boldsymbol{\omega} \times \dot{\mathbf{r}}_r$$

where

$$\ddot{\mathbf{R}} = 0$$
$$\boldsymbol{\alpha} = 0$$
$$\boldsymbol{\omega} = -5\mathbf{k}$$
$$\mathbf{r} = L\mathbf{e}_r = (4t + 2\cos 2t)\mathbf{e}_r$$
$$\dot{\mathbf{r}}_r = (4 - 4\sin 2t)\mathbf{e}_r$$
$$\ddot{\mathbf{r}}_r = -8\cos 2t\mathbf{e}_r$$

Therefore

$$\mathbf{a} = -(58\cos 2t + 100t)\mathbf{e}_r - (40 - 40\sin 2t)\mathbf{e}_\theta$$

When Newton's law is applied, we have

$$m\mathbf{a} = F_r\mathbf{e}_r + F_\theta\mathbf{e}_\theta - mg(\sin 60°\mathbf{e}_r + \cos 60°\mathbf{e}_\theta)$$

The resulting scalar equations are

$\mathbf{e}_r$ equation:

$$m(58\cos 2t + 100t) = mg\sin 60° - F_r$$

$\mathbf{e}_\theta$ equation:

$$m(40 - 40\sin 2t) = mg\cos 60° - F_\theta$$

where $t = 3$ s. Solving these equations we obtain

$$F_\theta = -46.27 \qquad \mathbf{F}_\theta = -46.27\mathbf{e}_\theta \text{ N}$$
$$F_r = -347.2 \qquad \mathbf{F}_r = -347.2\mathbf{e}_r \text{ N}$$

Answer

Example 3.4

A speed-governing device rotates in the horizontal plane at a constant angular speed of 20 rad/sec as shown in Fig. 3.4(a). There are pin connections at A and B and a mass of 0.5 slugs at C. Neglect the mass of the rods OA and AC. Determine (a) the spring constant k needed to hold the rod AC in the position shown, and (b) the force at A, if the spring is stretched by 0.03 ft. Let $r = 0.5$ ft and $b = 0.4$ ft.

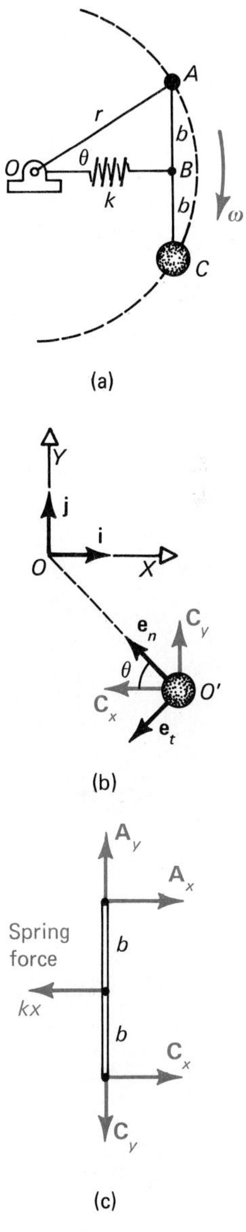

Choose the fixed reference frame at O as shown in Fig. 3.4(b), and the moving normal-tangential unit vectors at O'. Let C_x and C_y be the forces that are transmitted from the mass m to member AC. Since the mass follows a circular path at a constant angular speed ω, its acceleration is $r\omega^2$ in the normal direction. This can be derived using the general formula for $\mathbf{a}$, where it will be found that all terms except the normal acceleration term $\boldsymbol{\omega} \times (\boldsymbol{\omega} \times \mathbf{r})$ vanish. Applying Newton's law to the mass, and using the $\mathbf{i}$, $\mathbf{j}$ directions for convenience, we obtain

$$-C_x\mathbf{i} + C_y\mathbf{j} = mr\omega^2(-\cos\theta\mathbf{i} + \sin\theta\mathbf{j})$$

therefore

$$C_x = mr\omega^2 \cos\theta = 0.5(0.5)(400)\left(\frac{0.3}{0.5}\right) = 60 \text{ lb}$$

$$C_y = mr\omega^2 \sin\theta = 0.5(0.5)(400)\left(\frac{0.4}{0.5}\right) = 80 \text{ lb}$$

Since the rod AC is assumed massless, we sum moments about point A to find the spring force. Referring to Fig. 3.4(c),

$$\sum M_A = 0 \qquad 60(0.8) - kx(0.4) = 0$$

$$kx = 120 \text{ lb}$$

Since the spring is stretched by an amount $x = 0.03$ ft,

$$k = 4000 \text{ lb/ft} \qquad \textit{Answer}$$

The force at A is found by applying the equations of statics to the massless rod shown in Fig. 3.4(c).

$$\sum F_x = 0 \qquad A_x - kx + C_x = 0$$

$$A_x = 120 - 60 = 60 \text{ lb}$$

$$\sum F_y = 0 \qquad A_y - C_y = 0 \qquad A_y = 80 \text{ lb}$$

Consequently, the force at A is

$$\mathbf{A} = (60\mathbf{i} + 80\mathbf{j}) \text{ lb} \qquad \textit{Answer}$$

(a)

(b)

(c)

Figure 3.4

PROBLEMS

3.1 A particle of mass m moving along the X axis is acted upon by a force $\mathbf{F} = -c\dot{x}\mathbf{i}$. If the velocity is $\dot{x} = \dot{x}_0$ when $t = 0$, find the expression for $\dot{x}$ as a function of t.

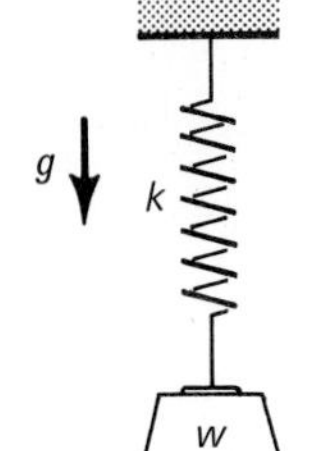

Figure P3.2

3.2 A weight w is suspended by a spring of modulus k (lb/ft) as seen in Fig. P3.2. (a) If the unstretched length of the spring is b ft, apply Newton's law to find the equation of motion. (b) If y ft is measured from the static equilibrium position, i.e., where the static spring force equals the weight, find the equation of motion in terms of y.

3.3 A particle undergoes rectilinear motion such that $F = F_0 e^{-\lambda t}$. If the initial velocity of the particle is V_0, find the expression for the velocity v as a function of t.

3.4 A particle of mass m is subjected to a force $\mathbf{F} = -F_0(r)\mathbf{e}_r$. Derive the equations of motion of the particle.

3.5 Obtain the spring constant k and reaction at A for $t = 3$ sec if $\alpha = 1.667$ rad/sec^2 in Example 3.4.

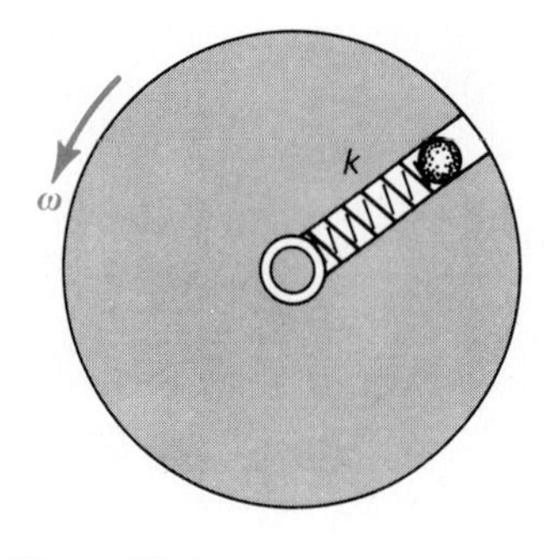

Figure P3.6

3.6 The horizontal disk shown in Fig. P3.6 rotates at a constant angular speed of 2 rad/sec. How much change is there in the length of the massless spring if the unstretched length is 0.2 ft? Let $m = 1$ slug and $k = 100$ lb/ft.

3.7 The rocket of mass m kg in Fig. P3.7, subjected to a thrust of T N, flies such that $\theta = (\theta_0 - kt)$ where t is time and k is constant. If $x \neq 0$, $y = y_0$, $\dot{x} = 0$, and $\dot{y} = \dot{y}_0$ at $t = 0$, obtain an equation for the motion of the rocket in the xy system. Assume that m and T are constant.

3.8 A sphere of mass m is attached to a massless rigid rod of length l that is restrained by a spring of modulus k. Obtain the equation of motion of the mass at the instant shown in Fig. P3.8. Assume θ, the angle between the rod and the horizontal static equilibrium position, is small so that the spring bending is neglected.

3.9 A sphere having a mass of 10 kg rests in a groove as shown in Fig. P3.9. The table rotates at a uniform angular velocity of 10 rad/s. Find the magnitudes of the forces acting on the sphere. Assume smooth surfaces.

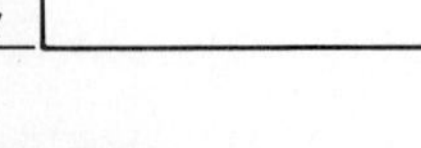
Figure P3.7

Figure P3.8

Figure P3.9

3.10 Calculate the magnitude of the force **P** for the 100-lb block shown in Fig. P3.10 to attain an acceleration of $0.2g$ to the right. The kinetic coefficient of friction between the block and the plane is 0.2 and $\theta = 30^\circ$.

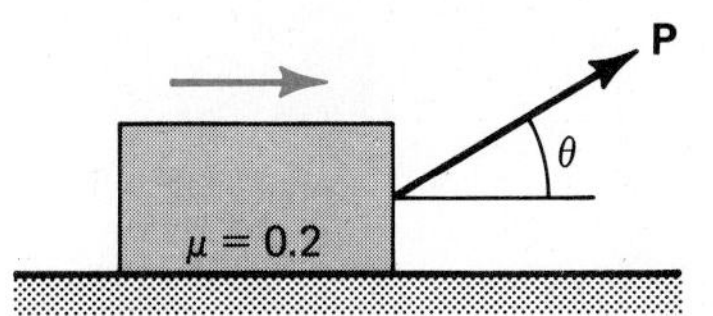

Figure P3.10

3.11 Obtain the magnitude of the force P for the block in Problem 3.10 to attain an acceleration of $0.2g$ to the right if the kinetic coefficient of friction is 0.5 and $\theta = 30^\circ$.

3.12 What is the angle θ for the block in Problem 3.10 that results in the largest horizontal acceleration of the block, assuming that the block remains in contact with the surface?

3.13 The tensile force in each rope of a swing reaches 250 N when a 20-kg child is at the lowest point on the circular arc of radius 3 m. What is the velocity of the child at this point?

3.14 A 50-kg boy stands on a scale inside an elevator. (a) When the elevator is accelerating upward at 4.905 m/s^2, what is the reading on the scale? (b) What is the scale reading when the elevator is decelerating upward at 4.905 m/s^2.

3.15 A 10-kg mass is attached by a rope to a moving support O' that is accelerating upwards as shown in Fig. P3.15. If the force necessary to break the rope is 150 N, what is the maximum angular velocity that the weight-rope pendulum can undergo? (*Hint*: The maximum force in the rope occurs at $\theta = 0^\circ$.)

3.16 Write the equations of motion for the spring-mass assembly shown in Fig. P3.16 assuming the motion is confined to the XY plane. Let l meters be the unstretched length of the spring.

3.17 The body in Fig. P3.17 has a mass of 1 kg and is traveling along the smooth guide at a constant speed of 2 m/s at the instant shown. If the spring is compressed 0.1 m and the radius of curvature is 3 m, find the reaction force from the guide on the mass. (*Hint*: Since the path is known, the tangential and normal directions can be readily used.)

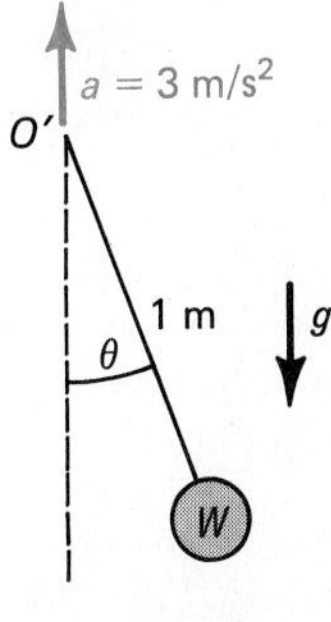

Figure P3.15

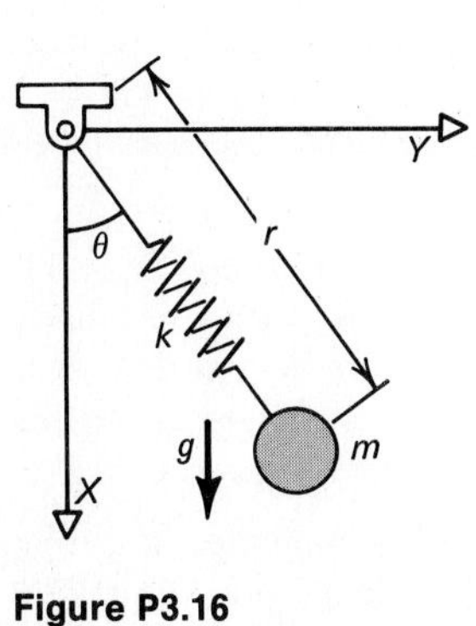

Figure P3.16

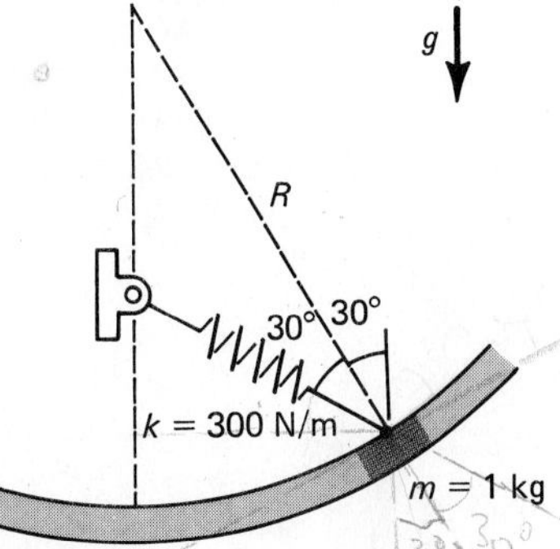

Figure P3.17

3.18 A ball is thrown over the structure as shown in Fig. P3.18. What is the minimum initial velocity to clear the structure?

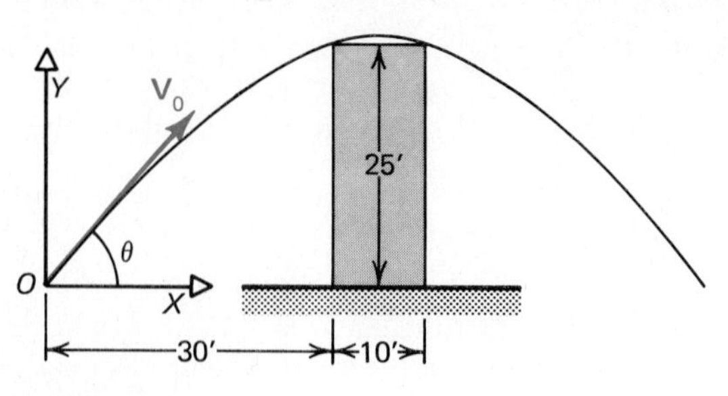

Figure P3.18

3.19 Where will the ball hit the wall in Problem 3.18 if the initial velocity is 50 ft/sec and $\theta = 40°$?

3.20 An airplane traveling at 200 mi/hr horizontally at an altitude of 1000 ft drops a package. Obtain the equation for the trajectory of the package and the force with which it strikes the ground if the package weighs 100 lb.

3.21 In Problem 3.20, if the safe vertical landing velocity of the package is 70 ft/sec, what is the maximum allowable height of the airplane for releasing the package?

3.22 Obtain the equation for the trajectory of the package discussed in Problem 3.20 assuming that a constant opposing force of 1 lb acts in the vertical direction due to atmospheric drag. Did neglecting the atmospheric drag allow a good approximation?

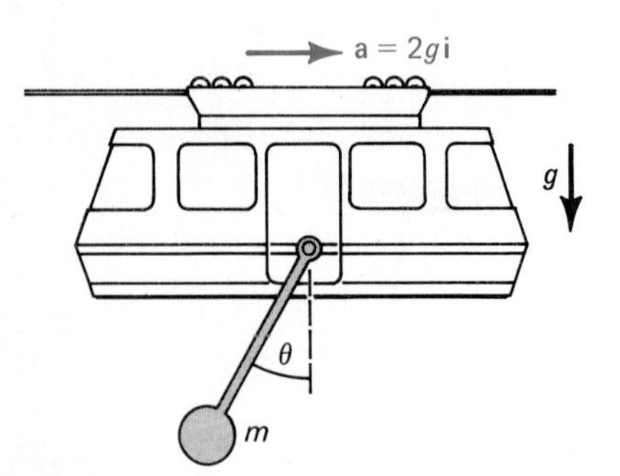

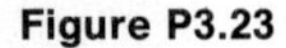
Figure P3.23

3.23 The vehicle in Fig. P3.23 traveling at an acceleration **a** to the right has a simple pendulum attached to it, making an angle θ from the vertical line. What is the equilibrium angle θ?

3.24 What is the maximum acceleration that the vehicle in Problem 3.23 can experience to assure that θ is less than 30°?

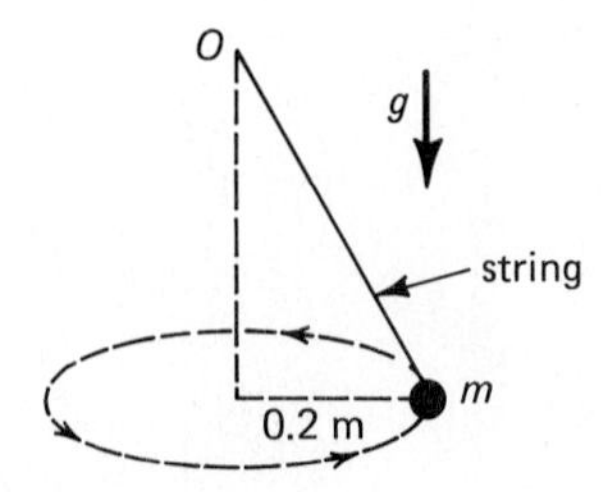

Figure P3.25

3.25 The 10-kg ball attached to a string shown in Fig. P3.25 travels around a circular path in the horizontal plane at a speed of 2 m/s. What is the tensile force in the string?

3.26 Find the acceleration of the 20-kg mass in the slider mechanism shown in Fig. P3.26 if the unstretched length of the spring is 0.5 m and $k = 50$ N/m. Assume that the kinetic coefficient of friction $\mu = 0.25$ between the mass and the rod.

3.27 A cantilever beam 8.5 ft long supports a 20-lb hoisting rig as shown in Fig. P3.27. When the hoisting rig is operating, a body is accelerating upward at a rate of $2g$. What are the reactions at the support of the beam for this condition? Assume the body weighs 100 lb and the cantilever beam weighs 300 lb.

3.28 A collapsible telescope is mounted on a turntable that is stationary, as shown in Fig. P3.28. For the instant shown, what shear and axial forces are transmitted to O'? Assume that the telescope is massless, $m = 10$ kg, $\theta = 30°$, the telescope is retracting at a constant rate $\dot{L} = 0.5$ m/s and $\dot{\theta} =$ 1 rad/s CW (constant).

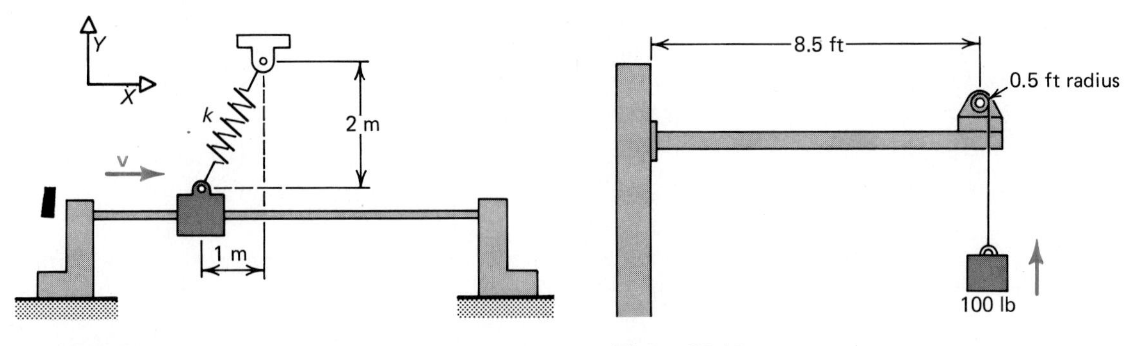

Figure P3.26 **Figure P3.27**

3.29 A flyball regulator, shown in Fig. P3.29 is used to control the speed of a steam engine. The position of the balls depends upon the angle θ which is governed by the rate of rotation ω. Neglecting gravitational effects, derive an expression for the force **F** acting on the fixed bearing V.

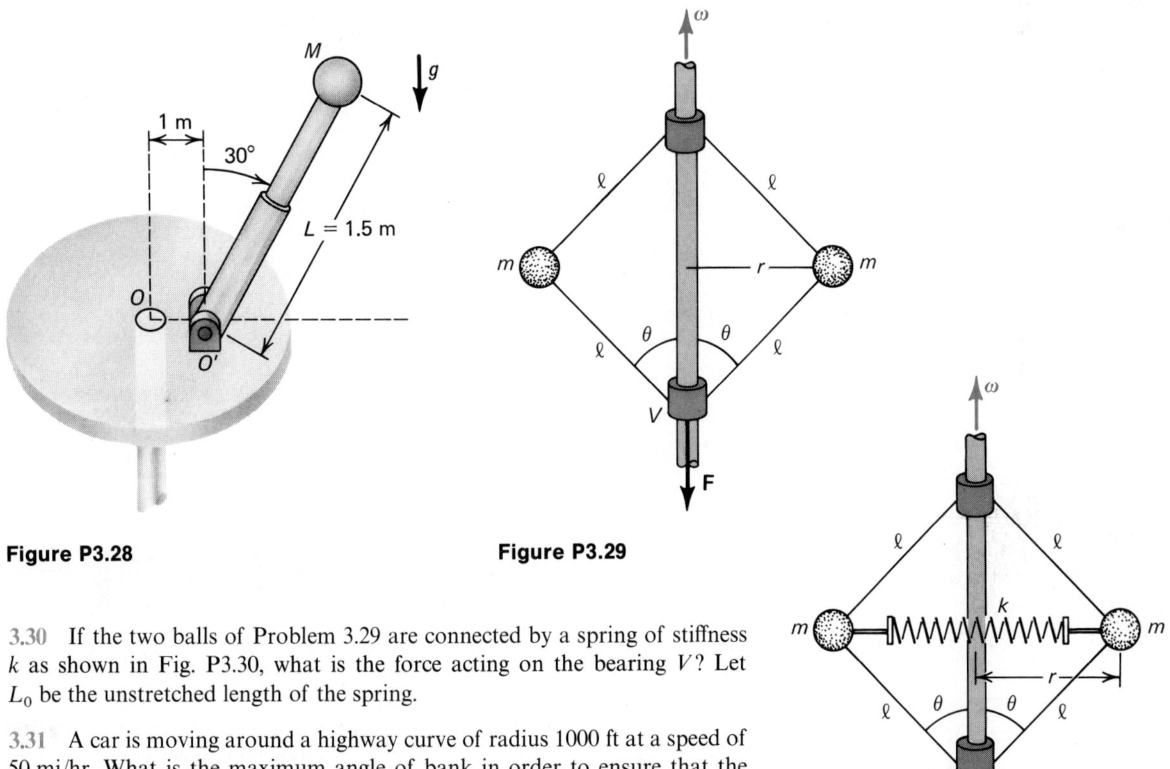

Figure P3.28 **Figure P3.29** **Figure P3.30**

3.30 If the two balls of Problem 3.29 are connected by a spring of stiffness k as shown in Fig. P3.30, what is the force acting on the bearing V? Let L_0 be the unstretched length of the spring.

3.31 A car is moving around a highway curve of radius 1000 ft at a speed of 50 mi/hr. What is the maximum angle of bank in order to ensure that the car does not slide?

3.32 The turntable in Fig. P3.32 with a block of mass 15 kg is rotating at a rate of ω rad/s. The block is restrained by a string that breaks if the tension exceeds 175 N. Assuming a coefficient of friction between the block and table top of 0.1, how fast can the table turn before the string breaks?

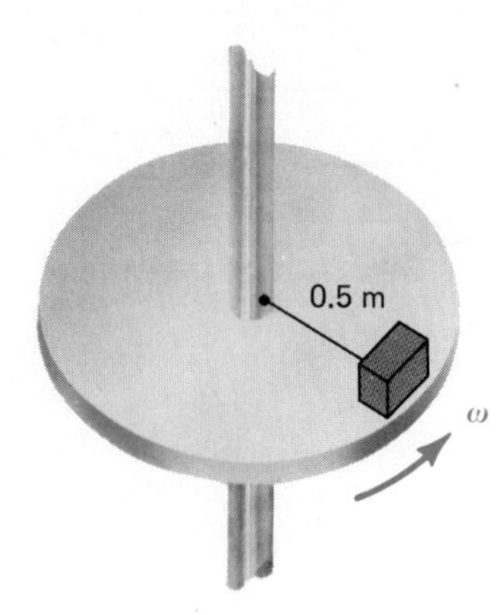

Figure P3.32

3.33 Assume that the turntable top of Problem 3.32 is 1 m from the floor and the string breaks such that the block velocity is $(18.4\mathbf{e}_r + 6.54\mathbf{e}_\theta)$ m/s. How far does the block land from the edge of the table? Compute the velocity when the block lands.

3.34 If the turntable in Problem 3.32 rotates at $0.5t$ rad/s, at what time does the block fly off?

3.35 A shaft supported at two bearings 7 m apart is unbalanced due to two masses attached as shown in Fig. P3.35. Neglecting static loading, that is, the weight of the shaft, what are the reactions at the supports when the shaft is rotating at 1800 rev/min? Let $m_1 = 0.3$ kg and $m_2 = 0.2$ kg.

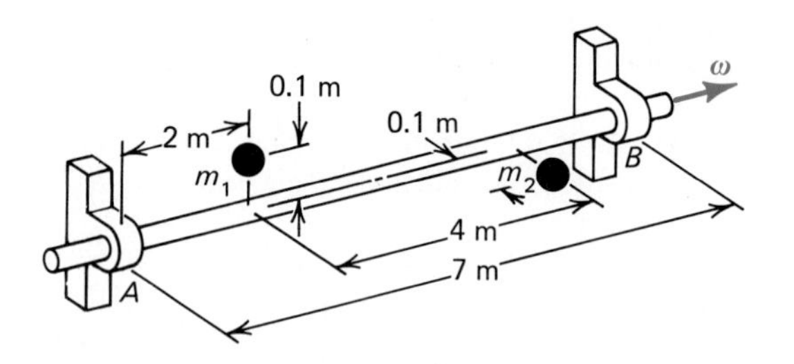

Figure P3.35

3.36 Obtain the reactions at the supports in Problem 3.35 if $m_1 = 0.5$ kg and $m_2 = 0.3$ kg.

3.2 Work and Energy of a Particle

Consider a force **F** acting on a particle as it moves from position A to B as shown in Fig. 3.5. The work done as the particle moves a distance $d\mathbf{r}$ is defined as

$$dW = \mathbf{F} \cdot d\mathbf{r}$$

The total work done by the force **F** is defined as the line integral

$$W = \int_A^B \mathbf{F} \cdot d\mathbf{r} \tag{3.7}$$

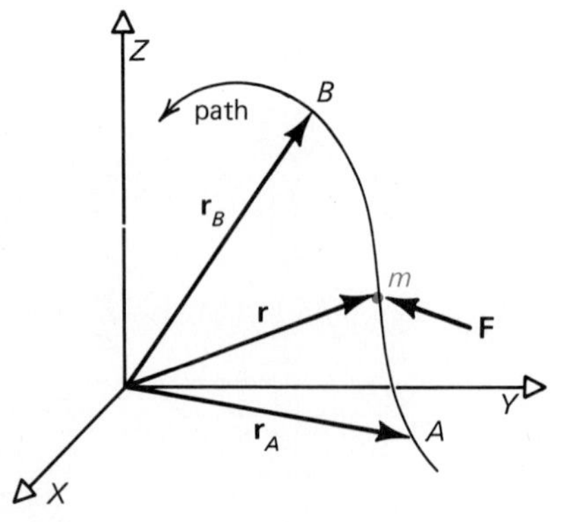

Figure 3.5

The units for work are usually N·m or lb·ft. Note that work is a scalar quantity.

A few examples will provide some physical insight into the mathematical definition. As a first case, suppose the block in Fig. 3.6(a) slides along a frictionless horizontal surface across a distance s under the action of the force of constant magnitude P and constant direction θ. The free-body diagram of the forces acting on the block is shown in Fig. 3.6(b). Thus,

$$\mathbf{F} = -mg\mathbf{j} + P(\cos\theta\mathbf{i} - \sin\theta\mathbf{j}) + N\mathbf{j}$$

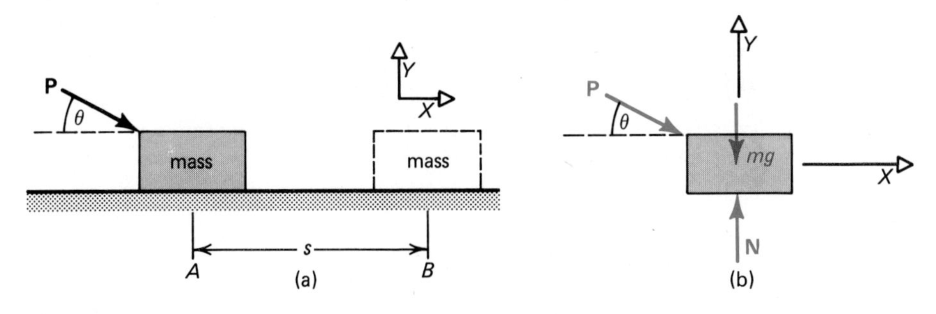

Figure 3.6

while the position vector is

$$\mathbf{r} = x\mathbf{i}$$

and therefore

$$d\mathbf{r} = dx\mathbf{i}$$

so that the work is

$$W = \int_A^B \mathbf{F} \cdot d\mathbf{r} = \int_0^s [-mg\mathbf{j} + P(\cos\theta\mathbf{i} - \sin\theta\mathbf{j}) + N\mathbf{j}] \cdot dx\mathbf{i}$$
$$= \int_0^s P\cos\theta\, dx = (P\cos\theta)s$$

From this example we can see that work is the product of the force component in the direction of travel and the distance traveled. Notice that the forces that remain perpendicular to the direction of travel do not contribute to the work.

As a second example, let us calculate the work done by the forces acting on the bead in Fig. 3.7(a) as it slides along the frictionless circular wire from position A at the top to position B at the bottom of the wire. In this case,

$$\mathbf{r} = r(\cos\theta\mathbf{i} + \sin\theta\mathbf{j})$$
$$d\mathbf{r} = r(-\sin\theta\, d\theta\mathbf{i} + \cos\theta\, d\theta\mathbf{j})$$

Figure 3.7(b) shows the free-body diagram of the forces acting on the bead.

$$\mathbf{F} = -mg\mathbf{j} + N(\cos\theta\mathbf{i} + \sin\theta\mathbf{j})$$

Here, the work is

$$W = \int_A^B \mathbf{F} \cdot d\mathbf{r}$$
$$= \int_{\pi/2}^{-\pi/2} [-mg\mathbf{j} + N(\cos\theta\mathbf{i} + \sin\theta\mathbf{j})] \cdot [-\sin\theta\mathbf{i} + \cos\theta\mathbf{j}]r\, d\theta$$

Note that the limits on the integral are expressed in terms of the θ values.

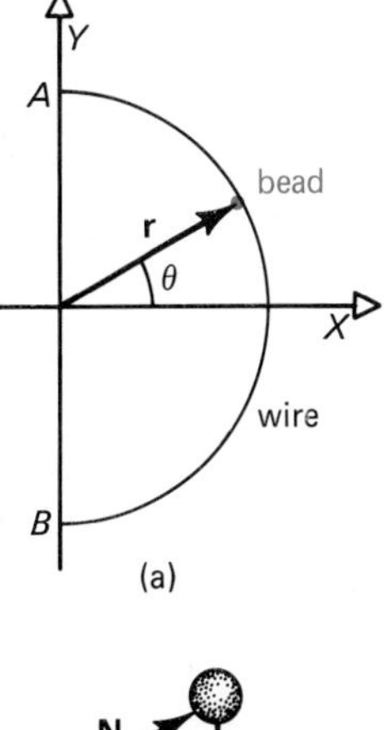

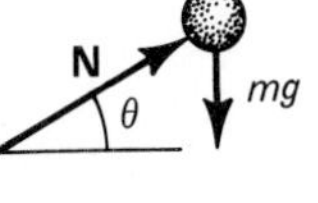

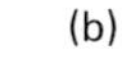

Figure 3.7

$$W = -\int_{\pi/2}^{-\pi/2} mgr \cos \theta \, d\theta = 2mgr$$

We observe that the gravitation force mg does the work; the reaction force N does zero work since it is always perpendicular to the path traveled by the bead.

Conservative Forces

A conservative force **F** is mathematically defined if it satisfies the following condition:

$$\int_A^B \mathbf{F} \cdot d\mathbf{r} \qquad \text{is independent of path}$$

This means that the value of the integral, namely the work, depends only on the upper and lower limits of the position vector **r**. To illustrate this, consider the previous example and suppose that the bead in Fig. 3.7 moves from A to B along the Y axis. Now

$$\mathbf{F} = -mg\mathbf{j}$$

$$\mathbf{r} = y\mathbf{j}$$

$$d\mathbf{r} = dy\mathbf{j}$$

and

$$W = -\int_r^{-r} mg\mathbf{j} \cdot dy\mathbf{j} = -mgy\Big]_r^{-r} = 2mgr$$

The work done by the gravitation force mg is the same whether the bead travels along the circular path or the Y axis. In fact, the work done by the gravitation force is the same regardless of which path we travel from point A to point B, that is, the work done by the gravitation force is independent of path. We shall return to the work done by a gravitation force later, but first we consider another example of a conservative force, viz., the force in a spring. Consider a linear spring of unstretched length L as shown in Fig. 3.8(a). Gradually apply a force **F** to the end of the spring so that it stretches by an amount x, as shown in Fig. 3.8(b). Linear springs are designed to provide a restoring elastic force that is proportional to the amount of stretching or compressing. This measure of resistance offered by a spring is denoted by the spring modulus k, whose units are N/m or lb/ft. The larger the magnitude of k, the stiffer the spring. Now, if we neglect the mass of the spring, the work done on the spring is

$$W = \int_A^B \mathbf{F} \cdot d\mathbf{r}$$

The force to stretch the spring is

$$\mathbf{F} = kx\mathbf{i} \tag{3.8}$$

as seen in the free-body diagram of the spring in Fig. 3.8(c).

The differential distance representing the deformation is $dx\mathbf{i}$. The work done becomes

$$W = \int_A^B kx\mathbf{i} \cdot dx\mathbf{i} = \int_{x_A}^{x_B} kx\,dx$$
$$= \tfrac{1}{2}kx_B^2 - \tfrac{1}{2}kx_A^2$$

which is positive. Note that the work done on the spring is positive whether the spring is being stretched or compressed.

Next, consider the work done on a block attached to the spring as it moves from position A to B on the frictionless surface as shown in Fig. 3.8(d). From the free-body diagram of Fig. 3.8(e) we have for the net horizontal force acting on the block,

$$\mathbf{F} = -kx\mathbf{i} \qquad d\mathbf{r} = dx\mathbf{i}$$

The work done by the spring on the mass is

$$W = \int_A^B \mathbf{F} \cdot d\mathbf{r} = -\int_{x_A}^{x_B} kx\mathbf{i} \cdot dx\mathbf{i}$$
$$= -(\tfrac{1}{2}kx_B^2 - \tfrac{1}{2}kx_A^2)$$

which is the negative of the work done on the spring.

The work done on a spring to stretch or compress it an amount x from its undeformed position is expressed as

$$W = \tfrac{1}{2}kx^2 \tag{3.9}$$

where x is the distance the spring is stretched or compressed from its undeformed position. This work is stored in the spring and is known as elastic potential energy or simply as potential energy and is denoted by V,

$$V = \tfrac{1}{2}kx^2 \tag{3.10}$$

We note that the potential energy stored in the spring is due to the action of a conservative force.

We can write expressions analogous to those written above for torsional springs that are used for resistance to angular deformation. The work done in unwinding or winding a torsional spring is given by

$$W = \int_A^B M\,d\theta \tag{3.11}$$

where the resisting moment M is given by

$$M = k_t\theta \tag{3.12}$$

The modulus k_t of a torsional spring has units of meter-newton per radian (m·N/rad) or foot-pound per radian (ft·lb/rad). Substituting for the moment, the work done on a spring becomes

$$W = \int_A^B k_t\theta\,d\theta = \tfrac{1}{2}k_t\theta_B^2 - \tfrac{1}{2}k_t\theta_A^2$$

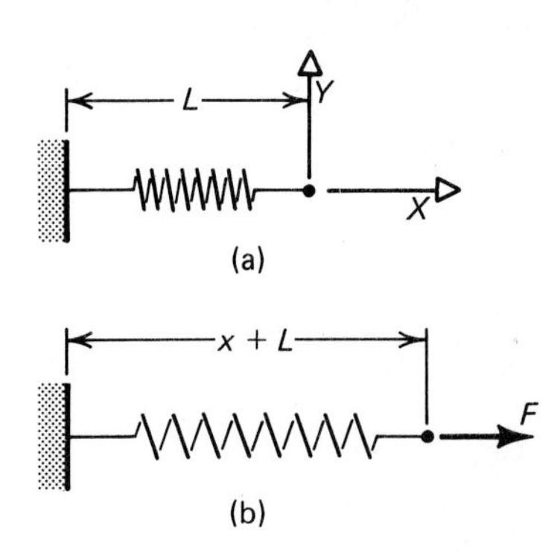

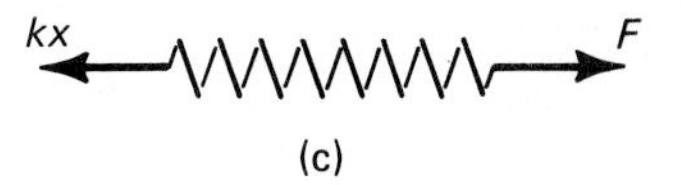

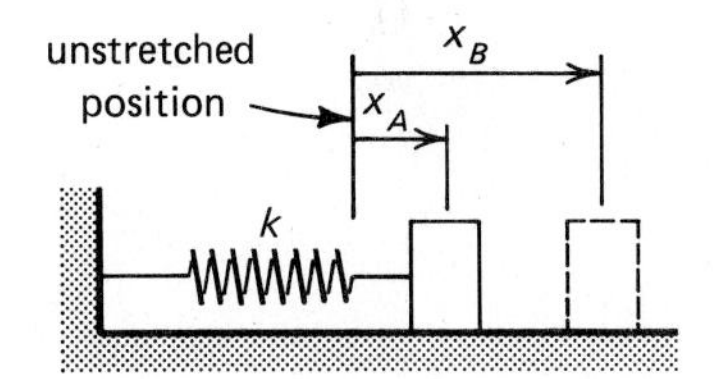

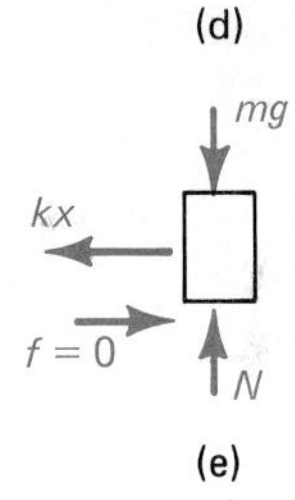

Figure 3.8

The potential energy stored in the spring as it winds or unwinds θ radians from its undeformed position becomes

$$V = \tfrac{1}{2}k_t\theta^2 \tag{3.13}$$

Now we note that this potential energy stored in the spring is due to the action of a conservative moment.

Potential Energy

Consider the work that is done by a conservative force such that the integral $\int_A^B \mathbf{F} \cdot d\mathbf{r}$ is independent of the path. This allows us to replace $\mathbf{F} \cdot d\mathbf{r}$ by an exact differential, $-dV$, the negative sign being chosen for convenience. Thus, for a conservative force $\mathbf{F}$,

$$W = \int_A^B \mathbf{F} \cdot d\mathbf{r} = -\int_A^B dV = -(V_B - V_A) \tag{3.14}$$

where V is the potential energy. Equation (3.14) reads as follows: The work done by a conservative force $\mathbf{F}$ on a body that moves from point A to point B is the negative of the difference in potential energy between B and A.

For the work done on the body attached to the spring shown in Fig. 3.8(d), we have

$$W = -(V_B - V_A) = -(\tfrac{1}{2}kx_B^2 - \tfrac{1}{2}kx_A^2)$$

so that $\quad V_B = \tfrac{1}{2}kx_B^2 \quad$ and $\quad V_A = \tfrac{1}{2}kx_A^2$

i.e., the work done by the spring force on the body as it moves from position A to position B is the negative of the difference in the potential energy stored in the spring in position B and position A.

Let us now return to the case of the gravitational force acting on a particle. Consider the free-body diagram of a particle at a distance h above the earth's surface, as shown in Fig. 3.9. Let h vary from h_A to h_B. Recall that the force varies with h as follows (see Chap. 1).

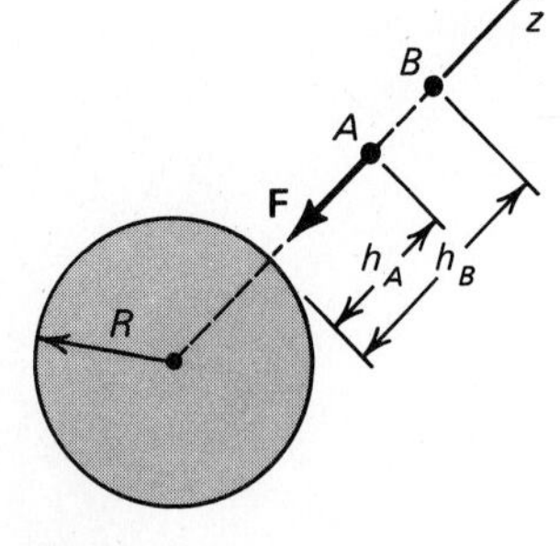

Figure 3.9

$$\mathbf{F} = -mg\frac{R^2}{(R+h)^2}\mathbf{k}$$
$$d\mathbf{r} = dz\mathbf{k} = dh\mathbf{k} \tag{3.15}$$

The work done by the gravitational force in moving the particle from h_A to h_B is

$$W = -\int_{h_A}^{h_B} mg\frac{R^2}{(R+h)^2}dh = mg\left(\frac{R^2}{R+h_B} - \frac{R^2}{R+h_A}\right)$$

which is seen to be a function of the end points. Using the potential energy notation,

$$W = -(V_B - V_A) = mg\left(\frac{R^2}{R + h_B} - \frac{R^2}{R + h_A}\right)$$

In general, then, if a particle is at a height h above the earth's surface, its potential energy is

$$V = -mg\frac{R^2}{R + h} \tag{3.16}$$

On the other hand, if a particle is close to the earth's surface, then the work done by the gravitational force as the particle moves a vertical distance dh is

$$dV = mg\left(\frac{R}{R + h}\right)^2 dh$$

If $R \gg h$ then $R^2/(R + h)^2$ goes to unity and

$$dV = mg\,dh$$

Integrating this expression yields

$$\int_{V_A}^{V_B} dV = mg \int_0^h dh$$

$$(V_B - V_A) = mgh$$

By convention, we let $V_A = 0$, so that $V_B = mgh$. Consequently, if a particle is a vertical distance h above a given reference point, its potential energy is

$$V = mgh \tag{3.17}$$

Note that if the body is a vertical distance h below the reference point, then $V = -mgh$.

In summary, the expressions for the potential energy associated with conservative forces are

Spring force: $V = \frac{1}{2}kx^2$

Gravitational force: $V = -mg\dfrac{R^2}{R + h}$

or $V = mgh$ (close to earth)

Example 3.5

A block is subjected to a force $\mathbf{F} = (50\mathbf{i} + 60\mathbf{j})$ lb and pushed along a smooth horizontal surface. If the distance the block is moved from the origin of the coordinate system is $(3\mathbf{i} + 2\mathbf{j})$ ft, how much work is done? Assume that the $\mathbf{i}, \mathbf{j}$ vectors lie in the horizontal plane.

The work done is given by

$$W = \int_A^B \mathbf{F} \cdot d\mathbf{r}$$

Now let

$$\mathbf{r} = x\mathbf{i} + y\mathbf{j}$$

so that

$$d\mathbf{r} = dx\mathbf{i} + dy\mathbf{j}$$

Substituting into the equation for the work yields

$$W = \int (50\mathbf{i} + 60\mathbf{j}) \cdot (dx\mathbf{i} + dy\mathbf{j})$$

$$= 50 \int_0^3 dx + 60 \int_0^2 dy$$

$$= 150 + 120 = 270 \text{ lb} \cdot \text{ft}$$ *Answer*

Example 3.6

A spring, having a spring modulus 10,000 N/m, is stretched by 0.2 m from its unstretched position. How much work is done? How much potential energy is stored in the spring?

The force to stretch the spring is kx. Therefore,

$$W = \int_0^{0.2} kx\,dx$$

$$= \frac{kx^2}{2}\Big]_0^{0.2}$$

$$= 200 \text{ N} \cdot \text{m}$$ *Answer*

This work stored in the spring equals the potential energy. Therefore,

$$V = W = 200 \text{ N} \cdot \text{m}$$ *Answer*

Example 3.7

A 500-kg particle is raised 100 m. What is the increase of potential energy of the particle measured (a) with reference to the surface of the earth and (b) with reference to 150 m above the surface of the earth?

The potential energy of a particle at a height h above the reference surface is

(a)

$$V = mgh = 500(9.81)(100)$$

$$= 490{,}500 \text{ N} \cdot \text{m}$$ *Answer*

(b) Using the reference point that is located 150 m above the earth's surface, h now equals -50 m. Therefore,

$$V = 500(9.81)(-50)$$
$$= -245{,}250\ \text{N}\cdot\text{m} \qquad \textit{Answer}$$

The Work-Energy Equation

Suppose $\mathbf{r}$ is the position vector of a particle measured from a fixed reference frame. Now $\dot{\mathbf{r}}$ is the absolute velocity of the particle while $\ddot{\mathbf{r}}$ is its absolute acceleration. Using this notation, Newton's second law takes the form

$$m\ddot{\mathbf{r}} = \mathbf{F} \tag{3.18}$$

We now form the scalar product by multiplying both sides of the equation by $\dot{\mathbf{r}}$.

$$m\ddot{\mathbf{r}} \cdot \dot{\mathbf{r}} = \mathbf{F} \cdot \dot{\mathbf{r}} \tag{3.19}$$

Next consider the identity

$$\frac{1}{2}\frac{d}{dt}(\dot{\mathbf{r}} \cdot \dot{\mathbf{r}}) = \ddot{\mathbf{r}} \cdot \dot{\mathbf{r}} \tag{3.20}$$

Substitute Eq. (3.20) into the left side of Eq. (3.19) to obtain

$$\frac{m}{2}\frac{d}{dt}(\dot{\mathbf{r}} \cdot \dot{\mathbf{r}}) = \mathbf{F} \cdot \frac{d\mathbf{r}}{dt}$$

The term dt cancels on both sides of the equation. Now integrating over the path of the particle from position A to position B, we obtain

$$\tfrac{1}{2}m \int_{v_A}^{v_B} d(\dot{\mathbf{r}} \cdot \dot{\mathbf{r}}) = \int_A^B \mathbf{F} \cdot d\mathbf{r} \tag{3.21}$$

Since $\dot{\mathbf{r}} = \mathbf{v}$ and $\dot{\mathbf{r}} \cdot \dot{\mathbf{r}} = v^2$, the integral on the left side of the equation becomes

$$\int_{v_A}^{v_B} d(\dot{\mathbf{r}} \cdot \dot{\mathbf{r}}) = \int_{v_A}^{v_B} d(v^2) = v^2 \Big]_{v_A}^{v_B} = (v_B^2 - v_A^2)$$

and Eq. (3.21) gives

$$\tfrac{1}{2}m(v_B^2 - v_A^2) = \int_A^B \mathbf{F} \cdot d\mathbf{r} \tag{3.22}$$

The term on the left side is the change in the kinetic energy; that on the right is the work done on the particle. Equation (3.22) reads as follows:

> *The work done by the forces acting on a particle as it moves from position A to position B is equal to the corresponding change in the kinetic energy.*

Equation (3.22) is rewritten in abbreviated form as follows:

$$W = \Delta T \tag{3.23}$$

where

$$W = \int_A^B \mathbf{F} \cdot d\mathbf{r} \tag{3.7}$$

$$T = \tfrac{1}{2}m\mathbf{v} \cdot \mathbf{v} = \tfrac{1}{2}mv^2 \tag{3.24}$$

We note that the work-energy equation is a scalar equation and the units of work and energy are usually given as newton-meters (N·m) or pound-feet (lb·ft).

If the forces acting on a body are all conservative forces, Eq. (3.22) is rewritten as follows:

$$\Delta T = T_B - T_A = \int_A^B \mathbf{F} \cdot d\mathbf{r}$$

Since the work is independent of path for this special case, the term $\mathbf{F} \cdot d\mathbf{r}$ is again replaced by the exact differential $-dV$. Thus,

$$T_B - T_A = -\int_A^B dV = -(V_B - V_A)$$

or

$$T_A + V_A = T_B + V_B \tag{3.25}$$

Equation (3.25) reads as follows:

The sum of the kinetic energy and the potential energy of a body at position A is equal to the sum of its kinetic energy and the potential energy at position B.

In other words, energy in the form of potential energy and kinetic energy is conserved as the body moves about under the influence of conservative forces.

Nonconservative Forces

Any force that is not a conservative force is called a nonconservative force. We have established that spring forces and gravitational forces are conservative forces. For our purposes, we will treat all other forces as nonconservative forces. In general, these nonconservative forces include friction forces and forces that vary with time or velocity.

Let us now consider the case where a particle is subjected to both conservative and nonconservative forces. Let

W_C = Work done by the conservative forces

W_{NC} = Work done by the nonconservative forces

The expression for the total work done by all the forces is

$$W = W_C + W_{NC}$$

and the equation for work done by the conservative forces $\mathbf{F}_C$ is

$$W_C = \int_A^B \mathbf{F}_C \cdot d\mathbf{r} = -(V_B - V_A) = -\Delta V$$

Thus,

$$W = -\Delta V + W_{NC}$$

Equating this to the left side of Eq. (3.23) we obtain

$$W_{NC} = \Delta T + \Delta V \tag{3.26}$$

from which we see that the work done by the nonconservative forces equals the change in the kinetic energy plus the change in the potential energy. Equation (3.26) automatically becomes the expression given by Eq. (3.25), if only conservative forces are present, and is a convenient form of the work-energy equation. Generally, the change in potential energy due to conservative forces is a straightforward calculation, so that attention is focused on calculating the work done by the nonconservative forces.

Power

Power is defined as the time rate of doing work. Since the differential form of work is

$$dW = \mathbf{F} \cdot d\mathbf{r}$$

then,

$$\text{Power} = P = \frac{dW}{dt} = \mathbf{F} \cdot \frac{d\mathbf{r}}{dt}$$

or

$$P = \mathbf{F} \cdot \mathbf{v} \tag{3.27}$$

Power is a scalar quantity with units N·m/s or lb·ft/sec. Often, mechanical power is given by the horsepower unit, where

$$1 \text{ hp} = 746 \text{ watts} = 550 \text{ lb·ft/sec}$$

where 1 watt = 1 N·m/s.

Example 3.8

A 100-kg particle has a velocity of $(20\mathbf{i} + 30\mathbf{j})$ m/s. What is the kinetic energy of this particle?

Referring to Eq. (3.24),

$$\begin{aligned} T &= \tfrac{1}{2}m\mathbf{v} \cdot \mathbf{v} \\ &= \tfrac{1}{2}(100)(20\mathbf{i} + 30\mathbf{j}) \cdot (20\mathbf{i} + 30\mathbf{j}) \\ &= 50(400 + 900) = 65{,}000 \text{ N·m} \end{aligned}$$

Answer

Example 3.9

A 10-slug block is traveling at 15 ft/sec to the left as shown in Fig. 3.10(a). The kinetic coefficient of friction between the block and horizontal surface is 0.2. At position B the block impacts with the spring. If the spring modulus $k = 1{,}000$ lb/ft, how much does the spring compress?

First consider the block moving from A to B. Now from Eq. (3.26)

$$\begin{aligned} W_{NC} &= \Delta T + \Delta V \\ &= (T_B - T_A) + \Delta V \end{aligned}$$

Since there is no change in potential energy from A to B, $\Delta V = 0$ so that

$$W_{NC} = T_B - T_A \qquad \text{(a)}$$

Figure 3.10(b) shows the free-body diagram of the block while traveling from A to B. The work done by the 64.4-lb friction force is

$$W_{NC} = -64.4(10) = -644 \text{ lb}\cdot\text{ft}$$

Substituting into Eq. (a),

$$\begin{aligned} -644 &= \tfrac{1}{2}mv_B^2 - \tfrac{1}{2}mv_A^2 \\ \tfrac{1}{2}mv_B^2 &= -644 + \tfrac{1}{2}(10)(15)^2 \\ &= 481 \text{ lb}\cdot\text{ft} \end{aligned}$$

Now apply Eq. (3.26) after impact between the block and the spring. The block moves from position B to position C, where it stops instantaneously so that $v_C = 0$. Thus,

$$W_{NC} = T_C - T_B + V_C - V_B$$

where

$$\begin{aligned} W_{NC} &= -64.4x \\ T_C &= 0 \\ T_B &= \tfrac{1}{2}mv_B^2 = 481 \\ V_C &= \tfrac{1}{2}kx^2 = 500x^2 \\ V_B &= 0 \quad \text{(spring is unstretched)} \end{aligned}$$

Now

$$-64.4x = -481 + 500x^2$$

Solving the quadratic equation,

$$x = 0.919 \qquad \text{or} \qquad x = -1.047$$

Choose the positive value to agree with the assumed positive displacement,

$$x = 0.919 \text{ ft} \qquad \textit{Answer}$$

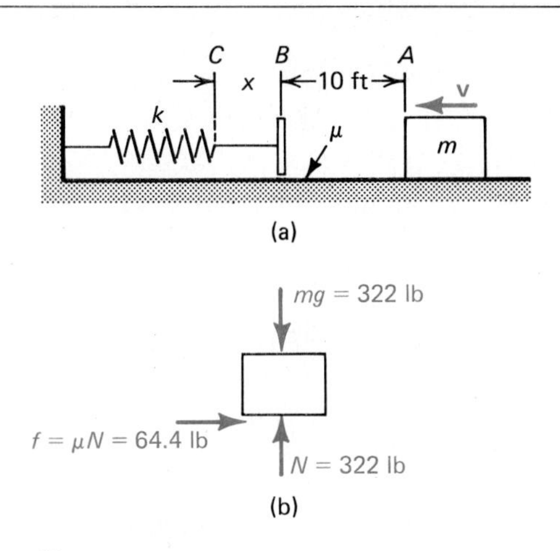

Figure 3.10

Example 3.10

A 2-kg particle slides down a frictionless chute as shown in Fig. 3.11(a). After leaving the chute the particle travels toward the wall. Does the particle reach the wall, go over the wall, or hit the wall? If it strikes the wall, how high above the floor does this occur?

Since there is no friction between the particle and the chute, we may use the principle of conservation of energy between points A and B to find the velocity when the particle leaves the chute.

$$V_B + T_B = V_A + T_A$$

Choose B as the reference point for the potential energy.

$$V_A = 2(9.81)5 = 98.1\ \text{N}\cdot\text{m} \qquad T_A = 0 \text{ (initially at rest)}$$

$$V_B = 0 \qquad T_B = \tfrac{1}{2}(2)v_B^2$$

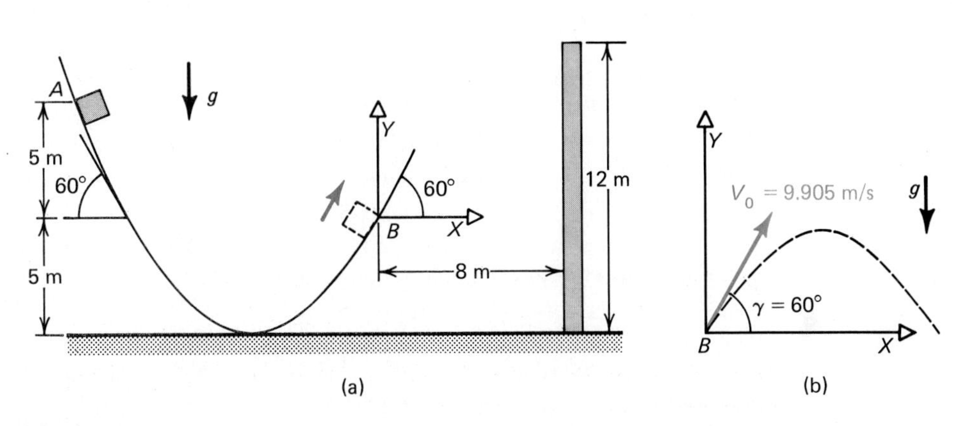

Figure 3.11

therefore $$v_B^2 = 98.1$$

$$v_B = 9.905 \text{ m/s}$$

Figure 3.11(b) shows the anticipated trajectory and the reference coordinate system. From the results of Example 3.1, we know that

$$\dot{y} = -gt + V_0 \sin\gamma \qquad \dot{x} = V_0 \cos\gamma$$

$$y = -\frac{gt^2}{2} + V_0 t \sin\gamma \qquad x = V_0 t \cos\gamma$$

where now $V_0 = v_B$ and $\gamma = 60^\circ$ at B.

Since the wall is higher than point A, we know from the conservation of energy that the particle can never go over the wall. However, it may strike the wall or never reach the wall before bouncing.

The time it takes to travel 8 m in the x direction is labeled t_1, so that

$$t_1 = \frac{x}{V_0 \cos\gamma} = \frac{8}{9.905 \cos 60^\circ} = 1.615 \text{ s}$$

The time it takes to travel -5 m in the y direction is labeled t_2, so that

$$y = -\frac{g}{2}t_2^2 + V_0 t_2 \sin\gamma$$

$$-5 = -4.905t_2^2 + 8.578t_2$$

Solving the quadratic for t_2,

$$t_2 = 2.210 \text{ s} \qquad \text{or} \qquad t_2 = -0.461 \text{ s}$$

We choose the positive value to agree with positive time having elapsed. Since $t_2 > t_1$, the particle strikes the wall. To obtain where the particle strikes the wall, we use $t = 1.615$ s in the expression for y.

$$y = -4.905(1.615)^2 + 8.578(1.615) = 1.06$$

Therefore, the particle strikes the wall 6.06 m above the floor.

Example 3.11

A particle riding on a frictionless circular wire of radius $a/2$ is attached to a spring as shown in Fig. 3.12. When the particle is at position A, the spring is unstretched and its velocity is $v_A = \sqrt{ag}$. Find the velocity of the body in position B.

Since the gravitational and spring forces are the only forces doing work, the conservation of energy prevails, i.e.,

$$T_B + V_B = T_A + V_A$$

Let us choose point 0 as a convenient reference point for writing the potential energy associated with gravity.

$$T_B = \tfrac{1}{2}mv_B^2$$

$$V_B = -\tfrac{3}{4}mga + \tfrac{1}{2}k(\tfrac{3}{4}a - \tfrac{1}{4}a)^2$$

$$T_A = \tfrac{1}{2}mga$$

$$V_A = \tfrac{1}{4}mga$$

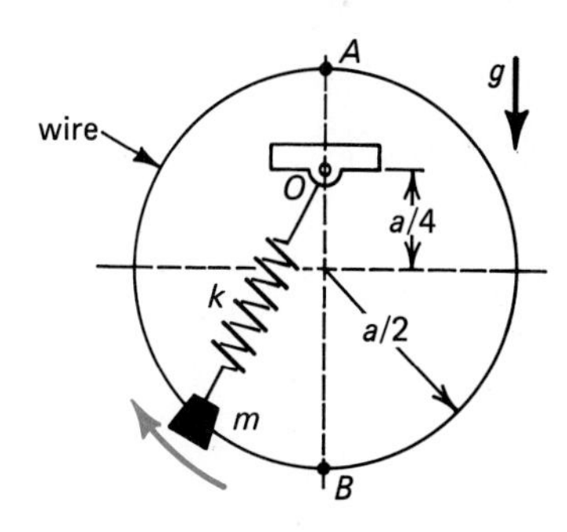

Figure 3.12

Upon substituting into the energy conservation equation we obtain

$$\tfrac{1}{2}mv_B^2 - \tfrac{3}{4}mga + \tfrac{1}{8}ka^2 = \tfrac{1}{2}mga + \tfrac{1}{4}mga$$

therefore

$$v_B = \sqrt{3ga - \frac{1}{4}\frac{ka^2}{m}} \qquad \textit{Answer}$$

Note that $v_B = 0$ if $k = 12mg/a$. What does it mean if $k > 12mg/a$?

Example 3.12

A body of mass m is initially at rest when a constant force of magnitude P is applied as shown in Fig. 3.13(a). The coefficient of kinetic friction equals μ between the body and the inclined plane. Assuming P is large enough to move the body up the plane, find the velocity of the body when it has been moved a distance of s meters.

The forces acting on the body in Fig. 3.13(b) can be separated as conservative and nonconservative forces. If this is done, Eq. (3.26) can be used. The only conservative force present is the force of gravity; the nonconservative force is expressed as:

$$\mathbf{F}_{NC} = P(\cos\theta\mathbf{i} - \sin\theta\mathbf{j}) + N\mathbf{j} - \mu N\mathbf{i}$$

We know that

$$W_{NC} = \Delta T + \Delta V$$

where

$$W_{NC} = \int_0^s \mathbf{F}_{NC} \cdot d\mathbf{r}$$

$$\Delta T = \tfrac{1}{2}mv_B^2, \quad \Delta V = mgs\sin\theta$$

$$N = mg\cos\theta + P\sin\theta$$

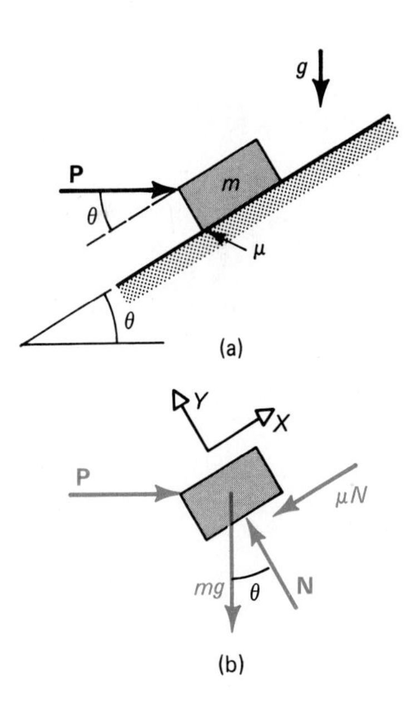

Figure 3.13

By substituting we have

$$\int_0^s [P(\cos\theta\mathbf{i} - \sin\theta\mathbf{j}) + N\mathbf{j} - \mu N\mathbf{i}] \cdot dx\mathbf{i} = \tfrac{1}{2}mv_B^2 + mgs\sin\theta$$

$$\int_0^s [P\cos\theta - \mu(mg\cos\theta + P\sin\theta)]\,dx = \tfrac{1}{2}mv_B^2 + mgs\sin\theta$$

so that

$$v_B = \sqrt{(2/m)[(P - \mu mg)s \cos\theta - (\mu P + mg)s \sin\theta]} \qquad \textit{Answer}$$

PROBLEMS

3.37 (a) What is the work done if a spring modulus $k = 5000$ lb/ft is stretched by 0.2 ft? (b) What is the additional work if it stretched 0.1 ft in addition to the original 0.2 ft stretching?

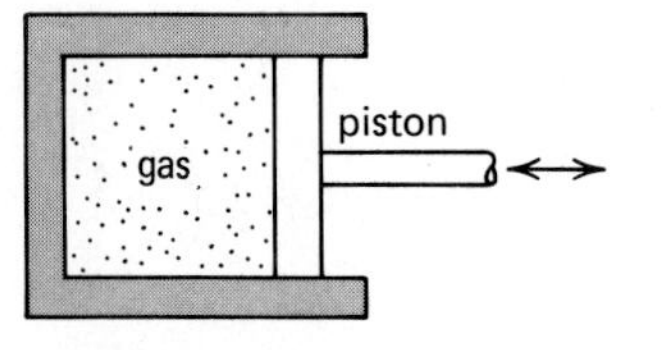

Figure P3.38

3.38 In the cylinder-piston arrangement shown in Fig. P3.38, a gas expands according to the relationship $pv^k =$ constant, where the pressure p is in lb/ft^2, the specific volume v is in ft^3/slug, and k is a constant. If the initial pressure and specific volume are p_1 and v_1, respectively, and the final pressure and specific volume are p_2 and v_2, respectively, show that the work per unit mass is

$$W = \frac{p_2 v_2 - p_1 v_1}{1 - k}$$

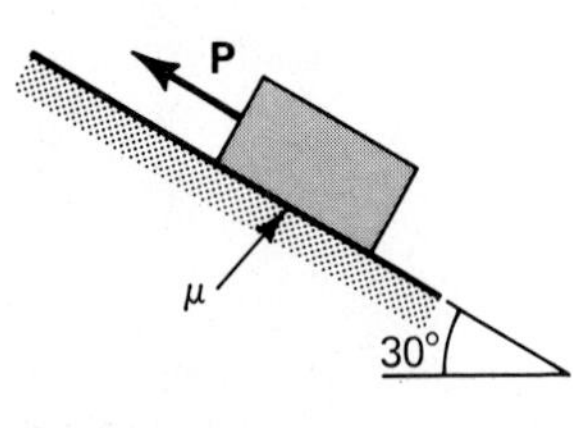

Figure P3.39

3.39 The 5-kg block in Fig. P3.39 slides 5 m up the inclined plane, which has a kinetic coefficient of friction equal to 0.1 between the block and the plane. Calculate the total work done if the magnitude of **P** is 50 N.

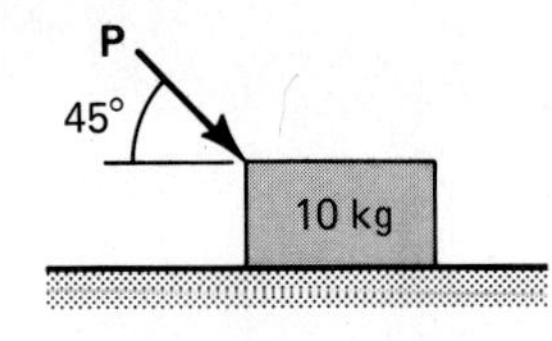

Figure P3.40

3.40 A 10-kg block is pushed across a horizontal surface, with a kinetic coefficient of friction equal to 0.2 between the block and the surface, by the force **P** as shown in Fig. P3.40. If 120 N·m of work is expended in moving the block through a distance of 10 m, find the magnitude of **P**.

3.41 A 100-lb body travels at a velocity of $\mathbf{v} = (2\mathbf{i} + 4\mathbf{j})$ ft/sec. What is the kinetic energy of the body?

3.42 The velocity of a 250-lb body is initially given by $\mathbf{v} = (5\mathbf{e}_r + 8\mathbf{e}_\theta)$ ft/sec. If the final velocity equals $\mathbf{v} = (2\mathbf{e}_r + 3\mathbf{e}_\theta)$ ft/sec, what is the change in kinetic energy of the body?

3.43 If the force acting on a particle is a function of time, i.e., if $\mathbf{F} = \mathbf{F}(t)$, and the velocity $\mathbf{v}$ of the particle is also known, how can the expression

$$W = \int_A^B \mathbf{F} \cdot d\mathbf{r}$$

be used to find the work during some time interval?

3.44 Using the results of Problem 3.43, find the work if

$$\mathbf{F} = 2\cos \pi t\,\mathbf{i} \text{ lb} \qquad \text{and} \qquad \mathbf{v} = 4 \sin \pi t\,\mathbf{i} \text{ ft/sec}$$

for the time interval from $t = 0$ to $t = \frac{1}{2}$ sec.

3.45 The position vector of a satellite in the plane of the orbit is $\mathbf{r} = r(\cos\theta\mathbf{i} + \sin\theta\mathbf{j})$, where $\theta = \omega t$ and r is constant. What is the kinetic energy of the satellite? Let m be the mass of the satellite.

3.46 An 8-kg particle is moving at constant acceleration of 2 m/s^2. If the initial velocity is 20 m/s, what is the change in kinetic energy after 30 s?

3.47 A 10-kg particle is subjected to a force defined by $\mathbf{F} = 2t\mathbf{i} + 3t\mathbf{j}$, where $\mathbf{F}$ is in newtons and t is in seconds. If the initial velocity is $20\mathbf{i}$ m/s, what is the change of kinetic energy after 10 s?

3.48 Using the information of Example 3.11, obtain the velocity at point B of the mass shown in Fig. 3.12 if $m = 10$ slugs, $k = 100$ lb/ft and $a = 0.5$ ft.

3.49 Compute the kinetic energy of the block in Problem 3.33 (a) when it flies off and (b) when it hits the ground 1 m below.

3.50 What is the kinetic energy of the balls of Problem 3.29 as they rotate at 1200 rev/min, if each ball has a mass of 0.2 kg and $r = 0.2$ m.

3.51 Compute the change in kinetic energy of the balls of Problem 3.29 as the angle changes by an increment of $\Delta\theta$.

3.52 What is the kinetic energy of the particle subjected to a radial force in Problem 3.4?

3.53 The acceleration of a 1-slug particle is given by $\mathbf{a} = (8/\pi)\sin \pi t\mathbf{i}$ ft/sec². It is known that when $t = 0$, $v = 6$ ft/sec. What is the kinetic energy of the particle after 5 sec?

3.54 The acceleration of a 50-lb particle is given by $\mathbf{a} = -Kv^2\mathbf{i}$ ft/sec². What is the kinetic energy at $t = 1$ sec? Assume that $v = 20$ ft/sec at $t = 0$, and $K = 0.2$.

3.55 A particle is rigidly attached to the edge of a massless drum at point A as shown in Fig. P3.55. The drum, rotating at a constant 10 rad/s, has a brake applied to it for 2 s, thus reducing the kinetic energy of the particle by one-half. What is the final angular velocity of the drum at the end of the braking period if the particle's mass equals 1.5 kg?

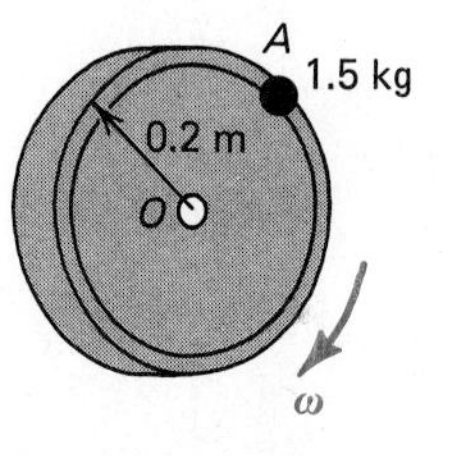

Figure P3.55

3.56 The block in Fig. P3.56 is released from rest and slides along a frictionless chute until it reaches the horizontal plane, at which point the kinetic coefficient of friction is $\mu_1 = 0.1$; 3 m further $\mu_2 = 0.2$. How far from point A does the block slide?

3.57 The block in Fig. P3.57 is released from rest and slides along a frictionless chute until it reaches the horizontal plane, at which point the kinetic coefficient of friction is $\mu = 0.1$. After 5 ft it slides up a smooth chute. How high does the block rise vertically?

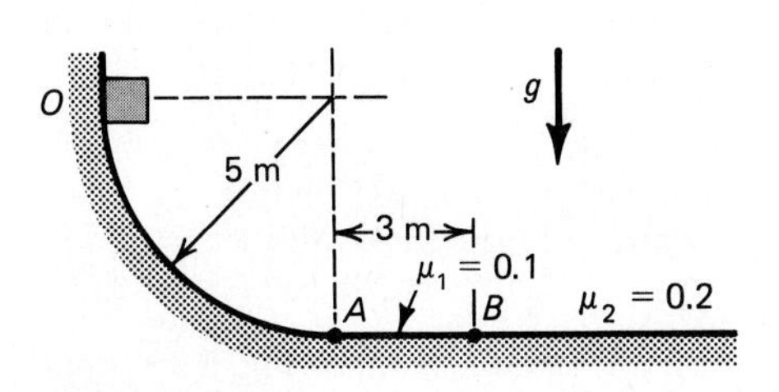

Figure P3.56

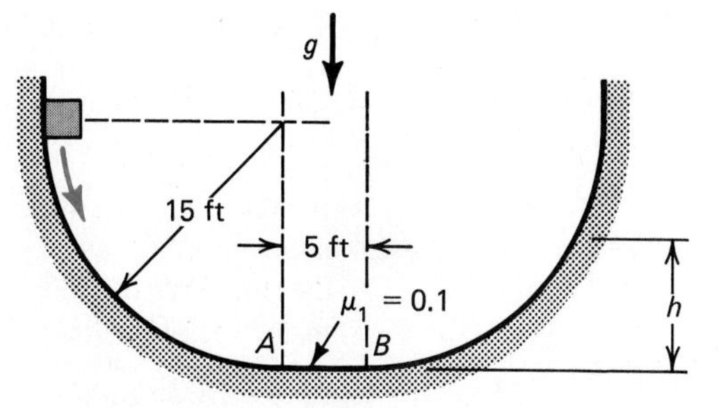

Figure P3.57

3.58 If the block in Problem 3.57 is allowed to slide back and forth in the chute after being released, how many times will the 5-ft section be traversed and where will the block stop?

3.59 A block is released with an initial velocity of 30 m/s and slides down a 20° inclined plane. Find the required kinetic coefficient of friction along the inclined plane if the block is to travel 100 m before coming to rest.

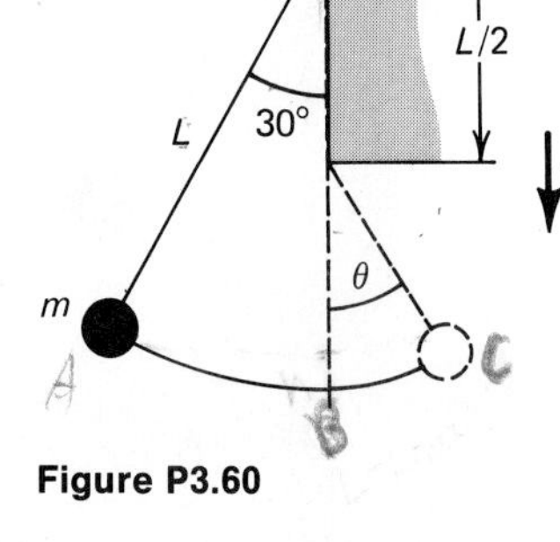

Figure P3.60

3.60 A mass m, attached to a string of length L, is released from rest as shown in Fig. P3.60. Find the maximum angle θ.

3.61 The 0.4-kg projectile in Fig. P3.61 is to be released such that it reaches a height of 1 km at a velocity of 20 m/s. It is known that the maximum distance the spring travels from its equilibrium position is 0.5 m. What must the spring constant be in order to meet the required criteria?

3.62 A 25,000-kg train is traveling at 2 km/h as it enters a railroad station. It comes to a complete stop as it collides with two restraining springs mounted in parallel on a concrete embankment. These springs are thereby compressed by 100 mm. How much potential energy is initially absorbed by the springs? What is the effective stiffness of the springs?

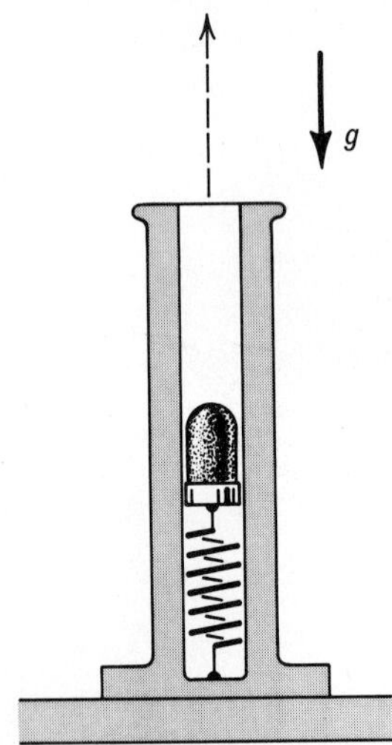

Figure P3.61

3.63 A 2,000-kg airplane is catapulted from the deck of a surface ship. The airplane must achieve a speed of 120 km/h at take-off. If the spring used for catapulting is stretched by 3 m, what is the stiffness of the spring?

3.64 What is the mass of the largest airplane that can be catapulted from the deck of a surface ship if the plane must achieve a speed of 100 km/h at take off? The spring used for catapulting has a stiffness of 2×10^7 N/m and it is stretched by 3m.

3.65 A 5-kg mass moves from a coordinate position (0, 8, 6) to (6, 0, 0) meters under the action of its own weight (gravity is in the negative y direction), that of a spring with one end attached at the origin and the other to the weight, and that of an externally applied force $\mathbf{P} = (6\mathbf{i} - 4\mathbf{j} - 2\mathbf{k})$ N. The spring modulus is $k = 5.4$ N/m and its unstretched length is 0.15 m. If the body starts from rest at the initial position, find the velocity of the mass when it reaches the coordinate (6, 0, 0).

3.66 The spring of modulus $k = 400$ lb/ft is initially compressed 0.1 ft as shown in Fig. P3.66. A 1-lb mass rests against the spring. How high will the mass rise when the spring is released?

3.67 If the 1-lb ball in Problem 3.66 strikes the spring after the pendulum is released from the horizontal position, how much will the spring compress? Assume $k = 400$ lb/ft and that the spring is undeformed when the ball is in the vertical position.

3.68 When a space vehicle is returning to earth from a trip to the moon, it has an absolute velocity of 6400 km/h when it is 17,600 km from the center of the earth. What is the absolute velocity of the vehicle when it is 6700 km from the earth's center?

3.69 A particle slides down the inclined surface shown in Fig. P3.69 from

A to B and then around the smooth semicircle to C. If the particle starts from rest at A, what is the velocity at C? Assume that all surfaces are smooth.

3.70 A 10-kg particle slides down the inclined plane from A to B, then around the smooth semicircle from B to C as shown in Fig. P3.69. The kinetic coefficient of friction between the particle and the inclined plane is $\mu = 0.2$. If the particle velocity at position A is 4 m/s downward, calculate the velocity at position C.

3.71 What should the distance be between A and B in Problem 3.70 if the particle is released from rest and its velocity at C is zero?

3.72 What is the maximum power (hp/lb) available from water flowing over a waterfall at a rate of 60 ft/sec?

3.73 If a machine delivers 10 hp for a period of 2 hours, how much work is performed during this time? Express your answer in N·m.

3.74 A spring is first stretched 0.5 m and then compressed 0.5 m over 10 s. If the stiffness is 10^5 N/m, what is the amount of power necessary to do this?

3.75 A tube with unequal cross-sectional areas is open at three ends as shown in Fig. P3.75. The fluid in the center column is pushed down by an amount x, causing a rise in the other columns. If the density of the fluid is ρ, obtain an expression for the change in potential energy of the system as a function of the cross-sectional areas A, the fluid density ρ, g, and x.

3.76 An electron of mass m with a velocity of magnitude v enters the region enclosed by two capacitor plates as shown in Fig. P3.76. The electric field between the plates is a constant E. Obtain an expression for the trajectory of the electron. How far can the electron travel before it hits a plate?

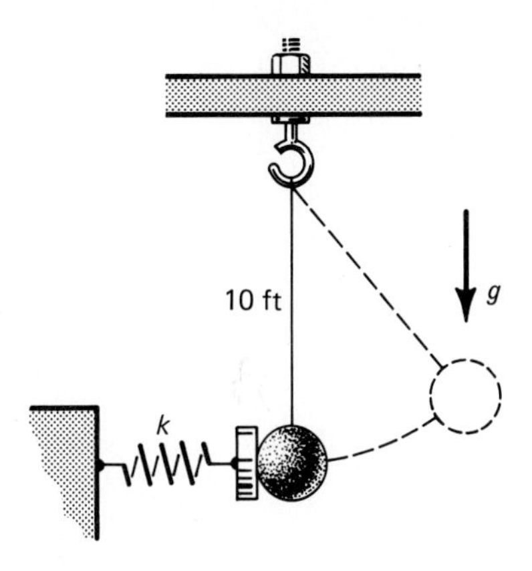

Figure P3.66

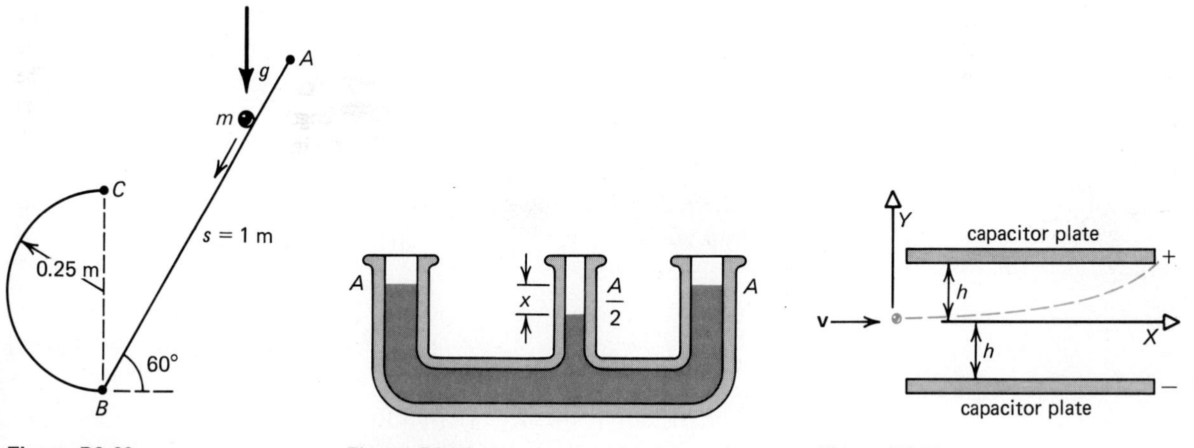

Figure P3.69

Figure P3.75

Figure P3.76

3.77 Compute the kinetic energy of the electron in Problem 3.76 as it hits a plate.

3.78 A 50-kg satellite is in a circular orbit about the earth at a distance

800 km above the earth's surface. Neglecting air resistance, find the energy needed to put the satellite into orbit.

3.79 A spring of modulus $k_1 = 8000$ lb/ft is compressed 0.1 ft as shown in Fig. P3.79. The 10-lb mass travels across the surface which has a kinetic coefficient of friction equal to 0.1. How much will the second spring be compressed if its modulus is $k_2 = 4000$ lb/ft? Neglect the dimensions of the mass.

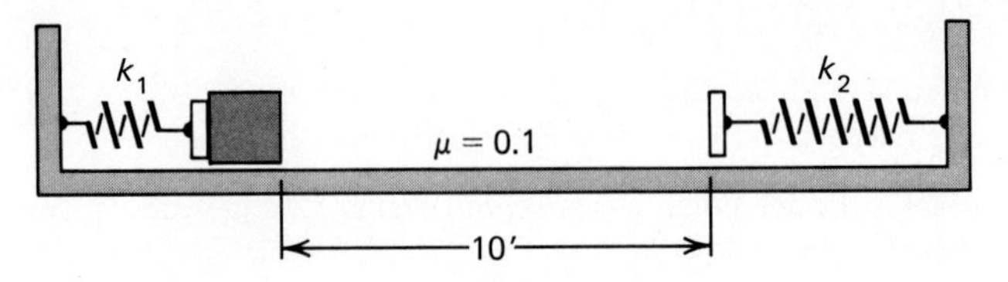

Figure P3.79

3.80 If the mass in Problem 3.79 has a velocity of 10 ft/sec just before it hits the left spring, then how many times will it hit the right spring? Assume $k_1 = 2k_2 = 8000$ lb/ft, and neglect any friction loss as the springs deform.

3.81 A force is applied to an 8-kg particle so that it moves along a plane curved path such that $y = x^2$. When $x = 0.1$ m, $\dot{x} = 2$ m/s and when $y = 9$ m, $\dot{y} = 6$ m/s; what is the change in the kinetic energy between these two positions?

3.3 Impulse and Momentum of a Particle

Having developed the work-energy relationship by spatial integration of Newton's second law, we shall now integrate the same equation over time in order to derive the impulse-momentum relationship of a particle.

Newton's second law can be written as follows:

$$\mathbf{F} = m\ddot{\mathbf{r}} = m\frac{d\mathbf{v}}{dt} = \frac{d}{dt}(m\mathbf{v}) \tag{3.28}$$

where the absolute velocity $\dot{\mathbf{r}} = \mathbf{v}$. We define the linear momentum as $\mathbf{p}$, so that

$$\mathbf{p} = m\mathbf{v} \tag{3.29}$$

$$\mathbf{F} = \frac{d\mathbf{p}}{dt} = \dot{\mathbf{p}} \tag{3.30}$$

This says that the net external force acting on the particle equals the time rate of change of the linear momentum of the particle. The units of linear momentum are kilogram-meter per second (kg·m/s) or pound-second (lb·sec). We note that N·s is equivalent to kg·m/s while lb·sec is equivalent to slug·ft/sec.

Consider the particle with linear momentum $m\mathbf{v}$ shown in

Fig. 3.14. For the instant under consideration, premultiply both sides of Eq. (3.28) by **r**.

$$\mathbf{r} \times \mathbf{F} = \mathbf{r} \times \left[\frac{d}{dt}(m\mathbf{v})\right] \tag{3.31}$$

Now $$\frac{d}{dt}(\mathbf{r} \times m\mathbf{v}) = \dot{\mathbf{r}} \times m\mathbf{v} + \mathbf{r} \times \left[\frac{d}{dt}(m\mathbf{v})\right] = \mathbf{r} \times \left[\frac{d}{dt}(m\mathbf{v})\right]$$

since $$\dot{\mathbf{r}} \times m\mathbf{v} = \mathbf{v} \times m\mathbf{v} = 0$$

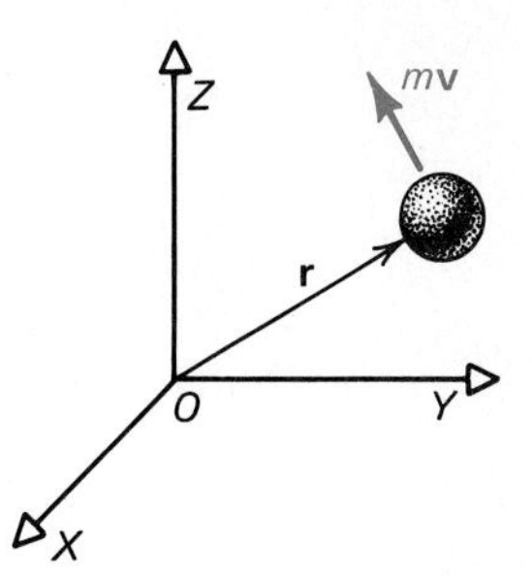

Figure 3.14

Equation (3.31) becomes

$$\mathbf{r} \times \mathbf{F} = \frac{d}{dt}(\mathbf{r} \times m\mathbf{v}) \tag{3.32}$$

But $\mathbf{r} \times \mathbf{F} = \mathbf{M}$, the moment of the external forces about the fixed reference point O. The term on the right side of Eq. (3.32) is the time derivative of the moment of the linear momentum. The term in the parenthesis on the right side is called the angular momentum and is denoted by **H**,

$$\mathbf{H} = \mathbf{r} \times m\mathbf{v} \tag{3.33}$$

The units of angular momentum are $\text{kg}\cdot\text{m}^2/\text{s}$ or lb·ft·sec. Substituting Eq. (3.33) into Eq. (3.32) yields

$$\mathbf{M} = \dot{\mathbf{H}} \tag{3.34}$$

This equation reads:

> *The moment about the fixed reference point O of the external forces acting on the particle equals the time rate of change of the angular momentum about O.*

This equation is particularly useful when dealing with rigid bodies as we will discover later. We note that if $\mathbf{M} = 0$ in Eq. (3.34), then $\dot{\mathbf{H}} = 0$ or **H** is a constant, i.e., the angular momentum is conserved about the point O. We conclude, therefore, that the angular momentum stays constant, or is conserved, when the resulting moment due to the forces is zero.

A relationship between linear impulse and linear momentum is obtained when Newton's second law is integrated over time. When a force **F** acts on a particle during the time interval t_1 to t_2, the **linear impulse** of the force is defined as

$$\mathscr{I} = \int_{t_1}^{t_2} \mathbf{F}\, dt \tag{3.35}$$

This can be related to the linear momentum by integrating Eq. (3.30) with respect to time,

$$\int_{t_1}^{t_2} \mathbf{F}\, dt = \mathbf{p}_2 - \mathbf{p}_1$$

so that

$$\mathscr{I} = \Delta\mathbf{p} \tag{3.36}$$

where $\Delta\mathbf{p} = \mathbf{p}_2 - \mathbf{p}_1$. The units of linear impulse and linear momentum are the same, i.e., kg·m/s or lb·sec.

Similarly the angular impulse and angular momentum can be related to each other. The angular impulse of a torque or moment **M** that acts on a body from time t_1 to time t_2 is defined as

$$\mathscr{M} = \int_{t_1}^{t_2} \mathbf{M}\, dt \tag{3.37}$$

This vector expression has units of newton-meter-second (N·m·s) or pound-feet-second (lb·ft·sec). The angular impulse can be related to the angular momentum by integrating Eq. (3.34) with respect to time,

$$\int_{t_1}^{t_2} \mathbf{M}\, dt = \mathbf{H}_2 - \mathbf{H}_1$$

so that

$$\mathscr{M} = \Delta\mathbf{H} \tag{3.38}$$

where $\Delta\mathbf{H} = \mathbf{H}_2 - \mathbf{H}_1$. Again the units of angular momentum and angular impulse are the same, i.e., kg·m^2/s or lb·ft·sec.

Notice that the impulse-momentum equations are vector equations. Also note that there is no limitation on the time interval for the impulse expressions, i.e., the impulse may cover a time interval measured in milliseconds or a time interval measured in hours. If the net force acting on the particle is zero during the time interval from t_1 to t_2, then $\Delta\mathbf{p} = 0$, or $\mathbf{p}_1 = \mathbf{p}_2$. This means there is a conservation of linear momentum during the time interval when the linear impulse is zero. Likewise, Eq. (3.38) says that the angular momentum about the reference point O is conserved if the net moment about O is zero during the time interval from t_1 to t_2.

Example 3.13

A hockey puck is subjected to a force as shown in Fig. 3.15 as it slides across a smooth horizontal sheet of ice. What is the linear impulse applied to the puck from $t = 0$ to $t = 3$ sec?

As the hockey puck slides across the smooth horizontal sheet of ice, the force is given by $\mathbf{F} = (4 - t)\mathbf{i}$ kip. The linear impulse of **F** between $t_1 = 0$ sec and $t_2 = 3$ sec is obtained by substituting into Eq. (3.35):

$$\mathscr{I} = \int_0^3 1000(4 - t)\mathbf{i}\, dt = 7500\mathbf{i} \text{ lb·sec} \qquad \textit{Answer}$$

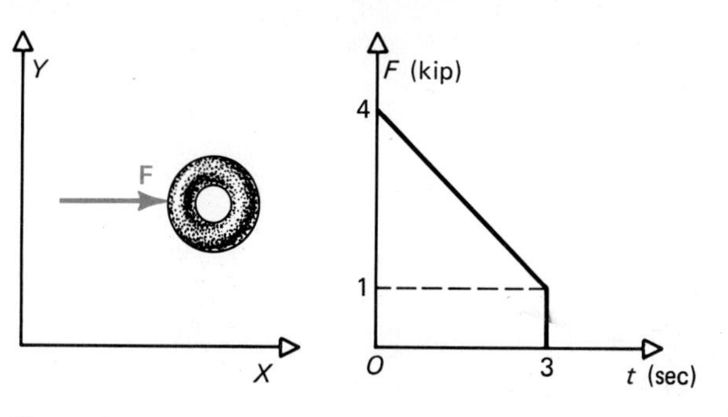

Figure 3.15

Example 3.14

A 100-slug body has a velocity of (1500**i** + 2000**j**) ft/sec at a particular instant when its position is given by (4**i** + 3**j**) ft. (a) What is the angular momentum at this instant? (b) If the velocity is (150**i** + 200**j**) ft/sec at a different time, what is the x coordinate of the body if $\mathbf{r} = x\mathbf{i} + 100\mathbf{j}$ ft? Assume that the angular momentum, expressed in the **i**, **j**, **k** system about the origin, is conserved.

Since the angular momentum is conserved,

$$\mathbf{H}_1 = \mathbf{H}_2$$

(a)

$$\begin{aligned} \mathbf{H}_1 &= \mathbf{r}_1 \times m\mathbf{v} \\ &= (4\mathbf{i} + 3\mathbf{j}) \times 100(1500\mathbf{i} + 2000\mathbf{j}) \\ &= 350{,}000\mathbf{k} \end{aligned}$$

Answer

(b)

$$\begin{aligned} \mathbf{H}_2 &= (x\mathbf{i} + 100\mathbf{j}) \times 100(150\mathbf{i} + 200\mathbf{j}) \\ &= (20{,}000x - 1.5 \times 10^6)\mathbf{k} \end{aligned}$$

From momentum conservation,

$$20{,}000x - 1.5 \times 10^6 = 350{,}000$$

$$x = 92.5 \text{ ft}$$

Answer

Example 3.15

A body of mass 5.1 kg is at rest when a force of magnitude $P = 25(1 + t)$ N is applied to it as seen in Fig. 3.16. The body slides up a frictionless rod. Calculate the speed of the body 2 s later.

Since the forces acting on the body are in the vertical direction, the scalar form of Eq. (3.36) is adequate. The weight of the body is

$$w = (5.1)(9.81) = 50 \text{ N}$$

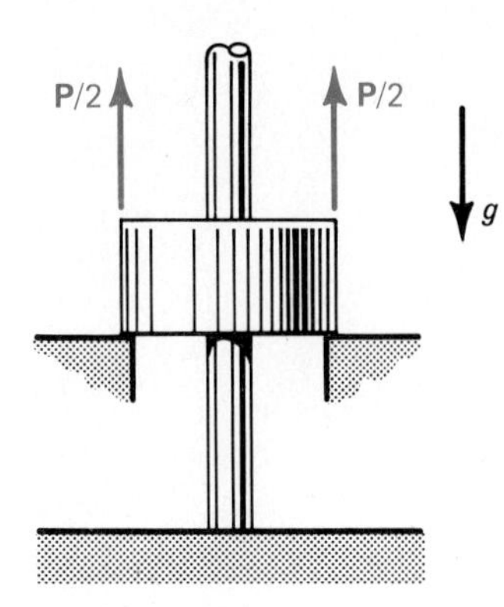

Figure 3.16

If t_1 is the time it takes for P to build up for overcoming the body weight, then

$$25(1 + t_1) = 50$$

$$t_1 = 1 \text{ s}$$

This means that motion is impending at time t_1. This becomes the lower limit in integrating Eq. (3.36). The forces acting on the body include both the force **P** and the weight of the body. Substituting into Eq. (3.36),

$$\int_1^2 [25(1 + t) - 50]\, dt = 5.1(v_2 - v_1)$$

Since the body is initially at rest, $v_1 = 0$ when $t_1 = 1$. Simplifying the above expression and rearranging terms,

$$v_2 = 4.90 \int_1^2 (t - 1)\, dt$$

$$= 4.90 \left[\frac{t^2}{2} - t \right]_1^2$$

$$v_2 = 2.45 \text{ m/s} \qquad \textit{Answer}$$

Example 3.16

A 1-kg body connected to a rigid rod of length 2 m slides in a circular path in the horizontal plane as shown in Fig. 3.17(a). When $\dot{\theta} = 4$ rad/s CW, a moment **M** is applied where $\mathbf{M} = 20t\mathbf{k}$ m·N and t is in seconds. If the kinetic coefficient of friction between the body and the plane is $\mu = 0.1$, find the angular velocity when $t = 1$ s. Neglect the mass of the rod.

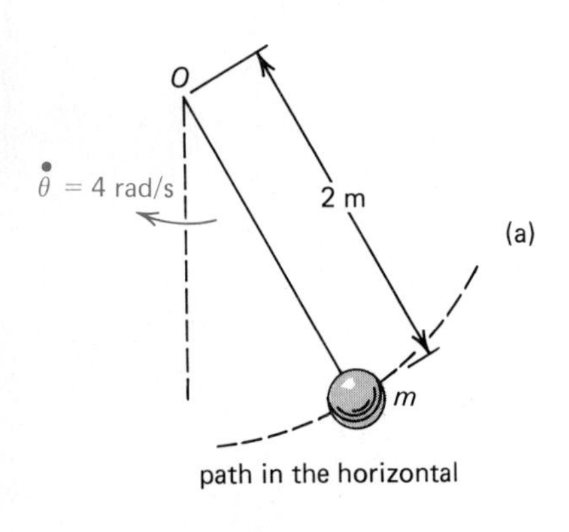

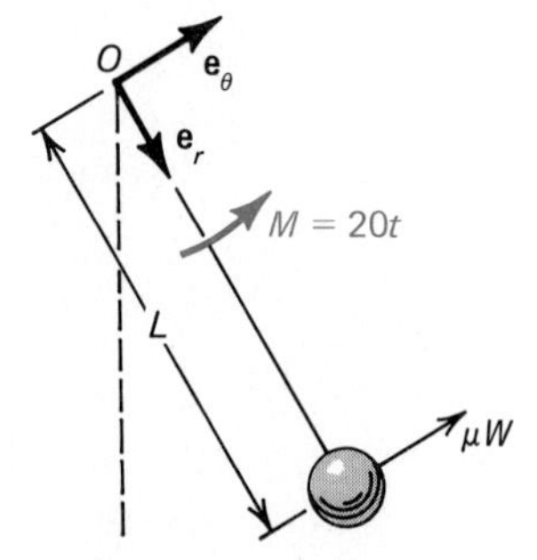

Figure 3.17

Figure 3.17(b) shows a moving reference frame $\mathbf{e}_r$, $\mathbf{e}_\theta$ attached to the rod, the friction force μW, and the applied torque. Note that $\mathbf{e}_r \times \mathbf{e}_\theta = \mathbf{k}$. Now

$$\mathbf{M} = (20\text{t} + \mu WL)\mathbf{k} = (20t + 1.962)\mathbf{k} \qquad \text{and} \qquad \mathbf{v} = L\dot{\theta}\mathbf{e}_\theta$$

therefore

$$\mathbf{H} = \mathbf{r} \times m\mathbf{v} = 2\mathbf{e}_r \times mL\dot{\theta}\mathbf{e}_\theta = 4m\dot{\theta}\mathbf{k} = 4\dot{\theta}\mathbf{k}$$

When $t_1 = 0$, $\dot{\theta} = -4$ rad/s so that

$$\mathbf{H}_1 = 4(-4)\mathbf{k} = -16\mathbf{k}$$

while at t_2,

$$\mathbf{H}_2 = 4\dot{\theta}_2\mathbf{k}$$

Equation (3.38) then becomes:

$$\left[\int_0^1 (20t + 1.962)\,dt\right]\mathbf{k} = (4\dot{\theta}_2 + 16)\mathbf{k}$$

Dropping the unit vector **k** and solving, we have

$$\dot{\theta}_2 = -1.01 \text{ rad/s}$$

so that

$$\boldsymbol{\omega}_2 = -1.01\mathbf{k} \text{ rad/s} \qquad \textit{Answer}$$

PROBLEMS

3.82 A particle of mass m is attached to a string of length l and rotates in the XY plane at an angular velocity of ω rad/s as shown in Fig. P3.82. What is (a) the kinetic energy of the particle, (b) the linear momentum of the particle, and (c) the angular momentum of the particle about the fixed point O?

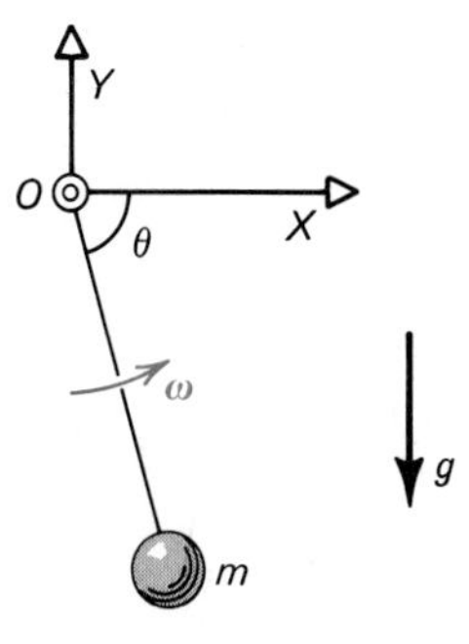

Figure P3.82

3.83 A rocket is subjected to thrust **P** as shown in Fig. P3.83. Calculate the linear impulse imparted to the rocket. The mass of the rocket is 100 kg.

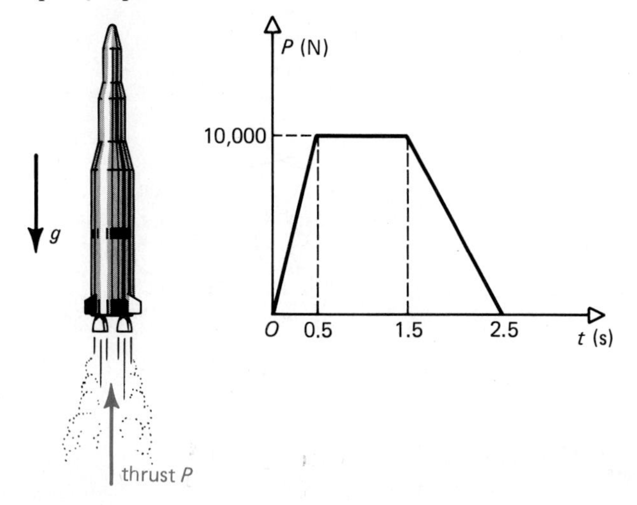

Figure P3.83

3.84 A particle is subjected to a force $\mathbf{F} = (3t^2\mathbf{i} + 4\mathbf{j})$ lb for 5 sec. Calculate the linear impulse that is imparted to the particle.

3.85 A force $\mathbf{F} = (3t^2\mathbf{i} + \sin 2\pi t\mathbf{j})$ lb acts on a particle. Find the linear impulse imparted to the particle by this force at the end of $\frac{1}{4}$ sec.

3.86 A bullet of mass m is traveling toward a suspended block of mass M at a speed v as shown in Fig. P3.86. The bullet hits the block and remains embedded in it. The bullet and block have an initial velocity of v' after impact has taken place. (a) Find the linear momentum of the bullet before impact. (b) Find the angular momentum of the bullet relative to O at the instant before impact. (c) Find the angular momentum of the bullet-block combination at the instant after impact.

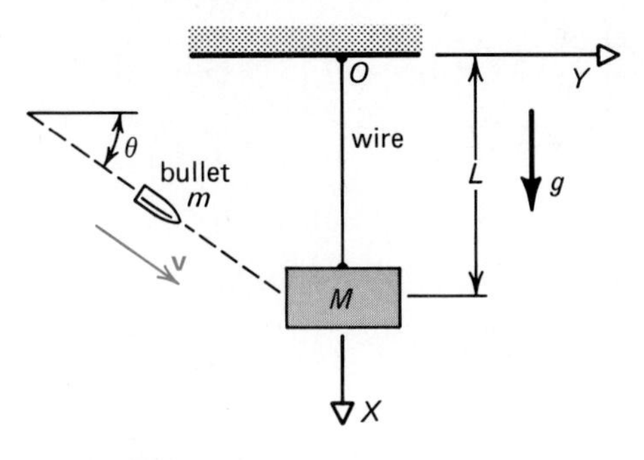

Figure P3.86

3.87 The angular velocity of the string in Problem 3.82 is changed in T seconds by applying a force of magnitude F to the mass. If this force is in the same direction as the linear velocity of the mass m, what angular impulse is imparted? Assume that $F = ae^{-bt}$.

3.88 A ball rolls along a horizontal surface and strikes a curb as shown in Fig. P3.88. The resultant force **F** due to the impact and the associated friction force μF are also shown. Considering only these impulsive forces, find the linear impulse and angular impulse for the time duration of **F** if $\mu = 0.1$. The radius of the ball is 2 in.

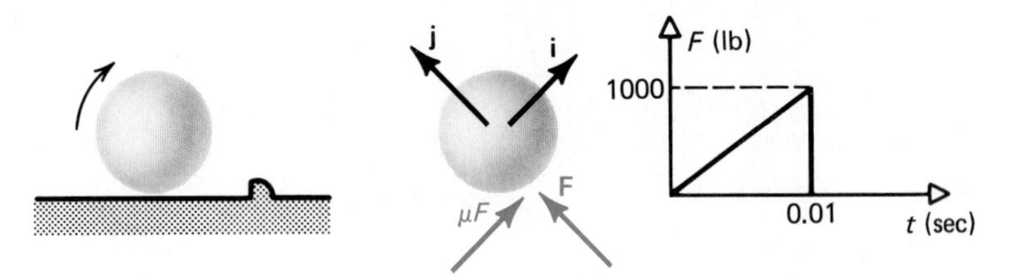

Figure P3.88

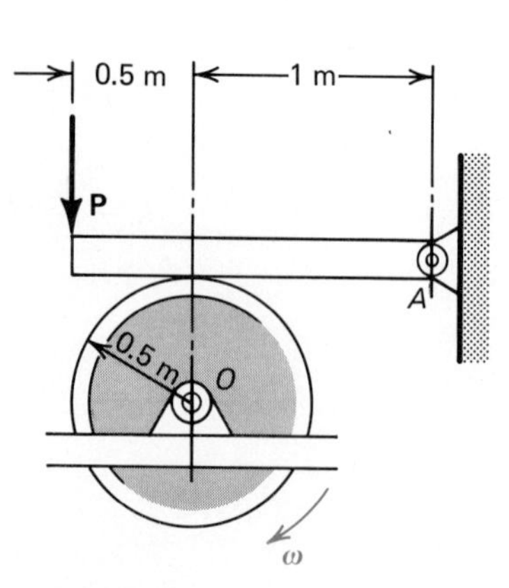

Figure P3.89

3.89 A brake is applied to a rotating drum as shown in Fig. P3.89. If $P =$ 50 N is applied for 10 s and the kinetic coefficient of friction between the brake and the drum is 0.3, find the angular impulse about O.

3.90 A particle of mass m travels along a curved path such that $x = 2 \sin \pi t$, $y = 3 \cos \pi t$, $z = 4t$, where t is in seconds and πt in radians. Find the angular momentum of the particle about the origin when $t = 2$ s.

3.91 A 10-kg particle travels along a parabolic path $y = 2x^2$. Find the linear momentum and angular momentum about the origin when $x = 3$ m and $\dot{x} = 4$ m/s.

3.92 A particle of mass m leaves an impeller at a velocity of

$$\mathbf{V} = V_1(t)\mathbf{i} + V_2(t)\mathbf{j} \qquad \text{located at } \mathbf{r} = r\mathbf{i}$$

What is (a) the angular momentum at the instant the particle leaves the impeller, and (b) the change of this angular momentum with respect to time?

3.93 In Example 3.16, obtain the time at which the angular velocity of the body becomes zero due to the applied moment **M**.

3.94 A particle is subjected to a force given by

$$\mathbf{F} = (e^t\mathbf{i} + t\mathbf{j}) \text{ lb}$$

for 2 sec. What is the linear impulse applied to the particle?

3.95 Compute the angular impulse applied to a body if the moment-time history is given by Fig. P3.95.

3.96 A 200-kg bucket is lifted up the inclined plane by the force $P = 1000(1 + t)$, in newtons, as shown in Fig. P3.96. Determine the velocity of the bucket when $t = 5$ s. Neglect friction losses and assume that the bucket is initially at rest.

3.97 A 400-kg rocket is subjected to the thrust shown in Fig. P3.97. Calculate the velocity of the rocket when $t = 2.5$ s.

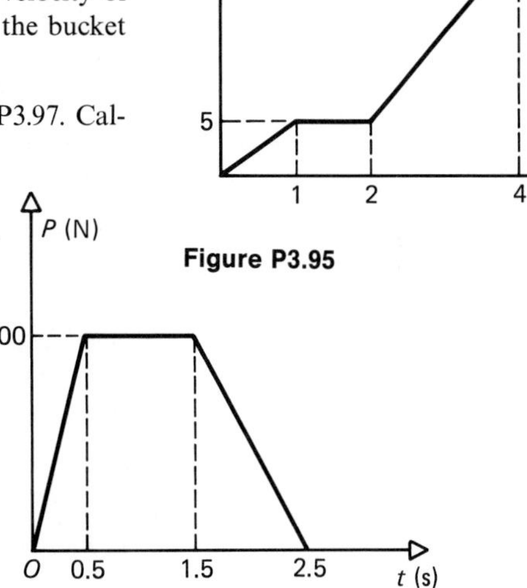

Figure P3.95

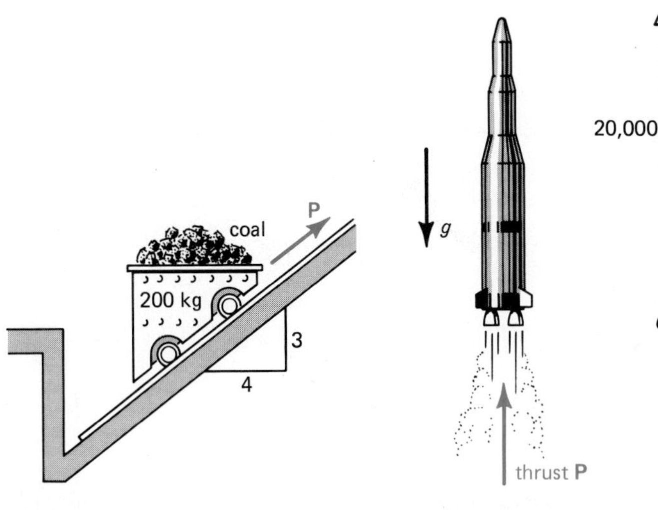

Figure P3.96 **Figure P3.97**

3.98 A 6-oz hockey puck is subjected to a 2-lb force. Find the velocity of the puck after 1 sec if the coefficient of friction between the puck and the ice is 0.02 and the puck is initially at rest.

3.99 The 5-kg mass is released from rest at position A as shown in Fig. P3.99. After traveling 5 m down the smooth inclined plane, a force of magnitude $P = 2t$, in newtons, is applied to the body, where t is measured in seconds. How much time elapses, after P is applied, for the body to come to rest?

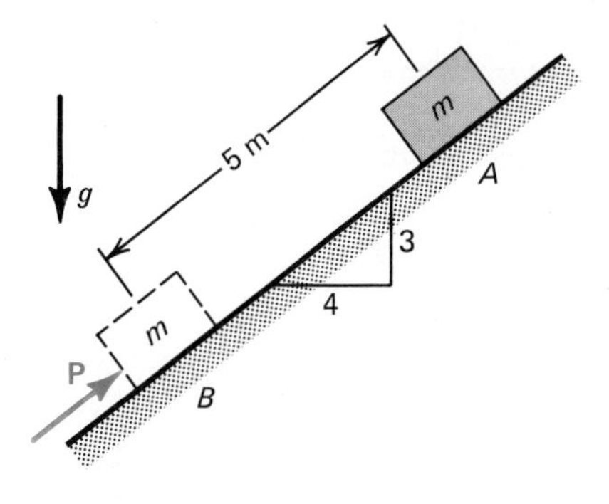

Figure P3.99

3.100 An airplane of 9000 kg flies through a sudden wind gust and the pilot notices that the velocity has decreased by 16 km/h during a 5-s interval. It is suspected that the net force causing the impulse is proportional to the velocity where the proportionality constant is 200 N·s/m. Find the distance through which this force is experienced.

3.101 In order to study the upper atmosphere, a 50-lb instrument package is to be launched by means of a balloon. The balloon is inflated until an upward buoyant force of 60 lb is achieved. At this point the balloon is released. Assuming constant gravity, obtain the velocity after 4 sec. Neglect air friction.

3.102 An instrument package of mass m kilograms is launched by a balloon that provides an upward buoyant force of F newtons. Find the maximum potential energy of the package if the variation of gravity is considered.

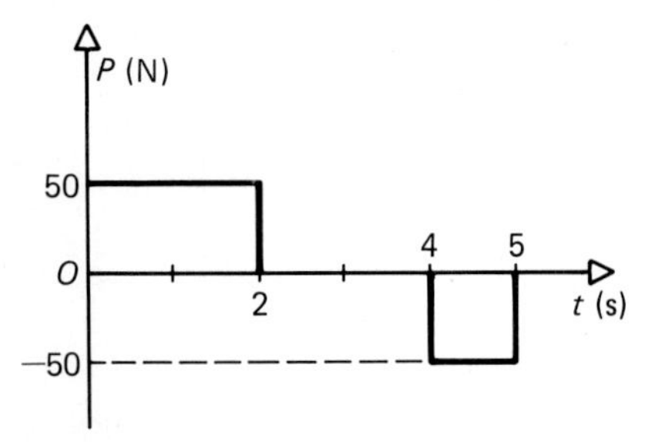

Figure P3.103

3.103 A 1-kg body subject to the force **P** shown in Fig. P3.103, moves along a smooth horizontal plane. If the body starts from rest, calculate the velocity of the body when $t = 1$, 3, and 5 s.

3.104 In the flyball governor of Problem 3.29, obtain the magnitude of the linear momentum of one ball and the magnitude of the angular momentum of one ball about the bearing V.

3.105 The 2-kg particle in Fig. P3.105 is attached to a massless rigid rod 0.1 m long. A constant force of 2 N acts perpendicular to the rod for 2 s. What is the magnitude of the change in the angular momentum of the mass relative to the origin if the mass was initially at rest and the rod turns in the smooth horizontal plane?

3.106 The particle in Fig. P3.106 is released from rest at point A and travels around the smooth looped path. What is the linear momentum of the particle at points B, C, and D, respectively?

3.107 A bead of mass m slides down a frictionless wire in the XY plane. Find the angular momentum of the bead relative to the origin O for the instant shown in Fig. P3.107. Assume the bead starts from rest at point A.

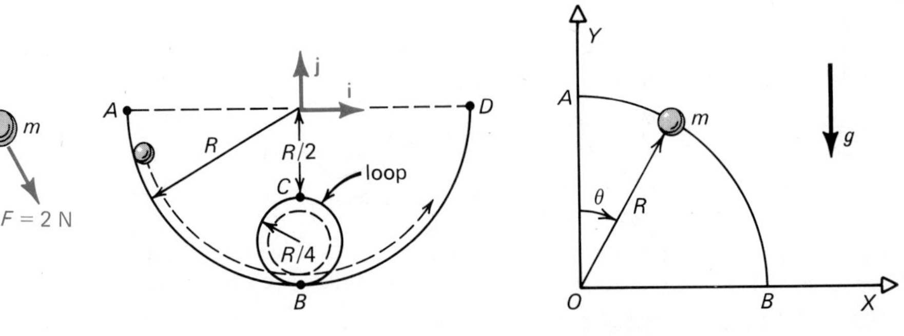

Figure P3.105 **Figure P3.106** **Figure P3.107**

3.108 A particle of mass m and velocity **v** is traveling up the inside wall of the right circular cone shown in Fig. P3.108. Point A is fixed on a vertical axis through the center of the cone. Find the angular momentum of the particle with respect to point A.

3.109 The simple pendulum rotating at a rate of $\dot{\theta}$ in Fig. P3.109 has a torsional spring that provides a resisting moment that is proportional to the angular displacement θ. The spring modulus is k_t. (a) What is the angular momentum of the mass about point O? (b) What is the net moment **M** about

O? (c) Substitute the results of parts (a) and (b) into Eq. (3.34). (d) Show that the result in part (c) could also be obtained using Eq. (3.1).

3.110 A particle of mass m slides outward along a rod at a rate of v relative to the rod as shown in Fig. P3.110. The rod is mounted on a table that is rotating at a uniform angular rate ω. (a) What is $\mathbf{H}$ about the point O? (b) Find $\dot{\mathbf{H}}$ about the point O.

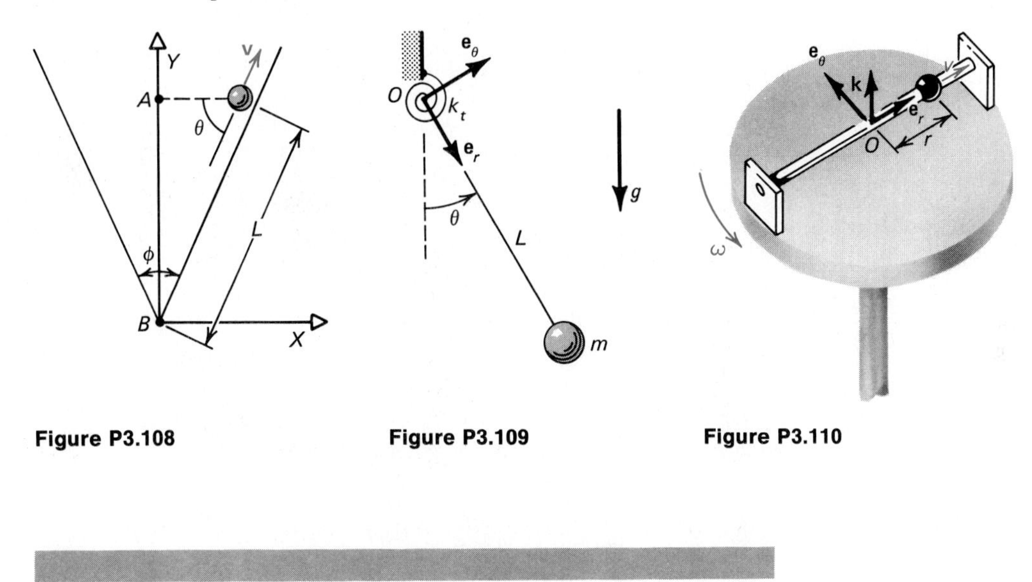

Figure P3.108 **Figure P3.109** **Figure P3.110**

3.4 Summary

A. The important *ideas*:

1. Newton's second law provides the basic relationship between the net force applied to a particle and the acceleration that force produces. This law is basically a differential equation that may be expressed in various forms.
2. The work-energy relationship is derived by integrating Newton's law spatially, i.e., with respect to position.
3. The elastic spring force as well as the gravitational force are examples of conservative forces. The work done by these forces is independent of the path of the body subjected to these forces.
4. When Eq. (3.26) is applied, W_{NC} represents the work done by the nonconservative forces acting on the particle. The work done by the conservative forces is accounted for in the ΔV term.
5. When Newton's law is integrated in the time domain we obtain the impulse-momentum relationship for a particle.

B. The important *equations*:

Newton's law: $$\mathbf{F} = m\mathbf{a} \tag{3.1}$$

$$\mathbf{F} = m\frac{d\mathbf{v}}{dt} \tag{3.2}$$

Work: $$W = \int_A^B \mathbf{F} \cdot d\mathbf{r} \tag{3.7}$$

Potential Energy:

Spring: $$V = \tfrac{1}{2}kx^2 \tag{3.10}$$

Gravity: $$V = -mgR^2/(R + h) \tag{3.16}$$

Gravity: $$V = mgh \qquad \text{(close to earth)} \tag{3.17}$$

Kinetic energy: $$T = \tfrac{1}{2}mv^2 \tag{3.24}$$

Conservation of energy:

$$T_A + V_A = T_B + V_B \tag{3.25}$$

Work-energy: $$W_{\mathrm{NC}} = \Delta T + \Delta V \tag{3.26}$$

Linear momentum: $$\mathbf{p} = m\mathbf{v} \tag{3.29}$$

Angular momentum: $$\mathbf{H} = \mathbf{r} \times m\mathbf{v} \tag{3.33}$$

Linear impulse: $$\mathscr{I} = \int_{t_1}^{t_2} \mathbf{F}\, dt \tag{3.35}$$

Angular impulse: $$\mathscr{M} = \int_{t_1}^{t_2} \mathbf{M}\, dt \tag{3.37}$$

Impulse-momentum:

Linear: $$\mathscr{I} = \Delta\mathbf{p} \tag{3.36}$$

Angular: $$\mathscr{M} = \Delta\mathbf{H} \tag{3.38}$$

4 Central Force Motion

When the force between two particles is directed along an imaginary line that connects the two particles, it is called a **central force.** Such a force on particle B due to particle A as shown in Fig. 4.1 is expressed as

$$\mathbf{F} = \frac{K}{r^2}\,\mathbf{e}_r \tag{4.1}$$

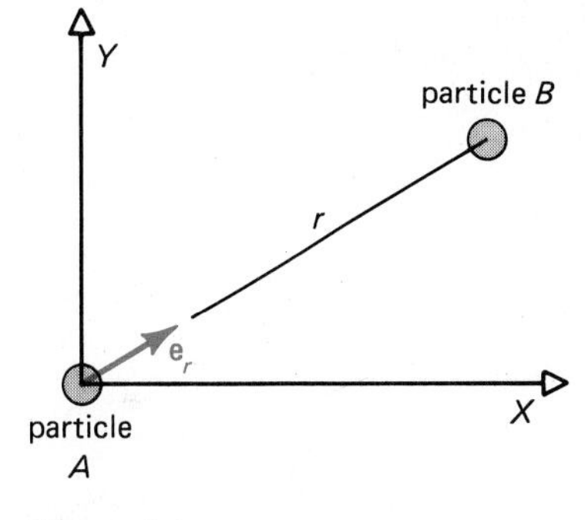

Figure 4.1

This is sometimes called the inverse-square law. For gravitational forces between the two particles, of mass m_A and m_B respectively, the constant K is given by

$$K = -Gm_Am_B \tag{4.2}$$

where

$$G = 6.675 \times 10^{-11}\,\frac{\text{N}\cdot\text{m}^2}{\text{kg}^2} \qquad \text{or} \qquad 3.44 \times 10^{-8}\,\frac{\text{lb}\cdot\text{ft}^2}{\text{slug}^2}$$

The negative sign in Eq. (4.2) indicates that the gravitational force is always attractive between the two particles. If the two particles are considered as point charges q_A and q_B, respectively, then

$$K = q_Aq_B$$

indicating that the electrostatic force between particles is attractive if q_A and q_B have opposite signs and repulsive if they have the same sign.

We are concerned here with the gravitational force and the motion that it causes. The motion described by satellites and planets is due to a central force, i.e., a force that is gravitational in nature. The three laws that govern such motion were first deduced by Kepler who based his work on data from astronomical observations of the planets. Kepler's first law states:

The orbit of each planet is an ellipse with the sun at one focus.

Figure 4.2 shows the sun at the focus of the ellipse, the position vector **r** from the sun to the planet, and the angle θ measured from the major axis of the ellipse.

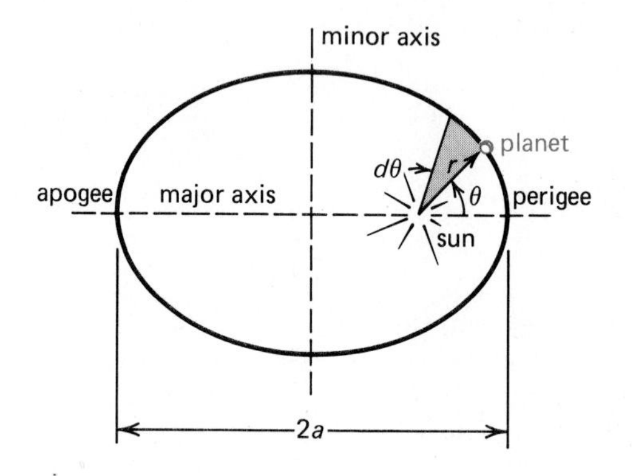

Figure 4.2

Kepler's second law states:

The radius vector drawn from the sun to a planet sweeps over equal areas in equal times.

Thus, during the time interval dt, the shaded area in Fig. 4.2 equals $dA = (\frac{1}{2}r)(r\,d\theta)$. The rate at which the area is swept by the radius vector is $\dot{A} = \frac{1}{2}r^2\dot{\theta}$. We shall show that the term $r^2\dot{\theta}$ is a constant.

Kepler's third law states:

The squares of the periods of the planets are proportional to the cubes of the semimajor axes of their respective orbits.

If τ is the period and $2a$ is the length of the major axis as shown in Fig. 4.2, then

$$\tau = \xi a^{3/2}$$

where ξ is a constant. If τ is measured in seconds and a in meters then ξ is equal to 3.147×10^{-7} s/m$^{3/2}$, and if τ is measured in hours and a in miles, then ξ is equal to 5.65×10^{-6} hr/mi$^{3/2}$.

In this chapter we shall obtain the mathematical representation of Kepler's laws. In deriving the equations we shall assume that the masses of the bodies in orbit are concentrated at a point.

4.1 Force Between Two Particles

Consider an isolated system consisting of two masses as shown in Fig. 4.3. The center of mass is defined such that

$$m_A r_A = m_B r_B \tag{4.3}$$

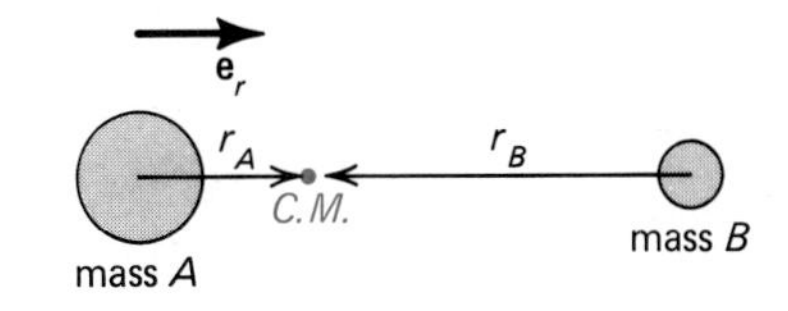

Figure 4.3

where r_A and r_B are the distances from the center of mass to each mass. Also, from Newton's gravitational law, the central force can be obtained. Using Eqs. (4.1) and (4.2), we have the force acting on m_B as follows:

$$\mathbf{F} = -\frac{Gm_A m_B}{(r_A + r_B)^2}\,\mathbf{e}_r$$

and if we substitute from Eq. (4.3),

$$\mathbf{F} = -\frac{Gm_A m_B}{r_B^2(1 + m_B/m_A)^2}\,\mathbf{e}_r \tag{4.4}$$

Now, if $m_A \gg m_B$, as is the case between the earth and a satellite, Eq. (4.4) reduces to

$$\mathbf{F} = -\frac{Gm_A m_B}{r_B^2}\,\mathbf{e}_r$$

Also, for this special case, r_A is insignificant so that

$$r_A + r_B \approx r_B \approx r \tag{4.5}$$

Therefore

$$\mathbf{F} = \frac{K}{r^2}\,\mathbf{e}_r = -\frac{Gm_A m_B}{r^2}\,\mathbf{e}_r \tag{4.6}$$

The distance r is the measured distance between a particle and the center of the earth, which is very close to the center of mass of the system under investigation. These assumptions reduce a two-body problem to a one-body problem which has to do with finding the motion of a particle attracted by a central force toward a fixed center.

The potential energy of a mass m_B in a gravitational field can be obtained by using Eqs. (3.16) and (1.23). This leads to the following:

$$V = -\frac{R^2 m_B g}{(r_A + r_B)} \qquad \text{and} \qquad gR^2 = Gm_A$$

from which we find that

$$V = -\frac{Gm_A m_B}{(r_A + r_B)} = \frac{K}{r_A + r_B}$$

If we use the assumption in Eq. (4.5), and require that r be greater than or equal to the earth's radius, we see that

$$V \approx \frac{K}{r} \tag{4.7}$$

Therefore, the potential energy varies inversely as the first power of the distance r. The variations of the magnitudes of V and F are shown in Fig. 4.4.

The acceleration of the particle B in the central force field due to the earth is obtained using Newton's second law, i.e.,

$$\mathbf{F} = m_B\mathbf{a}$$

If we substitute for $\mathbf{F}$ in Eq. (4.6), the equation for acceleration becomes

$$-\frac{Gm_A}{r^2} m_B\mathbf{e}_r = m_B\mathbf{a}$$

Let $m_A = M$ = mass of the earth.

$$\mathbf{a} = -\frac{GM}{r^2}\mathbf{e}_r \tag{4.8}$$

where the mass of the earth is approximately $M = 59.7 \times 10^{23}$ kg or 4.09×10^{23} slugs. The term GM/r^2 represents the magnitude of the acceleration of a particle in the earth's gravitational field.

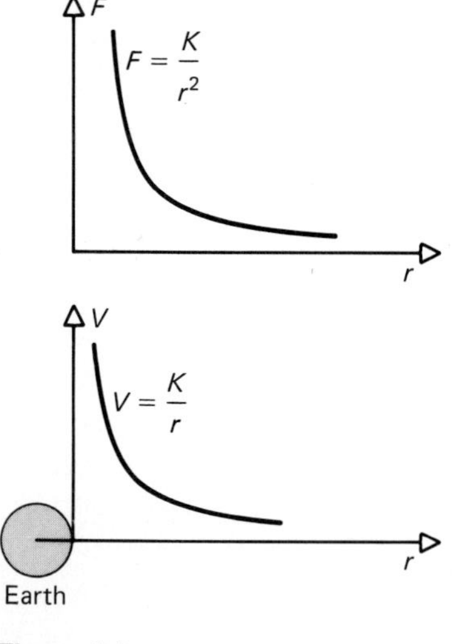

Figure 4.4

Example 4.1

A 200-kg satellite is in orbit 2000 km above the earth's surface. What is the magnitude of the central force acting on the satellite? What does the satellite weigh on the earth's surface?

The magnitude of the central force is obtained from Eq. (4.6) as

$$F = \frac{Gm_Am_B}{r^2}$$

where $G = 6.675 \times 10^{-11}\ \text{N}\cdot\text{m}^2/\text{kg}^2$

$m_A = 59.7 \times 10^{23}$ kg

$m_B = 200$ kg

The distance between the earth's center and the satellite equals the earth's radius of 6378 km plus the 2000-km elevation. That is,

$$r = 6378 + 2000 = 8378 \text{ km} = 8.378 \times 10^6 \text{ m}$$

The magnitude of the central force is

$$F = \frac{6.675 \times 10^{-11} \times 59.7 \times 10^{23} \times 200}{(8.378 \times 10^6)^2}$$

$$= 1135 \text{ N}$$ *Answer*

On the earth's surface, $r = 6.378 \times 10^6$ m, so that

$$F = \text{weight} = \frac{6.675 \times 10^{-11} \times 59.7 \times 10^{23} \times 200}{(6.378 \times 10^6)^2}$$

$$= 1959 \text{ N} \qquad \textit{Answer}$$

We observe that this latter force agrees with our earlier calculation for finding the weight of an object, i.e.,

$$w = mg = 200 \times 9.81 = 1962 \text{ N}$$

Example 4.2

Calculate the potential energy of the satellite in Example 4.1.

From Eq. (4.7),

$$V = \frac{K}{r} = -\frac{Gm_A m_B}{r}$$

$$= -\frac{6.675 \times 10^{-11} \times 59.7 \times 10^{23} \times 200}{8.378 \times 10^6}$$

$$= -9.513 \times 10^9 \text{ N}\cdot\text{m} \qquad \textit{Answer}$$

As an alternate approach, we can use Eq. (3.16) directly.

$$V = -mg\frac{R^2}{R+h}$$

$$= -200(9.81)\frac{(6.378 \times 10^6)^2}{8.378 \times 10^6}$$

$$= -9.526 \times 10^9 \text{ N}\cdot\text{m}$$

The slight difference is due to the approximate constants.

PROBLEMS

4.1 It can be shown that the magnitude of a conservative force is related to the potential energy by $F_r = -\partial V/\partial r$. Show that this relationship is satisfied for the force between two particles using Eq. (4.7).

4.2 Two isolated particles are 10 ft apart. If $m_A = 20$ slugs and $m_B = 10$ slugs, find the magnitude of the central force acting on m_B.

4.3 What is the potential energy of particle B relative to m_A in Problem 4.2?

4.4 From Eq. (4.8) calculate the acceleration due to gravity at sea level in m/s^2 and ft/sec^2.

4.5 What is the magnitude of the central force between two masses, 100 kg and 1 kg at a distance of 6 m?

4.6 At what altitude does the acceleration due to gravity become half of that at sea level?

4.7 If r is measured in earth radii, compute the constant GM in Eq. (4.8) so that the acceleration is given (a) in m/s^2, and (b) in km/h^2.

4.2 Motion in a Gravitational Field

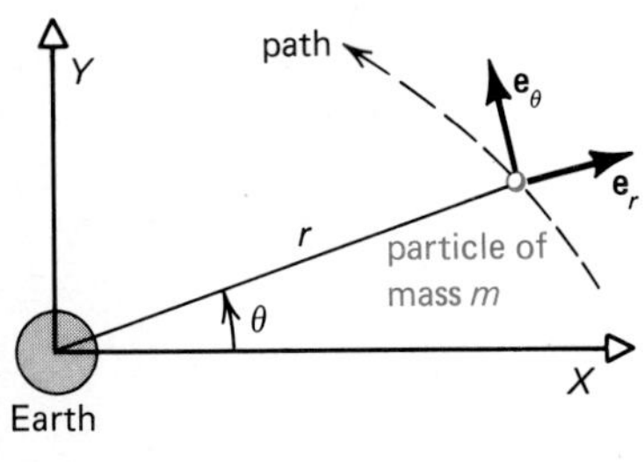

Figure 4.5

The acceleration of a particle of mass m moving in a gravitational field as seen in Fig. 4.5 can now be written as

$$\mathbf{a} = \ddot{\mathbf{r}} = -\frac{\mu}{r^2}\,\mathbf{e}_r$$

where $\mu = GM$. The components of $\mathbf{a}$ in polar coordinates were obtained in Chap. 2 and are given by Eq. (2.38),

$$\ddot{\mathbf{r}} = \mathbf{a} = (\ddot{r} - r\dot{\theta}^2)\mathbf{e}_r + (2\dot{r}\dot{\theta} + r\ddot{\theta})\mathbf{e}_\theta \tag{4.9}$$

Equating this to Eq. (4.8), we obtain

$$(\ddot{r} - r\dot{\theta}^2)\mathbf{e}_r + (2\dot{r}\dot{\theta} + r\ddot{\theta})\mathbf{e}_\theta = -\frac{\mu}{r^2}\,\mathbf{e}_r \tag{4.10}$$

Since this is a vector equation, we can write two equivalent scalar equations to obtain

$$\ddot{r} - r\dot{\theta}^2 = -\frac{\mu}{r^2} \tag{4.11}$$

$$2\dot{r}\dot{\theta} + r\ddot{\theta} = 0 \tag{4.12}$$

The second equation is equated to zero since there is no external force in the $\mathbf{e}_\theta$ direction as seen in Eq. (4.10). Rearranging Eq. (4.12), we obtain

$$r(2\dot{r}\dot{\theta} + r\ddot{\theta}) = \frac{d}{dt}(r^2\dot{\theta}) = 0$$

This expression can be integrated so that

$$r^2\dot{\theta} = h \qquad \text{a constant} \tag{4.13}$$

Therefore, as the particle moves about in a gravitational field, $r^2\dot{\theta}$ must remain constant. The fact that the radius vector sweeps out equal areas in equal times is Kepler's second law.

Equation (4.13) can be used to make another observation. The term $r^2\dot{\theta}$ can be viewed as the angular momentum of a particle of unit mass. Therefore, Eq. (4.13) states that the angular momentum of a particle in a gravitational orbit is constant. Since there is no force in the θ direction, this conclusion is not too surprising.

We now return to Eq. (4.11). This can be solved once it is put into a simpler form by redefining the variable r such that

$$w = \frac{1}{r} \tag{4.14}$$

and

$$r^2\dot{\theta} = \frac{\dot{\theta}}{w^2} = h$$

The derivatives of r then become,

$$\frac{dr}{dt} = \frac{dr}{dw}\frac{dw}{d\theta}\frac{d\theta}{dt} = -\frac{1}{w^2}\frac{dw}{d\theta}\dot{\theta} = -\frac{dw}{d\theta}h$$

$$\frac{d^2r}{dt^2} = \frac{d\dot{r}}{dt} = \frac{d\dot{r}}{d\theta}\frac{d\theta}{dt} = -\frac{d^2w}{d\theta^2}h\dot{\theta} = -\frac{d^2w}{d\theta^2}w^2h^2$$

Substituting these terms into Eq. (4.11), we obtain

$$-\frac{d^2w}{d\theta^2}w^2h^2 - w^3h^2 = -\mu w^2$$

or

$$\frac{d^2w}{d\theta^2} + w = \frac{\mu}{h^2} \tag{4.15}$$

This equation is a second-order linear differential equation with constant coefficients. Its solution is obtained in two parts. The first part is the homogeneous solution, which is $C\cos(\theta + \theta_0)$. The second part is the particular solution, which is μ/h^2. The complete solution becomes

$$w = \frac{1}{r} = C\cos\theta + \frac{\mu}{h^2} \tag{4.16}$$

where the integration constant θ_0 has been arbitrarily set to zero since we can choose any convenient reference for measuring θ. As we shall discover, Eq. (4.16) is the equation of a conic section, i.e., an ellipse, a circle, a parabola, or a hyperbola. The similarity between the equation governing the motion of a particle in a gravitational field and that of a conic section constitutes Kepler's first law.

From analytic geometry, the equation of a general conic is

$$\frac{1}{r} = \frac{1}{B\epsilon}(1 + \epsilon\cos\theta) \tag{4.17}$$

where θ is the angle between the radius vector and a fixed reference, ϵ is called the eccentricity, and B simply determines the relative scale of the figure. Comparing Eq. (4.16) with Eq. (4.17) we observe that

$$\frac{1}{\epsilon B} = \frac{\mu}{h^2} \qquad \frac{1}{B} = \frac{\epsilon\mu}{h^2} = C$$

Substitute these relationships into Eq. (4.16),

$$\frac{1}{r} = \frac{\mu}{h^2}(1 + \epsilon \cos \theta)$$

or

$$r = \frac{h^2/\mu}{1 + \epsilon \cos \theta} \tag{4.18}$$

The type of conic section depends upon the eccentricity as follows:

Circle	$\epsilon = 0$
Ellipse	$0 < \epsilon < 1$
Parabola	$\epsilon = 1$
Hyperbola	$\epsilon > 1$

Example 4.3

A satellite travels on an elliptic orbit about the earth such that $\epsilon = 0.1$. What is the distance between the satellite and the earth's center when (a) $\theta = 0^\circ$, and (b) $\theta = 90^\circ$ if $h^2/\mu = 1.52 \times 10^7$ m?

Referring to Eq. (4.18),

$$r = \frac{h^2/\mu}{1 + \epsilon \cos \theta}$$

(a)

$$r = \frac{1.52 \times 10^7}{1 + 0.1(1)} = 1.38 \times 10^7 \text{ m}$$ *Answer*

(b)

$$r = \frac{1.52 \times 10^7}{1 + 0.1(0)} = 1.52 \times 10^7 \text{ m}$$ *Answer*

PROBLEMS

4.8 The vector **r** is written as a complex number $re^{j\theta}$. Derive Eq. (4.10) by differentiating this expression for **r** twice.

4.9 Substitute Eq. (4.16) into Eq. (4.15) to show that it is the correct solution.

4.10 Using Eq. (4.18), plot r against θ for $\epsilon = 0, 0.1, 0.5$. Assume that $h^2/\mu = 1$.

4.11 Obtain the distance between a satellite and the earth's center if the satellite follows an elliptic path such that $\epsilon = 0.2$, $h^2/\mu = 1.52 \times 10^7$ m, and $\theta = 45^\circ$.

4.12 Review the geometric properties of an ellipse and sketch the ellipse showing the semi-latus rectum, directrix, semimajor and semiminor axes, and their relationships to the eccentricity.

4.13 For the ellipse shown in Fig. P4.13, derive a relationship between θ and ψ as a function of ϵ, a and r.

4.14 Derive the expression for the area of an ellipse having a semimajor axis a and semiminor axis b.

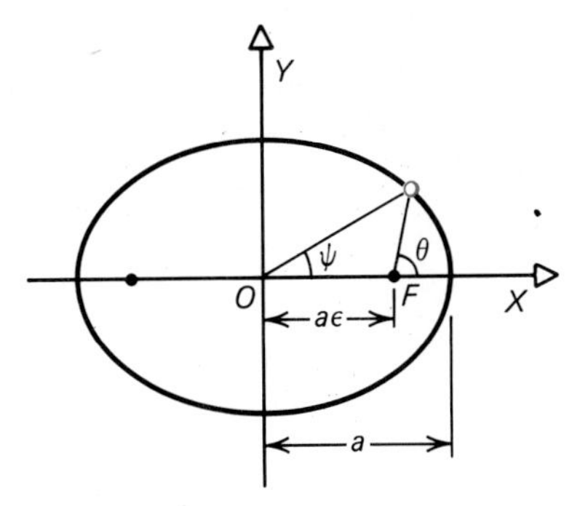

Figure P4.13

4.3 Elliptical and Circular Orbits

Since artificial satellites in orbit around the earth have orbits that are either elliptical or circular, we are concerned here with $0 \leq \epsilon < 1$. Orbits for different eccentricities are shown in Fig. 4.6(a). An elliptical orbit and its properties are shown in Fig. 4.6(b) where the area is πab. If x and y are measured from the center O, then the equation for an ellipse becomes

$$\frac{x^2}{a^2} + \frac{y^2}{b^2} = 1 \tag{4.19}$$

where a is the semimajor axis and b is the semiminor axis. The relationship between a and b with the eccentricity ϵ is given by

$$b = a\sqrt{1 - \epsilon^2}$$

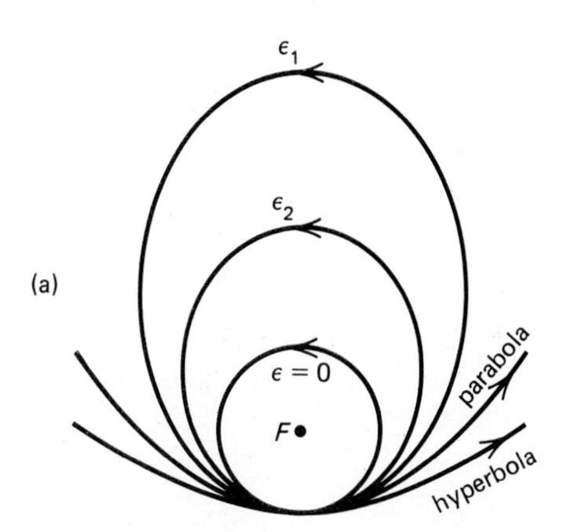

Orbits for different eccentricities: $\epsilon = 0$, circle; $0 < \epsilon < 1$, ellipse; $\epsilon = 1$, parabola; $\epsilon > 1$, hyperbola.

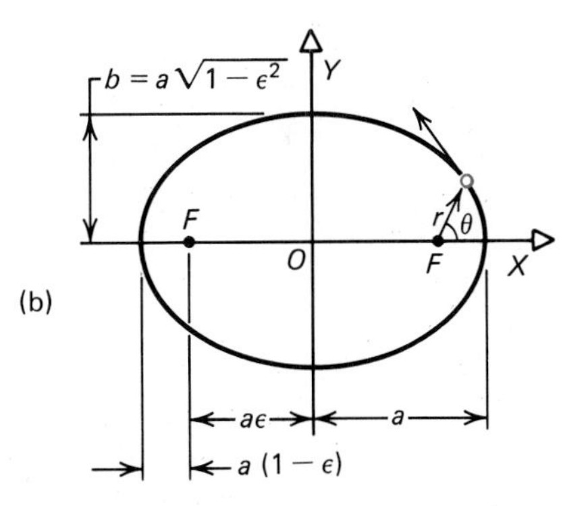

Ellipse with semimajor axis a and semiminor axis $b = a\sqrt{1 - \epsilon^2}$. The enclosed area is πab.

Figure 4.6

From Fig. 4.6(b) we can write the relationships,

$$x = a\epsilon + r\cos\theta \qquad y = r\sin\theta$$

If we substitute these into Eq. (4.19) and solve the resulting quadratic equation, we obtain the following positive root,

$$r = \frac{(1-\epsilon^2)a}{1+\epsilon\cos\theta} \tag{4.20}$$

Comparing this with Eq. (4.18), we obtain

$$\frac{h^2}{\mu} = (1-\epsilon^2)a$$

or

$$h = \sqrt{\mu a(1-\epsilon^2)} \tag{4.21}$$

This is a constant of the circular orbit or elliptical orbit of a particle.

Energy in Orbit

Consider now the equations that will describe the motion of particles in an elliptical orbit. The total energy of the particle as it moves in orbit is our first consideration. Since the gravitational field is conservative and we are not considering any dissipative forces, the total energy is conserved. By neglecting any dissipative forces, we are not only simplifying our derivation, but are obtaining answers that are excellent approximations for most cases. The conserved energy is, therefore, a natural constant of the system. The velocity of the particle in polar coordinates, from Eq. (2.37), is

$$\mathbf{v} = \dot{r}\mathbf{e}_r + r\dot{\theta}\mathbf{e}_\theta \tag{4.22}$$

and the kinetic energy per unit mass is

$$T = \tfrac{1}{2}\mathbf{v}\cdot\mathbf{v} = \tfrac{1}{2}(\dot{r}^2 + r^2\dot{\theta}^2) \tag{4.23}$$

The total energy per unit mass is

$$E = \tfrac{1}{2}\dot{r}^2 + \tfrac{1}{2}r^2\dot{\theta}^2 + PE \tag{4.24}$$

where PE is the potential energy per unit mass. Since the energy is conserved, we may choose to measure the energy at the maximum distance (apogee) or the minimum distance (perigee) from the origin. At perigee, $\dot{r} = 0$, $r = r_p$, and the energy becomes

$$E = \left(\frac{1}{2}r^2\dot{\theta}^2 + PE\right)_{r=r_p}$$

$$= \left(\frac{1}{2}\frac{h^2}{r^2} + PE\right)_{r=r_p}$$

Since the potential energy is $-\mu/r$, if we let $r = r_p$, we obtain

$$E = \frac{1}{2}\frac{h^2}{r_p^2} - \frac{\mu}{r_p} \tag{4.25}$$

From Eq. (4.20) we can find that at $\theta = 0$

$$r = r_p = a(1-\epsilon) \tag{4.26}$$

Substitution of Eqs. (4.21) and (4.26) into (4.25) yields

$$E = -\frac{1}{2}\frac{\mu}{a} \tag{4.27}$$

The energy constant is now used in the general equation.

Velocity in Orbit

If we denote the magnitude of the velocity as v, then

$$E = -\frac{1}{2}\frac{\mu}{a} = \frac{1}{2}v^2 - \frac{\mu}{r}$$

or

$$v^2 = \mu\left(\frac{2}{r} - \frac{1}{a}\right) \tag{4.28}$$

which may be used to compute the magnitude of the velocity anywhere in the orbit. Note that this is independent of direction.

The velocity at perigee is

$$v_p = \sqrt{\mu(2/r_p - 1/a)} = \sqrt{[(2a - r_p)/ar_p]\mu}$$

and since it can be shown (see Example 4.4) that

$$r_p + r_a = 2a$$

the velocity at perigee becomes

$$v_p = \sqrt{\frac{\mu}{a}\frac{r_a}{r_p}} \tag{4.29}$$

By similar arguments the velocity at apogee is

$$v_a = \sqrt{\frac{\mu}{a}\frac{r_p}{r_a}} \tag{4.30}$$

Period of Orbit

It is important to know the period, i.e., the time it takes for a particle to complete one full revolution, or an orbit. Recall that the area of an ellipse is πab. Also, we know from Kepler's second law that $\dot{A} = h/2$, so that

$$\int_0^\tau \frac{dA}{dt}\,dt = \int_0^\tau \frac{h}{2}\,dt \qquad \text{and} \qquad A = \pi ab$$

Since h is constant, we obtain $h\tau/2 = \pi ab$, or

$$\tau = \frac{2\pi ab}{h} \tag{4.31}$$

We have already found that

$$h = \sqrt{\mu a(1 - \epsilon^2)} \qquad \text{and} \qquad b = a\sqrt{1 - \epsilon^2}$$

and if we substitute these into Eq. (4.31), we find the period to be

$$\tau = \xi a^{3/2} \tag{4.32}$$

where $\xi = 2\pi/\sqrt{\mu}$. The period of an orbit is a function of the three-halves power of the semimajor axis. Equation (4.32) is the essential feature of Kepler's third law.

Example 4.4

Derive an equation for eccentricity in terms of the perigee and apogee distances.

From Eq. (4.20) we obtain

$$\text{at perigee} \quad \theta = 0 \quad \text{and} \quad r = r_p = a(1 - \epsilon)$$

$$\text{at apogee} \quad \theta = \pi \quad \text{and} \quad r = r_a = a(1 + \epsilon)$$

Solving these equations simultaneously gives us

$$\epsilon = \frac{r_a - r_p}{r_a + r_p} \qquad \textit{Answer}$$

(Note that $r_p + r_a = 2a$.)

Example 4.5

An object is fired from Cape Canaveral as shown in Fig. 4.7. When all the fuel burns out, the projectile velocity v_0 is perpendicular to the radius vector. (Such a launch is called a perigee launch.) What eccentricity does the orbit achieve?

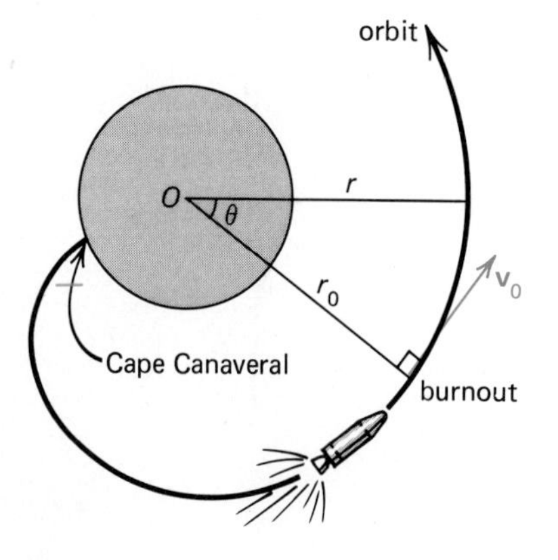

Figure 4.7

If we reference θ from the point of burnout, then the initial conditions become

$$r = r_0 \qquad v = v_0 \qquad \theta = 0$$

From Eqs. (4.20) and (4.28) we have

$$r_0 = \frac{(1 - \epsilon^2)a}{1 + \epsilon} = (1 - \epsilon)a \qquad \text{and} \qquad v_0^2 = \mu\left(\frac{2}{r_0} - \frac{1}{a}\right)$$

Rearranging the latter equation, we obtain

$$a = \frac{\mu r_0}{2\mu - v_0^2 r_0}$$

When we substitute this into the equation for r_0, we find that

$$\epsilon = \frac{v_0^2 r_0}{\mu} - 1 \qquad \textit{Answer}$$

This answer demonstrates that the eccentricity attained is independent of the mass of the body in orbit. In this case, it depends only upon the height achieved and the velocity at burnout.

Example 4.6

What must the projectile velocity be in mi/hr at burnout if the orbit is to be circular at a distance of 1040 mi above the earth's surface?

Since the orbit is circular, $\epsilon = 0$. Therefore, from the answer in the previous example we have,

$$v_0^2 r_0 = \mu \qquad v_0 = \sqrt{\mu/r_0}$$

The radius of the earth is 3960 mi, so that

$$r_0 = 3960 + 1040 = 5000 \text{ mi}$$

$$\mu = GM = 1.239 \times 10^{12} \text{ mi}^3/\text{hr}^2$$

Substituting,

$$v_0 = \sqrt{\frac{1.239 \times 10^{12}}{5 \times 10^3}}$$

$$v_0 = 15{,}740 \text{ mi/hr} \qquad \textit{Answer}$$

Example 4.7

The velocity of a rocket at burnout is 25,000 km/h. What perigee height can be achieved if an eccentricity of 0.1 is desired? What is the apogee height? What is the velocity at apogee?

From Example 4.5 we have

$$\epsilon = \frac{v_0^2 r_p}{\mu} - 1$$

This can be solved for the perigee height as follows:

$$r_p = \frac{(1 + \epsilon)\mu}{v_0^2}$$

Now $$v_0 = 25{,}000 \times \frac{1}{3600} \times 1000 = 6944 \text{ m/s}$$

And

$$\mu = GM = 6.675 \times 10^{-11} \times 59.7 \times 10^{23} = 3.984 \times 10^{14}\ \text{m}^3/\text{s}^2$$

Therefore,

$$r_p = \frac{(1 + 0.1)(3.984 \times 10^{14})}{(6944)^2} = 9.089 \times 10^6\ \text{m} \qquad \textit{Answer}$$

The height above the earth's surface is

$$(9.089 - 6.378)10^6 = 2.711 \times 10^6\ \text{m}$$

The apogee is obtained from the equation developed in Example 4.4, namely,

$$\epsilon = \frac{r_a - r_p}{r_a + r_p}$$

Solving for r_a and substituting numerical values, we obtain

$$r_a = r_p\left(\frac{1 + 0.1}{1 - 0.1}\right) = 1.22 r_p$$

$$= 1.109 \times 10^7\ \text{m} \qquad \textit{Answer}$$

Since the angular momentum of the rocket is constant, the velocity at apogee is

$$v_a = \frac{r_p}{r_a} v_p = \frac{9.089 \times 10^6}{1.109 \times 10^7}(25{,}000)$$

$$= 20{,}490\ \text{km/h} \qquad \textit{Answer}$$

Example 4.8

Obtain the period of the satellite discussed in the previous example.

The period is given by

$$\tau = \xi a^{3/2} \qquad \xi = 2\pi/\sqrt{\mu}$$

$$a = \frac{r_p + r_a}{2} = \frac{(9.089 + 11.09)10^6}{2} = 10.09 \times 10^6\ \text{m}$$

$$\xi = \frac{2\pi}{\sqrt{3.984 \times 10^{14}}} = 3.148 \times 10^{-7}\ s/m^{3/2}$$

Therefore,

$$\tau = 3.148 \times 10^{-7}(10.09 \times 10^6)^{3/2}$$

$$= 100.9 \times 10^2\ \text{s}$$

$$= 168\ \text{min} \qquad \textit{Answer}$$

Example 4.9

A satellite traveling around the earth has been observed to have a period of 2 hours. If the orbit is circular, how high is the satellite above the earth's surface?

The period of the satellite is given by

$$\tau = \xi a^{3/2}$$

Now $$\xi = \frac{2\pi}{\sqrt{\mu}} = \frac{2\pi}{\sqrt{1.239 \times 10^{12}}} = 5.65 \times 10^{-6}\ \text{hr/mi}^{3/2}$$

so that

$$a^{3/2} = \frac{\tau}{\xi} = \frac{2}{5.65 \times 10^{-6}}$$

$$a = 5004\ \text{mi}$$

Assuming the radius of the earth is 3960 mi,

$$h = a - R = 5004 - 3960$$

$$= 1044\ \text{mi} \qquad \textit{Answer}$$

Example 4.10

Obtain the velocity at which a particle can permanently escape from the attraction of the earth.

We can interpret this to mean that we want an orbit having a perigee of R and $a = \infty$. Therefore, using Eq. (4.28), we find

$$v^2 = \mu\left(\frac{2}{R} - 0\right) = \frac{2\mu}{R}$$

$$v = v_e = \sqrt{\frac{2\mu}{R}}$$

$$= \sqrt{\frac{2 \times 3.984 \times 10^{14}}{6.378 \times 10^6}} = 11{,}180\ \text{m/s}$$

$$= 40{,}240\ \text{km/h} \qquad \textit{Answer}$$

A particle leaving the surface of the earth at this velocity completely escapes the gravitational influence of the earth. This velocity is called the escape velocity.

PROBLEMS

4.15 Verify Eqs. (4.27), (4.30), and (4.32).

4.16 Derive a formula relating the periods and semimajor axes of planets as they move around the sun.

4.17 A 100-lb satellite is put into orbit with an apogee of 6000 mi and perigee of 4500 mi. (a) What is the eccentricity of the satellite? (b) What is the angular momentum of the satellite relative to the earth?

4.18 Derive the period of the satellite in Problem 4.17. What would be the semimajor axis if the satellite were in a circular orbit but with the same period?

4.19 What is the total energy of a 600-kg satellite in a circular orbit 6000 km above the earth's surface?

4.20 A satellite follows an elliptic orbit about the earth such that $\epsilon = 0.8$. If the semiminor axis $b = 2.7 \times 10^7$ ft, what is the satellite's velocity at apogee?

4.21 Calculate the period of the satellite in Problem 4.20.

4.22 What is the period and the eccentricity of the earth's orbit about the sun? Assume that the apogee is 1.52×10^8 km, the perigee is 1.47×10^8 km, and the sun's mass is 1.98×10^{30} kg.

4.23 Obtain the velocity at perigee and apogee of the earth as it moves around the sun. (Use the data in Problem 4.22.)

4.24 Estimate the distance between the centers of the moon and the earth knowing that the period of the moon's orbit is 28 days. As an approximation you may assume that the orbit is circular.

4.25 The apogee velocity of an earth satellite is one-half the perigee velocity. Obtain r_a and r_p as functions of the perigee velocity. Find the eccentricity ϵ.

4.26 Derive a formula for the eccentricity of an orbit if the burnout velocity of a satellite makes an angle of β_0 with the tangential velocity. The conditions are that $r = r_0$, $\theta = \theta_0$ at the instant shown in Fig. P4.26.

4.27 In the examples considered, the mass of the object put into orbit seems not to enter into our calculations. Why? When is this mass of importance?

4.28 What must the altitude of a satellite be in a circular orbit if it must rotate at the same rate as the earth? (These are the synchronous satellites.) Calculate the burnout velocity and the amount of energy the rocket must deliver if the orbit is to be circular and the satellite weighs 500 lb.

4.29 The perigee launch for a satellite was to have been such that r would be 8047 km and the burnout velocity would be 25,270 km/h. Although the burnout velocity and altitude were obtained, the angle between the velocity vector and radius vector was 100° instead of 90°. What was the eccentricity of the orbit?

4.30 The mean anomaly M is defined as the angular displacement from perigee of a line moving at an average angular rate of $2\pi/\tau$, so that

$$M = \frac{2\pi t}{\tau} = \psi - \epsilon \sin \psi$$

where t is the time measured from the perigee position as shown in Fig.

P4.30. Note that the area swept over by line FP' is constant. Find this area to get the above relationship which is called Kepler's equation.

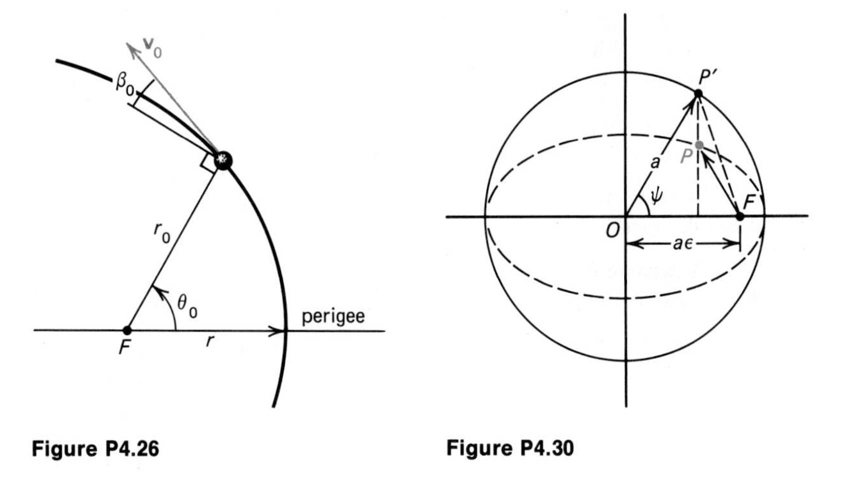

Figure P4.26

Figure P4.30

4.4 Summary

The motion of a particle subjected to a central force such as the gravitational force was described by the equation of a conic section. The particular form of the equation depends upon the eccentricity of the orbit in which the particle moves. Here we investigated the special case when the eccentricity varies between zero and one. Under this condition the particle moves in an elliptical orbit. The equations for the energy, velocity, period, as well as other parameters of the orbit were derived and applied to the motion of artificial satellites and planets.

A. The important *ideas*:

1. The angular momentum of a particle moving in a gravitational field remains constant.
2. The mathematical solution for the motion of a particle is equivalent to the equation of a conic section, i.e., an ellipse, a circle, a parabola, or a hyperbola.
3. Total energy is conserved for a particle in orbit.
4. Kepler's laws were verified mathematically in this chapter.

B. The important *equations*:

$$\mathbf{F} = \frac{K}{r^2}\mathbf{e}_r = -\frac{Gm_A m_B}{r^2}\mathbf{e}_r \qquad (4.6)$$

$$V = \frac{K}{r} \tag{4.7}$$

$$r^2\dot{\theta} = h \quad \text{a constant} \tag{4.13}$$

$$r = \frac{h^2/\mu}{1 + \epsilon \cos \theta} \tag{4.18}$$

$$r = \frac{(1 - \epsilon^2)a}{1 + \epsilon \cos \theta} \tag{4.20}$$

$$v^2 = \mu\left(\frac{2}{r} - \frac{1}{a}\right) \tag{4.28}$$

$$\tau = \xi a^{3/2} \tag{4.32}$$

5 Planar Kinematics of Rigid Bodies

The basic equations developed for studying the kinematics of a particle are extended to plane rigid-body motion in this chapter. The study of rigid bodies is important since we cannot, in many practical cases, ignore the dimensions of a body, and thereby treat it as a particle. The types of rigid-body motion that are briefly described here include rectilinear translation, curvilinear translation, rotation about a fixed line, rotation about a fixed point, and plane motion. This chapter however limits analysis to the case of plane motion. Even with this limitation we can solve a large variety of engineering problems. Problems in this chapter also include the motion of one rigid body relative to another. Finally, the concept of the instantaneous center of zero velocity for plane rigid-body motion is developed. Several examples are used to demonstrate the application of the planar kinematic equations to rigid-body motion.

5.1 Properties of Rigid Bodies

In Chapter 1 a rigid body was defined such that any two points of the body remain at a fixed distance relative to each other for all time. This definition implies that a rigid body has dimensions. In the case of a particle, dimensions are assumed to be negligible. Until this point, our study of the subject of dynamics has involved only particles; and the only inertia term to enter our calculations has been the particle mass. But since a rigid body has dimensions, the mass moment of inertia property must enter into our calculations when we study the motion of rigid bodies.

By way of reviewing the concept of mass moment of inertia, let us reconsider our earlier treatment of the angular momentum of a particle introduced in Chapter 3. We recall that the angular momentum of a particle about a fixed reference point is $\mathbf{H} = \mathbf{r} \times m\mathbf{v}$, where $\mathbf{r}$ is the position vector of the particle of mass m traveling at an absolute velocity $\mathbf{v}$. As an example, suppose the mass m is attached to one end of a rigid bar of negligible mass as shown in Fig. 5.1. The other end of the bar is pinned at the origin O and the body rotates in the XY plane at an angular speed ω. In this case, let $\mathbf{e}_r$ and $\mathbf{e}_\theta$ be attached to the bar, so that

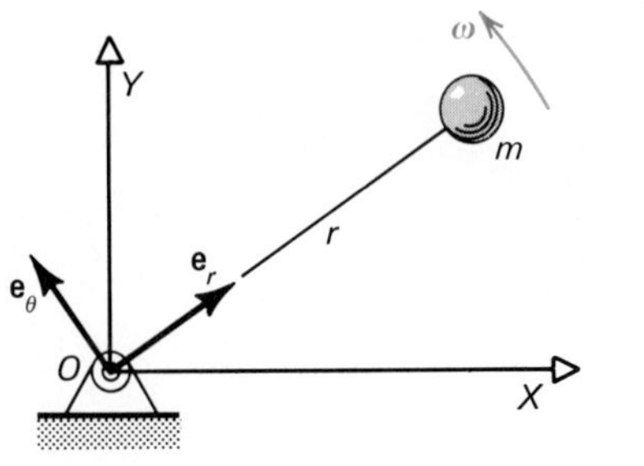

Figure 5.1

$$\mathbf{r} = r\mathbf{e}_r \qquad \boldsymbol{\omega} = \omega\mathbf{k} \qquad \mathbf{v} = \boldsymbol{\omega} \times \mathbf{r} = r\omega\mathbf{e}_\theta$$

and

$$\mathbf{H} = r\mathbf{e}_r \times mr\omega\mathbf{e}_\theta = mr^2\omega\mathbf{k}$$

If we use the notation $I_o = mr^2$, this equation becomes

$$\mathbf{H} = I_o\omega\mathbf{k} \tag{5.1}$$

where I_o is called the mass moment of inertia of the body about the point O. The units of I_o are given in $\text{kg}\cdot\text{m}^2$ or $\text{lb}\cdot\text{ft}\cdot\text{sec}^2$ (sometimes also written as $\text{slug}\cdot\text{ft}^2$). We note that it is the concept of angular momentum that introduces us to the idea of mass moment of inertia in the study of dynamics.

Next we consider a homogeneous rod whose mass is M. We wish to obtain the angular momentum of the rod. Figure 5.2 shows the rod broken down into a series of point masses as an approximate representation of the mass M of the rod. Thus,

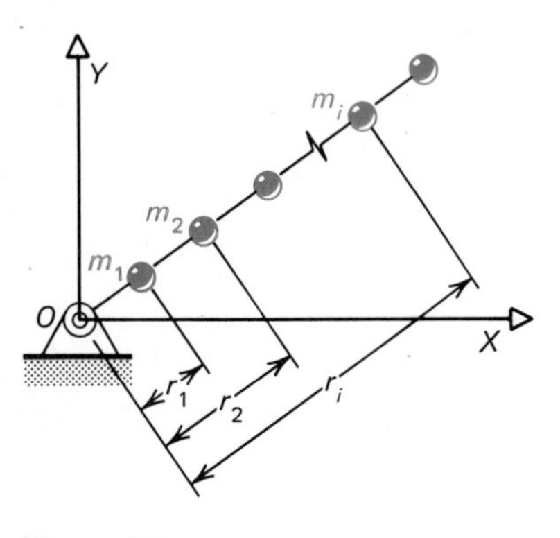

Figure 5.2

$$M = \sum_i m_i \tag{5.2}$$

From our previous example for calculating angular momentum about O, we see that

$$\begin{aligned}\mathbf{H} &= m_1 r_1^2\omega\mathbf{k} + m_2 r_2^2\omega\mathbf{k} + \cdots + m_j r_j^2\omega\mathbf{k} \\ &= \left(\sum_i m_i r_i^2\right)\omega\mathbf{k} = I_o\omega\mathbf{k}\end{aligned} \tag{5.3}$$

where I_o is again the mass moment of inertia of the body about O. If we were to increase the number of point masses, still adhering to the restriction of Eq. (5.2), the distance between particles would diminish. In the limit, as the particle spacing approaches zero, we obtain the expression

$$I_o = \lim_{\Delta r \to 0}\left(\sum m_i r_i^2\right) = \int r^2\, dm \tag{5.4}$$

Substituting this expression for I_o in Eq. (5.3) would give the exact result for the angular momentum of the rod about O.

The mass moment of inertia is also defined as

$$I_o = k^2 m$$

where k is the radius of gyration. Here, the mass moment of inertia of the body is equivalent to one in which the mass m is assumed to be concentrated at a distance k from O. In engineering applications, it is convenient to list the radii of gyration for bodies and shapes that are irregular or complicated.

Often it is necessary to obtain the mass moment of inertia of a body about a different axis but one which is parallel to the centroidal axis. This can be done by the use of the parallel-axis theorem which states that

$$I_o = I_c + md^2$$

where I_c is the centroidal mass moment of inertia and d is the distance between the axis through O and the centroidal axis.

The properties of the mass moments of inertia were developed in Volume 1, and the appendix in this volume summarizes the equations for the mass moments of inertia of rigid bodies typical of those we encounter in problems of dynamics that are considered here.

Example 5.1

Calculate the mass moment of inertia and angular momentum of the configuration shown in Fig. 5.3 about the fixed point O. The rod has a mass of 2 slugs. Assume the angular velocity is 5 rad/sec in the direction shown.

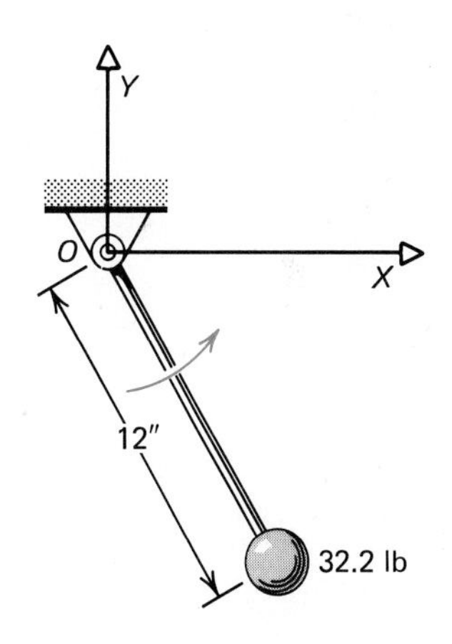

Figure 5.3

The mass moment of inertia about O is

$$I_o = (I_o)_{\text{rod}} + (I_o)_{\text{mass}}$$

To find $(I_o)_{\text{rod}}$, we employ the parallel-axis theorem.

$$(I_o)_{\text{rod}} = I_c + md^2$$

$$= \frac{1}{12}mL^2 + m\left(\frac{L}{2}\right)^2 = \frac{1}{3}mL^2$$

$$= \frac{1}{3}(2)(1)^2 = 0.667 \text{ lb}\cdot\text{ft}\cdot\text{sec}^2$$

Since no dimensions are given for the mass at the end of the rod, we treat it as a mass point.

$$(I_o)_{\text{mass}} = mL^2 = \left(\frac{32.2}{32.2}\right)(1)^2 = 1 \text{ lb}\cdot\text{ft}\cdot\text{sec}^2$$

$$I_o = 0.667 + 1 = 1.667 \text{ lb}\cdot\text{ft}\cdot\text{sec}^2 \qquad \textit{Answer}$$

$$\mathbf{H} = I_o\omega\mathbf{k} = 1.667(5)\mathbf{k}$$

$$= 8.335\mathbf{k} \text{ lb}\cdot\text{ft}\cdot\text{sec} \qquad \textit{Answer}$$

Example 5.2

The radius of gyration of a 3220-lb body about the axis of spin is 0.4 ft. If the body is spinning at 10 rad/sec, what is the magnitude of the angular momentum?

$$m = \frac{3220}{32.2} = 100 \text{ slugs}$$

$$I = mk^2 = 100(0.4)^2 = 16 \text{ lb} \cdot \text{ft} \cdot \text{sec}^2$$

$$H = I\omega = 16(10) = 160 \text{ lb} \cdot \text{ft} \cdot \text{sec} \qquad \textit{Answer}$$

Example 5.3

Obtain the mass moment of inertia about point O for the masses shown in Fig. 5.4. What is the radius of gyration of the system about O.

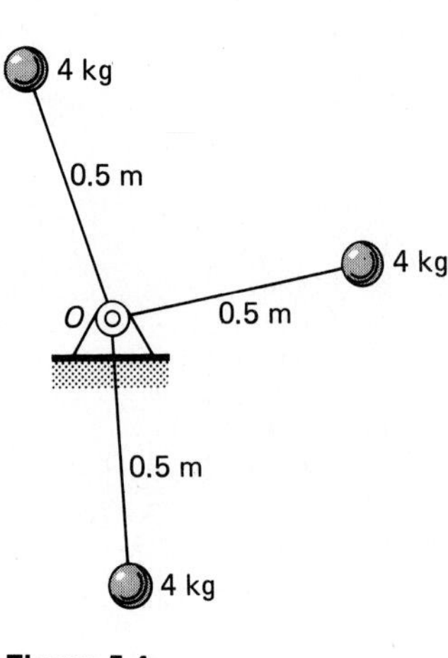

Figure 5.4

Since there are three equal masses each at a distance of 0.5 m from O,

$$I_o = 3(mr^2) = 3(4)(0.5)^2 = 3 \text{ kg} \cdot \text{m}^2 \qquad \textit{Answer}$$

The radius of gyration is obtained as follows:

$$k = \sqrt{I_o/m} = \sqrt{3/12} = 0.5 \text{ m} \qquad \textit{Answer}$$

This result is not surprising, since it means we could replace the given system by a mass equal to 12 kg located at a distance of 0.5 m.

PROBLEMS

5.1 Obtain the radius of gyration about point O for the rod and point mass in Fig. 5.3 using the results of Example 5.1.

5.2 An unbalanced 30-kg wheel has its center of mass located 0.2 m from the center of the wheel. If the radius of gyration about the center of mass is 0.1 m, find the mass moment of inertia about the center of the wheel. Assume all axes are perpendicular to the plane of the wheel.

5.3 A uniform rod is 0.3 ft long, has a mass of 4 slugs and rotates at a rate of 20 rad/sec. Attached to the end A of the rod is a point mass of 4 slugs. (a) Find the mass moment of inertia of the rod plus the mass point about the end O of the rod opposite the end A. (b) What is the magnitude of the angular momentum about the point O?

5.4 Figure P5.4 shows a pendulum consisting of two point masses. Find the mass moment of inertia of the system about O neglecting the mass of the slender rods.

5.5 Repeat Problem 5.4 to include the mass of the rods assuming that each rod has a mass of 1 slug.

5.6 A 4-kg cylinder has a 1.5-kg rod attached to it as shown in Fig. P5.6. Find the mass moment of inertia for the system (a) about O, and (b) about O'.

5.7 For Problem 5.6, obtain the radius of gyration about O and O'.

5.8 Obtain the radius of gyration about point O for the system shown in Fig. P5.8.

5.9 Find the angular momentum about O for the configuration of masses in Problem 5.8 which are rotating in the XY plane at 7 rad/sec CW.

5.10 If the body in Problem 5.6 is pinned at O' and rotates at 20 rad/s CCW, calculate the magnitude of the angular momentum about O'.

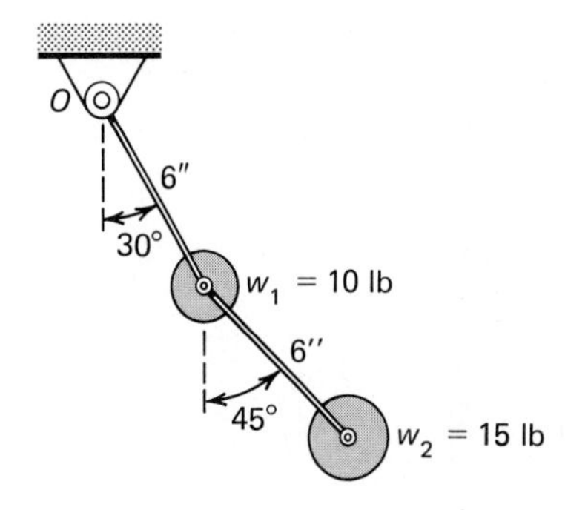

Figure P5.4

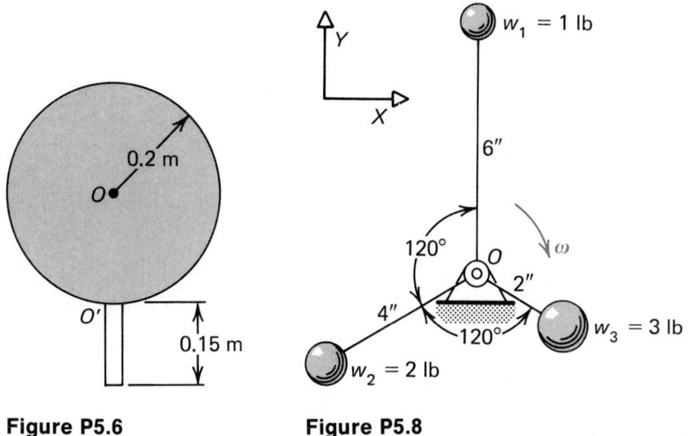

Figure P5.6

Figure P5.8

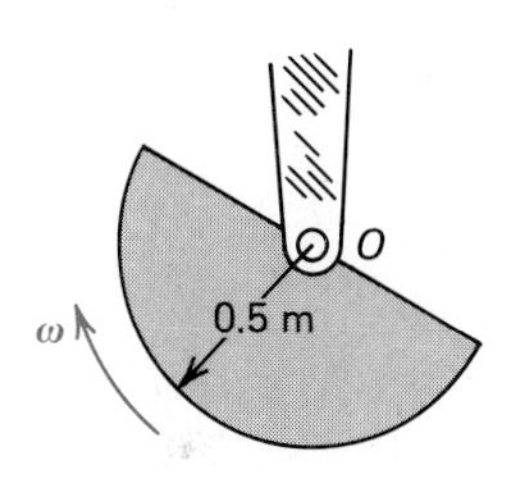

Figure P5.11

5.11 A 16-kg half cylinder is pinned at point O and rotates at an angular rate of 5 rad/s CW as shown in Fig. P5.11. Find the magnitude of the angular momentum about O.

5.12 The spacecraft in Fig. P5.12 has a uniform density of 10 slugs/ft^3 and is spinning about the Z axis at a rate of 100 rev/min. What is the magnitude of the angular momentum of the spacecraft?

5.13 If the spacecraft shown in Problem 5.12 has a radius of gyration of 0.4 ft about the spin axis, what is the magnitude of the angular momentum of the spacecraft if it is spinning at 120 rev/min?

5.14 Due to a launch error, the spacecraft of Problem 5.12 is put into orbit so that $\boldsymbol{\omega} = 100\mathbf{i} + 70\mathbf{j} + 5\mathbf{k}$ rad/sec. What is the magnitude of the angular momentum of the spacecraft? Assume that the mass moment of inertia about the resulting spin axis is the same as that obtained in Problem 5.12 and the spin axis is an axis of symmetry.

5.15 A 20-kg circular dish of radius 0.3 m is attached to a 10-kg rod, which in turn is attached to a 30-kg shaft as shown in Fig. P5.15. Find the angular

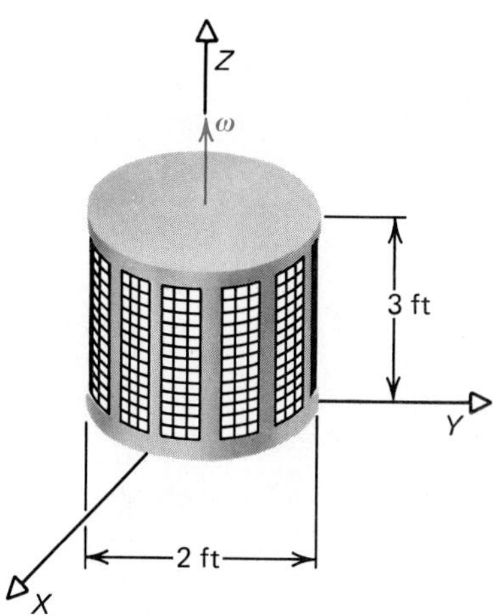

Figure P5.12

momentum of the assembly about the OO-axis if the angular velocity of the shaft is $\boldsymbol{\omega} = 10\mathbf{k}$ rad/s.

5.16 If the circular dish has an angular velocity by 120**k** rev/min in Problem 5.15, then what is the angular momentum of the dish about the OO-axis, assuming the remaining structure to be stationary?

5.17 The volume of the automobile tire shown in Fig. P5.17 is 0.1 m^3 when inflated and the effective density is 25 kg/m^3. If the radius of gyration about the spin axis is 0.25 m, (a) find the mass moment of inertia about the spin axis, and (b) find the magnitude of the angular momentum about the spin axis if the tire is spinning at 1000 rev/min.

5.18 An energy storage wheel is shown in Fig. P5.18. Its density is 10 $slugs/ft^3$ and its volume is 2 ft^3. What is the magnitude of the angular momentum of the wheel about the spin axis if the radius of gyration is 0.25 ft about the spin axis and it spins at a rate of 100 rev/sec?

5.19 If the energy storage wheel of Problem 5.18 has a radius of gyration of 0.20 ft about the spin axis, what is the percentage decrease in angular momentum that can be stored in the wheel?

5.20 A 40-kg beam supported by an elastic foundation is modeled by three point masses supported by a massless rigid rod in Fig. P5.20. What is the mass moment of inertia (a) about the center of mass of the model and (b) about one of the 15-kg masses?

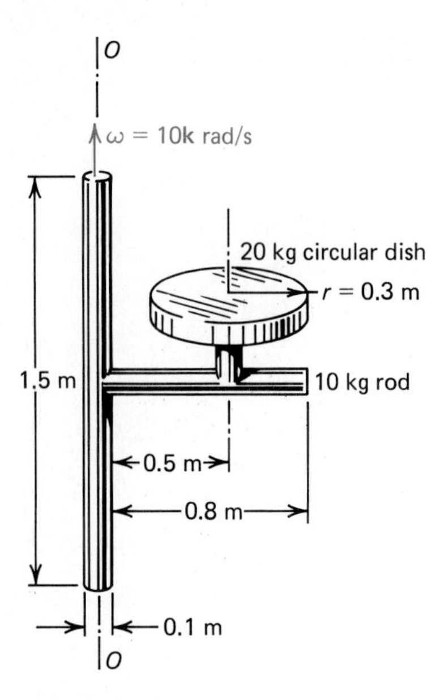

Figure P5.15

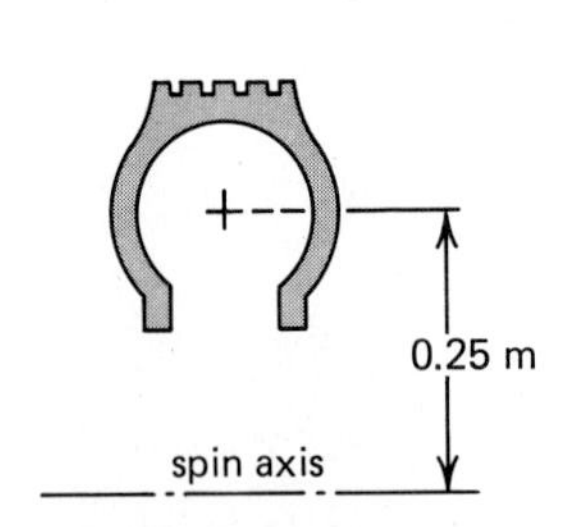

Figure P5.17

Figure P5.18

Figure P5.20

5.2 Types of Motion

The terms used to describe special types of rigid-body motion are summarized below.

Rectilinear Translation. All points in the rigid body move in a straight line. Figure 5.5(a) shows the change in position of the body with points A and B moving in a straight line to the right.

Curvilinear Translation. The orientation of all points in the rigid body remain fixed. Figure 5.5(b) shows that while the body moves along a curved path, the orientation of points A and B remains invariant. The body in Fig. 5.5(c) is an example of a rigid body undergoing this type of motion.

Rotation About a Fixed Line. All points of the rigid body move in circular motion about the fixed line. The fixed line in Fig. 5.6 is represented by the axis OO' and the arbitrary point P moves in a circle about the axis.

Rotation About a Fixed Point. All points of the rigid body move such that their respective distances from the fixed point are invariant. Figure 5.7 shows a spinning top, which is an example of this type of motion studied in advanced texts.

Plane Motion. Each point in the body moves in a path parallel to a fixed plane. Figure 5.8 shows a rolling cylinder as an example. For studying the motion of the cylinder, we can select any plane in the body parallel to some fictitious plane that is perpendicular to the axis OO'.

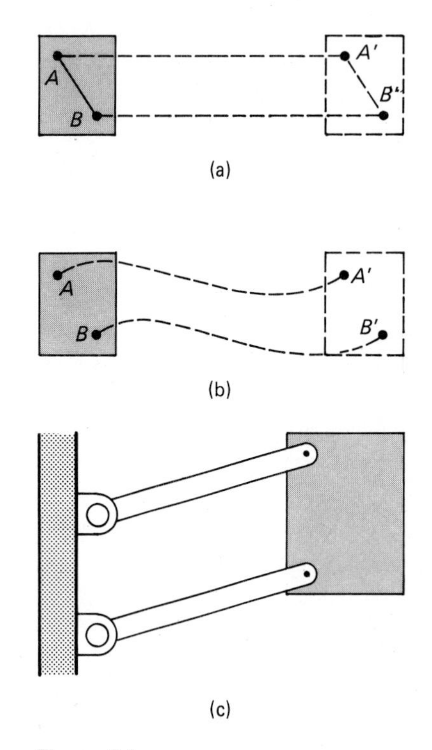

Figure 5.5

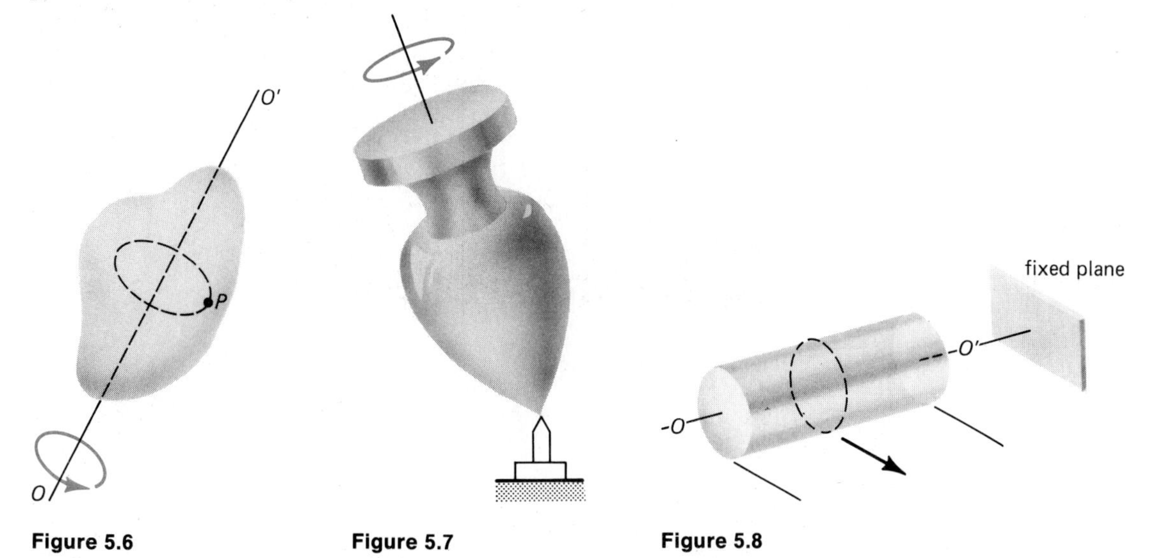

Figure 5.6 Figure 5.7 Figure 5.8

General Motion. The motion of a rigid body which is not restricted by any of the above special cases.

The present chapter is devoted to planar rigid-body motion, which is applicable to many engineering problems in dynamics. The case of general motion is more complicated and is normally treated as an advanced topic.

As a final comment on types of rigid-body motion, recall that in Chapter 2 we introduced the concept of infinitesimal translation and infinitesimal rotation. It can be shown that the order of this motion is immaterial. As an example, consider the motion of a bar moving from AB to $A'B'$ as shown in Fig. 5.9(a). In Fig. 5.9(a) the bar translates in the XY plane by an amount $d\mathbf{r}$ followed by a rotation $d\theta$, while Fig. 5.9(b) shows the order of motion reversed. In both cases the body rotated about the same pivot point. In going from AB to $A'B'$, any point in the body could have been used as the pivot point. For example, Fig. 5.9(c) shows point C as the pivot point for the infinitesimal rotation as the body moves from AB to $A'B'$. Note that $d\theta$ is the same as in Figs. 5.9(a) and (b). We refer to $d\theta$ and its time derivatives $\boldsymbol{\omega}$ and $\boldsymbol{\alpha}$ as free vectors. Consequently, when we talk about the angular motion of a body in planar motion undergoing translation and rotation, the angular velocity and angular acceleration of every line drawn on the body is the same.

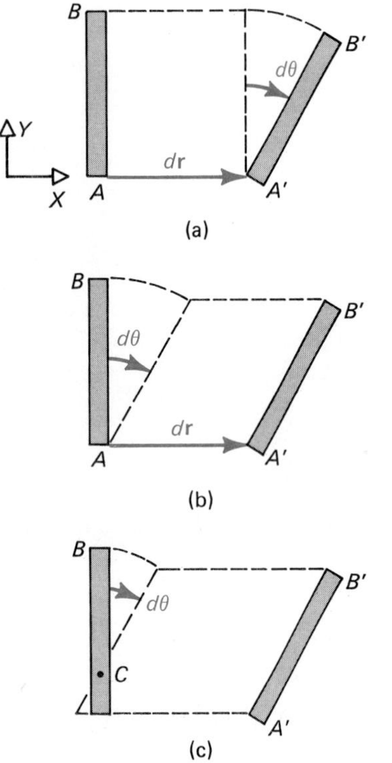

Figure 5.9

5.3 Plane Motion of Rigid Bodies

Before developing the equations for planar rigid-body motion, let us review the basic equations developed in Chapter 2. Figure 5.10(a) shows the fixed origin at O, with O' as the origin of the moving reference frame xyz. The equations for the absolute velocity and acceleration of point P are:

$$\mathbf{v} = \dot{\mathbf{R}} + \dot{\mathbf{r}}_r + \boldsymbol{\omega} \times \mathbf{r} \tag{5.5}$$

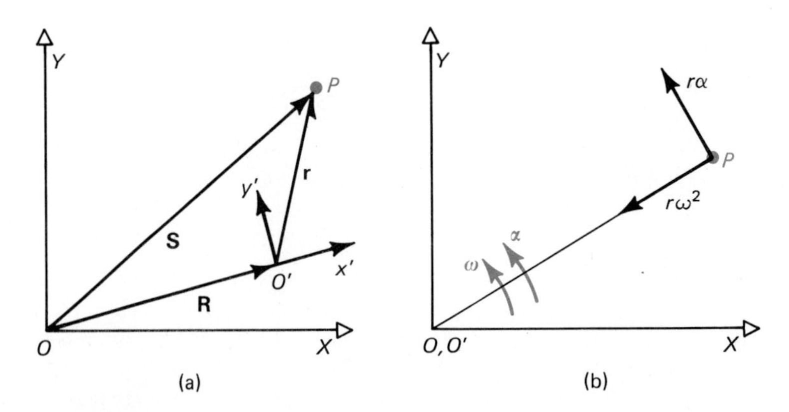

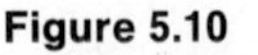
Figure 5.10

$$\mathbf{a} = \ddot{\mathbf{R}} + \ddot{\mathbf{r}}_r + \boldsymbol{\alpha} \times \mathbf{r} + \boldsymbol{\omega} \times (\boldsymbol{\omega} \times \mathbf{r}) + 2\boldsymbol{\omega} \times \dot{\mathbf{r}}_r \tag{5.6}$$

where $\dot{\mathbf{R}}$ = Velocity of point O' relative to O

$\dot{\mathbf{r}}_r$ = Velocity of point P as seen by an observer attached to the moving reference frame

$\boldsymbol{\omega}$ = Absolute angular velocity of the moving reference frame

$\ddot{\mathbf{R}}$ = Acceleration of point O' relative to O

$\ddot{\mathbf{r}}_r$ = Acceleration of point P as seen by an observer attached to the moving reference frame

$\boldsymbol{\alpha}$ = Absolute angular acceleration of the moving reference frame

In the case of rigid-body motion, if points O' and P are both on the rigid body, then the magnitude of $\mathbf{r}$ remains unchanged so that $\dot{\mathbf{r}}_r = \ddot{\mathbf{r}}_r = 0$. Consequently, Eqs. (5.5) and (5.6) are simplified to give us the governing equations for rigid-body motion,

$$\mathbf{v} = \dot{\mathbf{R}} + \boldsymbol{\omega} \times \mathbf{r} \tag{5.7}$$

$$\mathbf{a} = \ddot{\mathbf{R}} + \boldsymbol{\alpha} \times \mathbf{r} + \boldsymbol{\omega} \times (\boldsymbol{\omega} \times \mathbf{r}) \tag{5.8}$$

If the origin of the moving reference frame remains fixed so that O' remains at O, and O' and P are still on the rigid body, then $\dot{\mathbf{R}} = \ddot{\mathbf{R}} = 0$, and the kinematic equations are reduced to

$$\mathbf{v} = \boldsymbol{\omega} \times \mathbf{r} \tag{5.9}$$

$$\mathbf{a} = \boldsymbol{\alpha} \times \mathbf{r} + \boldsymbol{\omega} \times (\boldsymbol{\omega} \times \mathbf{r}) \tag{5.10}$$

The acceleration components of Eq. (5.10) are illustrated in Fig. 5.10(b).

Example 5.4

The wheel of a car of radius b rolls without slipping on a horizontal surface as shown in Fig. 5.11(a). Determine the velocity and acceleration of points A and C at the instant when the angular velocity and angular acceleration of the wheel are ω CW and α CW, respectively.

Figure 5.11(b) shows the displacement vector of A in moving to A' as $\mathbf{r}_A = s\mathbf{e}_1$. Since there is no slipping of the wheel, $s = b\theta$. Therefore $\mathbf{r}_A = b\theta\mathbf{e}_1$ and

$$\mathbf{v}_A = b\dot{\theta}\mathbf{e}_1 = b\omega\mathbf{e}_1 \qquad \mathbf{a}_A = b\dot{\omega}\mathbf{e}_1 = b\alpha\mathbf{e}_1 \qquad \textit{Answer}$$

To find the velocity of point C, use Eq. (5.7) and refer to Fig. 5.11(c). The fixed origin is at point O and the moving origin is at O'. Thus,

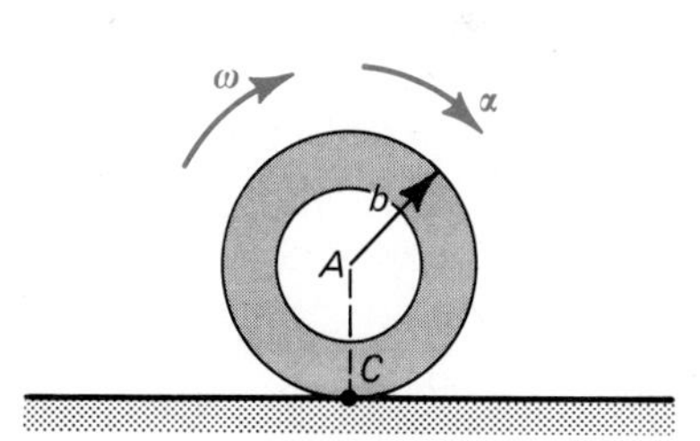

(a)

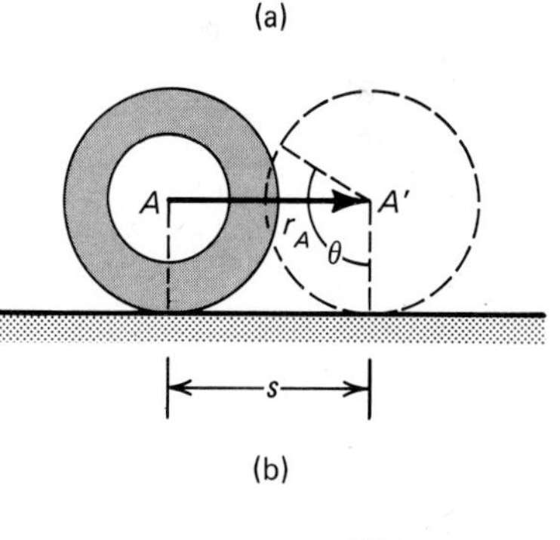

(b)

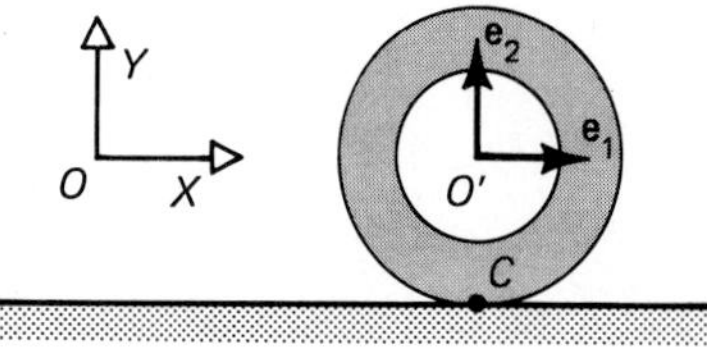

(c)

Figure 5.11

$$\mathbf{v}_C = \dot{\mathbf{R}} + \boldsymbol{\omega} \times \mathbf{r}$$

where $\dot{\mathbf{R}} = b\omega\mathbf{e}_1 \qquad \boldsymbol{\omega} = -\omega\mathbf{e}_3 \qquad \mathbf{r} = -b\mathbf{e}_2$

and therefore

$$\mathbf{v}_C = b\omega\mathbf{e}_1 - b\omega\mathbf{e}_1 = 0 \qquad \textit{Answer}$$

To find the acceleration of C, we turn to Eq. (5.8).

$$\mathbf{a}_C = \ddot{\mathbf{R}} + \boldsymbol{\alpha} \times \mathbf{r} + \boldsymbol{\omega} \times (\boldsymbol{\omega} \times \mathbf{r})$$

where $\ddot{\mathbf{R}} = b\alpha\mathbf{e}_1 \qquad \boldsymbol{\alpha} = -\alpha\mathbf{e}_3$

and therefore

$$\begin{aligned}\mathbf{a}_C &= b\alpha\mathbf{e}_1 - \alpha\mathbf{e}_3 \times (-b\mathbf{e}_2) - \omega\mathbf{e}_3 \times (-b\omega\mathbf{e}_1)\\ &= b\alpha\mathbf{e}_1 - b\alpha\mathbf{e}_1 + b\omega^2\mathbf{e}_2 = b\omega^2\mathbf{e}_2 \qquad \textit{Answer}\end{aligned}$$

We conclude that the point C, which is in contact with the surface, has zero velocity when the no-slippage condition exists; it does, however, experience an acceleration at the same instant.

Example 5.5

Obtain the velocity of points B and C of the four-bar linkage at the instant shown in Fig. 5.12(a). Let the angular velocity of member AB be 10 rad/sec CCW at the instant shown.

To find the velocity of B use A as the origin of the fixed system as shown in Fig. 5.12(b). Attach the moving unit vectors $\mathbf{e}_1$, $\mathbf{e}_2$, $\mathbf{e}_3$ on member AB. Now

$$\begin{aligned}\mathbf{v}_B &= \boldsymbol{\omega} \times \mathbf{r} = (10\mathbf{e}_3) \times (0.3\mathbf{e}_2)\\ &= -3\mathbf{e}_1 \text{ ft/sec} \qquad \textit{Answer}\end{aligned}$$

For obtaining the velocity of C, let point B be the origin of the moving coordinate system as shown in Fig. 5.12(c). Note that $\mathbf{e}_1$, $\mathbf{e}_2$, $\mathbf{e}_3$ are now attached to member BC but retain the same orientation that was used in Fig. 5.12(b). Now

$$\mathbf{v}_C = \dot{\mathbf{R}} + \boldsymbol{\omega} \times \mathbf{r}$$

where
$$\begin{aligned}\dot{\mathbf{R}} &= \mathbf{v}_B = -3\mathbf{e}_1\\ \mathbf{r} &= 0.5(\cos 15^\circ\mathbf{e}_1 + \sin 15^\circ\mathbf{e}_2)\\ \boldsymbol{\omega} &= \omega_{BC}\mathbf{e}_3\end{aligned}$$

The direction of ω_{BC} is arbitrarily assumed as counterclockwise. Substituting into the equation for $\mathbf{v}_C$ yields

$$\mathbf{v}_C = -3\mathbf{e}_1 + \omega_{BC}\mathbf{e}_3 \times 0.5(\cos 15^\circ\mathbf{e}_1 + \sin 15^\circ\mathbf{e}_2)$$

$$\mathbf{v}_C = -(3 + 0.129\omega_{BC})\mathbf{e}_1 + 0.483\omega_{BC}\mathbf{e}_2 \qquad \text{(a)}$$

To obtain a second expression for $\mathbf{v}_C$, we place $\mathbf{e}_1, \mathbf{e}_2, \mathbf{e}_3$ on member CD as shown in Fig. 5.12(d). Now the angular velocity $\boldsymbol{\omega} = \omega_{CD}\mathbf{e}_3$.

$$\mathbf{v}_C = \omega_{CD}\mathbf{e}_3 \times \mathbf{r}_{DC}$$

$$\mathbf{v}_C = \omega_{CD}\mathbf{e}_3 \times r(-\sin 15^\circ \mathbf{e}_1 + \cos 15^\circ \mathbf{e}_2)$$

We can compute r from the geometry in Fig. 5.12(a).

$$r = \frac{(0.3 + 0.5 \sin 15^\circ)}{\cos 15^\circ}$$

$$= 0.445 \text{ ft}$$

Substituting into the above expression for $\mathbf{v}_C$,

$$\mathbf{v}_C = -0.430\omega_{CD}\mathbf{e}_1 - 0.115\omega_{CD}\mathbf{e}_2 \qquad \text{(b)}$$

Equating Eqs. (a) and (b) we can solve for ω_{BC} and ω_{CD}.

$$-(3 + 0.129\omega_{BC})\mathbf{e}_1 + 0.483\omega_{BC}\mathbf{e}_2 = -0.430\omega_{CD}\mathbf{e}_1 - 0.115\omega_{CD}\mathbf{e}_2$$

This leads to two scalar equations

$$-0.430\omega_{CD} = -(3 + 0.129\omega_{BC})$$

$$-0.115\omega_{CD} = 0.483\omega_{BC}$$

Solving for ω_{BC} yields

$$\omega_{BC} = -1.55 \text{ rad/sec}$$

The velocity for point C can now be obtained as

$$\mathbf{v}_C = (-2.80\mathbf{e}_1 - 0.749\mathbf{e}_2) \text{ ft/sec} \qquad \textit{Answer}$$

Example 5.6

The cylinder shown in Fig. 5.13(a) rolls without slipping on the horizontal surface, such that, at the instant shown, its angular velocity is 3 rad/s CW. Find the velocity of points A and B.

To find the velocity of A, use O' as the origin of the moving reference frame attached to the cylinder as shown in Fig. 5.13(b). Applying Eq. (5.7),

$$\mathbf{v}_A = \dot{\mathbf{R}} + \boldsymbol{\omega} \times \mathbf{r}$$

where $\dot{\mathbf{R}} = b\omega\mathbf{e}_1 = 0.4(3)\mathbf{e}_1 = 1.2\mathbf{e}_1$

$\boldsymbol{\omega} = -3\mathbf{e}_3$

$\mathbf{r} = -0.3\mathbf{e}_1$

Therefore

$$\mathbf{v}_A = 1.2\mathbf{e}_1 - 3\mathbf{e}_3 \times (-0.3\mathbf{e}_1) = (1.2\mathbf{e}_1 + 0.9\mathbf{e}_2) \text{ m/s} \qquad \textit{Answer}$$

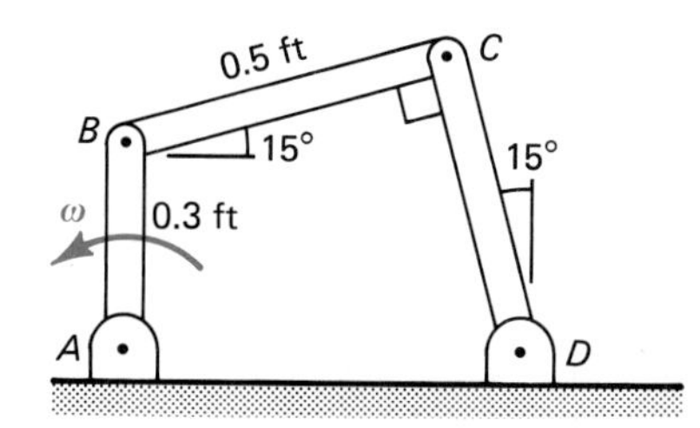

(a)

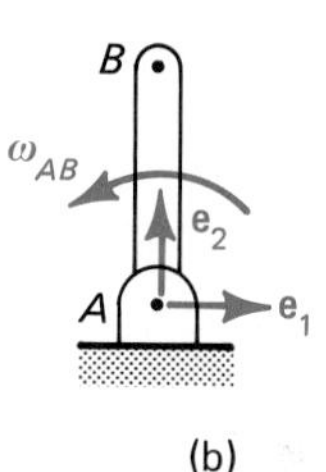

(b)

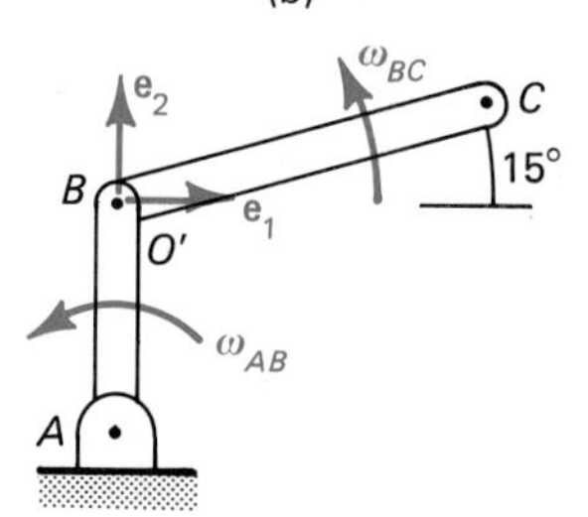

(c)

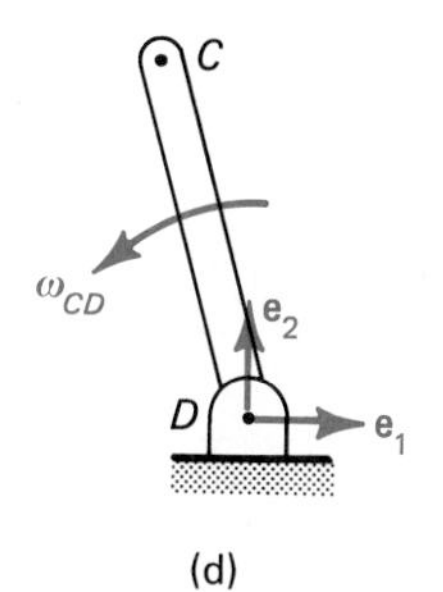

(d)

Figure 5.12

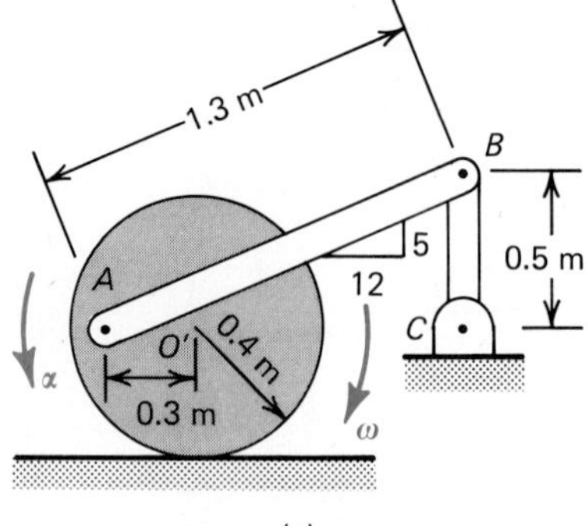

(a)

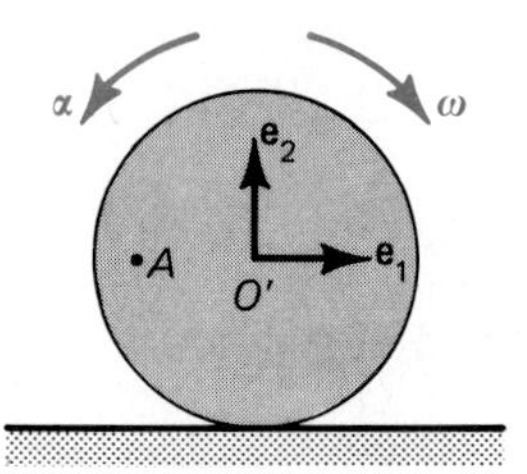

(b)

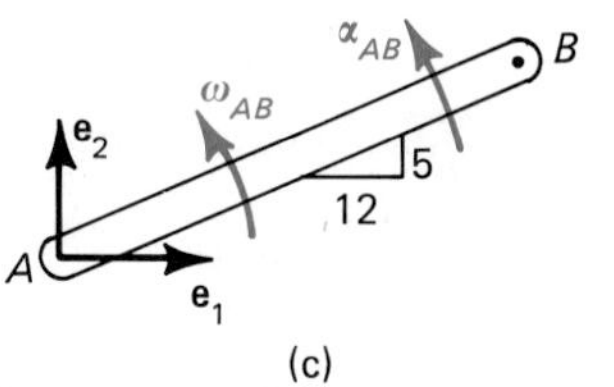

(c)

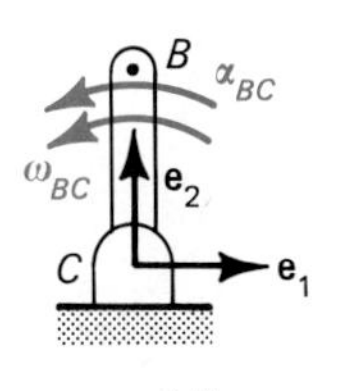

(d)

Figure 5.13

To find the velocity of B, choose point A as the origin of the moving reference point as shown in Fig. 5.13(c). Attach the moving reference frame to member AB. For convenience, orient the moving reference $\mathbf{e}_1$, $\mathbf{e}_2$, $\mathbf{e}_3$ similar to the reference frame used at point O'. Now, using Eq. (5.7),

$$\mathbf{v}_B = \dot{\mathbf{R}} + \boldsymbol{\omega} \times \mathbf{r}$$

where
$$\dot{\mathbf{R}} = \mathbf{v}_A = 1.2\mathbf{e}_1 + 0.9\mathbf{e}_2$$
$$\boldsymbol{\omega} = \omega_{AB}\mathbf{e}_3 \quad (\omega_{AB} \text{ is the angular speed of member } AB; \text{ the direction is assumed to be CCW})$$
$$\mathbf{r} = 1.2\mathbf{e}_1 + 0.5\mathbf{e}_2$$
$$\boldsymbol{\omega} \times \mathbf{r} = 1.2\omega_{AB}\mathbf{e}_2 - 0.5\omega_{AB}\mathbf{e}_1$$

We find that

$$\mathbf{v}_B = (1.2 - 0.5\omega_{AB})\mathbf{e}_1 + (0.9 + 1.2\omega_{AB})\mathbf{e}_2 \qquad \text{(a)}$$

Since Eq. (a) contains the angular velocity of bar AB, that is ω_{AB}, we need a second equation. Noting that point B is also a point on member BC, we turn again to Eq. (5.7), now with point C as the fixed reference point, and with the moving reference frame attached to member BC as seen in Fig. 5.13(d). Thus,

$$\mathbf{v}_B = \dot{\mathbf{R}} + \boldsymbol{\omega} \times \mathbf{r}$$

where
$$\dot{\mathbf{R}} = 0 \quad \text{(the reference point } C \text{ is fixed)}$$
$$\boldsymbol{\omega} = \omega_{BC}\mathbf{e}_3 \quad (\text{again we assume the direction of } \boldsymbol{\omega})$$
$$\mathbf{r} = 0.5\mathbf{e}_2$$

Therefore
$$\mathbf{v}_B = \omega_{BC}\mathbf{e}_3 \times 0.5\mathbf{e}_2 - -0.5\omega_{BC}\mathbf{e}_1 \qquad \text{(b)}$$

Before equating Eqs. (a) and (b), note that all vectors are expressed in terms of unit vectors oriented in the same direction.

$$(1.2 - 0.5\omega_{AB})\mathbf{e}_1 + (0.9 + 1.2\omega_{AB})\mathbf{e}_2 = -0.5\omega_{BC}\mathbf{e}_1$$

From the $\mathbf{e}_2$ equation,

$$\omega_{AB} = -0.75 \text{ rad/s}$$

From the $\mathbf{e}_1$ equation,

$$0.5\omega_{BC} = (0.5\omega_{AB} - 1.2)$$
$$\omega_{BC} = -3.15 \text{ rad/s}$$

The velocity of B becomes

$$\mathbf{v}_B = -0.5\omega_{BC}\mathbf{e}_1 = 1.575\mathbf{e}_1 \text{ m/s} \qquad \textit{Answer}$$

We note that in the above calculations, if a variable is calculated as a negative quantity, it means that we have assumed the wrong direction for this variable. However, we retain the negative sign in returning to our algebraic equations as we proceed with the computations.

Example 5.7

Obtain the acceleration of points A and B of the configuration studied in Example 5.6 if the angular acceleration of the cylinder is 2 rad/s^2 CCW.

To find the acceleration of A and B, we use the same approach as in Example 5.6. Referring to Fig. 5.13(b), apply Eq. (5.8) with the origin of the moving reference frame at O'. We have, for point A,

$$\mathbf{a}_A = \ddot{\mathbf{R}} + \boldsymbol{\alpha} \times \mathbf{r} + \boldsymbol{\omega} \times (\boldsymbol{\omega} \times \mathbf{r})$$

where we note that the acceleration of O' is to the left, so that

$$\ddot{\mathbf{R}} = b\alpha\mathbf{e}_1 = -0.4(2)\mathbf{e}_1 = -0.8\mathbf{e}_1$$

$$\boldsymbol{\alpha} = 2\mathbf{e}_3 \qquad \boldsymbol{\omega} = -3\mathbf{e}_3 \qquad \mathbf{r} = -0.3\mathbf{e}_1$$

Substituting into the expression for $\mathbf{a}_A$ yields

$$\begin{aligned}\mathbf{a}_A &= -0.8\mathbf{e}_1 + 2\mathbf{e}_3 \times (-0.3\mathbf{e}_1) \\ &\quad + (-3\mathbf{e}_3) \times [(-3\mathbf{e}_3 \times (-0.3\mathbf{e}_1)] \\ &= -0.8\mathbf{e}_1 - 0.6\mathbf{e}_2 + 2.7\mathbf{e}_1 \\ &= (1.9\mathbf{e}_1 - 0.6\mathbf{e}_2)\ \text{m/s}^2 \qquad \textit{Answer}\end{aligned}$$

To find $\mathbf{a}_B$, use point A as the moving reference point. Referring to Fig. 5.13(c) and again to Eq. (5.8), we have

$$\mathbf{a}_B = \ddot{\mathbf{R}} + \boldsymbol{\alpha} \times \mathbf{r} + \boldsymbol{\omega} \times (\boldsymbol{\omega} \times \mathbf{r})$$

where

$$\begin{aligned}\ddot{\mathbf{R}} &= \mathbf{a}_A = 1.9\mathbf{e}_1 - 0.6\mathbf{e}_2 \\ \boldsymbol{\alpha} &= \alpha_{AB}\mathbf{e}_3 \\ \boldsymbol{\omega} &= \boldsymbol{\omega}_{AB} = -0.75\mathbf{e}_3 \qquad \text{(see Example 5.6)} \\ \mathbf{r} &= 1.2\mathbf{e}_1 + 0.5\mathbf{e}_2\end{aligned}$$

Now

$$\begin{aligned}\boldsymbol{\alpha} \times \mathbf{r} &= 1.2\alpha_{AB}\mathbf{e}_2 - 0.5\alpha_{AB}\mathbf{e}_1 \\ \boldsymbol{\omega} \times (\boldsymbol{\omega} \times \mathbf{r}) &= -0.75\mathbf{e}_3 \times [(-0.75\mathbf{e}_3) \times (1.2\mathbf{e}_1 + 0.5\mathbf{e}_2)] \\ &= -0.675\mathbf{e}_1 - 0.281\mathbf{e}_2\end{aligned}$$

and therefore

$$\begin{aligned}\mathbf{a}_B &= 1.9\mathbf{e}_1 - 0.6\mathbf{e}_2 + 1.2\alpha_{AB}\mathbf{e}_2 - 0.5\alpha_{AB}\mathbf{e}_1 \\ &\quad - 0.675\mathbf{e}_1 - 0.281\mathbf{e}_2 \\ &= (1.225 - 0.5\alpha_{AB})\mathbf{e}_1 + (-0.881 + 1.2\alpha_{AB})\mathbf{e}_2 \qquad \text{(a)}\end{aligned}$$

As before in the computation of velocity in Example 5.6 we need a second equation to complete the solution. Using Eq. (5.8), with point C as the reference point, and referring to Fig. 5.13(d), we have

$$\mathbf{a}_B = \ddot{\mathbf{R}} + \boldsymbol{\alpha} \times \mathbf{r} + \boldsymbol{\omega} \times (\boldsymbol{\omega} \times \mathbf{r})$$

where
$$\ddot{\mathbf{R}} = 0$$
$$\boldsymbol{\alpha} = \alpha_{BC}\mathbf{e}_3$$
$$\mathbf{r} = 0.5\mathbf{e}_2$$
$$\boldsymbol{\omega} = \omega_{BC}\mathbf{e}_3 = -3.15\mathbf{e}_3 \qquad \text{(see Example 5.6)}$$

Now
$$\boldsymbol{\alpha} \times \mathbf{r} = -0.5\alpha_{BC}\mathbf{e}_1$$
$$\boldsymbol{\omega} \times (\boldsymbol{\omega} \times \mathbf{r}) = -4.96\mathbf{e}_2$$

and therefore
$$\mathbf{a}_B = -0.5\alpha_{BC}\mathbf{e}_1 - 4.96\mathbf{e}_2 \tag{b}$$

When we equate Eqs. (a) and (b) we have two scalar equations,

$$1.225 - 0.5\alpha_{AB} = -0.5\alpha_{BC}$$
$$-0.881 + 1.2\alpha_{AB} = -4.96$$

These yield

$$\alpha_{AB} = -3.40 \text{ rad/s}^2$$
$$\alpha_{BC} = -5.85 \text{ rad/s}^2$$

Substituting into Eq. (b), we obtain

$$\mathbf{a}_B = (2.925\mathbf{e}_1 - 4.96\mathbf{e}_2) \text{ m/s}^2$$ *Answer*

PROBLEMS

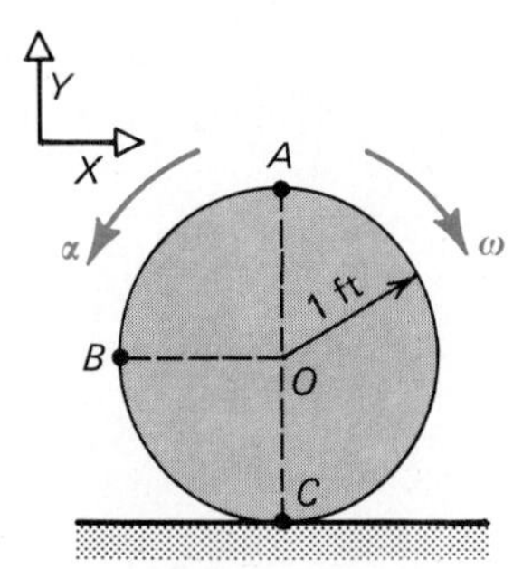

Figure P5.21

5.21 The cylinder in Fig. P5.21 rolls without slipping at an angular rate of 6 rad/sec CW and an angular acceleration of 3 rad/sec^2 CCW. Find the velocity and acceleration of points A and B.

5.22 Suppose the cylinder in Problem 5.21 is moving to the right such that

$$\mathbf{v}_O = 3\mathbf{i} \text{ ft/sec} \qquad \text{and} \qquad \mathbf{v}_A = 5\mathbf{i} \text{ ft/sec}$$

Obtain the velocity of point C. Is the cylinder slipping?

5.23 A cylinder undergoes motion as shown in Fig. P5.23 where

$$\mathbf{v}_C = 10\mathbf{i} \text{ ft/sec} \qquad \mathbf{a}_C = -3\mathbf{i} \text{ ft/sec}^2$$
$$\boldsymbol{\omega} = 8\mathbf{k} \text{ rad/sec} \qquad \boldsymbol{\alpha} = -1\mathbf{k} \text{ rad/sec}^2$$

Obtain the velocity and acceleration of point A.

5.24 Rework Problem 5.23 if $\boldsymbol{\omega} = -10\mathbf{k}$ rad/sec and $\boldsymbol{\alpha} = -3\mathbf{k}$ rad/sec^2.

5.25 Bar AB is pinned to the rolling cylinder at point A while point B slides along the horizontal plane. At the instant shown in Fig. P5.25, the angular speed of the cylinder is 10 rad/s CCW. Determine the magnitude of the velocity of points B and C.

5.26 An inverted pendulum is attached to a shaker table as shown in Fig. P5.26. The motion of the table and pendulum is given by

$$\mathbf{v}_{O'} = A \cos \omega t \mathbf{j}$$

$$\theta = B \sin \lambda t$$

What is the velocity of point P at the instant shown?

5.27 A cylinder is lifted using a rope as shown in Fig. P5.27. The rope is pulled so that the speed of B is a constant 5 m/s. What are the velocity and acceleration of points C, O, and A assuming no slippage?

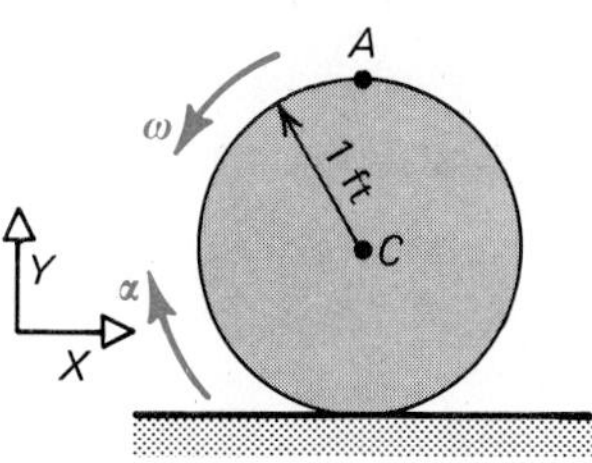

Figure P5.23

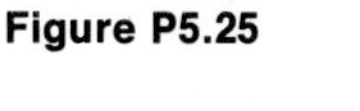

Figure P5.25

Figure P5.26

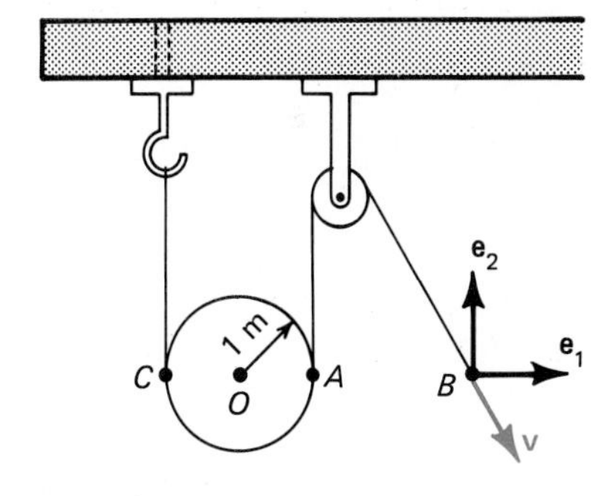

Figure P5.27

5.28 The crank OA shown in Fig. P5.28 has a constant angular rate of 700 rev/min CCW. Obtain the velocity of point C and the angular velocity of member AC at the instant when $\theta = 45°$.

5.29 For Problem 5.28, assume the crank OA to have an angular acceleration of 5 rad/sec^2 CW and an angular rate of 700 rev/min CCW. What are the velocity and acceleration magnitudes of point C when AC is perpendicular to the crank?

5.30 The velocity of the block B is 8 m/s downward at the instant shown in Fig. P5.30. Find the angular velocity of the cylinder, assuming it rolls without slipping.

5.31 The velocity of the block B in Problem 5.30 is 10 m/s upward at the instant shown. Find the angular velocity of the cylinder assuming it rolls without slipping.

5.32 In Fig. P5.32, the angular velocity of member AB is 2 rad/s CCW. Find the angular velocity of members BC and CD.

5.33 If the angular acceleration of member AB in Problem 5.32 is 3 rad/s^2 CW, find the angular accelerations of members BC and CD.

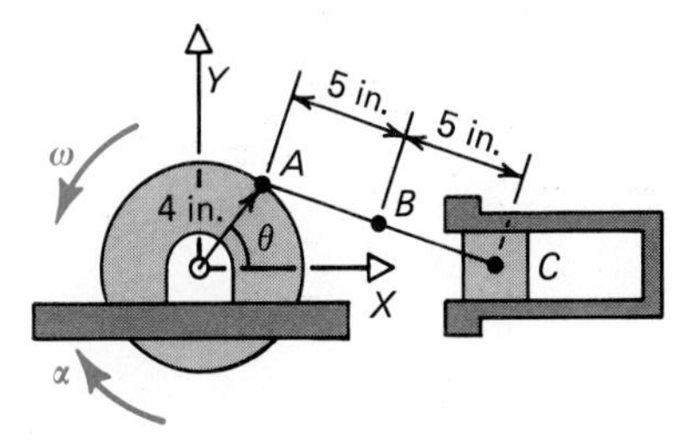

Figure P5.28

5.34 Using the data in Problem 5.33, find the velocity and acceleration of point E for the mechanism.

5.35 Find the velocity and acceleration of point B if the cylinder in Fig. P5.35 rotates at a constant angular rate of 7 rad/sec CCW.

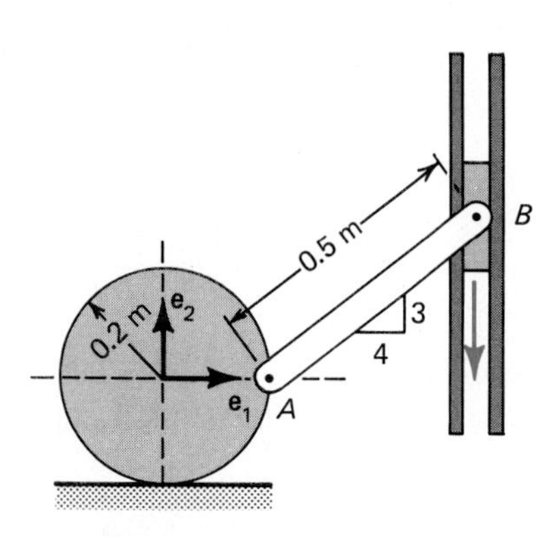

Figure P5.30

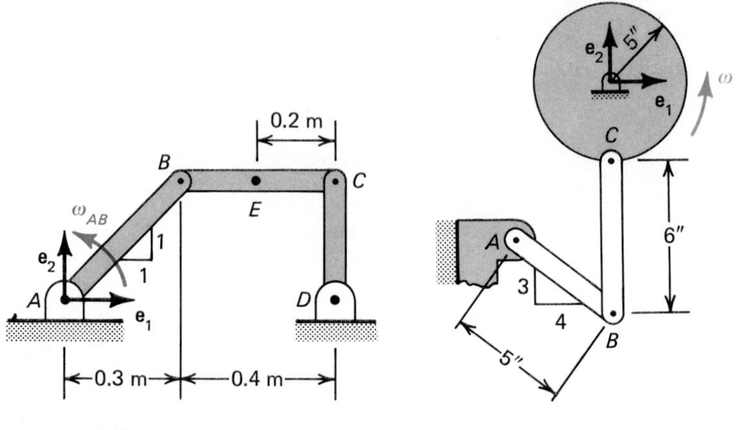

Figure P5.32 **Figure P5.35**

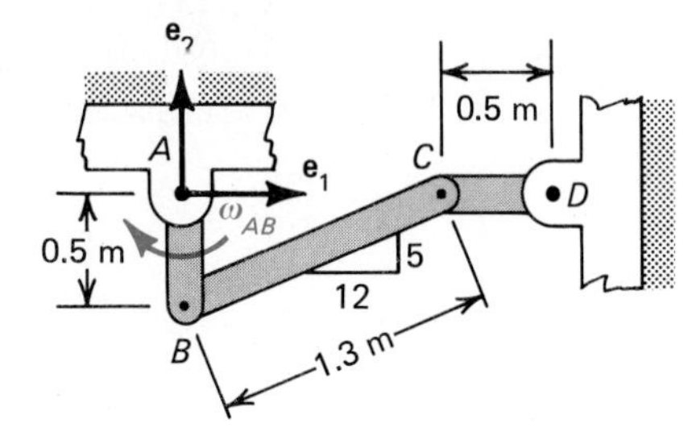

Figure P5.36

5.36 Link AB shown in Fig. P5.36 has an angular velocity of 12 rad/s CW at the instant shown. Find the angular velocities of members BC and CD.

5.37 If the angular acceleration of member AB in Problem 5.36 is 4 rad/s^2 CCW, find the angular acceleration of members BC and CD.

5.38 Two disks are joined by rod AB as shown in Fig. P5.38. At the instant shown, the left disk rotates at an angular rate of 1 rad/s CCW and an angular acceleration of 3 rad/s^2 CCW. Find the velocity and acceleration of point B.

5.39 Using the data in Problem 5.38 and Fig. P5.38, find the angular velocity and angular acceleration of the right disk.

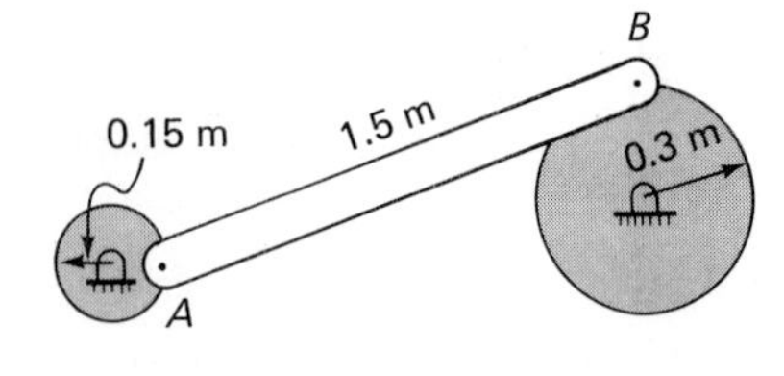

Figure P5.38

5.40 The ladder AB slides down the vertical wall at an angular rate of 1.0 rad/sec CW and an angular acceleration of 2 rad/sec^2 CW at the instant shown in Fig. P5.40. At the same time a box is sliding down the ladder at a uniform speed of 0.8 ft/sec relative to the ladder. Find the absolute velocity and acceleration of the box.

5.41 Bar AB slides up the inclined plane in Fig. P5.41 such that the velocity of point B is 6 ft/sec upward and the acceleration of point B is 4 ft/sec^2 downward. Find the velocity and acceleration of point A at the instant shown.

5.42 If a box slides down the bar AB in Problem 5.41 at a constant rate of 2 ft/sec relative to the bar, what is the velocity and acceleration of the box at the instant shown in Fig. P5.41?

5.43 The circular disk in Fig. P5.43 rolls without slipping along a curved surface whose radius of curvature is R. If the angular velocity and angular acceleration of the disk are ω CW and α CW, respectively, find the magnitude of the acceleration of point A on the disk.

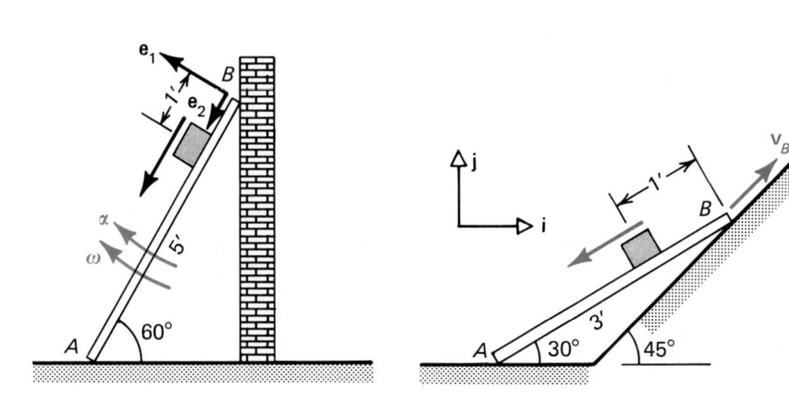

Figure P5.40

Figure P5.41

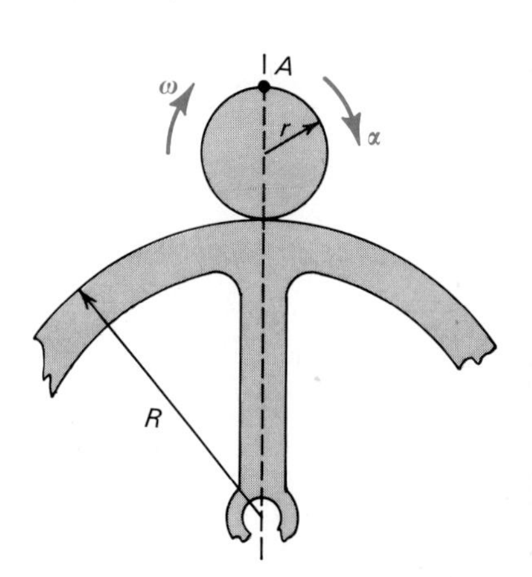

Figure P5.43

5.44 A slider crank mechanism rotates at a constant 200 rad/sec CW at the instant shown in Fig. P5.44. Find the magnitude of the velocity and acceleration of the piston B.

5.45 A cylinder rests on a flat car as shown in Fig. P5.45. Suddenly the flat car accelerates with an acceleration of 2 m/s^2 to the right and the cylinder center accelerates $\frac{2}{3}$ m/s^2 to the right. How long does it take for the cylinder to roll off the truck? Assume the cylinder rolls without slipping.

5.46 A large cylinder is supported by a smaller cylinder that runs on tracks as shown in Fig. P5.46. A constant acceleration of $a = 20$ ft/sec^2 is applied to the rim of the larger cylinder when both the cylinders have a constant rate of rotation of 10 rad/sec. What is the angular acceleration of the cylinders?

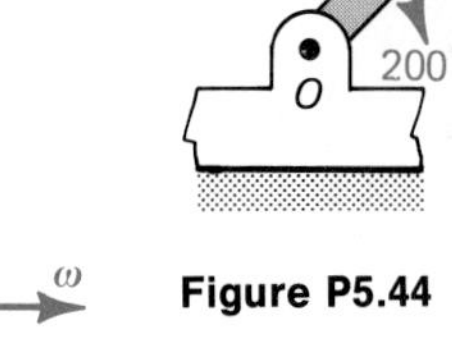

Figure P5.44

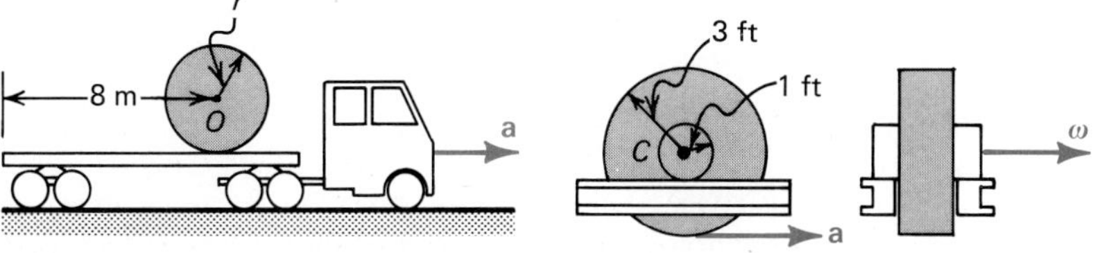

Figure P5.45

Figure P5.46

5.47 A man is climbing up a ladder as shown in Fig. P5.47 at a speed of 1 ft/sec when the ladder starts to slip so that $v_A = 0.5$ ft/sec. What is his velocity?

5.48 Shaft A of the gear system shown in Fig. P5.48 is used to drive the large wheel. If shaft A is rotating at 200 rad/s, what is the velocity of point C on the large wheel?

5.49 If the speed of point C on the wheel in Problem 5.48 is 100 m/s, how fast must shaft A rotate? Give your answer in revolutions per minute.

5.50 A weight w is used to hoist two slender bars as shown in Fig. P5.50

At the instant shown, the weight moves down at 4 ft/sec. What is the velocity of the bars at points B and A at this instant?

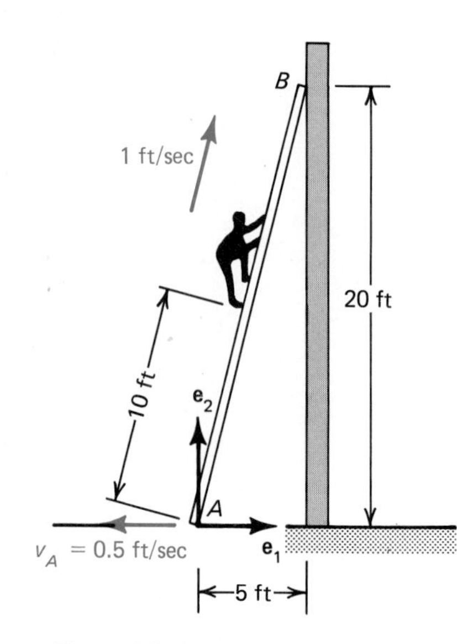

Figure P5.47

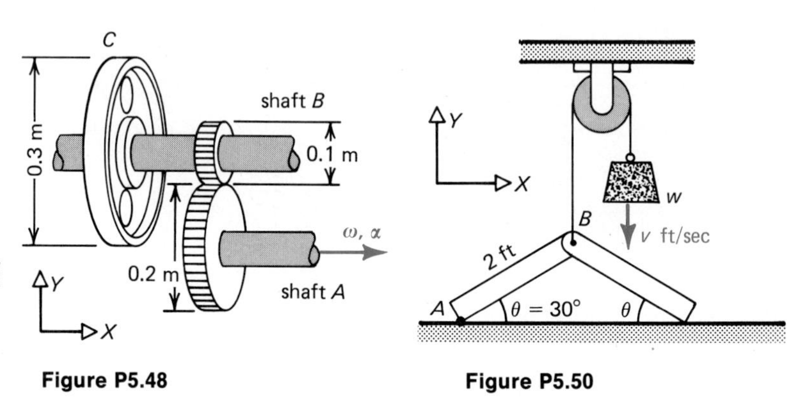

Figure P5.48

Figure P5.50

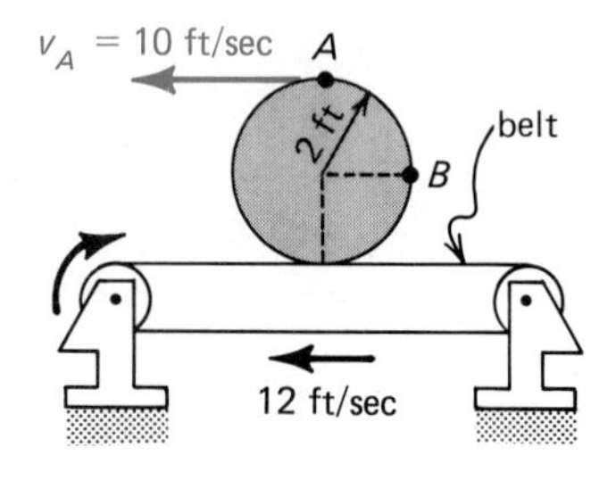

Figure P5.51

5.51 A conveyor belt travels to the right while a cylinder rolls without slipping as shown in Fig. P5.51. At the instant shown, the belt moves uniformly at 12 ft/sec while point A on the cylinder moves to the left at 10 ft/sec. Find the velocity and acceleration of point B.

5.52 Determine the angular velocity of the collapsible member AB at the instant shown in Fig. P5.52. Let the driving crank OA rotate at 20 rad/s CW and the velocity of the slider B move 0.18 m/s to the right.

5.53 What is the angular velocity of the collapsible member AB at the instant shown in Fig. P5.52 if member OA rotates at 10 rad/s CCW and the velocity of the slider B is 0.2 m/s to the left?

5.54 A cylinder is rolling without slipping down an incline as shown in Fig. P5.54. At a given instant the angular speed of the cylinder is constant at 5 rad/sec CCW and the incline is moving uniformly at 10 ft/sec. What is the velocity and acceleration of point A?

5.55 A force applied to the inclined block of Problem 5.54 causes it to accelerate to the left by 2 ft/sec^2. Find the magnitude of the acceleration of point A in Fig. P5.54 if, at the given instant, the angular speed of the cylinder is 5 rad/sec CCW and its angular acceleration is 2 rad/sec^2 CCW.

5.56 A spool of wire unwinds as shown in Fig. P5.56. The angular speed of the spool is 2 rad/s and its acceleration is 1 rad/s^2, both CCW. What is the velocity and acceleration of the center of the wheel? Neglect the thickness of the wire.

5.57 In Problem 5.56, what is the velocity and acceleration of point A viewed (a) as a point on the spool, and (b) as a point on the wire?

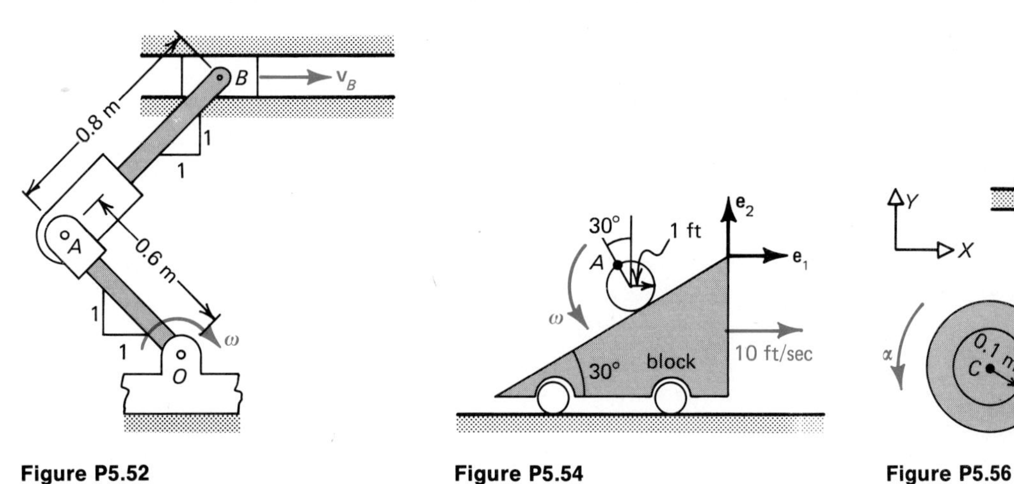

Figure P5.52 **Figure P5.54** **Figure P5.56**

5.58 A hollow half-cylinder is rotating without slipping at a constant rate of 3 rad/sec CCW as shown in Fig. P5.58. At the instant shown, a particle P is sliding at 4 ft/sec relative to the half-cylinder. Find the absolute velocity of the particle.

5.59 The gun on the tank in Fig. P5.59 rotates uniformly at $0.5\mathbf{k}$ rad/s CCW relative to the turret. At the instant shown, what is the acceleration of the point A if the tank is decelerating at 2 m/s^2?

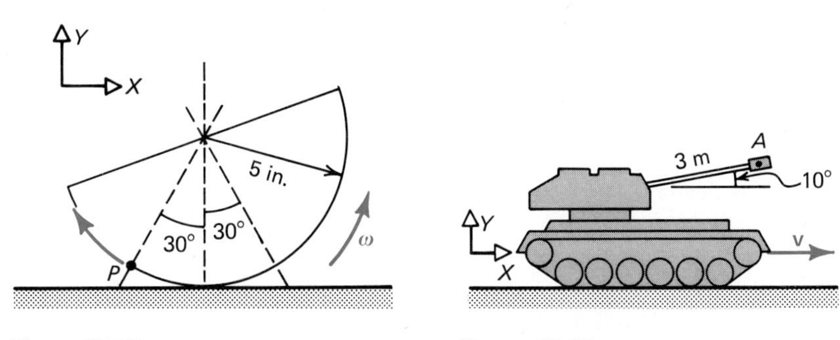

Figure P5.58 **Figure P5.59**

5.60 In Problem 5.59, assume that the tank is moving at a speed of 5 km/h to the right while the gun rotates at $0.5\mathbf{k}$ rad/s CCW relative to the turret. What is the velocity of the point A relative to the ground?

5.61 In Problem 5.60, assume the tank fires a mortar with a speed of 1000 m/s relative to the barrel. What is the velocity of the mortar relative to the ground?

5.4 Relative Motion Between Plane Rigid Bodies

In the previous section, Eqs. (5.5) and (5.6) were simplified to Eqs. (5.7) and (5.8) for cases where, because the moving origin O'

and the point P in question were *both* on the same rigid body, the terms $\dot{\mathbf{r}}_r$ and $\ddot{\mathbf{r}}_r$ vanished. There are, however, situations where we find it convenient to place the moving reference frame on one rigid body and work with $\dot{\mathbf{r}}_r$ and $\ddot{\mathbf{r}}_r$ as seen by the observer attached to the moving reference frame. For these situations, we will return to Eqs. (5.5) and (5.6).

Before we study some examples of relative motion between plane rigid bodies, a word about relative angular motion would be profitable. Consider Fig. 5.14, which shows a forearm B connected to an upper arm A. Suppose the absolute angular velocity and angular acceleration of the upper arm are $\mathbf{\Omega}$ and $\boldsymbol{\alpha}$, respectively. Furthermore, let us place a moving reference frame on body A at the elbow, O', so that the absolute angular velocity of the moving reference frame is $\mathbf{\Omega}$. Imagine that an observer at O' (actually the observer can be anywhere on body A), who is attached to the moving reference frame, measures the angular velocity of the forearm as ω_r. Thus, we can say that the absolute angular velocity of the forearm is

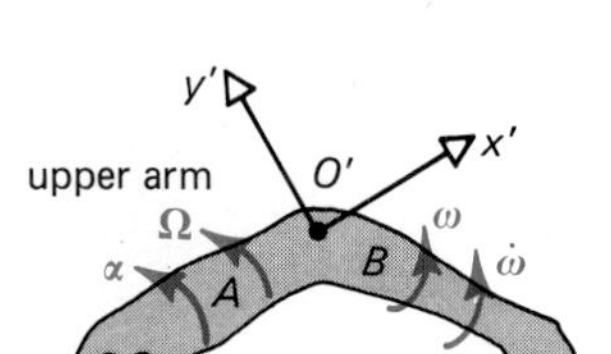

Figure 5.14

$$\omega = \mathbf{\Omega} + \omega_r \tag{5.11}$$

The absolute angular acceleration of the forearm is

$$\dot{\omega} = \dot{\mathbf{\Omega}} + \dot{\omega}_r \tag{5.12}$$

But

$$\dot{\mathbf{\Omega}} = \boldsymbol{\alpha} \tag{5.13}$$

Equation (5.12) then becomes

$$\dot{\omega} = \boldsymbol{\alpha} + \dot{\omega}_r \tag{5.14}$$

The terms for Eqs. (5.11) and (5.14) are summarized for convenience.

ω = Absolute angular velocity of body B

$\mathbf{\Omega}$ = Absolute angular velocity of body A; absolute angular velocity of the moving reference frame

ω_r = Angular velocity of body B as measured by an observer attached to body A

$\dot{\omega}$ = Absolute angular acceleration of body B

$\boldsymbol{\alpha}$ = Absolute angular acceleration of body A

$\dot{\omega}_r$ = Angular acceleration of body B as measured by an observer attached to body A

Example 5.8

The configuration in Fig. 5.15 consists of a circular wheel that rotates at an angular velocity of 6 rad/s CCW and an angular acceleration of 2 rad/s^2 CW. A simple pendulum is pinned to the wheel at O'. If the absolute angular velocity and acceleration of the pendulum are 6 rad/s CW and 1 rad/s² CW, respectively, find the angular velocity and acceleration of the pendulum as measured by an observer attached to the wheel.

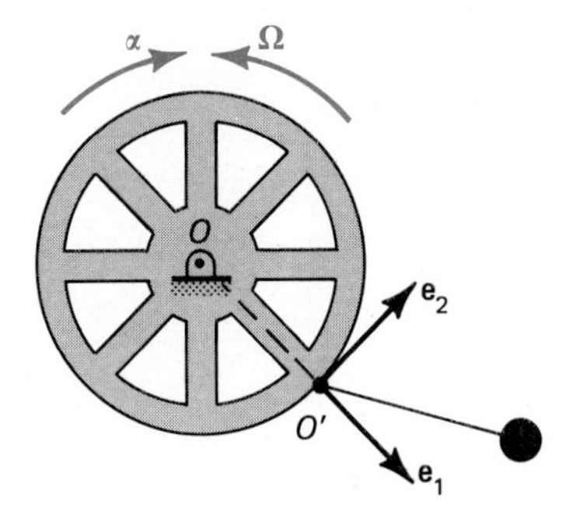

Figure 5.15

Attach the moving reference unit vectors $\mathbf{e}_1$, $\mathbf{e}_2$, $\mathbf{e}_3$ to the wheel. The following terms apply to Eqs. (5.11) and (5.14).

$$\boldsymbol{\omega} = -6\mathbf{e}_3 \qquad \dot{\boldsymbol{\omega}} = -1\mathbf{e}_3$$

$$\boldsymbol{\Omega} = 6\mathbf{e}_3 \qquad \boldsymbol{\alpha} = -2\mathbf{e}_3$$

$$\boldsymbol{\omega}_r = \omega_r\mathbf{e}_3 \qquad \dot{\boldsymbol{\omega}}_r = \dot{\omega}_r\mathbf{e}_3$$

Rearranging Eq. (5.11) to read $\boldsymbol{\omega}_r = \boldsymbol{\omega} - \boldsymbol{\Omega}$, we find that

$$\boldsymbol{\omega}_r = -6\mathbf{e}_3 - 6\mathbf{e}_3 = -12\mathbf{e}_3 \text{ rad/s} \qquad \textit{Answer}$$

Rearranging Eq. (5.14) to read $\dot{\boldsymbol{\omega}}_r = \dot{\boldsymbol{\omega}} - \boldsymbol{\alpha}$, we find that

$$\dot{\boldsymbol{\omega}}_r = -1\mathbf{e}_3 + 2\mathbf{e}_3 = 1\mathbf{e}_3 \text{ rad/s}^2 \qquad \textit{Answer}$$

Example 5.9

A 1-m radius wheel rotates at an angular velocity of 4 rad/s CCW and an angular acceleration of 2 rad/s² CCW as shown in Fig. 5.16. The 0.2-m radius wheel, pinned to the large wheel at O', rotates at a constant angular velocity of 2 rad/s CW measured relative to the large wheel. Find the acceleration of point A on the small wheel by placing the origin of the moving reference frame at O' (a) attached to the large wheel, and (b) attached to the small wheel.

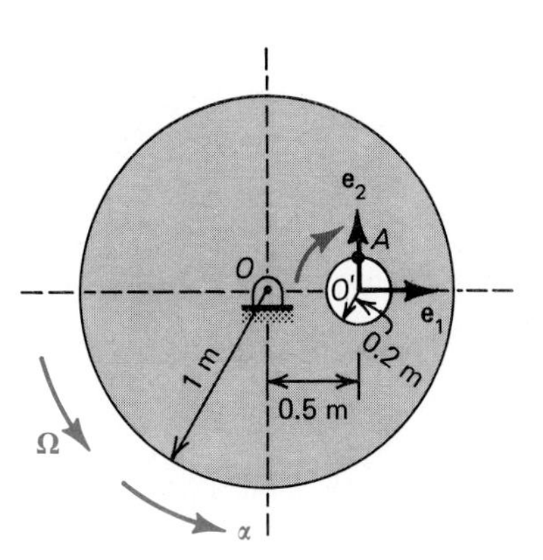

Figure 5.16

(a) When the moving reference frame is attached to the large wheel, point A moves relative to an observer attached to the moving reference frame. Hence, in accordance with Eq. (5.6),

$$\mathbf{a}_A = \ddot{\mathbf{R}} + \ddot{\mathbf{r}}_r + \boldsymbol{\alpha} \times \mathbf{r} + \boldsymbol{\omega} \times (\boldsymbol{\omega} \times \mathbf{r}) + 2\boldsymbol{\omega} \times \dot{\mathbf{r}}_r$$

where $\ddot{\mathbf{R}}$ = Acceleration of O' relative to O. Since O' moves on a circular path about O, its normal and tangential components are $-0.5(4)^2\mathbf{e}_1$ and $0.5(2)\mathbf{e}_2$. Thus,

$$\ddot{\mathbf{R}} = -8\mathbf{e}_1 + 1\mathbf{e}_2$$

$\ddot{\mathbf{r}}_r$ = Acceleration of A relative to the moving reference frame. Now A moves on a circular path relative to O' at a constant relative angular velocity, so that there is only a normal component of acceleration.

$$\ddot{\mathbf{r}}_r = -0.2(2)^2\mathbf{e}_2 = -0.8\mathbf{e}_2$$

and

$$\boldsymbol{\alpha} = 2\mathbf{e}_3 \qquad \mathbf{r} = 0.2\mathbf{e}_2 \qquad \boldsymbol{\omega} = 4\mathbf{e}_3 \qquad \dot{\mathbf{r}}_r = 0.4\mathbf{e}_1$$

$$\begin{aligned}\mathbf{a}_A &= -8\mathbf{e}_1 + 1\mathbf{e}_2 - 0.8\mathbf{e}_2 + (2\mathbf{e}_3 \times 0.2\mathbf{e}_2) \\ &\quad + 4\mathbf{e}_3 \times (4\mathbf{e}_3 \times 0.2\mathbf{e}_2) + 2(4\mathbf{e}_3 \times 0.4\mathbf{e}_1) \\ &= -8\mathbf{e}_1 + 1\mathbf{e}_2 - 0.8\mathbf{e}_2 - 0.4\mathbf{e}_1 - 3.2\mathbf{e}_2 + 3.2\mathbf{e}_2 \\ &= (-8.4\mathbf{e}_1 + 0.2\mathbf{e}_2)\ \text{m/s}^2 \qquad \textit{Answer}\end{aligned}$$

(b) By attaching the moving reference frame to the small wheel, the absolute angular velocity and acceleration of the reference frame are

$$\boldsymbol{\omega} = 4\mathbf{e}_3 - 2\mathbf{e}_3 = 2\mathbf{e}_3$$

$$\boldsymbol{\alpha} = 2\mathbf{e}_3$$

In this case, $\dot{\mathbf{r}}_r = \ddot{\mathbf{r}}_r = 0$ so that we can use Eq. (5.8).

$$\mathbf{a}_A = \ddot{\mathbf{R}} + \boldsymbol{\alpha} \times \mathbf{r} + \boldsymbol{\omega} \times (\boldsymbol{\omega} \times \mathbf{r})$$

where

$$\ddot{\mathbf{R}} = -8\mathbf{e}_1 + 1\mathbf{e}_2$$

$$\boldsymbol{\alpha} \times \mathbf{r} = 2\mathbf{e}_3 \times 0.2\mathbf{e}_2 = -0.4\mathbf{e}_1$$

$$\boldsymbol{\omega} \times (\boldsymbol{\omega} \times \mathbf{r}) = 2\mathbf{e}_3 \times (2\mathbf{e}_3 \times 0.2\mathbf{e}_2) = -0.8\mathbf{e}_2$$

and therefore

$$\begin{aligned}\mathbf{a}_A &= -8\mathbf{e}_1 + 1\mathbf{e}_2 - 0.4\mathbf{e}_1 - 0.8\mathbf{e}_2 \\ &= (-8.4\mathbf{e}_1 + 0.2\mathbf{e}_2)\ \text{m/s}^2 \qquad \textit{Answer}\end{aligned}$$

Example 5.10

Figure 5.17(a) shows member *BC* attached at *B* to a block that slides along member *AD*. At the instant shown, the angular velocity of *BC* is 2 rad/sec CCW. Find the velocity of point *B* as viewed by an observer attached to member *AD*, and the angular velocity of member *AD*.

Place the rotating reference frame on member *AD* at point *A*. Using this reference frame and Eq. (5.5) we have

$$\mathbf{v}_B = \dot{\mathbf{R}} + \dot{\mathbf{r}}_r + \boldsymbol{\omega} \times \mathbf{r}$$

where

$$\dot{\mathbf{R}} = 0 \qquad \boldsymbol{\omega} = \omega_{AD}\mathbf{e}_3$$

$$\dot{\mathbf{r}}_r = \dot{r}\mathbf{e}_1 \qquad \mathbf{r} = 1\mathbf{e}_1$$

and therefore

$$\mathbf{v}_B = \dot{r}\mathbf{e}_1 + \omega_{AD}\mathbf{e}_2 \qquad \text{(a)}$$

Since there are two unknowns on the right side of Eq. (a), a second equation is needed. Point *B* is a point on member *BC* as seen in Fig. 5.17(b). Note that the moving reference frame at *C* at the instant shown is oriented in the same way as that at point A. Now, from Eq. (5.9), we have

$$\mathbf{v}_B = \boldsymbol{\omega} \times \mathbf{r}$$

where

$$\boldsymbol{\omega} = 2\mathbf{e}_3 \qquad \text{and} \qquad \mathbf{r} = 0.6(\cos 60^\circ \mathbf{e}_1 - \sin 60^\circ \mathbf{e}_2)$$

and therefore

$$\begin{aligned} \mathbf{v}_B &= 2\mathbf{e}_3 \times 0.6(\cos 60^\circ \mathbf{e}_1 - \sin 60^\circ \mathbf{e}_2) \\ &= 0.6\mathbf{e}_2 + 1.04\mathbf{e}_1 \qquad \text{(b)} \end{aligned}$$

Substitute Eq. (b) into the left side of Eq. (a).

$$0.6\mathbf{e}_2 + 1.04\mathbf{e}_1 = \dot{r}\mathbf{e}_1 + \omega_{AD}\mathbf{e}_2$$

From the $\mathbf{e}_1$ equation we obtain $\dot{r} = 1.04$, so that

$$\dot{\mathbf{r}}_r = 1.04\mathbf{e}_1 \text{ m/s} \qquad \textit{Answer}$$

From the $\mathbf{e}_2$ equation we obtain $\omega_{AD} = 0.6$, therefore

$$\boldsymbol{\omega}_{AD} = 0.6\mathbf{e}_3 \text{ rad/s} \qquad \textit{Answer}$$

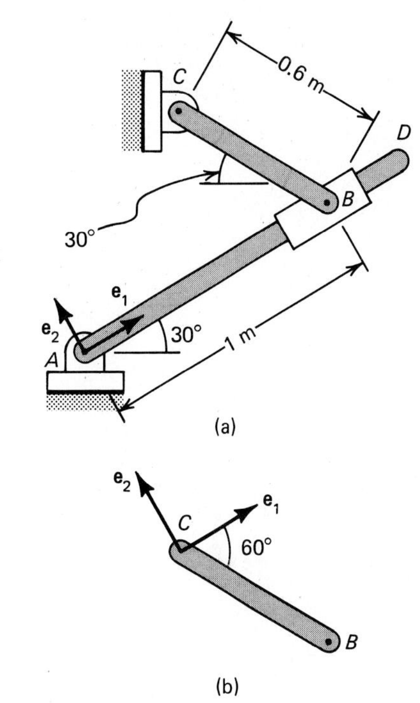

Figure 5.17

Example 5.11

Find the acceleration of point B in Fig. 5.17(a) as viewed by an observer attached to member AD. Assume the angular acceleration to be 1 rad/s^2 CCW and angular velocity of 2 rad/s CCW for member BC.

To find the acceleration of point B as viewed by the rotating observer on AD, attach the moving reference frame to member AD so that

$$\mathbf{a}_B = \ddot{\mathbf{R}} + \ddot{\mathbf{r}}_r + \boldsymbol{\alpha} \times \mathbf{r} + \boldsymbol{\omega} \times (\boldsymbol{\omega} \times \mathbf{r}) + 2\boldsymbol{\omega} \times \dot{\mathbf{r}}_r \qquad \text{(a)}$$

where

$$\begin{aligned} \ddot{\mathbf{R}} &= 0 \\ \ddot{\mathbf{r}}_r &= \ddot{r}\mathbf{e}_1 \\ \boldsymbol{\alpha} &= \alpha_{AB}\mathbf{e}_3 \\ \mathbf{r} &= 1\mathbf{e}_1 \\ \boldsymbol{\omega} &= 0.6\mathbf{e}_3 \qquad \text{(see Example 5.10)} \\ \dot{\mathbf{r}}_r &= 1.04\mathbf{e}_1 \qquad \text{(see Example 5.10)} \end{aligned}$$

Substituting into Eq. (a) yields

$$\begin{aligned} \mathbf{a}_B &= \ddot{r}\mathbf{e}_1 + \alpha_{AB}\mathbf{e}_2 - 0.36\mathbf{e}_1 + 1.248\mathbf{e}_2 \\ &= (\ddot{r} - 0.36)\mathbf{e}_1 + (\alpha_{AB} + 1.248)\mathbf{e}_2 \qquad \text{(b)} \end{aligned}$$

We now need to use the simplified Eq. (5.10) to obtain the acceleration of point B as a point on member BC. Again, we refer to Fig. 5.17(b).

$$\mathbf{a}_B = \boldsymbol{\alpha} \times \mathbf{r} + \boldsymbol{\omega} \times (\boldsymbol{\omega} \times \mathbf{r})$$

where

$$\boldsymbol{\alpha} = 1\mathbf{e}_3$$
$$\mathbf{r} = 0.6(\cos 60^\circ \mathbf{e}_1 - \sin 60^\circ \mathbf{e}_2)$$
$$\boldsymbol{\omega} = 2\mathbf{e}_3$$

so that

$$\begin{aligned}\mathbf{a}_B &= 0.3\mathbf{e}_2 + 0.52\mathbf{e}_1 - 1.2\mathbf{e}_1 + 2.08\mathbf{e}_2 \\ &= -0.68\mathbf{e}_1 + 2.38\mathbf{e}_2\end{aligned} \tag{c}$$

Equating the right side of Eq. (c) to the right side of Eq. (b),

$$-0.68\mathbf{e}_1 + 2.38\mathbf{e}_2 = (\ddot{r} - 0.36)\mathbf{e}_1 + (\alpha_{AB} + 1.248)\mathbf{e}_2$$

The $\mathbf{e}_1$ equation yields

$$\ddot{r} = -0.32$$

and therefore $\quad \ddot{\mathbf{r}}_r = -0.32\mathbf{e}_1 \text{ m/s}^2 \qquad$ *Answer*

Example 5.12

Bar AB is connected by pin C to the slider D that rides along the fixed vertical rod EF as shown in Fig. 5.18. Find the velocity and acceleration of the slider D if the bar AB rotates at a constant angular velocity of 10 rad/s CCW. Place O' on member AB at point A.

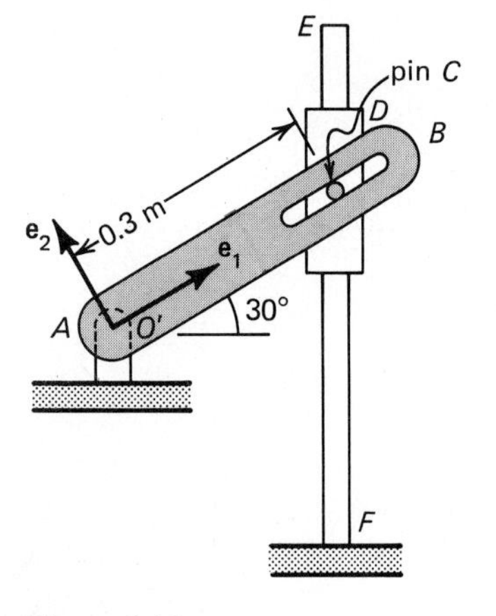

Figure 5.18

If the moving reference frame is attached to member AB at A, we turn to Eq. (5.5) to obtain the velocity of the slider $\mathbf{v}_D$.

$$\mathbf{v}_D = \dot{\mathbf{R}} + \dot{\mathbf{r}}_r + \boldsymbol{\omega} \times \mathbf{r}$$

where $\quad \dot{\mathbf{R}} = 0 \qquad \mathbf{r} = 0.3\mathbf{e}_1 \qquad \dot{\mathbf{r}}_r = \dot{r}\mathbf{e}_1 \qquad \boldsymbol{\omega} = 10\mathbf{e}_3$

and therefore

$$\mathbf{v}_D = 10\mathbf{e}_3 \times 0.3\mathbf{e}_1 + \dot{r}\mathbf{e}_1 = 3\mathbf{e}_2 + \dot{r}\mathbf{e}_1 \tag{a}$$

Since the slider moves vertically, we know the line of direction of the velocity and acceleration of the slider is along the line EF. Using the unit vectors associated with the moving reference frame at A, we can express the velocity of point D as follows:

$$\mathbf{v}_D = v_D(\sin 30^\circ \mathbf{e}_1 + \cos 30^\circ \mathbf{e}_2) = 0.5v_D\mathbf{e}_1 + 0.866v_D\mathbf{e}_2 \tag{b}$$

Substitute Eq. (b) into the left side of Eq. (a).

$$0.5v_D\mathbf{e}_1 + 0.866v_D\mathbf{e}_2 = 3\mathbf{e}_2 + \dot{r}\mathbf{e}_1$$

From the $\mathbf{e}_2$ equation and the $\mathbf{e}_1$ equation, respectively, we find

$$v_D = 3.46 \qquad \dot{r} = 1.73$$

Therefore, $\mathbf{v}_D = (1.73\mathbf{e}_1 + 3\mathbf{e}_2)$ m/s *Answer*

For the acceleration of point D, we refer first to Eq. (5.6) again with the origin of the moving reference frame at point A.

$$\mathbf{a}_D = \ddot{\mathbf{R}} + \ddot{\mathbf{r}}_r + \boldsymbol{\alpha} \times \mathbf{r} + \boldsymbol{\omega} \times (\boldsymbol{\omega} \times \mathbf{r}) + 2\boldsymbol{\omega} \times \dot{\mathbf{r}}_r$$

where

$$\ddot{\mathbf{R}} = 0$$
$$\ddot{\mathbf{r}}_r = \ddot{r}\mathbf{e}_1$$
$$\boldsymbol{\alpha} = 0$$
$$\mathbf{r} = 0.3\mathbf{e}_1$$
$$\boldsymbol{\omega} = 10\mathbf{e}_3$$

and

$$\boldsymbol{\omega} \times (\boldsymbol{\omega} \times \mathbf{r}) = 10\mathbf{e}_3 \times (10\mathbf{e}_3 \times 0.3\mathbf{e}_1) = -30\mathbf{e}_1$$

$$2\boldsymbol{\omega} \times \dot{\mathbf{r}}_r = 20\mathbf{e}_3 \times 1.73\mathbf{e}_1 = 34.6\mathbf{e}_2$$

so that

$$\mathbf{a}_D = (\ddot{r} - 30)\mathbf{e}_1 + 34.6\mathbf{e}_2 \qquad \text{(c)}$$

Then, knowing the direction of the slider motion, we can express the acceleration of point D in terms of the unit vectors associated with the moving reference frame at A.

$$\mathbf{a}_D = 0.5a_D\mathbf{e}_1 + 0.866a_D\mathbf{e}_2 \qquad \text{(d)}$$

Equating the right sides of Eqs. (c) and (d),

$$0.5a_D\mathbf{e}_1 + 0.866a_D\mathbf{e}_2 = (\ddot{r} - 30)\mathbf{e}_1 + 34.6\mathbf{e}_2$$

From the $\mathbf{e}_2$ equation

$$a_D = 39.95$$

and

$$\mathbf{a}_D = (20.0\mathbf{e}_1 + 34.6\mathbf{e}_2)\ \text{m/s}^2 \qquad \textit{Answer}$$

PROBLEMS

5.62 A truck carries a worker for checking electrical fixtures as shown in Fig. P5.62. At the instant shown, $\omega_{AB} = 1$ rad/sec CCW and the truck is at rest. The cockpit rotates at $-\omega_{AB}$ relative to member AB so that the worker stays upright. What is the velocity of point C?

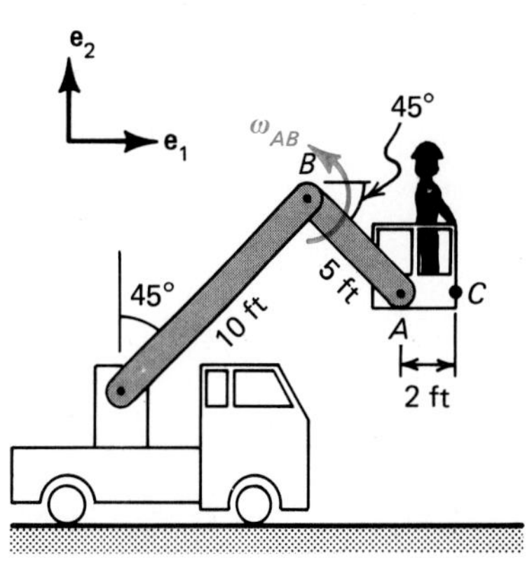

Figure P5.62

5.63 In Problem 5.62, assume that the truck has a velocity of $2\mathbf{e}_1$ ft/sec. What is the velocity of point C?

5.64 Member AB moves independently of member OA for the double pendulum shown in Fig. P5.64. Both members are moving in the plane as shown, and member OA turns at a uniform rate of 10 rad/s CCW while member AB turns at an angular velocity of 6 rad/s CW and at an angular acceleration of 4 rad/s^2 CW at the instant shown. What is the angular velocity and angular acceleration of member AB relative to an observer attached to member OA?

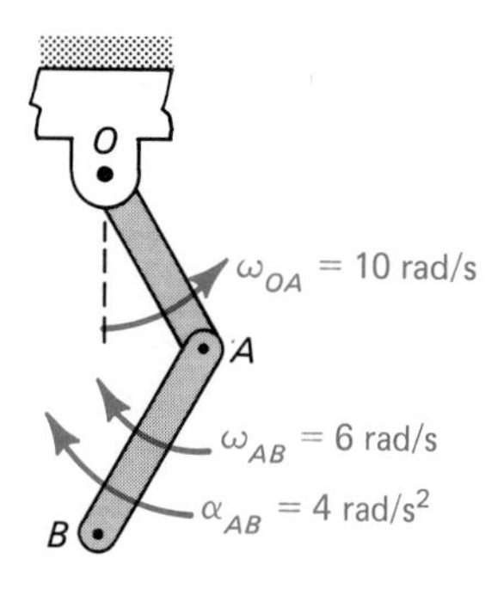

Figure P5.64

5.65 The motion of a satellite is controlled by a despinning device called a yo-yo, consisting of two masses that unwind from the satellite as shown in Fig. P5.65. At the instant shown the satellite is spinning at 4 rad/s CW and 2 rad/s^2 CW, while the masses are turning relative to the satellite at a uniform velocity of 1 rad/s CCW. What is the absolute angular velocity and angular acceleration of the despinning masses?

5.66 The circular wheel of radius 4 ft in Fig. P5.66 rotates at a uniform angular velocity of 4 rad/sec CW. A small wheel pinned at O' to the larger wheel rotates at a uniform velocity of 5 rad/sec CW relative to the large wheel. Find the absolute angular velocity of the small wheel.

5.67 If the large wheel of Problem 5.66 rotates at a uniform angular velocity of 4 rad/sec CCW, what is the absolute angular velocity of the small wheel?

5.68 In Problem 5.66, let the large wheel also have an angular acceleration of 6 rad/sec^2 CCW. Find the acceleration magnitude of point A on the small wheel, assuming the small wheel has zero angular acceleration.

5.69 The 3-m radius wheel in Fig. P5.69 is pinned at O and rotates at an angular velocity of 10 rad/s CCW and an angular acceleration of 5 rad/s^2 CCW. A 1-m radius wheel pinned at O' to the larger wheel rotates at an absolute uniform angular velocity of 10 rad/s CW. Find the angular velocity and angular acceleration of the small wheel relative to the large wheel.

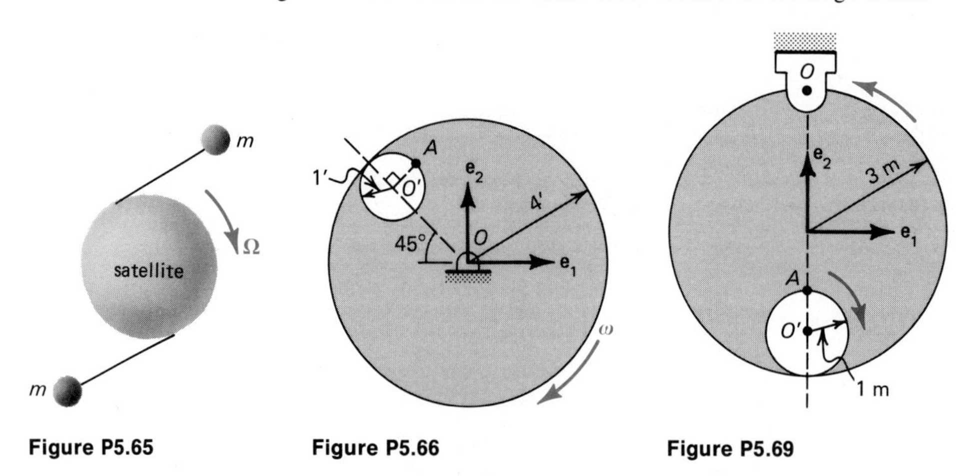

Figure P5.65 Figure P5.66 Figure P5.69

5.70 Using the data in Problem 5.69, find the acceleration of point A which is a point on the small wheel.

5.71 Find the velocity and acceleration of the sliding block C if the rod OA in Fig. P5.71 rotates at an angular velocity of 5 rad/sec CW and an angular acceleration of 10 rad/sec^2 CCW.

5.72 Find the velocity and acceleration of the sliding block C if the rod OA in Problem 5.71 rotates at an angular velocity of 10 rad/sec CCW and an angular acceleration of 5 rad/sec^2 CW.

5.73 A sliding collar at B attaches member AB to member CD as shown in

Fig. P5.73. If the angular velocity of member AB is uniform at 10 rad/sec CCW, find the angular velocity of member CD at the instant shown.

5.74 If member AB in Problem 5.73 has an angular acceleration of 2 rad/sec^2 CCW, obtain the angular acceleration of member CD at the instant shown in Fig. P5.73.

5.75 A crank shaper mechanism is shown in Fig. P5.75. The uniform angular velocity of member AC is 4 rad/s CCW. Find the velocity and acceleration of point B on member OB at the instant shown.

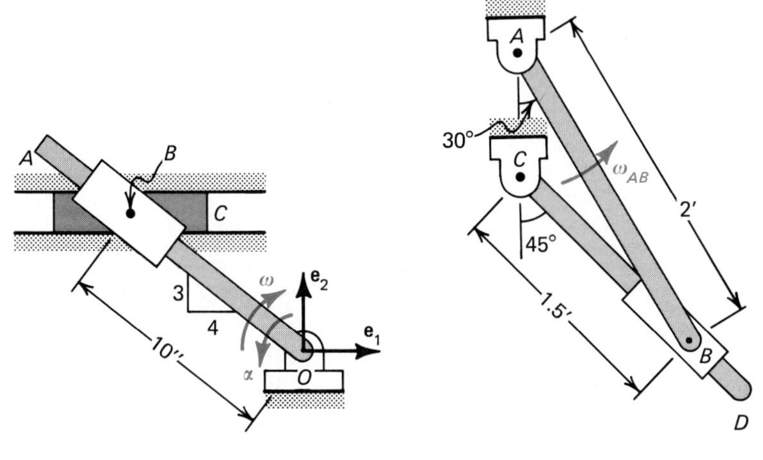

Figure P5.71

Figure P5.73

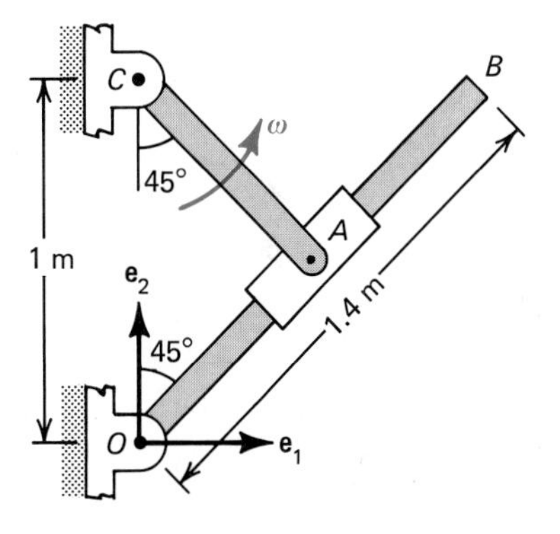

Figure P5.75

5.76 If the member AC also accelerates at 10 rad/s^2 CW, obtain the velocity of point B on member OB and the angular acceleration of OA in Problem 5.75.

5.5 Instantaneous Center of Zero Velocity

Our discussion on instantaneous centers of zero velocity for rigid bodies is concerned with the following simplified equation:

$$\mathbf{v} = \boldsymbol{\omega} \times \mathbf{r} \tag{5.9}$$

This equation is Eq. (5.7) simplified for the case where the origin of the moving reference frame, O', coincides with the fixed reference point O. Until now, we generally have used fixed-support conditions when we let O' coincide with O, so it has been apparent where a point of zero velocity existed. However, we recall that in Example 5.4 point C on the rolling circular cylinder had zero velocity, and in this case it was not readily apparent that point C had no velocity.

In general, a planar rigid body has a point of zero velocity associated with it which may be on the body or off the body. Such

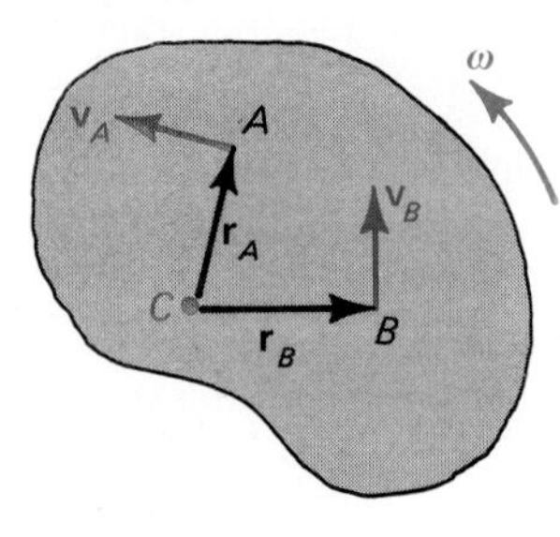

Figure 5.19

points are called instantaneous centers of zero velocity. Suppose point C of the body shown in Fig. 5.19 is a point of zero velocity. Then the velocity of point A is

$$\mathbf{v}_A = \boldsymbol{\omega} \times \mathbf{r}_A$$

when ω is the angular velocity of the body. We note that the vector $\mathbf{v}_A$ is of magnitude $r_A\omega$ and is perpendicular to $\mathbf{r}_A$. Suppose point B is some other point on the body. Then,

$$\mathbf{v}_B = \boldsymbol{\omega} \times \mathbf{r}_B$$

so that the velocity magnitude is $r_B\omega$ and the direction is perpendicular to $\mathbf{r}_B$.

It follows from this example that if we know the velocity of a point on the rigid body and the direction of the angular velocity of the body, we can locate the instantaneous center of zero velocity. Likewise, if we know the lines of direction of the velocity of two points on the body, we can construct perpendicular lines that will intersect at the instantaneous center.

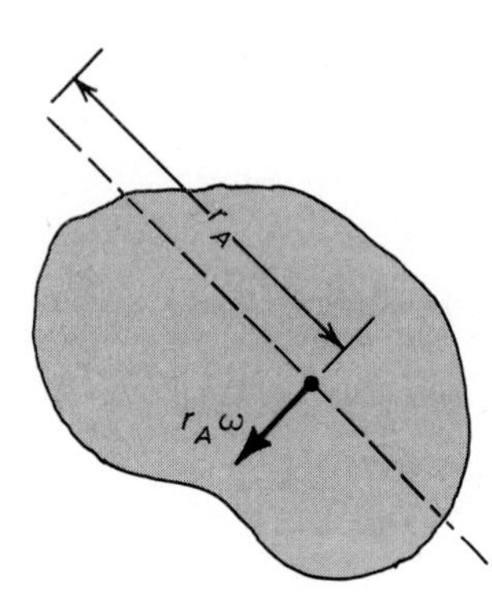

Figure 5.20

In the first case, suppose we know the angular velocity is clockwise for the body shown in Fig. 5.20. Knowing the magnitude and direction of $\mathbf{v}_A$, we locate point C on a line perpendicular to $\mathbf{v}_A$ as the instantaneous center of zero velocity. Figure 5.21 shows the case in which we know the line of direction of velocity for two points on the body. Here, we simply construct dashed lines perpendicular to the velocity vectors and locate the instantaneous center of zero velocity where these lines intersect. If the lines do not intersect, the instantaneous center of zero velocity is at infinity and the rigid body has no angular velocity at the instant under consideration. In other words, the body is moving in pure translation.

A few observations are in order at this time on instantaneous centers of zero velocity. We note that the problems of the kinematics of rigid bodies can be solved without the introduction of instantaneous centers of zero velocity. In some problems we may find that using instantaneous centers will significantly reduce the amount of work required to solve the problem. However, a word of caution is necessary. Instantaneous centers of zero velocity are exactly that—points of zero velocity. The danger here is that we may be tempted to call them points of zero acceleration, which they are not. Example 5.4 clearly showed that in Fig. 5.4 point C is a point of zero velocity having an acceleration equal to $b\omega^2\mathbf{e}_2$. Thus, point C is the instantaneous center of zero velocity for the rolling cylinder when there is no slippage present.

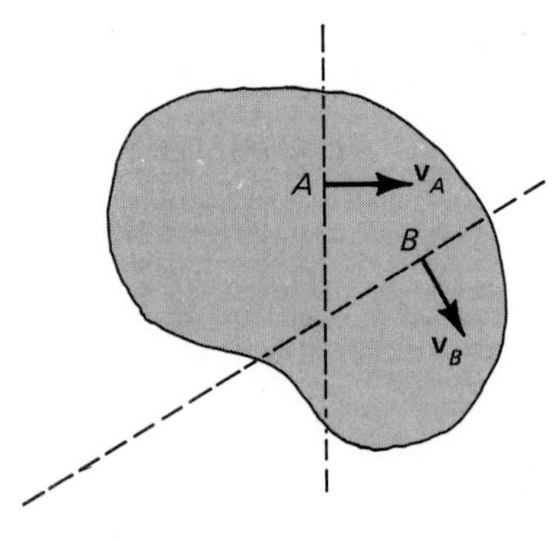

Figure 5.21

Example 5.13

Locate the instantaneous centers of zero velocity of members AB, BC, and CD for the four-link mechanism shown in Fig. 5.22. Also, find the angular velocity of BC if the angular velocity of CD is 3 rad/s CW.

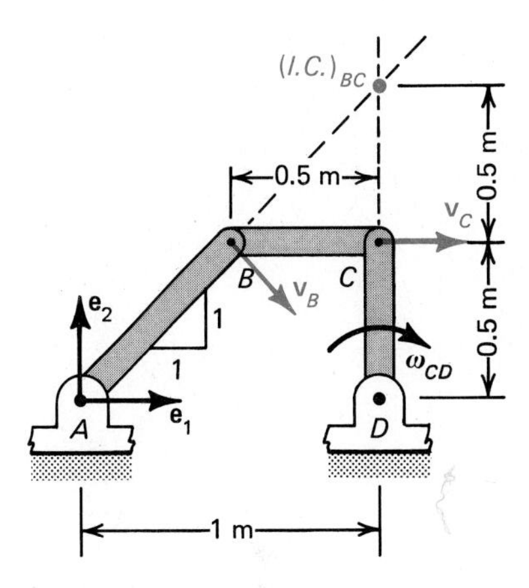

Figure 5.22

The instantaneous center of zero velocity of member AB is at A, while for member CD it is point D. We know the line of direction of the velocity of point B, since it is a point on member AB; and we know the direction (and magnitude) of the velocity of point C, since it is a point on member CD. Consequently, construction of the dashed lines shown in Fig. 5.22 locates the instantaneous center of zero velocity for member BC. This is labeled $(IC)_{BC}$ in the figure.

Having established the instantaneous center for BC, we can now calculate the angular velocity of BC. Considering member CD,

$$\mathbf{v}_C = \boldsymbol{\omega}_{CD} \times \mathbf{r}_{DC}$$
$$= -3\mathbf{e}_3 \times 0.5\mathbf{e}_2 = 1.5\mathbf{e}_1 \qquad \text{(a)}$$

Note that all unit vectors are referenced to the $\mathbf{e}_1$, $\mathbf{e}_2$, $\mathbf{e}_3$ unit vector system shown in Fig. 5.22. But point C is a point on the extended member BC (that is, we are including the $(IC)_{BC}$ as a point on the body). Now

$$\mathbf{v}_C = \boldsymbol{\omega}_{BC} \times \mathbf{r}_C$$

where

$$\boldsymbol{\omega}_{BC} = \omega_{BC}\mathbf{e}_3$$
$$\mathbf{r}_C = -0.5\mathbf{e}_2$$

so that

$$\mathbf{v}_C = \omega_{BC}\mathbf{e}_3 \times (-0.5\mathbf{e}_2) = 0.5\omega_{BC}\mathbf{e}_1 \qquad \text{(b)}$$

Equating the right sides of Eqs. (a) and (b), we obtain

$$\omega_{BC} = 3$$

or

$$\boldsymbol{\omega}_{BC} = 3\mathbf{e}_3 \text{ rad/s} \qquad \textit{Answer}$$

Example 5.14

The ladder in Fig. 5.23(a) slides down the smooth inclined plane. The speed of point A is v to the left. Using the instantaneous center of zero velocity of the ladder, find the angular velocity of the ladder and the speed of point B.

Figure 5.23(b) shows the construction for locating the instantaneous center of zero velocity of the ladder and the distances from

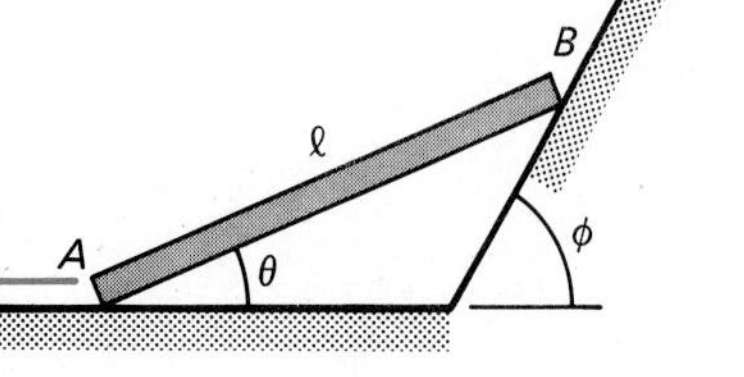

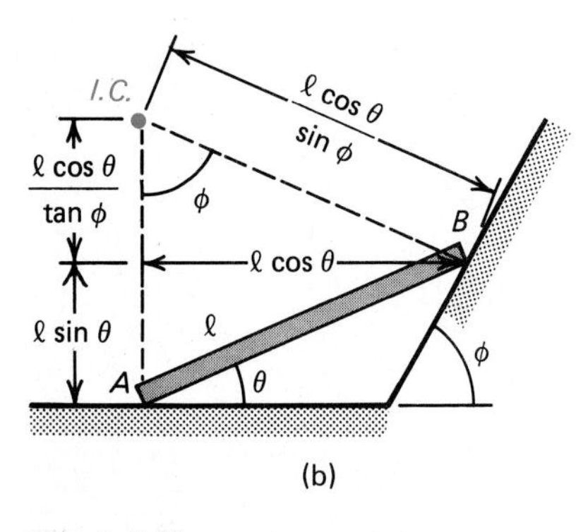

Figure 5.23

the instantaneous center to points A and B. Since we know the speed at point A, it follows that

$$\omega = \frac{v}{l \sin\theta + (l \cos\theta)/\tan\phi} \qquad \textit{Answer}$$

and

$$v_B = \frac{l \cos\theta}{\sin\phi}\,\omega$$

$$v_B = v\,\frac{\cos\theta}{\sin\theta\sin\phi + \cos\theta\cos\phi}$$

$$= v\,\frac{\cos\theta}{\cos(\theta - \phi)} \qquad \textit{Answer}$$

Note that when $\phi = 0°$, $v_B = v$, and when $\phi = 90°$, $v_B = v/\tan\theta$.

PROBLEMS

5.77 In Fig. 5.23, let $l = 10$ m, $\theta = 30°$, $\phi = 60°$ and assume that point B is slipping down the inclined wall at 4 m/s. Use the instantaneous center approach to obtain the velocity of point A.

5.78 Find the velocity of point A in Problem 5.23 using the instantaneous center method.

5.79 Solve Problem 5.28 using the instantaneous center method.

5.80 Solve Problem 5.32 using the instantaneous center method.

5.81 Find the velocity of point B in Problem 5.35 using the instantaneous center method.

5.82 Solve Problem 5.36 using the instantaneous center method.

5.83 Solve Problem 5.47 using the instantaneous center method.

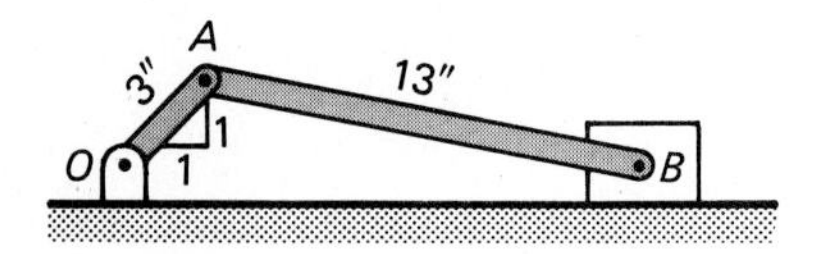

Figure P5.84

5.84 The slider crank mechanism shown in Fig. P5.84 rotates at 1500 rev/min CW. Using the instantaneous center of zero velocity, find (a) the angular velocity of member AB and (b) the velocity of B.

5.85 If the instantaneous velocity of B in Fig. P5.84 is 30 ft/sec to the right, use the instantaneous method to obtain the angular velocity of member OA.

5.86 The 0.1-m radius wheel in Fig. P5.86 rotates at 6 rad/s CW. Using the instantaneous center of zero velocity, find the angular velocity of (a) member AB and (b) the 0.25-m radius wheel.

5.87 Point B in Fig. P5.87 travels at a speed of 10 ft/sec to the left along the smooth floor. Find the magnitude of the velocity of point C using the method of instantaneous center of zero velocity.

5.88 Block B travels down the vertical wall at a speed of 8 ft/sec in Fig.

P5.88. Find the angular velocity of member *BC* using the method of instantaneous center of zero velocity.

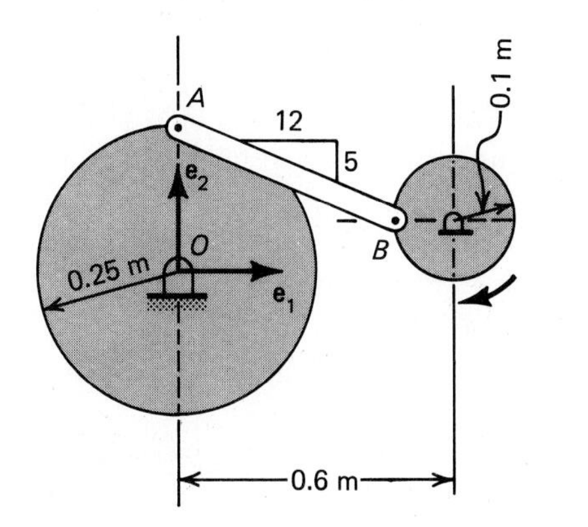

Figure P5.86

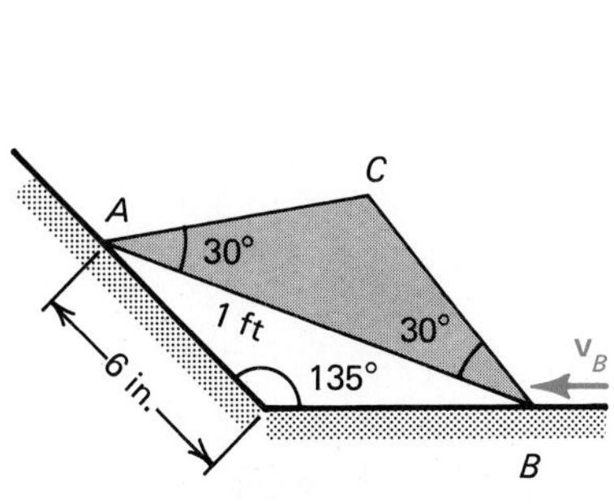

Figure P5.87

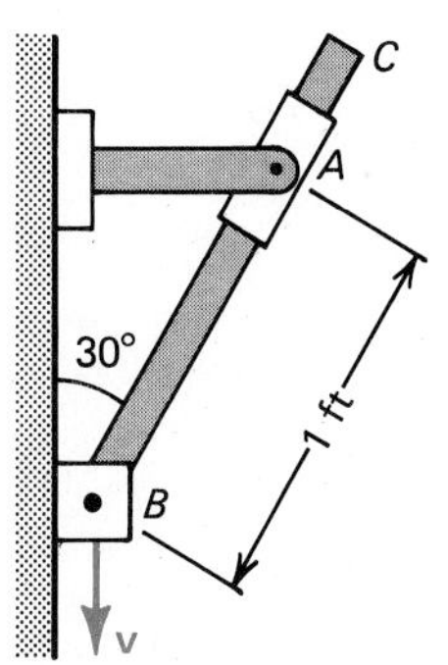

Figure P5.88

5.89 Locate the instantaneous centers of zero velocity for each member of the linkage mechanism in Fig. P5.89

5.90 Use the mechanism in Fig. P5.89 to find the angular velocity of member *AB* if member *BC* rotates 500 rev/min CW at the instant shown.

5.91 For the mechanism shown in Fig. P5.91, link *EB* has a clockwise rotation rate of 5 rad/s at the instant shown. Determine the magnitude of the velocity of point *A*.

5.92 For the mechanism shown in Fig. P5.91 let link *EB* have a counterclockwise rotation of 5 rad/s at the instant shown. Determine the angular velocity of member *CD*.

5.93 The body *B* shown in Fig. P5.93 rolls without slipping across the fixed half-cylinder *A*. At the instant shown, the block *D* moves at a speed of 5 m/s to the left. Using instantaneous centers, find the angular velocity of the rod *CD* which is 1.6 m long.

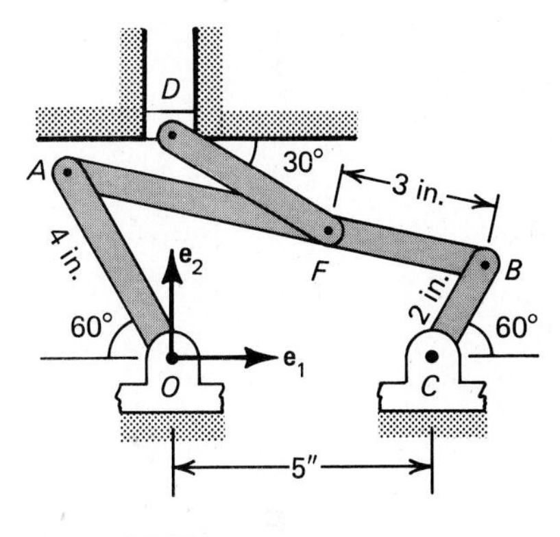

Figure P5.89

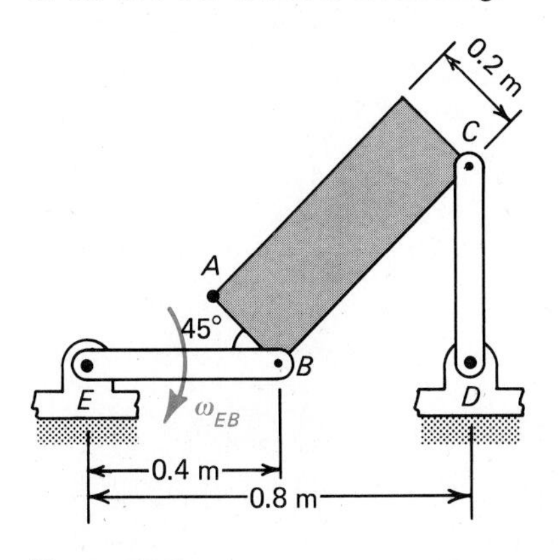

Figure P5.91

Figure P5.93

5.6 Summary

When a point on a rigid body has a velocity relative to another point on the rigid body, the body undergoes rotational motion. In general, a rigid body undergoes translational as well as rotational motion. If the motion is restricted to one plane, the rigid body is said to experience planar motion. Here, we derived the kinematic relationships of a rigid body in planar motion.

A. The important *ideas*:

1. When a point on a rigid body has a velocity relative to another point on that rigid body, the body undergoes rotational motion.
2. In studying the rotational motion of a rigid body, we introduced the relationship between the angular momentum and the mass moment of inertia.
3. The angular velocity and angular acceleration are free vectors.
4. The instantaneous center of zero velocity for a rigid body need not be a point on the body itself.
5. The acceleration at the instantaneous center of zero velocity is not necessarily zero.

B. The important *equations*:

Mass moment of inertia:

$$I_o = \int r^2\, dm \tag{5.4}$$

Plane motion of rigid bodies:

$$\mathbf{v} = \dot{\mathbf{R}} + \boldsymbol{\omega} \times \mathbf{r} \tag{5.7}$$

$$\mathbf{a} = \ddot{\mathbf{R}} + \boldsymbol{\alpha} \times \mathbf{r} + \boldsymbol{\omega} \times (\boldsymbol{\omega} \times \mathbf{r}) \tag{5.8}$$

Relative motion between plane rigid bodies:

$$\boldsymbol{\omega} = \boldsymbol{\Omega} + \boldsymbol{\omega}_r \tag{5.11}$$

$$\dot{\boldsymbol{\omega}} = \boldsymbol{\alpha} + \dot{\boldsymbol{\omega}}_r \tag{5.14}$$

6 Planar Kinetics of Rigid Bodies

From the study of planar kinematics of rigid bodies in the last chapter, we were able to describe the motion of rigid bodies without regard to the applied forces. We now consider the planar kinetics of rigid bodies. This means that in our analysis we shall include the forces acting on the rigid body as well as the resulting motion.

In this chapter we shall develop the equations of motion for both a system of finite particles and for plane rigid bodies. The equations include the direct application of Newton's second law, the work-energy relationships, and the impulse-momentum equations. We shall discover that while our approach to developing the planar kinetics of rigid bodies is similar to the kinetics of a particle that we studied in Chapter 3, additional equations become necessary in order to account for the rotational motion of the plane rigid body.

6.1 System of Particles

Equations of Motion

Suppose we have an isolated system composed solely of N particles and we wish to develop the equations of motion for the particles. Figure 6.1(a) shows a portion of the particles and includes a free-body diagram of the forces acting on the particle m_i. The force $\mathbf{F}_i$ represents the resultant external force acting on m_i. Forces $\mathbf{f}_{i1}$ and $\mathbf{f}_{i2}$ are internal forces acting between the particles. Let $\mathbf{f}_{ij}$ represent the general form of these internal forces where $\mathbf{f}_{ij}$ reads as the internal force acting on m_i due to the presence of m_j. These internal forces may be due to rigid or elastic connections between

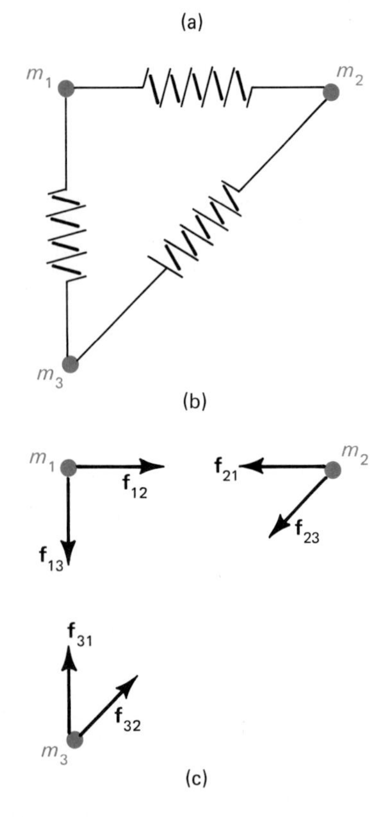

Figure 6.1

the particles, gravitational or magnetic forces between the particles, etc. We note from Newton's third law that $\mathbf{f}_{ij} = -\mathbf{f}_{ji}$; by definition $\mathbf{f}_{ii} = 0$; so that

$$\sum_{i=1}^{N} \sum_{j=1}^{N} \mathbf{f}_{ij} = 0$$

that is, the sum of all the internal forces is zero. As an example, consider the three masses connected by springs shown in Fig. 6.1(b). The free-body diagram is shown in Fig. 6.1(c). The sum of the internal forces yields

$$\sum_{i=1}^{3} \sum_{j=1}^{3} \mathbf{f}_{ij} = \mathbf{f}_{12} + \mathbf{f}_{21} + \mathbf{f}_{13} + \mathbf{f}_{31} + \mathbf{f}_{32} + \mathbf{f}_{23}$$

and since $\mathbf{f}_{ij} = -\mathbf{f}_{ji}$, the sum of the internal forces goes to zero.

Returning to Fig. 6.1(a), we now apply Newton's second law to particle m_i.

$$\mathbf{F}_i + \sum_{j=1}^{N} \mathbf{f}_{ij} = m_i \ddot{\mathbf{r}}_i \tag{6.1}$$

Since there is a similar equation for each mass, summing up all the N equations yields

$$\sum_{i=1}^{N} \mathbf{F}_i + \sum_{i=1}^{N} \sum_{j=1}^{N} \mathbf{f}_{ij} = \sum_{i=1}^{N} m_i \ddot{\mathbf{r}}_i \tag{6.2}$$

Since the second term is zero, we obtain

$$\sum_{i=1}^{N} \mathbf{F}_i = \sum_{i=1}^{N} m_i \ddot{\mathbf{r}}_i \tag{6.3}$$

We can write Eq. (6.3) in another form by introducing the center of mass for the system of particles. The center of mass $\mathbf{r}_c$ is defined by

$$\mathbf{r}_c = \sum_{i=1}^{N} \frac{m_i \mathbf{r}_i}{m} \tag{6.4}$$

where

$$m = \sum_{i=1}^{N} m_i$$

Figure 6.2 shows the position vector $\mathbf{r}_i$ of mass m_i relative to the origin of the XYZ coordinate system and the position vector $\mathbf{r}_c$ of the center of mass labeled c. The figure shows only the mass m_i for clarity. Rearranging and differentiating Eq. (6.4), we obtain

$$m\ddot{\mathbf{r}}_c = \sum_{i=1}^{N} m_i \ddot{\mathbf{r}}_i \tag{6.5}$$

If we substitute Eq. (6.5) into Eq. (6.3) and let

$$\mathbf{F} = \sum_{i=1}^{N} \mathbf{F}_i$$

then

$$\mathbf{F} = m\ddot{\mathbf{r}}_c \tag{6.6}$$

This equation states that the resultant of the external forces acting on a system of particles equals the product of the total mass of the particles and the acceleration of the center of mass of the particles. In effect, we can think of the particles as concentrated at the center of mass.

One important observation about Eq. (6.6) is that the resultant force vector $\mathbf{F}$ and the resultant vector $m\ddot{\mathbf{r}}_c$ are equal in magnitude and direction. However, $\mathbf{F}$ does not necessarily pass through the center of mass, so that, in general, the system of particles experiences rotational motion in addition to translational motion. Should the resultant force pass through the center of mass, the system of particles undergoes translational motion only.

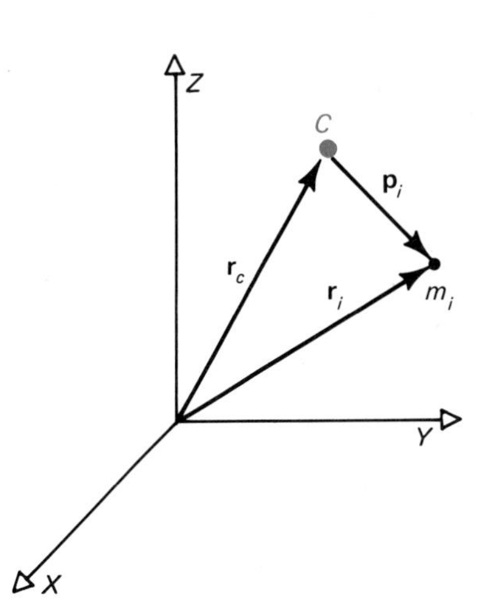

Figure 6.2

Work and Energy

The work-energy relationship as developed for a single particle has already been given:

$$\tfrac{1}{2}m(v_B^2 - v_A^2) = \int_A^B \mathbf{F} \cdot d\mathbf{r} \tag{3.22}$$

This equation can be written, with certain modifications, for each particle and then each equation can be summed to give the work-energy relationship for a set of N particles.

The first modification arises for the expression for work due to the presence of internal forces. For the ith particle, let the resultant internal force acting on particle m_i be expressed as

$$\mathbf{f}_i = \sum_{j=1}^{N} \mathbf{f}_{ij}$$

Then, the work done on m_i due to the external and internal forces is

$$W_i = \int_A^B (\mathbf{F}_i + \mathbf{f}_i) \cdot d\mathbf{r}_i \tag{6.7}$$

Summing over all of the particles, the total work is obtained as

$$W = \int_A^B \sum_{i=1}^{N} (\mathbf{F}_i + \mathbf{f}_i) \cdot d\mathbf{r}_i \tag{6.8}$$

This equation can be modified by introducing the location of m_i relative to the center of mass as shown in Fig. 6.2.

$$\mathbf{r}_i = \mathbf{r}_c + \boldsymbol{\rho}_i \tag{6.9}$$

This equation is represented in differential form as

$$d\mathbf{r}_i = d\mathbf{r}_c + d\boldsymbol{\rho}_i \tag{6.10}$$

Substituting Eq. (6.10) into Eq. (6.8), we find that

$$W = \int_A^B \sum(\mathbf{F}_i \cdot d\mathbf{r}_c + \mathbf{F}_i \cdot d\boldsymbol{\rho}_i + \mathbf{f}_i \cdot d\mathbf{r}_c + \mathbf{f}_i \cdot d\boldsymbol{\rho}_i)$$

Rearranging terms,

$$W = \int_A^B [\sum \mathbf{F}_i \cdot d\mathbf{r}_c + \sum \mathbf{f}_i \cdot d\mathbf{r}_c + \sum(\mathbf{F}_i + \mathbf{f}_i) \cdot d\boldsymbol{\rho}_i]$$

The first term under the integral reduces to $\mathbf{F} \cdot d\mathbf{r}_c$ while the second term goes to zero since

$$\sum \mathbf{f}_i \cdot d\mathbf{r}_c = d\mathbf{r}_c \cdot \sum \mathbf{f}_i = d\mathbf{r}_c \cdot \sum\sum \mathbf{f}_{ij} = 0$$

The expression for the total work becomes

$$W = \int_A^B [\mathbf{F} \cdot d\mathbf{r}_c + \sum(\mathbf{F}_i + \mathbf{f}_i) \cdot d\boldsymbol{\rho}_i] \tag{6.11}$$

The kinetic energy for a system of particles is the sum of the kinetic energy of each particle.

$$T = \sum_{i=1}^{N} \tfrac{1}{2} m_i v_i^2 \tag{6.12}$$

where v_i is the magnitude of the absolute velocity of the ith particle. Now

$$v_i^2 = \dot{\mathbf{r}}_i \cdot \dot{\mathbf{r}}_i = (\dot{\mathbf{r}}_c + \dot{\boldsymbol{\rho}}_i) \cdot (\dot{\mathbf{r}}_c + \dot{\boldsymbol{\rho}}_i)$$

$$\begin{aligned} v_i^2 &= \dot{r}_c^2 + 2\dot{\mathbf{r}}_c \cdot \dot{\boldsymbol{\rho}}_i + \dot{\rho}_i^2 \\ &= v_c^2 + 2\dot{\mathbf{r}}_c \cdot \dot{\boldsymbol{\rho}}_i + \dot{\rho}_i^2 \end{aligned} \tag{6.13}$$

where we let $v_c = \dot{r}_c$. Equation (6.12) is rewritten,

$$T = \sum \tfrac{1}{2} m_i v_c^2 + \sum \dot{\mathbf{r}}_c \cdot m_i \dot{\boldsymbol{\rho}}_i + \sum \tfrac{1}{2} m_i \dot{\rho}_i^2 \tag{6.14}$$

Since, by definition,

$$\begin{aligned} m\mathbf{r}_c &= \sum m_i \mathbf{r}_i = \sum m_i(\mathbf{r}_c + \boldsymbol{\rho}_i) \\ &= m\mathbf{r}_c + \sum m_i \boldsymbol{\rho}_i \end{aligned}$$

Consequently, $$\sum m_i \boldsymbol{\rho}_i = 0$$

Differentiating with respect to time,

$$\sum m_i \dot{\boldsymbol{\rho}}_i = 0$$

which allows us to eliminate the middle term in Eq. (6.14). The kinetic energy reduces to

$$T = \tfrac{1}{2}mv_c^2 + \sum\tfrac{1}{2}m_i\dot{\rho}_i^2 \tag{6.15}$$

Note that $\dot{\boldsymbol{\rho}}_i$ is the velocity of m_i as viewed by a nonrotating observer at the center of mass.

For a system of N particles, the work-energy relationship is

$$W = \Delta T = T_B - T_A$$

Now referring to Eqs. (6.11) and (6.15)

$$\int_A^B [\mathbf{F} \cdot d\mathbf{r}_c + \sum(\mathbf{F}_i + \mathbf{f}_i) \cdot d\boldsymbol{\rho}_i] = \tfrac{1}{2}mv_c^2\Big|_A^B + \sum\tfrac{1}{2}m_i\dot{\rho}_i^2\Big|_A^B \tag{6.16}$$

We observe that the first integral on the left side of Eq. (6.16) corresponds to the first term on the right side of the equation, namely, the work done by the resultant external force acting through the displacement of the center of mass equals the change in kinetic energy due to the total mass moving with the velocity of the center of mass. The second term on the left side represents the work done by the external and internal forces in moving through displacements relative to the center of mass and equals the change in kinetic energy of relative motion given by the second term on the right-hand side of the equation.

Impulse and Momentum

Equations (6.3) and (6.6) can be written in the linear momentum form as

$$\mathbf{F} = \sum_{i=1}^{N} \frac{d}{dt}(m_i\dot{\mathbf{r}}_i) = \sum\dot{\mathbf{p}}_i \tag{6.17}$$

and

$$\mathbf{F} = \frac{d}{dt}(m\dot{\mathbf{r}}_c) = \dot{\mathbf{p}}_c \tag{6.18}$$

where the linear momentum terms are defined as

$$\mathbf{p}_i = m_i\dot{\mathbf{r}}_i \qquad \text{and} \qquad \mathbf{p}_c = m\dot{\mathbf{r}}_c$$

We now wish to derive the moment–angular momentum expression for the system of particles. To begin, let us rewrite Eq. (6.1) as

$$\mathbf{F}_i + \mathbf{f}_i = \frac{d}{dt}(m_i\dot{\mathbf{r}}_i) \tag{6.19}$$

Cross multiply both sides of this equation by the position vector $\mathbf{r}_i$ and then sum on all the masses to obtain,

$$\sum\mathbf{r}_i \times (\mathbf{F}_i + \mathbf{f}_i) = \frac{d}{dt}\sum\mathbf{r}_i \times m_i\dot{\mathbf{r}}_i \tag{6.20}$$

Since the internal forces occur in equal pairs but are opposite in direction,

$$\sum_i \mathbf{r}_i \times \mathbf{f}_i = 0$$

By definition, let the moment of the external forces about the origin be

$$\mathbf{M} = \sum \mathbf{r}_i \times \mathbf{F}_i$$

and the angular momentum of the particles about the origin be

$$\mathbf{H} = \sum \mathbf{r}_i \times m_i \dot{\mathbf{r}}_i$$

Equation (6.20) then reduces to

$$\mathbf{M} = \dot{\mathbf{H}} \tag{6.21}$$

Equation (6.21) states that the sum of the moments about the origin O due to the external forces acting on the system of particles equals the time rate of change of the angular momentum of the system of particles about O.

While Eq. (6.21) was derived using the fixed reference point O at the origin of the XYZ coordinate system, we may derive a new relationship between moments and time derivatives of angular momenta using arbitrary points of reference. The center of mass is one convenient reference point. Let us define the angular momentum about the center of mass for the system of particles as

$$\mathbf{H}_c = \sum \boldsymbol{\rho}_i \times m_i \dot{\mathbf{r}}_i \tag{6.22}$$

Differentiating this equation, we obtain

$$\dot{\mathbf{H}}_c = \sum \dot{\boldsymbol{\rho}}_i \times m_i \dot{\mathbf{r}}_i + \sum \boldsymbol{\rho}_i \times m_i \ddot{\mathbf{r}}_i \tag{6.23}$$

Now $\dot{\boldsymbol{\rho}}_i = \dot{\mathbf{r}}_i - \dot{\mathbf{r}}_c$, so the first term on the right side of Eq. (6.23) is

$$\begin{aligned} \sum \dot{\boldsymbol{\rho}}_i \times m_i \dot{\mathbf{r}}_i &= \sum \dot{\mathbf{r}}_i \times m_i \dot{\mathbf{r}}_i - \dot{\mathbf{r}}_c \times \sum m_i \dot{\mathbf{r}}_i \\ &= -\dot{\mathbf{r}}_c \times m \dot{\mathbf{r}}_c = 0 \end{aligned}$$

Equation (6.23) reduces to

$$\dot{\mathbf{H}}_c = \sum \boldsymbol{\rho}_i \times m_i \ddot{\mathbf{r}}_i = \sum \boldsymbol{\rho}_i \times (\mathbf{F}_i + \mathbf{f}_i) \tag{6.24}$$

by means of Eq. (6.1). But $\sum \boldsymbol{\rho}_i \times \mathbf{f}_i = 0$, so that Eq. (6.24) becomes

$$\dot{\mathbf{H}}_c = \sum \boldsymbol{\rho}_i \times \mathbf{F}_i$$

Rearranging this expression and letting $\sum \boldsymbol{\rho}_i \times \mathbf{F}_i = \mathbf{M}_c$,

$$\mathbf{M}_c = \dot{\mathbf{H}}_c \tag{6.25}$$

This says that for a system of particles the moment of the external forces about the center of mass equals the time rate of change of the angular momentum about the center of mass. Note the similarity between this equation and Eq. (6.21).

Equations (6.17), (6.18), (6.21), and (6.25) are integrated over time to give the impulse-momentum relationships, i.e.,

$$\mathscr{I} = \int_{t_1}^{t_2} \mathbf{F}\, dt = \sum \mathbf{p}_i(t_2) - \sum \mathbf{p}_i(t_1) \tag{6.26}$$

$$\mathscr{I} = \int_{t_1}^{t_2} \mathbf{F}\, dt = \mathbf{p}_c(t_2) - \mathbf{p}_c(t_1) \tag{6.27}$$

$$\mathscr{M} = \int_{t_1}^{t_2} \mathbf{M}\, dt = \mathbf{H}(t_2) - \mathbf{H}(t_1) \tag{6.28}$$

$$\mathscr{M}_c = \int_{t_1}^{t_2} \mathbf{M}_c\, dt = \mathbf{H}_c(t_2) - \mathbf{H}_c(t_1) \tag{6.29}$$

where, for example, $\mathbf{H}_c(t_2)$ is the angular momentum about the center of mass at time t_2, $\mathbf{H}(t_1)$ is the angular momentum about the fixed reference point O at t_1, etc. These relationships are especially important when we study the planar motion of rigid bodies.

Example 6.1

Two particles connected by a spring of unstretched length 0.3 m with a modulus of 800 N/m move along a smooth plane. At the instant shown in Fig. 6.3(a), a force $\mathbf{F}_2$ acts on m_2. Find the acceleration of each mass and the acceleration of the center of mass. Let $m_1 = 1$ kg and $m_2 = 2$ kg.

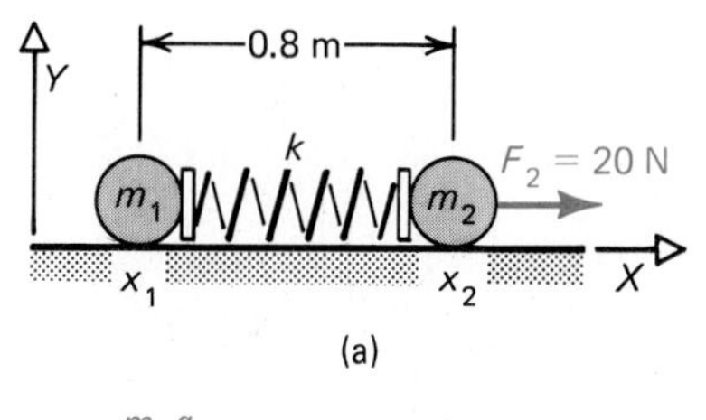

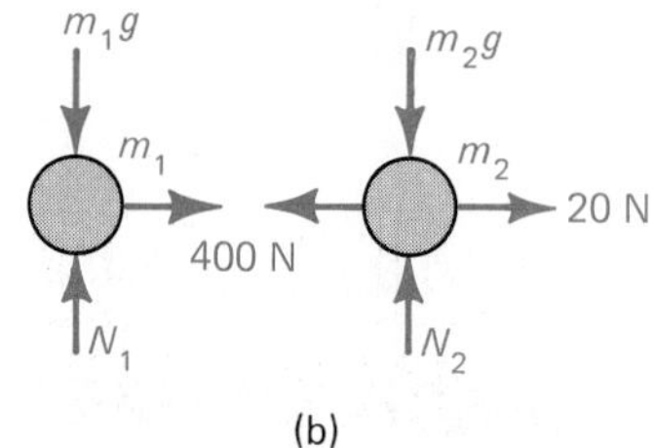

Figure 6.3

A free-body diagram of the forces acting on each particle is shown in Fig. 6.3(b). Note the direction of the spring force on each particle due to the stretching of the spring. Also, the magnitude of the spring force equals 800(0.8 − 0.3) = 400 N.

Mass m_1: $\sum F_x = m_1\ddot{x}_1 \qquad 400 = \ddot{x}_1$

$\therefore \ddot{\mathbf{x}}_1 = 400\mathbf{i}$ m/s^2 *Answer*

Mass m_2: $\sum F_x = m_2\ddot{x}_2 \qquad -400 + 20 = 2\ddot{x}_2$

$\therefore \ddot{\mathbf{x}}_2 = -190\mathbf{i}$ m/s^2 *Answer*

The only external force in the x direction is $\mathbf{F}_2$. Referring to Eq. (6.6),

$$\mathbf{F} = m\ddot{\mathbf{r}}_c$$

Apply this equation in the x direction, letting $\ddot{\mathbf{r}}_c = \ddot{x}_c\mathbf{i}$ since the motion is confined to the x direction only.

$$F_2\mathbf{i} = (1 + 2)\ddot{x}_c\mathbf{i}$$

or

$$20 = 3\ddot{x}_c$$

$$\ddot{x}_c = 6.67$$

$\ddot{\mathbf{r}}_c = 6.67\mathbf{i}$ m/s^2 *Answer*

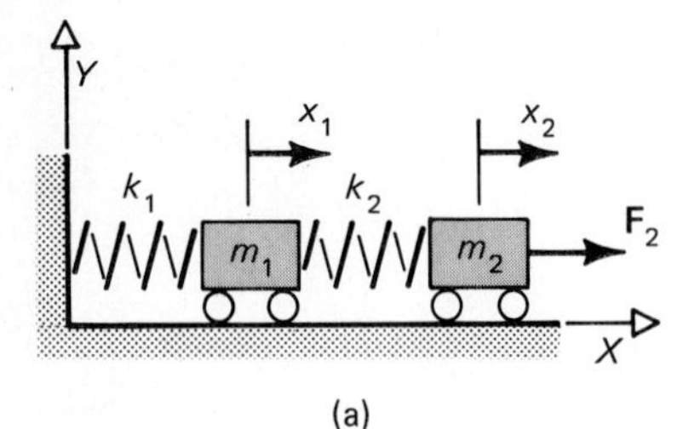

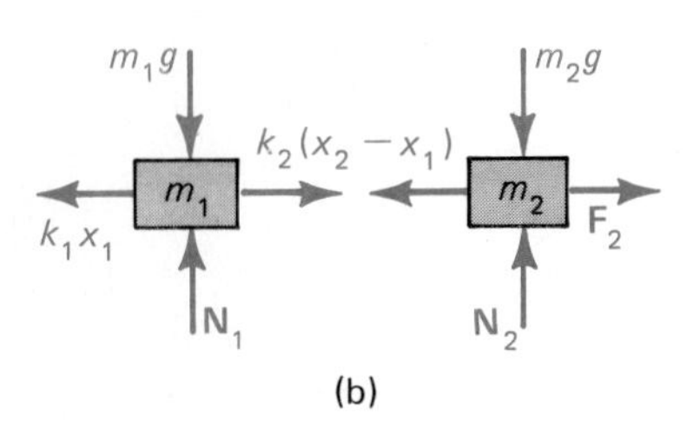

Figure 6.4

Example 6.2

The two particles shown in Fig. 6.4(a) oscillate on the smooth plane in the x direction. (a) Draw the free-body diagram of the forces acting on each mass. (b) Write the differential equation of motion for each mass. (c) Apply Eq. (6.6) to find the equation of motion for the center of mass. (d) Write the expressions for kinetic and potential energies of the system of particles.

(a) The free-body diagram is shown in Fig. 6.4(b). We allow each mass to be displaced a positive amount and arbitrarily assume $x_2 > x_1$. As a consequence of this assumption, the spring of modulus k_2 is in tension so that the internal force of magnitude $k_2(x_2 - x_1)$ acts on each mass as shown in Fig. 6.4(b).

(b) Mass m_1: $\sum F_x = m_1\ddot{x}_1 = -k_1x_1 + k_2(x_2 - x_1)$ *Answer*

Mass m_2: $\sum F_x = m_2\ddot{x}_2 = -k_2(x_2 - x_1) + F_2$ *Answer*

(c) $$(m_1 + m_2)\ddot{x}_c = F_2 - k_1x_1$$ *Answer*

Note that while the internal forces do not appear in this equation, the spring force of magnitude k_1x_1 does since it is an external force acting on m_1.

(d) $$T = \tfrac{1}{2}m_1\dot{x}_1^2 + \tfrac{1}{2}m_2\dot{x}_2^2$$

$$V = \tfrac{1}{2}k_1x_1^2 + \tfrac{1}{2}k_2(x_2 - x_1)^2$$ *Answer*

Notice that the potential energy consists of the potential energy stored in each spring. No potential energy due to gravity is included since each particle moves along the horizontal plane.

Example 6.3

A system of three masses is connected by massless rigid rods as shown in Fig. 6.5(a). If the three masses move 5 ft up the plane under the action of force **P**, find the work done on the system. Assume $P = 200$ lb, $m_1g = m_2g = 2m_3g = 100$ lb, $\mu = 0.1$, and that the maximum frictional force is developed between each mass and the surface.

The free-body diagram is shown in Fig. 6.5(b). Several observations are in order. First we note the orientation of the fixed XY coordinate system. This is done as a matter of convenience. Next, we see the internal forces have magnitudes f_{12} and f_{23} where $f_{12} = f_{21}$ and $f_{23} = f_{32}$. Finally, we observe the friction forces of magnitude f_1, f_2, and f_3 opposing the direction of motion of the system of particles.

Summing forces in the y direction yields:

$$N_1 = m_1g \cos 30^\circ \qquad N_2 = m_2g \cos 30^\circ \qquad N_3 = m_3g \cos 30^\circ$$

The magnitude of each friction force is now calculated.

$$f_1 = \mu(m_1 g)\cos 30^\circ = 0.1(100)\cos 30^\circ = 8.66 \text{ lb}$$

$$f_2 = \mu(m_2 g)\cos 30^\circ = 0.1(100)\cos 30^\circ = 8.66 \text{ lb}$$

$$f_3 = \mu(m_3 g)\cos 30^\circ = 0.1(50)\cos 30^\circ = 4.33 \text{ lb}$$

The work done on the system is given by Eq. (6.11).

$$W = \int_A^B [\mathbf{F}\cdot d\mathbf{r}_c + \sum(\mathbf{F}_i + \mathbf{f}_i)\cdot d\boldsymbol{\rho}_i]$$

$$= \int_A^B \mathbf{F}\cdot d\mathbf{r}_c$$

since $d\boldsymbol{\rho}_i = 0$ in this case, that is, there is no relative motion between the particles and the center of mass of the particles due to the rigid rods connecting the particles. Now let us sum the external forces acting on the system of particles in the x direction.

$$\mathbf{F} = \sum \mathbf{F}_i = [P - f_1 - f_2 - f_3 - (m_1 g + m_2 g + m_3 g)\sin 30^\circ]\mathbf{i}$$

$$= [200 - 8.66 - 8.66 - 4.33 - 250(0.5)]\mathbf{i} = 53.35\mathbf{i}$$

$$d\mathbf{r}_c = dx_c\mathbf{i}$$

Substituting into the equation for the work,

$$W = \int_0^5 (53.35\mathbf{i})\cdot(dx_c\mathbf{i})$$

$$= 266.8 \text{ lb}\cdot\text{ft} \qquad \textit{Answer}$$

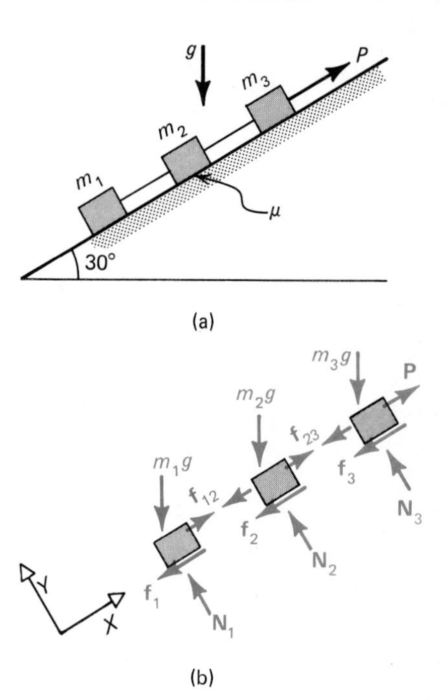

Figure 6.5

Example 6.4

A double pendulum, shown in Fig. 6.6(a), oscillates in the XY plane. At the instant shown $\omega_1 = 2$ rad/s CCW and $\omega_2 = 3$ rad/s CCW. What is $\mathbf{H}_o$ at this instant if $m_1 = m_2 = 1$ kg? Note that the lower pendulum is connected to mass m_1 by a pin joint and is free to rotate about this point.

To find the angular momentum about the origin O, recall that

$$\mathbf{H}_o = \sum \mathbf{r}_i \times m_i\dot{\mathbf{r}}_i = \sum \mathbf{r}_i \times m_i\mathbf{v}_i$$

$$\mathbf{H}_o = \mathbf{r}_1 \times m_1\mathbf{v}_1 + \mathbf{r}_2 \times m_2\mathbf{v}_2$$

where

$$\mathbf{r}_1 = 0.4(\cos 30^\circ\mathbf{i} + \sin 30^\circ\mathbf{j})$$

$$= 0.346\mathbf{i} + 0.2\mathbf{j}$$

$$\mathbf{v}_1 = 0.4(2)(-\sin 30^\circ\mathbf{i} + \cos 30^\circ\mathbf{j})$$

$$= -0.4\mathbf{i} + 0.693\mathbf{j}$$

$$\mathbf{r}_2 = \mathbf{r}_1 + 0.5(\cos 60^\circ\mathbf{i} + \sin 60^\circ\mathbf{j})$$

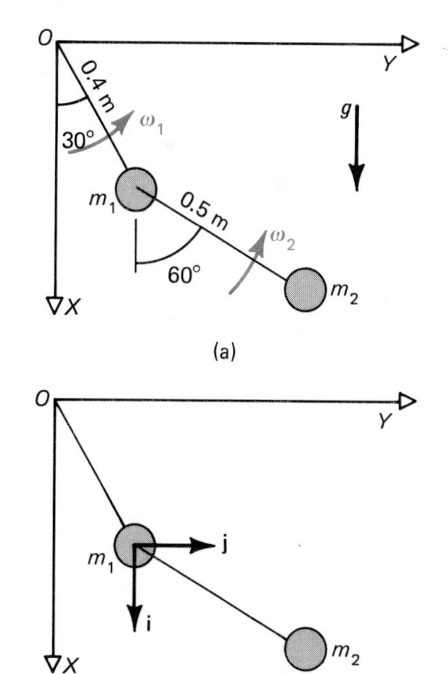

Figure 6.6

$$= 0.346\mathbf{i} + 0.2\mathbf{j} + 0.25\mathbf{i} + 0.433\mathbf{j}$$
$$= 0.596\mathbf{i} + 0.633\mathbf{j}$$

To find the absolute velocity of m_2, place the origin of the moving reference frame on the lower pendulum at m_1 as shown in Fig. 6.6(b). Then,

$$\mathbf{v}_2 = \dot{\mathbf{R}} + \dot{\mathbf{r}}_r + \boldsymbol{\omega} \times \mathbf{r}$$

where
$$\dot{\mathbf{R}} = \mathbf{v}_1 = -0.4\mathbf{i} + 0.693\mathbf{j}$$
$$\dot{\mathbf{r}}_r = 0$$
$$\boldsymbol{\omega} = 3\mathbf{k}$$
$$\mathbf{r} = 0.5(\cos 60^\circ \mathbf{i} + \sin 60^\circ \mathbf{j}) = 0.25\mathbf{i} + 0.433\mathbf{j}$$

Therefore,

$$\mathbf{v}_2 = -0.4\mathbf{i} + 0.693\mathbf{j} + 3\mathbf{k} \times (0.25\mathbf{i} + 0.433\mathbf{j})$$
$$= -1.70\mathbf{i} + 1.44\mathbf{j}$$

Substituting into the expression for angular momentum, we obtain the following:

$$\mathbf{H}_o = (0.346\mathbf{i} + 0.2\mathbf{j}) \times (-0.4\mathbf{i} + 0.693\mathbf{j})$$
$$+ (0.596\mathbf{i} + 0.633\mathbf{j}) \times (-1.7\mathbf{i} + 1.44\mathbf{j})$$
$$= (0.2398 + 0.08 + 0.8582 + 1.0761)\mathbf{k}$$

$$\mathbf{H}_o = 2.254\mathbf{k} \text{ kg}\cdot m^2/\text{s}$$ *Answer*

PROBLEMS

6.1 Two particles each of mass 2 kg are separated by 2 m. Assuming that the particles are isolated, so that the gravitational force of attraction between the particles is the only force present, find (a) the acceleration of each mass, and (b) the acceleration of the center of mass for the set of particles.

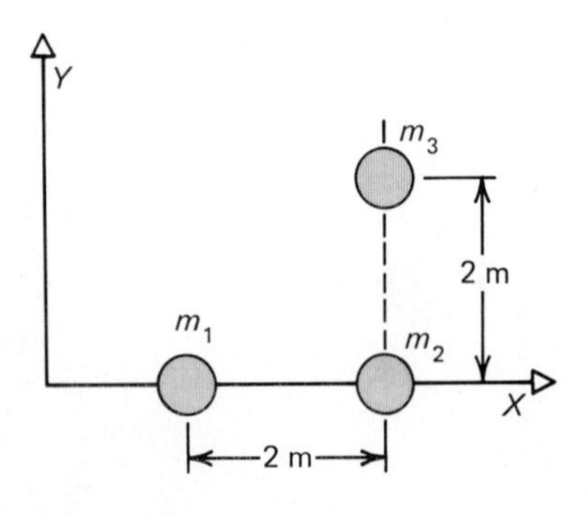

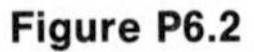
Figure P6.2

6.2 Three isolated particles lie on a smooth horizontal plane as shown in Fig. P6.2. Each particle has a mass of 1 kg. (a) What is the acceleration of mass m_1? (b) What external force $\mathbf{F}_3$ must be applied to m_3 for its acceleration to be zero at the instant shown?

6.3 In Example 6.2, it was assumed that $x_2 > x_1$ for the instant considered. Obtain the equations of motion assuming $x_1 > x_2$.

6.4 Find the tension in the massless wire that connects $w_1 = 16.1$ lb and $w_2 = 8.05$ lb as shown in Fig. P6.4. Let $\mathbf{P}$ be an applied force acting on w_2 and assume the kinetic coefficient of friction between each body and the horizontal plane is 0.2.

6.5 Two masses are connected to each other by a rope as shown in Fig. P6.5. At the instant shown, assume m_2 is released from rest and m_1 slides along the smooth inclined plane. If $m_1 = 1$ slug and $m_2 = 2$ slugs, find the tensile force in the rope.

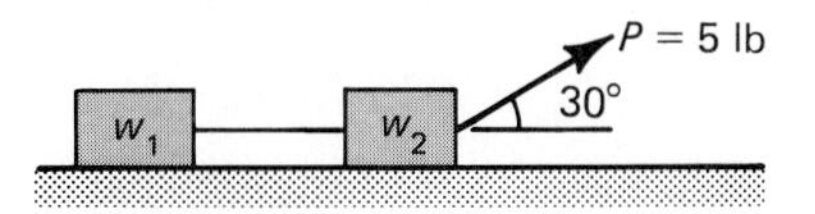

Figure P6.4

6.6 Suppose the system of particles in Problem 6.5 is again released from rest but that $m_1 = 2$ slugs and $m_2 = 1$ slug. Find the tensile force in the rope and the acceleration of m_1.

6.7 The masses shown in Fig. P6.7 are released from rest. Neglect the friction between the sliding rope and the curved surface. (a) Find the acceleration of each mass. (b) What is the velocity of the 100-kg mass after 10 s?

6.8 In Fig. P6.8, (a) what is the acceleration of the 50-kg mass at the instant shown and (b) what is its speed after 10 s?

Figure P6.5

Figure P6.7

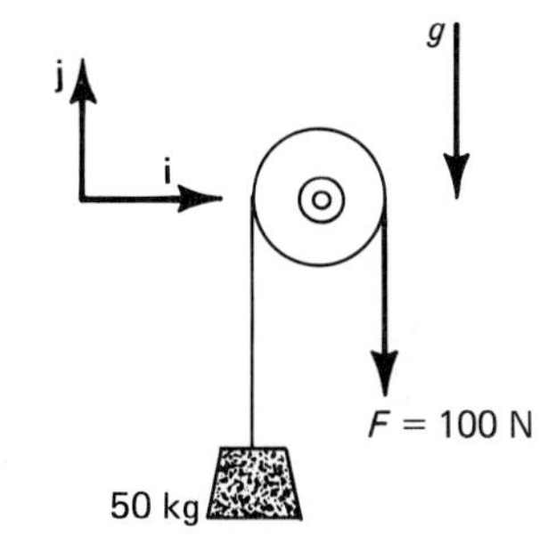

Figure P6.8

6.9 The 200-lb weight in Fig. P6.9 is to be lowered such that it accelerates at a constant rate of 16.1 ft/sec^2. What must the magnitude of the force **F** be to achieve this motion?

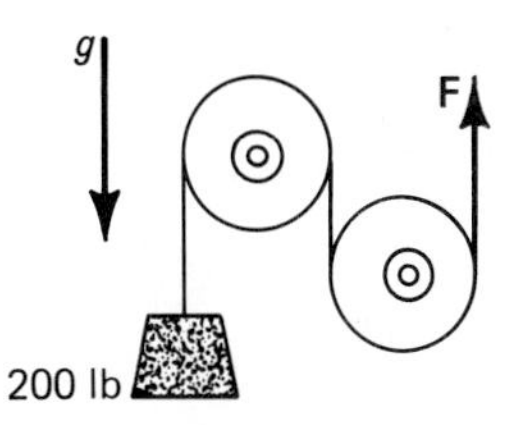

Figure P6.9

6.10 Find the work done on the system shown in Fig. 6.5 if the three masses are moved 10 m up the inclined plane by a force of $P = 400$ N. Let $\mu = 0.2$ and $m_1 = 10$ kg, $m_2 = 5$ kg, $m_3 = 5$ kg.

6.11 At the instant shown in Fig. P6.11, the velocity of m_1 is $\dot{\mathbf{x}}_1 = 4\mathbf{i}$ in/sec and the angular velocity of member AB is $\omega = 3\mathbf{k}$ rad/sec. Write the kinetic energy of the set of particles if $m_2 = m_1 = 2$ slugs.

6.12 If the unstretched length of the spring in Problem 6.11 is 8 in, write the potential energy for the system of particles. (*Note*: Use point A as the reference point for calculating the gravitational potential energy.)

6.13 Two masses, connected by a massless rigid rod AB, are released from rest at the instant shown in Fig. P6.13. What is the magnitude of the velocity of m_1 when it reaches point C and point D? Let each body have a mass of 10 kg. Assume that the spring is initially unstretched and $k = 400$ N/m.

6.14 What must be the minimum value of the spring modulus in Problem 6.13 for the velocity of m_1 to be zero at the instant when it reaches point C?

6.15 The massless rigid rod AB, having masses m_1 and m_2 attached to it, is released from rest as shown in Fig. P6.15. At the instant of release, the spring is unstretched. If each mass is 3 kg, what is the minimum spring modulus required to keep the rod from turning through 1° clockwise?

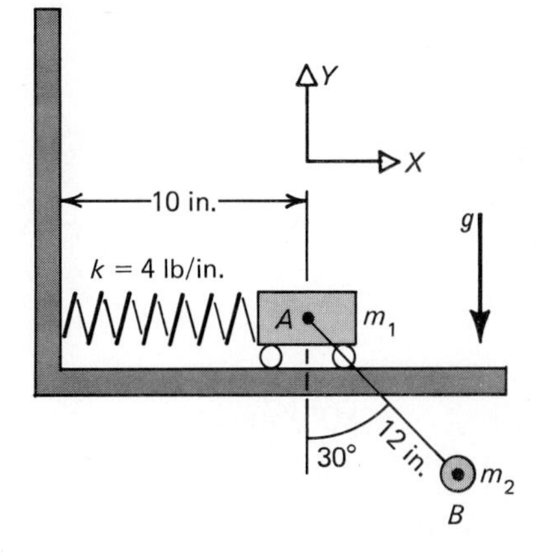

Figure P6.11

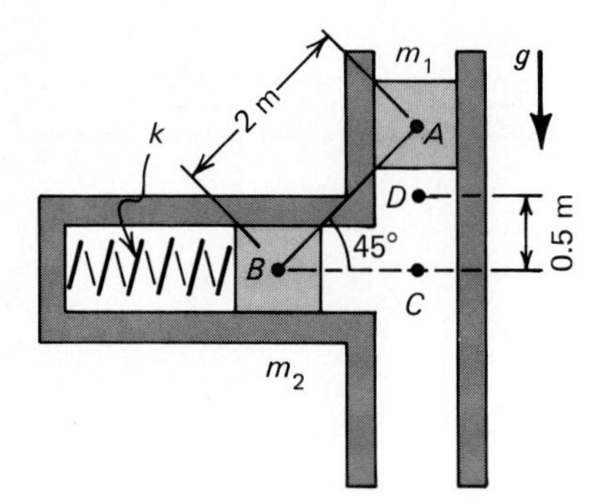

Figure P6.13

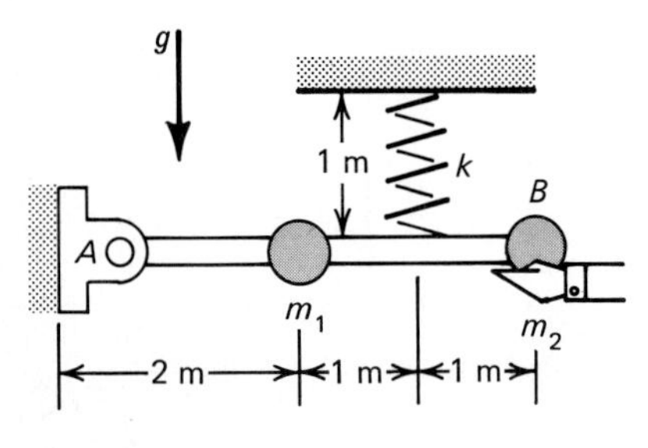

Figure P6.15

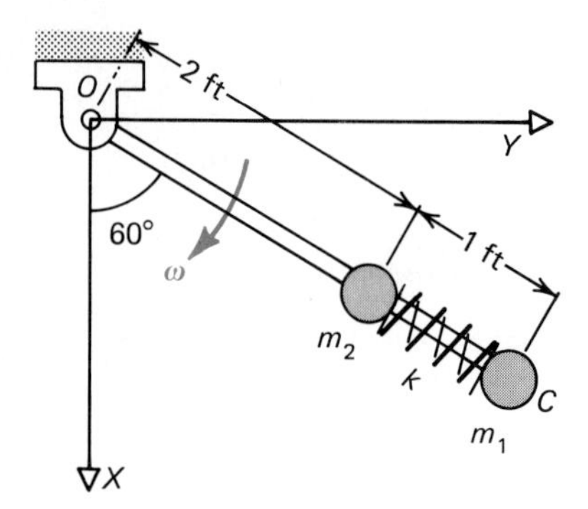

Figure P6.17

6.16 It is required that the spring shown in Fig. P6.15 have a modulus $k = 1000$ N/m. If $m_1 = 4$ kg, $m_2 = 5$ kg, then what is the maximum extension of the spring? How much energy is stored in the spring? Assume that the spring is unstretched in the position shown and assume small displacements.

6.17 The pendulum moving on the smooth horizontal plane in Fig. P6.17 consists of a mass $m_1 = 2$ slugs and a mass $m_2 = 1$ slug, where m_2 is free to slide along the frictionless rod OC and m_1 is fixed to the rod. At the instant shown the angular velocity of the rod is $\boldsymbol{\omega} = -3\mathbf{k}$ rad/sec, m_2 is moving relative to the rod at a rate of 3 ft/sec toward m_1, and the spring is compressed 1 in. Write the potential energy stored in the spring and the kinetic energy for the system.

6.18 Write the angular momentum for the set of particles in Problem 6.17 about the fixed point O.

6.19 Body A of mass $2m$ has three masses, each of mass m rotating at a constant rate of 2π rad/sec CCW, attached to it as shown in Fig. P6.19. If the system of masses is released from rest, find (a) the kinetic energy, (b) the linear momentum, and (c) the angular momentum about the point A one second after release. Let $m = \frac{1}{2}$ slug.

6.20 Obtain the linear momentum of the system shown in Fig. P6.19 3 sec after it is released from rest. Assume $\omega = 10$ rad/sec and $m = 1$ slug.

6.21 Two particles at rest, each of mass m, are connected by a spring that is unstretched at the instant shown in Fig. P6.21. The particles travel across the smooth cylindrical surface of radius R. What must the maximum spring modulus be for the masses to reach the horizontal position?

6.22 Referring to Problem 6.21, what is the angular momentum about O after the system is released from rest if the spring modulus is k?

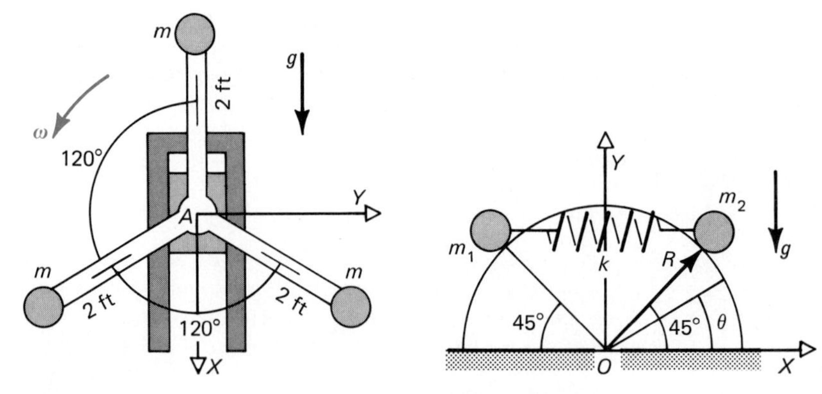

Figure P6.19

Figure P6.21

6.23 A massless rigid rod rotates about point O on a smooth horizontal plane at a uniform rate of 200 rev/min. Two masses are rigidly attached to the rod as shown in Fig. P6.23. Obtain the force in (a) the rod section OA and (b) the rod section AB, if $m_1 = 2$ kg and $m_2 = 1$ kg.

6.24 The mass $m = 2$ kg is free to slide along the massless rod that is turning on a smooth horizontal plane at a uniform rate of 10 rad/s as shown in Fig. P6.24. Mass m is connected to the spring of modulus $k = 800$ N/m whose unstretched length is 1 m. Find the amount by which the spring deforms.

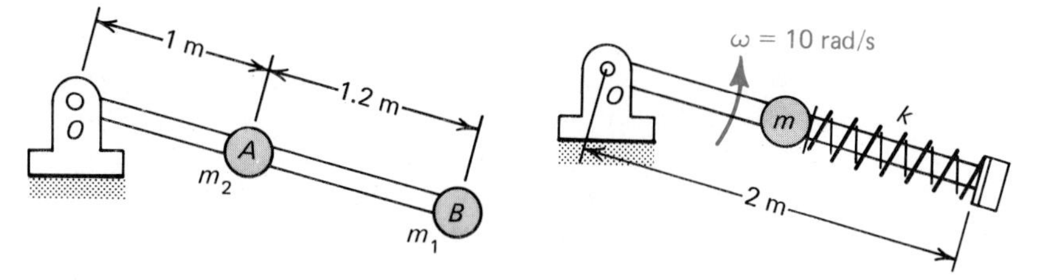

Figure P6.23 **Figure P6.24**

6.25 The mass shown in Fig. P6.24 is 5 kg and the spring modulus is 1000 N/m. If the spring is allowed to compress by 0.5 m, what is the maximum rate at which the rod can be turned? The unstretched length of the spring is 1 m.

6.2 Plane Rigid Body—Equations of Motion

A rigid body can be viewed as a system of infinite particles occupying a finite volume in which the distance between any two particles remains unchanged. Therefore, the equations derived in the preceding section for motion of a finite number of particles can be applied to a rigid body. Before applying these equations, let us consider a plane rigid body of mass m undergoing translation and rotation as shown in Fig. 6.7(a). At a given instant, the acceleration of the center of mass is $\mathbf{a}_c$, the body experiences an angular velocity $\boldsymbol{\omega}$ and an angular acceleration $\boldsymbol{\alpha}$, and external forces $\mathbf{F}_i$ and external moments $\mathbf{M}_i$ act on the body. Equation (6.6) can be directly applied to the rigid body, namely,

$$\mathbf{F} = m\mathbf{a}_c \tag{6.30}$$

where $\mathbf{F}$ is the resultant of the external forces acting on the body. Since we are working with plane rigid bodies, it is sometimes just as convenient to use scalar equations for Eq. (6.30). In cartesian coordinates, these are

$$\sum F_x = ma_{cx} \qquad \sum F_y = ma_{cy} \tag{6.31}$$

Note that a_{cx} and a_{cy} are the components of the absolute acceleration of the center of mass.

Due to the angular motion of the body, a third equation can be obtained directly from Eq. (6.25). However, a more convenient

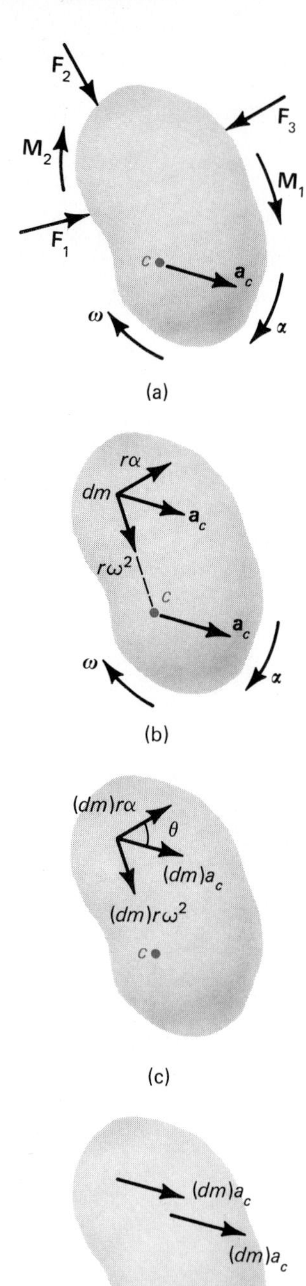

Figure 6.7

approach, and one that provides a closer understanding of the physical aspects of the problem, is to consider an element of mass dm of the rigid body as shown in Fig. 6.7(b). The acceleration of mass dm is given by

$$\mathbf{a} = \ddot{\mathbf{R}} + \boldsymbol{\alpha} \times \mathbf{r} + \boldsymbol{\omega} \times (\boldsymbol{\omega} \times \mathbf{r}) \tag{5.8}$$

where the origin of the moving reference frame is placed at the center of mass. Now

$$\ddot{\mathbf{R}} = \mathbf{a}_c$$

$\boldsymbol{\alpha} \times \mathbf{r} =$ Tangential component of acceleration of mass dm relative to the center of mass

$\boldsymbol{\omega} \times (\boldsymbol{\omega} \times \mathbf{r}) =$ Normal component of acceleration of mass dm relative to the center of mass

These components of acceleration are shown in Fig. 6.7(b). If we apply Newton's second law to the differential element of mass dm, we obtain

$$d\mathbf{F} = dm\mathbf{a} \tag{6.32}$$

Note that we are omitting the internal forces acting on dm in Eq. (6.32) since these forces will cancel out when we sum these forces over the rigid body. The resultant components on the right side of Eq. (6.32) are shown in Fig. 6.7(c). Figure 6.7(d) shows several forces of magnitude $a_c\,dm$ acting at arbitrary elements of mass throughout the rigid body. Note that the resultant of all the $a_c\,dm$ terms passes through the center of mass.

Returning to Fig. 6.7(c) and summing moments due to the resultant components about the center of mass yields

$$dM_c = r^2\alpha\,dm + ra_c \cos\theta\,dm \tag{6.33}$$

Note that the second term on the right side is due to the translation of the rigid body. When we sum all the moments for each differential element of mass we see that the sum of all the $(ra_c \cos\theta\,dm)$ terms is zero since their resultant passes through the center of mass. Dropping the last term of Eq. (6.33) and summing over all points of the body we obtain the following:

$$M_c = \int r^2\alpha\,dm = \alpha \int r^2\,dm$$

But by Eq. (5.4), this can be written

$$M_c = I_c\alpha \tag{6.34}$$

where I_c is the mass moment of inertia of the plane rigid body about its center of mass. The term on the left side of Eq. (6.34) represents the sum of the moments about the center of mass due to the external forces and moments.

Consequently, the basic instantaneous equations for the dynamics of a plane rigid body undergoing translation and rotation are summarized as follows:

$$\mathbf{F} = m\mathbf{a}_c \tag{6.35a}$$

$$M_c = I_c\alpha \tag{6.35b}$$

Figures 6.8(a) and (b) represent pictorially the meaning of Eq. (6.35) at a given instant in time. In Fig. 6.8(a) we show the external forces and moments that produce the instantaneous translational acceleration $\mathbf{a}_c$ and rotational acceleration α. Figure 6.8(b) shows the right side of Eqs. (6.35), namely ma_c and $I_c\alpha$.

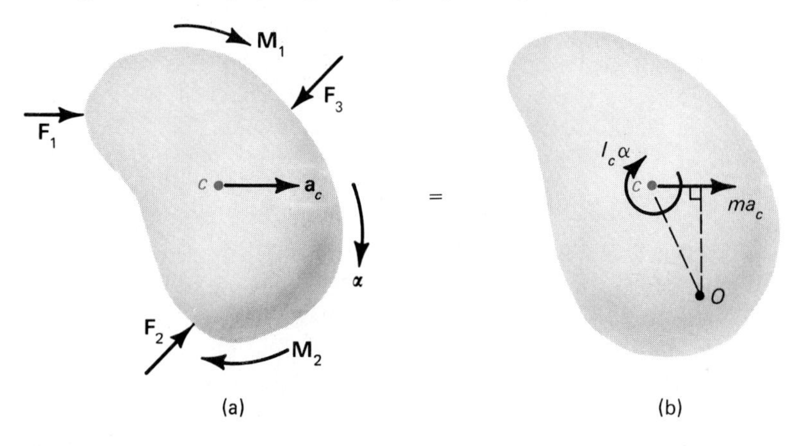

Figure 6.8

It is sometimes convenient to sum moments about an arbitrary point O rather than the center of mass. In this case we must modify Eq. (6.35b). The term on the left side becomes M_o, namely the sum of the moments due to the external forces and moments about point O. The term on the right side of the equation is obtained with the aid of Fig. 6.8(b). Summing moments about point O, we include the resultant moment $I_c\alpha$ and the contribution due to ma_c. Consequently, Eq. (6.35b) is modified as follows,

$$M_o = I_c\alpha + ma_cd \tag{6.36}$$

where d is the perpendicular distance between the ma_c vector and the point O. If point O is a *fixed point*, then the acceleration of the center of mass has a tangential component of acceleration (αd) and a normal component of accleration $(\omega^2 d)$. In this case, Eq. (6.36) becomes

$$M_o = I_c\alpha + \alpha md^2 = (I_c + md^2)\alpha \tag{6.37}$$

From the parallel-axis theorem for mass moments of inertia,

$$I_o = I_c + md^2$$

Therefore, Eq. (6.37) reduces to

$$M_o = I_o\alpha \tag{6.38}$$

Consequently, the moment equation may be taken about the center of mass as represented by Eq. (6.35b), or about an arbitrary point as given by Eq. (6.36). Should the arbitrary point be fixed, Eq. (6.38) may be used for convenience.

Example 6.5

A steel I beam is supported by a simple support at one end and a guy wire at the other end as shown in Fig. 6.9(a). The beam weighs 3220 lb and its radius of gyration from the center of mass at point c is 3 ft. Due to corrosion, the wire suddenly snaps. Find the magnitude of the reactions at the pinned support at the instant the wire snaps.

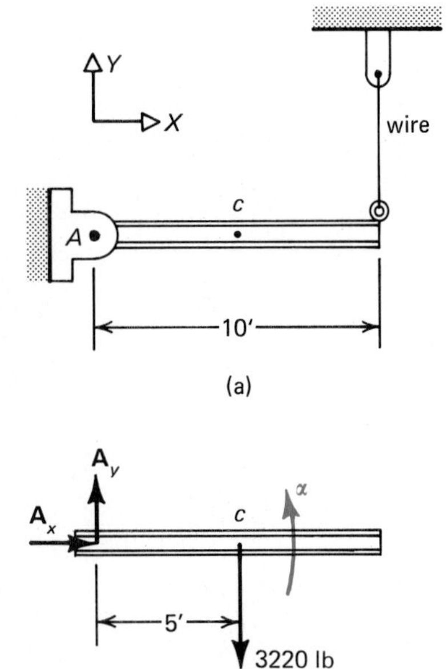

Figure 6.9

When the wire breaks, the initial angular velocity is zero. Figure 6.9(b) shows the free-body diagram of the beam. Note that the reactions $\mathbf{A}_x$ and $\mathbf{A}_y$ are the unknowns. First, apply the translational equation of motion in the x direction.

$$\sum F_x = ma_{cx} = 0$$

since $a_{cx} = 0$. Therefore,

$$A_x = 0 \qquad \textit{Answer}$$

Next, apply the translational equation of motion in the y direction,

$$\sum F_y = ma_{cy}$$

$$A_y - mg = ma_{cy} \tag{a}$$

Notice that we are assuming positive forces and the accleration are upward. Also positive moments and the angular acceleration are assumed in the counterclockwise direction.

$$M_c = I_c\alpha$$

$$-5A_y = m(3)^2\alpha = 100(9)\alpha = 900\alpha \tag{b}$$

From kinematics, we know

$$a_{cy} = r\alpha = 5\alpha \tag{c}$$

Equation (b) is rewritten

$$-5A_y = 900\left(\frac{a_{cy}}{5}\right)$$

or

$$A_y = -36a_{cy}$$

Substitute this into Eq. (a) for a_{cy}.

$$A_y - 3220 = 100\left(\frac{-A_y}{36}\right)$$

$$A_y = 852 \text{ lb} \qquad \textit{Answer}$$

Obtaining a positive value for A_y means our assumed direction for the reaction in Fig. 6.9(b) is correct.

Example 6.6

At a certain instant, an unbalanced drum is rotating at an angular velocity of 2 rad/s CW as shown in Fig. 6.10(a). The drum of mass 30 kg has an inextensible rope wrapped around its circumference supporting a 15-kg mass. The radius of gyration of the drum about the point O is 0.6 m. Find the reactions at the support O.

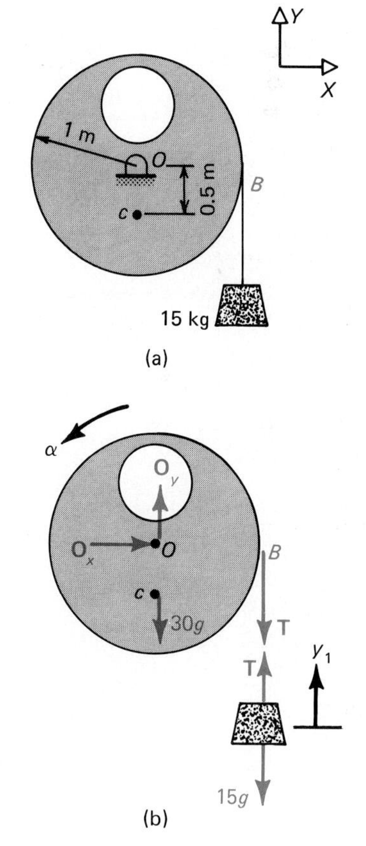

Figure 6.10

Figure 6.10(b) shows the free-body diagram of the drum and of the mass. Again we shall assume positive angular acceleration in the counterclockwise direction. We note that the acceleration of the 15-kg mass, $\ddot{\mathbf{y}}_1$, equals the tangential component of acceleration of point B on the drum, i.e., $(\mathbf{a}_B)_t$. Thus, $\ddot{\mathbf{y}}_1 = (\mathbf{a}_B)_t = r\alpha\mathbf{j} = \alpha\mathbf{j}$ since $r = 1$ m. The equation of motion in the y direction for the 15-kg mass is as follows:

$$\sum F_y = m\ddot{y}_1$$

$$T - mg = m\alpha$$

$$T = 15(9.81) + 15\alpha = 147.2 + 15\alpha \qquad \text{(a)}$$

For the drum, sum moments about the fixed point O.

$$M_o = I_o\alpha$$

$$-(1)T = 30(0.6)^2\alpha$$

$$T = -10.8\alpha \qquad \text{(b)}$$

Solving Eqs. (a) and (b) simultaneously yields

$$T = 61.6 \text{ N} \qquad \alpha = -5.70 \text{ rad/s}^2$$

The negative sign on α means that the angular acceleration is actually in the clockwise direction at the instant under consideration. To find the reactions,

$$\sum F_x = ma_{cx} \qquad \text{and} \qquad \sum F_y = ma_{cy}$$

We have the following relationship for the acceleration components of the center of mass.

$$a_{cx} = r\alpha = 0.5(-5.70) = -2.85$$

$$a_{cy} = r\omega^2 = (0.5)(2)^2 = 2$$

For the x direction

$$\sum F_x = ma_{cx} \qquad O_x = 30(-2.85) = -85.5$$

$$\mathbf{O}_x = -85.5\mathbf{i}\text{ N} \qquad \textit{Answer}$$

For the y direction

$$\sum F_y = ma_{cy} \qquad O_y - 30(9.81) - T = (30)2$$

$$O_y = 415.9$$

$$\therefore \mathbf{O}_y = 415.9\mathbf{j}\text{ N} \qquad \textit{Answer}$$

Example 6.7

A cylinder of mass m has a force **P** applied to it as shown in Fig. 6.11(a). Determine the minimum coefficient of friction μ for the cylinder to roll without slipping.

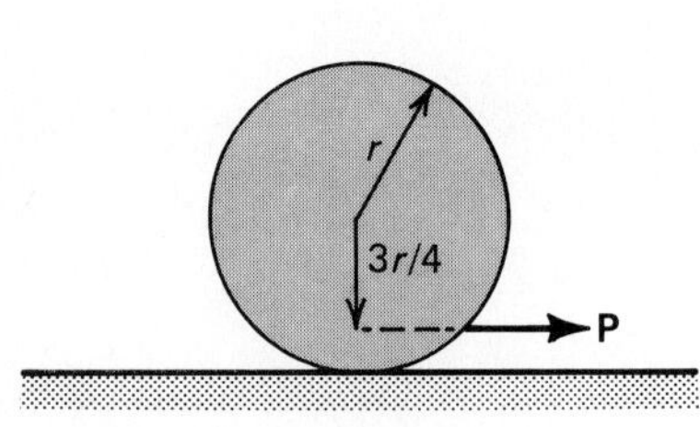

(a)

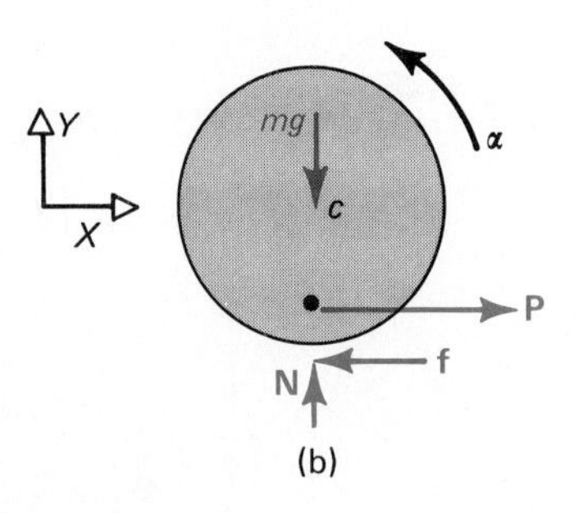

(b)

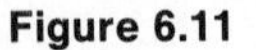
Figure 6.11

The free-body diagram of the cylinder is shown in Fig. 6.11(b). Once again, assume $\boldsymbol{\alpha}$ is counterclockwise so that the acceleration of the center of mass is $\mathbf{a}_{cx} = -r\alpha\mathbf{i}$ since there is no slippage occurring between the cylinder and the surface. Now

$$\sum F_x = ma_{cx} \qquad P - f = -mr\alpha$$

$$\alpha = \frac{f - P}{mr} \qquad \text{(a)}$$

Also,

$$M_c = I_c\alpha$$

$$\tfrac{3}{4}rP - fr = \tfrac{1}{2}mr^2\alpha$$

Substituting Eq. (a) for α, setting $f = \mu mg$ and solving for μ yields

$$\mu = \frac{5P}{6\,mg} \qquad \textit{Answer}$$

Example 6.8

A pendulum, initially at rest, is struck by a force **P** as shown in Fig. 6.12(a). A free-body diagram of the forces acting on the body is shown in Fig. 6.12(b). Find the distance q from the pinned support O, such that the magnitude of the reaction $O_x = 0$. Let k_c be the radius of gyration about the center of mass c.

Figure 6.12(c) shows the resultant components $I_c\alpha$ and $ma_c = mr\alpha$ acting on the pendulum. Let us apply Eq. (6.36) using point Q on the pendulum as the reference point. Now

$$M_Q = I_c\alpha + ma_c d$$

Let counterclockwise moments be positive. We observe from the free-body diagram in Fig. 6.12(b) that $\mathbf{O}_x$ is the only external force producing a moment about point Q. Therefore,

$$-O_x q = I_c\alpha - mr\alpha(q - r)$$

where $I_c = mk_c^2$. But we wish to find q such that $O_x = 0$. Therefore,

$$0 = mk_c^2\alpha - mr\alpha(q - r)$$

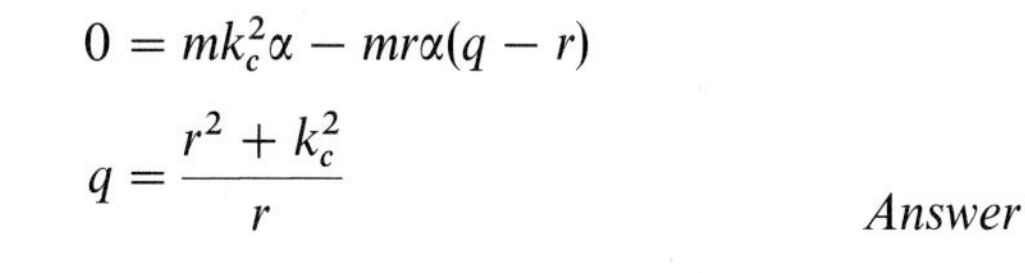

$$q = \frac{r^2 + k_c^2}{r} \qquad \textit{Answer}$$

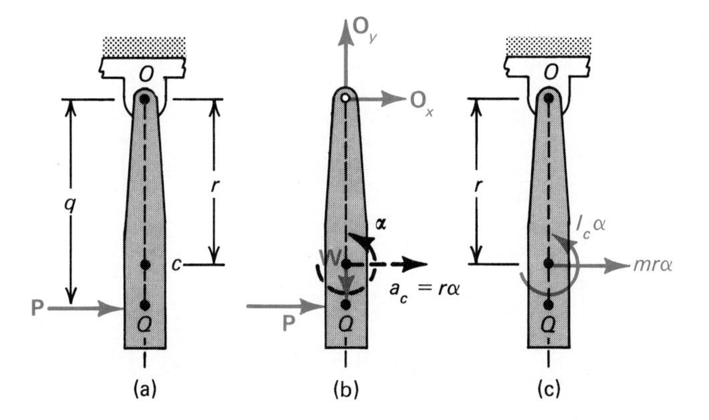

Figure 6.12

The point Q located along the line OC is called the **center of percussion.** In general, this is the point where the resultant of all the external forces passes. The center of percussion is a useful concept in designing such things as baseball bats and ballistic pendula.

PROBLEMS

6.26 A slender bar having a mass of 8 kg is struck by a hammer such that it rotates at an angular rate of 10 rad/s CW and an angular acceleration of 20 rad/s^2 CW at the instant shown in Fig. P6.26. Find the reactions at the support O.

6.27 If the angular acceleration of 20 rad/s^2 is CCW in Problem 6.26, while the angular velocity is 10 rad/s CW, then what are the reactions at the support O?

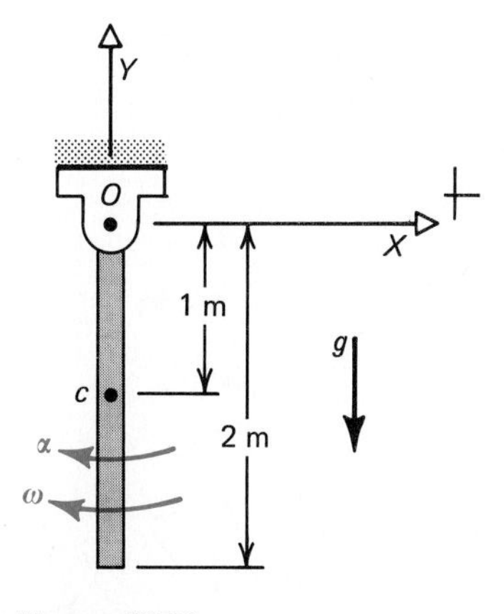

Figure P6.26

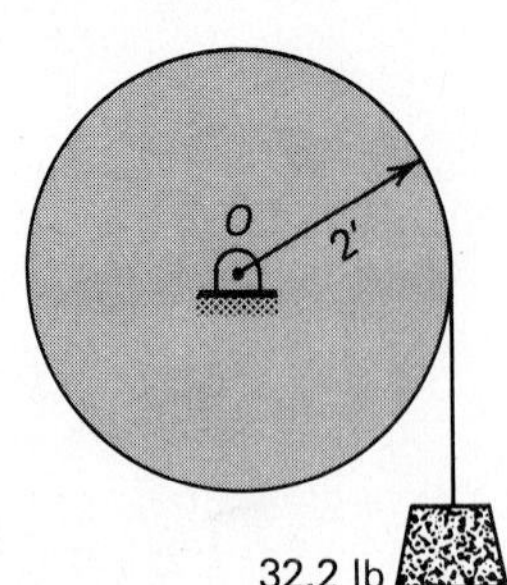

Figure P6.28

6.28 A 64.4-lb pulley has a 32.2-lb weight suspended from it by a rope as shown in Fig. P6.28. The radius of gyration of the pulley about O is 6 in. Find the angular acceleration of the pulley when the 32.2-lb weight is released from rest.

6.29 Repeat Problem 6.28 but replace the 32.2-lb weight by a downward force equal to 32.2 lb.

6.30 A cylinder of radius r and weight w rolls down a plane that is inclined at an angle θ with the horizontal. Assuming no slippage, find the expression for the friction force acting on the cylinder.

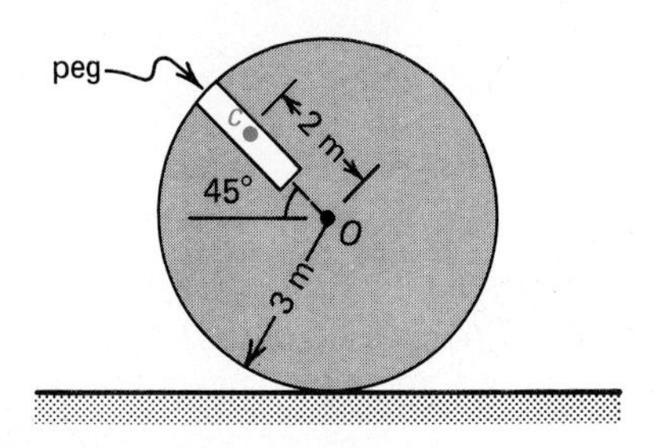

Figure P6.31

6.31 A thin hoop, supporting a 2-kg peg, is released from rest as shown in Fig. P6.31. Neglecting the mass of the hoop, find the initial angular acceleration of the hoop assuming no slippage.

6.32 A 10-kg sphere rolls and slips on the horizontal plane. The coefficient of friction is 0.4. The velocities of the center of mass c and the point A on the sphere are shown in Fig. P6.32. Find the angular acceleration of the sphere.

6.33 The 32.2-lb pulley in Fig. P6.33 supports two equal weights $w = 16.1$ lb that are released from rest. Find the acceleration of each weight w. The radius of gyration of the pulley is 1 ft about the center of mass.

6.34 A cylinder of mass m has a force of magnitude $P = mg$ applied to it as shown in Fig. P6.34. Determine the minimum coefficient of friction for the cylinder to roll without slipping.

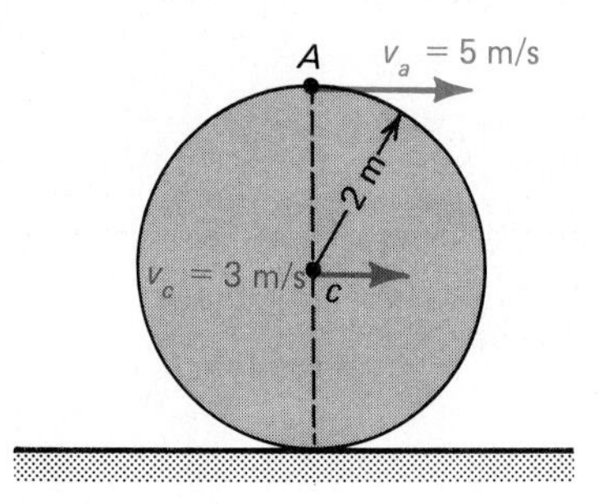

Figure P6.32

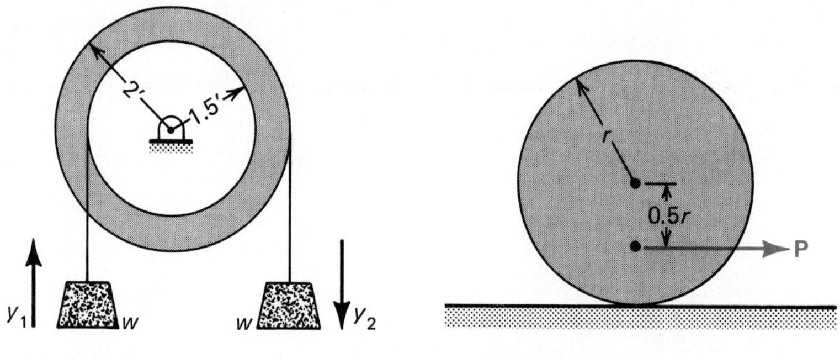

Figure P6.33

Figure P6.34

6.35 In Fig. P6.34, what is the frictional force if $P = 80$ N and the cylinder rolls without slipping?

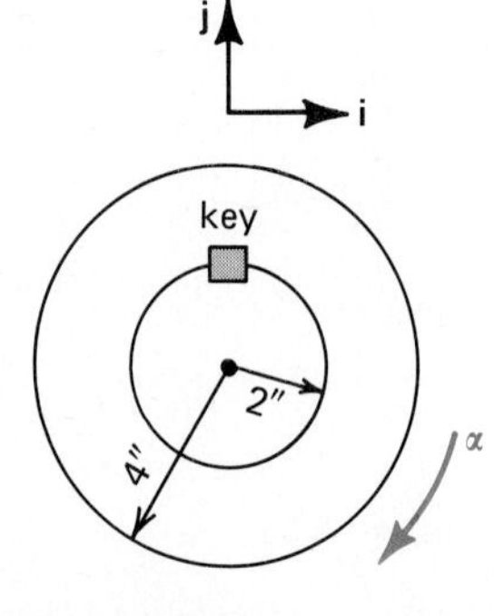

Figure P6.36

6.36 A flywheel is connected to a shaft by a key as shown in Fig. P6.36. A brake is applied to the shaft so that at the instant under consideration, the angular acceleration of the assembly is 100 rad/sec^2 CW. If the flywheel weighs 161 lb and its radius of gyration about the center of mass is 3 in, find the shear force transmitted across the key assuming no friction present between the shaft and the flywheel.

6.37 Rod OA is pinned to an accelerating car as shown in Fig. P6.37. The rod is initially vertical before the car accelerates. Find how much the spring deforms given that the weight of the rod is 10 lb, $a = 5$ ft/sec^2 and $k = 4$ lb/in. Assume a small deflection of the rod off its plumb position.

6.38 Figure P6.38 shows an activating device in its static equilibrium condition. The device is pinned at point A and restrained by the spring. The contacting mass is 1 kg and the spring modulus $k = 3600$ N/m. Find the magnitude of the acceleration **a** that causes contact to occur.

6.39 Where would you recommend a horizontal blow be applied to the bar in Problem 6.26 to minimize the reactions at the support?

6.40 If the rod in Problem 6.26 is to be struck by a horizontal blow at the position $3L/4$ from the support, how much mass would you recommend at the end of the rod in order to minimize the horizontal support reaction?

6.41 A spacecraft is spinning about the Z axis at 148 rev/min and is to be spun up to 200 rev/min in 100 min by applying a constant magnetic torque. If the angular acceleration is uniform and the mass moment of inertia about the Z axis is 200 slug·ft^2, calculate the torque that must be generated during the spin-up mode.

6.42 In Fig. P6.42 the energy storage wheel of density 3000 kg/m^3 is spun about the Y axis at a rate 1000 rev/min. It is then used to direct drive a small motor for 2 min during which it despins uniformly to 100 rev/min. What torque has it supplied the motor?

6.43 The crank shown in Fig. P6.43 has a uniform angular rotation of 100 rad/sec CCW. What pressure is generated on the piston at the instant shown? Assume that the effective mass of the gas is 0.015 slugs moving with the same acceleration as the piston. Consider only the force exerted on the piston from the gas.

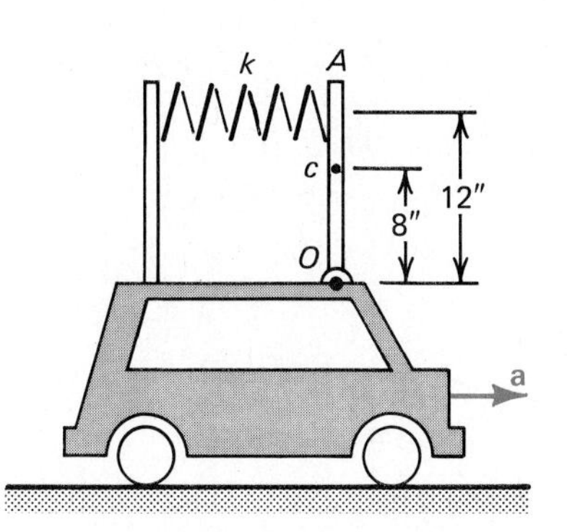

Figure P6.37

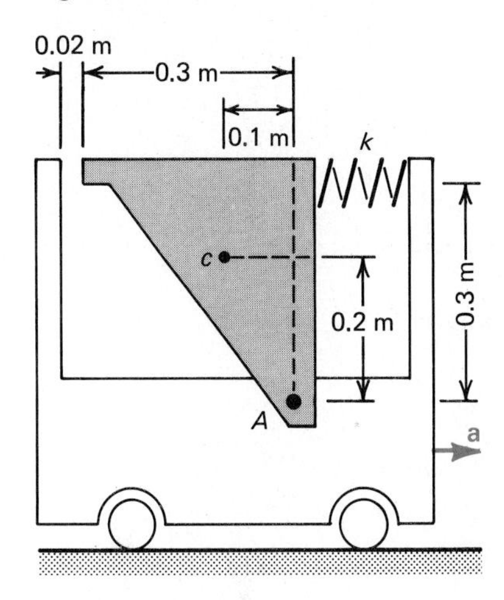

Figure P6.38

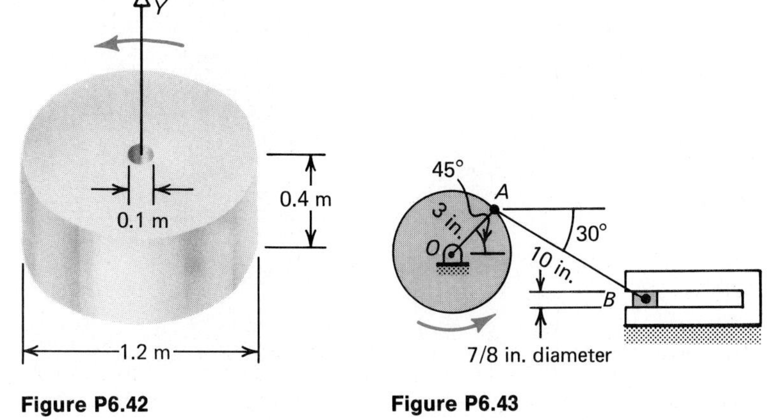

Figure P6.42

Figure P6.43

6.44 The truck shown in Fig. P5.62 is stationary while $\omega_{AB} = 8t\mathbf{k}$ rad/s for a short period of time. If the bar BA has a mass of 30 slugs and the driver plus cockpit have a mass of 200 slugs, what torque is generated at the joint B? Assume that the center of mass of the driver and cockpit is at A.

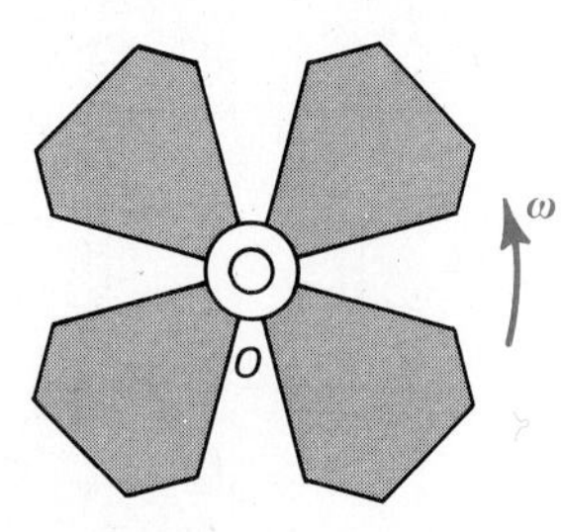

Figure P6.45

6.45 In Fig. P6.45 the fan is rotating at 200 rad/s when one blade having a length of 0.5 m becomes loose and flies off. Calculate the resultant force at the center O assuming that the blade has a mass of 0.5 kg. Assume a constant mass per unit length of blade.

6.46 In Fig. P6.46 a 200-lb flywheel is rotating at 500 rev/min and is to be stopped by a belt in 50 rev. The coefficient of friction between the flywheel and belt is 0.25 while the radius of gyration about the center of mass of the wheel is 1 ft. What must F be in order to bring the belt to rest uniformly? Let the radius of the flywheel be 2 ft.

6.47 A 20-lb ladder rests against a smooth wall as shown in Fig. P6.47. If the supporting block A is suddenly removed, calculate the initial angular acceleration by treating the ladder as a thin rod.

6.48 In Fig. P6.48 a spool of mass 200 kg has a radius of gyration about the center of mass equal to 0.1 m. At the instant shown, the spool is rolling without slipping under the applied force of magnitude $P = 100$ N. Find the angular acceleration of the body.

6.49 If a force of $P = 10$ N is applied as shown in Fig. P6.49, determine the angular acceleration of the 200-kg spool, which has a radius of gyration about the center C equal to 0.1 m.

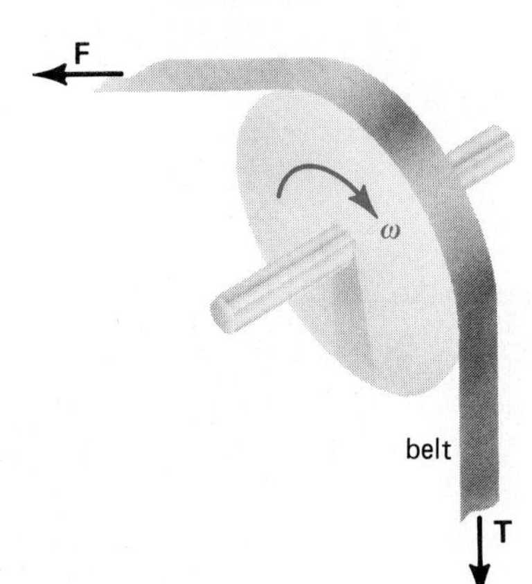

Figure P6.46

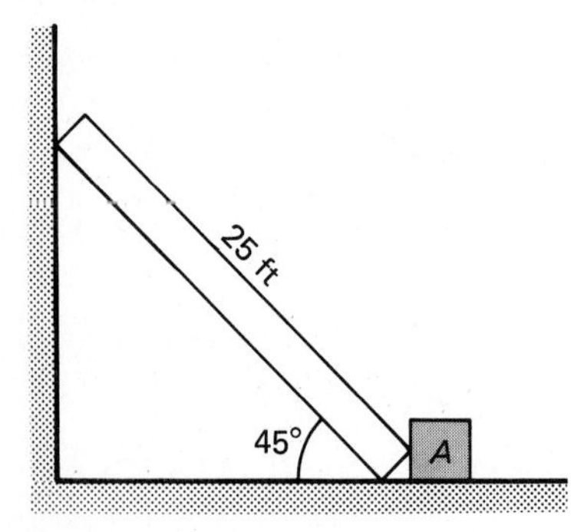

Figure P6.47

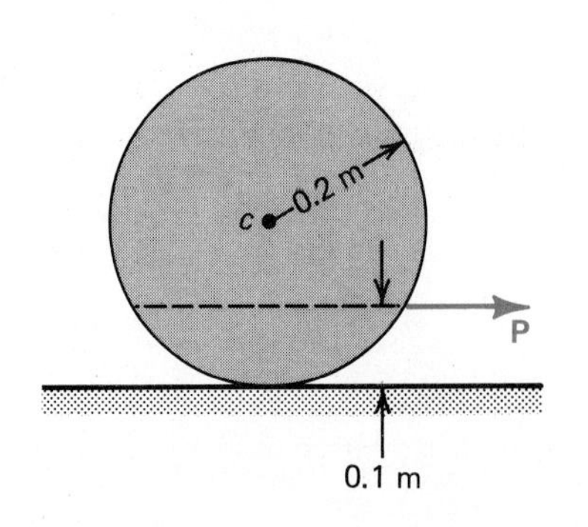

Figure P6.48

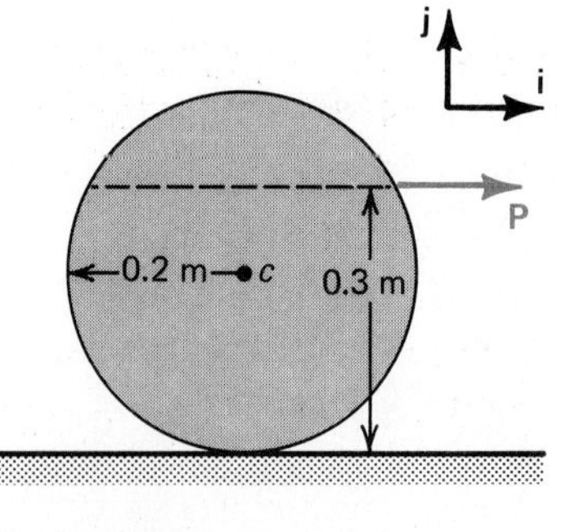

Figure P6.49

6.50 If a force of magnitude $P = 150$ N is applied to the spool shown in Fig. P6.49, calculate the acceleration of the center of mass C and the frictional force.

6.3 Work and Energy of a Plane Rigid Body

The work-energy relationship for a system of particles was based on the fundamental relationship

$$W = \Delta T$$

which led to the following equation:

$$\int_A^B [\mathbf{F} \cdot d\mathbf{r}_c + \sum(\mathbf{F}_i + \mathbf{f}_i) \cdot d\boldsymbol{\rho}_i] = \tfrac{1}{2}mv_c^2\Big|_A^B + \sum \tfrac{1}{2}m\dot{\rho}_i^2\Big|_A^B \qquad (6.16)$$

The kinetic energy is

$$T = \tfrac{1}{2}mv_c^2 + \sum \tfrac{1}{2}m_i\dot{\rho}_i^2 \qquad (6.15)$$

In the case of a rigid body, an element of mass dm has a velocity relative to a nonrotating observer at the center of mass equal to $r\omega$ as shown in Fig. 6.13. Consequently, for a plane rigid body let $m_i = dm$, $\dot{\rho}_i = r\omega$, and replace the summation symbol in Eq. (6.15) by an integral symbol. Doing this gives us the expression for the kinetic energy of a plane rigid body.

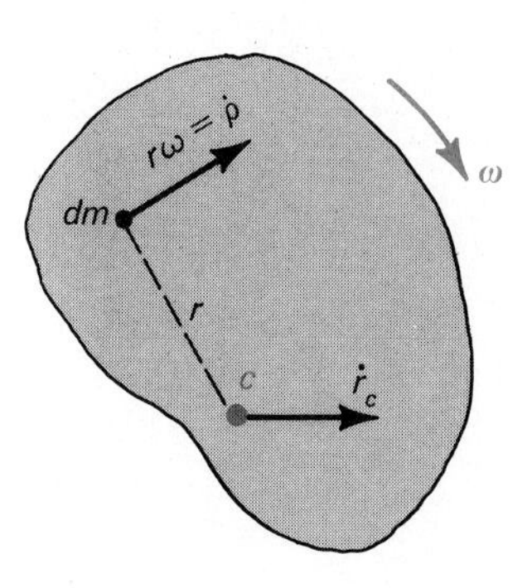

Figure 6.13

$$T = \tfrac{1}{2}mv_c^2 + \tfrac{1}{2}\omega^2 \int r^2\, dm$$
$$= \tfrac{1}{2}mv_c^2 + \tfrac{1}{2}I_c\omega^2 \qquad (6.39)$$

We see that the total kinetic energy consists of kinetic energy due to the translational motion, $\frac{1}{2}mv_c^2$, and kinetic energy due to the rotational motion, $\frac{1}{2}I_c\omega^2$, where I_c is the mass moment of inertia of the body about the center of mass. Note that v_c is the magnitude of the absolute velocity of the center of mass and ω is the magnitude of the absolute angular velocity of the rigid body.

If a point O on a body has zero velocity, such as the case of the cylinder rolling without slipping in Fig. 6.14, then $v_c = r\omega$. Consequently,

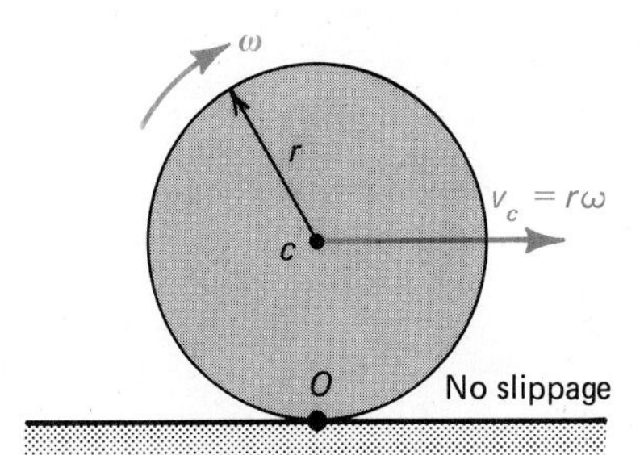

Figure 6.14

$$T = \tfrac{1}{2}mr^2\omega^2 + \tfrac{1}{2}I_c\omega^2 = \tfrac{1}{2}(I_c + mr^2)\omega^2$$
$$= \tfrac{1}{2}I_o\omega^2 \qquad (6.40)$$

where, by the parallel-axis theorem, I_o is the mass moment of inertia of the body about point O. Equation (6.40) is useful for a pinned connection of a rigid body and, on occasion, when the instantaneous center of zero velocity is available.

The basic form of the work-energy relationship of a system of particles was expressed as

$$W = \Delta T \qquad (6.41)$$

This work-energy relationship is directly applicable to a plane rigid body. We must now be careful to use Eq. (6.39) when calculating the kinetic energy of the body.

Suppose we have a set of N rigid bodies connected by nonrigid connections such as springs. It is important to realize that the expression for the work in Eq. (6.41) must include not only the work done by the external forces and moments acting on the set of rigid bodies, but also the internal forces and moments represented by the nonrigid connectors. This agrees with the left side of Eq. (6.16) that includes the contribution of the internal forces to the work expression for a set of particles.

An alternate form of Eq. (6.41) is to introduce the notion of potential energy. We recall that the work done by conservative forces on a particle equals the negative change in the potential energy, i.e.,

$$W_c = -\Delta V \tag{6.42}$$

Consequently, we can write Eq. (6.41) for a plane rigid body as

$$W_{NC} = \Delta T + \Delta V \tag{6.43}$$

where W_{NC} is the work done by all forces acting on the plane rigid body *except* those associated with gravity and spring forces. For these latter forces we have derived the following potential energy terms that we studied in Chapter 3.

Gravity (Body close to the earth's surface)

$$V_g = mgh$$

$$\Delta V_g = mg(h_B - h_A)$$

Note that h is measured from the reference plane to the center of mass of the rigid body.

Gravity (Body away from the earth's surface)

$$V_g = -mg\frac{R^2}{R+h}$$

$$\Delta V_g = -mg\left(\frac{R^2}{R+h_B} - \frac{R^2}{R+h_A}\right)$$

Elastic Spring

$$V_e = \tfrac{1}{2}kx^2$$

$$\Delta V_e = \tfrac{1}{2}k(x_B^2 - x_A^2)$$

Conservation of Mechanical Energy

Equation (6.43) is convenient for most problems when we wish to use the work-energy relationship. In the case of a conservative system, Eq. (6.43) reduces to the form for the conservation of energy, i.e.,

$$\Delta T + \Delta V = 0$$

or

$$T_A + V_A = T_B + V_B \tag{6.44}$$

Rolling Friction

Figure 6.15 shows the free-body diagram of a rolling body moving along a horizontal plane. If there is no slippage, the friction force **F**, whose magnitude is less than μN, does no work. This is because this force occurs at a given instant at the point of contact between the rolling body and the fixed surface. At the next instant a new set of stationary points of contact are interacting with each other producing the force **F**. Consequently, the force **F** shown in Fig. 6.15 does essentially zero work on the point of contact since it acts on a given point during an infinitesimal change in displacement. Another viewpoint of rolling friction work is to allow the body to translate a distance Δx and rotate such that a point on the circumference moves through a distance Δx. This two-step approach places the body in the same position it would be if pure rolling occurred. Now the work done by F during the translational part is $-F\,\Delta x$ while it is $+F\,\Delta x$ during the rotational part. Consequently, the net work due to the rolling friction is again zero. Note that if slippage occurs, $F = \mu N$ and the friction forces does negative work on the body.

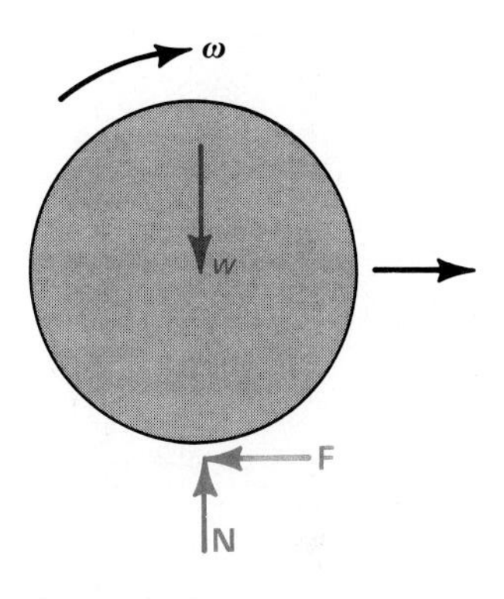

Figure 6.15

Power

It is sometimes important not only to obtain the work done on a plane rigid body but also the rate of doing this work. The work done per unit time is the definition of power. For rigid bodies, as in the case of particles, the instantaneous power due to the action of a force **F** is defined as

$$\text{Power} = P = \mathbf{F} \cdot \mathbf{v}_c$$

where $\mathbf{v}_c$ is the velocity of the center of mass.

In the case of a plane rigid body subjected to a moment of magnitude M, the element of work due to the moment producing the element of rotation $d\theta$ is

$$dW = M\,d\theta$$

The corresponding power is

$$P = \frac{dW}{dt} = M\frac{d\theta}{dt} = M\omega$$

The average power over the time interval Δt is

$$P_{avg} = \frac{\Delta W}{\Delta t}$$

The units for mechanical power are generally expressed in watts or kilowatts, where 1 watt = 1 N·m/s. Also, power in units of horsepower (hp) is common, where 1 hp = 746 W.

Example 6.9

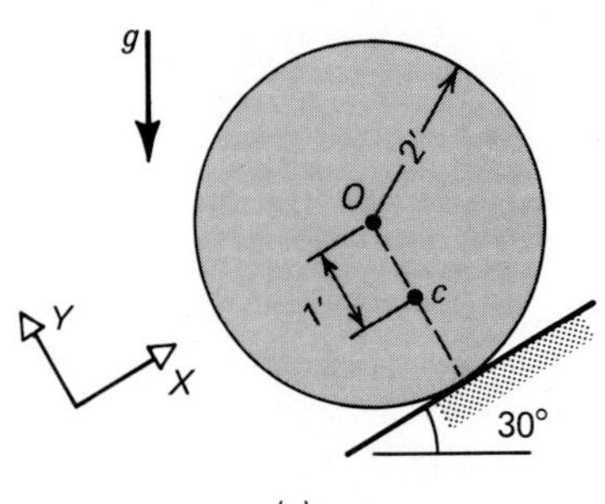

(a)

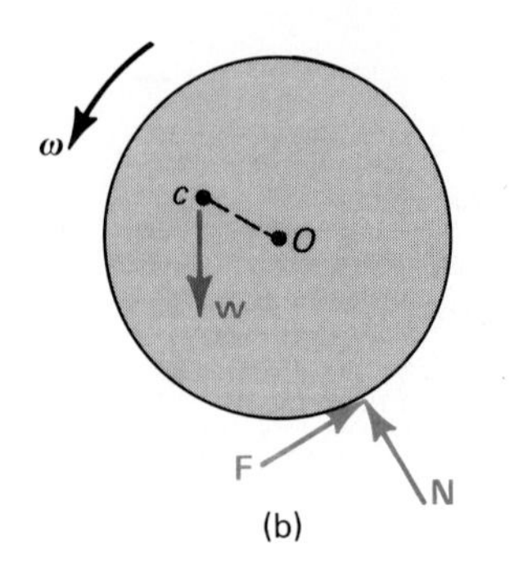

(b)

Figure 6.16

An unbalanced wheel is released from rest as shown in Fig. 6.16(a). The center of mass c is initially on a line perpendicular to the inclined plane. If the wheel weighs 32.2 lb and has a radius of gyration of 6 in about the center of mass, find the velocity of the center O after the wheel has rolled through two complete revolutions. Assume no slippage.

Figure 6.16(b) shows a free-body diagram of the forces acting on the wheel at some time during its downward travel. The normal reaction N always acts perpendicular to the path of motion so that it does no work on the system. The rolling friction force **F** also does zero work. Consequently, the conservation of energy holds in this example even though there is rolling friction present. Thus,

$$T_A + V_A = T_B + V_B$$

where A represents the initial position of the wheel and B is the final position. Now $T_A = 0$ since the body is released from rest and

$$T_B = \tfrac{1}{2}mv_c^2 + \tfrac{1}{2}I_c\omega^2$$

Since there is no slippage, the magnitude of the velocity of the center of mass is $v_c = r_c\omega$ where $r_c = 1$ ft = the distance from the instantaneous center of zero velocity to the center of mass.

$$\begin{aligned} T_B &= \frac{1}{2}\left(\frac{32.2}{32.2}\right)(1\omega)^2 + \frac{1}{2}\left(\frac{32.2}{32.2}\right)\left(\frac{6}{12}\right)^2 \omega^2 \\ &= 0.625\omega^2 \end{aligned}$$

For the potential energy terms, let the reference point be the center of mass at position B after two complete revolutions. Now

$$V_A = mgh_A = 32.2(4\pi)(2)\sin 30^\circ = 404.6 \text{ lb·ft}$$

$$V_B = 0$$

Substituting into the governing equation,

$$404.6 = 0.625\omega^2$$

$$\omega = 25.4 \text{ rad/sec}$$

Therefore

$$\mathbf{v}_o = -r\omega\mathbf{i} = -2(25.4)\mathbf{i} = -50.8\mathbf{i} \text{ ft/sec} \qquad \textit{Answer}$$

Example 6.10

A 10-kg mass, connected to a drum as shown in Fig. 6.17, starts from rest and moves down a distance of 20 m. What is the magnitude of the velocity of the 10-kg mass at this point? Assume that the drum is 100 kg and has a radius of gyration of 0.2 m about point O.

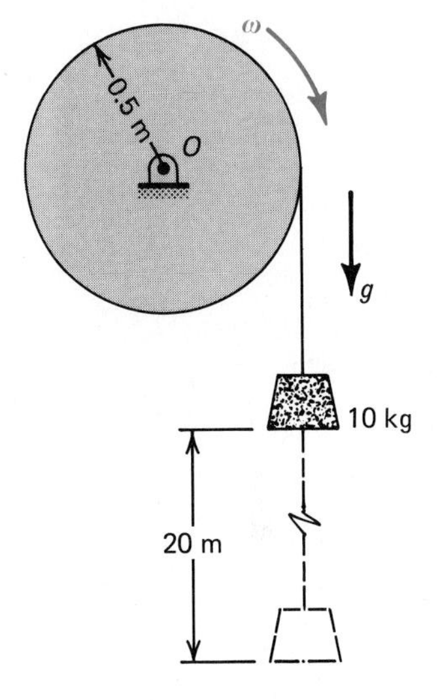

Figure 6.17

We have a conservative system since there are no nonconservative forces present.

$$T_A + V_A = T_B + V_B$$

Let position A be the position of the 10-kg mass when it is released from rest. After moving down 20 m, the system is at position B. Here the magnitude of the velocity of the 10-kg mass is $v = r\omega$. Now

$$T_A = 0$$

$$\begin{aligned} T_B &= (T_B)_{\text{mass}} + (T_B)_{\text{drum}} \\ &= \tfrac{1}{2}m(r\omega)^2 + \tfrac{1}{2}I_o\omega^2 \\ &= \tfrac{1}{2}(10)(0.5\omega)^2 + \tfrac{1}{2}(100)(0.2)^2\omega^2 \\ &= 3.25\omega^2 \end{aligned}$$

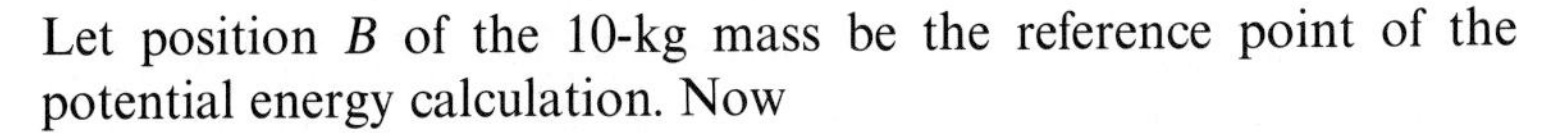

Let position B of the 10-kg mass be the reference point of the potential energy calculation. Now

$$V_A = mgh_A = 10(9.81)(20) = 1962$$

$$V_B = 0$$

Substituting into the governing equation,

$$1962 = 3.25\omega^2$$

$$\omega = 24.57 \text{ rad/s}$$

The magnitude of the velocity of the 10-kg mass is

$$\begin{aligned} v = r\omega &= 0.5(24.57) \\ &= 12.3 \text{ m/s} \qquad \textit{Answer} \end{aligned}$$

Example 6.11

Two identical bars each weighing 16.1 lb are connected by a spring as shown in Fig. 6.18. A force $\mathbf{P}$ acts downward at the pin connection O. $P = 100$ lb and the spring, as shown, is unstretched and has a spring modulus $k = 20$ lb/in. It is assumed that the horizontal surface is smooth. Determine the angular velocity of each bar at the instant when the rods become horizontal.

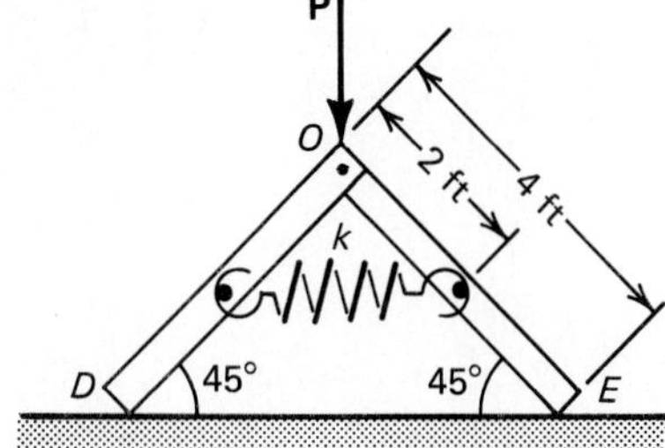

Figure 6.18

Equation (6.43) is applicable to this problem.

$$W_{NC} = \Delta T + \Delta V$$
$$= \Delta T + \Delta V_g + \Delta V_e$$

The applied force **P** produces W_{NC} as it acts through the distance traveled by point O.

$$W_{NC} = 100(4 \sin 45^\circ) = 282.8 \text{ lb}\cdot\text{ft}$$

For the kinetic energy,

$$\Delta T = T_B - T_A$$

Since the system starts from rest, $T_A = 0$.

When the bars are in the horizontal position, the end points D and E have zero velocity. Consequently, it is convenient to use Eq. (6.40) for each rod,

$$T_B = 2(\tfrac{1}{2}I_o\omega^2)$$

The mass moment of inertia of a rod about the end point is

$$I_o = \tfrac{1}{3}mL^2$$

so that

$$T_B = \frac{1}{3}mL^2\omega^2 = \frac{1}{3}\left(\frac{16.1}{32.2}\right)(4)^2\omega^2$$
$$= 2.667\omega^2$$

For the change in potential energy due to gravity for the two rods, we have

$$\Delta V_g = 2mg(h_B - h_A)$$

Let position B be the point of reference, so that $h_B = 0$ and $h_A = 2 \sin 45^\circ = 1.414$ ft.

$$\Delta V_g = 2(16.1)(0 - 1.414) = -45.53 \text{ lb}\cdot\text{ft}$$

The change in potential energy due to the stretching of the spring is found by

$$\Delta V_e = \tfrac{1}{2}k(x_B^2 - x_A^2)$$

At position A the spring is unstretched so that $x_A = 0$. At position B, $x_B = 4 - 2(2 \cos 45^\circ) = 1.172$ ft. It is important at this step in our solution to convert the units of the spring modulus from lb/in to lb/ft. Thus,

$$k = 20 \text{ lb/in} = 240 \text{ lb/ft}$$

Now

$$\Delta V_e = \tfrac{1}{2}(240)(1.172)^2 = 164.8 \text{ lb}\cdot\text{ft}$$

Substituting into the governing equation,

$$282.8 = 2.667\omega^2 - 45.53 + 164.8$$

$$\omega = 7.83 \text{ rad/sec} \qquad \textit{Answer}$$

PROBLEMS

6.51 The velocity of two points on a rolling and sliding sphere is shown in Fig. P6.51. Find the kinetic energy of the body if the mass of the sphere is 1 kg.

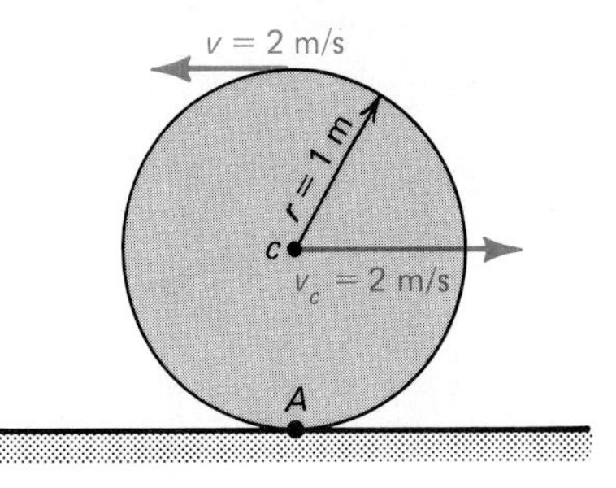

Figure P6.51

6.52 The two wheels in Fig. P6.52 are connected by a 2-kg bar AB. If the left wheel is rotating at 1200 rev/min CW, find the kinetic energy of the bar AB.

6.53 The 3.22-lb cylinder in Fig. P6.53 is released from rest and rolls without slipping along the circular surface. Find the maximum kinetic energy of the cylinder.

6.54 Under special conditions, an inverted pendulum attached to a shaker table can be made to oscillate about its unstable equilibrium position. If the pendulum weighs 2 lb, the table velocity $\dot{z} = 3$ in/sec at a given instant, $\theta = 10°$, and $\dot{\theta} = 2$ rad/sec CW as shown in Fig. P6.54, find the kinetic energy of the pendulum.

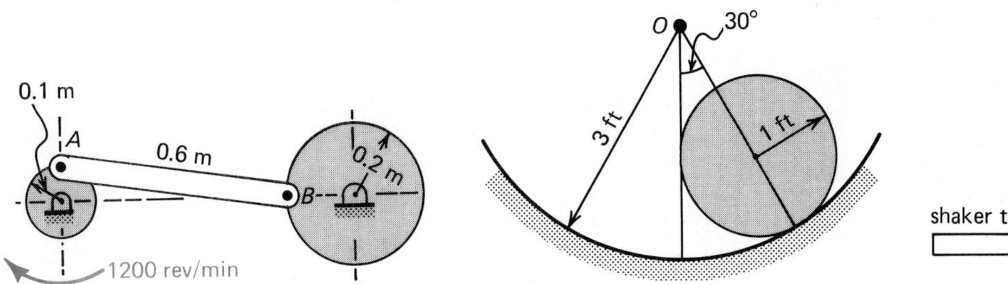

Figure P6.52

Figure P6.53

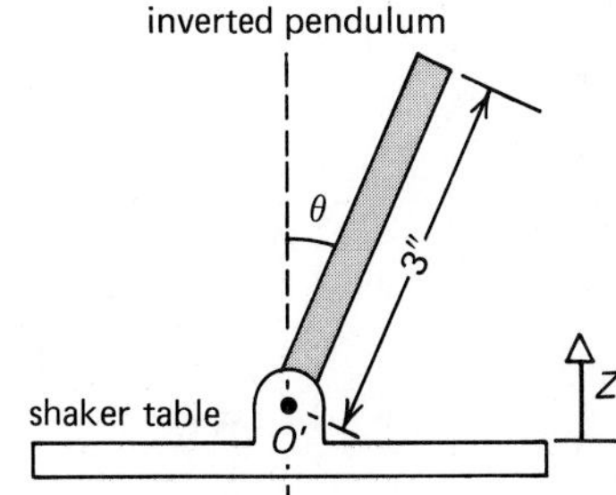

Figure P6.54

6.55 The 10-kg mass shown in Fig. P6.55 starts from rest and accelerates downward. The 16-kg drum has a radius of gyration of 0.5 m about point O. Compute the angular velocity of the drum after the mass has dropped 20 m.

6.56 Compute the angular velocity of the drum in Problem 6.55 if the 10-kg mass had an initial velocity of 8 m/s downward and again drops 20 m.

6.57 A $\frac{1}{2}$-lb pendulum is attached to a shaft in Fig. P6.57 to eliminate undesirable oscillations in the shaft. If $\dot{\theta} = 100$ rad/sec CCW, and $\dot{\phi} = 50$ rad/sec CCW at the instant shown, find the kinetic energy of the pendulum.

6.58 Find the angular velocity of the bar in Problem 6.26 when the bar reaches the horizontal position. Assume that the 8-kg bar has an angular velocity of 10 rad/s at the instant shown in Fig. P6.26.

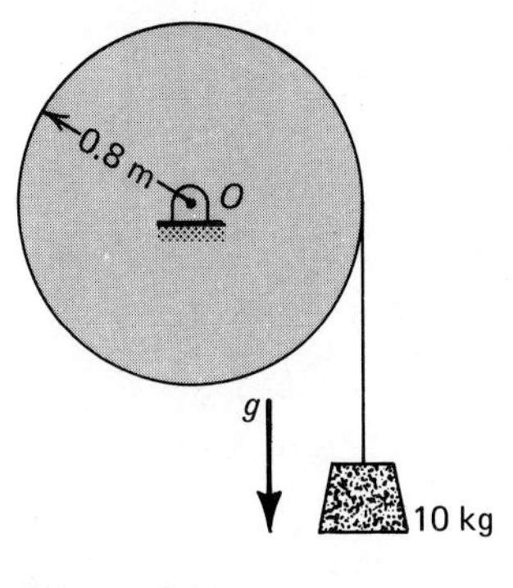

Figure P6.55

6.59 The brake in Fig. P6.59 is applied to a 10-kg drum rotating at 200 rev/min. If the brake force $P = 10$ N and the coefficient of friction between

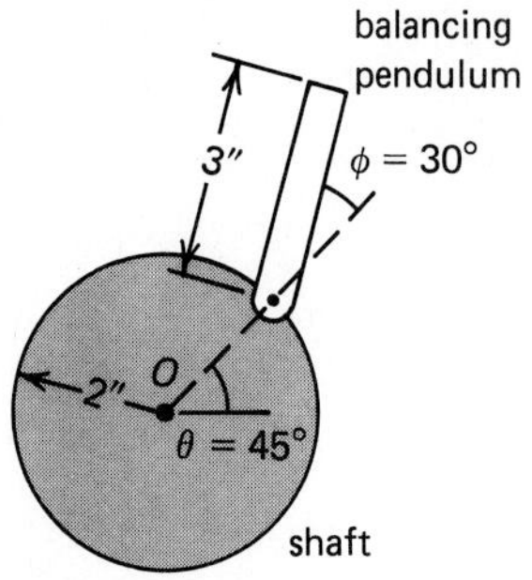

Figure P6.57

the brake and the drum is 0.2, determine the number of revolutions the drum experiences before coming to rest. Let the radius of gyration of the drum about an axis through O be 0.25 m.

6.60 What is the velocity of the weight in Problem 6.28 after it moves 10 ft downward having started from rest?

6.61 The 50-kg mass m in Fig. P6.61 is attached to a grooved 100-kg cylinder by means of a rope. If m is released from rest while the spring is unstretched, find the maximum drop of m if the spring constant $k = 4500$ N/m. (*Note*: The spring wraps around the cylinder.)

6.62 Gear B shown in Fig. P6.62 is to be spun up by gear A which is started from rest. If gear A is spun to 120 rad/sec in 100 revolutions, what constant moment must be applied to achieve this? Let $m_A = 1$ slug, $m_B = 2$ slugs, and the radii of gyration about the rotational axes are 0.5 in and 3 in, respectively.

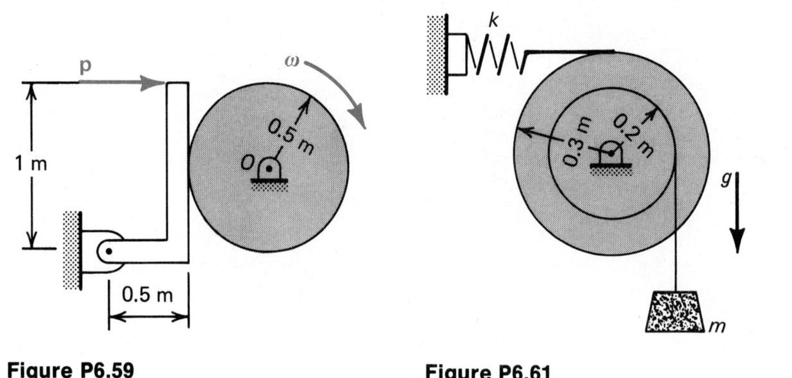

Figure P6.59

Figure P6.61

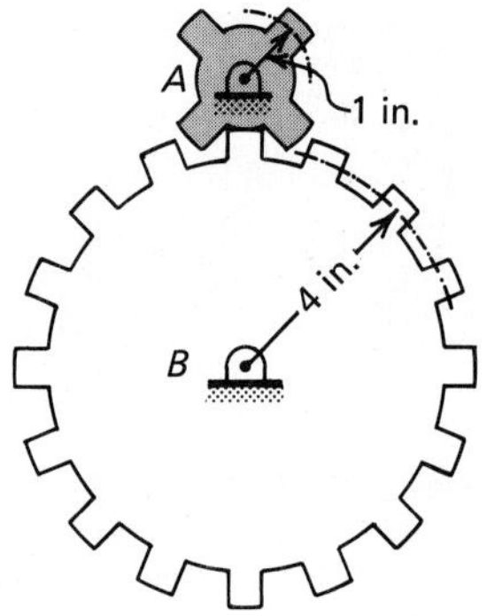

Figure P6.62

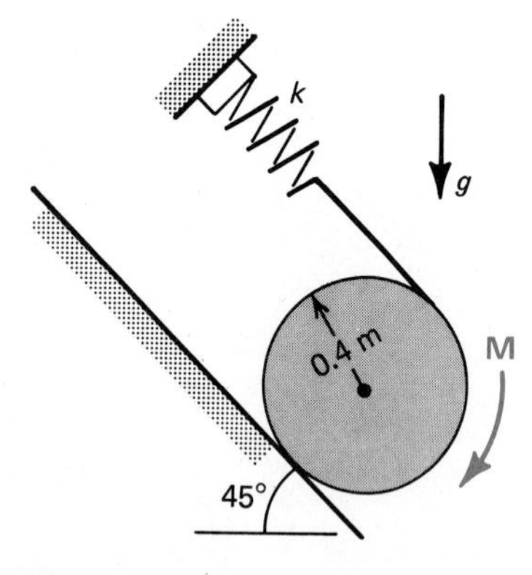

Figure P6.63

6.63 A constant torque of magnitude $M = 2$ m·N is applied to a spring-restrained rolling 20-kg cylinder as shown in Fig. P6.63. For the position shown, the spring, whose unstretched length is 0.3 m, is compressed 0.03 m, and the angular rate of the cylinder is 2 rad/s CW. Find the angular velocity of the cylinder after one-fourth of a revolution. Assume no slippage and let $k = 160$ N/m. (*Note*: The spring wraps around the cylinder.)

6.64 If the 10-kg drum of Problem 6.59 decelerates uniformly to rest, how much power is expended?

6.65 What is the velocity of each mass in Problem 6.7 after the right mass has dropped 1 m?

6.66 A coal car carries a load of 1.5 tons as shown in Fig. P6.66. Each of the four wheels weighs 100 lb, has a radius of 2 ft, and a radius of gyration about the rotational axis of 1.5 ft. The car is traveling at 4 mi/hr when it approaches a 2-ft drop at a switching station. Find the velocity of the car after the 2-ft drop. The weight of the empty car and the wheels is 1 ton.

6.67 A cylinder of mass m is raised from rest by the constant force **P** shown in Fig. P6.67. Derive a relationship between the velocity of the center of the cylinder and the amount the spring stretches.

6.68 A 0.5-kg restrained cylinder is raised from rest by a constant force **P** of magnitude $P = 20$ N as shown in Fig. P6.67. Find the spring modulus that causes the cylinder to come to rest after it is raised 1 m. Assume that the spring is initially unstretched.

6.69 A constant torque $M = 100$ ft·lb lifts a 25-lb gate shown in Fig. P6.69. The gate is initially at rest, and the restraining spring is initially unstretched. The center of mass of the gate is located at point C. Find the spring modulus k so that the gate comes to rest after being rotated through 90° CCW.

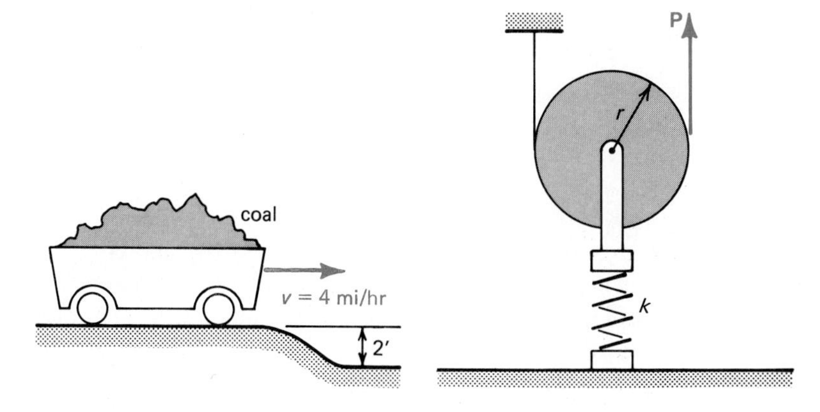

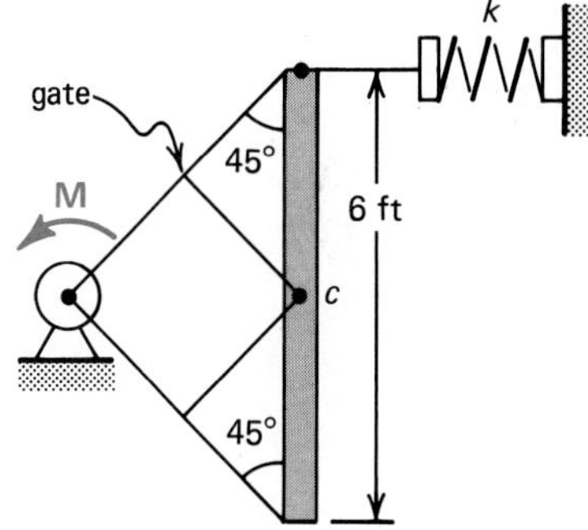

Figure P6.66

Figure P6.67

Figure P6.69

6.70 What is the average power that must be supplied to the gears in Problem 6.62 if gear A is spun up to 120 rad/s in 100 revolutions in 1 min?

6.71 How much average power is supplied into spinning the satellite of Problem 6.41 from 148 rev/min to 200 rev/min?

6.72 A 10,000-kg airplane lands at a ground speed of 150 km/h. At touchdown the jet engines are reversed so that the plane can be decelerated uniformly to a stop in 2 km. What average power must be generated to achieve this?

6.73 A helicopter is designed such that the main helicopter blades are accelerated to 500 rev/min uniformly in 1.25 min. The mass moment of inertia of each blade is 95 kg·m^2 about the vertical axis. (a) How much work is done in accelerating the two blades? (b) How much average power is needed? Express your answer in horsepower.

6.74 A child's toy car is propelled by a spinning energy wheel. This wheel has a mass moment of inertia about the spin axis of 0.5 kg·m^2 and is revved up to 100 rev/min. The car has a mass of 0.5 kg and travels along a horizontal plane 10 m before coming to rest. What is the average frictional force?

6.75 Assume the toy car in Problem 6.74 goes up an inclined plane of 30°. If the frictional force is a constant as found in Problem 6.74, how far up the incline does the car go?

6.4 Impulse and Momentum of Plane Rigid Bodies

The impulse and momentum relationships for a system of N particles, as developed in Section 6.1, can be applied directly to a plane rigid body. For example, the force-linear momentum change as given by Eq. (6.18) is

$$\mathbf{F} = \dot{\mathbf{p}}_c \tag{6.18}$$

where $\mathbf{F}$ is now the resultant external force acting on the plane rigid body and $\mathbf{p}_c$ is the linear momentum of the center of mass of the rigid body.

The equations for the angular momentum change about the center of mass were developed earlier as

$$\mathbf{M}_c = \dot{\mathbf{H}}_c \tag{6.25}$$

where $\mathbf{M}_c$ is now the moment about the center of mass c of the external forces and moments acting on the rigid body and $\mathbf{H}_c$ is the angular momentum of the body about c. In the case of plane rigid body motion, Eq. (6.25) is an equation whose unit vector is $\mathbf{k}$, that is, a vector perpendicular to the XY plane of motion. Consequently, we can rewrite Eq. (6.25) in scalar form as

$$M_c = \dot{H}_c \tag{6.45}$$

Consider the plane rigid body in Fig. 6.19 whose center of mass travels with a velocity $\mathbf{v}_c$ while the body rotates at ω rad/s CW. The components of the velocity of the element of mass dm consist of v_c and $\rho\omega$ as shown in Fig. 6.19. Consequently, the angular momentum of dm about the point C is

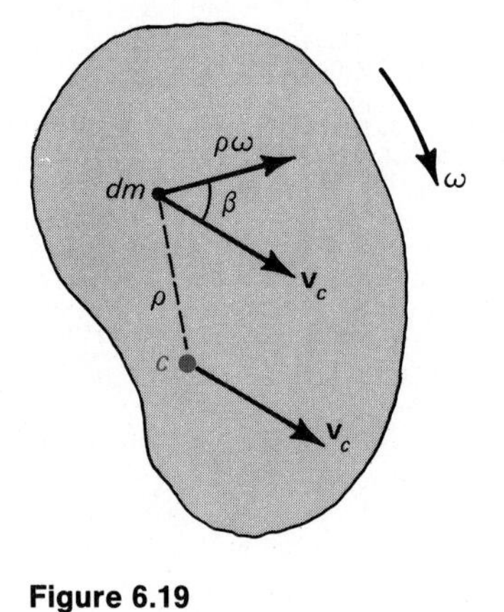

Figure 6.19

$$dH_c = (\rho\omega + v_c \cos\beta)\rho\, dm$$

Summing over the entire body,

$$H_c = \int (\rho\omega + v_c \cos\beta)\rho\, dm$$
$$= \int \rho^2 \omega\, dm$$

since the second term on the right side is zero. Now

$$H_c = I_c\omega \tag{6.46}$$

since

$$I_c = \int \rho^2\, dm$$

It is interesting to differentiate Eq. (6.46), i.e.,

$$\dot{H}_c = I_c\dot{\omega} = I_c\alpha$$

so that Eq. (6.45) reduces to

$$M_c = I_c\alpha \tag{6.47}$$

which is precisely Eq. (6.34) developed earlier.

Instead of using the center of mass as the reference point, the angular momentum about an arbitrary point O may be found by means of Fig. 6.20. Once again we observe the plane rigid body whose center of mass travels at velocity $\mathbf{v}_c$ while the body rotates ω rad/s CW. The momentum terms mv_c and $I_c\omega$ are shown as they act on the body. Now the total angular momentum about the point O is

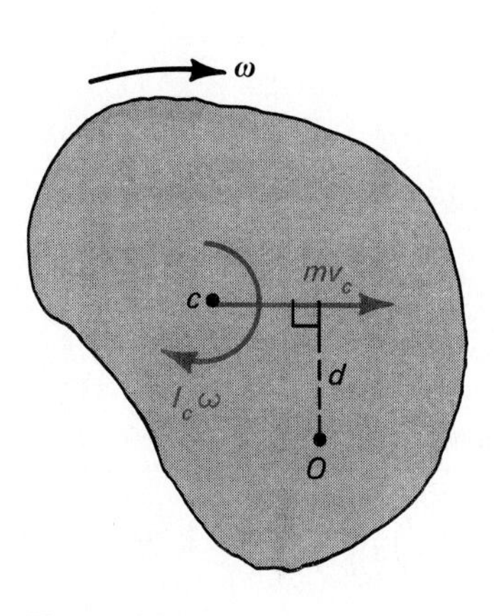

Figure 6.20

$$H_o = I_c\omega + mv_c d \tag{6.48}$$

where d is the lever arm between mv_c and point O. If point O is fixed, we can show that $H_o = I_o\omega$ where I_o is the mass moment of inertia of the body about point O.

The linear impulse–momentum relationship is obtained by integrating Eq. (6.18). Thus,

$$\mathscr{I} = \int_{t_1}^{t_2} \mathbf{F}\,dt = \mathbf{p}_{c2} - \mathbf{p}_{c1} \tag{6.49}$$

By integrating Eq. (6.45), we have the angular impulse–momentum relationship for a plane rigid body using the center of mass as the reference point.

$$\mathscr{M}_c = \int_{t_1}^{t_2} M_c\,dt = H_{c2} - H_{c1} \tag{6.50}$$

Equations (6.49) and (6.50) are particularly useful for cases where applied forces and moments are explicit functions of time. In this way we can calculate changes in velocities directly, without introducing accelerations.

In many instances we are able to assume that a plane rigid body experiences a conservation of linear momentum and/or a conservation of angular momentum. If the linear impulse is zero over a given time interval, then the linear momentum is conserved. That is,

$$\mathbf{p}_{c2} = \mathbf{p}_{c1} \tag{6.51}$$

In the case of a single rigid body, this equation reduces to

$$m\mathbf{v}_{c2} = m\mathbf{v}_{c1}$$

so that

$$\mathbf{v}_{c2} = \mathbf{v}_{c1}$$

Likewise, if the angular impulse is zero during a time interval, then the angular momentum of the plane rigid body is conserved, i.e.,

$$H_{c2} = H_{c1} \tag{6.52}$$

For a single rigid body this equation reduces to

$$I_c\omega_2 = I_c\omega_1$$

or

$$\omega_2 = \omega_1$$

The special class of problems dealing with bodies impacting with one another will be examined in the next chapter. There we shall see the important role played by the equations of impulse and momentum.

Example 6.12

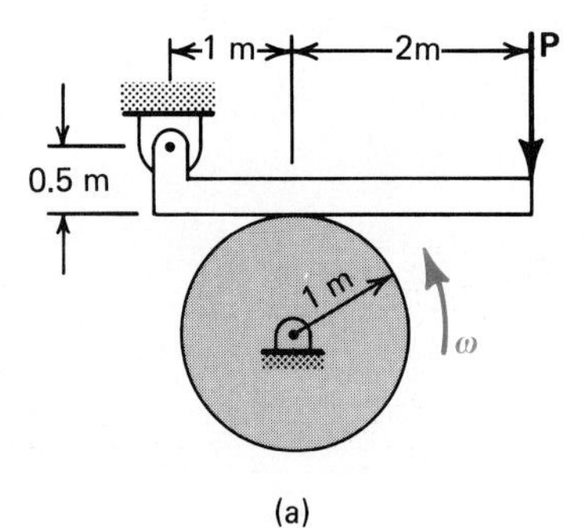

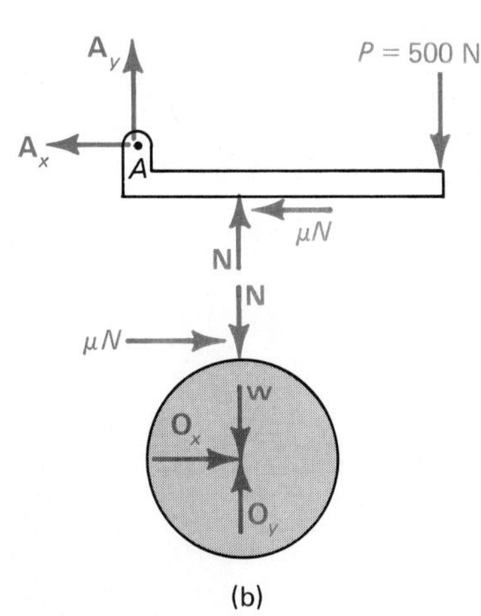

Figure 6.21

A constant force **P** is applied to the brake shown in Fig. 6.21(a) in order to bring the rotating flywheel to rest. The kinetic coefficient of friction between the brake and the wheel is 0.2. The flywheel has a mass of 300 kg and a radius of gyration of 0.5 m about the center of mass and is initially rotating at 1200 rev/min CCW. If $P = 500$ N, find the time it takes for the flywheel to come to rest.

The free-body diagram for each member is shown in Fig. 6.21(b). Since the brake is stationary, we may sum moments about point A letting clockwise moments be positive.

$$\sum M_A = 0 \qquad 500(3) + 0.2N(0.5) - N(1) = 0$$

$$N = 1667 \text{ N}$$

The mass moment of inertia of the wheel is

$$I_c = 300(0.5)^2 = 75 \text{ kg·m}^2$$

Applying the angular-momentum relationship of Eq. (6.50) to the flywheel and again letting clockwise moments be positive,

$$\int_0^t (1)\mu N\,dt = I_c[0 - (-\omega)]$$

$$0.2(1667)t = 75\omega$$

$$t = \frac{75(1200)(2\pi)}{0.2(1667)60}$$

$$t = 28.3 \text{ s} \qquad \textit{Answer}$$

Example 6.13

A satellite, spinning about its centroidal axis at 30 rad/sec, is accelerated to 50 rad/sec by firing jets. These jets generate a moment about a spin axis according to $M_c = 16t^3$, where t is in seconds, having units of lb·ft. If the satellite has a mass of 100 slugs and a radius of gyration of 0.5 ft about the spin axis, how much time does it take to spin up the satellite?

Using the angular impulse–momentum Eq. (6.50),

$$\int_0^t M_c\,dt = I_c(\omega_2 - \omega_1) = I_c(50 - 30)$$

$$\int_0^t 16t^3\,dt = 20I_c = 20(100)(0.5)^2$$

$$t^4 = 125$$

$$t = 3.34 \text{ sec} \qquad \textit{Answer}$$

Example 6.14

A 0.9-kg cue ball rolls without slipping along a pool table as shown in Fig. 6.22. At this instant the angular velocity of the mass is 4 rad/s CCW. Find the angular momentum about the center of mass and about the instantaneous center of zero velocity.

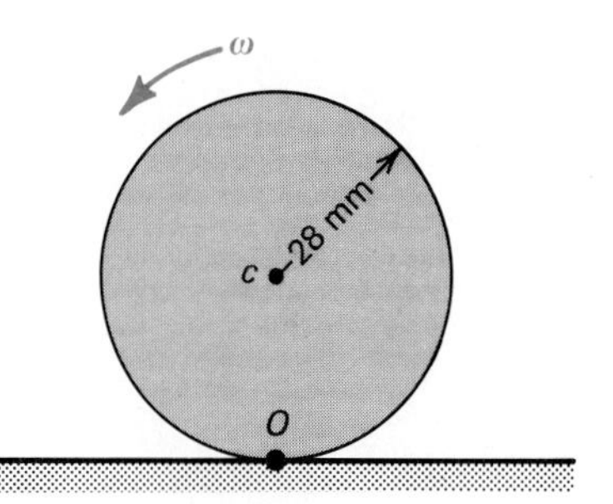

Figure 6.22

We start with the equation

$$H_c = I_c\omega$$

where for a sphere,

$$I_c = \tfrac{2}{5}mr^2 = \tfrac{2}{5}(0.9)(0.028)^2 = 2.82 \times 10^{-4}\ \text{kg}\cdot\text{m}^2$$

and therefore

$$H_c = 2.82 \times 10^{-4} \times 4 = 1.128 \times 10^{-3}\ \text{kg}\cdot\text{m}^2/\text{s CCW} \qquad \textit{Answer}$$

The angular momentum about the instantaneous center, labeled O in Fig. 6.22, is

$$H_o = I_c\omega + mv_c r = I_c\omega + mr^2\omega$$

$$= 1.128 \times 10^{-3} + 0.9(0.028)^2(4)$$

$$H_o = 3.95 \times 10^{-3}\ \text{kg}\cdot\text{m}^2/\text{s CCW} \qquad \textit{Answer}$$

PROBLEMS

6.76 What is the angular momentum about points C and A for the sphere in Problem 6.51?

6.77 What is the angular momentum about point O when the cylinder in Problem 6.53 is at its lowest point?

6.78 What is the angular momentum of the inverted pendulum in Problem 6.54 about the point O'?

6.79 What is the angular momentum of the pendulum in Problem 6.57 about the point O?

6.80 An unbalanced wheel of mass m kg rolls without slipping at an angular velocity of ω rad/s CW as shown in Fig. P6.80. If the radius of gyration about

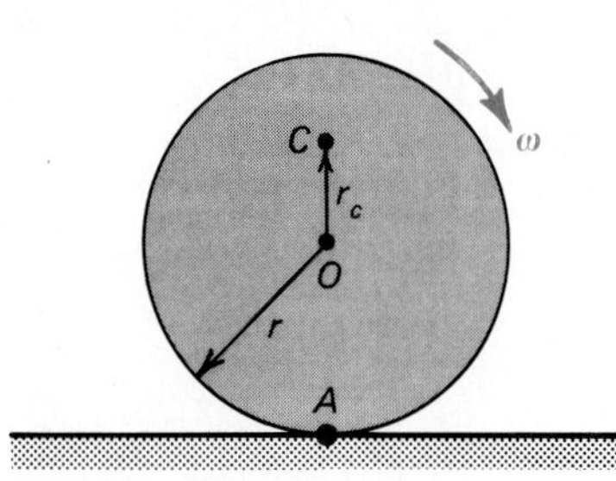

Figure P6.80

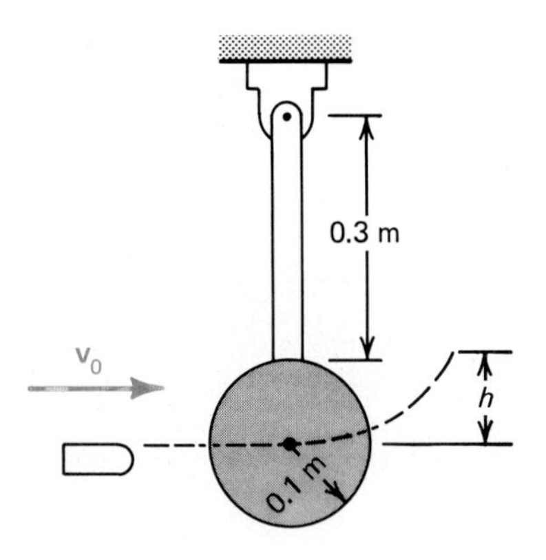

Figure P6.81

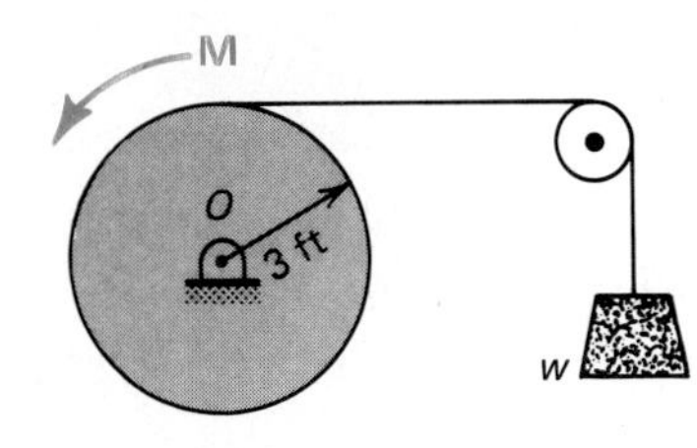

Figure P6.85

O is r_g, find the angular momentum about the center of mass at point C and the angular momentum about point O.

6.81 A 50-g bullet traveling at 15 m/s strikes a pendulum consisting of a 2-kg rod and a 1.5-kg sphere, as shown in Fig. P6.81. The bullet is embedded in the sphere after impact. Find how high the pendulum rises. (*Hint*: Conservation of angular momentum is used at impact followed by energy conservation after impact.)

6.82 Calculate the speed of the bullet necessary to raise the pendulum of Fig. P6.81 a distance 0.1 m. Assume the bullet is 0.05 kg and is embedded at the center of mass of the sphere. Neglect the mass of the rod and assume that the mass of the sphere is 3 kg.

6.83 Find the velocity of the falling mass in Problem 6.28 3 sec after the instant of release.

6.84 Find the angular momentum of the pulley and mass about point O in Problem 6.28 after the mass has moved down 10 ft starting from rest.

6.85 A torque of magnitude $M = 100t^2$ ft·lb in which t is measured in seconds is applied to a 64.4-lb wheel which hoists a 96.6-lb weight w as shown in Fig. P6.85. If the system starts from rest, find the speed of the weight w when $t = 4$ sec.

6.86 A cylinder of mass m kg rolls without slipping at a velocity $\mathbf{v}$ as shown in Fig. P6.86. What must be the minimum magnitude of the velocity $\mathbf{v}$ in order to just roll over the curb?

6.87 A despinning device used on satellites consists of equal masses unwinding as shown in Fig. P6.87. If the system in Fig. P6.87(a) is rotating at an angular velocity Ω_0 before releasing the masses, find the angular velocity Ω of the satellite as the masses unwind as shown in Fig. P6.87(b). Let I be the mass moment of inertia of the satellite alone and let each unwinding mass be m. (*Hint*: Use the principle of conservation of energy and the principle of conservation of angular momentum.)

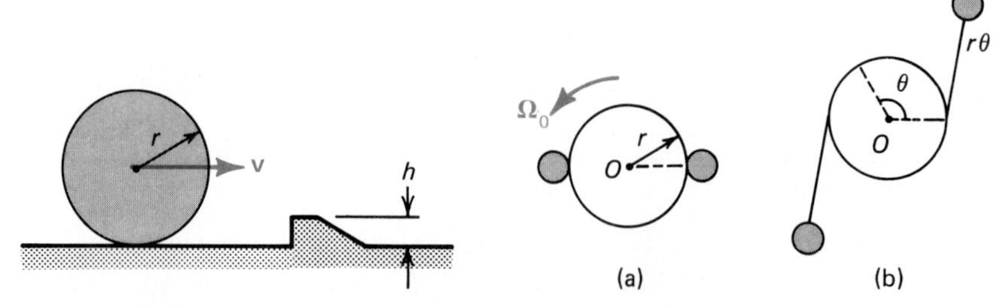

Figure P6.86

Figure P6.87

6.88 The satellite shown in Fig. P6.87(a) is spinning at 2 rad/s CCW and the system in Fig. P6.87(b) is spinning at 0.02 rad/s CCW. The satellite has $I_o = 500$ kg·m^2 and the two end masses are 1 kg each. How far should the masses be from the center of the satellite for the momentum to be conserved. Let $r = 0.5$ m.

6.89 Body B in Fig. P6.89 spins continuously about its axis at O' at a rate of $\dot{\theta}$ CCW when body A, to which it is pinned connected, is at rest. Find the time t for a motor to supply a constant torque M to body A to spin body A to an angular velocity Ω CCW. Let m_A, I_A and m_B, I_B be the masses and mass moments of inertia of each body about the respective axes, O and O'.

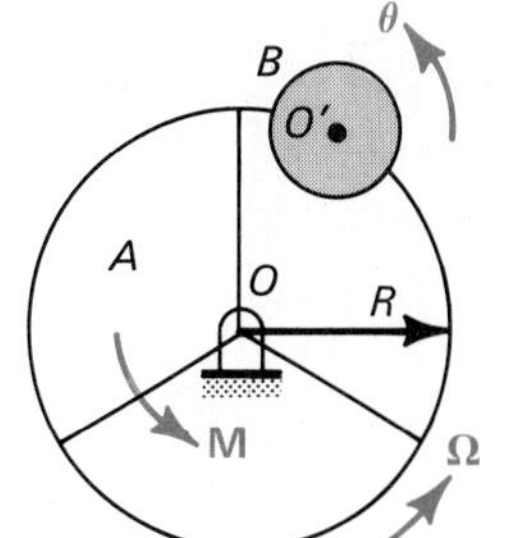

Figure P6.89

6.90 The satellite in Fig. P6.90 is spinning at 100 rev/min and has a mass moment of inertia about the spin axis of 200 slug·ft². The satellite must be spun up to 150 rev/min in 15 min. Assuming a constant force **F** is applied by a resisto-jet, compute the magnitude of **F**.

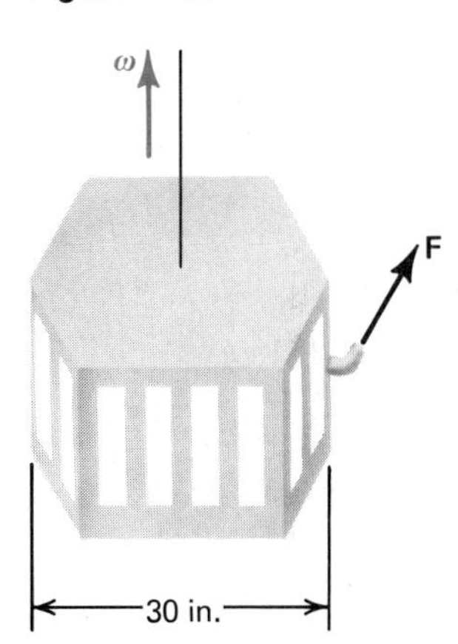

Figure P6.90

6.91 An angular impulse is generated by a moment $\mathbf{M} = 6t\mathbf{k}$ m·N and is applied from $t = 0$ to $t = 10$ s to a body whose mass moment of inertia is 100 kg·m². If the impulse changes the angular momentum of the body, compute the change of the angular speed.

6.92 A man with a 5-lb weight in each hand is standing on a turntable that is rotating at 10 rev/min as shown in Fig. P6.92. If the man suddenly puts his hands down, what is the new rate of rotation of the man? Assume the mass moment of inertia of the man and table is 4 slug-ft² about the spin axis. Neglect any frictional effects.

6.93 In Problem 6.92, calculate the angle through which the arms of the man must drop so that the angular velocity is increased to 12 rev/min.

6.94 A thin disk having a mass of 10 kg is rigidly connected to an 8-kg bar OA as shown in Fig. P6.94. A moment of magnitude $M = 10t$ m·N is applied to the vertical shaft. If the body starts from rest, what is the magnitude of its angular velocity after 5 s?

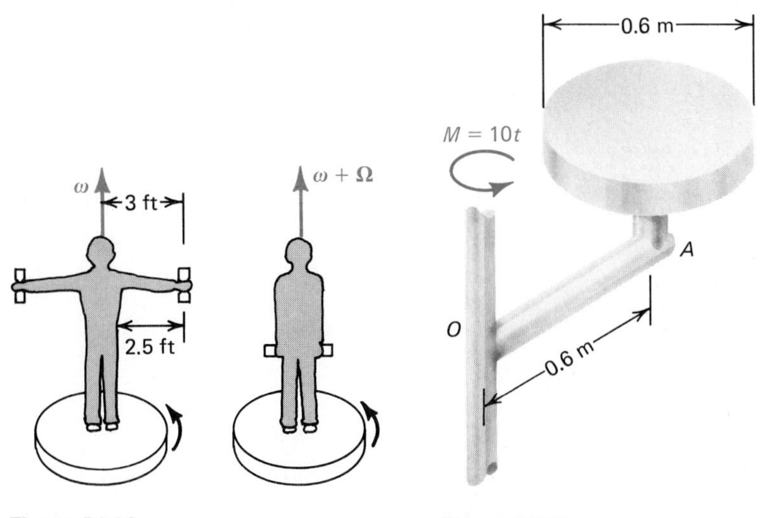

Figure P6.92 Figure P6.94

6.95 A particle of mass m is fired at speed v from each gun that is mounted on a horizontal turntable as shown in Fig. P6.95. The mass moment of inertia of the turntable is I about the rotation axis and it is assumed that the reaction

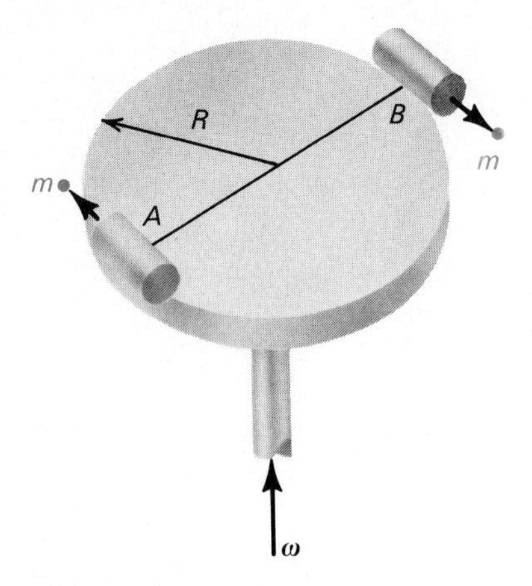

Figure P6.95

at the bearings is frictionless. If the guns are fired simultaneously, what is the angular velocity ω of the table? Assume that the mass of the table is much greater than the mass of each particle.

6.96 Describe the motion of the turntable in Problem 6.95 if gun B fires 2 s after gun A fires.

6.5 Summary

The planar kinetics of a rigid body is studied by applying Newton's second law to translational as well as rotational motion. By appropriately integrating this law, we derived relationships between work and energy as well as impulse and momentum.

A. The important *ideas*:

1. The forces that act on a system of particles are external forces and internal foces.
2. The equation for work done by the forces acting on a system of particles includes the work done by the internal forces acting on the particles.
3. For plane rigid-body motion, the equations of motion include the external moments acting on the body and the resulting angular acceleration. The inertia terms include the mass and the mass moment of inertia.
4. For plane rigid-body motion, the kinetic energy consists of kinetic energy due to translation plus the kinetic energy due to the rotational motion.

B. The important *equations*:

System of particles:

$$\mathbf{F} = m\ddot{\mathbf{r}}_c \tag{6.6}$$

$$\mathbf{F} = \dot{\mathbf{p}}_c \tag{6.18}$$

$$\mathbf{M} = \dot{\mathbf{H}} \tag{6.21}$$

$$\mathbf{M}_c = \dot{\mathbf{H}}_c \tag{6.25}$$

Plane rigid bodies:

$$\mathbf{F} = m\mathbf{a}_c \tag{6.35a}$$

$$M_c = I_c\alpha \tag{6.35b}$$

$$M_o = I_c\alpha + ma_c d \tag{6.36}$$

$$M_o = I_o\alpha \tag{6.38}$$

$$T = \tfrac{1}{2}mv_c^2 + \tfrac{1}{2}I_c\omega^2 \tag{6.39}$$

$$W_{\mathrm{NC}} = \Delta T + \Delta V \tag{6.43}$$

$$M_c = \dot{H}_c \tag{6.45}$$

$$\mathscr{I}_c = \int_{t_1}^{t_2} \mathbf{F}\, dt = \mathbf{p}_{c2} - \mathbf{p}_{c1} \tag{6.49}$$

$$\mathscr{M}_c = \int_{t_1}^{t_2} M_c\, dt = H_{c2} - H_{c1} \tag{6.50}$$

7 Impact

When we see a batter hitting a baseball, the operation of a pile driver at work, billiard balls hitting each other on a pool table, or the ramming of one car by another, we are witnessing the phenomenon of collision or **impact**. Generally, the forces at the contacting surfaces are large in magnitude and last for a short duration. Such forces are called **impulsive** forces.

Impulsive forces are very large in magnitude relative to other forces that may be present during a collision. Consequently, we assume that only impulsive forces act on the body during impact. As we shall see, this enables us to introduce the conservation of linear momentum of the impacting bodies. We shall also examine the loss of kinetic energy after impact due to the nature of the colliding bodies as measured by the coefficient of restitution.

7.1 Types of Impact

Impact in which the kinetic energy is conserved is called an **elastic impact**. If there is a loss of kinetic energy, the impact is called an **inelastic impact**.

A line perpendicular to the colliding surfaces of two bodies as shown in Fig. 7.1(a) is called the **line of impact** or the **line of action**. The line tangent to the colliding surfaces is called the **tangent line**. Direct impact occurs when the initial velocities of colliding objects are directed along the line of impact. Direct **central** impact, shown in Fig. 7.1(b), occurs when the mass centers of the objects are also on the line of impact. If the velocities of the mass center of two bodies are parallel but not collinear to the normal of the striking surfaces, then the impact is direct and eccentric as illustrated in Fig. 7.1(c).

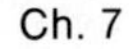

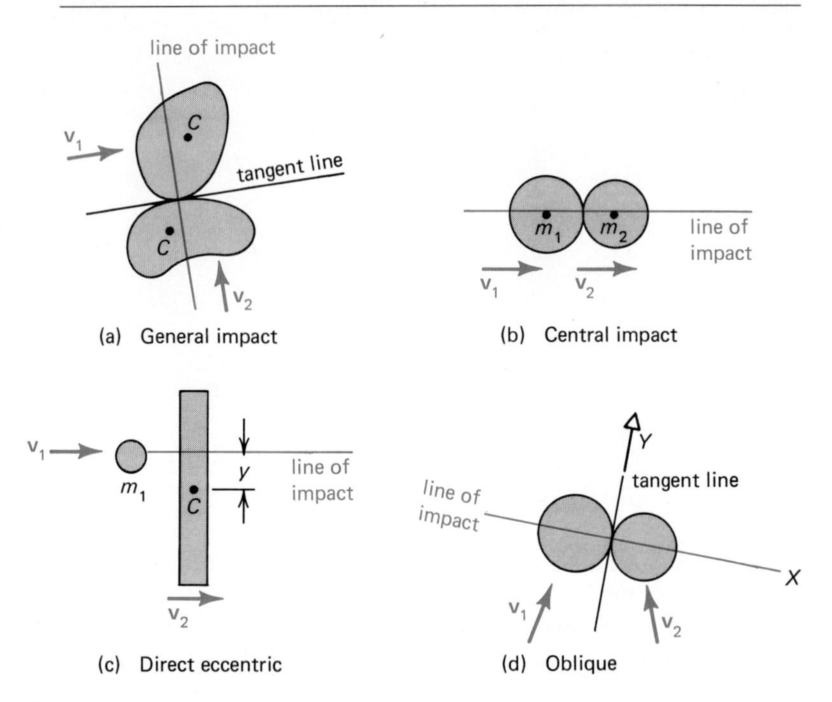

Figure 7.1

Finally, if the velocities are not directed along the line of impact, as shown in Fig. 7.1(d), then we speak of **oblique** impact. In general, during oblique impact, if the governing equations are written independently in the x and y directions, then the impact along the X axis is considered direct, whereas no impact occurs in the y direction. Since there is no impact in the y direction, the linear momentum is conserved and therefore the velocity components in the y direction remain unchanged.

7.2 Elastic and Perfectly Inelastic Direct Impact

Consider two spheres shown in Fig. 7.2 before, during, and after direct impact. The spheres have velocities $\mathbf{v}_1$ and $\mathbf{v}_2$ before impact and $\mathbf{u}_1$ and $\mathbf{u}_2$ after impact. Figure 7.3 shows the impulsive force acting on masses m_1 and m_2. Applying the linear impulse–momentum relationship to each mass, assuming positive forces and velocities are to the right, we obtain

$$\text{For mass } m_1: \quad -\int_{t_1}^{t_2} F\,dt = m_1(u_1 - v_1)$$

$$\text{For mass } m_2: \quad \int_{t_1}^{t_2} F\,dt = m_2(u_2 - v_2)$$

Adding these equations, we have the expression for the conservation of linear momentum.

$$m_1 v_1 + m_2 v_2 = m_1 u_1 + m_2 u_2 \tag{7.1}$$

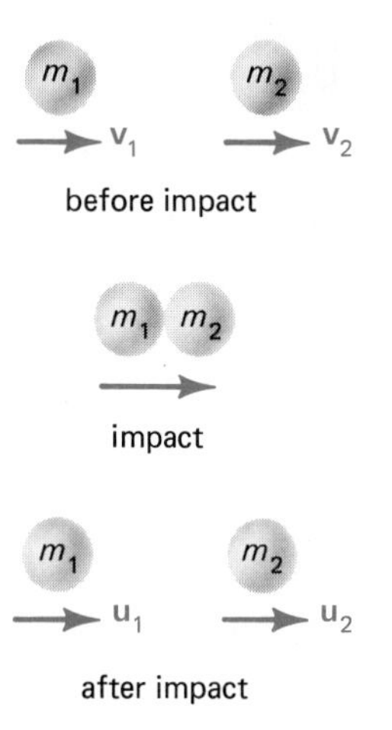

Figure 7.2

We are not using vector notation since it is clear that all the velocities are in the same direction. Note, however, that if v_2 were directed to the left in Fig. 7.2, we would take this into consideration in Eq. (7.1) by introducing a minus sign for v_2.

If the direct impact is elastic, then the kinetic energy is also constant. Equating the kinetic energies before and after impact,

$$\tfrac{1}{2}m_1 v_1^2 + \tfrac{1}{2}m_2 v_2^2 = \tfrac{1}{2}m_1 u_1^2 + \tfrac{1}{2}m_2 u_2^2 \tag{7.2}$$

We now rewrite Eq. (7.2) as

$$m_1 v_1^2 - m_1 u_1^2 = -m_2 v_2^2 + m_2 u_2^2$$

or

$$m_1(v_1 - u_1)(v_1 + u_1) = m_2(u_2 - v_2)(u_2 + v_2) \tag{7.3}$$

Figure 7.3

Equation (7.1) can also be written as

$$m_1(v_1 - u_1) = m_2(u_2 - v_2)$$

and substituting this into Eq. (7.3), we obtain

$$v_1 + u_1 = u_2 + v_2$$

or

$$(v_1 - v_2) = -(u_1 - u_2) \tag{7.4}$$

We call the relative velocity $(v_1 - v_2)$ the velocity of separation before impact, and the relative velocity $(u_1 - u_2)$ the velocity of separation after impact. Equation (7.4) shows that for the case of elastic impact, the velocity of separation before impact is the negative of the velocity of separation after impact. If we know the velocities before impact and the masses of the spheres, then the velocities after impact can be computed using Eqs. (7.1) and (7.4). Substituting Eq. (7.4) into Eq. (7.1) gives us

$$m_1 u_1 + m_2(v_1 - v_2 + u_1) = m_1 v_1 + m_2 v_2$$

Solving for u_1, we have

$$u_1 = \frac{v_1(m_1 - m_2) + 2m_2 v_2}{m_1 + m_2} \tag{7.5}$$

and substituting this into Eq. (7.4), we obtain

$$u_2 = \frac{v_2(m_2 - m_1) + 2m_1 v_1}{m_1 + m_2} \tag{7.6}$$

If the bodies that collide in direct impact are no longer elastic, then Eq. (7.2) is not valid, i.e., the kinetic energy is not conserved. As a matter of fact, if the collision is perfectly inelastic, as shown in

Figure 7.4

Fig. 7.4, the bodies stick to each other after impact and travel together at the same velocity, so that

$$u_1 = u_2 = u$$

Note that the velocity of separation after impact is zero for this case. Substituting this into the conservation of linear momentum as given by Eq. (7.1), we have

$$m_1 v_1 + m_2 v_2 = m_1 u_1 + m_2 u_2 = (m_1 + m_2)u$$

so that

$$u = \frac{m_1 v_1 + m_2 v_2}{m_1 + m_2} \tag{7.7}$$

There is a loss in kinetic energy for perfectly inelastic impact. To calculate this loss consider the kinetic energy T before impact,

$$T_1 = \tfrac{1}{2} m_1 v_1^2 + \tfrac{1}{2} m_2 v_2^2 \tag{7.8}$$

The kinetic energy after impact is

$$T_2 = \tfrac{1}{2}(m_1 + m_2)u^2 \tag{7.9}$$

Subtract Eq. (7.9) from (7.8) while substituting Eq. (7.7) into Eq. (7.9). This leads to

$$T_1 - T_2 = \frac{m_1 m_2}{2(m_1 + m_2)}(v_1 - v_2)^2 \tag{7.10}$$

So far, our discussion has assumed direct impact. If impact is oblique we simply apply the foregoing equations to the coordinate direction in which impact occurs. This approach for the oblique problem is fully developed in Section 7.4.

Example 7.1

A 5-kg sphere traveling at 20 m/s collides with a 10-kg sphere traveling at 10 m/s in the same direction, as shown in Fig. 7.5. If the impact is direct and central, what will the final velocities of the spheres be if the impact is (a) elastic, (b) perfectly inelastic? How much kinetic energy is lost in each case?

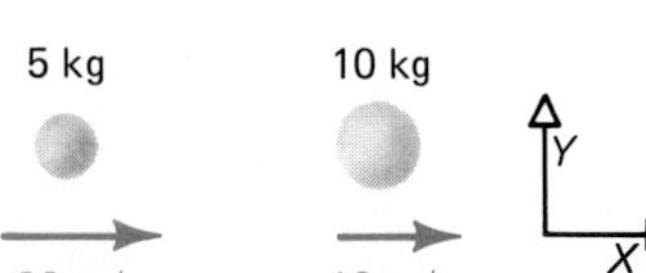
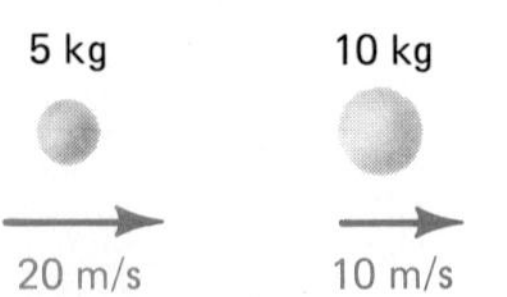

Figure 7.5

(a) From the equation for conservation of linear momentum we have

$$m_1 u_1 + m_2 u_2 = m_1 v_1 + m_2 v_2 \tag{a}$$

$$5u_1 + 10u_2 = 5(20) + 10(10)$$

This reduces to

$$u_1 + 2u_2 = 40 \tag{b}$$

Since the collision is elastic,

$$v_1 - v_2 = -(u_1 - u_2) \qquad \text{(c)}$$

Rearranging this equation and substituting the values for v_1 and v_2,

$$u_2 - u_1 = 20 - 10$$

or

$$-u_1 + u_2 = 10 \qquad \text{(d)}$$

Now add Eqs. (b) and (d) to obtain u_2,

$$3u_2 = 50$$

$$u_2 = 16.67$$

The positive sign on u_2 means that our assumed direction is correct, so that

$$\mathbf{u}_2 = 16.67\mathbf{i} \text{ m/s} \qquad \textit{Answer}$$

Recall Eq. (b) to find u_1.

$$u_1 + 2(16.67) = 40$$

$$u_1 = 6.66$$

$$\mathbf{u}_1 = 6.66\mathbf{i} \text{ m/s} \qquad \textit{Answer}$$

There is no change in kinetic energy since the impact is elastic.

(b) If the collision is perfectly inelastic, by Eq. (7.7),

$$u = \frac{5(20) + 10(10)}{5 + 10} = 13.33$$

or

$$\mathbf{u} = 13.33\mathbf{i} \text{ m/s} \qquad \textit{Answer}$$

The kinetic energy loss for inelastic impact is

$$\begin{aligned} \Delta T &= T_1 - T_2 \\ &= [\tfrac{1}{2}(5)(20)^2 + \tfrac{1}{2}(10)(10)^2] - [\tfrac{1}{2}(5 + 10)(13.33)^2] \\ &= 167.3 \text{ N}\cdot\text{m} \qquad \textit{Answer} \end{aligned}$$

Example 7.2

A 1-lb particle is released from rest as shown in Fig. 7.6. It strikes the stationary 0.8-lb particle. Assume the impact is direct and elastic. If the horizontal surface has a kinetic coefficient of friction $\mu = 0.3$, locate the final position of each mass from the origin of the X axis.

We can view the problem in three steps. First, we calculate the velocity of impact of the 1-lb particle after sliding down the smooth circular path by employing the conservation of energy

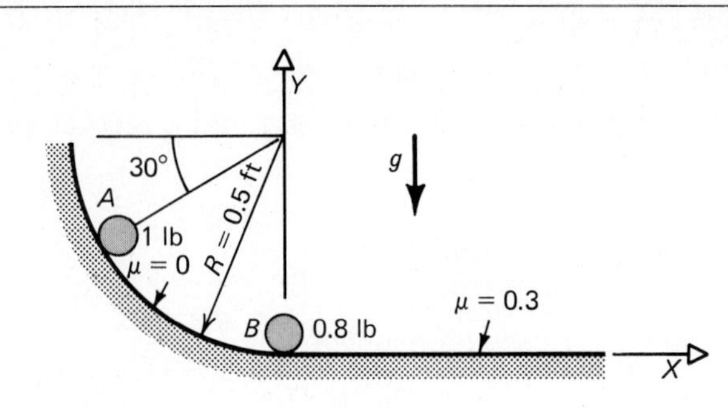

Figure 7.6

equation. Next, direct elastic impact occurs so that we use the conservation of linear momentum and Eq. (7.4) to find the velocities at the instant after impact. Finally, we make use of the work-energy equation for each particle to calculate how far each travels from the origin of the X axis.

Conservation of energy of the 1-lb particle:

$$V_A + T_A = V_B + T_B$$

Let position A be the initial position of the 1-lb particle and position B be the point of impact with the 0.8-lb particle. Also, use position B as the reference point to calculate the potential energy due to gravity.

$$V_A = mgh_A = 1(0.5 - 0.5 \sin 30^\circ)$$
$$= 0.25$$

$$T_A = 0$$

$$V_B = 0$$

$$T_B = \frac{1}{2} mv_B^2 = \frac{1}{2}\left(\frac{1.0}{32.2}\right) v_B^2 = 0.0155 v_B^2$$

Substituting into the conservation of energy equation,

$$v_B = 4.016 \text{ ft/sec}$$

Impact:

Let $$m_1 = \frac{1}{g} \qquad v_1 = v_B = 4.016$$

$$m_2 = \frac{0.8}{g} \qquad v_2 = 0$$

By the conservation of linear momentum,

$$\frac{1}{g}(4.016) + \frac{0.8}{g}(0) = \frac{1}{g} u_1 + \frac{0.8}{g} u_2$$

or
$$4.016 = u_1 + 0.8u_2$$

Also, by Eq. (7.4),
$$(4.016 - 0) = -(u_1 - u_2)$$

Solving these two equations simultaneously,
$$u_1 = 0.446 \text{ ft/sec} \qquad u_2 = 4.462 \text{ ft/sec}$$

Work-energy for the 1-lb particle: Recall Eq. (3.26),
$$W_{NC} = \Delta T + \Delta V$$
$$-0.3(1)x_1 = 0 - \frac{1}{2}\left(\frac{1}{32.2}\right)(0.446)^2$$
$$x_1 = 0.010 \text{ ft} \qquad \textit{Answer}$$

For the 0.8-lb particle:
$$W_{NC} = \Delta T + \Delta V$$
$$-0.3(0.8)x_2 = 0 - \frac{1}{2}\left(\frac{0.8}{32.2}\right)(4.462)^2$$
$$x_2 = 1.03 \text{ ft} \qquad \textit{Answer}$$

PROBLEMS

7.1 Calculate the velocity of each mass in Fig. P7.1 at the instant after impact. Assume the direct impact is elastic.

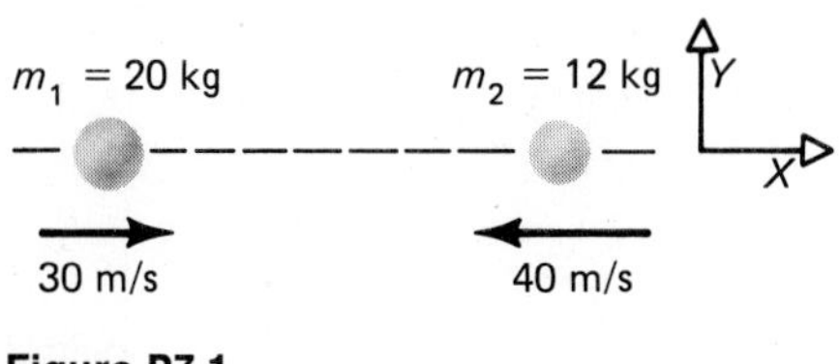

Figure P7.1

7.2 Assuming that the impact is perfectly inelastic in Problem 7.1, calculate the energy lost due to impact.

7.3 A 30,000-lb railroad car traveling at 10 mi/hr hits a 20,000-lb stationary car, and gets stuck to it. The two cars then travel together. (a) What is the new velocity of the two cars? (b) How much kinetic energy is lost?

7.4 A pile driver, shown in Fig. P7.4 is used to drive a 100-lb pile. It is noticed that upon impact the ram and pile stay together. If the ram weighs 500 lb (a) what is the velocity of the pile right after impact when the ram is dropped from 2 ft? (b) How much energy is transformed to the pile?

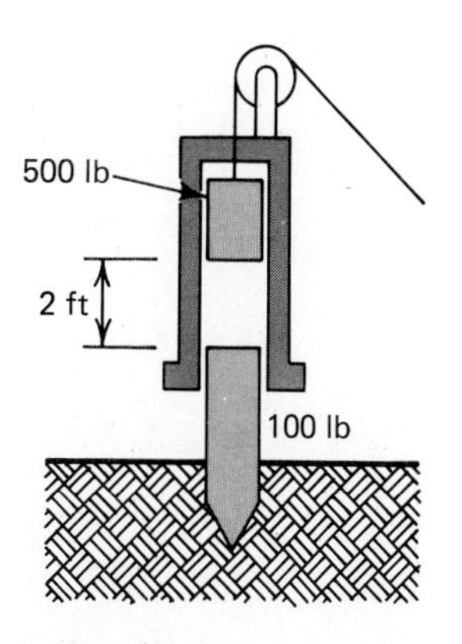

Figure P7.4

7.5 Three particles can slide along the smooth wire as shown in Fig. P7.5. The 5-kg particle m_1 moves to the right at 8 m/s while the remaining particles m_2 and m_3 are stationary. If the direct impact is perfectly inelastic between m_1 and m_2 and elastic between m_2 and m_3, what is the maximum compression of the spring whose modulus $k = 1500$ N/m?

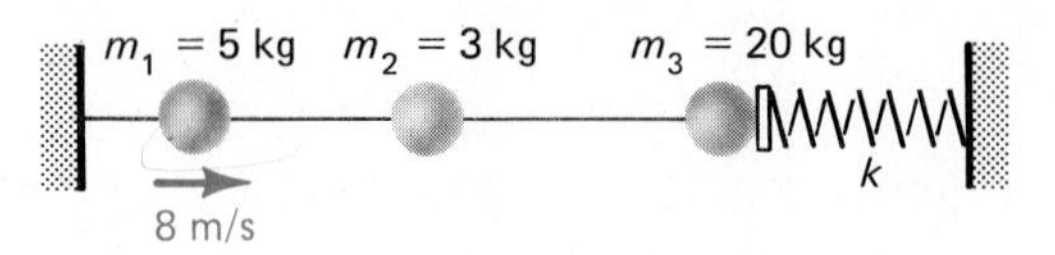

Figure P7.5

7.6 A 10-kg sphere traveling at 25 m/s collides with a 15-kg sphere traveling 20 m/s in the opposite direction. If the impact is direct and central, what will be the final velocities if the impact is (a) elastic, (b) perfectly inelastic? (c) How much kinetic energy is lost in each case?

7.7 In Problem 7.6 if the 15-kg sphere travels in the same direction as the 10-kg sphere what are the final velocities for (a) elastic and (b) perfectly inelastic impact?

7.8 Particles m_1 and m_2 sliding along the smooth wire in Fig. P7.8 collide such that the impact is perfectly inelastic. If the impact is elastic between m_2 and the stationary particle m_3, how long does it take before m_3 is struck if each particle has the same mass?

7.9 A 100-lb stationary mass is hit by a 5-lb mass traveling at 10 mi/hr. The collision is perfectly inelastic. If the kinetic coefficient of friction is 0.25 between the combined masses and the floor, then how far will the combined masses slide before stopping?

7.10 A 20-kg body is released from rest and slides down a smooth, curved chute to point B as shown in Fig. P7.10. At point B it travels along the horizontal surface with a coefficient of friction equal to 0.3. If the impact is elastic between the body and the wall at point C, where does the body come to rest?

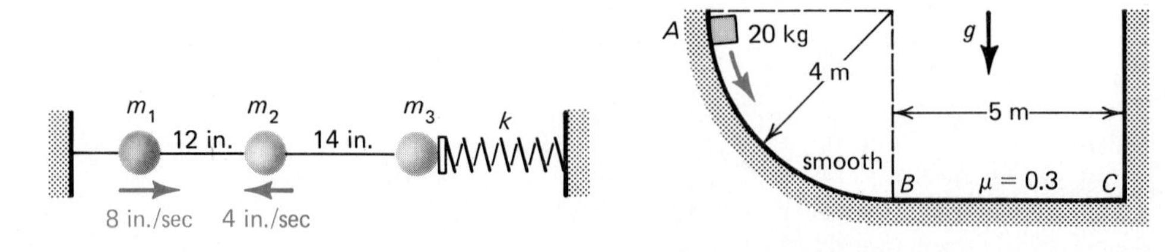

Figure P7.8

Figure P7.10

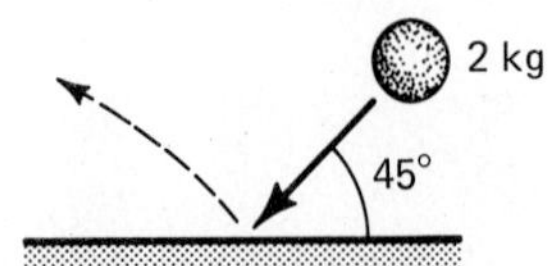

Figure P7.11

7.11 A 2-kg particle strikes a plane with a velocity of 20 m/s as shown in Fig. P7.11. How high does the particle travel above the point of impact if the impact between the particle and the plate is elastic? (*Hint*: Assume the impact is perpendicular to the surface.)

7.12 A 100-lb mass travels at a velocity of 1.98 ft/sec and impacts with a 200-lb plate that is supported by the spring modulus k, as shown in Fig.

P7.12. The impact is perfectly inelastic between the two masses. What must be the spring modulus to keep the spring from deforming more than 0.2 ft?

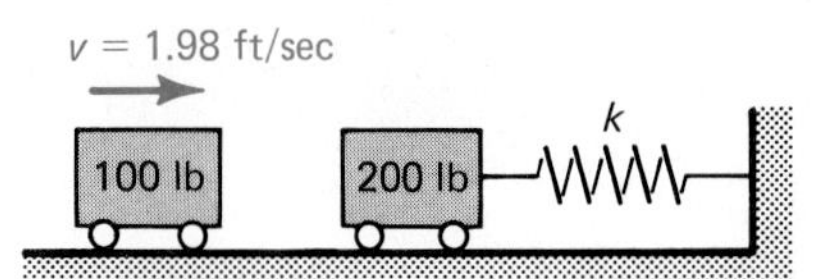

Figure P7.12

7.13 A 5-lb projectile is fired from a gun at a muzzle velocity of 1200 ft/sec as shown in Fig. P7.13. Find the maximum spring deformation from the motion of the gun if $k = 10{,}000$ lb/ft and the gun weighs 200 lb. Neglect friction.

7.14 A simple pendulum of mass m_1 is released from rest at A and strikes the mass m_2 as shown in Fig. P7.14. If all collisions are elastic, what must the coefficient of friction μ be for m_2 to strike the wall at C and just return to point B? Assume $m_1 = m_2$ are particles.

7.15 A 20-kg sandbag is suspended by a wire as shown in Fig. P7.15. A 0.03-kg bullet is shot into the bag which rises by 0.25 m. Assuming that the bullet stays in the bag, (a) what was the velocity of the bullet just before impact? (b) How much kinetic energy was lost due to impact?

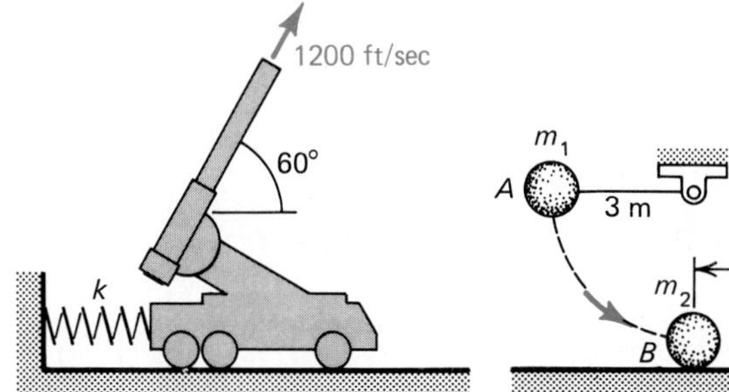

Figure P7.13

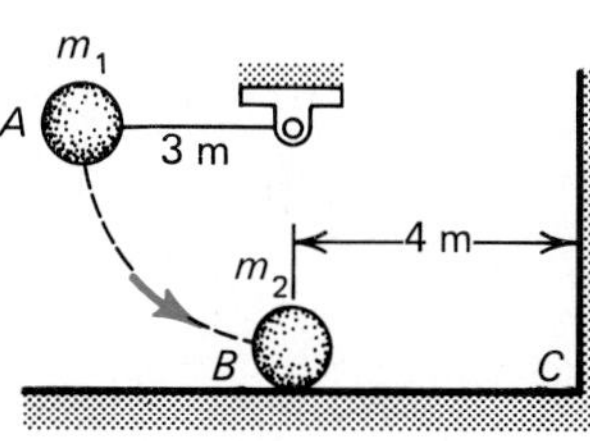

Figure P7.14

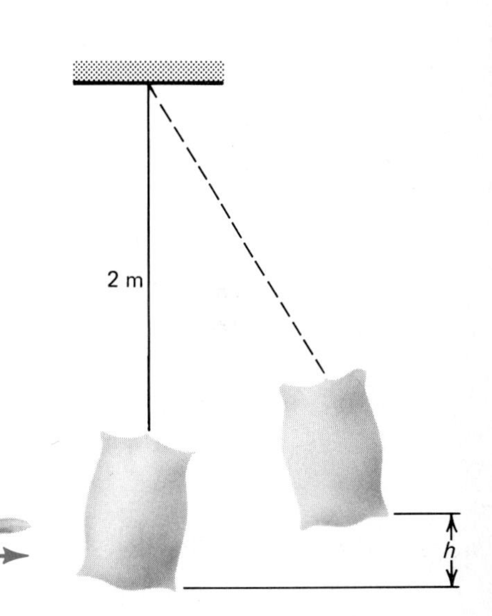

Figure P7.15

7.3 Coefficient of Restitution

Let us rewrite Eq. (7.4).

$$v_1 - v_2 = -(u_1 - u_2) \tag{7.4}$$

We recall that this equation states that the velocity of separation before an elastic impact is equal to the negative of the velocity of separation after impact. We also have established that the velocity of separation after a perfectly inelastic impact is zero. These two statements can be incorporated into the equation.

$$e(v_1 - v_2) = -(u_1 - u_2) \tag{7.11}$$

where $e = 0$ for perfectly inelastic impact and $e = 1$ for elastic impact. The term e, called the **coefficient of restitution**, depends on the shape and material properties of the impacting bodies. For practical impact problems the value of e will lie between 0 and 1.

Let us again consider two spheres of masses m_1 and m_2

having velocities v_1 and v_2 before impact, respectively. If the velocities after direct central impact are u_1 and u_2, the two equations used for computing u_1 and u_2 are based upon the conservation of linear momentum and the expression for the coefficient of restitution, respectively. Thus,

$$m_1u_1 + m_2u_2 = m_1v_1 + m_2v_2 \tag{7.1}$$

$$-(u_1 - u_2) = e(v_1 - v_2) \tag{7.11}$$

These are the two equations we need for calculating velocities for direct central impact of two bodies.

Example 7.3

Rework Example 7.1 assuming that $e = 0.8$ for the impacting spheres.

We begin by writing Eq. (7.1).

$$m_1v_1 + m_2v_2 = m_1u_1 + m_2u_2$$

$$5(20) + 10(10) = 5u_1 + 10u_2$$

$$u_1 + 2u_2 = 40$$

Now we introduce Eq. (7.11).

$$e(v_1 - v_2) = -(u_1 - u_2)$$

$$0.8(20 - 10) = -u_1 + u_2$$

$$u_2 = 8 + u_1$$

Substitute this last expression into the conservation equation.

$$u_1 + 2(8 + u_1) = 40$$

$$u_1 = 8 \qquad \text{so that} \qquad u_2 = 16$$

Therefore, $\mathbf{u}_1 = 8\mathbf{i}$ m/s *Answer*

$\mathbf{u}_2 = 16\mathbf{i}$ m/s *Answer*

The loss of kinetic energy is found as follows:

$$\Delta T = T_1 - T_2$$

$$\Delta T = [\tfrac{1}{2}(5)(20)^2 + \tfrac{1}{2}(10)(10)^2] - [\tfrac{1}{2}(5)(8)^2 + \tfrac{1}{2}(10)(16)^2]$$

$$= 1500 - 1440 = 60 \text{ N}\cdot\text{m}$$ *Answer*

We observe that the loss of energy is less than the energy loss for the case of perfectly inelastic impact.

Example 7.4

If the two spheres in the above example are moving in opposite directions, as shown in Fig. 7.7, what is the final velocity of each sphere if $e = 0.8$?

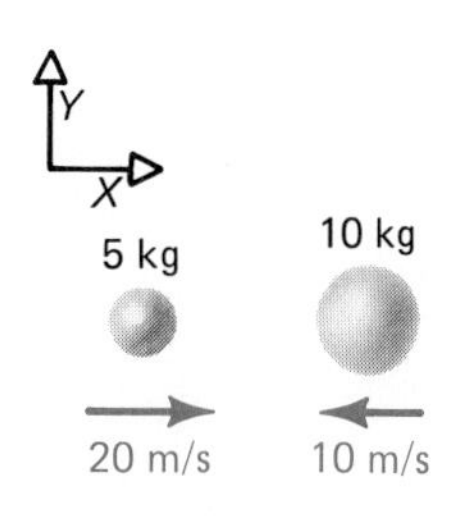

Figure 7.7

The equation for the coefficient of restitution is

$$-u_1 + u_2 = e(v_1 - v_2)$$

Now v_2 is negative since m_2 is moving to the left.

$$-u_1 + u_2 = 0.8(20 + 10) = 24$$

From the conservation of linear momentum we obtain

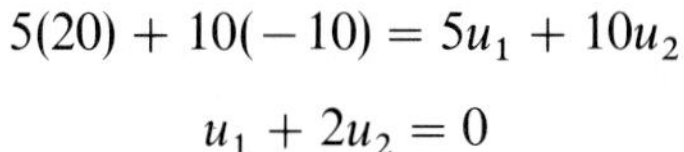

$$5(20) + 10(-10) = 5u_1 + 10u_2$$

or

$$u_1 + 2u_2 = 0$$

Solving the two equations for the velocities at the instant after impact, we find

$$\mathbf{u}_1 = -16\mathbf{i} \text{ m/s} \qquad \mathbf{u}_2 = 8\mathbf{i} \text{ m/s} \qquad \textit{Answer}$$

Notice that each sphere reverses its direction of travel after impact, i.e., m_1 travels to the left at 16 m/s while m_2 travels to the right at 8 m/s.

Example 7.5

Derive a formula for the distance the fifth sphere rises if the first sphere is raised by h_1 ft as shown in Fig. 7.8. Assume all spheres are of equal mass and the coefficient of restitution for all collisions is e.

If the first sphere is raised by h_1 ft and allowed to swing to its vertical position, then the velocity at impact is obtained by the conservation of energy.

$$V_A + T_A = V_B + T_B$$

$$mgh_1 + 0 = 0 + \tfrac{1}{2}mv_B^2$$

$$v_B = \sqrt{2gh_1}$$

Let $v_1 = v_B$ be the velocity of m_1 at the instant before impact while all of the other spheres have zero velocity. Therefore, applying the conservation of linear momentum between the first spheres,

$$mv_1 + m(0) = mu_1 + mu_2$$

so that

$$u_1 + u_2 = v_1 \qquad \text{(a)}$$

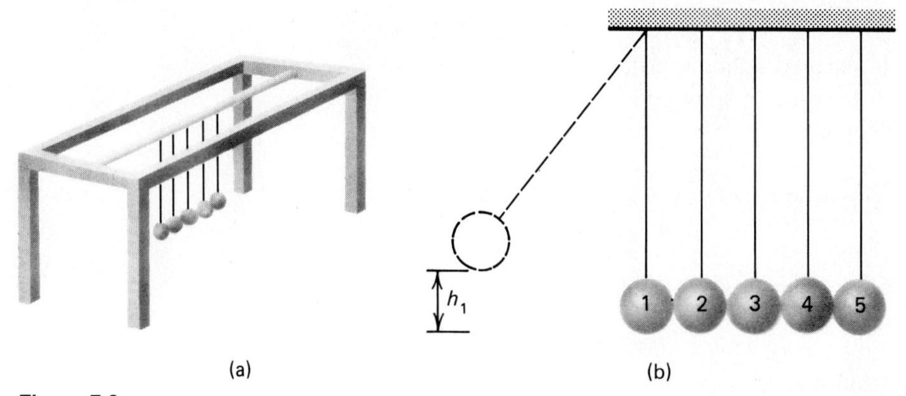

Figure 7.8

From the definition of the coefficient of restitution, we know that

$$-u_1 + u_2 = e(v_1 - 0)$$

or

$$-u_1 + u_2 = ev_1 \tag{b}$$

Solving Eqs. (a) and (b), we have

$$u_2 = \tfrac{1}{2}v_1(1 + e) \tag{c}$$

If we denote the velocity before impact as u_2 for sphere 2 and after impact as u_2', then we can write two equations of impact for spheres 2 and 3, that is,

$$mu_2 + m(0) = mu_2' + mu_3$$

Also,

$$-(u_2' - u_3) = e(u_2 - 0)$$

Solving for u_3 gives us

$$u_3 = \tfrac{1}{2}u_2(1 + e)$$

and, substituting for u_2 from Eq. (c), we finally have

$$u_3 = v_1\left(\frac{1 + e}{2}\right)^2$$

This procedure may be carried out to obtain the velocity of the fifth sphere when it is hit by the fourth sphere. The velocity of the fifth sphere is

$$u_5 = v_1\left(\frac{1 + e}{2}\right)^4 \tag{d}$$

If the fifth sphere swings up a distance h_5, then by the conservation of energy,

$$u_5 = \sqrt{2gh_5}$$

If we equate this to Eq. (d) and substitute for v_1, we obtain

$$\sqrt{2gh_5} = \sqrt{2gh_1}\left(\frac{1+e}{2}\right)^4$$

or

$$h_5 = h_1\left(\frac{1+e}{2}\right)^8 \qquad \textit{Answer}$$

Notice that if the spheres have a coefficient of restitution close to unity, then the fifth sphere will rise to approximately the same height as the first sphere. Since steel balls have a high e, they are used for the game in Fig. 7.8.

PROBLEMS

7.16 Rework Problem 7.1 assuming $e = 0.6$.

7.17 If the cars in Problem 7.3 collide assuming $e = 0.8$, what are the velocities of the cars after impact?

7.18 In Fig. P7.18, sphere B is shown at rest, while sphere A has a velocity of $10\mathbf{i}$ m/s. If $e = 0.8$, find the velocity of spheres A and B after impact. Assume that both spheres have the same density.

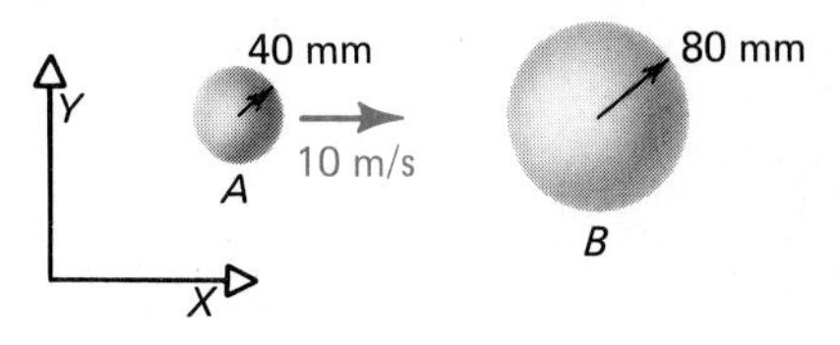

Figure P7.18

7.19 Repeat Problem 7.18 letting $\mathbf{v}_A = 10\mathbf{i}$ m/s and $\mathbf{v}_B = -15\mathbf{i}$ m/s before impact.

7.20 Rework Problem 7.5 assuming $e = 0.5$ between m_2 and m_3.

7.21 In Example 7.5, how much does the fifth ball rise if $h_1 = 1$ ft and $e = 0.95$?

7.22 Steel balls are dropped from a height of 4 ft onto an inclined surface, as shown in Fig. P7.22. After hitting the surface they follow the indicated trajectory. If they rise to a height of 2 ft (a) what is the coefficient of restitution? (b) Find the value of x.

7.23 In a billiard game, a player has called out ball B. If she plans to place the cue ball A and make the shot by rebounding as shown in Fig. P7.23, what must the distance x be to make the shot? Assume that $e = 0.75$.

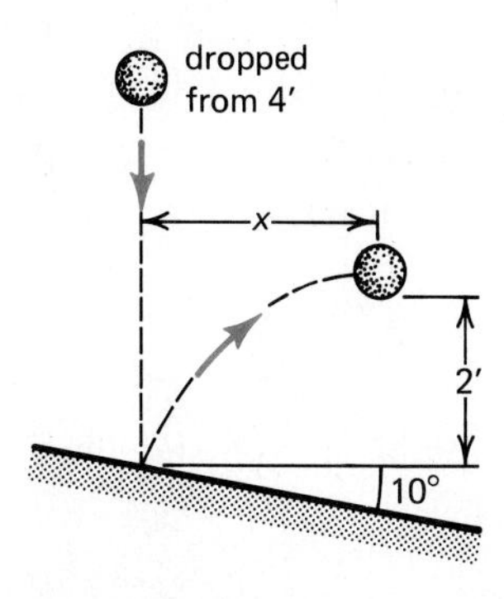

Figure P7.22

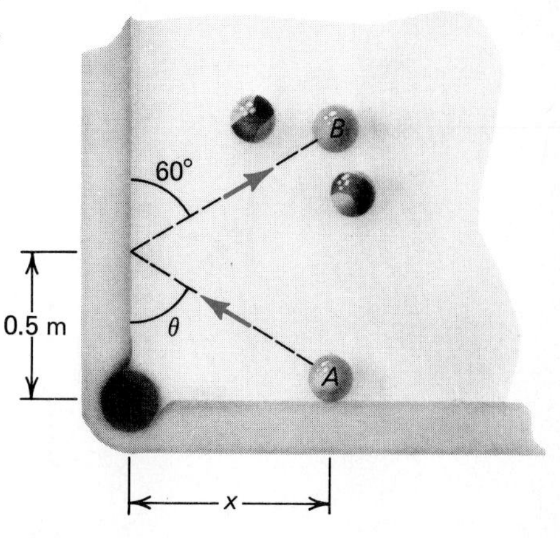

Figure P7.23

7.4 Oblique Impact

Earlier we defined oblique impact as occurring when the velocities of two colliding bodies are not directed along the line of impact. For such a case consider two spheres in Fig. 7.9 whose velocity directions before impact are measured by the θ angles off the line of impact and by the α angles after impact. Knowing v_1, v_2, θ_1, θ_2, and e, four unknowns remain, i.e., u_1, u_2, α_1, and α_2. Consequently, we need four equations to solve for the four unknowns. Equations (7.1) and (7.11) are two equations that apply along the line of impact, that is, we use the velocity components along the Y axis. The remaining two equations are obtained by recognizing that there is no linear impulse along the tangent line. This means that the linear momentum of each particle is conserved in the x direction.

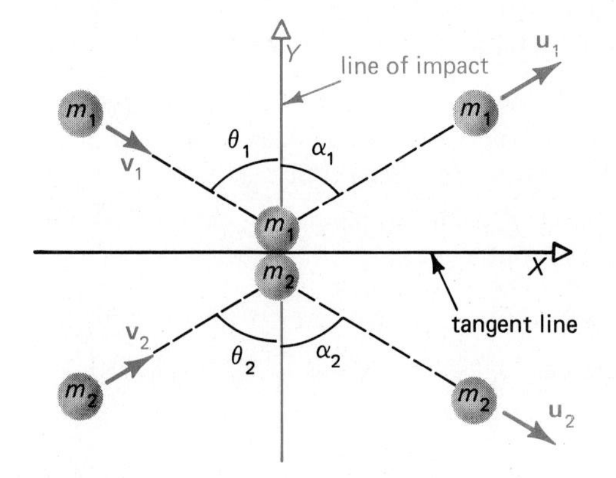

Figure 7.9

Equation (7.1) is applied to the particles in Fig. 7.9.

$$-m_1v_1 \cos \theta_1 + m_2v_2 \cos \theta_2 = m_1u_1 \cos \alpha_1 - m_2u_2 \cos \alpha_2$$

Notice the negative signs that take into account the velocity component directions. Applying Eq. (7.11),

$$e(-v_1 \cos \theta_1 - v_2 \cos \theta_2) = -(u_1 \cos \alpha_1 + u_2 \cos \alpha_2)$$

The conservation of linear momentum of each particle in the x direction leads to the following:

$$v_1 \sin \theta_1 = u_1 \sin \alpha_1$$

$$v_2 \sin \theta_2 = u_2 \sin \alpha_2$$

The above equations can be simplified by introducing the following notation:

$$v_{1x} = v_1 \sin \theta_1 \qquad v_{1y} = v_1 \cos \theta_1$$

$$v_{2x} = v_2 \sin \theta_1 \qquad v_{2y} = v_2 \cos \theta_1$$

$$u_{1x} = u_1 \sin \alpha_1 \qquad u_{1y} = u_1 \cos \alpha_1$$

$$u_{2x} = u_2 \sin \alpha_2 \qquad u_{2y} = u_2 \cos \alpha_1$$

The four equations for the problem of oblique impact reduce to the following:

$$-m_1 v_{1y} + m_2 v_{2y} = m_1 u_{1y} - m_2 u_{2y} \tag{7.12}$$

$$e(-v_{1y} - v_{2y}) = -(u_{1y} + u_{2y}) \tag{7.13}$$

$$v_{1x} = u_{1x} \tag{7.14}$$

$$v_{2x} = u_{2x} \tag{7.15}$$

It is important to remember that Eqs. (7.12) and (7.13) have been written for the velocity directions shown in Fig. 7.9 and for the case where the line of impact is along the Y axis.

Example 7.6

An experiment is conducted to determine the coefficient of restitution between the puck and the wall shown in Fig. 7.10. The mass of the puck is m slugs and travels at v ft/sec. After impact it rebounds at a different angle. Find the coefficient of restitution for (a) the general case, and (b) for $m = 0.02$ slugs, $v = 20$ ft/sec, $\beta = 60°$, and $\gamma = 45°$.

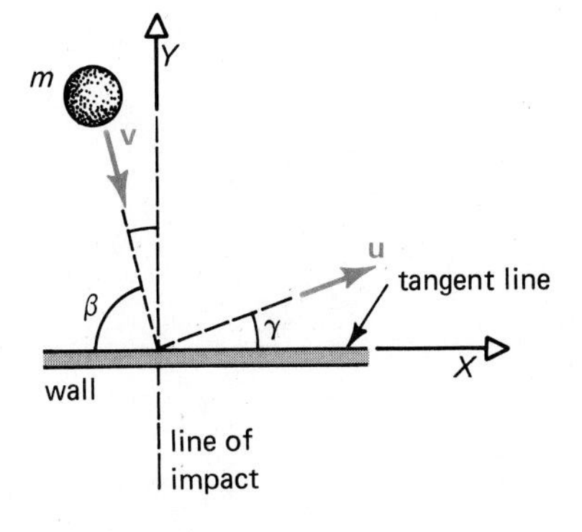

Figure 7.10

(a) We can refer to Eqs. (7.12–7.15) for this problem recognizing that the mass of the wall, m_2, remains stationary. This means that Eqs. (7.12) and (7.15) are not applicable. Referring to Eq. (7.14),

$$v_{1x} = u_{1x}$$

or

$$v \cos \beta = u \cos \gamma$$

$$u = \frac{\cos \beta}{\cos \gamma} v$$

From Eq. (7.13),

$$e(-v_{1y} - 0) = -(u_{1y} + 0)$$

$$-ev \sin \beta = -u \sin \gamma$$

$$= -\left(v \frac{\cos \beta}{\cos \gamma}\right) \sin \gamma$$

Therefore, the general expression for the coefficient of restitution is

$$e = \frac{\tan \gamma}{\tan \beta}$$

It is interesting to note that the coefficient of restitution is independent of the mass and velocity, but not the direction of the incoming and outgoing velocity.

(b) For $\beta = 60^\circ$, $\gamma = 45^\circ$, e becomes

$$e = \frac{\tan 45^\circ}{\tan 60^\circ} = 0.577 \qquad \textit{Answer}$$

Example 7.7

Consider $m_1 = 10$ kg and $m_2 = 25$ kg for the masses in Fig. 7.11. If they experience an oblique impact, find the velocity of each mass after the impact. Let $e = 0.4$ and assume impact occurs along the Y axis

Equations (7.12–7.15) apply directly to this problem.

$$-10(8 \cos 60^\circ) + 25(6 \cos 45^\circ) = 10u_{1y} - 25u_{2y}$$

$$0.4(-8 \cos 60^\circ - 6 \cos 45^\circ) = -(u_{1y} + u_{2y})$$

$$8 \sin 60^\circ = u_{1x} = 6.928$$

$$6 \sin 45^\circ = u_{2x} = 4.243$$

Rewriting the first two equations,

$$10u_{1y} - 25u_{2y} = 66.07$$

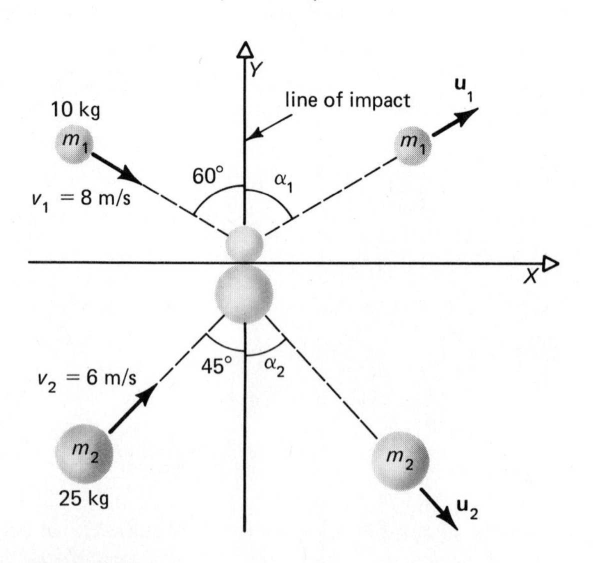

Figure 7.11

$$u_{1y} + u_{2y} = 3.297$$

Solving these equations simultaneously,

$$u_{1y} = 4.243 \qquad u_{2y} = -0.946$$

The negative sign on u_{2y} means that the assumed direction of m_2 in the y direction after impact is incorrect. Rather than traveling in the negative y direction as assumed, it actually travels in the positive y direction. Therefore, the velocities after impact are

$$\mathbf{u}_1 = 6.928\mathbf{i} + 4.243\mathbf{j} \text{ m/s}$$

$$\mathbf{u}_2 = 4.243\mathbf{i} + 0.946\mathbf{j} \text{ m/s}$$

PROBLEMS

7.24 Obtain the velocity after oblique impact for the two spheres in Fig. 7.11 if the impact is elastic, that is, $e = 1$.

7.25 Obtain the velocity after oblique impact for the two spheres in Fig. 7.11 if the impact is perfectly inelastic, that is, $e = 0$.

7.26 Two atomic particles A and B collide as shown in Fig. P7.26. Derive an expression for the velocities of the particles after impact assuming that they are spherical and $e = 0.99$.

7.27 A puck traveling 20 ft/sec strikes a wall as shown in Fig. P7.27. Assuming that the coefficient of restitution is 0.6, how long does it take the puck to reach the player after it strikes the wall? Let us also assume that the mass of the puck is 0.015 slugs and the coefficient of friction is 0.1 between the floor and the puck.

7.28 In the previous problem, (a) what is the kinetic energy of the puck as it reaches the player? (b) How much energy is lost due to friction?

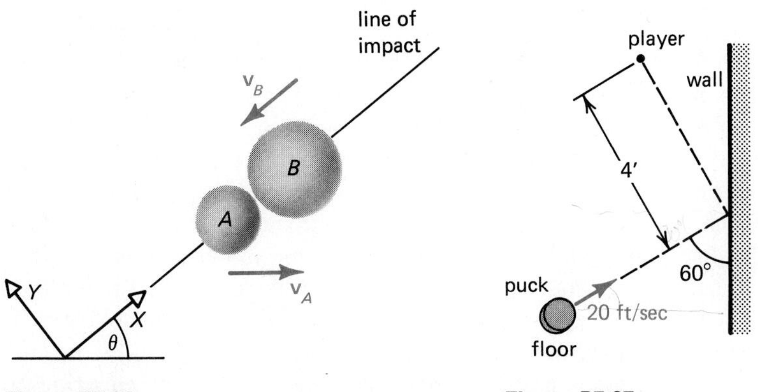

Figure P7.26

Figure P7.27

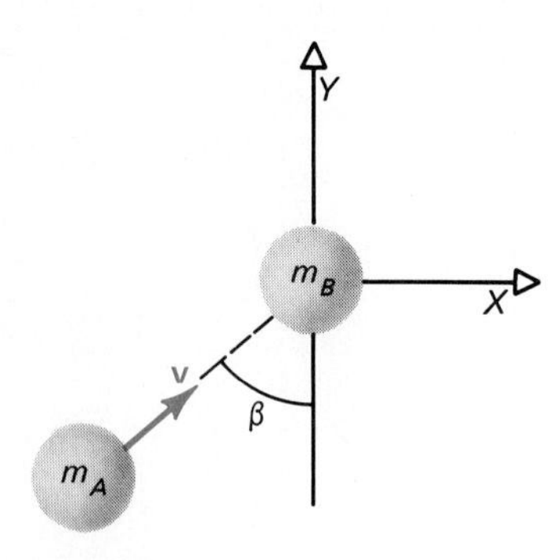

Figure P7.29

7.29 The 20-kg mass A strikes the stationary 15-kg mass B as shown in Fig. P7.29. Find the velocity of each mass after impact if $v = 30$ m/s, $\beta = 60°$, and $e = 0.6$.

7.30 Repeat Problem 7.29 for $v = 30$ m/s, $\beta = 40°$, $e = 1$, and $m_A = m_B = 20$ kg.

7.31 A ball is dropped as shown in Fig. P7.31. If the ball weighs 2 lb and $e = 0.5$, then derive an equation for the trajectory of the ball after impact. Let the velocity at impact be 40 ft/sec in the indicated direction. What is the maximum height reached by the ball?

7.32 A cue ball with a velocity v strikes a stationary ball which moves as shown in Fig. P7.32. What is the velocity, after impact, of each ball if $0 < e \leq 1$? Assume that both balls have the same mass and that impact is along the Y axis.

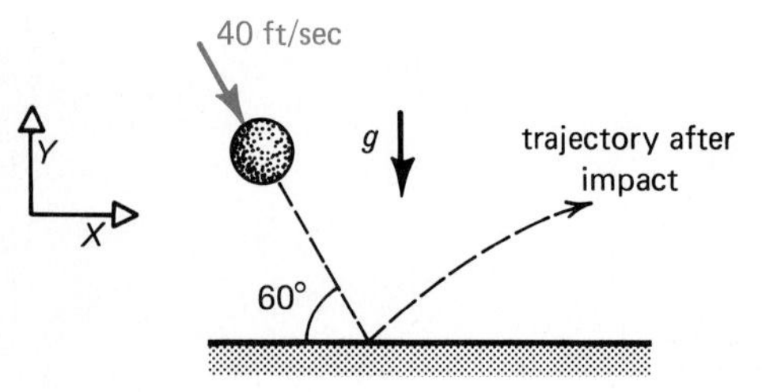

Figure P7.31

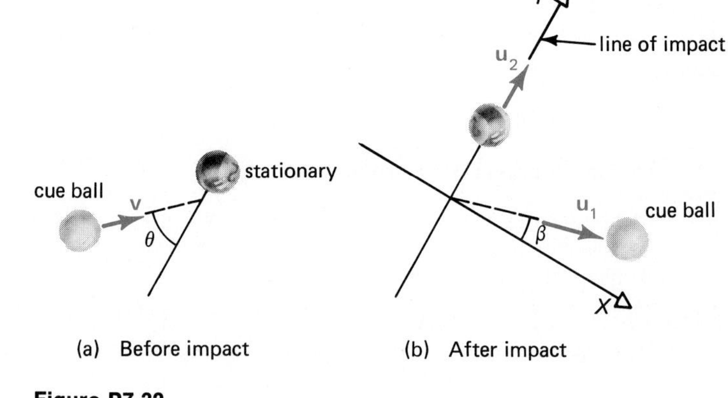

Figure P7.32

7.33 Solve Problem 7.32 for the case where $v = 15$ m/s, $e = 1$, and $\theta = 20°$.

7.5 Eccentric Impact

If the impact is eccentric, the angular momentum must be taken into account as well as the linear momentum. Consider, for example, a sphere of mass m_1 and velocity v_1 hitting a bar of mass m_2 whose center of mass is moving at a velocity v_2 while the bar is rotating at an angular rate of ω_1, as shown in Fig. 7.12(a). The impulsive force $\mathbf{F}$ acting on each mass during impact is shown for each body in Fig. 7.12(b). Figure 7.12(c) shows an equivalent force system acting on the rod in which the impulsive force is shifted to the center of mass, along with an impulsive moment of magnitude Fy. Figure 7.12(d) shows the assumed direction of the velocities at the instant after impact.

Now we apply the linear impulse–momentum equation to each body.

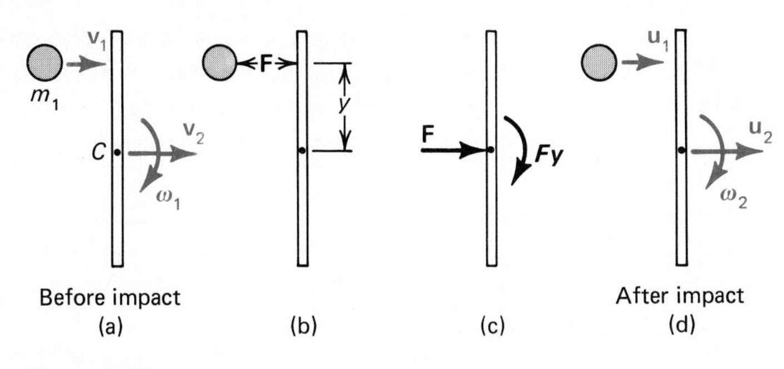

Figure 7.12

For m_1: $$-\int_0^t F\,dt = m_1(u_1 - v_1)$$

For m_2: $$\int_0^t F\,dt = m_2(u_2 - v_2)$$

Adding these two equations, we have the expression for the conservation of linear momentum.

$$m_1 v_1 + m_2 v_2 = m_1 u_1 + m_2 u_2 \tag{7.1}$$

Let us now apply the angular impulse–momentum relationship for the bar.

$$\int_0^t M_c\,dt = \Delta H_c$$

Substituting for the moment and momentum we obtain

$$\int_0^t Fy\,dt = I_c(\omega_2 - \omega_1)$$

Now substituting from the linear impulse–momentum equation yields

$$\int_0^t Fy\,dt = y\int_0^t F\,dt = ym_1(v_1 - u_1)$$

Therefore we have

$$ym_1(v_1 - u_1) = I_c(\omega_2 - \omega_1)$$

or $$m_1 v_1 y + I_c\omega_1 = m_1 u_1 y + I_c\omega_2 \tag{7.16}$$

where I_c is the mass moment of inertia of the bar about the center of mass. This equation represents the conservation of angular momentum of the system about the center of mass of the rod.

We now form the equation involving the coefficient of restitution. The initial velocity of the bar at the point of impact is $v_2 + y\omega_1$ and the final velocity of this point is $u_2 + y\omega_2$. Therefore, from the definition of the coefficient of restitution we obtain,

$$e(v_1 - v_2 - y\omega_1) = -(u_1 - u_2 - y\omega_2) \tag{7.17}$$

Equations (7.1), (7.16), and (7.17) allow us to find the three unknowns for eccentric impact, namely, u_1, u_2, and ω_2. The only word of caution concerns the sign convention. Assumed positive directions are shown in Fig. 7.12. If a velocity is directed opposite to that assumed in the derivation of the equations, negative values must be entered.

We again emphasize that the impact was direct. If the impact had been oblique, we would apply the above equations to compute velocities in the direction of the impact. The velocities in the tangential direction remain unchanged.

The previous equations are applicable *only* if there is translational and rotational motion of the bar. If the bar is pivoted, then no translation of the bar occurs. In this case, only the angular momentum is conserved about the pivot point. Therefore, the angular momentum and coefficient of restitution equations are used. This will be demonstrated in Example 7.9.

Let us say a word about notation before we consider some examples. When velocity components are computed along the line of action, vector notation is not used. If the final velocity has components in a direction other than line of impact, then we represent the final velocity in vector form.

Example 7.8

Consider a sphere striking a thin bar with the conditions shown in Fig. 7.13. What is the motion of the sphere and bar after impact? Assume that the line of impact coincides with the direction of the X axis. Also let $m_1 = 2$ slugs, $m_2 = 4$ slugs, $v_1 = 40$ ft/scc, $v_2 = 0$, $\theta = 60°$, $\omega_1 = 0$, and $e = 0.8$.

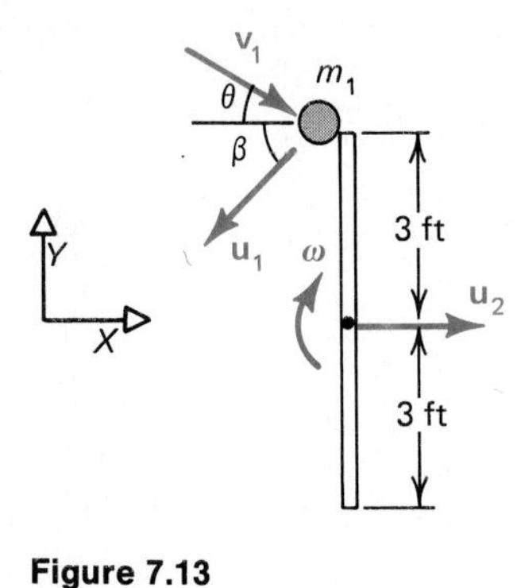

Figure 7.13

The mass moment of inertia about the center of mass of a thin rod is

$$I_c = \tfrac{1}{12}ml^2 = 12 \text{ slug}\cdot\text{ft}^2$$

We may substitute directly into Eqs. (7.1), (7.16), and (7.17), noting the correct signs for the assumed velocity directions after impact.

$$2(40 \cos 60°) + 4(0) = -2u_1 \cos \beta + 4u_2$$

$$2(40 \cos 60°)(3) + 12(0) = -2(u_1 \cos \beta)3 + 12\omega_2$$

$$0.8(40 \cos 60° - 0 - 0) = -(-u_1 \cos \beta - u_2 - 3\omega_2)$$

Simplifying these, we find that

$$-u_1 \cos \beta + 2u_2 = 20$$

$$-u_1 \cos \beta + 2\omega_2 = 20$$

$$u_1 \cos \beta + u_2 + 3\omega_2 = 16$$

Solving these equations, we obtain

$$u_1 \cos \beta = -8 \text{ ft/sec} \qquad u_2 = 6 \text{ ft/sec} \qquad \omega_2 = 6 \text{ rad/sec}$$

The magnitude of the velocity of the sphere in the y direction before and after impact is 40(sin 60°) or 34.6 ft/sec. Thus, the velocity of the sphere after impact is

$$\mathbf{u}_1 = 8\mathbf{i} - 34.6\mathbf{j} \text{ ft/sec} \qquad \textit{Answer}$$

The translation velocity of the bar after impact is

$$\mathbf{u}_2 = 6\mathbf{i} \text{ ft/sec} \qquad \textit{Answer}$$

and the rotational rate of the bar is

$$\boldsymbol{\omega}_2 = -6\mathbf{k} \text{ rad/sec} \qquad \textit{Answer}$$

Example 7.9

A sphere strikes a thin bar in oblique eccentric impact as shown in Fig. 7.14. (a) What is the motion of the bar and sphere after impact? (b) Solve for the velocities if $v_1 = 50$ m/s, $\theta = 60°$, $l = 6$ m, $l_1 = 4$ m, $m_1 = 2$ kg, $m_2 = 4$ kg, and $e = 0.8$. Assume that the bar is pinned at O and initially at rest.

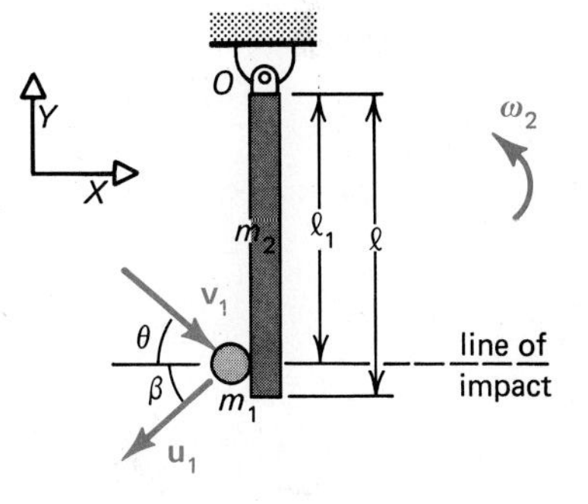

Figure 7.14

(a) We assume that the line of impact is perpendicular to the bar as shown in the figure. The velocity of the sphere in the y direction remains unchanged, and is $(-v \sin \theta)\mathbf{j}$. The angular momentum of the system about point O is conserved. Assuming counterclockwise moments and rotations are positive, we have

$$(m_1 v_1 \cos \theta) l_1 = -(m_1 u_1 \cos \beta) l_1 + I_o \omega_2 \qquad \text{(a)}$$

where I_o is the mass moment of inertia of the bar about point O. The definition of the coefficient of restitution yields the following:

$$e(v_1 \cos \theta - 0) = -(-u_1 \cos \beta - l_1 \omega_2)$$

or

$$\omega_2 l_1 + u_1 \cos \beta = e(v_1 \cos \theta) \qquad \text{(b)}$$

Solving Eqs. (a) and (b) simultaneously, we obtain

$$\omega_2 = \frac{(v_1 \cos \theta)(1 + e) m_1 l_1}{I_o + m_1 l_1^2} \qquad \textit{Answer} \quad \text{(c)}$$

and

$$-u_1 \cos \beta = (m_1 l_1^2 - e I_o) \frac{v_1 \cos \theta}{I_o + m_1 l_1^2}$$

Therefore, the velocity of the ball is

$$\mathbf{u}_1 = -(v_1 \sin \theta)\mathbf{j} + (m_1 l_1^2 - e I_o) \frac{v_1 \cos \theta}{I_o + m_1 l_1^2} \mathbf{i} \qquad \textit{Answer} \quad \text{(d)}$$

whereas the bar rotates at the rate given by Eq. (c).

(b) For the values given,

$$I_o = \tfrac{1}{3}m_2 l^2 = 48 \text{ kg}\cdot\text{m}^2$$

$$v_1 \cos\theta = 25 \text{ m/s} \qquad v_1 \sin\theta = 43.3 \text{ m/s}$$

$$m_1 l_1^2 = 32 \text{ kg}\cdot\text{m}^2 \qquad m_1 l_1 = 8 \text{ kg}\cdot\text{m}$$

Substituting into Eq. (c) and Eq. (d), we obtain

$$\mathbf{u}_1 = -2\mathbf{i} - 43.3\mathbf{j} \text{ m/s} \qquad \textit{Answer}$$

$$\boldsymbol{\omega}_2 = 4.5\mathbf{k} \text{ rad/s} \qquad \textit{Answer}$$

PROBLEMS

7.34 Rework Example 7.9 for the case where $e = 0$.

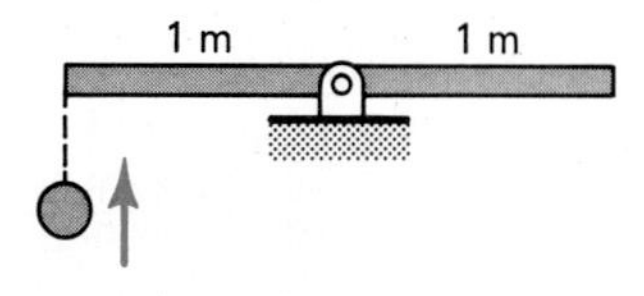

Figure P7.35

7.35 A 0.0243-kg coin traveling 5 m/s strikes a thin horizontal rod as shown in Fig. P7.35. Assuming that the mass of the rod is 2 kg and that it is initially at rest, what is the rate of rotation immediately after impact if $e = 0.7$?

7.36 A 1-slug sphere traveling 10 ft/sec strikes a 3-slug horizontal rod as shown in Fig. P7.36. Assuming the thin rod is initially at rest, what is the rate of rotation immediately after impact if $e = 0.5$?

7.37 The particle m_1 strikes the thin uniform bar of mass m_2 as shown in Fig. P7.37. Find the velocities of each body after impact if $e = 1$, $m_1 = 2$ kg, $m_2 = 10$ kg, $v_1 = 5$ m/s, $v_2 = 0$, $\omega_1 = 2$ rad/s.

7.38 Rework Problem 7.37 for $v_2 = 5$ m/s.

7.39 Rework Problem 7.37 for $m_1 = 4$ kg, $m_2 = 6$ kg, $v_1 = 12$ m/s, $v_2 = 4$ m/s, $\omega_1 = 0$, $e = 0.5$.

7.40 A mass m_1 is dropped from a height h onto a horizontal slender rod that is pivoted as shown in Fig. P7.40. Assuming that the coefficient of restitution is e, then: (a) What is the angular velocity right after impact? (b) How much does the spring deform? The mass of the rod m_2 and is initially in static equilibrium. The spring is also in its unstretched position before impact.

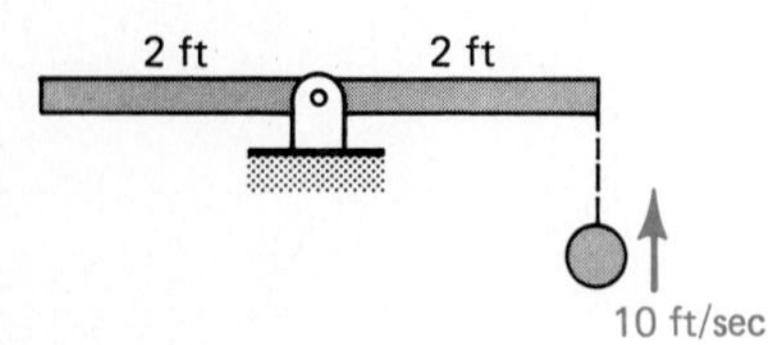

Figure P7.36

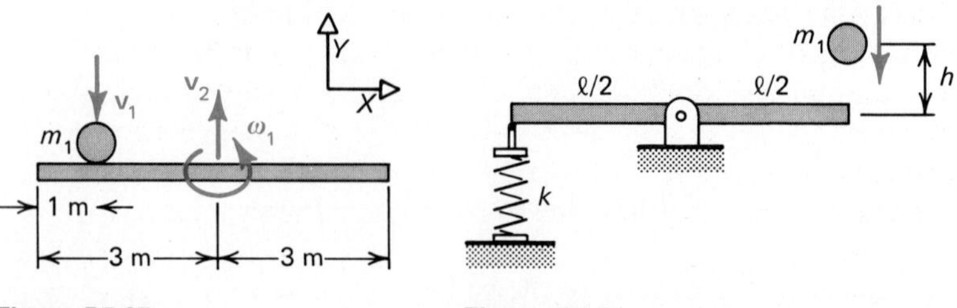

Figure P7.37 **Figure P7.40**

7.41 A 15-lb rod 2 ft long pinned at B is released from rest as shown in Fig. P7.41. The bar strikes a 5-lb body that slides across the horizontal surface. The coefficient of friction is zero for the first foot of travel and equals 0.2 thereafter. If $e = 0.8$ between the bodies, how far does the 5-lb body travel?

7.42 Suppose $e = 0$ in Problem 7.41. Describe the motion of the rod at the instant after impact and thereafter.

7.43 The 1-kg bar in Fig. P7.43 hits the ball as shown. If the initial velocity of the ball is zero and the initial angular velocity of the bar is 10 rad/s, how far does the ball land as measured by the coordinate x? Assume that the mass of the ball is 0.05 kg and $e = 0.8$.

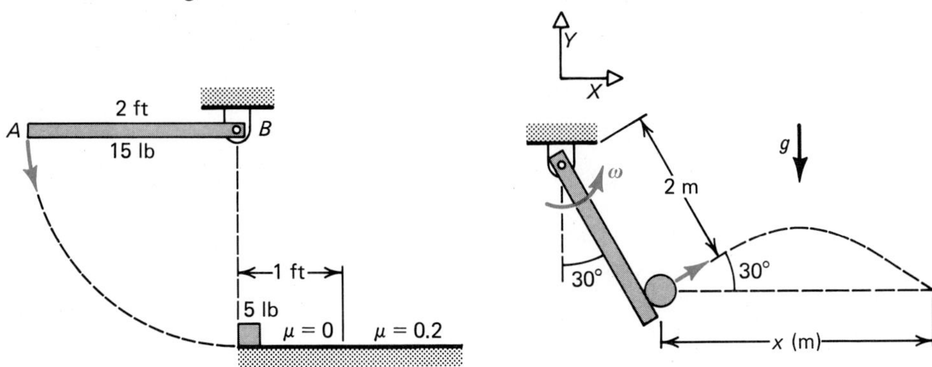

Figure P7.41 **Figure P7.43**

7.44 A 5-kg release mechanism is shown in Fig. P7.44 in its equilibrium position in the horizontal plane resting against a 2-kg particle A. If $e = 1$ between these bodies and we wish to impart a velocity of 20 m/s to the particle by impact with the rod, what must be the value of the torsional spring modulus if $\theta_{max} = 45°$?

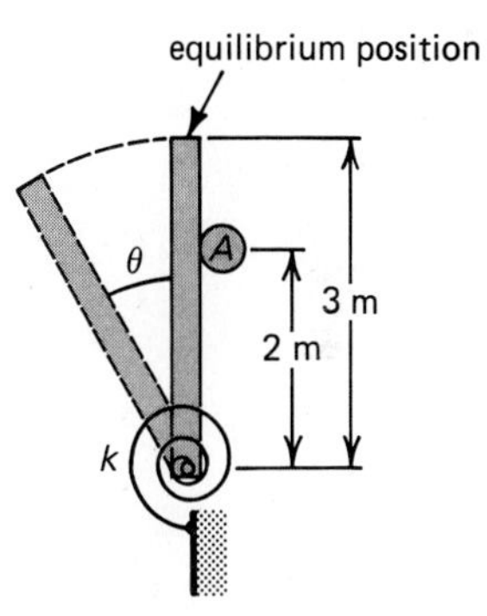

Figure P7.44

7.6 Summary

When two bodies collide with an impulsive force, i.e., a large force with a short time duration, we speak of these bodies as undergoing impact. During impact, the total momentum is conserved. The amount of lost kinetic energy depends on the coefficient of restitution, a property of the colliding bodies. Knowing this property and the motion of the bodies before impact, we developed equations useful for calculating the motion at the instant after impact.

A. The important *ideas*:

1. Impact occurs along a line of impact.
2. There is no velocity change in the direction perpendicular to the line of impact.

3. Linear momentum is conserved for the case of direct central impact.
4. If the impact involves rotational motion, as in the case of direct eccentric impact, then the angular momentum of the system about the center of mass of the rotating body is conserved.
5. The coefficient of restitution e is a measure of the elasticity of impact. If $e = 1$, the impact is elastic and there is no loss of kinetic energy. If $e = 0$, the impact is perfectly inelastic and there is a maximum loss of kinetic energy. In general, e varies between zero and one.

B. The important *equations*:

Conservation of linear momentum:

$$m_1 v_1 + m_2 v_2 = m_1 u_1 + m_2 u_2 \tag{7.1}$$

Definition of the coefficient of restitution

$$e(v_1 - v_2) = -(u_1 - u_2) \tag{7.11}$$

Conservation of angular momentum–eccentric impact

$$m_1 v_1 y + I_c \omega_1 = m_1 u_1 y + I_c \omega_2 \tag{7.16}$$

Coefficient of restitution–eccentric impact

$$e_1(v_1 - v_2 - y\omega_1) = -(u_1 - u_2 - y\omega_2) \tag{7.17}$$

8 Advanced Topics

Having studied Newton's second law and its various integral forms, we are now ready to broaden our problem-solving capabilities and delve more deeply into the subject of dynamics. We will therefore consider the concepts of generalized coordinates, coordinate transformations, virtual work, and generalized forces as preliminary steps to understanding the equations of Lagrange and Euler. The lagrangian approach to the study of motion is based on the total energy of the system; the eulerian approach begins with the angular momentum of the body under consideration. We shall see that in both cases proper identification of the coordinates as well as reference frames is very important.

8.1 Generalized Coordinates

So far, we have been solving problems using cartesian coordinates, sometimes called the ordinary coordinate system, or special coordinate systems such as polar coordinates and cylindrical coordinates. In many instances it is convenient to introduce generalized coordinates that specify the configuration of a system. These generalized coordinates may refer to the above-mentioned coordinate systems, or they may be composed of a variety of coordinates that serve to specify the configuration of the system.

For example, consider the simple pendulum that oscillates in the XY plane as shown in Fig. 8.1. Its motion is defined by the ordinary coordinates x and y, where x and y are related by the equation of constraint $x^2 + y^2 = L^2$. On the other hand, the angle θ serves as a convenient single generalized coordinate since the pendulum has only one degree of freedom. In general we may define the

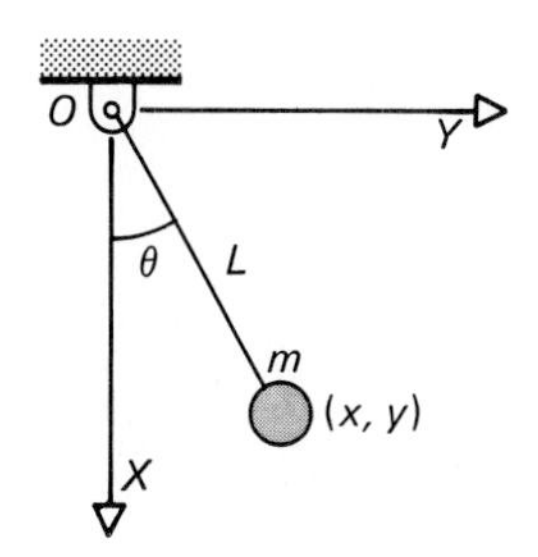

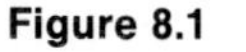
Figure 8.1

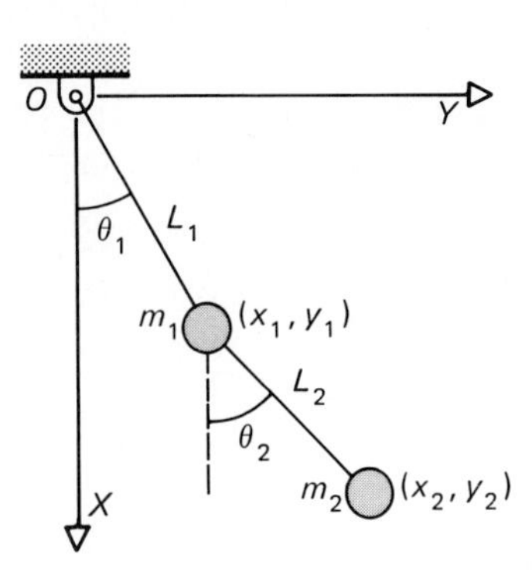

Figure 8.2

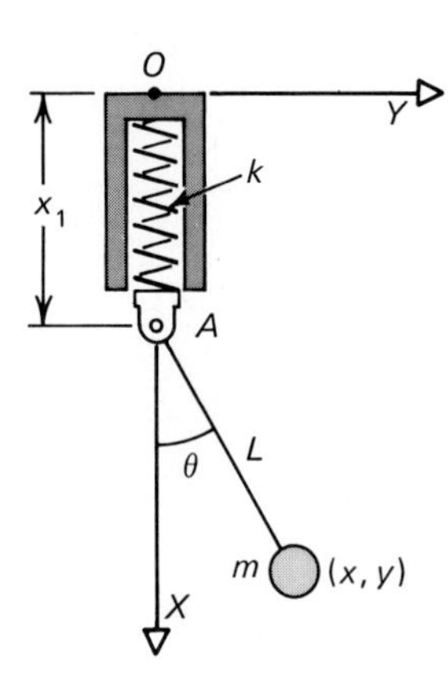

Figure 8.3

number of degrees of freedom as equal to the number of coordinates that are used to describe the configuration of the system minus the number of independent equations of constraint.

Figure 8.2 shows two pendulums pin-connected such that the lower pendulum is free to oscillate in the XY plane relative to the upper pendulum. Now there are four ordinary coordinates, x_1, y_1, x_2, and y_2 that locate the two masses m_1 and m_2. There are two independent equations of constraint, viz.

$$x_1^2 + y_1^2 = L_1^2$$

$$(x_2 - x_1)^2 + (y_2 - y_1)^2 = L_2^2$$

Consequently, the configuration in Fig. 8.2 has two degrees of freedom. The coordinates θ_1 and θ_2 serve as convenient generalized coordinates.

As a final variation on this example, consider the simple pendulum in series with a spring as shown in Fig. 8.3, where point A is constrained to move in the vertical direction. While the coordinates x and y could serve as the ordinary coordinates to locate the mass m, x_1 and θ prove to be convenient generalized coordinates for studying the two-degrees-of-freedom system.

In solving complicated problems of motion, the most important conditions are usually (1) the use of the proper number of degrees of freedom and (2) the assignment of the generalized coordinates. In this volume we are addressing ourselves only to those cases where there are as many generalized coordinates as there are degrees of freedom. We classify such problems as **holonomic**, and will refer to such a set of coordinates as an **independent set of generalized coordinates**.

Example 8.1

A double pendulum is attached to a cart as shown in Fig. 8.4(a). (a) How many degrees of freedom does the system have? (b) Choose an independent set of generalized coordinates.

(a) The cart of mass m_1 is described by the ordinary coordinate x. Each pendulum mass could be described by two ordinary coordinates so that there are a total of five ordinary coordinates. However, there are two equations of constraint for the length of each pendulum. Consequently, there are three degrees of freedom for the configuration.

(b) Three independent generalized coordinates are x, θ, ϕ. These are shown in Fig. 8.4(b).

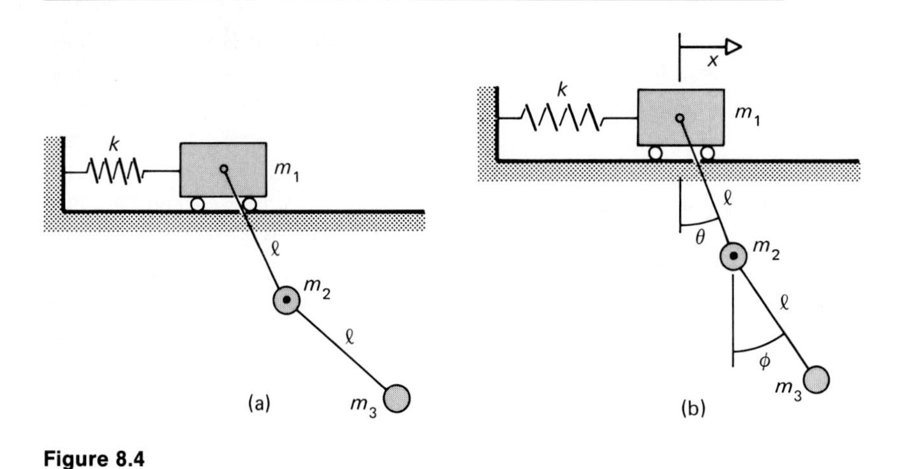

Figure 8.4

PROBLEMS

8.1 A mass m is attached to a spring of modulus k as shown in Fig. P8.1. L_0 is the unstretched length of the spring. (a) How many degrees of freedom does the mass have if its motion is confined to the XY plane? (b) Choose an independent set of generalized coordinates.

8.2 Mass m_1 is fixed to the end of the rod OA that rotates in the XY plane. Length L is the length of the rod. Mass m_2 is connected to m_1 by the spring of modulus k and is free to slide along the massless rod, as shown in Fig. P8.2. L_0 is the unstretched length of the spring. (a) How many degrees of freedom does the system of masses have? (b) Choose an independent set of generalized coordinates.

8.3 Masses m_2 and m_3 in Fig. P8.3 are free to slide along the massless rotating rod OA and are coupled by the springs of moduli k_1 and k_2 with unstretched lengths of L_1 and L_2, respectively. (a) How many degrees of freedom does the system of masses have if the rod rotates in the XY plane? (b) Choose an independent set of generalized coordinates.

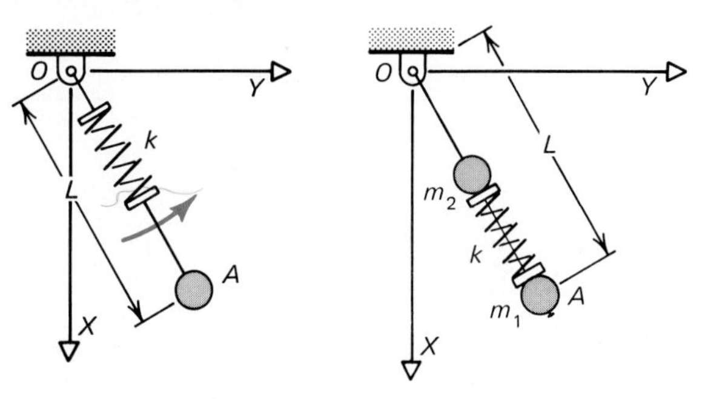

Figure P8.1 **Figure P8.2**

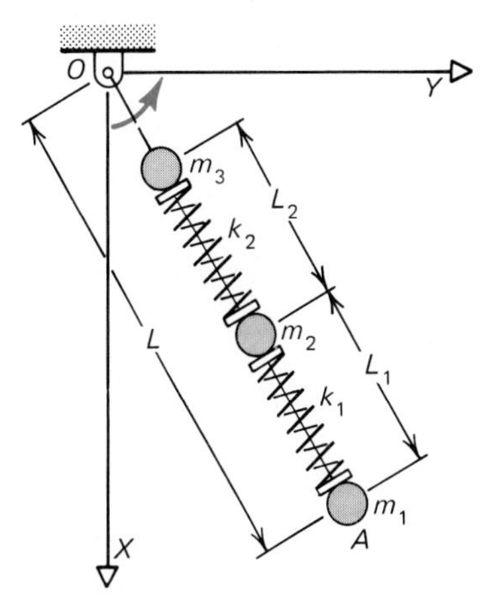

Figure P8.3

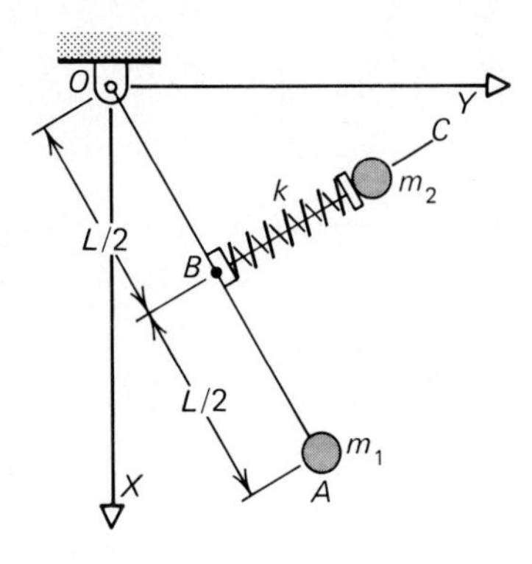

Figure P8.4

8.4 Mass m_1 is attached to the rotating massless rod OA while m_2 slides freely along the rod BC in the XY plane as shown in Fig. P8.4. A spring of modulus k and unstretched length L_0 connects m_2 to point B. (a) How many degrees of freedom does the system of masses have? (b) Choose an independent set of generalized coordinates.

8.5 A simple pendulum of mass m_2 is attached to m_1 as shown in Fig. P8.5. (a) How many degrees of freedom does the system have? (b) Choose an independent set of generalized coordinates.

8.6 An inverted pendulum of mass m is constrained by the spring of modulus k as shown in Fig. P8.6. (a) How many degrees of freedom does the mass have if its motion is restricted to the XY plane? (b) Choose an independent set of generalized coordinates.

8.7 A beam is represented by two masses in Fig. P8.7, each of mass m as shown in the static equilibrium position. The beam is supported by two springs, one of modulus k and the other of $2k$. (a) How many degrees of freedom does the beam have? (b) Choose an independent set of generalized coordinates.

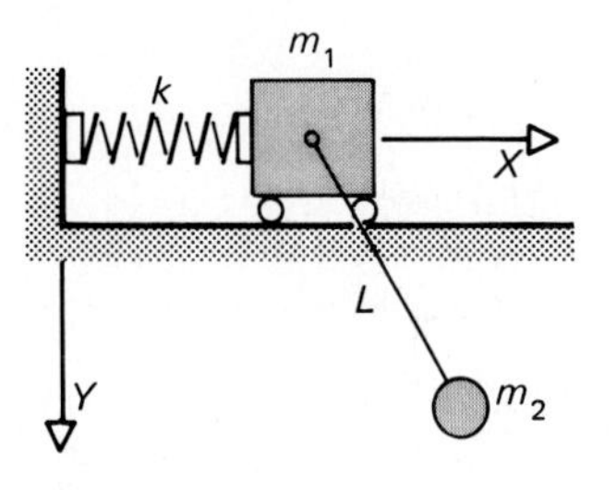

Figure P8.5

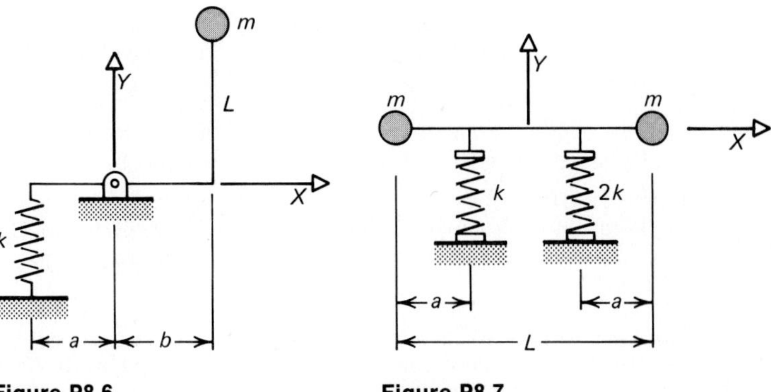

Figure P8.6

Figure P8.7

8.8 Determine the degrees of freedom and an independent set of generalized coordinates for the system shown in Fig. P8.8. All motion is confined to the XY plane.

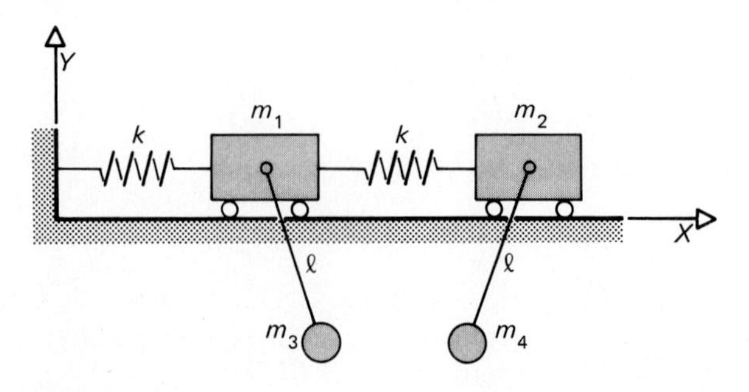

Figure P8.8

8.9 If mass m_1 is fixed in Problem 8.8, determine the degrees of freedom and an independent set of generalized coordinates.

8.10 A motor of mass m is modeled as shown in Fig. P8.10. It is connected to the foundation by springs of moduli k_1 and k_2. (a) How many degrees of freedom does the system have? (b) Choose a set of generalized coordinates.

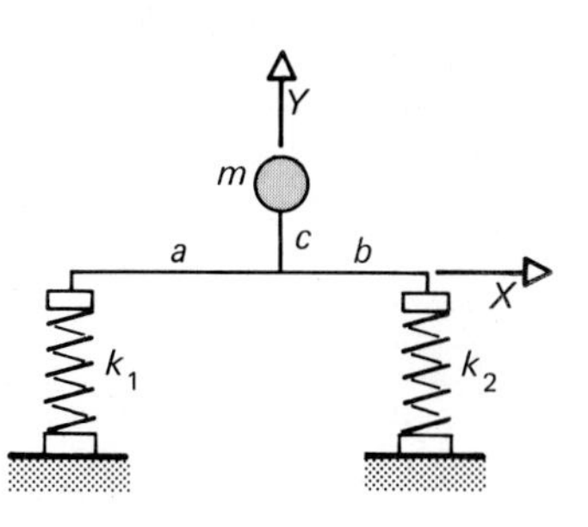

Figure P8.10

8.2 Coordinate Transformation

Coordinate transformation is the way by which we relate a set of generalized coordinates with another set of coordinates, usually ordinary coordinates. In Fig. 8.2, for example, we have the following relationships between the ordinary and the generalized coordinates:

$$\begin{aligned} y_1 &= L_1 \sin \theta_1 \\ x_1 &= L_1 \cos \theta_1 \\ y_2 &= L_1 \sin \theta_1 + L_2 \sin \theta_2 \\ x_2 &= L_1 \cos \theta_1 + L_2 \cos \theta_2 \end{aligned} \tag{8.1}$$

As a more general form, we could write these equations as

$$\begin{aligned} x_1 &= f_1(\theta_1) \\ y_1 &= f_2(\theta_1) \\ x_2 &= f_3(\theta_1, \theta_2) \\ y_2 &= f_4(\theta_1, \theta_2) \end{aligned} \tag{8.2}$$

where the f's refer to some function. Thus, x_1 is a function of θ_1, x_2 is a function of θ_1 and θ_2, etc.

Let the symbol q_i represent the ith generalized coordinate and x_i be the ith ordinary coordinate. Then, for the case where there are n generalized coordinates for a system with n degrees of freedom, we have the following general form for the coordinate transformations:

$$\begin{aligned} x_1 &= f_1(q_1, q_2, \ldots, q_n) \\ x_2 &= f_2(q_1, q_2, \ldots, q_n) \\ &\vdots \\ x_k &= f_k(q_1, q_2, \ldots, q_n) \end{aligned} \tag{8.3}$$

We note that the number of ordinary coordinates, k, does not necessarily equal the number of generalized coordinates, n. However, if there are l equations of constraint, then $(k - l) = n$. We see this was the case for the examples in Figs. 8.1 and 8.2. In the first case,

$k = 2$, $l = 1$, and $n = 1$. In the second case, $k = 4$, $l = 2$, and $n = 2$ where the two equations of constraint are formed from the length of each rod.

The derivative with respect to time of Eq. (8.3) is one of the important mathematical operations that we will be using in deriving Lagrange's equations. For example,

$$\frac{dx_1}{dt} = \dot{x}_1 = \frac{\partial x_1}{\partial q_1}\frac{dq_1}{dt} + \frac{\partial x_1}{\partial q_2}\frac{dq_2}{dt} + \cdots + \frac{\partial x_1}{\partial q_n}\frac{dq_n}{dt}$$

$$\dot{x}_1 = \frac{\partial x_1}{\partial q_1}\dot{q}_1 + \frac{\partial x_1}{\partial q_2}\dot{q}_2 + \cdots + \frac{\partial x_1}{\partial q_n}\dot{q}_n$$

$$\dot{x}_1 = \sum_{k=1}^{n} \frac{\partial x_1}{\partial q_k}\dot{q}_k \tag{8.4}$$

Derivatives for x_2, x_3, etc. can also be written in the form of Eq. (8.4). Consequently, we can write Eq. (8.4) in the general form

$$\dot{x}_j = \frac{dx_j}{dt} = \sum_{k=1}^{n} \frac{\partial x_j}{\partial q_k}\dot{q}_k \tag{8.5}$$

By way of example, let us apply Eq. (8.5) to the third expression in Eq. (8.1). Here $q_1 = \theta_1$ and $q_2 = \theta_2$, so that $\dot{q}_1 = \dot{\theta}_1$ and $\dot{q}_2 = \dot{\theta}_2$. Also, the coordinate y_2 is now replaced with the ordinary coordinate x_3.

$$y_2 = x_3 = L_1 \sin\theta_1 + L_2 \sin\theta_2$$

Thus,
$$\frac{\partial x_3}{\partial \theta_1} = L_1 \cos\theta_1 \qquad \frac{\partial x_3}{\partial \theta_2} = L_2 \cos\theta_2$$

and therefore

$$\dot{x}_3 = (L_1 \cos\theta_1)\dot{\theta}_1 + (L_2 \cos\theta_2)\dot{\theta}_2$$

Example 8.2

For the system shown in Fig. 8.4, write a set of transformation equations between the ordinary coordinates and the generalized coordinates. Include any equations of constraint and also apply Eq. (8.5) to the ordinary coordinates of mass m_2.

The generalized coordinates $q_1 = x$, $q_2 = \theta$, and $q_3 = \phi$ are shown in Fig. 8.4(b). For mass m_1, the relationship between the ordinary coordinate x_1 and the generalized coordinates is simply

$$x_1 = x$$

where x is measured from the unstretched spring position.

For mass m_2, let x_2 and x_3 be the ordinary coordinates. Now,

$$x_2 = x + l \sin \theta$$

$$x_3 = l \cos \theta$$

For mass m_3, let x_4 and x_5 be the ordinary coordinates. Thus,

$$x_4 = x_2 + l \sin \phi = x + l \sin \theta + l \sin \phi$$

$$x_5 = l \cos \theta + l \cos \phi$$

The equations of constraint are

$$(x_2 - x_1)^2 + x_3^2 = l^2$$

$$(x_4 - x_2)^2 + (x_5 - x_3)^2 = l^2$$

Applying Eq. (8.5) to x_2, x_3, we have the following results:

$$\dot{x}_2 = \dot{x} + \dot{\theta} l \cos \theta$$

$$\dot{x}_3 = -\dot{\theta} l \sin \theta$$

PROBLEMS

Using an independent set of generalized coordinates for the following problems, (a) write a set of transformation equations between the ordinary coordinates and the generalized coordinates; (b) include any equations of constraint; and (c) apply Eq. (8.5) to the equations found in (a):

8.11 For the system shown in Fig. P8.1.

8.12 For the system shown in Fig. P8.2.

8.13 For the system shown in Fig. P8.3.

8.14 For the system shown in Fig. P8.4.

8.15 For the system shown in Fig. P8.5.

8.16 For the system shown in Fig. P8.6.

8.17 For the system shown in Fig. P8.7.

8.18 For the system shown in Fig. P8.8.

8.19 For the system shown in Fig. P8.8 but for m_1 fixed.

8.20 For the system shown in Fig. P8.10.

8.3 Virtual Work and Generalized Forces

The concept of virtual work is normally developed in the study of statics. It is useful, however, to review the principle here. Let us begin by recalling that **virtual displacements** are fictitious displacements in that we do not consider the passage of time. Furthermore, they are arbitrarily small displacements that do not violate the constraints of the system. We normally represent a virtual displacement by δx rather than dx.

Suppose we have a system of particles whose positions are described by the ordinary coordinates $x_1, x_2, \ldots, x_k$ and the corresponding external forces are $F_1, F_2, \ldots, F_k$, each force being applied in the direction of the coordinate increasing in value. Let each coordinate experience a virtual displacement. The resulting work done by the forces is called the **virtual work**, i.e.,

$$\delta W = \sum_{i=1}^{k} F_i \delta x_i$$

Using the transformation of Eq. (8.3), we find that

$$\delta x_i = \sum_{j=1}^{n} \frac{\partial x_i}{\partial q_j} \delta q_j$$

Thus,

$$\delta W = \sum_{i=1}^{k} \sum_{j=1}^{n} F_i \frac{\partial x_i}{\partial q_j} \delta q_j$$

$$= \sum_{j=1}^{n} \left(\sum_{i=1}^{k} F_i \frac{\partial x_i}{\partial q_j} \right) \delta q_j \tag{8.6}$$

We define the *generalized force* Q_j as

$$Q_j = \sum_{i=1}^{k} F_i \frac{\partial x_i}{\partial q_j} \tag{8.7}$$

Equation (8.6) is then rewritten as

$$\delta W = \sum_{j=1}^{n} Q_j \delta q_j \tag{8.8}$$

This equation tells us that in order to find a generalized force, say Q_1, allow only δq_1 to occur, holding the remaining coordinates fixed, and calculate the virtual work. Then

$$Q_1 = \frac{\delta W}{\delta q_1}$$

The procedure may be repeated for each generalized coordinate to calculate all of the generalized forces. Although the computation of virtual work was thoroughly covered in statics, it would be helpful to review it again.

Example 8.3

Obtain the generalized forces for the double pendulum shown in Fig. 8.5. Let w_1 and w_2 be the weight of each point mass and let the length of each pendulum equal L.

Let the double pendulum experience a virtual displacement $\delta\theta_1$, while holding θ_2 fixed as shown in Fig. 8.5. Then

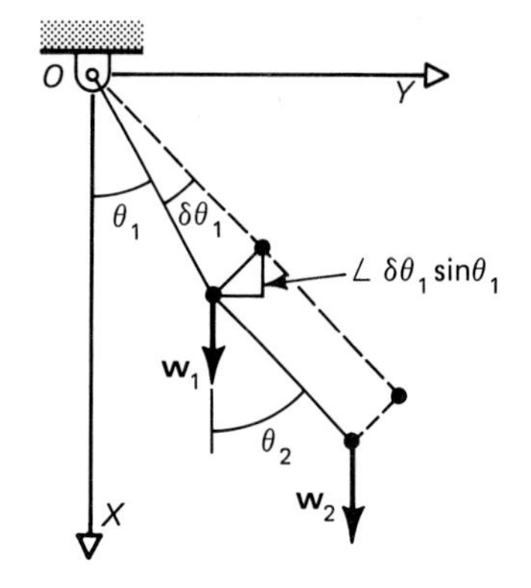

Figure 8.5

$$\delta W = -w_1 L \delta\theta_1 \sin\theta_1 - w_2 L\,\delta\theta_1 \sin\theta_1$$
$$= -(w_1 + w_2)L \sin\theta_1\, \delta\theta_1$$

Therefore,
$$Q_1 = \frac{\delta W}{\delta\theta_1} = -(w_1 + w_2)L \sin\theta_1 \qquad \textit{Answer}$$

To find the generalized force Q_2, hold θ_1 fixed and vary θ_2. The resulting virtual work is

$$\delta W = -w_2 L \delta\theta_2 \sin\theta_2$$

Therefore
$$Q_2 = \frac{\delta W}{\delta\theta_2} = -w_2 L \sin\theta_2 \qquad \textit{Answer}$$

We note from this particular example that the generalized forces may have units of a moment. Thus, while we speak of generalized forces, we actually include moments as well as forces under this term.

8.4 Lagrange's Equations

In order to simplify the algebra, Lagrange's equations will be derived for the case of a single particle of mass m described by the coordinates x_1, x_2, x_3 (instead of the usual x, y, z coordinates) acted upon by the components of the resultant force F_1, F_2, F_3. The kinetic energy of the particle is expressed as

$$T = \sum_{j=1}^{3} \frac{1}{2} m\dot{x}_j^2 \tag{8.9}$$

Let us assume the general form of the transformation equations is given by Eq. (8.3), where we assume a system described by n generalized coordinates. For this transformation, we perform the following mathematical operations,

$$\frac{\partial T}{\partial \dot{q}_i} = \sum_{j=1}^{3} m\dot{x}_j \frac{\partial \dot{x}_j}{\partial \dot{q}_i} = \sum_{j=1}^{3} m\dot{x}_j \frac{\partial x_j}{\partial q_i} \tag{8.10}$$

since $\partial\dot{x}_j/\partial\dot{q}_i = \partial x_j/\partial q_i$ by eliminating dt in the numerator and denominator.

Next we differentiate Eq. (8.10) with respect to time.

$$\frac{d}{dt}\left(\frac{\partial T}{\partial \dot{q}_i}\right) = \sum_{j=1}^{3} m\ddot{x}_j \frac{\partial x_j}{\partial q_i} + \sum_{j=1}^{3} m\dot{x}_j \frac{d}{dt}\left(\frac{\partial x_j}{\partial q_i}\right) \tag{8.11}$$

We wish to write the last term of Eq. (8.11) in terms of the kinetic energy. To do this, let us recall that

$$\dot{x}_j = \frac{dx_j}{dt} = \sum_{k=1}^{n} \frac{\partial x_j}{\partial q_k} \dot{q}_k \tag{8.5}$$

Thus,

$$\frac{d}{dt}\left(\frac{\partial x_j}{\partial q_i}\right) = \sum_{k=1}^{n} \frac{\partial^2 x_j}{\partial q_i \, \partial q_k} \dot{q}_k$$

Equation (8.11) now becomes

$$\frac{d}{dt}\left(\frac{\partial T}{\partial \dot{q}_i}\right) = \sum_{j=1}^{3} m\ddot{x}_j \frac{\partial x_j}{\partial q_i} + \sum_{j=1}^{3} m\dot{x}_j \sum_{k=1}^{n} \frac{\partial^2 x_j}{\partial q_i \, \partial q_k} \dot{q}_k \tag{8.12}$$

From Eq. (8.9) we see that

$$\frac{\partial T}{\partial q_i} = \sum_{j=1}^{3} m\dot{x}_j \frac{\partial \dot{x}_j}{\partial q_i} = \sum_{j=1}^{3} m\dot{x}_j \sum_{k=1}^{n} \frac{\partial^2 x_j}{\partial q_i \, \partial q_k} \dot{q}_k \tag{8.13}$$

by taking the partial derivative of Eq. (8.5) with respect to q_i. Substituting Eq. (8.13) into (8.12), we obtain

$$\frac{d}{dt}\left(\frac{\partial T}{\partial \dot{q}_i}\right) = \sum_{j=1}^{3} m\ddot{x}_j \frac{\partial x_j}{\partial q_i} + \frac{\partial T}{\partial q_i} \tag{8.14}$$

To reduce the above equation into the final desired form, we apply Newton's second law to the particle to obtain

$$F_j = m\ddot{x}_j \qquad j = 1, 2, 3 \tag{8.15}$$

and substitute Eq. (8.15) into (8.14) to get

$$\frac{d}{dt}\left(\frac{\partial T}{\partial \dot{q}_i}\right) - \frac{\partial T}{\partial q_i} = \sum_{j=1}^{3} F_j \frac{\partial x_j}{\partial q_i} \tag{8.16}$$

But, referring to Eq. (8.7) we see that the right side of Eq. (8.16) is the generalized force. Thus, Lagrange's equations for the motion of a particle become

$$\frac{d}{dt}\left(\frac{\partial T}{\partial \dot{q}_i}\right) - \frac{\partial T}{\partial q_i} = Q_i \tag{8.17}$$

where $i = 1, \ldots, n$. Of course $n \leq 3$ for this special case of a single particle. However, it can be shown that the same form of the equation would have been derived for a set of particles having n degrees of freedom. Equation (8.17) is but one form of Lagrange's equations.

An alternative form of Eq. (8.17) can be made for the case where all generalized forces are conservative forces and are therefore derivable from a potential energy function V. Let us represent V as a function of the generalized coordinates, i.e.,

$$V = V(q_1, q_2, \ldots, q_n) \tag{8.18}$$

For such a system we recall that the work done by these conservative forces equals the negative of the potential energy. In differential form, this says that $dW = -dV$, or, in virtual work form,

$$\delta W = -\delta V \tag{8.19}$$

Now, from Eq. (8.18),

$$\delta V = \sum_{j=1}^{n} \frac{\partial V}{\partial q_j} \delta q_j \tag{8.20}$$

Substitute Eq. (8.8) for the left side of Eq. (8.19) and Eq. (8.20) for the right side.

$$\sum_{j=1}^{n} Q_j \delta q_j = -\sum_{j=1}^{n} \frac{\partial V}{\partial q_j} \delta q_j$$

Matching both sides of this equation, we have

$$Q_j = -\frac{\partial V}{\partial q_i}$$

Lagrange's equations can now be written

$$\frac{d}{dt}\left(\frac{\partial T}{\partial \dot{q}_i}\right) - \frac{\partial T}{\partial q_i} = -\frac{\partial V}{\partial q_i}$$

or

$$\frac{d}{dt}\left(\frac{\partial T}{\partial \dot{q}_i}\right) - \frac{\partial (T - V)}{\partial q_i} = 0 \tag{8.21}$$

We now introduce the lagrangian function $\mathscr{L} = T - V$ and assume that V is not a function of the $\dot{q}$'s. Equation (8.21) becomes

$$\frac{d}{dt}\left(\frac{\partial \mathscr{L}}{\partial \dot{q}_i}\right) - \frac{\partial \mathscr{L}}{\partial q_i} = 0 \qquad i = 1, 2, \ldots, n \tag{8.22}$$

From the development of Eq. (8.22), we can now write Lagrange's equations of motion in a second form, namely,

$$\frac{d}{dt}\left(\frac{\partial \mathscr{L}}{\partial \dot{q}_i}\right) - \frac{\partial \mathscr{L}}{\partial q_i} = Q_i \qquad i = 1, 2, \ldots, n \tag{8.23}$$

where the Q_i's on the right side are obtained from those forces that are not derivable from a potential function. Examples of such forces include friction forces and time-variant forcing functions. Of the two general forms of Lagrange's equations, Eq. (8.23) is usually the most convenient to use. In applying this set of equations, we must perform the following steps:

1. Establish the number of degrees of freedom for the system.
2. Select the generalized coordinates.

3. Write the total kinetic energy for the system.
4. Write the total potential energy for the system.
5. Form the lagrangian function $\mathscr{L}$.
6. Perform the differentiation of $\mathscr{L}$.
7. Using the procedure discussed in Section 8.3, find the generalized forces for those forces not derivable from a potential function.
8. Substitute the results of steps 6 and 7 into Eq. (8.23). We should obtain as many equations of motion as there are degrees of freedom in the system.

Example 8.4

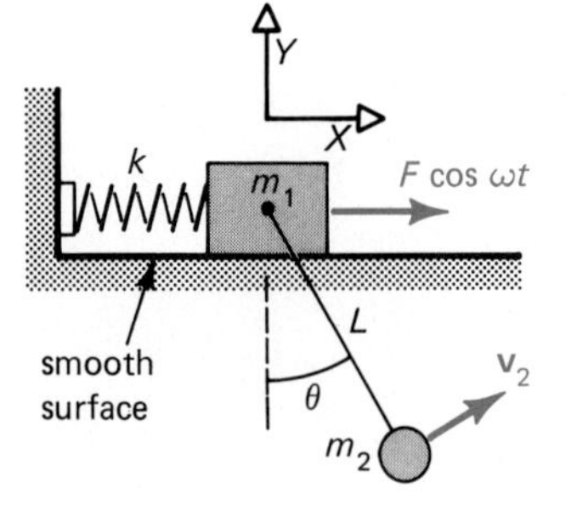

Figure 8.6

Figure 8.6 shows a simple pendulum of length L and mass m_2 pinned to a mass m_1 which is driven by an applied force $F \cos \omega t$. The motion of m_1 is constrained by the spring of modulus k. Find Lagrange's equations of motion for this system using the eight steps outlined above.

1. The mass m_1 is constrained to move in one direction while mass m_2 behaves like a simple pendulum. Hence, there are two degrees of freedom for the system.
2. Let $q_1 = x$ and $q_2 = \theta$, where x is measured from the unstretched position of the spring.
3. $$T = \tfrac{1}{2}m_1\dot{x}^2 + \tfrac{1}{2}m_2 v_2^2$$

   where v_2 is the absolute velocity of m_2. We wish to express v_2 in terms of the generalized coordinates. Using a moving reference frame attached to m_1, we have

   $$\begin{aligned}\mathbf{v}_2 &= \dot{x}\mathbf{i} + L\dot{\theta}(\cos\theta\mathbf{i} + \sin\theta\mathbf{j})\\ &= (\dot{x} + L\dot{\theta}\cos\theta)\mathbf{i} + L\dot{\theta}\sin\theta\mathbf{j}\end{aligned}$$

   therefore

   $$\begin{aligned}v_2^2 = \mathbf{v}_2 \cdot \mathbf{v}_2 &= (\dot{x} + L\dot{\theta}\cos\theta)^2 + (L\dot{\theta}\sin\theta)^2\\ &= \dot{x}^2 + L^2\dot{\theta}^2 + 2L\dot{x}\dot{\theta}\cos\theta\end{aligned}$$

   The kinetic energy is now expressed in terms of the generalized coordinates.

   $$T = \tfrac{1}{2}m_1\dot{x}^2 + \tfrac{1}{2}m_2(\dot{x}^2 + L^2\dot{\theta}^2 + 2L\dot{x}\dot{\theta}\cos\theta)$$

4. The potential energy includes the potential energy stored in the spring and the potential energy of mass m_2 due to gravity.

   $$V = \tfrac{1}{2}kx^2 + m_2 gL(1 - \cos\theta)$$

5.
$$\mathcal{L} = \tfrac{1}{2}m_1\dot{x}^2 + \tfrac{1}{2}m_2(\dot{x}^2 + L^2\dot{\theta}^2 + 2L\dot{x}\dot{\theta}\cos\theta) - \tfrac{1}{2}kx^2 - m_2 gL(1 - \cos\theta)$$

6.
$$\frac{\partial \mathcal{L}}{\partial \dot{x}} = m_1\dot{x} + m_2\dot{x} + m_2 L\dot{\theta}\cos\theta$$

$$\frac{\partial \mathcal{L}}{\partial x} = -kx$$

$$\frac{\partial \mathcal{L}}{\partial \dot{\theta}} = m_2 L^2\dot{\theta} + m_2 L\dot{x}\cos\theta$$

$$\frac{\partial \mathcal{L}}{\partial \theta} = -m_2 L\dot{x}\dot{\theta}\sin\theta - m_2 gL\sin\theta$$

7. The applied force is not derivable from a potential function. Consequently, calculate the virtual work due to this force, first allowing δx to occur keeping θ fixed.

$$\delta W = (F\cos\omega t)\,\delta x$$

$$\frac{\delta W}{\delta x} = F\cos\omega t = Q_1 = Q_x$$

Now vary θ, and hold x fixed. The resulting virtual work $\delta W = 0$ for the applied force. Therefore,

$$Q_2 = Q_\theta = 0$$

8. Lagrange's equations of motion are

$$\frac{d}{dt}(m_1\dot{x} + m_2\dot{x} + m_2 L\dot{\theta}\cos\theta) + kx = F\cos\omega t$$

$$\frac{d}{dt}(m_2 L^2\dot{\theta} + m_2 L\dot{x}\cos\theta) + m_2 L\dot{x}\dot{\theta}\sin\theta + m_2 gL\sin\theta = 0$$

Answer

Example 8.5

The spring-pendulum combination of mass m shown in Fig. 8.7 has a force $F\cos\omega t$ applied at A and moment $M\cos\lambda t$ applied to the pendulum. Point A is constrained to move in the vertical direction. Derive Lagrange's equations of motion using the eight steps outlined previously.

1. The pendulum mass m has two degrees of freedom.
2. Let $q_1 = x$, $q_2 = \theta$ and L_0 equal the unstretched length of the spring.

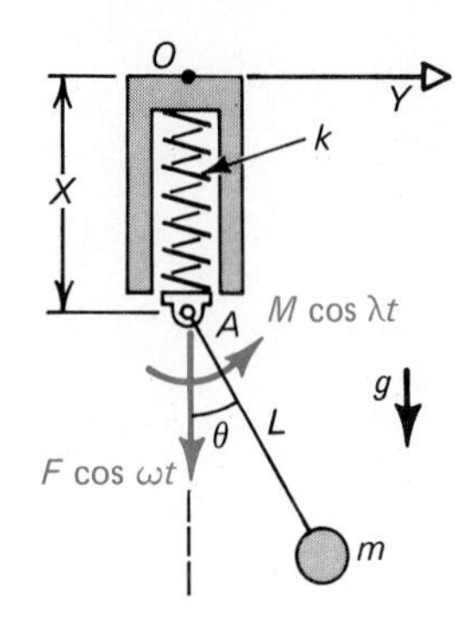

Figure 8.7

3. $T = \frac{1}{2}mv^2$, where

$$\mathbf{v} = \dot{x}\mathbf{i} + L\dot{\theta}[(-\sin\theta)\mathbf{i} + (\cos\theta)\mathbf{j}]$$

$$v^2 = \mathbf{v}\cdot\mathbf{v} = \dot{x}^2 + L^2\dot{\theta}^2 - 2L\dot{x}\dot{\theta}\sin\theta$$

The kinetic energy is now expressed in terms of the generalized coordinates.

$$T = \tfrac{1}{2}m(\dot{x}^2 + L^2\dot{\theta}^2 - 2L\dot{x}\dot{\theta}\sin\theta)$$

4. The potential energy includes the potential energy stored in the spring and the gravitational potential energy of the mass m.

$$V = \tfrac{1}{2}k(x - L_0)^2 - mg(x + L\cos\theta)$$

Note that the gravitational potential energy is taken relative to the fixed reference point at O.

5. $$\mathscr{L} = \tfrac{1}{2}m(\dot{x}^2 + L^2\dot{\theta}^2 - 2L\dot{x}\dot{\theta}\sin\theta) - \tfrac{1}{2}k(x - L_0)^2 + mg(x + L\cos\theta)$$

6. $$\frac{\partial\mathscr{L}}{\partial\dot{x}} = m\dot{x} - mL\dot{\theta}\sin\theta$$

$$\frac{\partial\mathscr{L}}{\partial x} = -k(x - L_0) + mg$$

$$\frac{\partial\mathscr{L}}{\partial\dot{\theta}} = mL^2\dot{\theta} - mL\dot{x}\sin\theta$$

$$\frac{\partial\mathscr{L}}{\partial\theta} = -mL\dot{x}\dot{\theta}\cos\theta - mgL\sin\theta$$

7. Allow δx to occur while holding θ fixed. The virtual work due to the nonconservative force $F\cos\omega t$ is $\delta W = (F\cos\omega t)\,\delta x$. Therefore,

$$Q_1 = Q_x = F\cos\omega t$$

Allow $\delta\theta$ to occur with x fixed. $\delta W = (M\cos\lambda t)\,\delta\theta$, and therefore

$$Q_2 = Q_\theta = M\cos\lambda t$$

8. Lagrange's equations of motion are:

$$\frac{d}{dt}(m\dot{x} - mL\dot{\theta}\sin\theta) + k(x - L_0) - mg = F\cos\omega t$$

$$\frac{d}{dt}(mL^2\dot{\theta} - mL\dot{x}\sin\theta) + mL\dot{x}\dot{\theta}\cos\theta + mgL\sin\theta = M\cos\lambda t$$

Answer

PROBLEMS

8.21 Obtain Lagrange's equations of motion for Problem 8.1.

8.22 Obtain Lagrange's equations of motion for Problem 8.2.

8.23 Obtain the lagrangian function for Problem 8.3.

8.24 Obtain Lagrange's equations of motion for Problem 8.4.

8.25 Obtain Lagrange's equations of motion for Problem 8.5.

8.26 Obtain Lagrange's equations of motion for Problem 8.6. Simplify for small displacements.

8.27 Obtain Lagrange's equations of motion for Problem 8.7. Simplify for small displacements.

8.28 Obtain the lagrangian function for Problem 8.10.

8.29 Obtain the lagrangian function for the system shown in Fig. P8.8.

8.30 Two masses are connected in series by springs as shown in Fig. P8.30. An applied force $F_0 \cos \lambda t$ acts on mass m_2. Assuming no friction forces acting on the masses, find Lagrange's equations of motion.

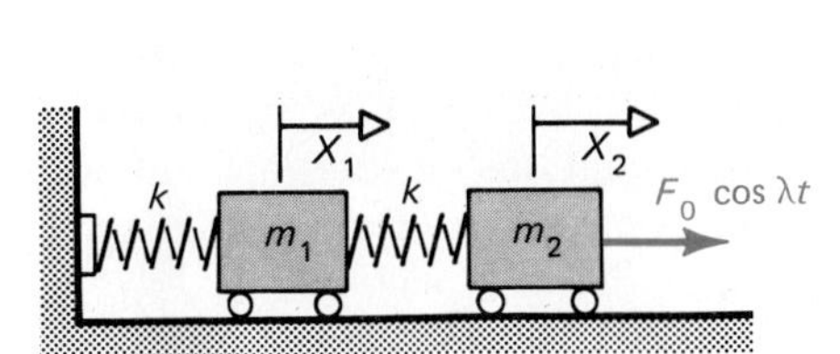

Figure P8.30

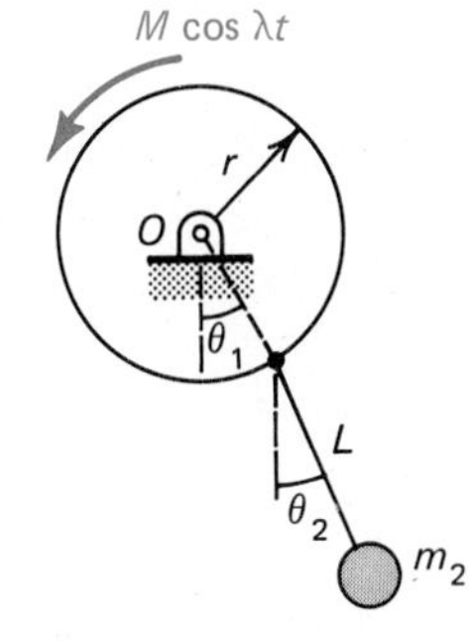

Figure P8.31

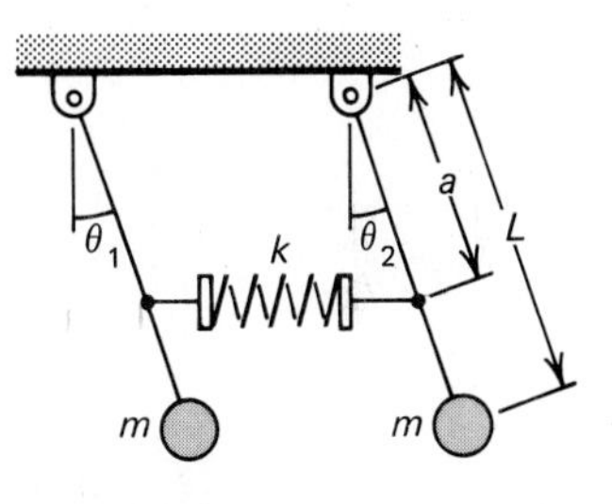

Figure P8.32

8.31 Find the lagrangian function for the system in Fig. P8.31. The configuration consists of a drive wheel of mass m_1 and mass moment of inertia I about the rotational axis through O, and an attached simple pendulum of mass m_2.

8.32 Write Lagrange's equations of motion for the coupled pendula shown in Fig. P8.32. Assume the spring is unstretched in the static equilibrium position and assume small displacements.

8.33 A mass m_1 is driven by an external force and slides on a smooth horizontal surface as shown in Fig. P8.33. Attached is a simple pendulum of mass m_2. Write the equations of motion for the system.

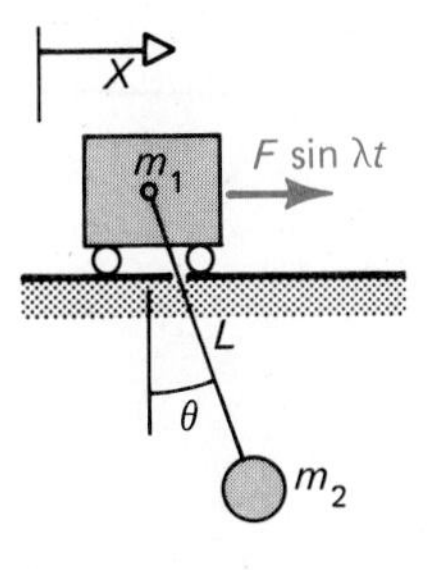

Figure P8.33

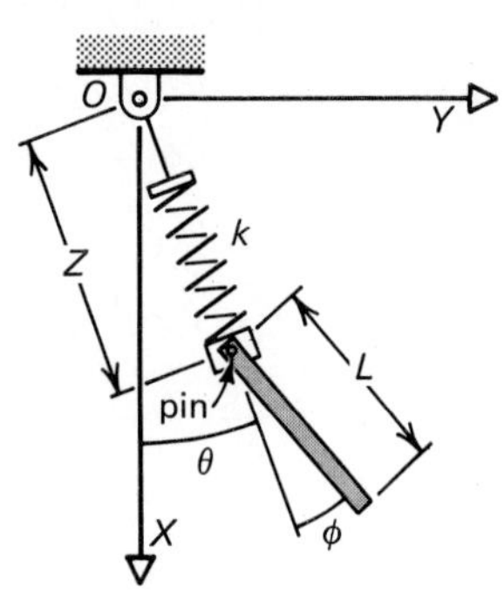

Figure P8.34

8.34 A thin rod of mass m and length L is attached to a spring of modulus k and unstretched length L_0, as shown in Fig. P8.34. Write the lagrangian function for motion confined to the XY plane.

8.5 Euler's Equations

In the previous section we developed Lagrange's equations for the motion of a particle and used these to find equations for the motion of a system of particles. It should be noted that these equations can also be used to study the motion of rigid bodies. For more involved problems, this approach usually incorporates a coordinate system called the Euler angle coordinates, which will not be presented here due to the complexities involved. However, Euler's fundamental equations of motion, which are introduced below, are adequate to the study of a large number of problems concerned with the rotational motion of rigid bodies.

As a start, let us recall the basic equations governing the motion of a rigid body.

$$\mathbf{F} = m\ddot{\mathbf{r}}_c \tag{8.24}$$

$$\mathbf{M} = \dot{\mathbf{H}} \tag{8.25}$$

where $\mathbf{F}$ is the resultant external force acting on the body of mass m, $\ddot{\mathbf{r}}_c$ is the absolute acceleration of the center of mass, $\mathbf{M}$ is the resultant moment of the external forces about a fixed point (or the center of mass), and $\dot{\mathbf{H}}$ is the time rate of change of the angular momentum of the body about the fixed point (or the center of mass). Since Euler's equations are a set of scalar equations for rotational motion of a rigid body about a fixed point or the center of mass of the body, they are essentially Eq. (8.25) written out in explicit detail.

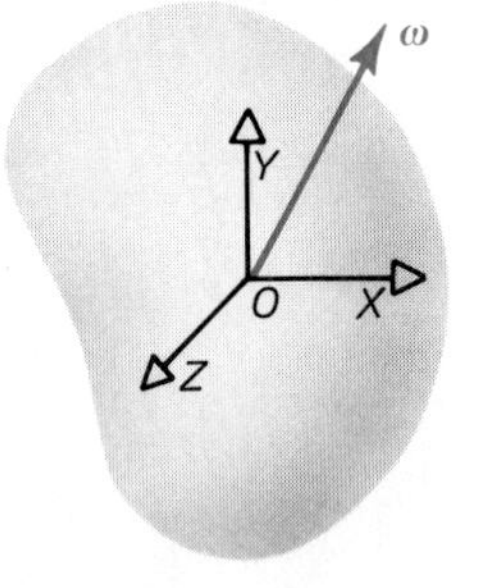

Figure 8.8

Suppose the rigid body shown in Fig. 8.8 has an absolute angular velocity $\boldsymbol{\omega}$ and the origin O of the coordinate system is either fixed or represents the center of mass of the rigid body. Further, assume that the coordinate system forms the principal axes of the body and that it is attached to the rigid body so that its angular velocity is $\boldsymbol{\omega}$. Because of this rotating reference, the time derivative of $\mathbf{H}$ is

$$\dot{\mathbf{H}} = \dot{\mathbf{H}}_r + \boldsymbol{\omega} \times \mathbf{H} \tag{8.26}$$

where

$$\mathbf{H} = H_x\mathbf{i} + H_y\mathbf{j} + H_z\mathbf{k}$$

$$\dot{\mathbf{H}}_r = \dot{H}_x\mathbf{i} + \dot{H}_y\mathbf{j} + \dot{H}_z\mathbf{k} \tag{8.27}$$

$$\boldsymbol{\omega} = \omega_x\mathbf{i} + \omega_y\mathbf{j} + \omega_z\mathbf{k}$$

Therefore,

$$\boldsymbol{\omega} \times \mathbf{H} = \begin{vmatrix} \mathbf{i} & \mathbf{j} & \mathbf{k} \\ \omega_x & \omega_y & \omega_z \\ H_x & H_y & H_z \end{vmatrix}$$

$$= (H_z\omega_y - H_y\omega_z)\mathbf{i} + (H_x\omega_z - H_z\omega_x)\mathbf{j} + (H_y\omega_x - H_x\omega_y)\mathbf{k} \tag{8.28}$$

Substituting Eqs. (8.27) and (8.28) into (8.26), gives us

$$\dot{\mathbf{H}} = (\dot{H}_x + H_z\omega_y - H_y\omega_z)\mathbf{i} + (\dot{H}_y + H_x\omega_z - H_z\omega_x)\mathbf{j} + (\dot{H}_z + H_y\omega_x - H_x\omega_y)\mathbf{k}$$

Consequently, Eq. (8.25) can be written as three scalar equations:

$$M_x = \dot{H}_x + H_z\omega_y - H_y\omega_z$$

$$M_y = \dot{H}_y + H_x\omega_z - H_z\omega_x$$

$$M_z = \dot{H}_z + H_y\omega_x - H_x\omega_y$$

But $\qquad H_x = I_x\omega_x \qquad H_y = I_y\omega_y \qquad H_z = I_z\omega_z$

since XYZ are the principal axes of the rigid body.

Therefore

$$M_x = I_x\dot{\omega}_x + (I_z - I_y)\omega_y\omega_z$$

$$M_y = I_y\dot{\omega}_y + (I_x - I_z)\omega_x\omega_z \tag{8.29}$$

$$M_z = I_z\dot{\omega}_z + (I_y - I_x)\omega_x\omega_y$$

Equations (8.29) are in the usual form of Euler's equations of motion. The main points to remember in using these equations are:

1. The origin is fixed or is the center of mass of the rigid body.
2. The XYZ axes are the principal axes of the rigid body.
3. The XYZ axes are fixed to the body (sometimes called a body-axis coordinate system).
4. ω_x, ω_y, ω_z are the components of the absolute angular velocity of the body.
5. M_x, M_y, M_z are the moments due to external forces acting on the body about their respective axes.
6. I_x, I_y, I_z are the principal mass moments of inertia of the rigid body.

One of the shortcomings of Euler's equations is that the integrals of ω_x, ω_y, and ω_z do not give the angular orientation of the body. Therefore, a complete solution to the problem requires use of the Euler angles, mentioned earlier, along with Lagrange's equations. This approach is developed in advanced texts.

Example 8.6

A thin rod of mass m and length l is attached to a massless bar AB that is attached to a rotating table as shown in Fig. 8.9(a). The table rotates at a constant angular speed Ω rad/s, while the rod rotates about AB at a constant rate p rad/s as shown in the figure. Find the moments transmitted to the bearings at the supports A and B.

Figure 8.9(b) shows the X axis as a principal axis along the length of the rod while the Z axis is along the bar AB. The absolute angular velocity components projected onto the body-axis system will include the angular velocity of the table Ω as well as p. Referring to Fig. 8.9(c), we see that

$$\omega_x = \Omega \sin pt \qquad \omega_y = \Omega \cos pt \qquad \omega_z = p$$

$$\dot{\omega}_x = \Omega p \cos pt \qquad \dot{\omega}_y = -\Omega p \sin pt \qquad \dot{\omega}_z = 0$$

$$I_x = 0 \qquad I_y = I_z = \tfrac{1}{12} mL^2 = I$$

Substituting into Eqs. (8.29), we have

$$M_x = 0$$

$$M_y = -I\Omega p \sin pt - I\Omega p \sin pt = -2I\Omega p \sin pt$$

$$M_z = I\Omega^2 \sin pt \cos pt \qquad \textit{Answer}$$

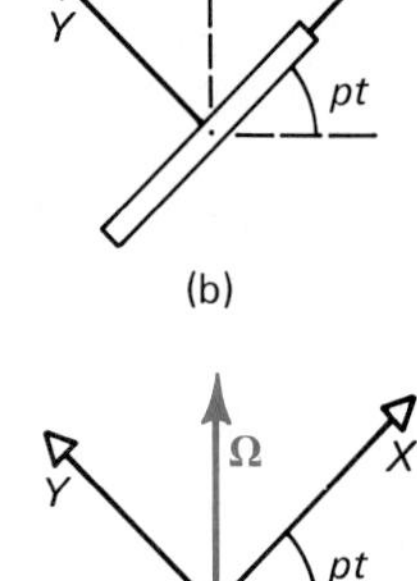

Figure 8.9

Example 8.7

A rigid body is spinning freely about its X axis at Ω rad/s. The body is symmetric so that $I_z = I_y = I$ and the applied moments are zero so that $M_x = M_y = M_z = 0$. Obtain an expression for the angular velocity of the rigid body. Assume $\omega_y = 0$ at $t = 0$ and $\omega_z = \omega_o$ at $t = 0$ as initial conditions.

Substituting $I_z = I_y = I$ and $M_x = M_y = M_z = 0$ into Eqs. (8.29) yields

$$I_x \dot{\omega}_x = 0$$

$$I\dot{\omega}_y + (I_x - I)\omega_x \omega_z = 0 \qquad \text{(a)}$$

$$I\dot{\omega}_z + (I - I_x)\omega_x \omega_y = 0$$

From the first equation

$$\omega_x = \text{constant} = \Omega$$

Defining $[(I_x/I) - 1] = \beta$ and substituting into the last two of Eqs. (a) yields,

$$\begin{aligned} \dot{\omega}_y + \beta\Omega\omega_z &= 0 \\ \dot{\omega}_z - \beta\Omega\omega_y &= 0 \end{aligned} \qquad \text{(b)}$$

These are two coupled first-order differential equations. They can be solved by taking the derivative of the first equation and substituting into the second equation to yield,

$$\ddot{\omega}_y + (\beta\Omega)^2\omega_y = 0$$

This is a second-order ordinary differential equation having the solution

$$\omega_y = A \sin(\beta\Omega t + \phi)$$

where A and ϕ are constants established by the initial conditions. Since $\omega_y = 0$ at $t = 0$, then $\phi = 0$ and

$$\omega_y = A \sin(\beta\Omega t)$$

Differentiating this result and substituting into the first of Eqs. (b) gives,

$$\omega_z = -A \cos(\beta\Omega t)$$

Using $\omega_z = \omega_o$ at $t = 0$ gives $A = -\omega_o$. Substituting this leads to $\omega_z = \omega_o \cos(\beta\Omega t)$. The angular velocity of the body is

$$\begin{aligned} \boldsymbol{\omega} &= \omega_x\mathbf{i} + \omega_y\mathbf{j} + \omega_z\mathbf{k} \\ &= \Omega\mathbf{i} - \omega_o \sin(\beta\Omega t)\mathbf{j} + \omega_o \cos(\beta\Omega t)\mathbf{k} \end{aligned} \qquad \textit{Answer}$$

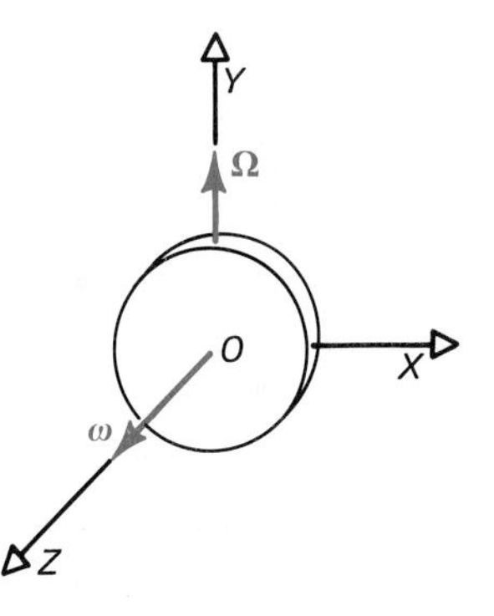

Figure P8.36

PROBLEMS

8.35 A rigid body, having one point fixed and no external torque about this point, has equal principal axes of inertia about the X and Y axes. What conclusion can you draw concerning its angular velocity about the Z axis?

8.36 A thin disk of mass m and radius r rotates at an angular rate ω rad/s about the Z axis as shown in Fig. P8.36. The X, Y, and Z axes are fixed to the disk. At the same time the X axis rotates slowly about the Y axis at an angular rate Ω rad/s. Compute the torques about the axes OX, OY, and OZ.

8.37 A sphere of mass m and radius r in Fig. P8.37 is mounted by a thin massless rod to a forked apparatus that is rotating at a constant angular rate Ω rad/s. At the same time, the sphere is spinning about its axis at a constant angular rate $\dot{\theta}$ rad/s. Neglecting the mass of the fork, find the torques transmitted to the rod.

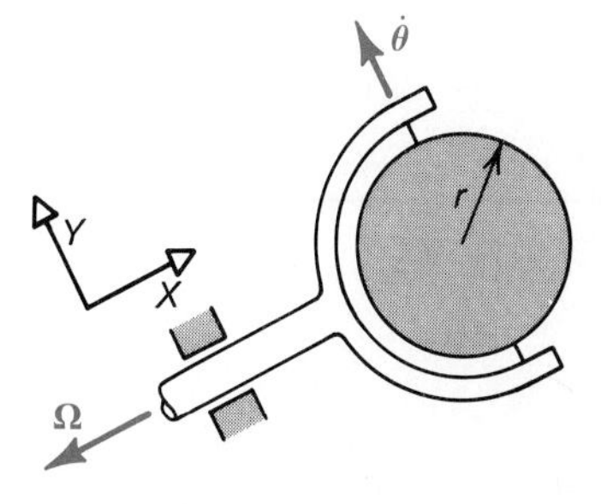

Figure P8.37

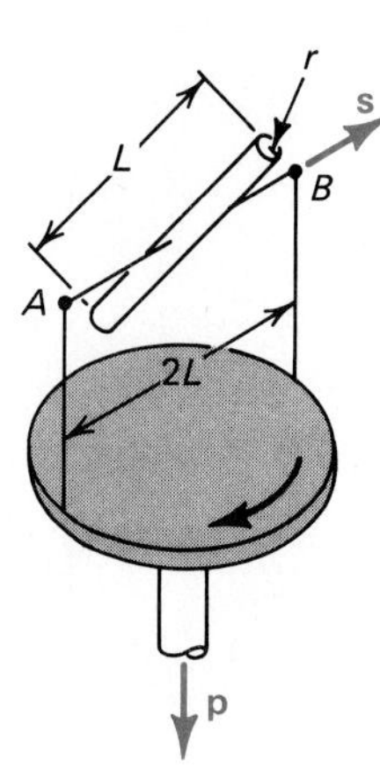

Figure P8.38

8.38 A cylinder of mass m is attached to a massless shaft AB which in turn is attached to a rotating table as shown in Fig. P8.38. The table rotates at a variable angular rate of p rad/s while the shaft AB rotates independently at a constant rate s rad/s as shown in the figure. Find the moments transmitted to the bearings at the supports A and B.

8.39 A sphere of mass m and radius r is attached to point O and is free to oscillate about O as shown in Fig. P8.39. However, point O is pinned to a vertical shaft that rotates at a constant angular rate Ω rad/s. Obtain expressions for the moments about point O.

8.40 Rework Example 8.7 for the initial conditions $\omega_y = \omega_0$ at $t = 0$ and $\dot{\omega}_y = 0$ at $t = 0$.

8.41 A fork oscillates in the plane of the page while a sphere is spinning about the shaft AB at a constant rate of p rad/s as shown in Fig. P8.41. Derive expressions for the moments about the X, Y, and Z axes.

8.42 The two masses shown in Fig. P8.42 move so that $\dot{\theta} = Bt$ rad/s, while the collar to which they are attached rotates at a constant rate ω rad/s. Derive the rotational equations of motion for one mass using the eulerian approach. Neglect the dimension a.

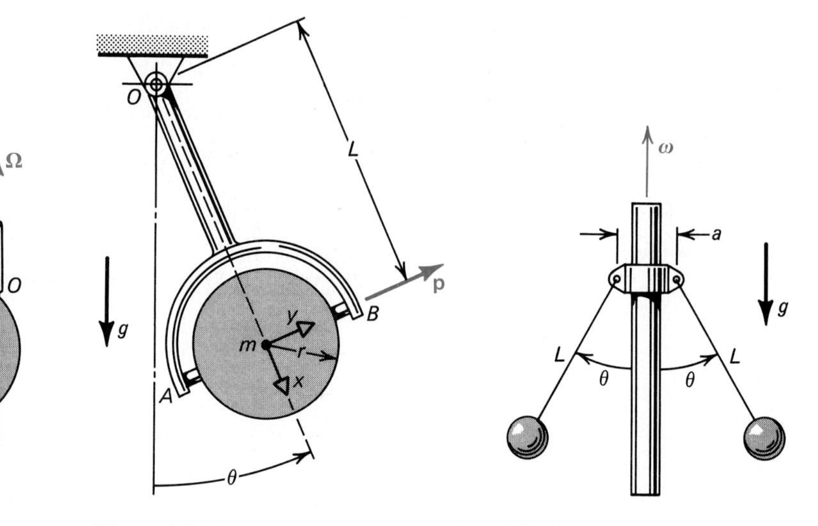

Figure P8.39 Figure P8.41 Figure P8.42

8.43 Check the equations derived in Problem 8.42 by using the lagrangian approach.

8.44 A nonsymmetrical rigid body is spinning about its X axis at Ω rad/s. Assuming that ω_y and $\omega_z \ll \Omega$ and that $I_x \gg I_y$ and $I_x \gg I_z$, obtain expressions for the three angular velocities. (*Hint*: Linearize your equations by using the above inequalities and first show that ω_x stays constant. Then use the approach of Example 8.7. Assume zero applied moments.)

8.6 Summary

The equations of particle and rigid body motion have been derived using the newtonian approach. Sometimes it is desirable to derive the equations of motion by considering the total energy or the momentum of the particles or bodies. This can be done via the lagrangian and eulerian approaches developed here.

A. The important *ideas*:
   1. Generalized coordinates are any convenient set of coordinates that describe the motion of moving bodies.
   2. Generalized forces include both forces and moments acting on a body.
   3. Conservative forces are derivable from a potential energy function.
   4. Lagrange's equations are equations of translational as well as rotational motion derived for systems by considering their kinetic energy, potential energy, and generalized forces.
   5. Euler's equations are a set of scalar equations for rotational motion of a rigid body about a fixed point or the center of mass of the body. The body axis coordinate system used in Euler's equations is fixed in the body and represents the principal axes of the rigid body.

B. The important *equations*:

Lagrange's equations:

$$\frac{d}{dt}\left(\frac{\partial \mathscr{L}}{\partial \dot{q}_i}\right) - \frac{\partial \mathscr{L}}{\partial q_i} = Q_i \tag{8.23}$$

$$\mathscr{L} = T - V$$

$$Q_i = \frac{\delta W}{\delta q_i}$$

Euler's equations:

$$\begin{aligned} M_x &= I_x\dot{\omega}_x + (I_z - I_y)\omega_y\omega_z \\ M_y &= I_y\dot{\omega}_y + (I_x - I_z)\omega_x\omega_z \\ M_z &= I_z\dot{\omega}_z + (I_y - I_x)\omega_x\omega_y \end{aligned} \tag{8.29}$$

9 Some Special Problems

The purpose of this chapter is to apply the fundamental concepts of dynamics to a wider range of problems than those treated in the first eight chapters. These problems are designed to generate an awareness that the solutions to many engineering problems are closely related. After we understand the physical forces and motions present in a problem, the notation, the units, the mathematics, and the analysis are frequently common in reaching a solution.

The concepts of the previous eight chapters will be used to investigate four different problems—particle motion due to electric and magnetic fields, equations of an electrical network, the vibration of a mass that is coupled to a spring and a damper, and the motion of rigid bodies. Fundamental notions will be reviewed when necessary and each problem will be developed in sufficient detail to emphasize its central idea.

9.1 Motion in Electric and Magnetic Fields

In investigating the dynamics of particles in an electric and/or magnetic field, we shall assume that the velocities are low enough so that the motion may be considered nonrelativistic. Recall that the force on a particle at rest, of charge q, due to an electrostatic field $\mathbf{E}$ is

$$\mathbf{F}_e = q\mathbf{E}$$

where q is in coulombs (C) and $\mathbf{E}$ is in newtons/coulomb (N/C). If the particle is moving in a magnetic field at a velocity $\mathbf{v}$, then there is an additional force

$$\mathbf{F}_m = q\mathbf{v} \times \mathbf{B}$$

where $\mathbf{B}$ is the magnetic field intensity in teslas (T); where 1 tesla = 1 kg/s·C. The total force on the particle of charge q is

$$\mathbf{F} = q\mathbf{E} + q\mathbf{v} \times \mathbf{B} \tag{9.1}$$

The force $\mathbf{F}$ is called the Lorentz force. From Newton's law we already know that

$$\mathbf{F} = m\mathbf{a} \tag{9.2}$$

so that the acceleration $\mathbf{a}$ of a charged particle of mass m is obtained by equating Eq. (9.1) and Eq. (9.2):

$$\mathbf{a} = \frac{1}{m}(q\mathbf{E} + q\mathbf{v} \times \mathbf{B}) \tag{9.3}$$

Notice that we have omitted the gravitational force. This is because the gravitational force is much weaker than the electrostatic and magnetic forces. The mass of an electron is 9.1×10^{-31} kg, while the mass of a proton and neutron is approximately 1.67×10^{-27} kg.

Finally, we recall that the current i is related to the charge, q, by

$$i = \frac{dq}{dt} \tag{9.4}$$

The unit for current is the ampere (A), where one ampere equals the quantity of electricity in coulombs that is transported in a conductor in one second.

Example 9.1

A particle of mass m moves in a magnetic field such that $\mathbf{B} = B\mathbf{k}$. Obtain an equation for the general motion of the particle. Assume that the velocity of the particle is $\mathbf{v} = v_x\mathbf{i} + v_y\mathbf{j} + v_z\mathbf{k}$

The acceleration can be obtained by substitution into Eq. (9.3):

$$m\mathbf{a} = q(v_x\mathbf{i} + v_y\mathbf{j} + v_z\mathbf{k}) \times B\mathbf{k}$$

or

$$(a_x\mathbf{i} + a_y\mathbf{j} + a_z\mathbf{k}) = \frac{q}{m}(v_yB\mathbf{i} - v_xB\mathbf{j})$$

Equating each component, we obtain

$$a_x = \frac{d^2x}{dt^2} = \frac{q}{m}Bv_y \tag{a}$$

$$a_y = \frac{d^2y}{dt^2} = -\frac{q}{m}Bv_x \tag{b}$$

$$a_z = \frac{d^2z}{dt^2} = 0 \qquad \text{(c)}$$

The last equation states that

$$\frac{dz}{dt} = A, \qquad \text{a constant}$$

or

$$z = At + A_0$$

and, if $z = 0$ at $t = 0$, then $A_0 = 0$ so that

$$z = At \qquad \textit{Answer} \quad \text{(d)}$$

Equations (a) and (b) are two coupled first-order linear differential equations that can be written as

$$\frac{d^2x}{dt^2} = \frac{dv_x}{dt} = \frac{q}{m} Bv_y \qquad \text{(e)}$$

$$\frac{d^2y}{dt^2} = \frac{dv_y}{dt} = -\frac{q}{m} Bv_x \qquad \text{(f)}$$

For $v_y = v_1$ at $t = 0$, the solution of Eqs. (e) and (f) has the form

$$v_x = v_1 \sin \omega t \qquad \textit{Answer} \quad \text{(g)}$$

$$v_y = v_1 \cos \omega t \qquad \textit{Answer} \quad \text{(h)}$$

where $\omega = qB/m$ and is known as the gyro frequency, and v_1 is the initial velocity in the y direction. The linear displacements $x(t)$ and $y(t)$ may be obtained by integrating Eqs. (g) and (h). The trajectory of this motion is shown in Fig. 9.1 where the velocity along the Z axis is constant and the trajectory looks like a helix.

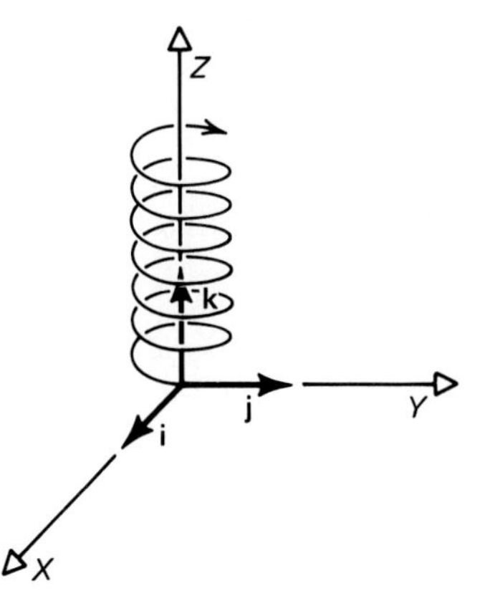

Figure 9.1

Example 9.2

Obtain the change of kinetic energy of a particle traveling through a constant magnetic field.

The kinetic energy T is given by

$$T = \tfrac{1}{2} m\mathbf{v} \cdot \mathbf{v}$$

so that the change in T becomes

$$\frac{dT}{dt} = \tfrac{1}{2} m[(\dot{\mathbf{v}} \cdot \mathbf{v}) + (\mathbf{v} \cdot \dot{\mathbf{v}})] = m\mathbf{v} \cdot \dot{\mathbf{v}}$$

Since there is only the magnetic field, Eq. (9.3) becomes

$$\dot{\mathbf{v}} = \mathbf{a} = \frac{q}{m} \mathbf{v} \times \mathbf{B}$$

so that
$$\frac{dT}{dt} = m\mathbf{v} \cdot \left(\frac{q}{m}\mathbf{v} \times \mathbf{B}\right) = 0 \qquad \textit{Answer}$$

Therefore, there is no change in the kinetic energy of a particle traveling through a constant magnetic field.

Example 9.3

Calculate the gyro frequency of a particle in a 1.5 T magnetic field. Assume the particle to be (a) an electron and (b) a proton.

The gyro frequency was obtained in Example 9.1 as

$$\omega = \frac{qB}{m}$$

where the elementary charge $q = 1.6 \times 10^{-19}$ C.

(a) For an electron,

$$\omega = \frac{1.6 \times 10^{-19} \times 1.5}{9.1 \times 10^{-31}}$$
$$= 2.64 \times 10^{11} \text{ rad/s} \qquad \textit{Answer}$$

(b) For a proton, the gyro frequency is

$$\omega = \frac{1.6 \times 10^{-19} \times 1.5}{1.67 \times 10^{-27}}$$
$$= 1.44 \times 10^{8} \text{ rad/s} \qquad \textit{Answer}$$

PROBLEMS

9.1 If the force acting on an electron is $2 \times 10^{-8}\mathbf{i}$ N, what is the electrostatic field?

9.2 Calculate the force acting on an electron moving in a magnetic field of $2\mathbf{i}$ T if the velocity $\mathbf{v} = 6 \times 10^6\mathbf{j}$ m/s.

9.3 If a charged particle is accelerated in an electrostatic field such that $\mathbf{E} = E \sin \omega t\mathbf{i}$, show that the displacement $x(t)$ is given by

$$x(t) = -\frac{q}{m\omega^2} E \sin \omega t + \frac{qE}{m\omega} t + x_0$$

where $x = x_0$ at $t = 0$ and $\dot{x} = 0$ at $t = 0$.

9.4 Obtain the equations of the trajectory of a particle of charge q if it is moving under a constant magnetic field $\mathbf{B}$.

9.5 An electron entering at a velocity $\dot{x}\mathbf{i}$ at $x = 0$ is deflected by a transverse constant electric field as shown in Fig. P9.5. Obtain the angle θ.

9.6 Calculate the kinetic energy of the particle in Problem 9.5 as it moves through a constant electrostatic field.

9.7 Verify the solution in Example 9.1 given by Eqs. (g) and (h) by substituting into Eqs. (e) and (f).

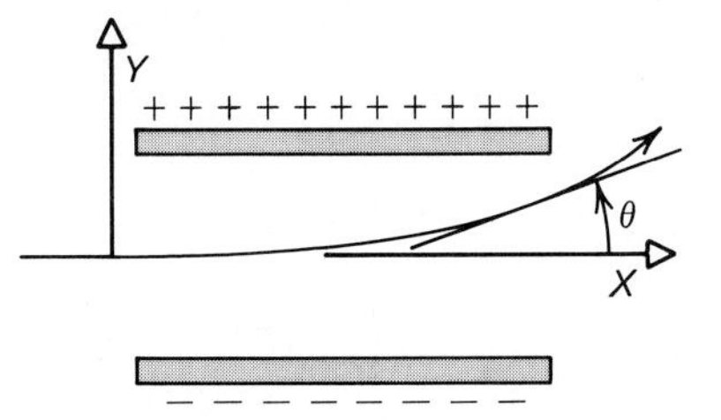

Figure P9.5

9.2 Mechanical Vibrations

Mechanical vibrating systems are frequently modeled by a combination of lumped masses, dampers, and springs. The simple mass-damper-spring model is the fundamental model that we wish to consider at this time. In this case, the mass has one degree of freedom and, as it vibrates, energy is stored in the spring while energy is dissipated by the damper as heat.

The spring shown in Fig. 9.2(a) provides a restoring force that is proportional to the relative displacement of the end terminals. For the case shown, we are assuming $x_1 > x_2$. The magnitude of the spring force is $k|x_1 - x_2|$.

The damper is shown in Fig. 9.2(b) in which the damper force is proportional to the relative velocity of the terminal points. Here, we are assuming $\dot{x}_1 > \dot{x}_2$. The magnitude of the damping force is $c|\dot{x}_1 - \dot{x}_2|$ where the units on c are usually N·s/m or lb·sec/ft.

Let us obtain the deflection x measured from the static equilibrium position for the mass-damper-spring model shown in Fig. 9.3(a). We shall assume that the mass is set into motion by an initial displacement x_0 or an initial velocity V_0. The free-body diagram is shown in Fig. 9.3(b). Since we are measuring the displacement of the mass from the static equilibrium position, the initial force in the spring cancels the weight mg. Hence, mg is not shown in the free-body diagram. Using Newton's second law,

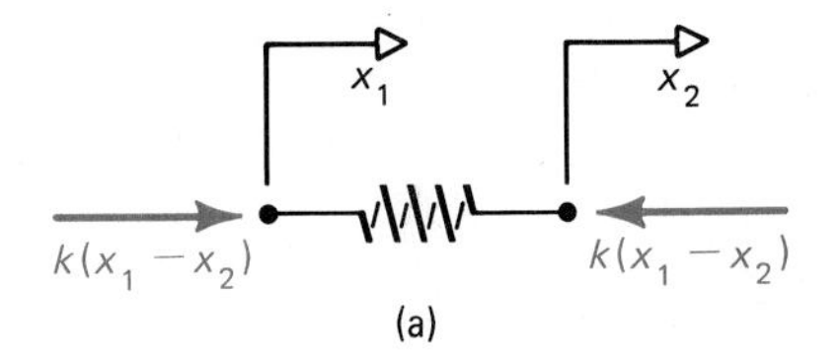

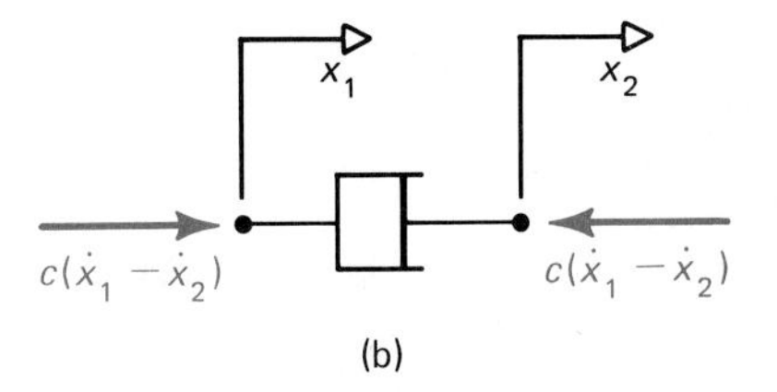

Figure 9.2

$$m\ddot{x} = -c\dot{x} - kx$$

or

$$\ddot{x} + \frac{c}{m}\dot{x} + \frac{k}{m}x = 0$$

With the definitions

$$\omega_n^2 - \frac{k}{m} \quad \text{and} \quad \xi = \frac{c}{2\sqrt{mk}}$$

this equation becomes

$$\ddot{x} + 2\xi\omega_n\dot{x} + \omega_n^2 x = 0 \tag{9.5}$$

where ω_n is the undamped natural frequency and ξ is the damping ratio. The auxiliary equation associated with Eq. (9.5) is

$$D^2 + 2\xi\omega_n D + \omega_n^2 = 0$$

which has the roots

$$-\xi\omega_n + \omega_n\sqrt{\xi^2 - 1} \qquad -\xi\omega_n - \omega_n\sqrt{\xi^2 - 1}$$

Naturally, the form of the solution to Eq. (9.5) depends upon the value of $\sqrt{\xi^2 - 1}$. The three possible situations are classified as follows:

(a)

$\sqrt{\xi^2 - 1}$ is real,	$\xi^2 > 1$, system is overdamped
$\sqrt{\xi^2 - 1}$ is zero,	$\xi^2 = 1$, system is critically damped
$\sqrt{\xi^2 - 1}$ is negative,	$\xi^2 < 1$, system is underdamped

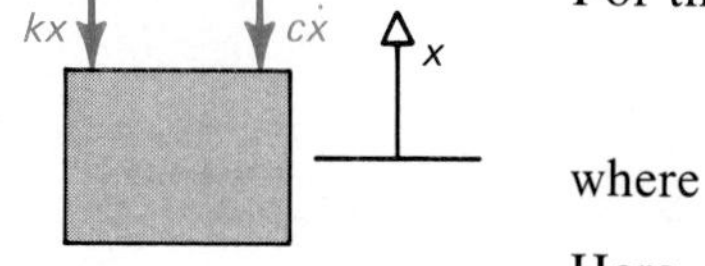

(b)

Figure 9.3

For the underdamped case, the roots become

$$-\xi\omega_n + j\omega_d \qquad -\xi\omega_n - j\omega_d$$

where

$$j = \sqrt{-1} \qquad \omega_d = \omega_n\sqrt{1 - \xi^2}$$

Here, ω_d is called the damped natural frequency. The solution for this case can be written as

$$x(t) = e^{-\xi\omega_n t}(A_1 \cos \omega_d t + A_2 \sin \omega_d t) \qquad \textit{Answer} \quad (9.6)$$

where A_1 and A_2 are determined from initial conditions. If,

$$\text{at } t = 0 \quad \begin{cases} x = 0 \\ \dot{x} = V_0 \end{cases}$$

then Eq. (9.6) becomes

$$x(t) = \frac{V_0}{\omega_d} e^{-\xi\omega_n t} \sin \omega_d t \qquad (9.7)$$

Physically, this means that the mass moves up and down about its static equilibrium position with decreasing amplitude. This amplitude decreases exponentially and the rate of decay is a function of the damping ratio ξ. A plot of $x(t)$ versus t is shown in Fig. 9.4(a). For critical damping and the overdamped case, $x(t)$ has the form shown in Fig. 9.4(b).

Equation (9.5) is called a homogenous differential equation. It is used in studying the motion of the mass about its equilibrium position when perturbed by an initial displacement or initial velocity. In many cases the mass is not only perturbed about its equilibrium position but a forcing function $F(t)$ is applied to the mass so that the right side of Eq. (9.5) becomes $F(t)/m$. The solution of such an equation is equal to the sum of the homogenous solution that was derived and a particular solution which is a function of $F(t)/m$.

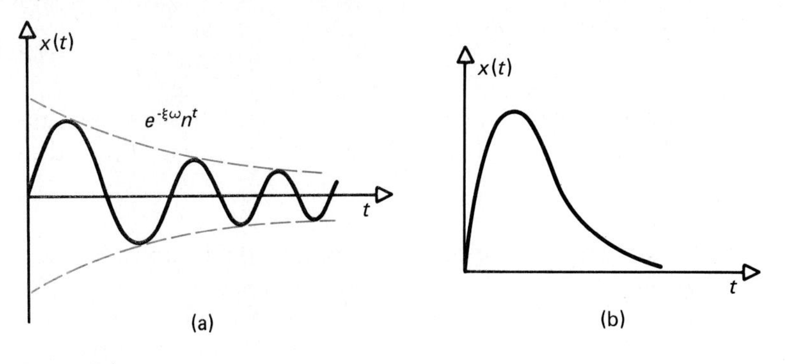

Figure 9.4

Example 9.4

What is the maximum displacement for a vibrating mass if $m = 25$ kg, $k = 100$ N/m, and $c = 10$ N·s/m; and the initial condition is $\dot{x} = V_0 = 0.1$ m/s?

We first compute ξ.

$$\xi = \frac{10}{2\sqrt{25 \times 100}} = 0.1$$

Since $\xi < 1$, the system is underdamped. Evaluating ω_n and ω_d, we have

$$\omega_n = \sqrt{\frac{k}{m}} = \sqrt{\frac{100}{25}} = 2$$

$$\omega_d = \omega_n\sqrt{1 - \xi^2} = 2\sqrt{0.99} = 1.990$$

The maximum displacement can be obtained from Eq. (9.7),

$$x(t) = \frac{V_0}{\omega_d} e^{-\xi\omega_n t} \sin \omega_d t$$

and occurs when $dx/dt = 0$. This leads to

$$\tan \omega_d t = \frac{\omega_d}{\xi\omega_n} = \frac{1.99}{0.1(2)} = 9.95$$

$$\omega_d t = 1.471$$

Substituting, we have the time when the maximum displacement occurs.

$$t = \frac{1.471}{1.990} = 0.739 \text{ s}$$

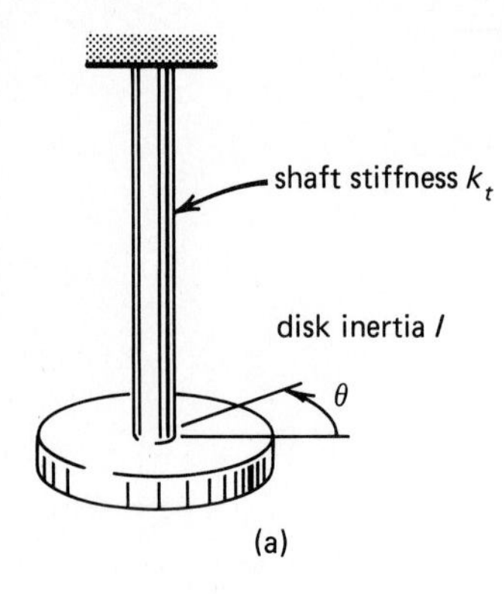

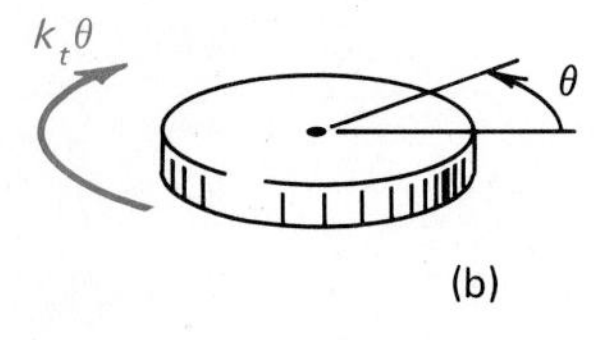

Figure 9.5

If we now substitute into $x(t)$, at $t = 0.739$,

$$x(t) = \frac{0.1}{1.990} e^{-0.1(2)(0.739)} \sin(1.471) = 0.0431 \text{ m}$$

Answer

Example 9.5

Figure 9.5(a) shows a torsional vibration model in which the disk has a mass moment of inertia of I about the longitudinal axis of the shaft, and the massless shaft provides a restoring moment expressed in terms of a torsional spring modulus k_t. Find the natural frequency of free vibrations for the system.

Figure 9.5(b) shows the free-body diagram of the disk. Let θ represent the angle of rotation of the disk about the longitudinal axis so that the restoring moment has a magnitude of $k_t\theta$ and a direction opposing the rotational motion. Now,

$$I\ddot{\theta} = -k_t\theta$$

or

$$I\ddot{\theta} + k_t\theta = 0$$

The solution for this undamped case is

$$\theta = A_1 \cos \omega_n t + A_2 \sin \omega_n t$$

where the undamped natural frequency is

$$\omega_n = \sqrt{\frac{k_t}{I}}$$

Answer

We notice that this expression is similar to the mass-spring model in that k_t is analogous to k, and I is analogous to m.

PROBLEMS

9.8 Derive Eq. (9.5) using Lagrange's equation.

9.9 Show that for the mass-damper-spring system in Fig. 9.3

$$x(t) = (A_1 + A_2 t)e^{-\omega_n t}$$

if the system is critically damped.

9.10 Show that for the system in Fig. 9.3.

$$x(t) = e^{-\xi\omega_n t}(A_1 e^{-\omega_m t} + A_2 e^{\omega_m t})$$

if the system is overdamped. We define ω_m as $\omega_m = \omega_n\sqrt{\xi^2 - 1}$.

9.11 Find the response $x(t)$ for the mass-damper-spring system in Fig. 9.3 if at $t = 0$, $x = x_0$ and $\dot{x} = v_0$. Assume an underdamped system.

9.12 A package of mass m is protected from mechanical shock by polyurethane. The mathematical model of the system is shown in Fig. P9.12. Obtain the equations of motion for the system vibrating freely.

9.13 Obtain the equations of motion for the two-mass system shown in Fig. P9.13. Neglect the friction at the wheels.

9.14 An automobile suspension system is modeled as shown in Fig. P9.14. The automobile, of mass m, has a mass moment of inertia of I about the center of mass. The tires and suspension system behave as springs of moduli k_1 and k_2. Obtain the equations of motion for small displacements about the static equilibrium position.

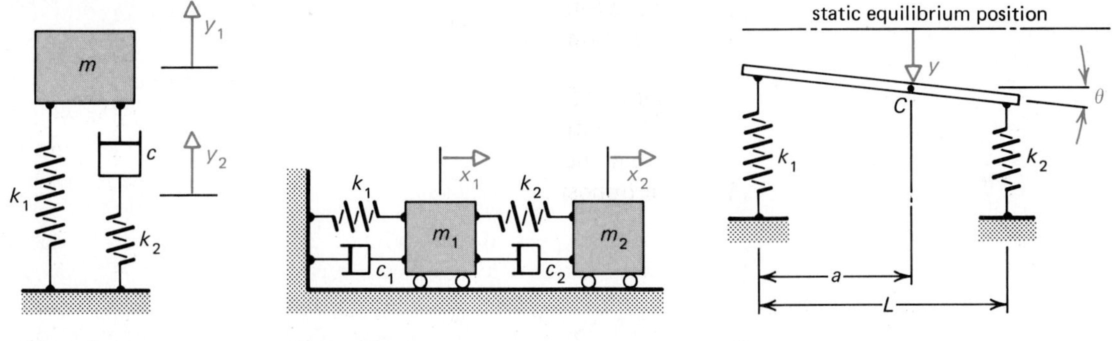

Figure P9.12 Figure P9.13

Figure P9.14

9.15 The mass-damper-spring system shown in Fig. P9.15 is subjected to a moving base described by the function $z(t)$. Obtain the equation of the motion of the mass.

9.16 Suppose the period of free vibration for the torsional model in Fig. 9.5 is measured at 0.2 s. If $k_t = 20$ m·N/rad, calculate the mass moment of inertia of the disk.

9.17 A torsional vibrating system consists of two disks connected by massless shafts as shown in Fig. P9.17. The shafts act as torsional springs each of modulus k_t (m·N/rad). If the mass moment of inertia of each disk about the longitudinal axis is I, obtain the equations of motion.

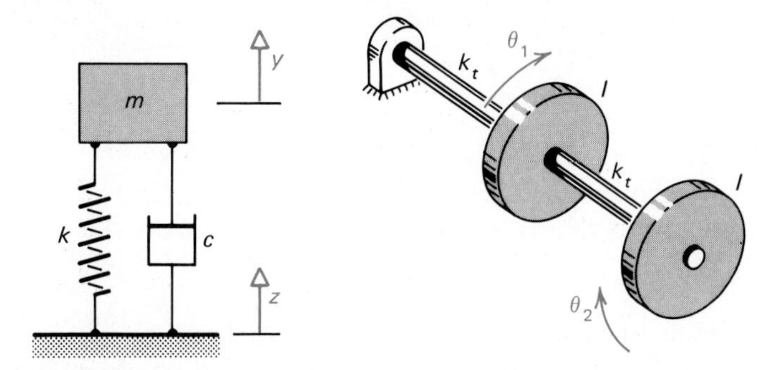

Figure P9.15 Figure P9.17

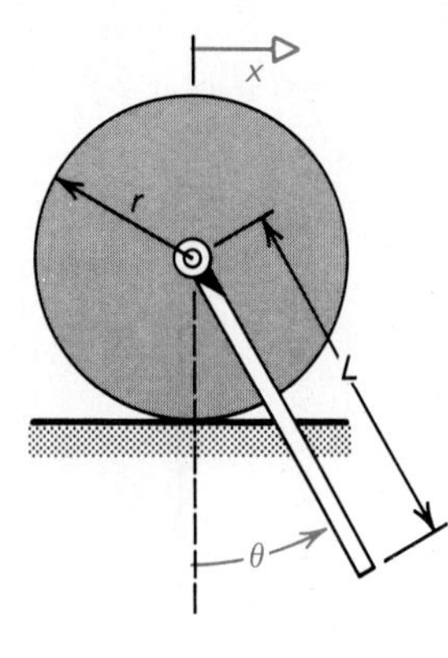

Figure P9.18

9.18 A cylinder of mass M and radius r has a rod of mass m and length L pinned to it as shown in Fig. P9.18. If the cylinder rolls without slipping, obtain the equations of motion using Lagrange's equations.

9.3 Electrical Networks

The lagrangian method developed in the previous chapter, although derived for mechanical problems, can be used to derive equations for electrical networks and may, if the network is complex, save time over the loop or the node method.

In Chapter 8 Lagrange's equations were shown to be

$$\frac{d}{dt}\left(\frac{\partial \mathscr{L}}{\partial \dot{q}_i}\right) - \frac{\partial \mathscr{L}}{\partial q_i} = Q_i \tag{8.23}$$

where q_i is a generalized coordinate, Q_i is a generalized force, and $\mathscr{L} = T - V$. Recall that in obtaining $\mathscr{L}$ we included only systems that did not contain any dissipation terms. Since energy dissipation occurs in most real systems, one can take this into account in the lagrangian formulation by the generalized forces Q_i or by a dissipation function. If the damping constant (this can be a mechanical damper or an electrical resistor) is denoted by R, then we define a dissipation function D such that

$$D = \tfrac{1}{2}R\dot{q}_i^2 \tag{9.8}$$

Sometimes D is called the Rayleigh dissipation function. Lagrange's equations for systems that contain this dissipation function can be written

$$\frac{d}{dt}\left(\frac{\partial \mathscr{L}}{\partial \dot{q}_i}\right) - \frac{\partial \mathscr{L}}{\partial q_i} + \frac{\partial D}{\partial \dot{q}_i} = Q_i \tag{9.9}$$

Although this can be rigorously derived, we shall content ourselves with its application.

Example 9.6

Derive the governing equation for the circuit shown in Fig. 9.6. Let i be the current, l the inductance, C the capacitance, and R the resistance.

While there are several methods available to obtain the governing equations, we shall apply Eq. (9.9). Since the inductor stores kinetic energy, the kinetic energy may be written in terms of the electrical charge q as

$$T = \tfrac{1}{2} l\dot{q}^2$$

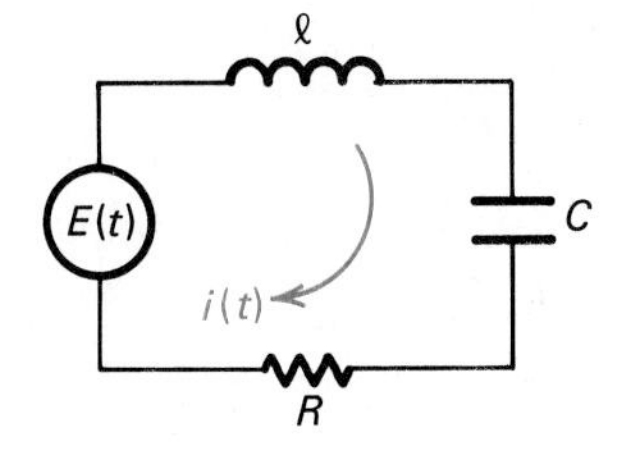

Figure 9.6

The capacitor stores potential energy, so that

$$V = \frac{1}{2C} q^2$$

Since the resistance dissipates energy as heat,

$$D = \tfrac{1}{2} R\dot{q}^2$$

The governing equation is

$$\frac{d}{dt}\left(\frac{\partial \mathscr{L}}{\partial \dot{q}}\right) - \frac{\partial \mathscr{L}}{\partial q} + \frac{\partial D}{\partial \dot{q}} = Q$$

where $Q = E(t)$, and

$$\mathscr{L} = \frac{1}{2} l\dot{q}^2 - \frac{1}{2C} q^2$$

Now $\qquad \dfrac{\partial \mathscr{L}}{\partial \dot{q}} = l\dot{q} \qquad \dfrac{\partial \mathscr{L}}{\partial q} = -\dfrac{q}{C} \qquad \dfrac{\partial D}{\partial \dot{q}} = R\dot{q}$

The governing equation becomes

$$l\ddot{q} + R\dot{q} + \frac{1}{C} q = E \qquad \textit{Answer}$$

We observe that the governing differential equation for the electrical charge of the basic circuit in Fig. 9.6 is analogous to the mechanical system shown in Fig. 9.3. In the analogy between the electrical and mechanical system, l is related to m, R to c, $1/C$ to k, q to x, and E to an applied force F.

Example 9.7

Obtain the governing equation for the circuit shown in Fig. 9.7.

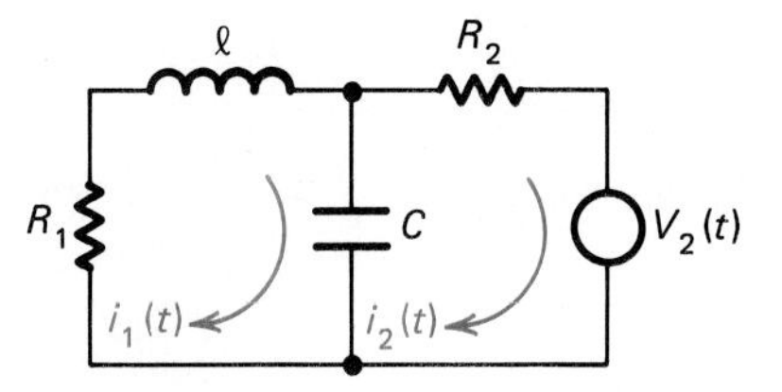

Figure 9.7

The inductor stores kinetic energy, the capacitor stores potential energy, and the resistances dissipate energy as heat. Therefore, in terms of the electric charges

$$T = \tfrac{1}{2} l\dot{q}_1^2$$

$$V = \frac{1}{2C}(q_1 - q_2)^2$$

$$D = \tfrac{1}{2} R_1\dot{q}_1^2 + \tfrac{1}{2} R_2\dot{q}_2^2$$

If the generalized coordinates are q_1 and q_2 then the two equations are

$$\frac{d}{dt}\left(\frac{\partial \mathscr{L}}{\partial \dot{q}_1}\right) - \frac{\partial \mathscr{L}}{\partial q_1} + \frac{\partial D}{\partial \dot{q}_1} = Q_1 \tag{a}$$

$$\frac{d}{dt}\left(\frac{\partial \mathscr{L}}{\partial \dot{q}_2}\right) - \frac{\partial \mathscr{L}}{\partial q_2} + \frac{\partial D}{\partial \dot{q}_2} = Q_2 \tag{b}$$

where $Q_1 = 0$ and $Q_2 = V_2(t)$ and the lagrangian $\mathscr{L}$ is given by

$$\mathscr{L} = \frac{1}{2} l\dot{q}_1^2 - \frac{1}{2C}(q_1 - q_2)^2$$

Substituting into Eqs. (a) and (b) gives us

$$l\ddot{q}_1 + \frac{1}{C}(q_1 - q_2) + R_1\dot{q}_1 = 0$$

$$-\frac{1}{C}(q_1 - q_2) + R_2\dot{q}_2 = V_2(t) \qquad \textit{Answer}$$

As an alternate form we can substitute Eq. (9.4) to obtain the governing equations in terms of the current i.

$$l\frac{di_1}{dt} + \frac{1}{C}\int(i_1 - i_2)\,dt + R_1 i_1 = 0 \tag{c}$$

$$-\frac{1}{C}\int(i_1 - i_2)\,dt + R_2 i_2 = V_2(t) \tag{d}$$

PROBLEMS

9.19 Write the lagrangian function for the network shown in Fig. P9.19.

9.20 Using Lagrange's equations and the Rayleigh dissipation function, find the equations of motion for the system shown in Fig. P9.20.

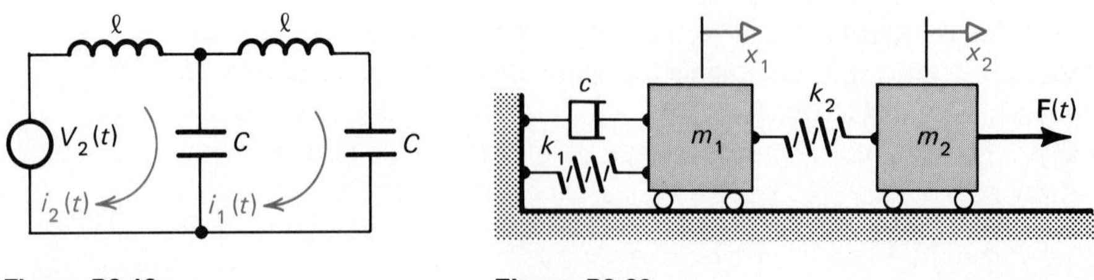

Figure P9.19

Figure P9.20

9.21 Write the equations of motion for the systems shown in Fig. P9.21. Are these two systems dynamically equivalent?

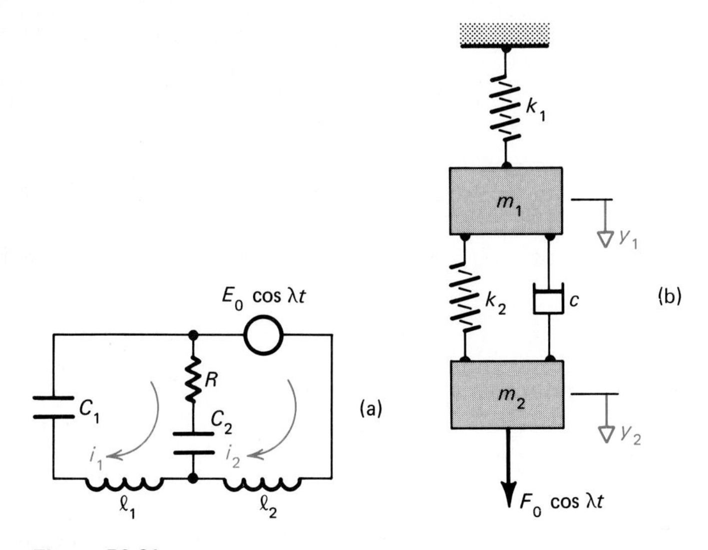

Figure P9.21

9.22 Obtain the equations of motion of the system in Fig. P9.22 using the lagrangian function. Assume translational motion only and a Rayleigh dissipation function between the masses. The friction c is developed by mass m moving in the casing.

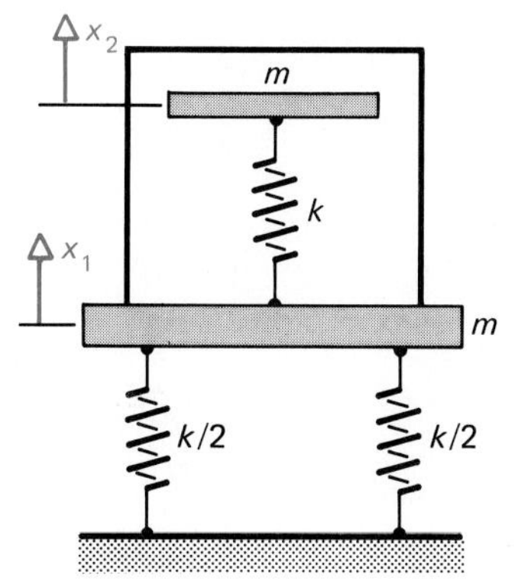

Figure P9.22

9.23 Verify Eqs. (c) and (d) in Example 9.7 using the node or loop method of analysis.

9.4 General Motion of a Rigid Body

In general, a rigid body can have three degrees of freedom due to translational motion and three additional degrees of freedom due to rotational motion. Quite often these motions are coupled and the governing equations are nonlinear. Even when rigid bodies undergo purely rotational or oscillatory motions, e.g., spinning tops, gyroscopes, satellites, etc., the dynamic equations are generally nonlinear and nontrivial to analyze. The study of the general motion of rigid bodies is, therefore, quite appropriately relegated to advanced texts. Nevertheless, there are some problems which, if carefully posed, are amenable to analysis with the tools we have developed so far. The rigid-body problems treated in earlier chapters had relatively straightforward solutions. Here, we consider two problems in which we develop their governing equations using the lagrangian and eulerian approaches.

Example 9.8

Consider a bar of mass m connected by a frictionless pin rotating at a constant rate of $\dot{\Phi}$ rad/s as shown in Fig. 9.8. (a) Obtain the equa-

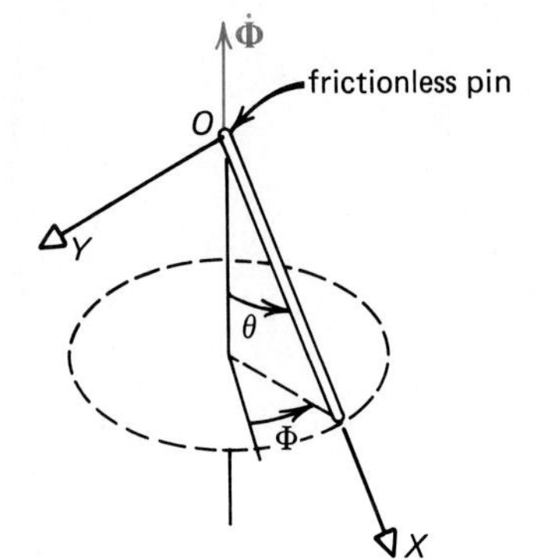

Figure 9.8

tion of motion of the bar using Lagrange's equations. (b) Assuming that θ is small, solve for $\theta(t)$. Assume the bar to be very slender with uniform mass distribution.

(a) For the bar we select a coordinate system such that X and Z are in the plane containing θ. The Y axis is perpendicular to this plane. The absolute rotational velocity of this coordinate system is $\boldsymbol{\omega}$.

$$\begin{aligned}\boldsymbol{\omega} &= \omega_x\mathbf{i} + \omega_y\mathbf{j} + \omega_z\mathbf{k} \\ &= (-\dot{\Phi}\cos\theta)\mathbf{i} + \dot{\theta}\mathbf{j} + (-\dot{\Phi}\sin\theta)\mathbf{k}\end{aligned} \tag{a}$$

The mass moments of inertia about the origin O are

$$\begin{aligned} I_x &= 0 \\ I_y &= I_z = \tfrac{1}{3}mL^2 \end{aligned} \tag{b}$$

Since the XYZ axes are principal axes, the kinetic energy is

$$T = \tfrac{1}{2}(I_x\omega_x^2 + I_y\omega_y^2 + I_z\omega_z^2) \tag{c}$$

Substituting Eqs. (a) and (b) into Eq. (c) we obtain

$$T = \tfrac{1}{2}I_y\dot{\theta}^2 + \tfrac{1}{2}I_z\dot{\Phi}^2\sin^2\theta \tag{d}$$

The potential energy of the bar may be expressed as

$$V = \frac{mgL}{2}(1 - \cos\theta) \tag{e}$$

The lagrangian function is $\mathscr{L} = T - V$, and substituting Eqs. (d) and (e) yields

$$\mathscr{L} = \frac{1}{2}I_y\dot{\theta}^2 + \frac{1}{2}I_z\dot{\Phi}^2\sin^2\theta - \frac{mgL}{2}(1 - \cos\theta) \tag{f}$$

The only generalized coordinate is θ, since $\dot{\Phi}$ is constant, so that the Lagrange equations reduce to

$$\frac{d}{dt}\left(\frac{\partial\mathscr{L}}{\partial\dot{\theta}}\right) - \frac{\partial\mathscr{L}}{\partial\theta} = 0 \tag{g}$$

Substituting Eq. (f) into Eq. (g) yields

$$I_y\ddot{\theta} - I_z\dot{\Phi}^2\sin\theta\cos\theta + \frac{mgL}{2}\sin\theta = 0$$

When we substitute for I_y and I_z from Eq. (b) and simplify, we have the required equation of motion.

$$\ddot{\theta} + \left(\frac{3g}{2L} - \dot{\Phi}^2\cos\theta\right)\sin\theta = 0 \qquad \textit{Answer} \tag{h}$$

(b) Now if θ is small,

$$\sin\theta \approx \theta \qquad \cos\theta \approx 1$$

If we substitute this into Eq. (h), we obtain

$$\ddot{\theta} + \left(\frac{3g}{2L} - \dot{\Phi}^2\right)\theta = 0 \tag{i}$$

The equation of motion is now a linearized equation. When a frequency ω is defined such that

$$\omega^2 = \left(\frac{3g}{2L} - \dot{\Phi}^2\right)$$

Eq. (i) becomes

$$\ddot{\theta} + \omega^2\theta = 0$$

This equation is similar to the equation of motion of a pendulum having a solution that is harmonic. Assuming $\theta(0) = \theta_0$ and $\dot{\theta}(0) = 0$,

$$\theta(t) = \theta_0 \cos\omega t \qquad \textit{Answer}$$

Notice that since ω^2 must be positive for harmonic motion, we require that $3g/2L > \dot{\Phi}^2$. If $\dot{\Phi} = 0$, then the bar acts as a simple pendulum oscillating in a plane.

Example 9.9

A rigid body, having principal mass moments of inertia I_x, I_y, and I_z kg·m² is rotating such that

$$\boldsymbol{\omega} = \dot{\theta}_x\mathbf{i} + \dot{\theta}_y\mathbf{j} + \dot{\theta}_z\mathbf{k} \text{ rad/s}$$

where $\dot{\theta}_x$ is large but $\dot{\theta}_y$ and $\dot{\theta}_z$ are very small. The moment applied to the body is

$$\mathbf{M} = \gamma\theta_y\mathbf{j} \text{ m·N}$$

where γ is a constant.* Obtain the governing equations of motion for the rigid body and solve for the motion as a function of time.

Here, we use Euler's equations derived in Chapter 8. Referring to Eq. (8.29) and replacing the ω terms by $\dot{\theta}$ terms and $\dot{\omega}$ terms by $\ddot{\theta}$ terms, we have

$$I_x\ddot{\theta}_x + (I_z - I_y)\dot{\theta}_y\dot{\theta}_z = M_x$$

* The situation described is approximately similar to a satellite that is spinning about its X axis and is in orbit around the earth. The torque $\gamma\theta_y$ is due to gravity.

$$I_y\ddot{\theta}_y + (I_x - I_z)\dot{\theta}_x\dot{\theta}_z = M_y$$

$$I_z\ddot{\theta}_z + (I_y - I_x)\dot{\theta}_x\dot{\theta}_y = M_z$$

where M_x, M_y, and M_z are the moments about the three principal axes. Let us assume that the rigid body under question is rotating very rapidly around the X axis at a rate of A rad/s. We may then consider $\dot{\theta}_x$ to be equal to A plus $\dot{x}$ which is a small variation in A. Let us also assume that under normal conditions $\dot{\theta}_y$ and $\dot{\theta}_z$ are very small and represented by $\dot{y}$ and $\dot{z}$. We can then write

$$\dot{\theta}_x = A + \dot{x}$$

$$\dot{\theta}_y = \dot{y}$$

$$\dot{\theta}_z = \dot{z}$$

We can now substitute this in the above three Euler's equations to obtain

$$I_x\ddot{x} + (I_z - I_y)\dot{y}\dot{z} = M_x = 0$$

$$I_y\ddot{y} + (I_x - I_z)(A\dot{z} + \dot{x}\dot{z}) = M_y = \gamma\theta_y$$

$$I_z\ddot{z} + (I_y - I_x)(A\dot{y} + \dot{x}\dot{y}) = M_z = 0$$

Notice that if $\dot{x}$, $\dot{y}$, and $\dot{z}$ are small, then their products $\dot{x}\dot{y}$, $\dot{x}\dot{z}$, and $\dot{y}\dot{z}$ may be neglected. This linearizes the differential equations. With this in mind and noting that $\theta_y = y$, the above equations become

$$I_x\ddot{x} = 0 \tag{a}$$

$$I_y\ddot{y} + (I_x - I_z)A\dot{z} = \gamma y \tag{b}$$

$$I_z\ddot{z} + (I_y - I_x)A\dot{y} = 0 \tag{c}$$

Equation (a) simply states that

$$\ddot{x} = 0 \qquad \text{or} \qquad \dot{x} = \text{Constant}$$

i.e., the spin about the X axis stays constant as a first approximation.

If we now define

$$\lambda_1 = \frac{I_x - I_z}{I_y}A \qquad \lambda_2 = \frac{I_x - I_y}{I_z}A \qquad C = \frac{\gamma}{I_y}$$

then Eqs. (b) and (c) become

$$\ddot{y} + \lambda_1\dot{z} = Cy \tag{d}$$

$$\ddot{z} - \lambda_2\dot{y} = 0 \tag{e}$$

Let us solve these coupled equations in the following way. We take the derivative of Eq. (d)

$$\dddot{y} + \lambda_1\ddot{z} = C\dot{y}$$

and substitute into Eq. (e) to obtain

$$\dddot{y} + (\lambda_1\lambda_2 - C)\dot{y} = 0 \tag{f}$$

This is an odd-looking equation, based on our experience so far. Therefore, we write it in a slightly different form. Let us define

$$p = \dot{y} \qquad \lambda^2 = \lambda_1\lambda_2 - C$$

so that Eq. (f) becomes

$$\ddot{p} + \lambda^2 p = 0$$

which we recognize as an equation having a harmonic solution. Let us select

$$p = p_0 \cos \lambda t$$

or

$$\dot{y} = \dot{y}_0 \cos \lambda t \tag{g}$$

so that the rate of rotation about the Y axis is oscillatory with an amplitude $\dot{y}_0$ applied at $t = 0$. From Eq. (e), we see that

$$\ddot{z} = \lambda_2 \dot{y}_0 \cos \lambda t$$

Integrating,

$$\dot{z} = \frac{\lambda_2 \dot{y}_0}{\lambda} \sin \lambda t \tag{h}$$

where $\dot{z}(0) = 0$. From this result we see that the rate of rotation about the Z axis is sinusoidal. This is out of phase with the Y axis motion by $\pi/2$.

We conclude that from an analysis of the linearized dynamic equations, the rigid body moves according to

$$\dot{\theta}_x = \text{constant}$$

$$\dot{\theta}_y = \dot{\theta}_{y0} \cos \lambda t \qquad \textit{Answer}$$

$$\dot{\theta}_z = \frac{\lambda_2 \dot{\theta}_{y0}}{\lambda} \sin \lambda t$$

A plot of this is shown in Fig. 9.9.

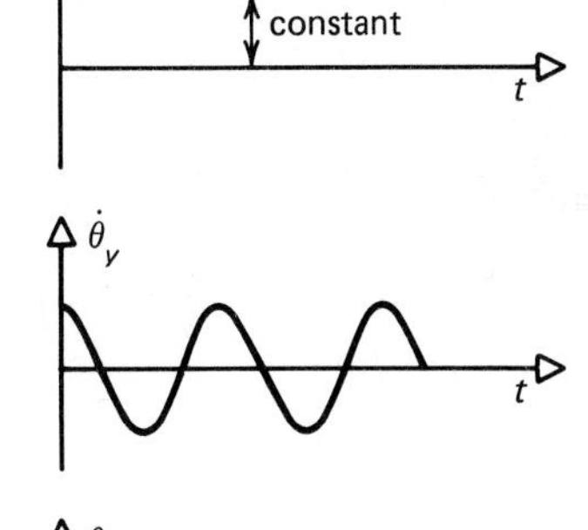

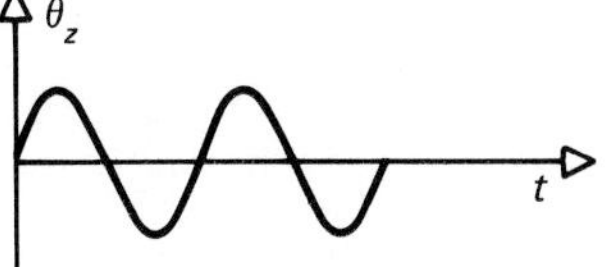

Figure 9.9

PROBLEMS

9.24 Verify Eqs. (d) and (h) in Example 9.8.

9.25 Verify Eqs. (g) and (h) in Example 9.9 by substituting into Eqs. (d) and (e).

9.26 The thin rod AB of mass m is connected to a rotating massless bar OA as shown in Fig. P9.26. There is a torsional spring of modulus k_t (m·N/rad) at A. Obtain the equation of motion of the rod using Euler's equations of motion.

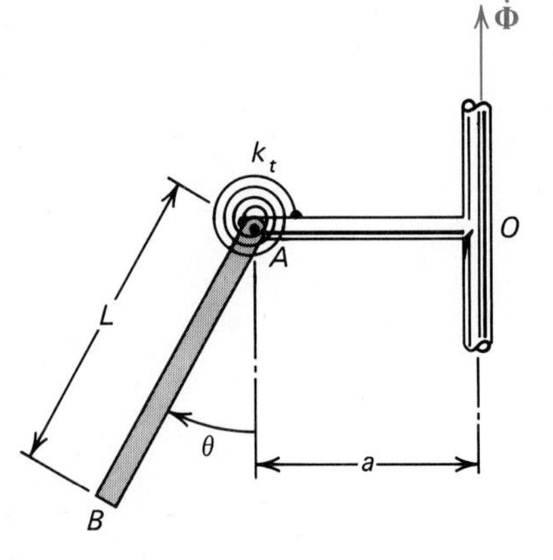

Figure P9.26

9.27 Solve Problem 9.26 using Lagrange's equations of motion.

9.28 Find the equations of motion for the bar in Fig. 9.8 under the conditions stated in Example 9.8, except that $\dot{\Phi}$ is no longer constant.

9.29 A gravity-gradient satellite is rotating around the earth as shown in Fig. P9.29. The mass moments of inertia are given by I_x, I_y, and I_z kg·m^2 and are principal mass moments of inertia. The motion of the satellite is such that the absolute velocity is given by

$$\boldsymbol{\omega} = (\dot{\theta}_x\mathbf{i} + \dot{\theta}_y\mathbf{j} + \dot{\theta}_z\mathbf{k}) \text{ rad/s}$$

where $\dot{\theta}_x = \dot{x}$, $\dot{\theta}_y = \dot{y}$, $\dot{\theta}_z = \dot{z} + \omega$, and $\dot{x}$, $\dot{y}$, $\dot{z}$ are small, while ω is the rate of rotation about the earth. If the moment due to gravity is given by

$$\mathbf{M} = 4\omega^2(I_z - I_y)\theta_x\mathbf{i} + \omega^2(I_z - I_x)\theta_y\mathbf{j} + 3\omega^2(I_x - I_y)\theta_z\mathbf{k} \text{ m·N}$$

derive the equations of motion using Euler's equations. Assume ω is constant. Solve for the actual motion about the Z axis, if $I_x = 0$ and $I_y = I_z$, and if at $t = 0$, $z = 0$ and $\dot{z} = 0.1$ rad/s.

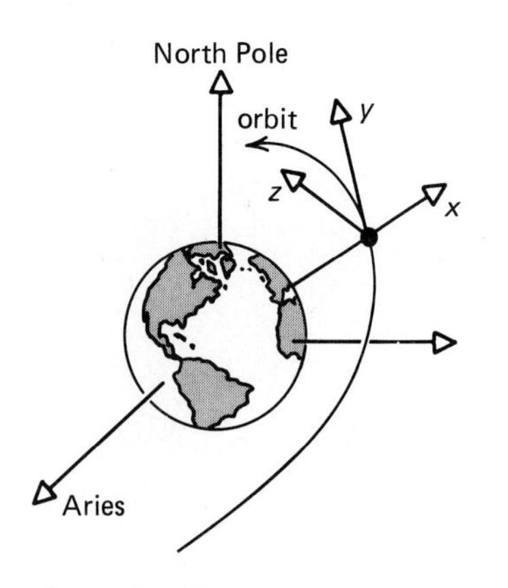

Figure P9.29

Answers to Even-Numbered Problems

Chapter 1

1.2 (a) $2\mathbf{j} + 25\mathbf{k}$
(b) $-14\mathbf{j} + 5\mathbf{k}$
(c) $28\mathbf{j} + 80\mathbf{k}$
(d) $48\mathbf{j} - 30\mathbf{k}$

1.4 0

1.6 95

1.8 $-0.535\mathbf{i} + 0.267\mathbf{j} + 0.802\mathbf{k}$

1.10 $20\mathbf{i} - 18\mathbf{j} - 10\mathbf{k}$

1.12 $-64\mathbf{i} + 40\mathbf{j} + 352\mathbf{k}$

1.16 28.7

1.18 $-2\mathbf{i} - 2\mathbf{j} - 2\mathbf{k}$

1.20 45.11×10^6 m, $0.707\mathbf{i} + 0.424\mathbf{j} + 0.566\mathbf{k}$

1.22 $7\mathbf{k}$

1.24 $[(V_2 \sin \phi_2 \sin \theta_2 - V_1 \sin \phi_1 \sin \theta_1)^2 + (V_2 \cos \phi_2 - V_1 \cos \phi_1)^2 + (V_2 \sin \phi_2 \cos \theta_2 - V_1 \sin \phi_1 \cos \theta_1)^2]^{1/2}$

1.28 $20\mathbf{j}$

1.32 $-(V_r \cos \theta + V_\theta \sin \theta)\mathbf{i} + (-V_r \sin \theta + V_\theta \cos \theta)\mathbf{j}$

1.34 $(4t + 6t^2 - 2e^t)\mathbf{i} + (4t^2 - 4t + 3e^t)\mathbf{j}$

1.36 $e^t(3t + 3)\mathbf{i} + e^t\mathbf{j} + e^t(t^2 + 2t)\mathbf{k}$, for moving coordinate system include $d\mathbf{i}/dt$, $d\mathbf{j}/dt$, $d\mathbf{k}/dt$

1.38 $\dot{\theta}\mathbf{e}_\theta$

1.44 $21.08\mathbf{i}$ m/s^2

1.46 $\left(\frac{F_1 t}{m} + C_1\right)\mathbf{i} + \left(\frac{F_2 t}{m} + C_2\right)\mathbf{j}$

1.48 $10\mathbf{i} + 20\mathbf{j} + 35\mathbf{k}$

1.50 $m(12 \cos \theta - \pi^2 \sin \pi t \sin \theta)\mathbf{e}_r - m(12 \sin \theta + \pi^2 \sin \pi t \cos \theta)\mathbf{e}_\theta$

1.52 8.11 m/s^2, 2.45 m/s^2, 1.09 m/s^2

1.54 3000 lb, 186.3 slugs

Chapter 2

2.2 1.27 m/s

2.4 60 rad/sec, 60 rad/sec^2

2.6 $te^{-t}(2 \sin t - t \sin t + t \cos t)\mathbf{k}$ rad/s^2

2.8 $\omega \sin \theta(\cos \phi\mathbf{i} - \sin \phi\mathbf{j})$, 2 m/s

2.10 $0.88\mathbf{k}$ rad/sec

2.18 114 s

2.20 $14 - \frac{8 \cos \pi t}{\pi}$ ft/sec, 76 ft

2.22 0.247 m/s^2, 90 s

2.24 $\sqrt{3k/4m}\,x_0$, $\sqrt{k/m}\,x_0$

2.26 -1.05 m/s^2

2.28 1.686 hr

2.32 $\omega_0 e^{ct}$

2.34 3770 rad/min^2, 1200 rev

2.36 8.33 rad/min

2.38 119.4 rad/sec^2

2.42 -60 rad, 100 rad

2.44 $-9{,}600\pi$ rad, $12{,}000\pi$ rad

2.46 4.25 rad, 12.59 rad

2.48 $-2.83\mathbf{i} - 2.83\mathbf{j}$ m/s
2.50 0.12 m/s
2.52 $-3\mathbf{e}_3$ rad/s
2.54 $-5\mathbf{e}_1 - 1.5\mathbf{e}_2$ m/s
2.56 $\sqrt{v^2 + \omega^2}$ ft/sec
2.58 $15.66\mathbf{e}_1 + 2.5\mathbf{e}_2$ m/s
2.60 0, $1.57\mathbf{i}$ ft/sec
2.62 4.795 m/s
2.64 $(\sin^2\theta + 2\cos\theta - \cos^2\theta)\dot{\theta}\mathbf{j}$
2.66 $-2\mathbf{e}_1 + 7\mathbf{e}_2$ in/sec
2.68 $-30\mathbf{e}_1 + 40\mathbf{e}_2$ km/h
2.70 $-5.65\mathbf{e}_1 - 4.35\mathbf{e}_2$ mi/hr
2.72 $+1.41\mathbf{e}_1 + 2.51\mathbf{e}_2$ m/s²
2.74 $1\mathbf{e}_1$ m/s²
2.76 $-800\mathbf{e}_1 + 100\mathbf{e}_2$ in/sec², $100\mathbf{e}_2$ in/sec²
2.78 $L\omega^2\mathbf{e}_1 - 2\omega v\mathbf{e}_2$
2.80 $-8\mathbf{e}_1 - 882\mathbf{e}_2$ m/s²
2.82 7.33 m/s²
2.84 $30\mathbf{e}_2 - 4.5\mathbf{e}_1$ m/s²
2.86 $-0.6\mathbf{i} + 59.2\mathbf{j}$ in/sec²
2.88 $[\ddot{\theta}(\sin^2\theta + 2\cos\theta - \cos^2\theta) + 2\dot{\theta}^2(2\cos\theta - 1)\sin\theta]\mathbf{j}$
2.90 $-1.5\mathbf{e}_2$ m/s²
2.92 $6\mathbf{e}_\theta - 0.9\mathbf{e}_r$ m/s²
2.94 $10e^{2t}(\cos\theta - \sin\theta)\mathbf{i} + 10e^{2t}(\cos\theta + \sin\theta)\mathbf{j}$ ft/sec
2.96 $23\mathbf{e}_r - 8.48\mathbf{e}_\theta$ m/s
2.98 $1511\mathbf{e}_r + 756\mathbf{e}_\theta$ in/sec,
$4.75 \times 10^5\mathbf{e}_r + 6.33 \times 10^5\mathbf{e}_\theta$ in/sec²
2.100 $[R\dot{\theta} + r(\dot{\theta} + \dot{\phi})\cos\phi]\mathbf{e}_\theta - [(\dot{\theta} + \dot{\phi})r\sin\phi]\mathbf{e}_r$
2.102 $7.6\mathbf{e}_r + 8.8\mathbf{e}_\theta$ m/s
2.104 $v\mathbf{e}_\tau, \dfrac{24\sin t\cos t}{v}\mathbf{e}_\tau + \dfrac{5}{v}\mathbf{e}_n$
where $v = \sqrt{24\sin^2 t + 1}$
2.106 $\sqrt{36t^4 + 25}\mathbf{e}_\tau, \dfrac{t}{\sqrt{36t^4 + 25}}(72t^2\mathbf{e}_\tau + 60\mathbf{e}_n)$
2.108 $8t\mathbf{e}_\tau$ ft/sec, $\left[8\mathbf{e}_\tau + \dfrac{128t^2\mathbf{e}_n}{(1 + 4x^2)^{3/2}}\right]$ ft/sec²
2.110 $2\mathbf{e}_\tau + 1.61\mathbf{e}_n$ ft/sec²
2.112 $129.6\mathbf{e}_n + 16\mathbf{e}_\tau$ ft/sec², 58.7 ft
2.114 $-20\mathbf{e}_1 + 10\mathbf{e}_2 - 4\mathbf{e}_3$ m/s, $-242\mathbf{e}_1 - 395\mathbf{e}_2$ m/s²
2.116 $5\mathbf{e}_1 + 80\mathbf{e}_2 - 40\mathbf{e}_3$ in/sec
$-1000\mathbf{e}_1 + 100\mathbf{e}_2 - 50\mathbf{e}_3$ in/sec²
2.118 $-(\omega_2^2 + \omega_3^2)R\mathbf{i} + (\alpha_3 + \omega_1\omega_2)R\mathbf{j} + (\omega_1\omega_3 - \alpha_2)R\mathbf{k}$

Chapter 3

3.2 (a) $m\ddot{x} = -k(x - b) + mg$, (b) $m\ddot{y} = -ky$
3.4 $m(\ddot{r} - r\dot{\theta}^2) = -F_0$, $m(2\dot{r}\dot{\theta} + r\ddot{\theta}) = 0$
3.6 0.1 in
3.8 $ml\dot{\theta}^2 = O_n,\ ml\ddot{\theta} = mg - \dfrac{kl}{2}\theta + O_\tau$
3.10 41.4 lb
3.12 11.3°
3.14 (a) 735.8 N, (b) 245.3 N
3.16 $-k(r - l) + mg\cos\theta = m(\ddot{r} - r\dot{\theta}^2)$,
$(2\dot{r}\dot{\theta} + r\ddot{\theta}) = -g\sin\theta$
3.18 49.15 ft/sec at $\theta = 55.6°$
3.20 $y = 1000 - 1.87 \times 10^{-4}x^2$, 3220 lb
3.22 $y = 1000 - 1.85 \times 10^{-4}x^2$, good approximation
3.24 5.66 m/s²
3.26 0.459 m/s²
3.28 59.1 N, 70 N
3.30 $[-ml\omega^2\cos\theta + k\cot\theta(2l\sin\theta - L_0)]\mathbf{j}$
3.32 5.03 rad/s
3.34 10.1 s
3.36 $\mathbf{A} = 152.3\mathbf{i} + 1269\mathbf{j}$ N
$\mathbf{B} = 913.7\mathbf{i} + 507.8\mathbf{j}$ N
3.40 55.9 N
3.42 −295 lb·ft
3.44 1.273 lb·ft
3.46 24,000 N·m
3.48 6.9 ft/sec
3.50 126.3 N·m
3.52 $\frac{1}{2}m(\dot{r}^2 + r^2\dot{\theta}^2)$
3.54 12.42 lb·ft
3.56 26.5 m
3.58 30 passes, ends at point A
3.60 42.9°
3.62 3,858 N·m, 385,500 N/m
3.64 233,400 kg
3.66 2 ft
3.68 8,764 m/s
3.70 4.61 m/s
3.72 0.109 hp/lb
3.74 3750 watts
3.76 $y = \dfrac{qE}{2mV^2}x^2,\ x = V\sqrt{2mh/qE}$

3.78 1.737×10^9 N·m

3.80 25 times

3.82 $T = \frac{1}{2}m\omega^2 l^2$, $\mathbf{p} = m\omega l(\sin\theta\mathbf{i} + \cos\theta\mathbf{j})$, $\mathbf{H} = m\omega l^2\mathbf{k}$

3.84 $125\mathbf{i} + 20\mathbf{j}$ lb·sec

3.86 $\mathbf{p} = mv(\sin\theta\mathbf{i} + \cos\theta\mathbf{j})$, $\mathbf{H} = mLv\cos\theta\mathbf{k}$, $\mathbf{H}' = (M + m)Lv'\mathbf{k}$

3.88 $0.5\mathbf{i} + 5\mathbf{j}$ lb·sec, $0.083\mathbf{k}$ lb·ft·sec

3.90 $m(12\mathbf{i} + 15.3\mathbf{j} - 18.8\mathbf{k})$

3.92 $mrV_2\mathbf{k}$, $mr\dfrac{dV_2}{dt}\mathbf{k}$

3.94 $6.39\mathbf{i} + 2\mathbf{j}$ lb·sec

3.96 58.24 m/s

3.98 171.1 ft/sec

3.100 200 m

3.102 $-R\sqrt{Fmg_e}$

3.104 $ml\omega\sin\theta$, $ml^2\omega\sin\theta$

3.106 $m\sqrt{2gR}\mathbf{i}$, $-m\sqrt{gR}\mathbf{i}$, 0

3.108 $\dfrac{mLv}{2}\sin\phi\mathbf{k}$

3.110 $mr^2\omega\mathbf{k}$, $mr(2\dot{r}\omega + r\alpha)\mathbf{k}$

Chapter 4

4.2 6.89×10^{-8} lb

4.4 9.795 m/s², 32.18 ft/sec²

4.6 2638 km

4.14 πab

4.16 $\dfrac{\tau_A^2}{\tau_B^2} = \dfrac{a_A^3}{a_B^3}$

4.18 127.3 min, same semimajor axis.

4.20 5893 ft/sec

4.22 365.6 days, 0.01672

4.24 389,400 km

4.26 $\epsilon^2 = \left(\dfrac{r_0 V_0^2}{\mu} - 1\right)^2 \cos^2\beta_0 + \sin^2\beta_0$

4.28 22,283 mi, 10,076 ft/sec, 7.88×10^{18} lb·ft

4.30 $M = \psi - \epsilon\sin\psi$

Chapter 5

5.2 1.5 kg·m²

5.4 0.5355 slug·ft²

5.6 0.196 kg·m², 0.251 kg·m²

5.8 3.65 in

5.10 5.02 kg·m²/s

5.12 493.5 slug·ft²/sec

5.14 5757 slug·ft²/sec

5.16 $87.34\mathbf{k}$ kg·m²/s

5.18 785.4 slug·ft²/sec

5.20 (a) 120 kg·m²; (b) 320 kg·m²

5.22 $1\mathbf{i}$ ft/sec, cylinder is slipping

5.24 $20\mathbf{i}$ ft/sec, $-100\mathbf{j}$ ft/sec²

5.26 $\lambda BL\cos\lambda t\cos\theta\mathbf{i} + (A\cos\omega t - \lambda BL\cos\lambda t\sin\theta)\mathbf{j}$

5.28 $-268\mathbf{i}$ in/sec, $-21.6\mathbf{k}$ rad/sec

5.30 $120\mathbf{k}$ rad/s

5.32 $\omega_{BC} = -1.5\mathbf{k}$ rad/s, $\omega_{CD} = 2\mathbf{k}$ rad/s

5.34 $(-0.6\mathbf{i} + 0.3\mathbf{j})$ m/s, $(-0.75\mathbf{i} - 1.65\mathbf{j})$ m/s²

5.36 $\omega_{BC} = -12\mathbf{e}_3$ rad/s, $\omega_{CD} = 28.8\mathbf{e}_3$ rad/s

5.38 $0.0306\mathbf{i}$ m/s, $(-0.0031\mathbf{j} - 0.0735\mathbf{i})$ m/s²

5.40 $(-0.25\mathbf{e}_1 + 2.965\mathbf{e}_2)$ m/s, $(-1.065\mathbf{e}_1 + 7.08\mathbf{e}_2)$ m/s²

5.42 $(3.32\mathbf{e}_1 + 1.83\mathbf{e}_2)$ m/s, $(2.98\mathbf{e}_1 - 7.54\mathbf{e}_2)$ m/s²

5.44 141.4 ft/sec, 28,300 ft/sec²

5.46 $10\mathbf{k}$ rad/sec²

5.48 $-60\mathbf{k}$ m/s

5.50 $\mathbf{V}_B = 4\mathbf{j}$ ft/sec, $\mathbf{V}_A = 2.31\mathbf{i}$ ft/sec

5.52 $\omega_{AB} = -0.159\mathbf{k}$ rad/s

5.54 $(1.34\mathbf{e}_1 - 5\mathbf{e}_2)$ ft/sec, $(12.5\mathbf{e}_1 - 21.65\mathbf{e}_2)$ ft/sec²

5.56 $-0.2\mathbf{j}$ m/s, $-0.1\mathbf{j}$ m/s²

5.58 $(-5.47\mathbf{i} - 5.5\mathbf{j})$ in/sec

5.60 $(1.13\mathbf{i} + 1.48\mathbf{j})$ m/s

5.62 $3.535(\mathbf{e}_1 + \mathbf{e}_2)$ ft/sec

5.64 16 rad/s CW, 4 rad/s² CW

5.66 $-9\mathbf{e}_3$ rad/sec

5.68 110 ft/sec²

5.70 $(25\mathbf{e}_1 + 400\mathbf{e}_2)$ m/s²

5.72 -166.7 in/sec, -4362 in/sec²

5.74 $-51.76\mathbf{k}$ rad/s²

5.76 $V_B = 0$, $\alpha = 16$ rad/s² CCW

5.78 $2\mathbf{i}$ ft/sec

5.84 26.02 rad/s, 387.8 in/sec

5.86 $\omega_{AB} = 1.2\mathbf{e}_3$ rad/sec, $\omega_{OA} = 1\mathbf{e}_3$ rad/sec

5.88 4 rad/sec CCW

5.90 $-14.96\mathbf{e}_3$ rad/sec

5.92 $-5\mathbf{k}$ rad/s

Chapter 6

6.2 (a) $(2.26\mathbf{i} + 0.59\mathbf{j})10^{-11}$ m/s^2
(b) $(0.59\mathbf{i} + 2.26\mathbf{j})10^{-11}$ N

6.4 3.22 lb

6.6 36.65 lb, $-4.45\mathbf{i}$ ft/sec^2

6.8 (a) $-7.81\mathbf{j}$ m/s^2, (b) 78.1 m/s

6.10 2679 N·m

6.12 55.1 lb·ft

6.14 808 N/m

6.16 0.183 m, 16.77 N·m

6.18 $-66\mathbf{k}$ lb·ft·sec

6.20 $483\mathbf{i}$ lb·sec

6.22 0

6.24 0.33 m

6.26 $(-160\mathbf{i} + 878.5\mathbf{j})$ N

6.28 14.3 rad/sec^2 CW

6.30 $f = \dfrac{W}{3} \sin\theta$

6.32 $-4.905\mathbf{k}$ rad/s^2

6.34 0.667

6.36 187.5 lb

6.38 108 m/s^2

6.40 1.33 kg

6.42 191.9 N·m

6.44 66,470 ft·lb

6.46 41.7 lb

6.48 1 rad/s^2 CW

6.50 $0.9\mathbf{i}$ m/s^2, $30\mathbf{i}$ N

6.52 1896 N·m

6.54 0.00385 lb·ft

6.56 21.8 rad/s CW

6.58 9.23 rad/s CW

6.60 23.92 ft/sec

6.62 0.1095 ft·lb

6.64 17.45 watts

6.66 12.56 ft/sec

6.68 70.2 N/m

6.70 1.146 lb·ft/sec

6.72 9.04×10^4 watts

6.74 2.74 N

6.76 $\mathbf{H}_c = 1.6$ kg·m^2/s CCW, $\mathbf{H}_A = 0.4$ kg·m^2/s CW

6.78 $-.00225\mathbf{k}$ slug·ft^2/sec

6.80 $\mathbf{H}_c = m(r_g^2 - r_c^2)\omega$ CW, $\mathbf{H}_o = m\omega(r_g^2 + rr_c)$ CW

6.82 85.8 m/s

6.84 53.82 slug·ft^2/sec CW

6.86 $\dfrac{6r}{3r - 2h}\sqrt{\dfrac{gh}{3}}$

6.88 157.4 m

6.90 0.9307 lb

6.92 16.66 rev/min

6.94 24.95 rad/s

6.96 $\omega = \dfrac{2mVR}{I}$

Chapter 7

7.2 18,375 N·m

7.4 (a) 9.46 ft/sec, (b) 139 lb·ft

7.6 (a) $-29\mathbf{i}$ m/s, $16\mathbf{i}$ m/s; (b) $-2\mathbf{i}$ m/s; (c) 0, 6075 N·m

7.8 10 s

7.10 $x = 3.33$ m

7.12 304.4 lb/ft

7.14 0.375

7.16 $-12\mathbf{i}$ m/s, $30\mathbf{i}$ m/s

7.18 $\mathbf{u}_A = -6\mathbf{i}$ m/s, $\mathbf{u}_B = 2\mathbf{i}$ m/s

7.20 0.247 m

7.22 (a) 0.76; (b) 1.7 ft

7.24 $(6.93\mathbf{i} + 7.78\mathbf{j})$ m/s, $(4.24\mathbf{i} - 0.47\mathbf{j})$ m/s

7.26
$$\mathbf{u}_A = \left(-1.99 m_B v_B + m_A(v_A \cos\theta)\left(1 - 0.99\frac{m_B}{m_A}\right)\right)\frac{\mathbf{i}}{m_A + m_B} - v_A \sin\theta\mathbf{j}$$
$$\mathbf{u}_B = \left(1.99 m_A v_A \cos\theta - m_B v_B\left(1 - 0.99\frac{m_A}{m_B}\right)\right)\frac{\mathbf{i}}{m_A + m_B}$$

7.28 (a) 1.36 lb·ft, (b) 0.193 lb·ft

7.30 $\mathbf{u}_A = 0$, $\mathbf{u}_B = (23\mathbf{i} + 19.3\mathbf{j})$ m/s

7.32 $V\dfrac{(1-e)}{3}\cos\theta\mathbf{i} + V\sin\theta\mathbf{j}$, $V\dfrac{(1+e)}{2}\cos\theta\mathbf{i}$

7.34 $\omega_2 = 2.5$ rad/s, $\mathbf{u}_1 = -43.3\mathbf{j} - 10\mathbf{i}$ m/s

7.36 3.75 rad/s CCW

7.38 $\mathbf{u}_1 = 3.18\mathbf{j}$ m/s, $\mathbf{u}_2 = 3.364\mathbf{j}$ m/s, $\boldsymbol{\omega}_2 = 3.90\mathbf{k}$ rad/s

7.40 $\omega = \dfrac{6(1+e)m_1\sqrt{2gh}}{l(m_2+3m_1)}$

$x = \dfrac{6(1+e)m_1\sqrt{m_2gh}}{(m_2+3m_1)\sqrt{6k}}$

7.42 $\omega_2 = 3.48$ rad/sec

7.44 1432 N·m/rad

Chapter 8

8.2 (a) 2; (b) r, θ

8.4 (a) 2; (b) z, θ

8.6 (a) 1; (b) θ

8.8 Four degrees of freedom, two translation and two rotational

8.10 (a) 2; (b) Y, θ

8.12 $\dot{x}_1 = -L\dot{\theta}\sin\theta,\ \dot{x}_2 = L\dot{\theta}\cos\theta,$
$\dot{x}_3 = -(L-z)\dot{\theta}\sin\theta - \dot{z}\cos\theta,$
$\dot{x}_4 = (L-z)\dot{\theta}\cos\theta - \dot{z}\sin\theta$

8.14 $\dot{x}_1 = -L\dot{\theta}\sin\theta,\ \dot{x}_2 = L\dot{\theta}\cos\theta,$

$\dot{x}_3 = \left(-\dfrac{L}{2}\sin\theta - z\cos\theta\right)\dot{\theta} - \dot{z}\sin\theta$

$\dot{x}_4 = \left(\dfrac{L}{2}\cos\theta - z\sin\theta\right)\dot{\theta} + \dot{z}\cos\theta$

8.16 $\dot{x}_1 = -\dot{\theta}(b\sin\theta + L\cos\theta),$
$\dot{x}_2 = \dot{\theta}(b\cos\theta - L\sin\theta)$

8.18 $\dot{y}_1 = \dot{x}_1,\ \dot{y}_2 = \dot{x}_1 + l\dot{\theta}_1\cos\theta_1,\ \dot{y}_3 = l\dot{\theta}_1\sin\theta_1$
$\dot{y}_4 = \dot{x}_2,\ \dot{y}_5 = \dot{x}_2 + l\dot{\theta}_2\cos\theta_2,\ \dot{y}_6 = l\dot{\theta}_2\sin\theta_2$

8.20 $\dot{x}_1 = -c\dot{\theta}\sin\theta,\ \dot{x}_2 = \dot{y} - c\dot{\theta}\sin\theta$

8.22 $\dfrac{d}{dt}[m_1L^2\dot{\theta} + m_2(L-z)^2\dot{\theta}] + m_1gL\sin\theta$

$+ m_2g(L-z)\sin\theta = 0$

$\dfrac{d}{dt}(m_2\dot{z}) + m_2(L-z)\dot{\theta}^2 + m_2g\cos\theta + k(z-L_0) = 0$

8.24 $\dfrac{d}{dt}\left[m_1L^2\dot{\theta} + m_2\left(\dfrac{L^2}{4}\dot{\theta} + \dfrac{L}{2}\dot{z} + z^2\dot{\theta}\right)\right]$

$+ m_1gL\sin\theta + m_2g\left(\dfrac{L}{2}\sin\theta + z\cos\theta\right) = 0$

$\dfrac{d}{dt}\left[m\left(\dfrac{L}{2}\dot{\theta} + \dot{z}\right)\right] - m_2z\dot{\theta}^2 + m_2g\sin\theta$
$+ k(z-L_0) = 0$

8.26 $m(b^2+L^2)\ddot{\theta} + ka^2\theta = 0$

8.28 $\dfrac{m}{2}[(\dot{y}\sin\theta - c\dot{\theta})^2 + (\dot{y}\cos\theta)^2] - \dfrac{k_1}{2}(y - a\sin\theta)^2$

$-\dfrac{k_2}{2}(y + b\sin\theta)^2$

8.30 $m_1\ddot{x}_1 + k_1x_1 - k(x_2-x_1) = 0$
$m_2\ddot{x}_2 + k(x_2-x_1) = F_0\cos\lambda t$

8.32 $mL^2\ddot{\theta}_2 + mgL\theta_2 + ka^2(\theta_2-\theta_1) = 0$
$mL^2\ddot{\theta}_1 + mgL\theta_1 - ka^2(\theta_2-\theta_1) = 0$

8.34 $\dfrac{m}{2}\left\{(\dot{z}\cos\phi + z\dot{\theta}\sin\phi)^2\right.$

$\left.+\left[-\dot{z}\sin\phi + z\dot{\theta}\cos\phi + \dfrac{L}{2}(\dot{\theta}+\dot{\phi})\right]^2\right\}$

$+\dfrac{1}{24}mL^2(\dot{\theta}+\dot{\phi})^2 - \dfrac{1}{2}k(z-L_0)^2$

$+ mg\left[\dot{z}\cos\theta + \dfrac{L}{2}\cos(\theta+\phi)\right]$

8.36 $M_x = \frac{1}{4}mr^2\omega\Omega,\ M_y = \frac{1}{4}mr^2\dot{\Omega},\ M_z = \frac{1}{2}mr^2\dot{\omega}$

8.38 $M_x = \frac{1}{2}mr^2(ps\sin st - \dot{p}\cos st)$

$M_y = -\dfrac{m}{12}(3r^2+L^2)(\dot{p}\sin st + ps\cos st)$

$+\left(\dfrac{mr^2}{4} - \dfrac{mL^2}{12}\right)sp\cos st$

$M_z = \left(\dfrac{mL^2}{12} - \dfrac{mr^2}{4}\right)p^2\sin st\cos st$

8.40 $\Omega\mathbf{i} + \omega_0\cos(\beta\Omega t)\mathbf{j} + \omega_0\sin(\beta\Omega t)\mathbf{k}$

8.42 $mL^2\ddot{\theta} - mL^2\omega^2\sin\theta\cos\theta + mgL\sin\theta = 0$

8.44 $\Omega\mathbf{i} + \dfrac{\alpha}{\beta}A_1\cos(\alpha\beta t + \phi_1)\mathbf{j} + A_1\sin(\alpha\beta t + \phi_1)\mathbf{k}$
where A_1 and ϕ_1 are constants and $\alpha^2 = I_x\Omega/I_y$ and $\beta^2 = I_x\Omega/I_z$

Chapter 9

9.2 $-19.2\times10^{-13}\mathbf{k}$ N

9.4 $a_x = \dfrac{q}{mc}(B_zV_y - B_yV_z),\ a_y = \dfrac{q}{mc}(B_xV_z - B_zV_x)$

$a_z = \dfrac{q}{mc}(B_yV_x - B_xV_y)$

9.6 $\frac{1}{2}m\left[\dot{x}^2 + \left(\frac{qEt}{m}\right)^2\right]$

9.12 $m\ddot{y}_1 + c(\dot{y}_1 - \dot{y}_2) + k_1 y_1 = 0$
$k_2 y_2 - c(\dot{y}_1 - \dot{y}_2) = 0$

9.14 $m\ddot{x} = -k_1(x - a\theta) - k_2(x + \theta(L - a))$
$I\ddot{\theta} = -k_1 a(x - a\theta) + k_2[x + \theta(L - a)](L - a)$

9.16 0.0203 kg·m^2

9.18 $\frac{d}{dt}\left(\frac{3}{2}M\dot{x} + m\dot{x} + \frac{mL}{2}\dot{\theta}\cos\theta\right) = 0$

$$\frac{d}{dt}\left(\frac{mL}{2}\left(\dot{x}\cos\theta + \frac{L}{2}\dot{\theta}\right) + \frac{mL^2}{12}\dot{\theta}\right)$$

$$+\frac{m\dot{x}L\dot{\theta}}{2}\sin\theta + \frac{mgL}{2}\sin\theta = 0$$

9.20 $m_1\ddot{x}_1 + k_1 x_1 - k_2(x_2 - x_1) + c\dot{x}_1 = 0$
$m_2\ddot{x}_2 + k_2(x_2 - x_1) = F(t)$

9.22 $m\ddot{x}_1 + 2kx_1 - kx_2 - c(\dot{x}_2 - \dot{x}_1) = 0$
$m\ddot{x}_2 + k(x_2 - x_1) + c(\dot{x}_2 - \dot{x}_1) = 0$

9.26 $\frac{d}{dt}(\dot{\phi}\sin\theta) + \dot{\phi}\dot{\theta}\cos\theta = 0$

$$\ddot{\theta} - \dot{\phi}^2\sin\theta\cos\theta + \frac{k_t\theta}{I} + \frac{O_y L}{2I} = 0$$

9.28 $\ddot{\theta} - \dot{\phi}^2\sin\theta\cos\theta + \frac{3g}{2L}\sin\theta = 0$

$$\frac{d}{dt}(I_z\,\dot{\phi}\sin^2\theta) = 0$$

Index

Numbers in regular type refer to pages in *Engineering Mechanics: Statics*.
Numbers in bold italics refer to pages in *Engineering Mechanics: Dynamics*.

L

M

N

U

V

W

Z

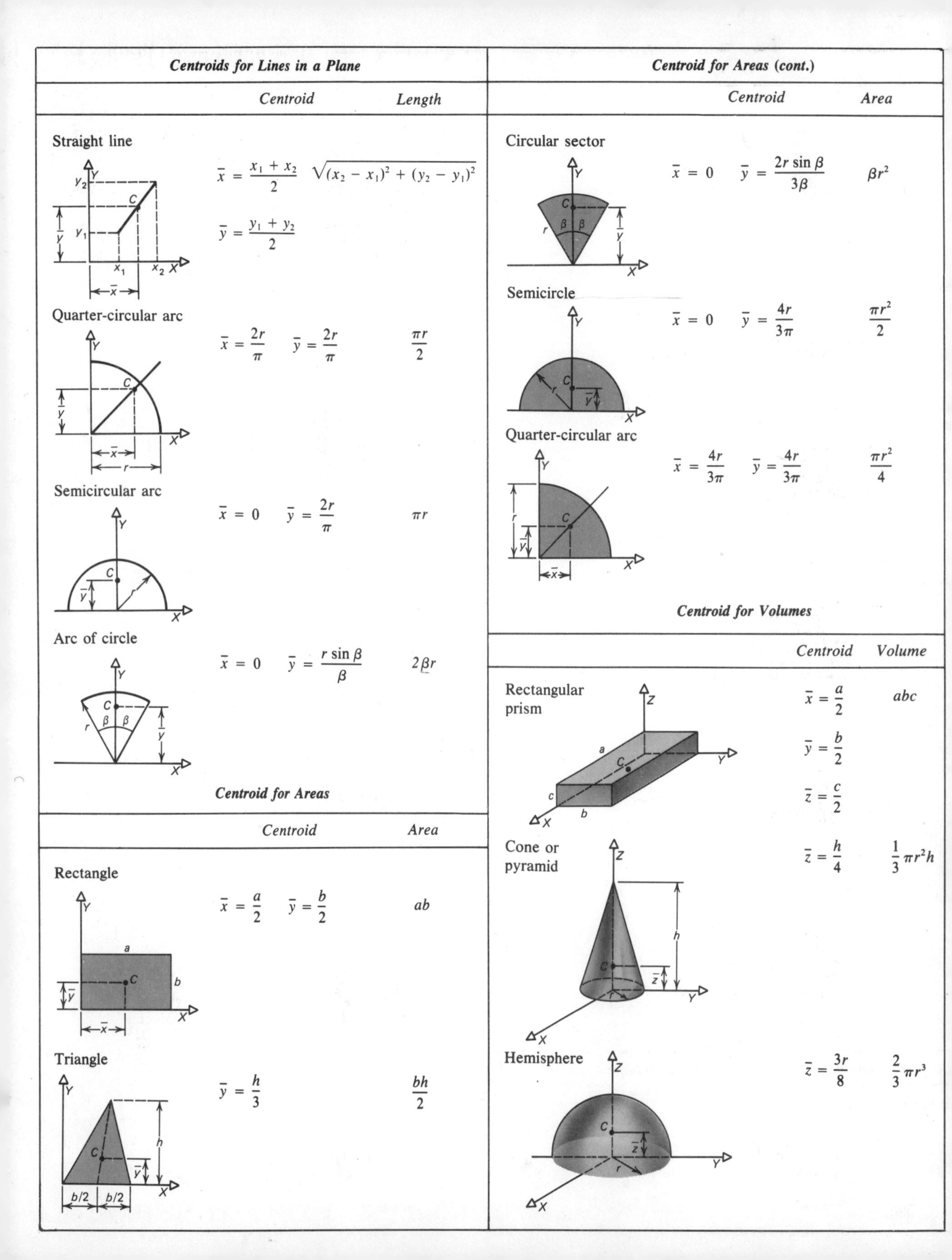

Centroids for Lines in a Plane

	Centroid		Length
Straight line	$\bar{x} = \dfrac{x_1 + x_2}{2}$	$\bar{y} = \dfrac{y_1 + y_2}{2}$	$\sqrt{(x_2 - x_1)^2 + (y_2 - y_1)^2}$
Quarter-circular arc	$\bar{x} = \dfrac{2r}{\pi}$	$\bar{y} = \dfrac{2r}{\pi}$	$\dfrac{\pi r}{2}$
Semicircular arc	$\bar{x} = 0$	$\bar{y} = \dfrac{2r}{\pi}$	πr
Arc of circle	$\bar{x} = 0$	$\bar{y} = \dfrac{r \sin \beta}{\beta}$	$2\beta r$

Centroid for Areas

	Centroid		Area
Rectangle	$\bar{x} = \dfrac{a}{2}$	$\bar{y} = \dfrac{b}{2}$	ab
Triangle		$\bar{y} = \dfrac{h}{3}$	$\dfrac{bh}{2}$

Centroid for Areas (cont.)

	Centroid		Area
Circular sector	$\bar{x} = 0$	$\bar{y} = \dfrac{2r \sin \beta}{3\beta}$	βr^2
Semicircle	$\bar{x} = 0$	$\bar{y} = \dfrac{4r}{3\pi}$	$\dfrac{\pi r^2}{2}$
Quarter-circular arc	$\bar{x} = \dfrac{4r}{3\pi}$	$\bar{y} = \dfrac{4r}{3\pi}$	$\dfrac{\pi r^2}{4}$

Centroid for Volumes

	Centroid	Volume
Rectangular prism	$\bar{x} = \dfrac{a}{2}$, $\bar{y} = \dfrac{b}{2}$, $\bar{z} = \dfrac{c}{2}$	abc
Cone or pyramid	$\bar{z} = \dfrac{h}{4}$	$\dfrac{1}{3}\pi r^2 h$
Hemisphere	$\bar{z} = \dfrac{3r}{8}$	$\dfrac{2}{3}\pi r^3$